Comprehensive Applied Mathematical Modeling in the Natural and Engineering Sciences

David J. Wollkind · Bonni J. Dichone

Comprehensive Applied Mathematical Modeling in the Natural and Engineering Sciences

Theoretical Predictions Compared with Data

David J. Wollkind
Department of Mathematics
Washington State University
Pullman, WA
USA

Bonni J. Dichone
Department of Mathematics
Gonzaga University
Spokane, WA
USA

ISBN 978-3-030-08804-0 ISBN 978-3-319-73518-4 (eBook)
https://doi.org/10.1007/978-3-319-73518-4

Mathematics Subject Classification (2010): 35Q35, 34E15, 34A34, 35B36, 35K57, 76E06, 78A60, 92D40, 93A30

© Springer International Publishing AG, part of Springer Nature 2017
Softcover re-print of the Hardcover 1st edition 2017
This work is subject to copyright. All rights are reserved by the Publisher, whether the whole or part of the material is concerned, specifically the rights of translation, reprinting, reuse of illustrations, recitation, broadcasting, reproduction on microfilms or in any other physical way, and transmission or information storage and retrieval, electronic adaptation, computer software, or by similar or dissimilar methodology now known or hereafter developed.
The use of general descriptive names, registered names, trademarks, service marks, etc. in this publication does not imply, even in the absence of a specific statement, that such names are exempt from the relevant protective laws and regulations and therefore free for general use.
The publisher, the authors and the editors are safe to assume that the advice and information in this book are believed to be true and accurate at the date of publication. Neither the publisher nor the authors or the editors give a warranty, express or implied, with respect to the material contained herein or for any errors or omissions that may have been made. The publisher remains neutral with regard to jurisdictional claims in published maps and institutional affiliations.

Printed on acid-free paper

This Springer imprint is published by the registered company Springer International Publishing AG part of Springer Nature
The registered company address is: Gewerbestrasse 11, 6330 Cham, Switzerland

For my parents, Monte and Norma Wollkind.
–David J. Wollkind

In loving memory of my daddy, George Kealy.
–Bonni J. Dichone

"It is a capital mistake to theorize before one has data. Insensibly one begins to twist facts to suit theories, instead of theories to suit facts."

Sherlock Holmes from A Scandal in Bohemia [1]

[1] Doyle, A.C.: The Adventures of Sherlock Holmes (1891)

Preface

Scope

Our book is about comprehensive applied mathematical modeling in the natural and engineering sciences. The basic theme of such modeling is that its theoretical predictions are compared with observational or experimental data from the phenomenon under investigation. The book's main purpose is to demonstrate the process of comprehensive applied mathematical modeling by introducing various case studies. These are arranged for the most part in increasing order of the complexity of the mathematical methods required to analyze those models with the methods being developed in parallel to the models as "pastoral interludes" to preserve the goal of a self-contained presentation. Further, these models are deduced from scientific first principles. To reinforce and supplement the material introduced, problems are provided involving closely related case studies.

From 1970 to 2015, David Wollkind developed five classes relevant to the teaching of comprehensive applied mathematical modeling in the Department of Mathematics at Washington State University (WSU): Namely, a two-semester sequence that introduced the basic concepts of continuum mechanics in the context of stability methods and perturbation techniques, predominantly to analyze fluid flow phenomena; and three single-semester courses that concentrated on the partial differential equations of mathematical physics, on advanced topics in comprehensive mathematical modeling, and on applied analysis as related both to solving linear systems and to obtaining asymptotic expansions for integral representations of special functions. These courses were as self-contained as possible and, rather than using textbooks, provided the students with carefully constructed lecture notes instead. To reinforce and supplement the material introduced, original problems sets were assigned involving case studies closely related to ones presented in class and designed expressly for that purpose. In the Spring of 2007, Robert E. O'Malley, Jr., from the University of Washington gave the Twenty-Sixth Annual Theodore G. Ostrom Lecture at WSU on the history of the development of boundary-layer theory. During this visit, he had the occasion to examine these lecture notes and dovetailed problem sets from the courses described above and strongly endorsed their publication in book form given that, in his opinion, when taken collectively they represented a unique perspective on comprehensive applied mathematical modeling. Hence the genesis of this book was that endorsement.

It has been organized using the advanced topics course in comprehensive applied mathematical modeling as a template, while interspersing related material consisting of both models and methodologies from the other four courses. This organization is in the spirit of David Wollkind's thesis adviser Lee A. Segel, who was renowned as a leading scholar in Comprehensive Applied Mathematical Modeling. His book with C.C. Lin and its sequel have been the gold standard in the teaching of modeling, continuum mechanics, and asymptotic methods. That book "Mathematics Applied to Deterministic Problems in the Natural Sciences" was a charter selection for the SIAM classics series, and its sequel "Mathematics Applied to Continuum Mechanics" appears as the 52nd entry in this series. Originally trained as an applied mathematical modeler in the physical and engineering sciences with an emphasis on fluid mechanics and materials science, Lee Segel was also a pioneer in modern Mathematical Biology. His work spanning three and one-half decades in this field has had a huge impact. He was one of the earliest promoters of the need for close collaborative contact between theoretical and experimental biology and fostered the appreciation of modeling by biologists and of biology by mathematicians. Lee Segel wrote many papers in mathematical biology and was the author of a handful of books based on his pioneering teaching in that subject at the Weizmann Institute in Rehovot, Israel, and Rensselaer Polytechnic Institute (RPI) in Troy, New York. His work led to the creation of new mathematical and computational tools for investigating the riches of biological behavior. His research on pattern formation and morphogenesis has been seminal in launching a burgeoning field at the intersection of developmental biology, applied mathematics, and numerical computation. He was a forefather of the field now known as theoretical immunology. As the editor of the *Bulletin of Mathematical Biology*, Lee Segel introduced the idea that theoretical mathematical modeling predictions of biological phenomena should be compared with relevant experimental or observational data in order to be eligible for publication. It is in this spirit that Wollkind and his co-author Bonni J. Dichone have organized this book. Dichone, currently a faculty member at Gonzaga University, completed her doctoral dissertation under Wollkind's direction, on ecological pattern formation of vegetation in arid flat environments. In the Fall Semester of 2007, soon after Bob O'Malley's Ostrom Lecture, she entered the Ph.D. program in Mathematics at WSU, began taking the applied mathematics classes described above, and has been working closely together with Wollkind ever since. Hence this book is a natural extension of that research and her educational experience in comprehensive applied mathematical modeling as a graduate student, postdoctoral scholar, and faculty member. Indeed, the completion of our book is as much dependent upon her encouragement of and collaborative effort with Wollkind as its genesis was on Bob O'Malley's initial endorsement.

Content

The book consists of twenty-two chapters organized in the following fashion: Chapter 1; Part I: Chapters 2–8; Part II: Chapters 9–14; Part III: Chapters 15–18; Part IV: Chapters 19–21; and Chapter 22. Let us explain our rationale for grouping these chapters together in this way.

Chapter 1 serves as an introduction by defining comprehensive applied mathematical modeling and explaining the rationale for basing a textbook on this subject. Chapter 2 examines a prototype projectile problem related to determining the escape velocity of the Earth and serves as a template for the multistep process of comprehensive applied mathematical modeling described in Chapter 1, while the rest of Part I concentrates on modeling phenomena involving nonmoving continua using predominantly linear stability theory techniques although global and nonlinear stability methods are discussed. In particular, a predator–prey mite system on fruit trees, soap films, heat conduction in finite and infinite bars, slime mold aggregation, and chemical Turing patterns are investigated in Chapters 3–8, respectively. The pastoral interludes in this part include the introduction of the concepts of regular and singular perturbation theory, the Calculus of Variations, linear and global stability methods applied to dynamical systems, vector integral relations, similarity solutions, Laplace transforms, and Fourier series and integrals.

Part II concerns the modeling of moving continua. Hence the governing equations and boundary conditions appropriate for describing such continua in fluid mechanics are developed in Chapters 9 and 10, respectively, while Chapters 11 and 12 concentrate on modeling two inviscid (subsonic sound waves and potential flow past a cylinder) and 13 and 14, two viscous (Couette–Poiseuille flows and Blasius flow past a flat plate) fluid situations, respectively. In particular, the pastoral interludes in this part introduce characteristic coordinates and Leibniz's Rule of Differentiation, develop Legendre polynomials, and revisit the concepts of the Calculus of Variations and singular perturbation theory.

Part III is concerned with the application of nonlinear stability theory to four models appropriate to investigating pattern formation for a buoyancy-driven viscous fluid convection layer, heat conduction in a finite bar with a source through terms of fifth-order, laser light in an optical gaseous ring cavity, and vegetation distributed over a flat arid environment, these case studies comprising Chapters 15–18, respectively.

Part IV consists of a variety of topics offered for the purpose of tying up some loose ends. Chapter 19 revisits the Calculus of Variations for constrained optimization as well as dynamical systems and examines gamma and Bessel functions some of which are included in pastoral interludes. Chapter 20 uses Laplace and Fourier transform methods to solve heat and wave equation problems solved in earlier chapters by other means while introducing the Dirac delta function in the last pastoral interlude.

Asymptotic series and techniques for finding such representations for integrals with a large parameter as applied to approximating the special functions appearing in solutions is a recurring theme throughout Chapters 2–20. Although a lion's share of the chosen mathematical models deals with the continuum mechanical treatment of natural and engineering science phenomena in these chapters and in five of the six capstone problems in concluding Chapter 22, some discrete-type models of life, behavioral, and social science phenomena are also introduced both for the sake of completeness in Chapter 21 and in the last capstone problem in Chapter 22 and out of necessity when appropriate (*e.g.*, recursion relations often arising in the analysis of continuous-type models are actually finite difference equations and numerical methods applied to such models involve equations of this sort as well).

The book concludes with the six capstone problems of Chapter 22 mentioned above that are related to the application of comprehensive applied mathematical modeling techniques to phenomena in the areas of current interest of virology, cosmology, chemically reactive flow, complexity, quantitative finance, and ecological conservation, respectively, each of which synthesizes several concepts presented in earlier chapters. Finally, certain topics initially appear in a simplified intuitive manner (*e.g.*, perturbation techniques, the Calculus of Variations, and stability methods) and are only later developed in greater theoretical detail as required for more complicated applications. The advanced topics class in mathematical modeling included four slide shows based on research results. Here the main emphasis was on the comparison of theoretical model-based predictions with experimental or field data for the phenomena under examination. Chapters 3, 15, 17, and 18 contain more complicated material introduced in the spirit of these slide shows for the purpose of enrichment.

The case studies presented and their mode of presentation draw liberally from what Wollkind learned from those five individuals who have most influenced him professionally: These are Lee Segel, Richard C. DiPrima, George H. Handelman, and Donald S. Cohen, from his time at RPI and his postdoctoral supervisor Harry L. Frisch at the State University of New York at Albany. Lee Segel's contribution in this regard is catalogued above, while that of the other four are as follows: George Handelman in his Foundations of Applied Mathematics course showed that mathematics could be used effectively to describe real-world phenomena particularly in the field of elasticity; Richard DiPrima in his Topics in Applied Mathematics course dovetailed perturbation techniques and stability methods with the governing equations of fluid mechanics to model viscous flow problems realistically; Don Cohen in his Advanced Methods in Applied Mathematics course introduced the concepts of Green's functions, distributions, and the use of asymptotic methods to analyze the special functions of mathematical physics; and Harry Frisch demonstrated the power of employing a variety of mathematical techniques to understand physical chemistry from first principles.

Audience

The book is intended to be used as a text in a two-semester sequence for advanced undergraduate and beginning graduate students, requiring rudimentary knowledge of advanced calculus and differential equations, along with a basic understanding of some simple physical and biological scientific principles. In particular, the mathematical methods developed in the "pastoral interludes" assume familiarity, for example, with Edwards [33] and Boyce and DiPrima [11] as prerequisites. The students, who took Wollkind's WSU courses described above as providing the original source material for this book, were from mathematics, physics, forestry, chemistry, geology, metallurgy, natural resources, entomology, biology, agriculture, veterinary medicine and biological systems, civil, electrical, mechanical, and computer engineering departments. They employed methodologies learned in those classes for their other advanced courses and dissertation research, much of which included applications of comprehensive applied mathematical modeling techniques.

The book could also be used as a text in a one-semester such course by an instructor picking and choosing the appropriate topics. In particular, exclusive of the slide shows just mentioned Wollkind's one-semester advanced comprehensive modeling course included Chapters 2, 3, 4, 5, 8, 9, 10.1, 10.2, 15, 16, and 21; his one-semester partial differential equations of mathematical physics course, Chapters 5, 6, 9 (restricted to inviscid fluids), 10.1, 11, 12, 16, 19, and 20; the second part of his applied analysis course, Chapters 12.6, 12.7, 19.5, 19.6, 19.7, 19.8, 19.9, and 20.2 while its first part came predominantly from Friedman [37] and Greenberg [45]; and finally the two-semester continuum mechanics course, Chapters 2, 3.2, 5, 6 (applied to Rayleigh impulsive flow of a viscous fluid in a semi-infinite channel driven by a lower plate moving at a constant velocity), 7, 9, 10, 11, 12, 13, 14, 15, and 16.

Acknowledgements

We would like to acknowledge the role Professor Richard A. Cangelosi has played in the preparation of this book. Rick, who also completed his Ph.D. dissertation at WSU under Wollkind's direction on mussel bed pattern formation in a quiescent marine layer containing an algae food source and is currently a faculty colleague of Dichone in the Department of Mathematics at Gonzaga University, was instrumental in the selection and completion of some of its figures and problems. Further, he has distributed certain of its chapters to the students taking his applied mathematics courses at Gonzaga University and in so doing gave the book a dry run in that pedagogical regard.

Pullman, WA, USA David J. Wollkind
Spokane, WA, USA Bonni J. Dichone

Contents

Part IV Chapters 19–22

Chapter 1
Introduction

1.1 What is Comprehensive Applied Mathematical Modeling?

Fifty years ago the Society for Industrial and Applied Mathematics (SIAM) sponsored a Conference on Education in Applied Mathematics that was held in Aspen, Colorado, from 24–27 May 1966 and hosted by the University of Denver. The proceedings of that conference were published in *SIAM Review* (1967) **9**, pp. 289–415. Chia-Chiao Lin of MIT, in his contribution to that proceedings, defined what he termed Comprehensive Applied Mathematicians as individuals who have a thorough knowledge, understanding, and appreciation of the proper relationship between mathematics and all the sciences and who professionally devote themselves to the totality of such phenomenological modeling. These people are neither pure mathematicians with an interest in certain specializations of the theoretical sciences, nor are they theoretical scientists with an acquired expertise in a given subdiscipline of mathematics closely related to their particular field. Rather, their goal is to elucidate scientific concepts and describe specific phenomena through the use of mathematics. Lin divided this process of using mathematics for increasing scientific understanding into the following sequential efforts: (a) the formulation of the scientific problems in mathematical terms, (b) the "solution" of the resultant mathematical problems, and (c) the discussion, interpretation, and evaluation of the results of that analysis, including its empirical verification in scientific terms. Here, the term "solution" is being employed to describe either an approximate or exact solution of the original mathematical model or a simplified version of the latter. Indeed it is often necessary to introduce simplifications into the original mathematical model system in order to make the resulting system more tractable. This procedure of reducing a model to a skeletal form, that still preserves the salient features of the original system in terms of making predictions relevant to the observational or experimental data of the phenomenon under examination, is a delicate operation. In fact, all three of these efforts as described by C. C. Lin are more of an art than they are a science, as pointed out by George F. Carrier of Harvard University in his contribution to those

© Springer International Publishing AG, part of Springer Nature 2017

D. J. Wollkind and B. J. Dichone, *Comprehensive Applied Mathematical Modeling in the Natural and Engineering Sciences*,
https://doi.org/10.1007/978-3-319-73518-4_1

proceedings. Carrier went on to say that, above all, such an endeavor involved the judgment to permit an intelligent compromise between the adequacy of the mathematical model adopted and the tractability of the mathematical problems so posed, which leads eventually to a result that can be interpreted usefully in the context of the original question. Raymond D. Mindlin of Columbia University described this complete cycle as: The confrontation of the physical problem, the construction of the mathematical model, the examination of the model to produce the solution of a particular question, the prediction of a critical result, and a satisfactory laboratory test of that result. It is not being claimed here that C. C. Lin's proposing this multi-step process to investigate a particular phenomenon was especially new in that Pólya [97] described a similar four-step process for solving complex problems as early as 1945 (see Section 21.2). What was novel about the concept of comprehensive applied mathematical modelers as coined by Lin is that since it refers to individuals concerned with the totality of such phenomenological modeling across a large number of scientific and engineering disciplines, the term comprehensive is being used in both the horizontal and vertical sense.

A decade later, the American Mathematical Society (AMS) Committee on Employment and Education Policy sponsored a panel discussion entitled "The Role of Applications in Ph.D. Programs in Mathematics" on 21 August 1975 at its Summer Meetings in Kalamazoo, Michigan, hosted by Western Michigan University. The texts of the panelists' presentations were published in *AMS Notices* (1976) **23**, pp. 158–164. In his presentation, Henry O. Pollak of Bell Laboratories introduced the schematic Venn diagram depicted in Fig. 1.1 of what he meant by applications of mathematics: Take a region and label it "Mathematics" (by which is meant Stephen Wolfram's definition of that term as encompassing all studies of abstract systems or Keith Devlin's as the science of patterns both real and imagined). Within that region, there is a certain subset corresponding to "Classical Applied Mathematics" – *i.e.*, those parts of mathematics traditionally recognized as applied – *e.g.*, mathematical physics. There is another subset of "Mathematics" that Pollak termed "Applicable Mathematics" for which significant practical applications are known. When included in the Venn diagram, "Applicable Mathematics" has a fair amount of overlap with "Classical Applied Mathematics," although the former by no means contains all of the latter and it includes a lot of other mathematics as well. For instance, the whole subject area of differential equations may be classified as "Applicable Mathematics" while some of its existence and uniqueness theorems are not part of "Classical Applied Mathematics," nor are some of the phenomenological interpretations of the partial differential equations of mathematical physics included in "Classical Applied Mathematics" generally considered a part of "Applicable Mathematics." Pollak then stated that both of these subsets are used as definitions of applied mathematics which is part of the reason that there are so many arguments about what is meant by this term. Now draw another region disjoint from "Mathematics" and label it "Rest of the World." Pollak stated that there was a third definition of applied mathematics: Namely, starting out with a problem in the "Rest of the World" and making your way over to those subsets of "Mathematics" described above, you construct and analyze a mathematical model relevant to that problem, and finally, having developed some

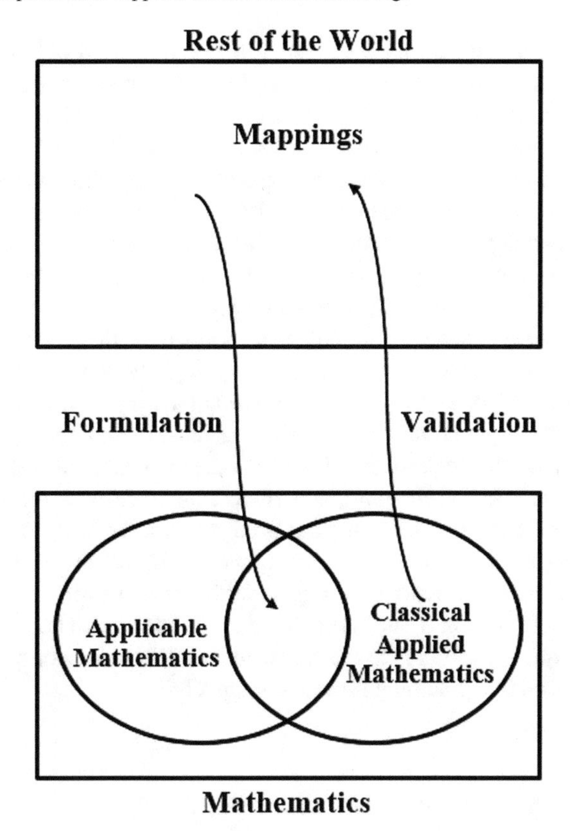

Fig. 1.1 Schematic Venn diagram representing H. O. Pollak's idea of the processes involved in comprehensive applied mathematical modeling.

understanding of the original problem from this mathematical analysis, return to the "Rest of the World." Applied mathematics in this sense not only is a subset of "Mathematics," but also includes these two mappings, one from and one back to the "Rest of the World." Pollak next asserted that this is the definition many university instructors teaching applied mathematics courses try to use, although they often do a very poor job of explaining the two mappings. Such a hypothetical instructor might begin by mouthing a few words from another discipline and then say "consider the following partial differential equation" without explaining how one got from the real world to that equation, what was kept or thrown away, and what the resulting solution means physically; hence, in effect, ignoring these mappings to and from the real world. Pollak lastly posed a fourth definition of applied mathematics, that he actually preferred, which was an iterative one using this process sequentially. That is, you start out with a situation in the real world, map it into a mathematical model, analyze this model, and come back to the real world only to find out that the resulting predictions are nonsense – *i.e.*, they don't make sense in terms of the original physical, life, or engineering science situation. So you keep trying by going around the loop between the real world and a modified mathematical model as many times as it takes to obtain good agreement between the resulting predictions and the observational or experimental data of the relevant phenomenon.

In summary, a comprehensive applied mathematician starts with a particular scientific phenomenon, abstracts to a mathematical model using fundamental first principles, introduces whatever simplifications are necessary to make the model tractable, analyzes that modified model by employing a variety of analytical or numerical techniques, and then compares the resulting theoretical predictions of that analysis with existing observational or experimental data relevant to the phenomenon under examination. Hence these phenomenological observables are used both to formulate and validate the model. The optimal situation is that the theoretical modeling predictions provide very good qualitative and quantitative agreement with those observables. Sometimes, there may also be modeling predictions for which no relevant observational data exists. The comprehensive applied mathematician can then use these unexpected modeling predictions to suggest further observation or experimentation that might be performed in order to provide their verification. The main purpose for obtaining such quantification is to reach a deeper understanding of the phenomenon under investigation by explaining it more completely.

1.2 What is the Rationale for this Book?

At the Aspen conference mentioned above, there was a general consensus that the best way to expose students to this sort of comprehensive applied mathematics would be to present a series of phenomenological modeling case studies. There was also a feeling that the best approach for doing so would be one that concentrated on continuum mechanics. C. C. Lin stated that continuum mechanics had a favored place in comprehensive applied mathematical modeling for the following three reasons: (i) the physical concepts are simple; (ii) the mathematical problems involved are interesting and of wide applicability; and (iii) a large number of phenomena in nature fall within its scope. He asserted that the simplicity of physical concept had two advantages: Namely, it is easier for students to learn and the interesting observed phenomena could be more directly associated with mathematical reasoning. According to Lin, nonlinearity is perhaps one of the chief mathematical properties that characterize the interesting problems of continuum mechanics. Elliott W. Montroll of the University of Rochester and Julian D. Cole of the California Institute of Technology suggested that the application of stability methods and perturbation techniques, respectively, to such nonlinear systems was a good vehicle for presenting comprehensive applied mathematical modeling concepts. What was needed in order to accomplish these educational goals in Montroll's opinion were a set of books, or probably better, clearly written lecture notes that included such case studies where applications of this sort had been made.

This book was designed with that goal in mind. As described in the preface, it consists of twenty-one chapters in addition to this introductory one with a lion's share of the case studies chosen involving mathematical models, after C. C. Lin's suggestion, dealing with the continuum mechanical treatment of natural and engineering science phenomena (Chapters 2–20 and five of the capstone problems of Chapter 22) although some discrete-type models of life, behavioral, and social sciences are also introduced in both Chapter 21 and the sixth capstone problem of Chapter 22. We have tried to make this presentation relatively self-contained by developing the required mathematical methods as pastoral interludes introduced in parallel with the models, which have been deduced from scientific first principles, as much as possible. Further, to reinforce and supplement this material, we have provided problems closely related to those case studies. Moreover, each of these chapters opens with a paragraph containing signposts to point the way toward what it includes and how this dovetails with material presented in other chapters, as well as to offer a rationale for that inclusion. We have also included more algebraic detail in our developments than is usually retained in traditional textbooks. Those intermediate steps have been provided to aid both prospective instructors and their students of advanced modeling courses in following these developments. From our pedagogical experience, we have found that it is far easier for students to learn and replicate material presented in this manner rather than if a more concise but esoteric approach had been employed instead. In writing our book, we have endeavored to promote comprehensive applied mathematical modeling as an exciting voyage of discovery that follows in the footsteps of its original pioneers and hopefully will help others add to its continuing legacy in the future. The book is intended to be used as a text in a two-semester sequence for advanced undergraduate and beginning graduate students in the natural and engineering sciences, requiring rudimentary knowledge of advanced calculus and differential equations along with a basic understanding of some simple physical and biological scientific principles. It could also be used as a text in a one-semester such course by an instructor picking and choosing the appropriate topics as indicated in the preface.

Part I
Chapters 2–8

Chapter 2
Canonical Projectile Problem: Finding the Escape Velocity of the Earth

In this chapter, the escape velocity of the Earth is calculated using an idealized projectile model that allows for the determination of projectile velocity as a function of its altitude. In order to obtain an approximate solution for projectile altitude as a function of time which cannot be determined as an exact solution, the concept of regular perturbation theory in ordinary differential equations is introduced as a pastoral interlude. The proper nondimensionalization of the governing projectile equation and initial conditions produces a formulation involving a single dimensionless parameter that is small whenever the projectile initial velocity is much less than that escape velocity. Then a regular perturbation expansion in this parameter is performed on the model to obtain the desired asymptotic solution of altitude as a function of time under that condition. An energy argument, making use of the fact that gravity acts as a conservative force for this canonical model, is also introduced to examine that phenomenon in more detail. The problems extend these analyses to the rest of the solar system planets and to two other canonical projectile problems that are nonconservative.

2.1 Newton's Second Law and the Basic Governing Projectile Equation of Motion

Consider a projectile of mass m that is shot vertically upward from the "surface" of the Earth with an initial velocity V_0 and assumed to be acted upon only by the force of gravity (see Fig. 2.1). Now let $t \equiv$ time and $x(t) \equiv$ altitude of the projectile. Then from Newton's second law of motion it follows that

$$m\boldsymbol{a}(t) = \boldsymbol{G}(t), \ t > 0, \tag{2.1.1a}$$

where

$$\boldsymbol{a}(t) = \frac{d^2x(t)}{dt^2}\boldsymbol{k} \equiv \text{ acceleration and } \boldsymbol{G}(t) = G(t)\boldsymbol{k} \equiv \text{ gravitational force} \tag{2.1.1b}$$

© Springer International Publishing AG, part of Springer Nature 2017

D. J. Wollkind and B. J. Dichone, *Comprehensive Applied Mathematical Modeling in the Natural and Engineering Sciences*,
https://doi.org/10.1007/978-3-319-73518-4_2

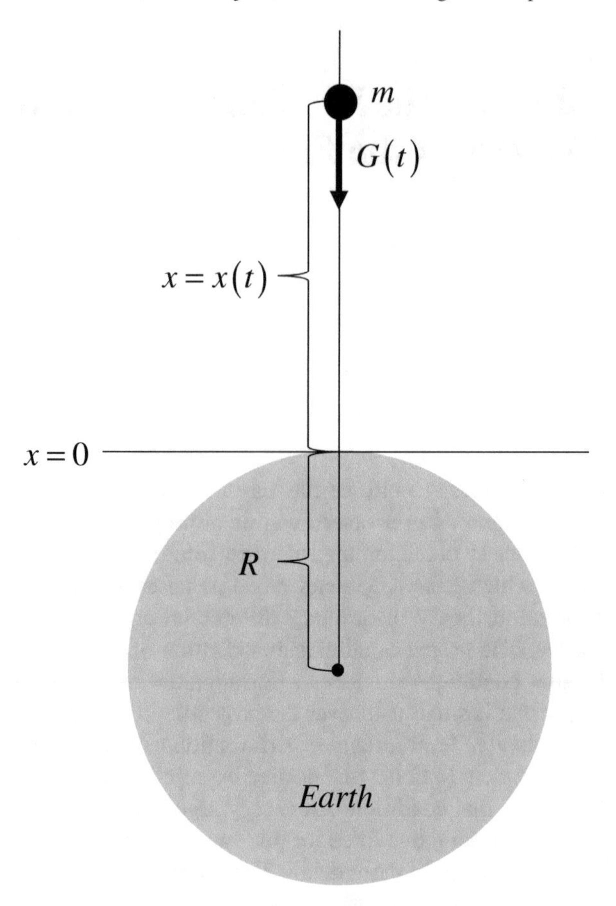

Fig. 2.1 Schematic plot of the projectile problem where m and $x = x(t)$ are the projectile mass and altitude, respectively; $R \equiv$ Earth's radius; and $G(t) \equiv$ vertical gravitation force while $t \equiv$ time.

for $\boldsymbol{k} \equiv$ a unit vector in the vertical direction. Here $\boldsymbol{v}(t) = v(t)\boldsymbol{k} = \frac{dx(t)}{dt}\boldsymbol{k} \equiv$ velocity. Hence our projectile problem has initial conditions

$$x(0) = 0 \text{ and } v(0) = V_0. \tag{2.1.1c}$$

(a) For comparison purposes to the actual case of the variation of the gravitational force with altitude, we shall first assume that this force is a constant. That is we take

$$G(t) \equiv -mg \tag{2.1.2a}$$

where

$$g \equiv \text{ acceleration due to gravity at sea level.} \tag{2.1.2b}$$

Thus our constant gravity case reduces to solving the second-order nonlinear ordinary differential equation initial value problem

$$\frac{d^2x(t)}{dt^2} = \frac{dv(t)}{dt} = -g;\ x(0) = 0,\ \frac{dx(0)}{dt} = v(0) = V_0. \tag{2.1.3}$$

To deduce a relationship between g and the so-called gravitational constant $G_0 = 6.67 \times 10^{-11}$ m^3/(sec^{2}kg) we apply Newton's universal law of gravitation to a body of mass m on the surface of a spherical Earth with mass M and radius R obtaining

$$mg = G_0 m \frac{M}{R^2} \tag{2.1.4a}$$

or

$$g = G_0 \frac{M}{R^2}. \tag{2.1.4b}$$

Since M is related to the mean density of the Earth ρ_0 by

$$M = \rho_0 \left(\frac{4\pi R^3}{3} \right), \tag{2.1.4c}$$

substitution of (2.1.4c) into (2.1.4b) yields

$$g = 4\pi G_0 \rho_0 \frac{R}{3}. \tag{2.1.4d}$$

Employing the appropriate values for G_0 and

$$\rho_0 = 5.5 \times 10^3 \text{ kg/m}^3,\ R = 6,378 \text{ km} \tag{2.1.5a}$$

from Table 2.1 [108] in (2.1.4d), one arrives at the traditional acceleration due to gravity at sea level for the Earth

$$g = 9.8 \text{ m/sec}^2 = 32 \text{ ft/sec}^2. \tag{2.1.5b}$$

Table 2.1 The values for mean density and equatorial radius of the 8 planets of the solar system

Planet	Mercury	Venus	Earth	Mars	Jupiter	Saturn	Uranus	Neptune
ρ_0 (10^3 kg/m^3)	5.4	5.2	5.5	3.9	1.3	0.7	1.2	1.7
R (10^3 km)	2.440	6.052	6.378	3.394	71.400	60.000	25.900	24.750

Solving the constant gravitational force problem of (2.1.3) by direct integration we find that

$$v(t) = \frac{dx(t)}{dt} = -gt + V_0 \text{ and } x(t) = \frac{-gt^2}{2} + V_0 t. \tag{2.1.6}$$

Thus

$$v(t_m) = 0 \Rightarrow t_m = \frac{V_0}{g} \text{ and } x_m = x(t_m) = \frac{V_0^2}{2g} \tag{2.1.7a}$$

while

$$x(T) = 0 \text{ for } T > 0 \Rightarrow T = 2t_m \text{ and } v(T) = -V_0. \tag{2.1.7b}$$

Hence

$$\max_{0\le t\le T} |x(t)| = \frac{V_0^2}{2g}, \max_{0\le t\le T}\left|\frac{dx(t)}{dt}\right| = V_0, \max_{0\le t\le T}\left|\frac{d^2x(t)}{dt^2}\right| = g. \tag{2.1.8}$$

Finally, from (2.1.6) we can deduce the relationship that

$$v^2(t) - V_0^2 = -2gx(t). \tag{2.1.9}$$

We could also derive this result directly as follows: First invert $x = x(t)$ to obtain $t = t(x)$ and then introduce the change of variables

$$V(x) = v[t(x)]. \tag{2.1.10}$$

Since

$$\frac{dv[t(x)]}{dt} = \frac{dV(x)}{dt} = \left\{\frac{dV(x)}{dx}\right\}\left\{\frac{dx[t(x)]}{dt}\right\} = \left\{\frac{dV(x)}{dx}\right\}V(x) \tag{2.1.11a}$$

and

$$V(x)\frac{dV(x)}{dx} = \frac{1}{2}\frac{dV^2(x)}{dx} \tag{2.1.11b}$$

problem (2.1.3) now transforms into the first-order linear differential equation for $V^2 = V^2(x)$

$$\frac{dV^2(x)}{dx} = -2g, x > 0; V^2(0) = V_0^2. \tag{2.1.12}$$

Upon integration, this yields the solution

$$V^2(x) = -2gx + V_0^2, \tag{2.1.13}$$

which is equivalent to the result of (2.1.9).

(b) We shall now explicitly take the variation of gravitational force with altitude into account by assuming that

$$G(t) = F[x(t)] \tag{2.1.14a}$$

where from Newton's universal law of gravitation (see Fig. 2.1)

$$F(x) = -G_0 m\frac{M}{(x+R)^2} \tag{2.1.14b}$$

or, employing (2.1.4b) in (2.1.14b),

$$F(x) = -mg\frac{R^2}{(x+R)^2}. \tag{2.1.14c}$$

Thus our variation of gravitational force with altitude case reduces to solving the second-order nonlinear ordinary differential equation initial value problem

$$\frac{d^2x(t)}{dt^2} = \frac{dv(t)}{dt} = -g\frac{R^2}{[x(t)+R]^2}; \; x(0) = 0, \; \frac{dx(0)}{dt} = v(0) = V_0. \tag{2.1.15}$$

2.2 Exact Solution Method Involving Velocity as a Function of Altitude

Given the nature of the nonlinearity appearing in the differential equation of (2.1.15), it cannot be solved to obtain a closed-form expression for either $v(t)$ or $x(t)$ as was done for the constant gravity case of (2.1.3). Hence in order to obtain the desired escape velocity of the Earth we must now introduce the change of variables of (2.1.10) and transform (2.1.15) into the first-order separable differential equation for $V^2 = V^2(x)$

$$\frac{dV^2(x)}{dx} = -2g\frac{R^2}{(x+R)^2}, \; x > 0; \; V^2(0) = V_0^2. \tag{2.2.1}$$

Note, that whereas this change of variables was introduced into the problem of (2.1.3) for the sake of convenience in obtaining (2.1.13), it is being introduced in this case out of necessity. Solving the variation of gravitational force with altitude problem of (2.2.1) by direct integration yields

$$V^2(x) = 2g\frac{R^2}{x+R} + V_0^2 - 2gR. \tag{2.2.2}$$

Observe that for

$$V_0^2 \geq 2gR \text{ or } V_0 \geq \sqrt{2gR} = V_e, \tag{2.2.3}$$

$V^2(x) > 0$ for all $x > 0$ and hence there exists no x_m such that $V^2(x_m) = 0$. Thus the projectile does not return to Earth. The smallest value for V_0 satisfying this condition is termed the escape velocity and has been denoted by V_e in (2.2.3). Let us calculate the Earth's V_e. Although the metric system will predominantly be employed elsewhere throughout the book, here we shall use the British engineering system of units by choosing that version for g of (2.1.5b) and taking

$$R = 4,000 \text{ mi}, \tag{2.2.4}$$

since this yields the following algebraically convenient form for

$$V_e = \sqrt{2gR} = \sqrt{64\frac{\text{ft}}{\text{sec}^2}4,000 \text{ mi}\frac{1\text{mi}}{5,280 \text{ ft}}} \cong \sqrt{\frac{(64)(400)}{529}}\frac{\text{mi}}{\text{sec}} = \frac{160}{23}\frac{\text{mi}}{\text{sec}} \cong 6.96\frac{\text{mi}}{\text{sec}}. \tag{2.2.5}$$

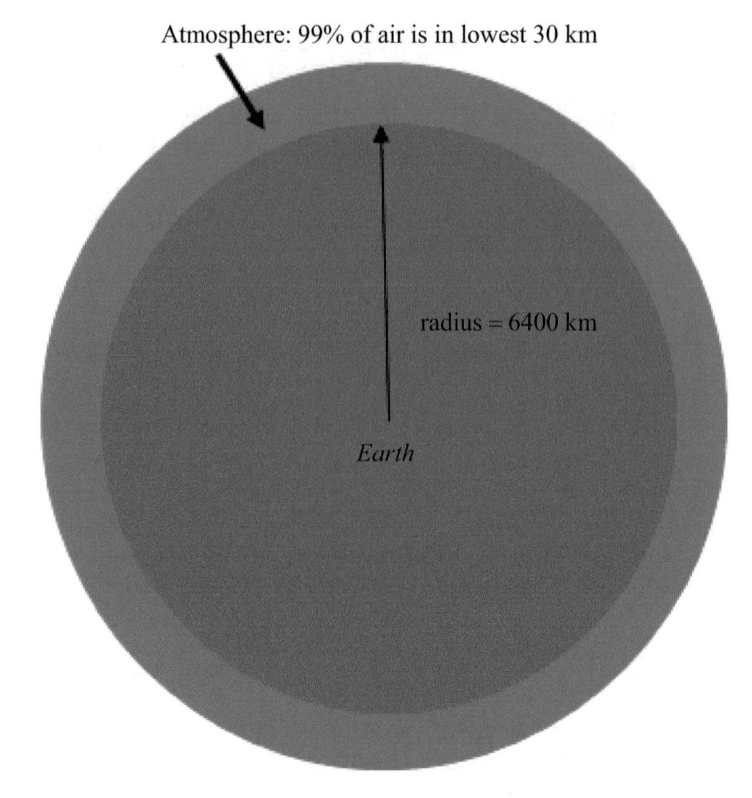

Fig. 2.2 Depiction of the proto-Earth consisting of the terrestrial planet plus its atmosphere based upon a Lyndon State College Atmospheric Sciences numerical example of the atmospheric depth versus the radius of the Earth. In this context note that 4,000 mi = 6437 km.

The actual escape velocity of the Earth is $V_e \cong 7$ mi/sec = 11.3 km/sec. Therefore, in spite of all the simplifications introduced in the model – *e.g.*, the neglect of air resistance and rocketry effects, it does an extremely good job of predicting the escape velocity of the Earth. To see why this is true, we need only examine our prediction within the framework of the model. For a real rocket with several stages each one provides an acceleration until the final stage reaches the top of the atmosphere and achieves its terminal velocity of V_0 after which the rocket is shut off. There is no air resistance at that point nor any rocketry effects, and the R in question is the effective radius of a proto-Earth consisting of the terrestrial planet plus its atmosphere (see Fig. 2.2). Since the R employed in Table 2.1 for the Earth is equivalent to 3,963 miles, our value of 4,000 miles employed in (2.2.4) corresponds to an atmospheric thickness of 37 miles which is about 6 miles above the top of the stratosphere. This explains the accuracy of our prediction given that it is standard operating procedure for the purpose of computing the escape velocity to consider g as having the constant value of (2.1.5b) in the whole atmospheric column. Note in this context that the employment of such an effective radius basically entails viewing the four terrestrial planets consisting of Mercury, Venus, Earth, and Mars in the same way that the four Jovian ones consisting of the two gas giants Jupiter and Saturn and the two ice giants Uranus and Neptune are viewed when determining the proper R-value for them in Table 2.1. The use of an effective radius in this computation is why we placed the word "surface" of the Earth in quotation marks when describing our projectile problem in Section 2.1.

Finally, should

$$V_0^2 < 2gR = V_e^2, \tag{2.2.6}$$

we can always find a value x_m which corresponds to the maximum projectile height such that

$$V^2(x_m) = 2g\frac{R^2}{x_m + R} + V_0^2 - 2gR = 0 \tag{2.2.7a}$$

$\Rightarrow$

$$\frac{V_e^2}{1 + x_m/R} = V_e^2 - V_0^2. \tag{2.2.7b}$$

Then, defining the dimensionless parameter

$$0 < \varepsilon = \frac{V_0^2}{V_e^2} = \frac{V_0^2}{2gR} < 1 \tag{2.2.7c}$$

we find that

$$x_m = \frac{R\varepsilon}{1 - \varepsilon}. \tag{2.2.7d}$$

For the special case of

$$0 < \varepsilon << 1, \tag{2.2.8a}$$

this reduces to

$$x_m \cong R\varepsilon = \frac{V_0^2}{2g} \tag{2.2.8b}$$

in agreement with the constant gravity result of (2.1.7a). This correspondence type principle indicates the variation of gravitational force with altitude does not have to be taken into account should (2.2.8a) be valid for a particular projectile problem.

2.3 Selection of Scale Factors and Introduction of Nondimensional Variables

For the purpose of this section we first represent the differential equation problem of (2.1.15) in the following form where $x^* = x^*(t^*) \equiv$ dimensional projectile altitude and $t^* \equiv$ dimensional time

$$\frac{d^2x^*}{dt^{*2}} = -g\frac{R^2}{(x^*+R)^2},\ t^* > 0;\ x^*(0) = 0,\ \frac{dx^*(0)}{dt^*} = V_0. \tag{2.3.1}$$

Consider a general second-order ordinary differential equation with dependent variable $x^*(t^*)$ where the independent variable $t^* \in I^*$, I^* being a finite, semi-infinite, or infinite interval. We wish to select scale factors T_0, X_0 and introduce *nondimensional variables*

$$t = \frac{t^*}{T_0},\ x(t) = \frac{x^*(t^*)}{X_0}, \tag{2.3.2a}$$

such that

$$\max_{t\in I}|x(t)| = 1,\ \max_{t\in I}\left|\frac{dx(t)}{dt}\right| \le 1,\ \max_{t\in I}\left|\frac{d^2x(t)}{dt^2}\right| \le 1 \tag{2.3.2b}$$

where $t \in I$ if an only if $t^* \in I^*$. Then

$$x^*(t^*) = X_0 x(t) \text{ and } |x^*(t^*)| = X_0|x(t)| \tag{2.3.3a}$$

$\Rightarrow$

$$\max_{t^*\in I^*}|x^*(t^*)| = X_0 \max_{t\in I}|x(t)| = X_0; \tag{2.3.3b}$$

while

$$\frac{dx^*(t^*)}{dt^*} = X_0\frac{dx(t)}{dt^*} = X_0\frac{dx(t)}{dt}\frac{dt}{dt^*} = \frac{X_0}{T_0}\frac{dx(t)}{dt} \tag{2.3.4a}$$

$\Rightarrow$

$$\max_{t^*\in I^*}\left|\frac{dx^*(t^*)}{dt^*}\right| = \frac{X_0}{T_0}\max_{t\in I}\left|\frac{dx(t)}{dt}\right| \le \frac{X_0}{T_0} \tag{2.3.4b}$$

$\Rightarrow$

$$T_0 \le T_c^{(1)} = \frac{X_0}{\max_{t^*\in I^*}|dx^*(t^*)/dt^*|} \tag{2.3.4c}$$

and

$$\frac{d^2x^*(t^*)}{dt^{*2}} = \frac{X_0}{T_0}\frac{d}{dt^*}\frac{dx(t)}{dt} = \frac{X_0}{T_0^2}\frac{d^2x(t)}{dt^2} \tag{2.3.5a}$$

$\Rightarrow$

$$\max_{t^*\in I^*}\left|\frac{d^2x^*(t^*)}{dt^{*2}}\right| = \frac{X_0}{T_0^2}\max_{t\in I}\left|\frac{d^2x(t)}{dt^2}\right| \le \frac{X_0}{T_0^2} \tag{2.3.5b}$$

$\Rightarrow$

$$T_0 \le T_c^{(2)} = \sqrt{X_0\Big/\left[\max_{t^*\in I^*}\left|\frac{d^2x^*(t^*)}{dt^{*2}}\right|\right]}. \tag{2.3.5c}$$

Now, for the inequalities of (2.3.4c) and (2.3.5c) to hold simultaneously, we must have

$$T_0 \le T_c = \min\{T_c^{(1)}, T_c^{(2)}\}. \tag{2.3.6a}$$

In order to make this estimate as sharp as possible we take the largest value of T_0 that satisfies (2.3.6a) and hence select

$$T_0 = T_c. \tag{2.3.6b}$$

We now wish to select scale factors X_0 and T_0 for our projectile problem of (2.3.1) and introduce nondimensional variables as defined by (2.3.2). Observe that formulae (2.3.3b) and (2.3.6b) for these scale factors involve precisely the solution $x^*(t^*)$ we cannot evaluate. Hence in what follows we shall assume that

$$V_0^2 << V_e^2 = 2gR. \tag{2.3.7}$$

Under this condition which is equivalent to our special case of (2.2.8) it is permissible for us to use the solution of the constant gravity problem in order to find X_0 and T_0. Thus substituting the results of (2.1.8) into (2.3.3b), (2.3.4c), and (2.3.5c) yields

$$X_0 = \frac{V_0^2}{2g} \tag{2.3.8a}$$

and

$$T_c^{(1)} = \frac{[V_0^2/(2g)]}{V_0} = \frac{V_0}{2g}, T_c^{(2)} = \sqrt{\frac{[V_0^2/(2g)]}{g}} = \frac{V_0}{\sqrt{2}g} \tag{2.3.8b}$$

while (2.3.6) then implies that

$$T_0 = T_c = \min\left\{\frac{V_0}{2g}, \frac{V_0}{\sqrt{2}g}\right\} = \frac{V_0}{2g}. \tag{2.3.8c}$$

Now employing these scale factors of (2.3.8) in (2.3.2a), we introduce the change of variables

$$t = \frac{t^*}{[V_0/(2g)]}, x(t) = \frac{x^*(t^*)}{[V_0^2/(2g)]}. \tag{2.3.9a}$$

Upon noting from (2.3.4a) and (2.3.5a) that then

$$\frac{dx^*(t^*)}{dt^*} = \frac{[V_0^2/(2g)]}{[V_0/(2g)]}\frac{dx(t)}{dt} = V_0\frac{dx(t)}{dt} \tag{2.3.9b}$$

and

$$\frac{d^2x^*(t^*)}{dt^{*2}} = \frac{[V_0^2/(2g)]}{[V_0/(2g)]^2}\frac{d^2x(t)}{dt^2} = 2g\frac{d^2x(t)}{dt^2}, \tag{2.3.9c}$$

this change of variables transforms (2.3.1) into the dimensionless problem for $x = x(t;\varepsilon)$

$$2\ddot{x} = -(1+\varepsilon x)^{-2}, t > 0, 0 \le \varepsilon = \frac{V_0^2}{2gR} << 1; x(0;\varepsilon) = 0, \dot{x}(0;\varepsilon) = 1; \tag{2.3.9d}$$

where we are employing the notation $(\dot{-}) \equiv d(-)/dt$ to reflect the historical interchangeability of Newton's fluxion dot and Leibniz's double d notation for time derivatives.

Observe that (2.3.9d) only involves the single quantity ε which is a nondimensional combination of V_0, g, and R. Also note that this parameter which arises naturally from our proper selection of scale factors and introduction of nondimensional variables is exactly the same as the one defined in (2.2.7c), which played such a fundamental role for the correspondence principle argument of the previous section. In what follows, we shall ultimately be seeking a regular perturbation solution for this problem through terms of order ε^2 of the form

$$x(t;\varepsilon) \sim x_0(t) + \varepsilon x_1(t) + \varepsilon^2 x_2(t) \text{ as } \varepsilon \to 0. \tag{2.3.9e}$$

As a prelude to that development we next present our first so-called "pastoral interlude" dealing with this subject.

2.4 Pastoral Interlude: Regular Perturbation Theory of Ordinary Differential Equations

One says $f(\varepsilon) = O(g(\varepsilon))$ as $\varepsilon \to 0$ provided $\lim_{\varepsilon\to 0}[f(\varepsilon)/g(\varepsilon)] = C$, a constant. If $C = 0$, one says $f(\varepsilon) = o(g(\varepsilon))$ as $\varepsilon \to 0$ while if $C = 1$, one says $f(\varepsilon) \sim g(\varepsilon)$ as $\varepsilon \to 0$. Given an $f(\varepsilon)$, and an asymptotic sequence $\{\varphi_n(\varepsilon)\}_{n=0}^N$ – *i.e.*, $\varphi_{n+1}(\varepsilon) = o(\varphi_n(\varepsilon))$ as $\varepsilon \to 0$ – and a set of constants $\{a_n\}_{n=0}^N$ such that

$$\lim_{\varepsilon\to 0}\left[\frac{f(\varepsilon) - \sum_{n=0}^N a_n\varphi_n(\varepsilon)}{\varphi_N(\varepsilon)}\right] = 0,$$

we say that

$$f(\varepsilon) = \sum_{n=0}^{N} a_n\varphi_n(\varepsilon) + o(\varphi_N(\varepsilon)) \text{ as } \varepsilon \to 0$$

or

$$f(\varepsilon) \sim \sum_{n=0}^{N} a_n\varphi_n(\varepsilon) \text{ as } \varepsilon \to 0.$$

Further

$$f(\varepsilon) = \sum_{n=0}^{m} a_n\varphi_n(\varepsilon) + O(\varphi_{m+1}(\varepsilon)) \text{ as } \varepsilon \to 0 \text{ for } m = 0, 1, 2, \ldots, N-1.$$

Given an $x = x(t;\varepsilon)$ with $t \in I$ and $0 < \varepsilon << 1$, an asymptotic sequence $\{\varphi_n(\varepsilon)\}_{n=0}^N$, and a set of functions $\{x_n(t)\}_{n=0}^N$ such that $\lim_{\varepsilon\to 0}\{[x(t;\varepsilon) - \sum_{n=0}^N x_n(t)\varphi_n(\varepsilon)]/\varphi_N(\varepsilon)\} = 0$, then we say $\sum_{n=0}^N x_n(t)\varphi_n(\varepsilon)$ is an $(N+1)$-term asymptotic series for $x(t;\varepsilon)$ as first proposed by Henri Poincaré [96] or

$$x(t;\varepsilon) \sim \sum_{n=0}^{N} x_n(t)\varphi_n(\varepsilon) \text{ as } \varepsilon \to 0.$$

If that limit is uniform in t – *i.e.*, given an arbitrary $\alpha > 0$, there exists a $\delta = \delta(\alpha, N)$ such that

$$\left| \frac{x(t;\varepsilon) - \sum_{n=0}^{N} x_n(t)\varphi_n(\varepsilon)}{\varphi_N(\varepsilon)} \right| < \alpha$$

whenever $0 < \varepsilon < \delta(\alpha, N)$ – then this Poincaré-type asymptotic series is said to be uniformly valid for all $t \in I$.

A *regular perturbation problem* in ordinary differential equations involves an equation with an independent variable $t \in I$, a small parameter $0 < \varepsilon << 1$, and a dependent variable $x(t;\varepsilon)$ such that its solution possesses an asymptotic series of the Poincaré-type which is uniformly valid for all $t \in I$ as $\varepsilon \to 0$. The rule of thumb for deciding whether a particular perturbation problem exhibits such regularity is to set $\varepsilon = 0$ in it and see if the equation retains its qualitative behavior. For our problem of (2.3.9d) defining $x_0(t) = x(t;0)$, we find that

$$2\ddot{x}_0(t) = -1,\ t > 0;\ x_0(0) = 0,\ \dot{x}_0(0) = 1;$$

which represents the constant gravity case and retains the same qualitative behavior as for $\varepsilon > 0$. Hence our problem is regular and we can seek a solution for it of the form of (2.3.9e). In that event the higher-order terms merely represent refinements of $x_0(t)$. Should a problem not be regular in this sense then it is said to be singular. As an example of such a *singular perturbation problem* we shall consider the following canonical singularly perturbed boundary value problem for $y = y(x;\varepsilon)$ in the next chapter:

$$\varepsilon \frac{d^2y}{dx^2} + 2\frac{dy}{dx} + y = 0,\ 0 < x < 1,\ 0 < \varepsilon << 1;$$
$$y(0;\varepsilon) = 0,\ y(1;\varepsilon) = 1.$$

Note that $y_0(x) = y(x;0)$ satisfies a first-order differential equation

$$2\frac{dy_0}{dx} + y_0 = 0$$

and hence setting $\varepsilon = 0$ does not preserve the qualitative behavior of the original second-order differential equation involved when $\varepsilon > 0$.

2.5 Approximate Solution to the Projectile Problem Involving Regular Perturbation Theory

Returning to (2.3.9d) we observe that although this problem cannot be solved exactly for $x(t;\varepsilon)$, it is possible using the concept of regular perturbation theory in the case of sufficiently small ε to find its approximate asymptotic solution to any degree of accuracy desired. Toward that end we seek a regular perturbation solution of this problem with $\varphi_n(\varepsilon) = \varepsilon^n$ and $N = 2$ of the form

$$x(t;\varepsilon) = x_0(t) + \varepsilon x_1(t) + \varepsilon^2 x_2(t) + O(\varepsilon^3) \text{ as } \varepsilon \to 0. \tag{2.5.1}$$

Substituting this series into our basic system written in the form

$$2\ddot{x}(t;\varepsilon) + [1 + \varepsilon x(t;\varepsilon)]^{-2} = 0; \; x(0;\varepsilon) = 0, \; \dot{x}(0;\varepsilon) = 1; \tag{2.5.2}$$

we shall obtain and solve a differential equation and initial conditions for each problem proportional to ε^n for $n = 0$, 1, and 2. Since considering each term in (2.5.2) separately yields

$$2\ddot{x}(t;\varepsilon) = 2[\ddot{x}_0(t;\varepsilon) + \varepsilon\ddot{x}_1(t;\varepsilon) + \varepsilon^2\ddot{x}_2(t) + O(\varepsilon^3)] \tag{2.5.3a}$$

and, recalling the binomial series

$$(1+J)^\alpha = \sum_{n=0}^{\infty} \binom{\alpha}{n} J^n, \; |J| < 1,$$

where

$$\binom{\alpha}{n} = \frac{\alpha(\alpha-1)\cdots(\alpha-n+1)}{n!},$$

or

$$(1+J)^\alpha = 1 + \alpha J + \frac{\alpha(\alpha-1)}{2} J^2 + O(J^3)$$

for $J = \varepsilon x(t;\varepsilon)$ and $\alpha = -2$,

$$\begin{aligned}
[1 + \varepsilon x(t;\varepsilon)]^{-2} &= 1 - 2[\varepsilon x(t;\varepsilon)] + 3[\varepsilon x(t;\varepsilon)]^2 + O(\varepsilon^3) \\
&= 1 - 2\varepsilon[x_0(t) + \varepsilon x_1(t) + O(\varepsilon^2)] + 3\varepsilon^2[x_0(t) + O(\varepsilon)]^2 + O(\varepsilon^3) \\
&= 1 - 2\varepsilon x_0(t) + \varepsilon^2[3x_0^2(t) - 2x_1(t)] + O(\varepsilon^3);
\end{aligned} \tag{2.5.3b}$$

with initial conditions

$$x(0;\varepsilon) = x_0(0) + \varepsilon x_1(0) + \varepsilon^2 x_2(0) + O(\varepsilon^3) = 0 \tag{2.5.3c}$$

and

$$\dot{x}(0;\varepsilon) = \dot{x}_0(0) + \varepsilon\dot{x}_1(0) + \varepsilon^2\dot{x}_2(0) + O(\varepsilon^3) = 1; \tag{2.5.3d}$$

we obtain

$$O(1): \ddot{x}_0(t) = -1/2; x_0(0) = 1, \dot{x}_0 = 0; \tag{2.5.4a}$$

$$O(\varepsilon): \ddot{x}_1(t) = x_0(t); x_1(0) = \dot{x}_1 = 0; \tag{2.5.4b}$$

$$O(\varepsilon^2): \ddot{x}_2(t) = x_1(t) - (3/2)x_0^2(t); x_2(0) = \dot{x}_2(0) = 0. \tag{2.5.4c}$$

Now, solving these problems of (2.5.4) sequentially, we find

$$\dot{x}_0(t) = 1 - t/2; x_0(t) = t - t^2/4; \tag{2.5.5a}$$

$$\ddot{x}_1(t) = t - t^2/4, \dot{x}_1(t) = t^2/2 - t^3/12, x_1(t) = t^3/6 - t^4/48; \tag{2.5.5b}$$

$$\begin{aligned}\ddot{x}_2(t) = -3t^2/2 + 11t^3/12 - 11t^4/96,&\\ \dot{x}_2(t) = -t^3/2 + 11t^4/48 - 11t^5/480,&\\ x_2(t) = -t^4/8 + 11t^5/240 - 11t^6/2{,}880.&\end{aligned} \tag{2.5.5c}$$

Thus

$$x(t;\varepsilon) \sim t - t^2/4 + \varepsilon(t^3/6 - t^4/48) + \varepsilon^2(-t^4/8 + 11t^5/240 - 11t^6/2{,}880) \text{ as } \varepsilon \to 0. \tag{2.5.6}$$

Further

$$\dot{x}(t;\varepsilon) \sim 1 - t/2 + \varepsilon(t^2/2 - t^3/12) + \varepsilon^2(-t^3/2 + 11t^4/48 - 11t^5/480) \text{ as } \varepsilon \to 0. \tag{2.5.7}$$

We wish to determine a $t_m(\varepsilon)$ such that $\dot{x}(t_m;\varepsilon) = 0$ or

$$1 - t_m/2 + \varepsilon(t_m^2/2 - t_m^3/12) + \varepsilon^2(-t_m^3/2 + 11t_m^4/48 - 11t_m^5/480) + O(\varepsilon^3) = 0. \tag{2.5.8}$$

Now, taking the limit of (2.5.8) as $\varepsilon \to 0$, we can deduce that $t_m(0) = 2$, which is consistent with $t_m^* = (V_0/(2g))t_m$ of (2.3.9a) and $\lim_{\varepsilon\to 0} t_m^* = V_0/g$ of (2.1.7a). Thus, we assume that

$$t_m(\varepsilon) = 2[1 + a_1\varepsilon + a_2\varepsilon^2 + O(\varepsilon^3)]. \tag{2.5.9}$$

Then substituting (2.5.9) into (2.5.8) and collecting terms, we find that

$$\begin{aligned}1 - [1 + a_1\varepsilon + a_2\varepsilon^2 + O(\varepsilon^3)] + \varepsilon(2[1 + a_1\varepsilon + O(\varepsilon^2)]^2&\\ - (2/3)[1 + a_1\varepsilon + O(\varepsilon^2)]^3) + \varepsilon^2[-4 + 11/3 - 11/15 + O(\varepsilon)] + O(\varepsilon^3)&\\ = -[a_1\varepsilon + a_2\varepsilon^2 + O(\varepsilon^3)] + \varepsilon(2[1 + 2a_1\varepsilon + O(\varepsilon^2)]&\\ - (2/3)[1 + 3a_1\varepsilon + O(\varepsilon^2)]) + \varepsilon^2[-16/15 + O(\varepsilon)] + O(\varepsilon^3)&\\ = \varepsilon(4/3 - a_1) + \varepsilon^2(2a_1 - a_2 - 16/15) + O(\varepsilon^3) = 0,&\end{aligned} \tag{2.5.10}$$

which implies that

$$a_1 = 4/3, a_2 = 2a_1 - 16/15 = 8/3 - 16/15 = 8/5. \tag{2.5.11}$$

Then substituting (2.5.11) into (2.5.9) yields

$$t_m(\varepsilon) = 2[1+4\varepsilon/3+8\varepsilon^2/5+O(\varepsilon^3)]. \tag{2.5.12}$$

Next we examine $x_m = x(t_m;\varepsilon)$:

$$\begin{aligned}
x_m &= t_m - t_m^2/4 + \varepsilon(t_m^3/6 - t_m^4/48) + \varepsilon^2(-t_m^4/9 + 11t_m^5/240 - 11t_m^6/2,880) + O(\varepsilon^3)\\
&= 2[1+4\varepsilon/3+8\varepsilon^2/5+O(\varepsilon^3)] - [1+4\varepsilon/3+8\varepsilon^2/5+O(\varepsilon^3)]^2\\
&+\varepsilon(\{4/3\}[1+4\varepsilon/3+O(\varepsilon^2)]^3 - \{1/3\}[1+4\varepsilon/3+O(\varepsilon^2)]^4)\\
&+\varepsilon^2[-2+22/15-11/45+O(\varepsilon)]+O(\varepsilon^3)\\
&= 2[1+4\varepsilon/3+8\varepsilon^2/5+O(\varepsilon^3)] - [1+8\varepsilon/3+16\varepsilon^2/5+16\varepsilon^2/9+O(\varepsilon^3)]\\
&+\varepsilon(\{4/3\}[1+4\varepsilon+O(\varepsilon^2)] - \{1/3\}[1+16\varepsilon/3+O(\varepsilon^2)])\\
&+\varepsilon^2[-7/9+O(\varepsilon)]+O(\varepsilon^3)\\
&\sim 2-1+\varepsilon(8/3-8/3+4/3-1/3)+\varepsilon^2(16/5-16/5-16/9+16/3-16/9-7/9)
\end{aligned}$$

which implies that

$$x_m = x_m(\varepsilon) \sim 1+\varepsilon+\varepsilon^2 \text{ as } \varepsilon \to 0. \tag{2.5.13}$$

In order to check this calculation we solve (2.5.2) in exactly the same manner as employed earlier to find the escape velocity for its dimensional form by inverting $x = x(t;\varepsilon)$ to obtain $t = t(x;\varepsilon)$ and defining $V(x,t) = \dot{x}[t(x;\varepsilon);\varepsilon]$. Then, recalling that

$$2\ddot{x}[t(x;\varepsilon);\varepsilon] = \frac{dV^2(x;\varepsilon)}{dx},$$

this problem transforms into

$$\frac{dV^2(x;\varepsilon)}{dx} = \frac{-1}{(1+\varepsilon x)^2},\ x>0;\ V^2(0)=1. \tag{2.5.14}$$

Now integrating (2.5.14), we find that

$$V^2(x;\varepsilon)-1 = \frac{1}{\varepsilon}\left(\frac{1}{1+\varepsilon x}-1\right) \tag{2.5.15}$$

Then, since for the case $0<\varepsilon<1$,

$$V^2(x_m;\varepsilon)=0, \tag{2.5.16a}$$

it follows from (2.5.15) that x_m satisfies the relation

$$1-\varepsilon = \frac{1}{1+\varepsilon x_m} \tag{2.5.16b}$$

or solving (2.5.16b) we obtain the explicit formula

$$x_m = x_m(\varepsilon) = \frac{1}{\varepsilon}\left(\frac{1}{1-\varepsilon}-1\right) = \frac{1}{1-\varepsilon}. \tag{2.5.16c}$$

Finally, recalling the geometric series result that

$$\frac{1}{1-\varepsilon} = \sum_{n=0}^{\infty} \varepsilon^n < 1 \text{ for } |\varepsilon| < 1,$$

it follows that

$$x_m = x_m(\varepsilon) = \frac{1}{1-\varepsilon} = 1 + \varepsilon + \varepsilon^2 + O(\varepsilon^3) \sim 1 + \varepsilon + \varepsilon^2 \text{ as } \varepsilon \to 0 \qquad (2.5.16d)$$

in agreement with our asymptotic expansion (2.5.13).

We close this section by reiterating a few observations about our regular perturbation asymptotic expansion of the solution to the projectile problem made at the end of Section 2.4. The $O(1)$ terms in our expansions represent the constant gravity situation. Indeed the problem for $x_0(t)$ can be obtained by taking the limit of problem (2.5.2) as $\varepsilon \to 0$ and defining

$$x_0(t) = \lim_{\varepsilon \to 0} x(t; \varepsilon).$$

The $O(\varepsilon)$ and $O(\varepsilon^2)$ terms in our expansions represent the variation of gravitational force with altitude corrections to this constant gravity situation. Since we have only retained such terms through second order the accuracy of our truncated expansions depends upon the relative size of ε. In the binomial series expansion of (2.5.3b) we have implicitly assumed that $|\varepsilon x| < 1$. Let us examine this restriction in more detail. For the sake of the accuracy of our truncated expansions we require that $0 < \varepsilon << 1$ while the choice of scale factors for the nondimensional variables allows us in this case to conclude that $|x| \leq 1$. Hence it follows that $|\varepsilon x| << 1$. Note that because these correction terms are positive in both the expansions for $t_m(\varepsilon)$ and $x_m(\varepsilon)$ of (2.5.12) and (2.5.13), respectively, the variation of gravity with altitude causes the maximum projectile altitude and the time required to reach that height to increase when compared with the corresponding values for the constant gravity case which is to be expected since gravitational force decreases with altitude. Further, note that there is nothing sacred about retaining only terms through $N = 2$ in these expansions. We could in principle carry out such expansions to any desired higher-order $N > 2$ although in practice the calculations associated with such extensions become increasingly more complicated. Also observe that the higher the order of the terms retained, the relatively larger ε can be and still maintain the same degree of accuracy since the terms neglected are of $O(\varepsilon^{N+1})$. Finally let us reiterate for the purpose of emphasis that the most important thing about this procedure is it allows one to obtain an asymptotic expansion of $x(t; \varepsilon)$ valid in the limit as $\varepsilon \to 0$ to any degree of accuracy desired even though a closed-form solution to the original problem cannot be found exactly. Although such nonlinear problems can be solved numerically, the computation must be performed sequentially for each different value of ε. The advantage of this regular perturbation approach is that it deduces a representation involving both t and ε.

2.6 Energy Method of Solution

As a rationale for our investigation of this section demonstrating that $T = 2t_m$, we examine

$$\begin{aligned}
x(2t_m;\varepsilon) &= 2t_m - t_m^2 + \varepsilon(4t_m^3/3 - t_m^4/3) + \varepsilon^2(-2t_m^4 + 22t_m^5/15 - 11t_m^6/48) + O(\varepsilon^3) \\
&= 4[1 + 4\varepsilon/3 + 8\varepsilon^2/5 + O(\varepsilon^3)] - 4[1 + 4\varepsilon/3 + 8\varepsilon^2/5 + O(\varepsilon^3)]^2 \\
&\quad + \varepsilon(\{32/3\}[1 + 4\varepsilon/3 + O(\varepsilon^2)]^3 - \{16/3\}[1 + 4\varepsilon/3 + O(\varepsilon^2)]^4) \\
&\quad + \varepsilon^2[-32 + \{22/15\}32 - \{22/45\}32 + O(\varepsilon)] + O(\varepsilon^3) \\
&= 4[1 + 4\varepsilon/3 + 8\varepsilon^2/5 + O(\varepsilon^3)] - 4[1 + 8\varepsilon/3 + 16\varepsilon^2/5 + 16\varepsilon^2/9 + O(\varepsilon^3)] \\
&\quad + \varepsilon(\{32/3\}[1 + 4\varepsilon + O(\varepsilon^2)] - \{16/3\}[1 + 16\varepsilon/3 + O(\varepsilon^2)]) \\
&\quad + \varepsilon^2[-32/45 + O(\varepsilon)] + O(\varepsilon^3) \\
&= 4 - 4 + \varepsilon(16/3 - 32/3 + 32/3 - 16/3) \\
&\quad + \varepsilon^2(32/5 - 64/5 - 64/9 + 128/3 - 256/9 - 32/45) + O(\varepsilon^3) \\
&= 0 + 0\varepsilon + 0\varepsilon^2 + O(\varepsilon^3) \\
&= 0 + O(\varepsilon^3) \text{ as } \varepsilon \to 0
\end{aligned} \tag{2.6.1a}$$

and

$$\begin{aligned}
\dot{x}(2t_m;\varepsilon) &= 1 - t_m + \varepsilon(2t_m^2 - 2t_m^3/3) + \varepsilon^2(-4t_m^3 + 11t_m^4/3 - 11t_m^5/15) + O(\varepsilon^3) \\
&= 1 - 2[1 + 4\varepsilon/3 + 8\varepsilon^2/5 + O(\varepsilon^3)] \\
&\quad + \varepsilon(8[1 + 4\varepsilon/3 + O(\varepsilon^2)]^2 - \{16/3\}[1 + 4\varepsilon/3 + O(\varepsilon^2)]^3) \\
&\quad + \varepsilon^2[\{-2\}16 + \{11/3\}16 - \{22/15\}16 + O(\varepsilon)] + O(\varepsilon^3) \\
&= 1 - 2[1 + 4\varepsilon/3 + 8\varepsilon^2/5 + O(\varepsilon^3)] \\
&\quad + \varepsilon(8[1 + 8\varepsilon/3 + O(\varepsilon^2)]) - \{16/3\}[1 + 4\varepsilon + O(\varepsilon^2)] \\
&\quad + \varepsilon^2[16/5 + O(\varepsilon)] + O(\varepsilon^3) \\
&= 1 - 2 + \varepsilon(-8/3 + 8 - 16/3) + \varepsilon^2(-16/5 + 64/3 - 64/3 + 16/5) + O(\varepsilon^3) \\
&= -1 + 0\varepsilon + 0\varepsilon^2 + O(\varepsilon^3) \\
&= -1 + O(\varepsilon^3) \text{ as } \varepsilon \to 0.
\end{aligned} \tag{2.6.1b}$$

Given these results through terms of third order one might suspect that

$$x(T;\varepsilon) = 0 \text{ for } T > 0 \Rightarrow T = 2t_m \text{ and } \dot{x}(T;\varepsilon) = -1 \tag{2.6.2}$$

as in the constant gravity case. In order to demonstrate this conclusively, we first consider a general so-called conservative dynamical system involving a point mass m being acted upon by a force $\boldsymbol{F}[\boldsymbol{r}(t)] = -\nabla\Omega[\boldsymbol{r}(t)]$ derivable as the gradient of a scalar potential function Ω that is dependent only on its position vector $\boldsymbol{r}(t)$ where

$t \equiv$ time and all variables are measured in dimensional units. Then from Newton's second law we have the governing equation of motion

$$m\ddot{\boldsymbol{r}}(t) = \boldsymbol{F}[\boldsymbol{r}(t)] = -\nabla\Omega[\boldsymbol{r}(t)],\ t > 0; \tag{2.6.3a}$$

with initial conditions

$$\boldsymbol{r}(0) = \boldsymbol{r_0},\ \dot{\boldsymbol{r}}(0) = \boldsymbol{V_0} \tag{2.6.3b}$$

We now define the total energy of the system $E(t)$ by

$$E(t) = \frac{m}{2}\dot{\boldsymbol{r}}(t) \bullet \dot{\boldsymbol{r}}(t) + \Omega[\boldsymbol{r}(t)] \tag{2.6.4}$$

which is the sum of its kinetic and potential energies. Note that $\dot{\boldsymbol{r}}(t) \bullet \dot{\boldsymbol{r}}(t) = |\dot{\boldsymbol{r}}(t)|^2$. We next compute

$$\dot{E}(t) = m\dot{\boldsymbol{r}}(t) \bullet \ddot{\boldsymbol{r}}(t) + \nabla\Omega[\boldsymbol{r}(t)] \bullet \dot{\boldsymbol{r}}(t) = \{m\ddot{\boldsymbol{r}}(t) + \nabla\Omega[\boldsymbol{r}(t)]\} \bullet \dot{\boldsymbol{r}}(t) = \boldsymbol{0} \bullet \dot{\boldsymbol{r}}(t) = 0 \tag{2.6.5}$$

which implies that

$$E(t) \equiv E(0) \tag{2.6.6a}$$

or

$$\frac{m}{2}\dot{\boldsymbol{r}}(t) \bullet \dot{\boldsymbol{r}}(t) + \Omega[\boldsymbol{r}(t)] \equiv \frac{m}{2}|\boldsymbol{V_0}|^2 + \Omega[\boldsymbol{r_0}]. \tag{2.6.6b}$$

Particularizing this development to our variation of gravitational force with altitude projectile problem of Section 2.1 in dimensional form, we make the identifications from (2.1.15) that

$$\boldsymbol{r}(t) = x(t)\boldsymbol{k},\ \boldsymbol{F}[\boldsymbol{r}(t)] = F[x(t)]\boldsymbol{k} = -\Omega'[x(t)]\boldsymbol{k},\ \boldsymbol{r_0} = \boldsymbol{0},\ \boldsymbol{V_0} = V_0\boldsymbol{k} \tag{2.6.7a}$$

where

$$\Omega'(x) = mg\frac{R^2}{(x+R)^2}. \tag{2.6.7b}$$

Thus

$$\Omega(x) = -mg\frac{R^2}{x+R} + mgR \tag{2.6.8}$$

Hence, employing these results in (2.6.6) we obtain

$$E(t) = \frac{m}{2}|\dot{x}(t)|^2 - mg\frac{R^2}{x(t)+R} + mgR \equiv \frac{m}{2}V_0^2 \tag{2.6.9}$$

since $\Omega(0) = 0$. Finally, solving (2.6.9) for $|\dot{x}(t)|^2$ yields

$$|\dot{x}(t)|^2 = V_0^2 + 2g\frac{R^2}{x(t)+R} - 2gR = V_e^2\left[\frac{1}{1+x(t)/R} + \varepsilon - 1\right]. \tag{2.6.10}$$

Consider the schematic plot of the projectile's ascent $0 \le t_1 \le t_m$ and descent $t_m \le t_2 \le T$ depicted in Fig. 2.3 where

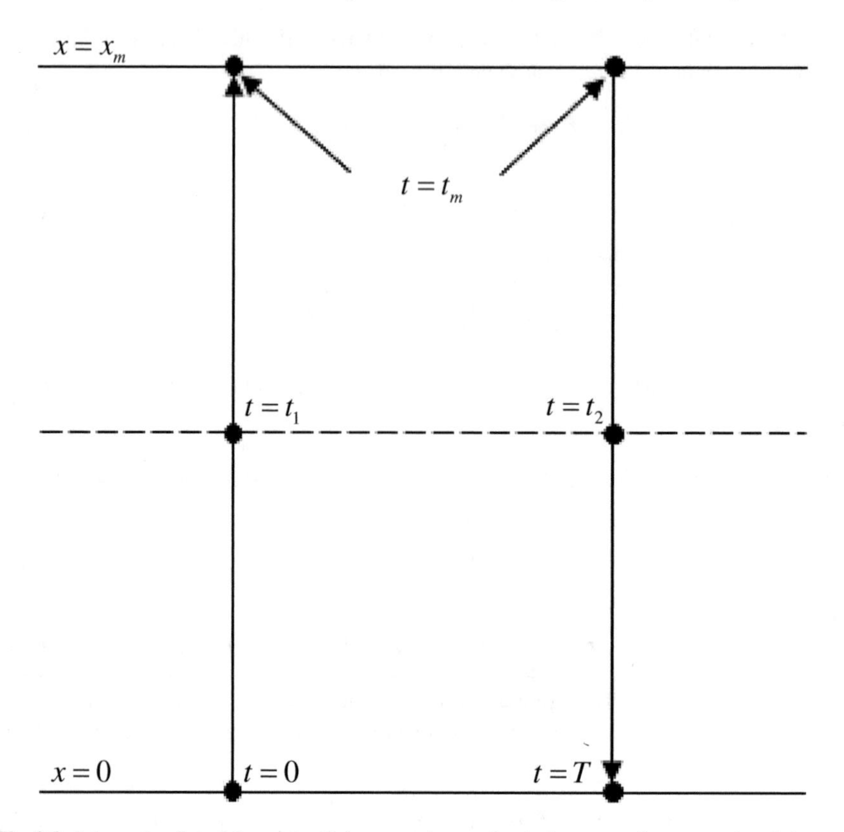

Fig. 2.3 Schematic plot of the projectile's ascent (upward pointing arrow $0 \leq t_1 \leq t_m$) and descent (downward pointing arrow $t_m \leq t_2 \leq T$) where $x = x(t) \equiv$ its altitude and $t \equiv$ time.

$$x(t_1) = x(t_2) \tag{2.6.11a}$$

and

$$x(0) = x(T) = 0, x(t_m) = x_m. \tag{2.6.11b}$$

We can conclude from (2.6.10) and (2.6.11a) that

$$|\dot{x}(t_1)| = |\dot{x}(t_2)|. \tag{2.6.12}$$

Hence the speed of the projectile at any given altitude is the same on the way down as it was on the way up at that altitude. Then, since the distance traveled x_m going up is obviously the same as that traveled going down, the time of ascent must equal the time of descent. Thus

$$t_m - 0 = T - t_m \Rightarrow T = 2t_m, \tag{2.6.13}$$

demonstrating the desired result. Further observe that the escape velocity for $\varepsilon \geq 1$ and the maximum altitude $x_m = x(t_m)$ where $\dot{x}(t_m) = 0$ for $0 \leq \varepsilon < 1$ could also be determined from the energy relation of (2.6.10). Finally note that should the dissipative force of air resistance be included in a problem (see Problems 2.2 and 2.3) then

$$\dot{E}(t) < 0 \Rightarrow |\dot{x}(t_1)| > |\dot{x}(t_2)| \Rightarrow t_m - 0 < T - t_m \Rightarrow T > 2t_m. \quad (2.6.14)$$

Problems

2.1. Compute g and V_e for all the other planets in both metric and British engineering units using the values in Table 2.1. Note that there are 3,280 ft in a km. Compare these values to that for Earth.

2.2. Consider a projectile of mass m that is shot vertically upward from the surface of the Earth with an initial velocity V_0. Assume that the gravitational force acts downward at a constant acceleration g while the force of air resistance has a magnitude proportional to the square of the velocity with proportional constant $k > 0$ and acts to resist motion. Let $x = x(t)$ denote the altitude of the projectile at time t and $v(t) = dx(t)/dt$ be its velocity.

(a) Explain why the governing equation of motion for the projectile is given by

$$\begin{cases} m\frac{dv}{dt} = -kv^2 - mg, & v > 0 \\ m\frac{dv}{dt} = kv^2 - mg, & v < 0 \end{cases} \text{ for } t > 0;$$

$$x(0) = 0,\ v(0) = V_0.$$

(b) Introduce the change of variables $V(x) = v[t(x)]$ and define $V_1(x) = V^2(x)$ for $V(x) > 0$ and $V_2(x) = V^2(x)$ for $V(x) < 0$. Show that $V_1(x)$ and $V_2(x)$ satisfy the linear first-order ordinary differential equations

$$\frac{dV_1}{dx} + 2\frac{k}{m}V_1 = -2g,\ V_1(0) = V_0^2;$$
$$\frac{dV_2}{dx} - 2\frac{k}{m}V_2 = -2g,\ V_2(x_m) = 0;$$

where x_m is defined implicitly by $V_1(x_m) = 0$.

(c) Solve these equations explicitly for $V_1(x)$ and $V_2(x)$ and show that

$$V_2(0) = \frac{V_0^2}{1 + kV_0^2/(mg)} < V_0^2 = V_1(0).$$

Discus this result in the context of (2.6.14).

2.3. In the projectile problem, when air resistance is taken into account but the variation of gravitational force with altitude is neglected, the governing dimensional ordinary differential equation and initial conditions are

$$m\frac{d^2x^*}{dt^{*2}} + k\frac{dx^*}{dt^*} = -mg;\, x^*(0) = 0,\, \frac{dx^*(0)}{dt^*} = V_0,$$

where $x^* = x^*(t^*) \equiv$ its altitude and $t^* \equiv$ time should air resistance be assumed to exert a force proportional to the speed of the projectile with proportionality constant $k > 0$ while acting in a direction to oppose motion.

(a) When the effect of air resistance is small, the change of variables

$$t = \frac{t^*}{V_0/(2g)},\, x(t) = \frac{x^*(t^*)}{V_0^2/(2g)}$$

is still appropriate (why does this follow from the concepts of scaling and introduction of nondimensional variables?). Show that this change of variables transforms the governing equation and initial conditions into the dimensionless form for $x = x(t;\beta)$

$$\ddot{x} + \beta\dot{x} = -\frac{1}{2};\, 0 < \beta = k\frac{V_0}{2mg} << 1,\, t > 0;\, x(0;\beta) = 0,\, \dot{x}(0;\beta) = 1;$$

where $\dot{(-)} \equiv d(-)/dt$.

(b) Seek a regular perturbation expansion of the solution to this dimensionless problem of the form

$$x(t;\beta) = x_0(t) + \beta x_1(t) + \beta^2 x_2(t) + O(\beta^3).$$

Proceed as in the corresponding problem treated in Section 2.5 to find

$$t_m = 2[1 + b_1\beta + b_2\beta^2 + O(\beta^3)],\, x_m = x(t_m;\beta) \text{ to } O(\beta^2).$$

Also compute T such that $x(T;\beta) = 0$ to that order and compare t_m with $T - t_m$. Discuss this result in the context of (2.6.14).

(c) Inverting $x = x(t;\beta)$ to obtain $t = t(x;\beta)$ and defining $V(x;\beta) = \dot{x}[t(x;\beta);\beta]$, demonstrate that this change of variables transforms the nondimensional problem into

$$V\frac{dV}{dx} + \beta V = -\frac{1}{2};\, V(0;\beta) = 1.$$

Solve this separable first-order ordinary differential equation and obtain an implicit relationship satisfied by V and x.

(d) Given that a power series can be integrated term-by-term in its interval of convergence, deduce from the geometric series

$$\frac{1}{1+\varepsilon} = \sum_{n=0}^{\infty}(-1)^n\varepsilon^n \text{ for } |\varepsilon| < 1,$$

that

$$\ln(1+\varepsilon) \sim \varepsilon - \frac{\varepsilon^2}{2} + \frac{\varepsilon^3}{3} - \frac{\varepsilon^4}{4} \text{ for } |\varepsilon| < 1.$$

(e) Determine an explicit formula for x_m as a function of β from the relation of part (c) where $V(x_m;\beta) = 0$. Check the result of part (b) by using the Maclaurin series of part (d) to obtain a power series for x_m from this formula through terms of $O(\beta^2)$ assuming β "small" in some sense. Discuss this restriction in the context of the result of part (b) with respect to the behavior of T when compared to $2t_m$.

Chapter 3
Of Mites and Models

This chapter analyzes a simple ordinary differential equations-based model for a predator–prey mite interaction on apple tree foliage ([21, 139]). That model is a specific Kolmogorov-type exploitation ordinary differential equation system assembled by May [80] from predator and prey components developed by Leslie [67] and Holling [49], respectively. It is a composite model in that its parameters are chosen appropriately for a temperature-dependent mite interaction on fruit tree leaves by curve-fitting to the relevant data. In particular, the proper temperature-rate relationship for arthropods is deduced by the knowledge of the results of singular perturbation theory applied to ordinary differential equations, which is introduced in a pastoral interlude. Its linear behavior is examined by developing a standard linear stability analysis for a general such system, in another pastoral interlude, and then applying that result to this model. Its nonlinear behavior is examined by means of Kolmogorov's theorem for dynamical systems and by the numerical bifurcation code AUTO. Since these results involve limit cycles, that concept is developed in the final pastoral interlude. The predictions of this model are then compared with general ecological field results and particular laboratory experimental data. The problems extend singular perturbation-type analyses to the investigation of a cubic polynomial, the specific temperature-rate relationship for the prey mite species to the predaceous mite, and linear stability analyses to a competing species model and a predator–prey model employing a Holling type I functional response as opposed to the type II response included in the May mite model.

3.1 Temperature-Rate Phenomena in Arthropods

Mites and ticks are arthropods in the taxon *Acarina* that possess one body part and eight legs. They are insect relatives, often classified as a suborder of *Arachnids* which predominantly include spiders, and have been typically studied by entomologists. Indeed the prey species in this interaction the McDaniel spider mite *Tetranychus mcdanieli* (McGregor), a pest on apple or pear tree foliage, weaves webs on the

© Springer International Publishing AG, part of Springer Nature 2017

D. J. Wollkind and B. J. Dichone, *Comprehensive Applied Mathematical Modeling in the Natural and Engineering Sciences*,
https://doi.org/10.1007/978-3-319-73518-4_3

undersides of fruit tree leafs and protected by that silken layer sucks the liquid from these leafs. Its economic impact on apple orchards is that the diameter of the average apple in the crop produced by trees so infested can be markedly decreased. The predaceous species in this interaction is a soil mite *Metaseiulus occidentalis* (Nesbitt). It is a voracious predator and has been introduced into apple tree orchards to provide biological control of McDaniel spider mite infestations (see Fig. 3.1). The major pest on apple trees in the interior valleys of the Wenatchee area of Washington State is the codling moth whose larvae feed on the apple itself. Its economic impact is so severe that it must be treated by pesticides. The comprehensive control program established in these orchards of the central valleys of Washington is one that integrates the chemical control of this insect pest with the biological control of the McDaniel spider mite.

Fig. 3.1 The artist W. G. Wiles' impression of the attack by the predacious mite *Metaseiulus occidentalis* on the spider mite pest *Tetranychus mcdanieli* occurring upon an apple tree leaf.

Of all exogenous – *i.e.*, external – variables affecting mite populations on apple tree leafs temperature is by far the most important, say, as compared to humidity. The McDaniel spider mite attained pest status in 1951, and hence, its life history became of significant interest. The critical importance of temperature to the rate of development of these populations has long been recognized, and entomologists have expended a great deal of experimental effort determining the temperature dependence of their life history parameters. In particular, the effect of temperature on the rate of development for this arthropod species measured in the number of mites produced per day can be divided into two phases: Namely, Phase I, which occurs from some base to optimum temperature and is characterized by a monotone increasing slope; and Phase II, which occurs should this optimum temperature ever be exceeded and is characterized by a rapid, often precipitous, decline in growth rate to zero at its lethal maximum temperature – *i.e.*, a temperature at or above which life processes cannot be maintained for any appreciable length of time. This standard arthropod behavior is represented schematically in Fig. 3.2.

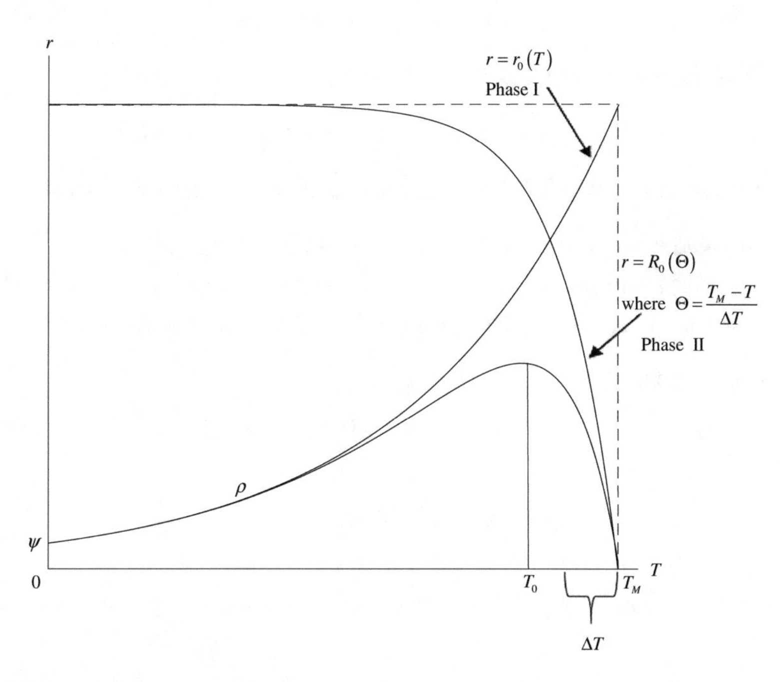

Fig. 3.2 A schematic representation of rate-temperature phenomena in arthropods with $r_0(T)$ for Phase I and $R_0(\Theta)$ for Phase II as defined in Section 3.3.

Although one could deduce empirical formulations to describe the rate of development in these two phases separately, it is advantageous to determine a single analytical expression for rate of development over the whole relevant temperature domain involving biologically meaningful and hence measurable parameters. In order to motivate that determination we next present our second "pastoral interlude" dealing with *singular perturbation theory* of ordinary differential equations given that the mere knowledge of the form of such solutions can then be used to derive the required relationship.

3.2 Pastoral Interlude: Singular Perturbation Theory of Ordinary Differential Equations

(a) As a prelude to examining singularly perturbed boundary value problems in ordinary differential equations [88], we shall consider the following canonical quadratic for $m = m(\varepsilon)$:

$$\varepsilon m^2 + 2m + 1 = 0, \; 0 < \varepsilon << 1. \tag{3.2.1}$$

There exists one root of (3.2.1) of the form

$$m^{(1)}(\varepsilon) = m_0 + \varepsilon m_1 + \varepsilon^2 m_2 + \varepsilon^3 m_3 + O(\varepsilon^4) \text{ as } \varepsilon \to 0. \tag{3.2.2}$$

Upon substituting (3.2.2) into (3.2.1) and collecting terms, we find that

$$\begin{aligned} 0 &= \varepsilon[m_0 + \varepsilon m_1 + \varepsilon^2 m_2 + O(\varepsilon^3)]^2 + 2[m_0 + \varepsilon m_1 + \varepsilon^2 m_2 + \varepsilon^3 m_3 + O(\varepsilon^4)] + 1 \\ &= \varepsilon[m_0^2 + 2m_0\{\varepsilon m_1 + \varepsilon^2 m_2\} + \varepsilon^2 m_1^2 + O(\varepsilon^3)] + 2[m_0 + \varepsilon m_1 + \varepsilon^2 m_2 + \varepsilon^3 m_3 + O(\varepsilon^4)] + 1 \\ &= 2m_0 + 1 + \varepsilon(2m_1 + m_0^2) + 2\varepsilon^2(m_2 + m_0 m_1) + \varepsilon^3(2m_3 + 2m_0 m_2 + m_1^2) + O(\varepsilon^4), \end{aligned}$$

which implies

$$O(1): 2m_0 + 1 = 0 \Rightarrow m_0 = -\frac{1}{2}; \tag{3.2.3a}$$

$$O(\varepsilon): 2m_1 + m_0^2 = 0 \Rightarrow m_1 = -\frac{m_0^2}{2} = -\frac{1}{8}; \tag{3.2.3b}$$

$$O(\varepsilon^2): m_2 + m_0 m_1 = 0 \Rightarrow m_2 = -m_0 m_1 = -\frac{1}{16}; \tag{3.2.3c}$$

$$O(\varepsilon^3): 2m_3 + 2_0 m_2 + m_1^2 = 0 \Rightarrow m_3 = -\frac{m_1^2}{2} - m_0 m_2 = -\frac{5}{128}. \tag{3.2.3d}$$

Hence

$$m^{(1)}(\varepsilon) \sim -\frac{1}{2} - \frac{\varepsilon}{8} - \frac{\varepsilon^2}{16} - \frac{5\varepsilon^3}{128} \text{ as } \varepsilon \to 0. \tag{3.2.4}$$

We wish to find the form of the second root which is unbounded in the limit as $\varepsilon \to 0$. Let

$$m^{(2)}(\varepsilon) = \frac{\xi(\varepsilon)}{\delta(\varepsilon)} \tag{3.2.5a}$$

where

$$\delta(\varepsilon) > 0 \text{ and } \lim_{\varepsilon \to 0} \delta(\varepsilon) = 0 \tag{3.2.5b}$$

while

$$\xi(\varepsilon) = O(1) \text{ but } \xi(\varepsilon) \neq o(1) \text{ as } \varepsilon \to 0. \tag{3.2.5c}$$

In what follows we shall employ the notation $\xi(\varepsilon) = O^*(\varepsilon)$ as $\varepsilon \to 0$ to represent this situation.

Then substituting (3.2.5) into our quadratic of (3.2.1) and multiplying by $\delta^2(\varepsilon)/\varepsilon$ we obtain

$$\xi^2(\varepsilon)+2\left[\frac{\delta(\varepsilon)}{\varepsilon}\right]\xi(\varepsilon)+\left[\frac{\delta^2(\varepsilon)}{\varepsilon}\right]=0. \tag{3.2.6}$$

Since $\xi(\varepsilon)=O^*(1)$ as $\varepsilon\to 0$, we do not wish it to go to ∞ or 0 as $\varepsilon\to 0$. Thus neither of the bracketed terms in (3.2.6) can go to ∞ as $\varepsilon\to 0$ and both cannot go to 0 simultaneously in that limit. Hence without loss of generality it is only necessary for us to consider the following two cases:

(i) $\delta^2(\varepsilon)/\varepsilon=1\Rightarrow\delta(\varepsilon)=\varepsilon^{1/2}\Rightarrow\delta(\varepsilon)/\varepsilon=1/\varepsilon^{1/2}\to\infty$ as $\varepsilon\to 0$ and must be rejected.

(ii) $\delta(\varepsilon)/\varepsilon=1\Rightarrow\delta(\varepsilon)=\varepsilon\Rightarrow\delta^2(\varepsilon)/\varepsilon=\varepsilon\to 0$ as $\varepsilon\to 0$ and is the correct determination.

Therefore employing $\delta(\varepsilon)=\varepsilon$ in (3.2.5) and (3.2.6) yields

$$m^{(2)}(\varepsilon)=\frac{\xi(\varepsilon)}{\varepsilon}\ \text{where}\ \xi^2(\varepsilon)+2\xi(\varepsilon)+\varepsilon=0, \tag{3.2.7}$$

respectively. We now seek a solution of (3.2.7) of the form

$$\xi(\varepsilon)=\xi_0+\varepsilon\xi_1+\varepsilon^2\xi_2+O(\varepsilon^3)\ \text{as}\ \varepsilon\to 0, \tag{3.2.8a}$$

where

$$\xi_k=O(1)\ \text{as}\ \varepsilon\to 0\ \text{for}\ k=0,1\ \text{and}\ 2\ \text{while}\ \xi_0\neq 0, \tag{3.2.8b}$$

by virtue of (3.2.5c). Upon substituting (3.2.8a) into (3.2.7) and again collecting terms, we find that

$$\begin{aligned}0&=[\xi_0+\varepsilon\xi_1+\varepsilon^2\xi_2+O(\varepsilon^3)]^2+2[\xi_0+\varepsilon\xi_1+\varepsilon^2\xi_2+O(\varepsilon^3)]+\varepsilon\\&=\xi_0^2+2\xi_0(\varepsilon\xi_1+\varepsilon^2\xi_2)+\varepsilon^2\xi_1^2+O(\varepsilon^3)+2[\xi_0+\varepsilon\xi_1+\varepsilon^2\xi_2+O(\varepsilon^3)]+\varepsilon\\&=\xi_0(\xi_0+2)+\varepsilon[2(\xi_0+1)\xi_1+1]+\varepsilon[2(\xi_0+1)\xi_2+\xi_1^2]+O(\varepsilon^3),\end{aligned}$$

which implies

$$O(1):\xi_0(\xi_0+2)=0\ \Rightarrow\ \xi_0=-2\ \text{since}\ \xi_0\neq 0; \tag{3.2.9a}$$

$$O(\varepsilon):2(\xi_0+1)\xi_1+1=0\ \Rightarrow\ \xi_1=-\frac{1}{2(\xi_0+1)}=\frac{1}{2}; \tag{3.2.9b}$$

$$O(\varepsilon^2):2(\xi_0+1)\xi_2+\xi_1^2=0\ \Rightarrow\ \xi_2=-\frac{\xi_1^2}{2(\xi_0+1)}=\frac{1}{8}. \tag{3.2.9c}$$

Hence

$$\xi(\varepsilon)\sim -2+\frac{\varepsilon}{2}+\frac{\varepsilon^2}{8}\ \text{as}\ \varepsilon\to 0 \tag{3.2.10a}$$

and

$$m^{(2)}(\varepsilon) = \frac{\xi(\varepsilon)}{\varepsilon} \sim -\frac{2}{\varepsilon} + \frac{1}{2} + \frac{\varepsilon}{8} \text{ as } \varepsilon \to 0. \tag{3.2.10b}$$

In order to check these asymptotic expansions of (3.2.4) and (3.2.10) for the roots $m^{(1,2)}(\varepsilon)$ of (3.2.1), we next solve this quadratic exactly and obtain

$$m^{(1,2)}(\varepsilon) = \frac{-1 \pm (1-\varepsilon)^{1/2}}{\varepsilon}. \tag{3.2.11a}$$

Employing the binomial series formula of Chapter 2 for $(1+J)^{\alpha}$ with $J = -\varepsilon$ and $\alpha = 1/2$, we find that

$$(1-\varepsilon)^{1/2} = 1 - \frac{\varepsilon}{2} - \frac{\varepsilon^2}{8} + O(\varepsilon^3) \text{ as } \varepsilon \to 0 \tag{3.2.11b}$$

which upon substitution of (3.2.11b) into (3.2.11a) yields

$$m^{(1)}(\varepsilon) = \frac{-1 + (1-\varepsilon)^{1/2}}{\varepsilon} \sim -\frac{1}{2} - \frac{\varepsilon}{8} + O(\varepsilon^2) \text{ as } \varepsilon \to 0 \tag{3.2.12a}$$

and

$$m^{(2)}(\varepsilon) = \frac{-1 - (1-\varepsilon)^{1/2}}{\varepsilon} \sim -\frac{2}{\varepsilon} + \frac{1}{2} + \frac{\varepsilon}{8} + O(\varepsilon^2) \text{ as } \varepsilon \to 0 \tag{3.2.12b}$$

serving as a partial check on these asymptotic expansions when compared with (3.2.4) and (3.2.10), respectively.
Finally let us deduce the one-term asymptotic representations for these roots by proceeding as follows: Consider our canonical quadratic in the form

$$\textcircled{1} + \textcircled{2} + \textcircled{3} = 0 \text{ where } \textcircled{1} \equiv \varepsilon m^2,\ \textcircled{2} \equiv 2m, \text{ and } \textcircled{3} \equiv 1 \text{ with } 0 < \varepsilon << 1. \tag{3.2.13}$$

We shall sequentially retain two of these terms, solve the truncated equation, and compare the resulting size of the neglected term with those retained. If it is small for $0 < \varepsilon << 1$ our result will be valid but if it is not then that result must be rejected. There are three possible cases:

(i) Retain ② & ③ and neglect ①: $2m + 1 = 0 \Rightarrow m = -1/2$

$$\Rightarrow 0 < \varepsilon m^2 = \varepsilon/4 << 1.$$

(ii) Retain ① & ② and neglect ③: $\varepsilon m^2 + 2m = 0 \Rightarrow m = -2/\varepsilon$

$$\Rightarrow \varepsilon m^2 = 4/\varepsilon >> 1.$$

(iii) Retain ① & ③ and neglect ②: $\varepsilon m^2 + 1 = 0 \Rightarrow m = \pm i/\varepsilon^{1/2}$

$$\Rightarrow |2m| = 2/\varepsilon^{1/2} >> 1.$$

Observe that cases (i) and (ii) are valid corresponding to the lead terms of $m^{(1,2)}(\varepsilon)$, respectively, while case (iii) must be rejected. In this context observe that the $m = 0$ root for case (ii) corresponds to the root of case (i) while case (iii) corresponds to $\delta(\varepsilon) = \varepsilon^{1/2}$ from our previous development which also had to be rejected. Here $i \equiv \sqrt{-1}$ and the notation $|\ldots|$ is to be interpreted as the absolute value of a complex quantity where, for $z = x + iy$ with $x, y \in \mathbb{R} \equiv$ reals, $|z| \equiv \sqrt{x^2 + y^2}$.

(b) We shall now after Lin and Segel [69] consider the following canonical singularly perturbed boundary value problem for $y = y(x;\varepsilon)$:

$$\varepsilon \frac{d^2y}{dx^2} + 2\frac{dy}{dx} + y = 0,\ 0 < x < 1,\ 0 < \varepsilon << 1; \tag{3.2.14a}$$

$$y(0;\varepsilon) = 0,\ y(1;\varepsilon) = 1; \tag{3.2.14b}$$

and begin our investigation of its behavior by first assuming there exists an interval I such that

$$y, \frac{dy}{dx}, \frac{d^2y}{dx^2} = O(1) \text{ as } \varepsilon \longrightarrow 0 \text{ for } x \in I \subset [0,1]. \tag{3.2.15}$$

Then defining

$$y_0(x) = \lim_{\varepsilon \to 0} y(x;\varepsilon) \text{ for } x \in I$$

and taking the limit of (3.2.14a) as $\varepsilon \longrightarrow 0$, we obtain

$$2y_0'(x) + y_0(x) = 0 \Rightarrow y_0(x) = c_0 e^{-x/2}. \tag{3.2.16}$$

Note that since this naïve approximation is a solution to a first-order equation it cannot satisfy both the boundary conditions (BC's) of (3.2.14b). We next sequentially consider the possibilities of either 0 or 1 being included in I (see Fig. 3.3).

(i) Let $0 \in I$. Hence, we retain BC's $y(0;\varepsilon) = 0 \Rightarrow y_0(0) = 0 \Rightarrow y_0(x) \equiv 0$;
(ii) Let $1 \in I$. Hence, we retain BC's $y(1;\varepsilon) = 1 \Rightarrow y_0(1) = 1 \Rightarrow y_0(x) = e^{(1-x)/2}$.

In order to decide which if any of these two possible naïve approximations to our actual solution is correct, we shall solve the original problem (3.2.1) exactly and examine its asymptotic behavior as $\varepsilon \longrightarrow 0$. Let $y(x;\varepsilon) = e^{m(\varepsilon)x}$ where $m(\varepsilon)$ satisfies our quadratic (3.2.1), upon substitution of this expression into (3.2.14a), and thus

$$m(\varepsilon) = m^{(1,2)}(\varepsilon)$$

as defined by (3.2.11a). Hence, since $m^{(1)}(\varepsilon) \neq m^{(2)}(\varepsilon)$, we have the general solution

$$y(x;\varepsilon) = c_1(\varepsilon)e^{m^{(1)}(\varepsilon)x} + c_2(\varepsilon)e^{m^{(2)}(\varepsilon)x}. \tag{3.2.17a}$$

Then applying the BC's (3.2.14b) yields

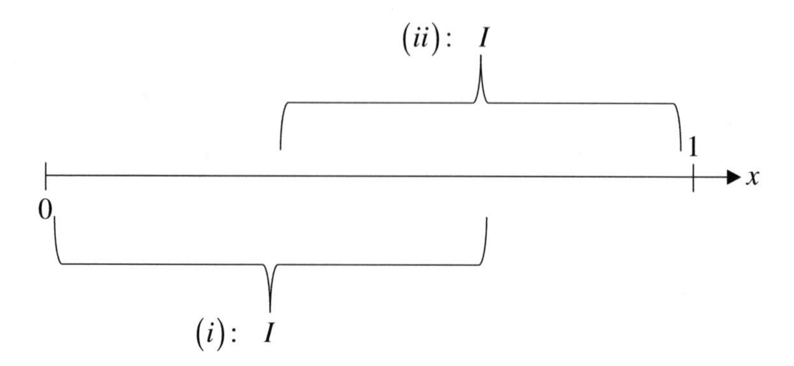

Fig. 3.3 The two possible choices for I defined in (3.2.15) where $0 \in I$ in (i) and $1 \in I$ in (ii).

$$c_1(\varepsilon) = -c_2(\varepsilon) = \frac{1}{e^{m^{(1)}(\varepsilon)} - e^{m^{(2)}(\varepsilon)}}. \tag{3.2.17b}$$

Therefore our exact solution is given by

$$y(x;\varepsilon) = \frac{e^{m^{(1)}(\varepsilon)x} - e^{m^{(2)}(\varepsilon)x}}{e^{m^{(1)}(\varepsilon)} - e^{m^{(2)}(\varepsilon)}}. \tag{3.2.17c}$$

Recalling that to lowest order

$$m^{(1)}(\varepsilon) \sim -\frac{1}{2},\ m^{(2)}(\varepsilon) \sim -\frac{2}{\varepsilon} \text{ as } \varepsilon \to 0, \tag{3.2.17d}$$

we can deduce from (3.2.17c, 3.2.17d) that

$$y(x;\varepsilon) \sim \frac{e^{-x/2} - e^{-2x/\varepsilon}}{e^{-1/2} - e^{-2/\varepsilon}} \sim e^{1/2}(e^{-x/2} - e^{-2x/\varepsilon}) = y_u^{(0)}(x, x/\varepsilon) \text{ as } \varepsilon \to 0. \tag{3.2.17e}$$

(i) Consider $x = O^*(\varepsilon)$ as $\varepsilon \to 0$ and introduce $\xi = x/\varepsilon = O^*(1)$ as $\varepsilon \to 0$. Then from (3.2.17e)

$$y_u^{(0)}(x, x/\varepsilon) = e^{1/2}(e^{-\varepsilon\xi/2} - e^{-2\xi}) \sim e^{1/2}(1 - e^{-2\xi}) = Y_0(\xi) \not\equiv 0 \text{ as } \varepsilon \to 0. \tag{3.2.18a}$$

(ii) Consider $x = O^*(1)$ as $\varepsilon \to 0$. Then from (3.2.17e)

$$y_u^{(0)}(x, x/\varepsilon) \sim e^{1/2}e^{-x/2} = y_0(x) \text{ as } \varepsilon \to 0. \tag{3.2.18b}$$

Hence the results of this analysis demonstrate that case (ii) is the correct naïve representation and what the proper asymptotic behavior of the solution actually is near the origin. That is

$$y(x;\varepsilon) \sim y_u^{(0)}(x;x/\varepsilon) \sim \begin{cases} y_0(x) = e^{(1-x)/2}, & \text{for } x = O^*(1) \\ Y_0(x/\varepsilon) = e^{1/2}(1-e^{-2x/\varepsilon}), & \text{for } x = O^*(\varepsilon) \end{cases} \text{ as } \varepsilon \to 0. \tag{3.2.19}$$

These results are plotted in Fig. 3.4. Note, in this context that

$$y_0(0) = \lim_{\xi\to\infty} Y_0(\xi) = e^{1/2}.$$

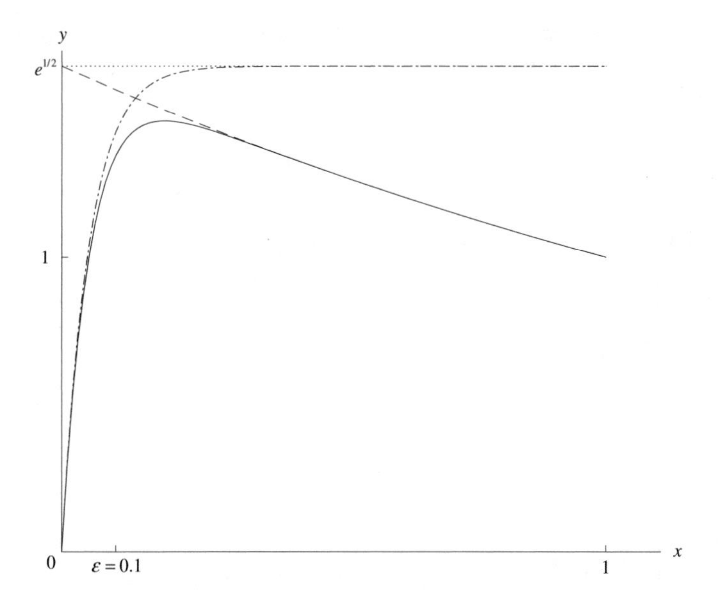

Fig. 3.4 A plot of $y(x;\varepsilon)$ of (3.2.19) with $\varepsilon = 0.1$ where the dashed curves having y-intercept and horizontal asymptote, respectively, of $e^{1/2}$ denote its one-term outer $y_0(x)$ and inner $Y_0(x/\varepsilon)$ expansions.

Finally the question naturally arises about whether there are any circumstances under which case (i) actually represents the correct naïve approximation. Toward that end, consider problem (3.2.14) with $|\varepsilon| << 1$ but $\varepsilon < 0$. Then defining $\alpha = -\varepsilon > 0$ and $u(x;\alpha) = y(x;\varepsilon)$ that problem transforms into

$$\alpha\frac{d^2u}{dx^2} - 2\frac{du}{dx} - u = 0,\ 0 < x < 1,\ 0 < \alpha << 1;\ u(0;\alpha) = 0,\ u(1;\alpha) = 1. \tag{3.2.20}$$

This problem has exact solution

$$u(x;\alpha) = \frac{e^{n^{(1)}(\alpha)x} - e^{n^{(2)}(\alpha)x}}{e^{n^{(1)}(\alpha)} - e^{n^{(2)}(\alpha)}} \tag{3.2.21a}$$

where $n^{(1,2)}(\alpha)$ are the root of the canonical quadratic $\alpha n^2 - 2n - 1 = 0$. Thus

$$n^{(1)}(\alpha) \sim -\frac{1}{2},\ n^{(2)}(\alpha) \sim \frac{2}{\alpha} \text{ as } \alpha \to 0. \tag{3.2.21b}$$

Then for $\eta = (1-x)/\alpha$,

$$u(x;\alpha) \sim \frac{e^{-x/2} - e^{2x/\alpha}}{e^{-1/2} - e^{2/\alpha}} \sim U_0(\eta) = e^{-2\eta} \text{ as } \alpha \to 0. \tag{3.2.21c}$$

Here the naïve approximation $u_0(x) \equiv 0$ is the valid asymptotic representation everywhere except for $1 - x = O^*(\alpha)$ as $\alpha \to 0$ and $u_0(1) = \lim_{\eta\to\infty} U_0(\eta) = 0$ (see Fig. 3.5).

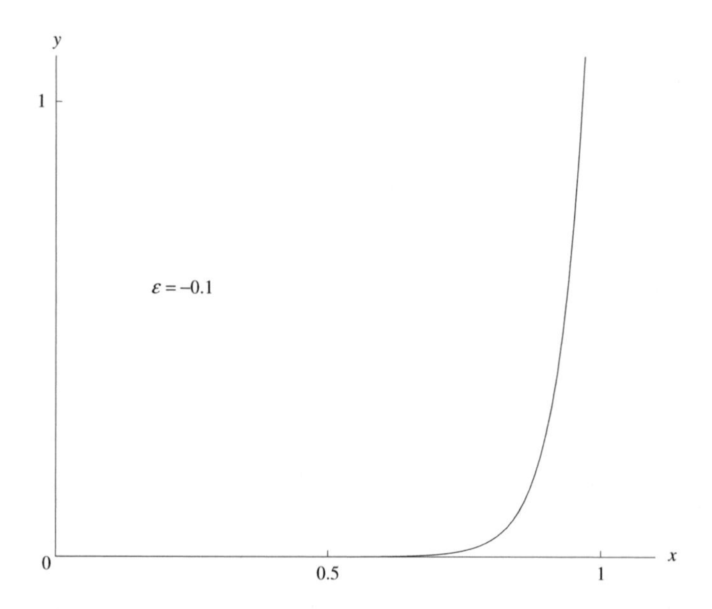

Fig. 3.5 A plot of $y(x;\varepsilon) = u(x;\alpha)$ of (3.2.21) with $\alpha = -\varepsilon = 0.1$.

(c) We would like to develop a procedure called the Method of Matched Asymptotic Expansions which allows us to deduce asymptotic behavior (3.2.19) without having to resort to solving (3.2.14) explicitly since in many instances such singularly perturbed boundary value problems cannot be solved exactly. Toward that end consider Fig. 3.6 and proceed as follows:

   (i) The outer solution: For $x \in I$, which is an interval flanking $x = 1$, assume that

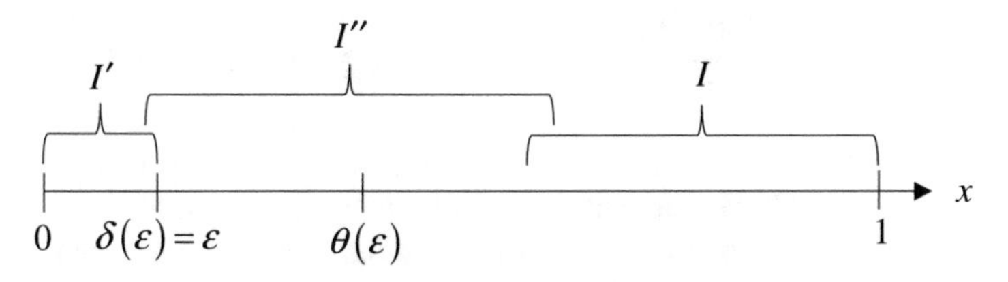

Fig. 3.6 A graphical depiction of the intervals I, I', and I'' as well as the scale factors $\delta(\varepsilon) = \varepsilon$ and $\theta(\varepsilon)$ for the case of the boundary layer located at $x = 0$.

$$y, \frac{dy}{dx}, \frac{d^2y}{dx^2} = O^*(1) \text{ as } \varepsilon \to 0. \tag{3.2.22}$$

(We shall ultimately examine what transpires should it have been assumed that I flanks $x = 0$ instead.) Then

$$y_0(x) = \lim_{\varepsilon \to 0} y(x;\varepsilon) \text{ for } x \in I$$

is such that

$$\frac{d^n y_0}{dx^n}(x) = O(1) \text{ as } \varepsilon \to 0 \text{ for } n = 0, 1, \text{ and } 2.$$

Hence taking the limit of (3.2.14a) and the BC of (3.2.14b) at $x = 1$ as $\varepsilon \to 0$, we obtain

$$2y_0'(x) + y_0(x) = 0,\ y_0(1) = 1 \Rightarrow y_0(x) = e^{(1-x)/2} \tag{3.2.23}$$

which, not surprisingly, was our correct naïve approximation.

(ii) The inner or boundary-layer solution: We wish to select a scale factor $\delta(\varepsilon) > 0$ where $\lim_{\varepsilon \to 0} \delta(\varepsilon) = 0$ and introduce the rescaled variables

$$\xi = \frac{x}{\delta(\varepsilon)},\ Y(\xi;\delta) = y(x;\varepsilon) \tag{3.2.24a}$$

such that for $x \in I'$, an interval where $x = O^*(\delta)$ or $\xi = O^*(1)$ as $\varepsilon \to 0$ (see Fig. 3.6),

$$Y, \frac{dY}{d\xi}, \frac{d^2Y}{d\xi^2} = O^*(1) \text{ as } \varepsilon \to 0. \tag{3.2.24b}$$

Note, in this context, that while for this problem it is not necessary to rescale the dependent variable for problems that are nonlinear and/or nonhomogeneous it is often necessary to rescale both independent and dependent variables (see Chapter 14). Now substituting these variables of (3.2.24) into (3.2.14a) and that BC of (3.2.14b) at $x = 0$, we obtain

$$\frac{d^2Y}{d\xi^2}+2\left[\frac{\delta(\varepsilon)}{\varepsilon}\right]\frac{dY}{d\xi}+\left[\frac{\delta^2(\varepsilon)}{\varepsilon}\right]Y=0,\ \xi>0;\ Y(0;\delta)=0. \tag{3.2.25}$$

Since in particular $d^2Y/d\xi^2 = O^*(1)$ as $\varepsilon \to 0$, neither of the bracketed terms in (3.2.25) can become infinite in that limit and both cannot go to zero simultaneously. Hence employing the same reasoning as introduced earlier in conjunction with our canonical quadratic we may conclude that

$$\delta(\varepsilon)=\varepsilon,\ Y(\xi;\varepsilon)=y(x;\varepsilon) \tag{3.2.26a}$$

which from (3.2.25) yields the *boundary-layer equation*

$$\frac{d^2Y}{d\xi^2}+2\frac{dY}{d\xi}+\varepsilon Y=0,\ \xi>0;\ Y(0;\varepsilon)=0. \tag{3.2.26b}$$

Then

$$Y_0(\xi)=\lim_{\varepsilon\to 0}Y(\xi;\varepsilon)\text{ for }x\in I'$$

is such that

$$\frac{d^nY(\xi)}{d\xi^n}=O^*(1)\text{ as }\varepsilon\to 0\text{ for }n=0,1,\text{ and }2.$$

Thus taking the limit of the boundary-layer equation as $\varepsilon \to 0$ we obtain

$$Y_0''(\xi)+2Y_0'(\xi)=0,\ Y_0(0)=0\ \Rightarrow\ Y_0(\xi)=C_0(1-e^{-2\xi}). \tag{3.2.27}$$

(iii) Intermediate limit technique of matching: Our one-term outer solution $y_0(x)$ is asymptotically valid for $x = O^*(1)$ as $\varepsilon \to 0$ in I while our one-term inner solution $Y_0(\xi)$ is asymptotically valid for $x = O^*(\varepsilon)$ as $\varepsilon \to 0$ in I'. Let $x \in I''$ such that $\gamma = x/\theta(\varepsilon) = O^*(1)$ as $\varepsilon \to 0$ where $\theta(\varepsilon) > 0$ (see Fig. 3.6) and

$$\lim_{\varepsilon\to 0}\theta(\varepsilon)=\lim_{\varepsilon\to 0}\frac{\varepsilon}{\theta(\varepsilon)}=0.$$

Since such $x \in I''$ are intermediate between $x \in I$ and $x \in I'$ we can assume that both the one-term outer and inner solutions are valid for $x \in I''$ as $\varepsilon \to 0$ and hence equal in this limit. Thus

$$\lim_{\varepsilon\to 0}[y_0(x)]_{x=\gamma\theta(\varepsilon)}=\lim_{\varepsilon\to 0}[Y_0(\xi)]_{\xi=\gamma\theta(\varepsilon)/\varepsilon}$$

from which we can deduce the so-called one-term matching rule

$$y_0(0)=\lim_{\xi\to\infty}Y_0(\xi)\ \Rightarrow\ e^{1/2}=C_0.$$

Therefore synthesizing the results of (ii) and (iii) yields the final determination of the one-term inner solution

$$Y_0(\xi) = e^{1/2}(1 - e^{-2\xi}) \text{ for } \xi = \frac{x}{\varepsilon} \tag{3.2.28}$$

in agreement with (3.2.19).

Note that $y_0(x)$ is an asymptotically valid expansion for $y(x;\varepsilon)$ except near $x = 0$. In fact

$$\lim_{\varepsilon\to 0} y(x;\varepsilon) = y_0(x) \text{ for all } x \in (0,1]$$

but that limit is not uniform in x. Observe that

$$\lim_{x\to 0}\left[\lim_{\varepsilon\to 0} y(x;\varepsilon)\right] = \lim_{x\to 0} y_0(x) = e^{1/2} \neq \lim_{\varepsilon\to 0}\left[\lim_{x\to 0} y(x;\varepsilon)\right] = \lim_{\varepsilon\to 0} y(0;\varepsilon) = 0,$$

which is a consequence of this nonuniformity. Indeed,

$$\lim_{\varepsilon\to 0} y(x;\varepsilon) = y_0(x) \text{ uniformly for all } x \in [a,1] \text{ where } 0 < a < 1,$$

but it is not uniform for $x \in (0,b]$ where $0 < b < 1$.

It would be convenient to offer an asymptotic representation for $y(x;\varepsilon)$ that was uniformly valid for all $x \in [0,1]$ as $\varepsilon \to 0$. Toward that end we shall construct the so-called uniformly valid additive composite obtained by adding the inner expansion to the outer and subtracting the common part. That is

$$y_u^{(0)}(x, x/\varepsilon) = y_0(x) + Y_0(x/\varepsilon) - y_0(0). \tag{3.2.29a}$$

Let us examine its behavior in the three intervals depicted in Fig. 3.6:

(i) For $x \in I$:

$$y_u^{(0)}(x, x/\varepsilon) \sim y_0(x) + \lim_{\xi\to\infty} Y_0(\xi) - y_0(0) = y_0(x) \text{ as } \varepsilon \to 0;$$

(ii) For $x \in I'$:

$$y_u^{(0)}(x, x/\varepsilon) \sim \lim_{x\to 0} y_0(x) + Y_0(\xi) - y_0(0) = Y_0(\xi) \text{ as } \varepsilon \to 0;$$

(iii) For $x \in I''$:

$$y_u^{(0)}(x, x/\varepsilon) \sim \lim_{x\to 0} y_0(x) + \lim_{\xi\to\infty} Y_0(\xi) - y_0(0) = y_0(0) \text{ as } \varepsilon \to 0;$$

as required. For our problem

$$y_u^{(0)}(x, x/\varepsilon) = e^{(1-x)/2} + e^{1/2}(1 - e^{-2x/\varepsilon}) - e^{1/2} = e^{1/2}(e^{-x/2} - e^{-2x/\varepsilon}) \tag{3.2.29b}$$

in agreement with (3.2.17e). Hence by deducing all the asymptotic results of the previous subsection we have now come full circle.

Finally we examine what would have occurred had it been assumed that the interval I flanked $x = 0$ – *i.e.*, that the boundary layer were located at $x = 1$. Then we would have found that

$$y(x;\varepsilon) \sim \begin{cases} y_0(x) \equiv 0, & \text{for } x \in I \text{ such that } 0 \in I \\ Y_0(\eta) = C_0 + (1 - C_0)e^{2\eta}, & \text{for } \eta = (1-x)/\varepsilon = O^*(1) \end{cases} \quad \text{as } \varepsilon \to 0,$$

while C_0 must be selected to satisfy the one-term matching rule

$$y_0(1) = \lim_{\eta \to \infty} Y_0(\eta).$$

This would require that we choose C_0 to satisfy

$$C_0 + (1 - C_0) \lim_{\eta \to \infty} e^{2\eta} = 0 \Rightarrow C_0 = 0 \text{ and } 1,$$

simultaneously, which, of course, is impossible. It is characteristic of the Method of Matched Asymptotic Expansions that if one assumes the boundary layer to be located at the wrong end of the interval then the matching condition cannot be satisfied.

Note, in this context, that problem (3.2.20) which does have its boundary layer located at $x = 1$ would yield the following "outer" and boundary-layer one-term asymptotic representations upon employment of the Method of Matched Asymptotic Expansions

$$u(x;\alpha) \sim \begin{cases} u_0(x) = 0, & \text{for } x \in I \text{ such that } 0 \in I \\ U_0(\eta) = e^{-2\eta}, & \text{for } \eta = (1-x)/\alpha = O^*(1) \end{cases}$$

which satisfies the one-term matching rule

$$u_0(1) = \lim_{\eta \to \infty} U_0(\eta) = 0.$$

Generalizing the behavior of these two prototype examples, we may conclude that given the singularly perturbed boundary value problem for $y(x;\varepsilon)$

$$\varepsilon \frac{d^2y}{dx^2} + a_0 \frac{dy}{dx} + b_0 y = 0,\ 0 < \varepsilon << 1,\ b_0 > 0,\ 0 < x < 1; \tag{3.2.30a}$$

$$y(0;\varepsilon) = 0,\ y(1,\varepsilon) = 1; \tag{3.2.30b}$$

the location of the boundary layer is determined by the sign of a_0 being located at $x = 0$ should $a_0 > 0$ and at $x = 1$ should $a_0 < 0$. Indeed the latter case can be transformed into the former one by introducing the change of variables

$$z = 1 - x,\ u(z;\varepsilon) = y(x;\varepsilon);$$

since then

$$\frac{dy}{dx} = -\frac{du}{dz} \text{ while } \frac{d^2y}{dx^2} = \frac{d^2u}{dz^2}.$$

3.3 Closed-Form Temperature-Rate Relationships for Mites Using Singular Perturbation Theory

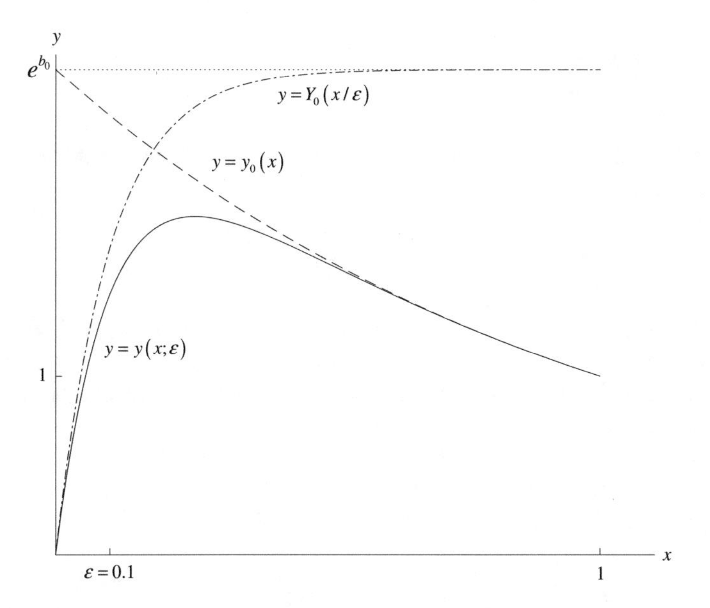

Fig. 3.7 A schematic representation of the solution $y(x;\varepsilon)$ of (3.2.30) with $a_0 \equiv 1$ and $b_0 \equiv$ constant where $y_0(x)$ and $Y_0(x/\varepsilon)$ are as defined by (3.3.1).

One of the most perplexing problems encountered in the natural and engineering sciences concerns itself with the deduction of an optimal closed-form representation which provides the best fit for a given set of data points. It is the aim of this section to demonstrate how the mere knowledge of the singular perturbation results developed in the last section can be used to accomplish the determination of the appropriate analytic expression for rate-temperature phenomena in arthropods. Returning to this topic as described in Section 3.1 and depicted schematically in Fig. 3.2, we can observe from the latter that Phase II exhibits boundary-layer type behavior at $T = T_M$. Motivated both by that occurrence in the high-temperature region and by the form of the solution $y(x;\varepsilon)$ to the singularly perturbed boundary value problem (3.2.30) with $a_0 \equiv 1$ and $b_0 \equiv$ constant given by (see Fig. 3.7 and Problem 3.2)

$$y(x;\varepsilon) \sim \begin{cases} y_0(x) = e^{b_0(1-x)}, & \text{for } x = O^*(1) \\ Y_0(x/\varepsilon) = e^{b_0}(1-e^{-x/\varepsilon}), & \text{for } x = O^*(\varepsilon) \end{cases} \text{ as } \varepsilon \to 0, \tag{3.3.1}$$

it is possible to deduce an analytic expression for population growth rate as a function of temperature which will provide an extremely good fit to our mite data [74]. In particular, we consider Phase I to be an "outer" solution and Phase II to be a boundary layer one. Then denoting the intrinsic growth rate of either mite species by $r(T)$ where T is temperature measured in °C above the given base temperature of 10°C, we define a boundary-layer variable $\Theta = (T_M - T)/\Delta T$ and assume a representation of the form

$$r(T) \sim \begin{cases} r_0(T) = \psi e^{\rho T}, & \text{for } T \text{ in Phase I} \\ R_0(\Theta) = \psi e^{\rho T_M}(1-e^{-\Theta}), & \text{for } T \text{ in Phase II} \end{cases} \text{ as } \frac{\Delta T}{T_M} \to 0, \tag{3.3.2}$$

which follows from (3.3.1) upon introduction of the change of variables and parameters (see Problem 3.2)

$$x = 1 - \frac{T}{T_M},\ y(x;\varepsilon) = \frac{r(T)}{\psi},\ \varepsilon = \frac{\Delta T}{T_M},\ b_0 = \rho T_M, \tag{3.3.3}$$

where $\psi \equiv$ the rate of population growth at the base temperature $T = 0$, $\rho \equiv$ the constant rate of increase to the optimum temperature, $T_M \equiv$ the lethal maximum temperature, and $\Delta T \equiv$ width of the high-temperature boundary layer. Implicit to the formulation of (3.3.2) is the satisfaction of the one-term matching rule $r_0(T_M) = \lim_{\Theta\to\infty} R_0(\Theta) = \psi e^{\rho T_M}$ denoted by the dashed lines in Fig. 3.2. Seeking a representation that will be uniformly valid for all $T \in [0, T_M]$, we form in the usual way the additive composite relevant to (3.3.2)

$$r(T) = r_0(T) + R_0\left(\frac{T_M - T}{\Delta T}\right) - r_0(T_M) = \psi[e^{\rho T} - e^{\rho T_M - (T_M - T)/\Delta T}] \tag{3.3.4a}$$

from which the optimal temperature T_0 such that $r'(T_0) = 0$ can be shown to satisfy (see Problem 3.2)

$$T_0 = T_M\left[1 + \varepsilon\frac{\ln(\rho\Delta T)}{1 - \rho\Delta T}\right]. \tag{3.3.4b}$$

Equation (3.3.4) has several desirable characteristics. It is analytic over its entire temperature domain and is described by biologically meaningful parameters. Here the relevance of ψ and T_M in this context follows directly from their definitions while ρ and ΔT can be interpreted as a composite temperature coefficient for the critical biochemical reaction associated with population growth in Phase I and the length of the temperature interval for which thermal breakdown due to desiccation becomes the overriding influence in Phase II, respectively. The appropriate values of these parameters for each of our mite species can be determined by a least squares fit to generated data points, which incorporate fecundity and life history information [141]. In particular, Fig. 3.8 is a plot of the results of such a procedure for the McDaniel spider mite where

$$\psi = 0.048,\ \rho = 0.103,\ T_M = 28.033,\ \Delta T = 2.710. \tag{3.3.5}$$

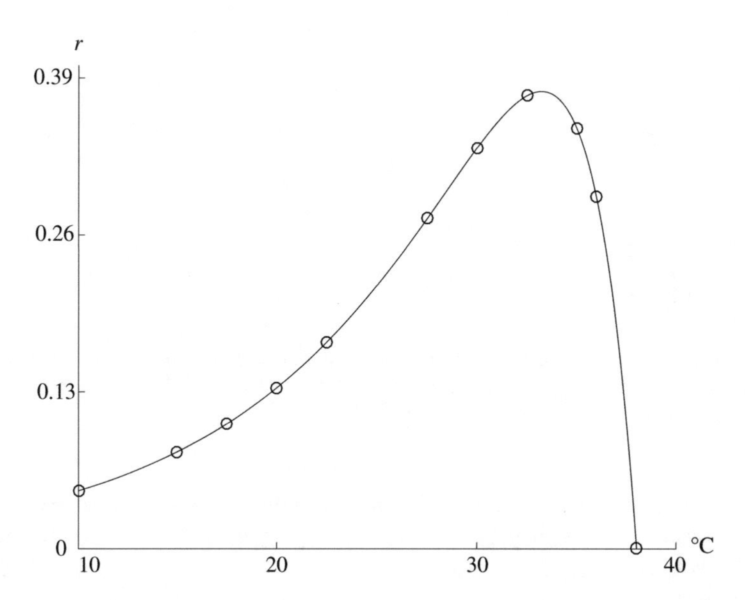

Fig. 3.8 A plot of the intrinsic growth rate per day for the McDaniel spider mite versus temperature measured in °C. The circled points represent the generated data while the curve is a graph of the derived formulation of (3.3.6a) for $r_1(T)$ obtained from (3.3.4a) by the parameter identification of (3.3.5). Here the base temperature is 10°C and the optimal temperature measured in °C above it as determined by (3.3.4b) satisfies $T_0 = 23.24$ and $r_1(T_0) = 0.38$.

These parameter identifications then yield the following explicit temperature-dependent per capita growth rates per day for the two mite species, *T. mcdanieli* and *M. occidentalis*, respectively, given by

$$r_1(T) = 0.048(e^{0.103T} - e^{0.369T-7.457}) \tag{3.3.6a}$$

and (see Problem 3.2)

$$r_2(T) = 0.089(e^{0.055T} - e^{0.483T-11.648}) \tag{3.3.6b}$$

where

$$T \equiv {}^\circ\mathrm{C} - 10^\circ\mathrm{C}, \tag{3.3.6c}$$

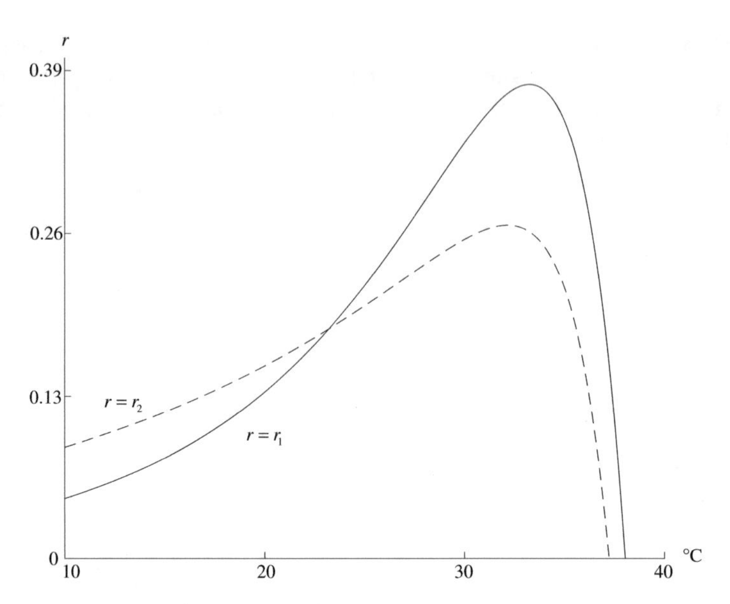

Fig. 3.9 A comparison of the simulated $r_{1,2}$ values as functions of temperature for the two mite species. The solid line is $r_1(T)$ for *T. mcdanieli* as given in Fig. 3.8 while the broken line is $r_2(T)$ for *M. occidentalis* of (3.3.6b) where T =°C−10°C.

which are plotted in Fig. 3.9. Finally we pose the question of whether a procedure of the sort illustrated here can provide further biological insights. Specifically problem (3.2.30) with $a_0 \equiv 1$ under our transformation of (3.3.3) has dimensional form

$$r'(T) = \rho r(T) + \Delta T r''(T),\ 0 < T < T_M; \quad (3.3.7a)$$

$$r(0) = \psi,\ r(T_M) = 0; \quad (3.3.7b)$$

which has the biological interpretation that the change of growth rate with temperature is composed of a term (ρr) due to the effect on critical enzyme-catalyzed biochemical reactions and another ($\Delta T r''$) due to the dissociation of the epicuticle monomolecular wax layer resulting in subsequent desiccation. Fundamentally we are asking the question of whether it is possible to use such a process to deduce governing biological laws where only empiricism existed before. It should be noted that were the results of this procedure to prove fruitful, the usual order of steps standardly taken when one quantifies natural phenomena mathematically would have been exactly reversed.

3.4 Composite May Predator–Prey Mite Model

One of the common interactions between individual organisms most frequently studied by the techniques of biological population dynamics is that of exploitation wherein a victim species is exploited by a predator species. The simplest example of this sort involving two animal populations is a predator–prey interaction involving a carnivore feeding on a herbivore such as the mite biological control situation under investigation. This predator–prey mite ecosystem represents one of the few two-species interactions between terrestrial arthropod populations that can be effectively modeled by differential equations. The generally high biotic potential of mite populations allows them quite rapidly to achieve a stable age distribution with overlapping generations, the latter being a necessary prerequisite for the employment of a differential equations representation [80]. Difference equations models are more appropriate to situations where the generations are nonoverlapping [47]. If, in addition, these populations are relatively numerous with the mean distance of separation between individuals relatively small when compared to the characteristic length scale of their habitat, then they may be represented by differential functions of continuous variables. Further should those populations be distributed uniformly in space then they may be considered to depend upon time alone.

Under these conditions we define $H(t) \equiv$ population density of the prey (*T. mcdanieli*) and $P(t) \equiv$ population density of the predator (*M.occidentalis*), measured in numbers of mites per apple tree leaf, where $t \equiv$ time is measured in days. One of the modeling techniques of biological population dynamics for a scenario of this sort is to represent the rate of growth of these populations by the following system of coupled nonlinear autonomous ordinary differential equations of the form

$$\frac{dH}{dt} = HF(H,P),\ t > 0;\ H(0) = H_0; \tag{3.4.1a}$$

$$\frac{dP}{dt} = PG(H,P),\ t > 0;\ P(0) = P_0; \tag{3.4.1b}$$

where the specific per capita growth rate functions F and G satisfy the exploitation conditions

$$\frac{\partial F}{\partial P} < 0, \tag{3.4.1c}$$

$$\frac{\partial G}{\partial H} > 0, \tag{3.4.1d}$$

and are defined on a case-by-case basis. Toward that end we adopt those particular per capita growth rate functions first employed for the purpose of modeling exploitation interactions by May [80]

$$F(H,P) = r_1\left(1 - \frac{H}{K}\right) - \frac{aP}{H+b}, \tag{3.4.1e}$$

$$G(H,P) = r_2\left(1 - \frac{P}{\gamma H}\right), \tag{3.4.1f}$$

who assembled them from dynamical components originally developed by Holling [49] and Leslie [67], respectively. Here $r_1, r_2 \equiv$ prey and predator growth rates; $a \equiv$ maximal predator per capita consumption rate; $b, K \equiv$ prey densities necessary to achieve one-half that rate and to be at carrying capacity where $b < K$; and $\gamma \equiv$ food quality coefficient of the prey for conversion into predator births. Note that this formulation implicitly employs a Holling [49] type II functional response $f_2(H) = aH/(H+b)$ and a variable predator carrying capacity of γH.

After Danby [25], we are referring to system (3.4.1) as the May model. When Caughley [16] employed this system to model the biological control of prickly pear cactus by the moth *Cactoblastis cactorum*, he generated the theoretical prediction that within two years initially extensive prickly pear stands of density 500 plants per acre would be reduced to a highly *stable equilibrium* value of 11 plants per acre in impressive agreement with actual field data. For other parameter ranges, the May model is capable of generating stable *limit cycles* or isolated periodic orbits possessing both maximum amplitudes and periods independent of initial conditions [80]. Perhaps the most salient feature that can be discovered upon examination of population data collected during laboratory, greenhouse, or field experiments involving a single predacious species feeding upon a single herbivorous one is the occurrence of periodic oscillations or limit cycles. Such population cycles can be anticipated by arguing that once the predator has overexploited its prey, the density level of the former will decline drastically because of lack of food. Then, if this, in turn, gives the prey population a sufficient chance to recover so that the surviving predators can be provided with food soon enough to enable them to increase until they overexploit their prey again, a self-perpetuating predator–prey cycle will have been established. Some of the most compelling evidence in support of such an interpretation for population fluctuations was obtained by Huffaker et al. [51], who studied the interaction between *M. occidentalis* and its prey the six-spotted spider mite feeding on oranges in a carefully controlled environment, which was an ecosystem very similar to the one under examination. Another occurrence which predator–prey model systems have been developed to describe is that of *outbreak phenomenon* wherein population levels can typically increase by several orders of magnitude. Since mean temperature is a biologically meaningful parameter for mite predator–prey systems in seasonal climates and outbreaks also occur naturally in such communities [50], we wish to develop a temperature-dependent so-called *composite* (see below) May model that is unified in the sense it can be used to represent both periodic oscillations and outbreak occurrences in populations simultaneously.

It is important not to confuse the concepts of biomass and number density in this model. The mite populations are being measured in terms of number rather than biomass density so no conclusions should be drawn between the prey biomass consumed by the predator from the functional response $f_2(H)$ measured in number of prey per leaf per day per predator and that converted to predator biomass (numerical response) from its carrying capacity γH measured in number of predators per leaf.

Indeed, since the biomass of one individual prey is so much greater than that of one individual predator and the predators are so much more active than the prey due to their size differential, the determination of such a relationship is a very complicated process. In this context the May model with the appropriate assignment of its parameter values has been used to reproduce successfully many laboratory and field data sets involving two-species exploitation interactions such as that of Caughley [16] described above where $J = 1/\gamma$ was interpreted as a constant of proportionality relating the number of plants needed to sustain a single herbivore.

One of the criticisms frequently levelled against models such as (3.4.1) is that of being overly simplistic. In order to help mitigate this concern we wish to complete our formulation by selecting the model parameters in such a way as to add biological credibility to our system. Toward this end we first assume

$$r_1 = r_1(T),\ r_2 = r_2(T) \tag{3.4.2}$$

as defined in (3.3.6) and then take

$$a = a(T) = a_0 \frac{r_2^2(T)}{r_1(T)} \tag{3.4.3a}$$

with

$$a_0 = 16. \tag{3.4.3b}$$

Since the data points used for the parameter identifications in (3.3.6) were generated by means of a discrete time dynamical systems approach, with the various life stages of mite metamorphosis as state variables and the age-specific deterministic transfer mechanisms from a specific such stage to the next as information flows, while a as defined by (3.3.3) depends only on these representations, our model is actually a *composite* or hybrid one consisting of an analytical framework containing simulation components [72]. The idea behind such a composite model is to capture the descriptive capabilities of an analytical model and the predictive capabilities of a simulation one by combining the best features of these two methods through the linkage just described [95].

Finally, we close this section with a brief discussion of the biological rationale for our choice of a as defined by (3.4.3). In doing so we have keyed a, the maximal predator consumption rate, on r_2, the predator growth rate, by taking

$$a = A_0\left(\frac{r_2}{r_1}\right) r_2 \text{ where the linear function } A_0\left(\frac{r_2}{r_1}\right) \equiv a_0 \frac{r_2}{r_1} \tag{3.4.4}$$

and then a has the biologically reasonable form of being directly proportional to the square of r_2 and inversely proportional to r_1 while the value of its proportionality constant a_0 has been chosen in (3.4.3b) to provide the maximal amount of correlation between the resulting expression of (3.4.3) and the linear interpolation employed by Logan [71]

$$a_\ell(T) = 0.29(8.520 + 0.125T) \tag{3.4.5}$$

over the latter's domain of validity, $11.1 \leq T \leq 22.2$, where the two end points in question were determined by growth chamber experiments at those fixed temperatures. These functions are plotted in Fig. 3.10.

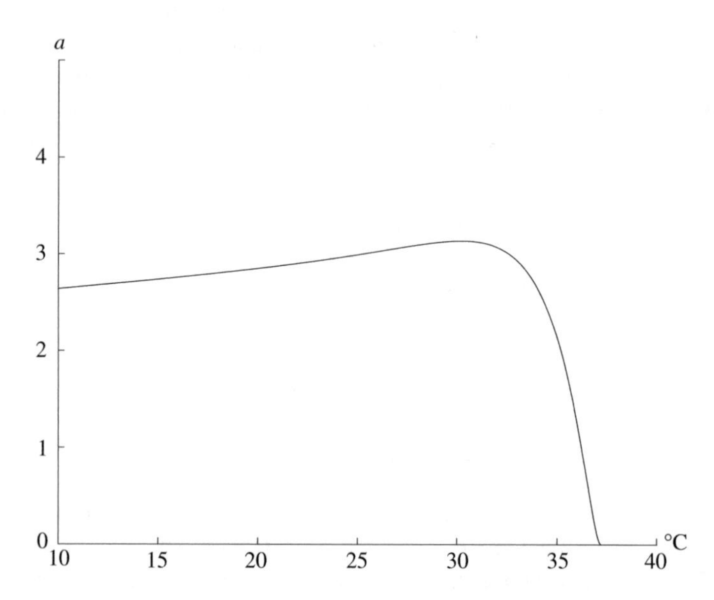

Fig. 3.10 A plot of the maximal consumption rate $a(T)$ of (3.4.3) versus temperature. Note that it virtually coincides with the linear interpolation $a_\ell(T)$ of (3.4.5) over the latter's temperature domain of validity from 21.1°C to 32.2°C.

In order to produce a nondimensionalized version of the May model system, we now introduce the following dimensionless variables and parameters

$$\tau = r_1 t,\ \mathcal{H} = \frac{H}{K},\ \mathcal{P} = \frac{P}{\gamma K},\ D = \frac{b}{K},\ \theta = \frac{r_2}{r_1},\ \varphi = \frac{a\gamma}{r_1}, \tag{3.4.6}$$

which transforms system (3.4.1) into the form

$$\frac{d\mathcal{H}}{d\tau} = \mathcal{H}\mathcal{F}(\mathcal{H},\mathcal{P}),\ \frac{d\mathcal{P}}{d\tau} = \mathcal{P}\mathcal{G}(\mathcal{H},\mathcal{P}),\ \tau > 0; \tag{3.4.7a}$$

$$\mathcal{H}(0) = \mathcal{H}_0,\ \mathcal{P}(0) = \mathcal{P}_0; \tag{3.4.7b}$$

where

$$\mathcal{F}(\mathcal{H},\mathcal{P}) = 1 - \mathcal{H} - \varphi\frac{\mathcal{P}}{\mathcal{H}+D},\ \mathcal{G}(\mathcal{H},\mathcal{P}) = \theta\left(1 - \frac{\mathcal{P}}{\mathcal{H}}\right); \tag{3.4.7c}$$

with

$$\mathcal{H}_0 = \frac{H_0}{K},\ \mathcal{P}_0 = \frac{P_0}{\gamma K}. \tag{3.4.7d}$$

3.5 Pastoral Interlude: Linear Stability Analysis of the Community Equilibrium Point for a General Predator–Prey Model Involving Slopes of the Isoclines and Exploitation Parameters

In this section we return to the general system (3.4.1a)–(3.4.1d) and develop linear stability criteria for its *community equilibrium point*. For such a system a critical point

$$H(T) \equiv H_e,\ P(t) \equiv P_e \text{ satisfying } F(H_e, P_e) = G(H_e, P_e) = 0, \tag{3.5.1}$$

where $H_e, P_e > 0$ is said to be a community equilibrium point. One of the ways to represent a point or points of this sort graphically is to plot the *zero isoclines* of the system in the first quadrant of the H-P plane (see Fig. 3.11)

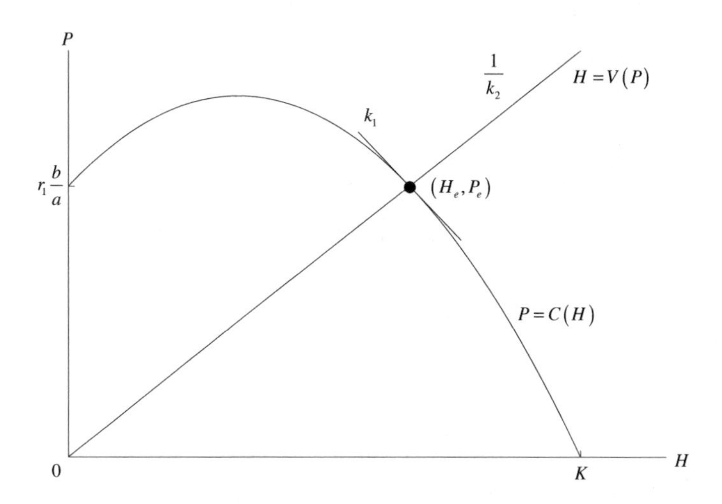

Fig. 3.11 Plots of the zero isoclines for the predator–prey May model system of (3.4.1).

$$F(H,P) \equiv 0 \Rightarrow P = C(H) \tag{3.5.2a}$$

and

$$G(H,P) \equiv 0 \Rightarrow H = V(C) \tag{3.5.2b}$$

and then their point(s) of intersection determines the community equilibrium point(s)

$$P_e = C(H_e) \tag{3.5.2c}$$

and

$$H_e = V(P_e). \tag{3.5.2d}$$

To examine the linear stability of this equilibrium point we investigate the long-time behavior of its initially infinitesimal perturbations. That is considering solutions to (3.4.1a) and (3.4.1b) of the form

$$H(t) = H_e + \varepsilon x(t) + O(\varepsilon^2),\ P(t) = P_e + \varepsilon y(t) + O(\varepsilon^2) \text{ where } |\varepsilon| << 1, \tag{3.5.3}$$

substituting (3.5.3) into (3.4.1a) and (3.4.1b), neglecting terms of $O(\varepsilon^2)$, and cancelling the resulting common ε-factor, we obtain a linear homogeneous system of ordinary differential equations involving the perturbation quantities $x(t)$ and $y(t)$. We shall demonstrate this procedure explicitly for (3.4.1a) after recalling the two-dimensional Taylor series expansion for $F(H,P)$ about the point (H_e, P_e)

$$\begin{aligned} F(H,P) &= F(H_e,P_e) + \frac{\partial F}{\partial H}(H_e,P_e)(H-H_e) + \frac{\partial F}{\partial P}(H_e,P_e)(P-P_e) + O(\varepsilon^2) \\ &= 0 + \frac{\partial F}{\partial H}(H_e,P_e)(\varepsilon x) + \frac{\partial F}{\partial P}(H_e,P_e)(\varepsilon y) + O(\varepsilon^2) \\ &= \varepsilon[(F_1)_e x + (F_2)_e y] + O(\varepsilon^2). \end{aligned}$$

That is,

$$\begin{aligned} \frac{dH}{dt} &= \varepsilon\frac{dx}{dt} + O(\varepsilon^2) = HF(H,P) \\ &= [H_e + \varepsilon x + O(\varepsilon^2)][\varepsilon\{(F_1)_e x + (F_2)_e y\} + O(\varepsilon^2)] \\ &= \varepsilon H_e[(F_1)_e x + (F_2)_e y] + O(\varepsilon^2), \end{aligned}$$

which yields

$$\frac{dx}{dt} = H_e(F_1)_e x + H_e(F_2)_e y \tag{3.5.4a}$$

and, upon extrapolating this result to (3.4.1b),

$$\frac{dy}{dt} = P_e(G_1)_e x + P_e(G_2)_e y. \tag{3.5.4b}$$

We wish to represent the partial derivatives appearing in this perturbation system in terms of the slopes of the isoclines and exploitation parameters of system (3.4.1a)–(3.4.1d) at the community equilibrium point. We can accomplish this goal by employing the following isocline analysis:

Prey: Equation (3.5.2a) implies that

$$f(H) = F[H, C(H)] \equiv 0 \Rightarrow f'(H) = F_1[H, C(H)] + F_2[H, C(H)]C'(H) \equiv 0.$$

Evaluating this result at $H = H_e$, using (3.5.2c), and defining $k_1 = C'(H_e)$, we obtain

$$f'(H_e) = (F_1)_e + (F_2)_e k_1 = 0 \Rightarrow H_e(F_1)_e + H_e(F_2)_e k_1 = 0. \tag{3.5.5}$$

Defining the prey exploitation parameter at equilibrium

$$\alpha = -H_e(F_2)_e > 0, \tag{3.5.6a}$$

by virtue of (3.4.1c), it follows from (3.5.5) that

$$H_e(F_1)_e = \alpha k_1. \tag{3.5.6b}$$

Predator: Equation (3.5.2b) implies that

$$g(P) = G[V(P), P] \equiv 0 \Rightarrow g'(P) = G_1[V(P), P]V'(P) + G_2[V(P), P] \equiv 0.$$

Evaluating this result at $P = P_e$, using (3.5.2d), and defining $k_2 = V'(P_e)$, we obtain

$$g'(P_e) = (G_1)_e k_2 + (G_2)_e = 0 \Rightarrow P_e(G_1)_e k_2 + P_e(G_2)_e = 0. \tag{3.5.7}$$

Defining the predator exploitation parameter at equilibrium

$$\beta = P_e(G_1)_e > 0, \tag{3.5.8a}$$

by virtue of (3.4.1d), it follows from (3.5.7) that

$$P_e(G_2)_e = -\beta k_2. \tag{3.5.8b}$$

Now substitution of (3.5.6) into (3.5.4a) and (3.5.8) into (3.5.4b) yields the desired formulation

$$\frac{dx}{dt} = \alpha(k_1 x - y), \quad \frac{dy}{dt} = \beta(x - k_2 y). \tag{3.5.9}$$

Besides exploitation the other two-species *inter*specific interactions are *mutualism* in which both the partial derivatives of (3.4.1c) and (3.4.1d) are positive and competition in which they are negative. One can also characterize *intra*specific interactions in a similar manner. In this context there is intraspecific prey mutualism at equilibrium when $(F_1)_e > 0$ or $k_1 > 0$ and competition when $(F_1)_e < 0$ or $k_1 < 0$ while there is intraspecific predator mutualism at equilibrium when $(G_2)_e > 0$ or $k_2 < 0$ and competition when $(G_2)_e < 0$ or $k_2 > 0$.

We wish to determine the long-time behavior of these perturbation quantities $x(t)$ and $y(t)$ which is the basic tenet of linear stability theory. That is, wishing to ascertain the fate of disturbances to our community equilibrium solution which are initially infinitesimal, we proceed as usual to seek a solution to (3.5.9) of the form $x(t) = x_1 e^{\lambda t}$

and $y(t) = y_1 e^{\lambda t}$, which upon cancellation of the common exponential factor, yields the following linear homogeneous system of equations for the constants x_1 and y_1:

$$(\lambda - \alpha k_1)x_1 + \alpha y_1 = 0, \quad -\beta x_1 + (\lambda + \beta k_2)y_1 = 0. \tag{3.5.10}$$

Observe that $x_1 = y_1 = 0$ is a solution to (3.5.10). Hence this is an *eigenvalue problem* for λ and we wish to calculate those values of λ that allow nontrivial solutions satisfying $|x_1|^2 + |y_1|^2 \neq 0$. To guarantee the existence of such nontrivial solutions, we require, by Cramer's rule, that the determinant of the coefficients of x_1 and y_1 in (3.5.10) vanishes or

$$\begin{vmatrix} \lambda - \alpha k_1 & \alpha \\ -\beta & \lambda + \beta k_2 \end{vmatrix} = \lambda^2 + (\beta k_2 - \alpha k_1)\lambda + \alpha\beta(1 - k_1 k_2) = 0. \tag{3.5.11}$$

Noting that, for $\lambda = \mu + i\nu$ where $\mu, \nu \in \mathbb{R}$, $e^{\lambda t} = e^{\mu t}[\cos(\nu t) + i \sin(\nu t)]$ and

$$|e^{\lambda t}| = \sqrt{e^{2\mu t}[\cos^2(\nu t) + \sin^2(\nu t)]} = \sqrt{e^{2\mu t}} = e^{\mu t} = e^{\mathrm{Re}(\lambda)t}, \tag{3.5.12a}$$

we may conclude that

$$|e^{\lambda t}| \to \begin{cases} 0 & \text{for } \mathrm{Re}(\lambda) < 0 \\ 1 & \text{for } \mathrm{Re}(\lambda) = 0 \\ \infty & \text{for } \mathrm{Re}(\lambda) > 0 \end{cases} \text{ as } t \to \infty. \tag{3.5.12b}$$

Thus we can say that our community equilibrium point is stable, neutrally stable, or unstable to initially infinitesimal disturbances depending upon whether the Re(λ) is less than, equal to, or greater than, zero, respectively. Given the form of (3.5.11) the desired stability criteria can be deduced from the fact that for $\lambda_{1,2}$ satisfying the quadratic

$$\lambda^2 + 2c\lambda + d = (\lambda - \lambda_1)(\lambda - \lambda_2) = 0, \ \mathrm{Re}(\lambda_{1,2}) < 0 \text{ if and only if (iff) } c, d > 0, \tag{3.5.13}$$

which is demonstrated as follows: Here $\lambda_{1,2} = -c \pm \sqrt{c^2 - d}$ and there are two cases to consider

(i) $c^2 - d < 0$: $\mathrm{Re}(\lambda_{1,2}) = -c < 0$ iff $c > 0 \Rightarrow d > c^2 > 0$;
(ii) $c^2 - d \geq 0$: $\mathrm{Re}(\lambda_{1,2}) = \lambda_{1,2}$; since $\lambda_1 + \lambda_2 = -2c$, $\lambda_1\lambda_2 = d \Rightarrow \lambda_{1,2} < 0$ iff $c, d > 0$.

For (3.5.12), the criteria of (3.5.13) yield the stability region plotted in the k_2-k_1 plane of Fig. 3.12

$$k_1 < \frac{\beta k_2}{\alpha}, \ k_1 k_2 < 1. \tag{3.5.14}$$

The boundaries of that stability region are composed of the union of the curves

$$k_1 = \frac{\beta k_2}{\alpha} \text{ where } k_1 k_2 < 1 \text{ and } k_1 k_2 = 1 \text{ where } k_1 < \frac{\beta k_2}{\alpha}, \tag{3.5.15}$$

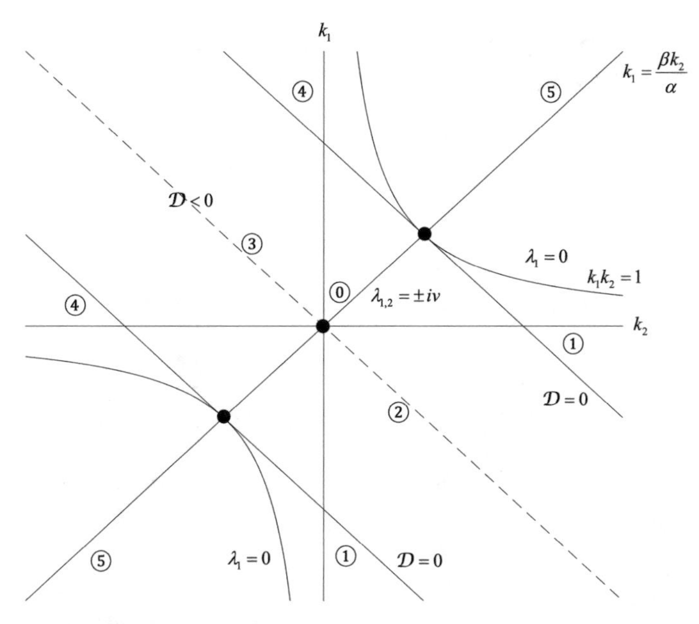

Fig. 3.12 Plots of the stability regions in the k_2-k_1 plane relevant to (3.5.11). Here ⓪ ≡ center, ① ≡ stable node, ② ≡ stable focus, ③ ≡ unstable focus, ④ ≡ unstable node, ⑤ ≡ saddle point.

which have intersection points $\pm(\sqrt{\alpha/\beta}, \sqrt{\beta/\alpha})$, correspond to $\lambda_{1,2} = \pm i\nu$ and $\lambda_1 = 0$, respectively, and comprise the neutral stability locus for this figure. Given the two cases considered above we now examine the sign of the discriminant $\mathcal{D}$ for the quadratic (3.5.11) in λ. Since

$$\mathcal{D} = (\beta k_2 - \alpha k_1)^2 - 4\alpha\beta(1 - k_1 k_2) = (\beta k_2 + \alpha k_1)^2 - 4\alpha\beta, \tag{3.5.16a}$$

$$\mathcal{D} \begin{cases} < 0 & \text{provided } |\beta k_2 + \alpha k_1| < 2\sqrt{\alpha\beta} \\ = 0 & \text{provided } \beta k_2 + \alpha k_1 = \pm 2\sqrt{\alpha\beta}\,. \\ > 0 & \text{provided } |\beta k_2 + \alpha k_1| > 2\sqrt{\alpha\beta} \end{cases} \tag{3.5.16b}$$

The loci corresponding to $\mathcal{D} = 0$ are also plotted in Fig. 3.12. Observe that the intersection points of the portions of the marginal curve defined in (3.5.15) lie on these loci which are tangent to their $k_1 k_2 = 1$ portions. The region corresponding to $\mathcal{D} < 0$ is identified in this figure as well.

We now investigate the linear stability of the community equilibrium point by cataloguing the dynamical behavior of $x = y = 0$ for perturbation system (3.5.9) in the x-y plane relevant to all the k_2-k_1 regions identified by circled numbers in Fig. 3.12.

These six qualitatively different cases are plotted in Fig. 3.13. We begin with the three cases involving $\lambda_{1,2} \in \mathbb{R}$ for which $\mathcal{D} > 0$. Then

$$\boldsymbol{X}(t) = \begin{bmatrix} x(t) \\ y(t) \end{bmatrix} = c_1 \boldsymbol{\xi}_1 e^{\lambda_1 t} + c_2 \boldsymbol{\xi}_2 e^{\lambda_2 t}.$$

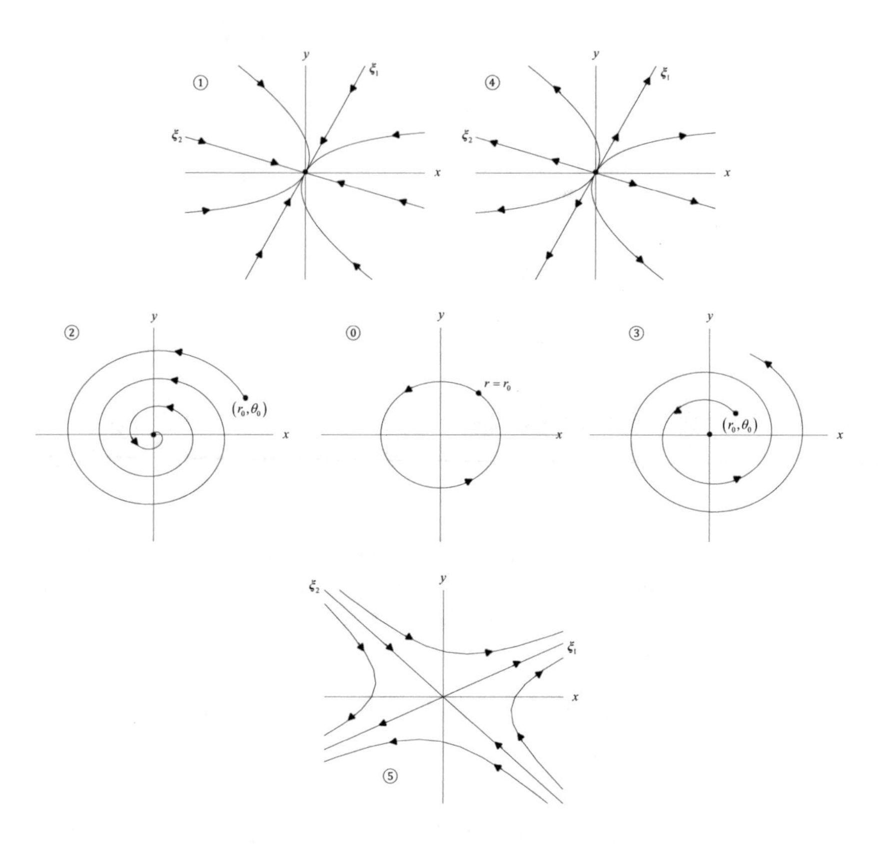

Fig. 3.13 Phase-plane portraits relevant to the stability regions of Fig. 3.12.

①: $\lambda_2 < \lambda_1 < 0$: $\lim_{t\to\infty} \boldsymbol{X}(t) = \boldsymbol{0} \Rightarrow x = y = 0$ is a *stable node*.

④: $\lambda_1 > \lambda_2 > 0$: $\lim_{t\to\infty} |\boldsymbol{X}(t)| \to \infty \Rightarrow x = y = 0$ is an *unstable node*.

⑤: $\lambda_1 > 0, \lambda_2 < 0$: $\lim_{t\to\infty} |\boldsymbol{X}(t)|$ is unbounded as $t \to \infty \Rightarrow x = y = 0$ is a *saddle point*.

We continue with the three cases involving $\lambda_{1,2} = \mu \pm i\nu$ where $\mu \in \mathbb{R}$ and $\nu > 0$ for which $\mathcal{D} < 0$ or $\beta k_2 + \alpha k_1 = \delta$ where $|\delta| < 2\sqrt{\alpha\beta}$. Since the behavior in these regions is qualitatively identical independent of the value of δ, we explicitly consider $\delta = 0$ or $\alpha k_1 = -\beta k_2 = \mu$ (see the dotted line in Fig. 3.13) with symmetric exploitation parameters $\alpha = \beta = \nu$ for ease of exposition. Under these conditions (3.5.9) and (3.5.11) become

$$\frac{dx}{dt} = \mu x - \nu y, \ \frac{dy}{dt} = \nu x + \mu y \tag{3.5.17a}$$

and

$$(\lambda - \mu)^2 + \nu^2 = 0, \tag{3.5.17b}$$

respectively, which is consistent with $\lambda_{1,2} = \mu \pm i\nu$. In order to solve (3.5.17a), we introduce the polar coordinate change of variables

$$x = r\cos(\theta), \ y = r\sin(\theta) \text{ where } r = r(t), \ \theta = \theta(t), \tag{3.5.18a}$$

which transforms (3.5.17a) into

$$\frac{dr}{dt} = \mu r, \ r\frac{d\theta}{dr} = \nu r. \tag{3.5.18b}$$

Note that (3.5.18b) admits the trivial solution $r \equiv 0$. Then, for $r \not\equiv 0$,

$$r(t) = r_0 e^{\mu t}, \ \theta(t) = \nu t + \theta_0.$$

(2): $\mu < 0$: $\lim_{t\to\infty} r(t) = 0 \Rightarrow x = y = 0$ is a *stable focus* or *spiral point*.

(0): $\mu = 0$: $r(t) \equiv r_0 \Rightarrow x = y = 0$ is a *center*.

(3): $\mu > 0$: $\lim_{t\to\infty} r(t) \to \infty \Rightarrow x = y = 0$ is an *unstable focus* or *spiral point*.

We close with a discussion of the two degenerate cases corresponding to $\mathcal{D} = 0$ for which $\lambda_{1,2} \in \mathbb{R}$ where $\lambda_1 = \lambda_2 \neq 0$. Then

$$\boldsymbol{X}(t) = c_1\boldsymbol{\xi_1}e^{\lambda_1 t} + c_2(\boldsymbol{\xi_1}te^{\lambda_1 t} + \boldsymbol{\eta}e^{\lambda_1 t}).$$

For these cases $x = y = 0$ can be classified as an *improper node* which is stable for $\lambda_1 < 0$ and unstable for $\lambda_1 > 0$. Fig. 3.14 contains phase portraits of these two cases from which it can be observed that each appears not surprisingly to be a cross between a node and a spiral point since $\mathcal{D} = 0$ serves as the boundary separating the two regions characteristic of such behavior, respectively.

We conclude this section with the observation that the stability region of (3.5.13) consists of stable nodes in (1), stable improper nodes on $\mathcal{D} = 0$, and stable foci or spiral points in (2). Note that the whole fourth quadrant of the k_2-k_1 plane is contained in that region, and hence, intraspecific competitive interaction for both species guarantees linear stability for the community equilibrium point. In addition the onset of instability when crossing the $k_1k_2 = 1$ portion of the marginal curve occurs from a stationary state while this onset occurs from an oscillatory state when

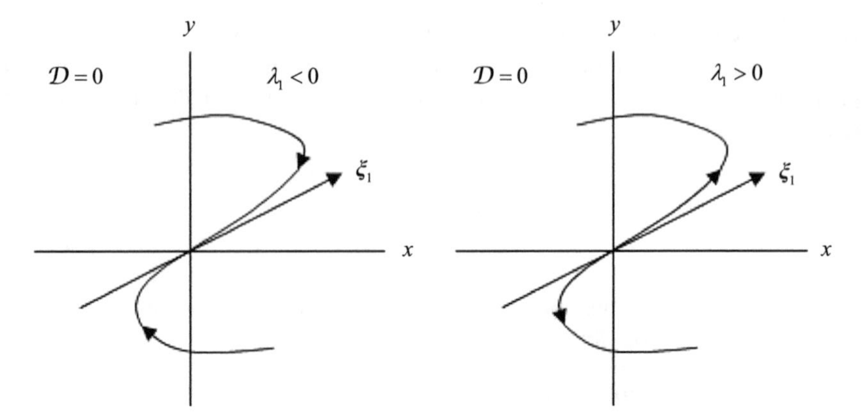

Fig. 3.14 Phase-plane portraits relevant to the $\mathcal{D}=0$ lines of Fig. 3.12 where $\lambda_1<0$ and $\lambda_1>0$ correspond to stable and unstable improper nodes, respectively.

crossing the $k_1=\beta k_2/\alpha$ portion, the former occurrence being called an exchange of stabilities and the latter, a Hopf bifurcation.

3.6 Linear Stability Analysis of the Temperature-Dependent Composite May Predator–Prey Mite Model

We wish to apply the results of the linear stability analysis of the last section to our temperature-dependent composite May predator–prey mite model developed in Section 3.4. Toward that end we first calculate α, β, k_1, and k_2 relevant to system (3.4.1). For the specific per capita growth rate functions F and G of (3.4.1e) and (3.4.1f) the prey and predator isoclines are given by

$$P=C(H)=\frac{r_1}{a}\left[-\frac{H^2}{K}+\left(1-\frac{b}{K}\right)H+b\right],\ H=V(P)=\frac{P}{\gamma},\tag{3.6.1}$$

respectively. Hence the community equilibrium point satisfies

$$P_e=C(H_e)=\gamma H_e\ \Rightarrow\ \mathcal{H}_e^2+(\varphi+D-1)\mathcal{H}_e-D=0 \text{ where } \mathcal{H}_e=\frac{H_e}{K}\tag{3.6.2}$$

or, taking the positive root of (3.6.2), we obtain

$$2\mathcal{H}_e=1+\Delta-(\varphi+D) \text{ with } \Delta>0\tag{3.6.3a}$$

where

$$\Delta^2 = (\varphi + D - 1)^2 + 4D = (\varphi + D)^2 + 2(D - \varphi) + 1 = \varphi^2 + 2(D-1)\varphi + (D+1)^2. \tag{3.6.3b}$$

Now, for the specific F of (3.4.1), $(\partial F/\partial P)(H,P) = -a/(H+b)$; hence

$$\alpha = -H_e(F_2)_e = a\frac{H_e}{H_e+b} = a\frac{2\mathcal{H}_e}{2\mathcal{H}_e+2D} = \frac{2a}{1+\Delta+D+\varphi} \tag{3.6.4a}$$

given that

$$\frac{2\mathcal{H}_e}{2\mathcal{H}_e+2D} = \frac{1+\Delta-(\varphi+D)}{1+\Delta+D-\varphi} = \frac{2}{1+\Delta+D+\varphi}$$

since

$$\begin{aligned}[1+\Delta-(\varphi+D)](1+\Delta+\varphi+D) &= (1+\Delta)^2-(\varphi+D)^2\\ &= \Delta^2+2\Delta+1-(\varphi+D)^2 = 2(D-\varphi)+1+2\Delta+1\\ &= 2(1+\Delta+D-\varphi);\end{aligned}$$

while, for the $C(H)$ of (3.6.1), $C'(H) = (r_1/a)(-2H/K + 1 - b/K)$; hence

$$k_1 = C'(H_e) = \frac{r_1}{a}(-2\mathcal{H}_e + 1 - D) = \frac{r_1}{a}(\varphi - \Delta). \tag{3.6.4b}$$

Further, for the specific G of (3.4.1f), $(\partial G/\partial H)(H,P) = r_2P/(\gamma H^2)$; hence

$$\beta = P_e(G_1)_e = r_2\frac{P_e^2}{\gamma H_e^2} = r_2\frac{(\gamma H_e)^2}{\gamma H_e^2} = r_2\gamma; \tag{3.6.5a}$$

while for the $V(P)$ of (3.6.1), $V'(P) \equiv 1/\gamma$; hence

$$k_2 = V'(P_e) = \frac{1}{\gamma}. \tag{3.6.5b}$$

We next consider the so-called secular equation (3.5.11) in λ and rewrite it in terms of σ, a dimensionless linear growth rate relevant to our nondimensional time $\tau = r_1 t$ of (3.4.6). The relationship between λ and σ can most easily be determined by examining

$$e^{\lambda t} = e^{(\lambda/r_1)\tau} = e^{\sigma\tau}, \tag{3.6.6a}$$

from which it becomes obvious we should define

$$\sigma = \frac{\lambda}{r_1}. \tag{3.6.6b}$$

Then, dividing (3.5.11) by r_1^2 and employing (3.6.6b), we obtain

$$\sigma^2 + \left(\frac{\beta k_2 - \alpha k_1}{r_1}\right)\sigma + \alpha\beta\frac{1-k_1k_2}{r_1^2} = 0. \tag{3.6.7}$$

In order to particularize this result to our system of Section 3.4, we calculate the coefficients of (3.6.7) for the parameter values of (3.6.4) and (3.6.5):

$$\frac{\beta k_2 - \alpha k_1}{r_1} = \theta - \theta_c \text{ where } \theta_c = 2\frac{\varphi - \Delta}{1 + \Delta + D + \varphi}; \tag{3.6.8a}$$

$$\frac{\alpha\beta(1 - k_1 k_2)}{r_1^2} = \frac{2\theta\varphi[1 - (\varphi - \Delta)/\varphi]}{1 + \Delta + D + \varphi} = \frac{2\theta\Delta}{1 + \Delta + D + \varphi}. \tag{3.6.8b}$$

Substitution of (3.6.8) into (3.6.7) yields the explicit secular equation for our system

$$\sigma^2 + (\theta - \theta_c)\sigma + \frac{2\theta\Delta}{1 + \Delta + D + \varphi} = 0, \tag{3.6.9}$$

from which we can conclude, upon application of (3.5.13), that there will be linear stability provided both the coefficients in (3.6.9) are positive. Since coefficient (3.6.8b) is always positive, we have deduced the linear stability criterion

$$\theta > \theta_c = \frac{2(\varphi - \Delta)}{1 + \Delta + D + \varphi}. \tag{3.6.10}$$

In Fig. 3.15 we plot the curves θ and θ_c of (3.6.10) versus T upon employment of (3.3.6), (3.4.2), (3.4.3), and (3.4.6) and upon assignment of the typical parameter values ([73]; [36])

$$\gamma = 0.15,\ K = 300 \text{ mites per leaf},\ b = 1 \text{ mite per 25 leafs} \Rightarrow D \cong 0.00013. \tag{3.6.11}$$

This figure represents graphically the stability criterion (3.6.10) from which we can deduce that the community equilibrium point is locally stable for both $0 \le T < T_1$ and $T_2 < T < T_M$ while it is locally unstable for $T_1 < T < T_2$. Here $T_M = 27.21$ denotes the predator lethal maximum temperature while $T_1 = 20.89$ and $T_2 = 25.56$ denote the temperatures at which Hopf bifurcation occurs. The main fruit of linear stability theory is the determination of the critical conditions for the onset of instability to initially small disturbances, and its results are valid for times sufficiently short so that the nonlinear terms can be neglected but as these disturbances grow those terms must be taken into account.

3.7 Pastoral Interlude: Global Stability Behavior of Kolmogorov-Type Predator–Prey Systems and a Limit Cycle Example

There are two major deficiencies of linear stability theory as applied to dynamical systems such as our predator–prey model: (i): It does not determine the proper long-time behavior of those growing disturbances that are predicted to be linearly unstable. (ii): It does not ascertain the effect of finite amplitude disturbances that are not initially infinitesimal. Hence linear theory alone cannot be used to deduce

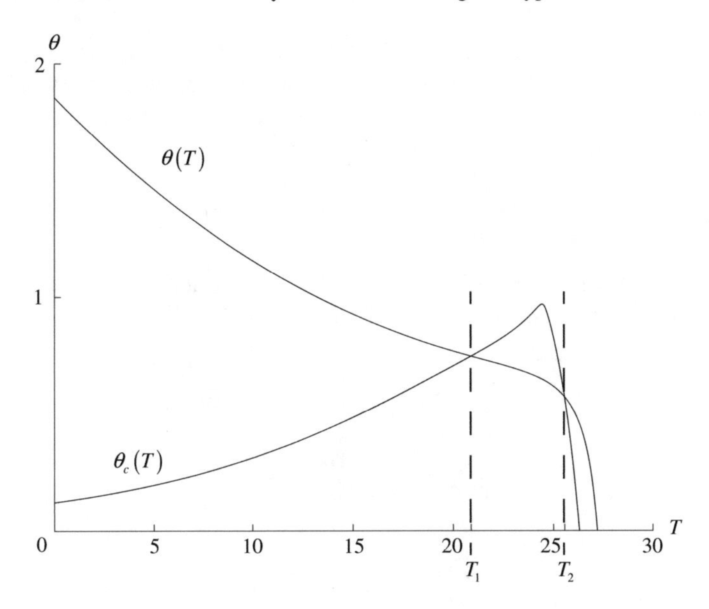

Fig. 3.15 A plot of θ and θ_c of (3.6.10) as functions of $T \equiv$°C −10°C for the mite and maximal consumption rates as given in Figs. 3.9 and 3.10 with $\gamma = 0.15$ and $D = 0.00013$, showing their intersection points which occur at $T_1 = 20.89$ and $T_2 = 25.56$. Note that the composite May mite model has a linearly stable community equilibrium point when $\theta > \theta_c$.

the global stability behavior of such dynamical systems. One of the methods that can be used to predict the global behavior of ordinary differential equation systems of the general form (3.4.1a)–(3.4.1b) is Kolmogorov's Theorem which is a special application of Poincaré–Bendixson theory (see [11]) as pointed out by May [80]. If that system satisfies the following properties, (1) $\partial F/\partial P < 0$, $\partial G/\partial P < 0$, (2) $H\ \partial F/\partial H + P\ \partial F/\partial P < 0$, $H\ \partial G/\partial H + P\ \partial G/\partial P > 0$, (3) $F(0,0) > 0$, $F(0,A) = 0$ with $A > 0$, (4) $F(K,0) = 0$, $G(C,0) = 0$ with $K > C > 0$, one can conclude that there exists a single community equilibrium point which is either globally stable or surrounded by a closed integral curve which is globally stable from the outside. Various possibilities can occur inside it including the existence of a bilaterally stable limit cycle (see below). For a predator–prey system these conditions have the biological implications: (1): The prey and predator per capita rates of increase are decreasing functions of the predator population size (note the former condition is always satisfied for exploitation interactions). (2): For any given ratio between the species the rate of increase of the prey species is a decreasing function of population size while that of the predator is an increasing function. (3): When both populations are small the prey have a positive rate of increase and there can be a predator population size that can stop further prey increase even for very few prey. (4): There exists a prey carrying capacity beyond which they cannot increase even in the absence of the predator and a critical prey density that stops predator increase even for low predator densities while the system will crash unless the carrying capacity exceeds that critical density [80]. This theorem can also hold for specific models when certain of the above equations are equalities rather than inequalities. In particular, it is valid for the May predator–prey model even though [80]

$$H\frac{\partial G}{\partial H} + P\frac{\partial G}{\partial P} = 0 \text{ and } G(C,C) = 0 \text{ with } C \to 0.$$

Let us demonstrate the satisfaction of those conditions for this May predator–prey model in its nondimensionalized form of system (3.4.7). Then

$$\frac{\partial \mathcal{F}}{\partial \mathcal{H}} = -1 - \varphi\frac{\mathcal{P}}{(\mathcal{H}+D)^2},\ \frac{\partial \mathcal{F}}{\partial \mathcal{P}} = -\frac{\varphi}{\mathcal{H}+D};$$

$$\frac{\partial \mathcal{G}}{\partial \mathcal{H}} = \theta\frac{\mathcal{P}}{\mathcal{H}^2},\ \frac{\partial \mathcal{G}}{\partial \mathcal{P}} = -\frac{\theta}{\mathcal{H}}.$$

Thus

$$\textcircled{1}: \frac{\partial \mathcal{F}}{\partial \mathcal{P}} < 0,\ \frac{\partial \mathcal{G}}{\partial \mathcal{P}} < 0;$$

$$\textcircled{2}: \mathcal{H}\frac{\partial \mathcal{F}}{\partial \mathcal{H}} + \mathcal{P}\frac{\partial \mathcal{F}}{\partial \mathcal{P}} = -\mathcal{H} + \varphi\frac{\mathcal{H}\mathcal{P}}{(\mathcal{H}+D)^2} - \varphi\frac{\mathcal{P}}{\mathcal{H}+D} = -\mathcal{H} - \varphi\frac{D\mathcal{P}}{(\mathcal{H}+D)^2} < 0,$$

$$\mathcal{H}\frac{\partial \mathcal{G}}{\partial \mathcal{H}} + \mathcal{P}\frac{\partial \mathcal{G}}{\partial \mathcal{P}} = 0;$$

$$\textcircled{3}: \mathcal{F}(0,0) = 1 > 0;\ \mathcal{F}(0,\mathcal{A}) = 1 - \varphi\frac{\mathcal{A}}{D} = 0 \Rightarrow \mathcal{A} = \frac{D}{\varphi} > 0;$$

$$\textcircled{4}: \mathcal{F}(\mathcal{K},0) = 1 - \mathcal{K} = 0 \Rightarrow \mathcal{K} = 1;\ \mathcal{G}(\mathfrak{C},\mathfrak{C}) = \theta\left(1 - \frac{\mathfrak{C}}{\mathfrak{C}}\right) = 0 \text{ with } 0 < \mathfrak{C} << 1 = \mathcal{K}.$$

Hence, from Kolmogorov's Theorem, we may conclude, as stated above, that the single community equilibrium point of the predator–prey May model is either globally stable or surrounded by a closed integral curve that is globally stable from the outside. We shall investigate the global stability behavior of our temperature-dependent mite composite version of this May predator–prey model more completely in the next section. Since that behavior is crucially dependent upon the occurrence of limit cycles we close this section by offering an example of a model dynamical system which can give rise to such stable periodic solutions.

Let $x_1 = x_1(t)$, $x_2 = x_2(t)$ satisfy the coupled autonomous homogeneous nonlinear system [4]

$$\frac{dx_1}{dt} = F_1(x_1,x_2) = \mu x_1 - x_2 - x_1(x_1^2 + x_2^2), \tag{3.7.1a}$$

$$\frac{dx_2}{dt} = F_2(x_1,x_2) = x_1 + \mu x_2 - x_2(x_1^2 + x_2^2); \tag{3.7.1b}$$

where the *bifurcation* parameter $\mu \in \mathbb{R}$. Consider

$$x_1(t) \equiv x_e,\ x_2(t) \equiv y_e \text{ such that } F_1(x_e, y_e) = F_2(x_e, y_e) = 0. \tag{3.7.2a}$$

We shall demonstrate that

$$x_e = y_e = 0 \tag{3.7.2b}$$

as follows: First note that since (3.7.1) is autonomous $F_1(0,0) = F_2(0,0) = 0$. Then assume that $x_e, y_e \neq 0$. Thus (3.7.2a) implies that $\mu - (x_e^2 + y_e^2) = y_e/x_e = -x_e/y_e$ or $x_e^2 + y_e^2 = 0$, which yields no nontrivial solutions. Hence the origin (3.7.2a) is the only critical point of (3.7.1).

We next perform a linear stability analysis of that critical point by seeking a solution of (3.7.1) of the form

$$x_1(t) = x_e + \varepsilon x(t) + O(\varepsilon^2),\ x_2(t) = y_e + \varepsilon y(t) + O(\varepsilon^2) \text{ where } x_e = y_e = 0,\ |\varepsilon| << 1. \tag{3.7.3}$$

Then upon substitution of (3.7.3) into (3.7.1), neglect of terms of $O(\varepsilon^2)$, and cancellation of the resulting common ε-factor, we obtain the following linear homogeneous system of ordinary differential equations in the perturbation quantities $x(t)$ and $y(t)$:

$$\frac{dx}{dt} = \mu x - y,\ \frac{dy}{dt} = x + \mu y. \tag{3.7.4a}$$

This is of the same form as (3.5.17) with $\nu = 1$. Hence, letting

$$x(t) = x_{11}e^{\lambda t},\ y(t) = y_{11}e^{\lambda t} \text{ where } |x_{11}|^2 + |y_{11}|^2 \neq 0, \tag{3.7.4b}$$

we would find that

$$\lambda_{1,2} = \mu \pm i. \tag{3.7.4c}$$

Then employment of the polar coordinate change of variables of (3.5.18) would yield the solution

$$x = r\cos(\theta),\ y = r\sin(\theta) \text{ where } r = r_0 e^{\mu t},\ \theta = t + \theta_0. \tag{3.7.4d}$$

Hence as in Section 3.5 and the corresponding plots of Fig. 3.13 the origin can be classified as (2): A stable spiral point if $\mu < 0$; (0): A center if $\mu = 0$; (3): An unstable spiral point if $\mu > 0$.

In order to determine the global behavior of this system we introduce the polar coordinate transformation

$$x_1 = r\cos(\theta),\ x_2 = r\sin(\theta) \text{ where } r = r(t), \theta = \theta(t) \tag{3.7.5a}$$

directly into system (3.7.1) and find that r and θ satisfy the following ordinary differential equations

$$\frac{dr}{dr} = (\mu - r^2)r,\ r\frac{d\theta}{dt} = r \text{ or } \frac{d\theta}{dt} = 1 \text{ for } r \not\equiv 0. \tag{3.7.5b}$$

Thus

$$\theta(t) = t + \theta_0 \tag{3.7.6}$$

while multiplication of the equation for dr/dt by r and making the change of variables

$$\tau = 2t, \ \rho(\tau) = r^2(t) \tag{3.7.7a}$$

yields the Bernoulli equation

$$-\frac{d\rho}{d\tau} + \mu\rho = \rho^2, \tag{3.7.7b}$$

which, upon division by ρ^2 and introduction of

$$u(\tau) = \frac{1}{\rho(\tau)}, \tag{3.7.8a}$$

becomes

$$\frac{du}{d\tau} + \mu u = 1. \tag{3.7.8b}$$

Solving (3.7.8) and employing (3.7.7a) we find that

$$u(2t) = \frac{1}{\rho(2t)} = \frac{1}{r^2(t)} = \begin{cases} 1/\mu + ce^{-2\mu t}, & \text{for } \mu \neq 0 \\ 2t + c_0, & \text{for } \mu = 0 \end{cases}, \tag{3.7.9a}$$

where

$$c = \frac{1}{r_0^2} - \frac{1}{\mu} \text{ and } c_0 = \frac{1}{r_0^2}. \tag{3.7.9b}$$

Since, from (3.7.9) it follows that

$$\frac{1}{\lim_{t\to\infty} r^2(t)} \begin{cases} \to \infty, & \text{for } \mu \leq 0 \\ 1/\mu, & \text{for } \mu > 0 \end{cases},$$

we can conclude

$$\lim_{t\to\infty} r(t) = \begin{cases} 0, & \text{for } \mu \leq 0 \\ \mu^{1/2}, & \text{for } \mu > 0 \end{cases}.$$

Hence the origin is a globally stable spiral point for system (3.7.1) when $\mu \leq 0$ while it is surrounded by the bilaterally stable limit cycle solution to system (3.7.1)

$$x_1(t) = \mu^{1/2}\cos(t + \theta_0), x_2(t) = \mu^{1/2}\sin(t + \theta_0) \tag{3.7.10a}$$

when $\mu > 0$ which represents a circle centered at the origin with radius $\mu^{1/2}$ traversed in the counterclockwise direction and therefore satisfying the equation

$$x_1^2 + x_2^2 = \mu. \tag{3.7.10b}$$

The phase portraits relevant to these two cases are plotted in the x_1-x_2 plane of Fig. 3.16. Note in this context that (3.7.10a) is a solution to system (3.7.1) by virtue of (3.7.10b) and

$$\frac{dx_1(t)}{dt} = -\mu^{1/2}\sin(t+\theta_0) = -x_2(t),\ \frac{dx_2(t)}{dt} = \mu^{1/2}\cos(t+\theta_0) = x_1(t). \quad (3.7.10c)$$

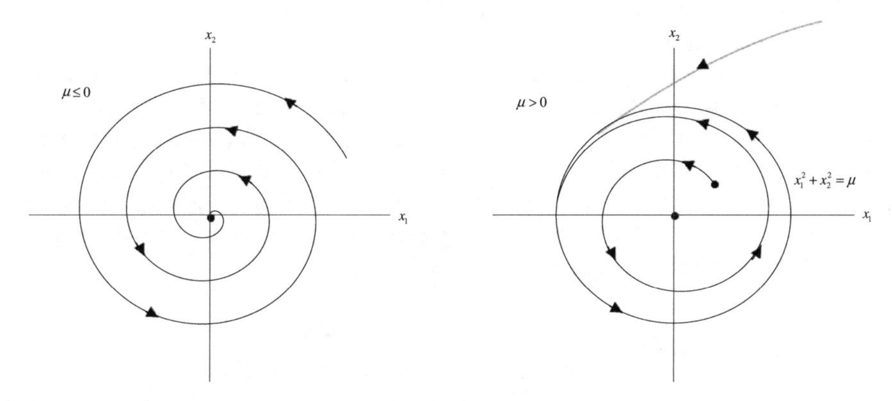

Fig. 3.16 Phase-plane plots in the x_1-x_2 plane of the solutions to system (3.7.1) showing that the origin is a globally stable equilibrium point for $\mu \le 0$ and is surrounded by the bilaterally stable limit cycle $x_1^2 + x_2^2 = \mu$ traversed in the counterclockwise direction for $\mu > 0$.

Thus upon inclusion of the nonlinear terms of system (3.7.1) its spiral point that is locally stable for $\mu < 0$ becomes globally stable, its center which exists for $\mu = 0$ is converted to a globally stable spiral point, and its spiral point that is locally unstable for $\mu > 0$ re-equilibrates to a bilaterally stable limit cycle.

3.8 Global Stability Behavior of the Temperature-Dependent Composite May Predator–Prey Mite Model

Since the predator–prey May model satisfies Kolmogorov's Theorem as demonstrated in the last section, we know that the community equilibrium point $\mathcal{H}_e = \mathcal{P}_e$ of (3.6.3) for its nondimensional composite mite version (3.4.3) is either globally stable or surrounded by a closed integral curve which is globally stable from the outside. From our linear theory results for the composite May mite model depicted in Fig. 3.15, we concluded that this equilibrium point was locally stable for $T \in [0, T_1) \cup (T_2, T_M)$ while it was locally unstable for $T \in (T_1, T_2)$. Given the behavior of our limit cycle example just presented one might suspect that the community equilibrium point of this system would be globally stable for $T \in [0, T_1) \cup (T_2, T_M)$ while for $T \in (T_1, T_2)$ there would be a bilaterally stable limit cycle. In order to test this conjecture we determine the global stability behavior with temperature of the composite May mite model by applying AUTO, a numerical continuation and bifurcation software package for autonomous ordinary differential equations [30], to our system. The code AUTO traces a bifurcation diagram for such a system as a given parameter is varied. It computes equilibrium points and determines their stability for both stationary and Hopf bifurcation behavior. In the latter case, it automatically follows branches of periodic solutions, investigates the orbital stability properties of these limit cycles, and measures the periods and maximum amplitudes of oscillation. Further AUTO also has the capability of identifying stability regions of interest in a two-parameter space.

We begin with a one-parameter bifurcation analysis in which all other parameters except T are fixed by taking γ and D to have their values assigned in (3.6.10) – *i.e.*, $\gamma = 0.15$ and $D = 0.00013$. Using AUTO we then determine how the solution structure of our composite May mite model depends on T. The results of this computation are summarized in the bifurcation diagram of Fig. 3.17. In this diagram T is on the horizontal axis while the vertical axis represents the maximum value of the prey population and has been designated by max $\mathcal{H}$. Here

$$\max \mathcal{H} = \begin{cases} \mathcal{H}_e, & \text{for stationary solutions} \\ \max_{\tau\in\Omega} \mathcal{H}(\tau), & \text{for periodic solutions.} \end{cases}$$

where Ω is the period of oscillation. There are two such branches, one of each type in Fig. 3.17. Here lines are being used to denote steady-state behavior on the stationary solution branch while the periodic solution branch consists of circles designating limit cycle behavior. Solid squares represent Hopf bifurcation points. Stable stationary solutions are drawn with solid lines and unstable ones with dashed. Solid circles represent stable periodic orbits while open ones denote unstable oscillations. That particular circle designated by a half-moon symbol in Fig. 3.17 serves as a so-called limit point for the periodic solution branch where it undergoes a change of direction which can, in this instance, be characterized by the occurrence of a vertical tangent.

Although Fig. 3.17 has only been drawn for $19 \le T \le 26$, our AUTO calculations were actually carried out for the same interval $0 \le T \le 27.21$ as employed in Fig. 3.15. We can recover our linear theory results completely from the stationary solution branch of Fig. 3.17 in regard to both the stability and steady-state values of the community equilibrium point. Hence we see that our Hopf bifurcations occur at $T = T_1 = 20.89$ and $T = T_2 = 25.56$, consistent with Fig. 3.15. In addition there is a low-temperature-low-population stable equilibrium for $19 \le T \le 20.89$ (or for $0 \le T \le 20.89$ from our extended AUTO results) and a high-temperature-high-population stable equilibrium for $25.56 < T \le 26$ (or for $25.56 < T \le 27.21$ from our extended AUTO results) while there is instability of the stationary solution branch for $20.89 < T < 25.56$. We now turn our attention to the periodic solution branch of Fig. 3.17. The limit point on that branch occurs at $T = T_{-1} = 19.86 \equiv$ the lowest temperature at which the system exhibits periodic solutions and separates its upper portion consisting of stable limit cycles from its lower portion consisting of unstable ones. The limit point itself corresponds to a semi-stable limit cycle. Unlike the bilaterally symmetric behavior characteristic of stable (attracting) or unstable

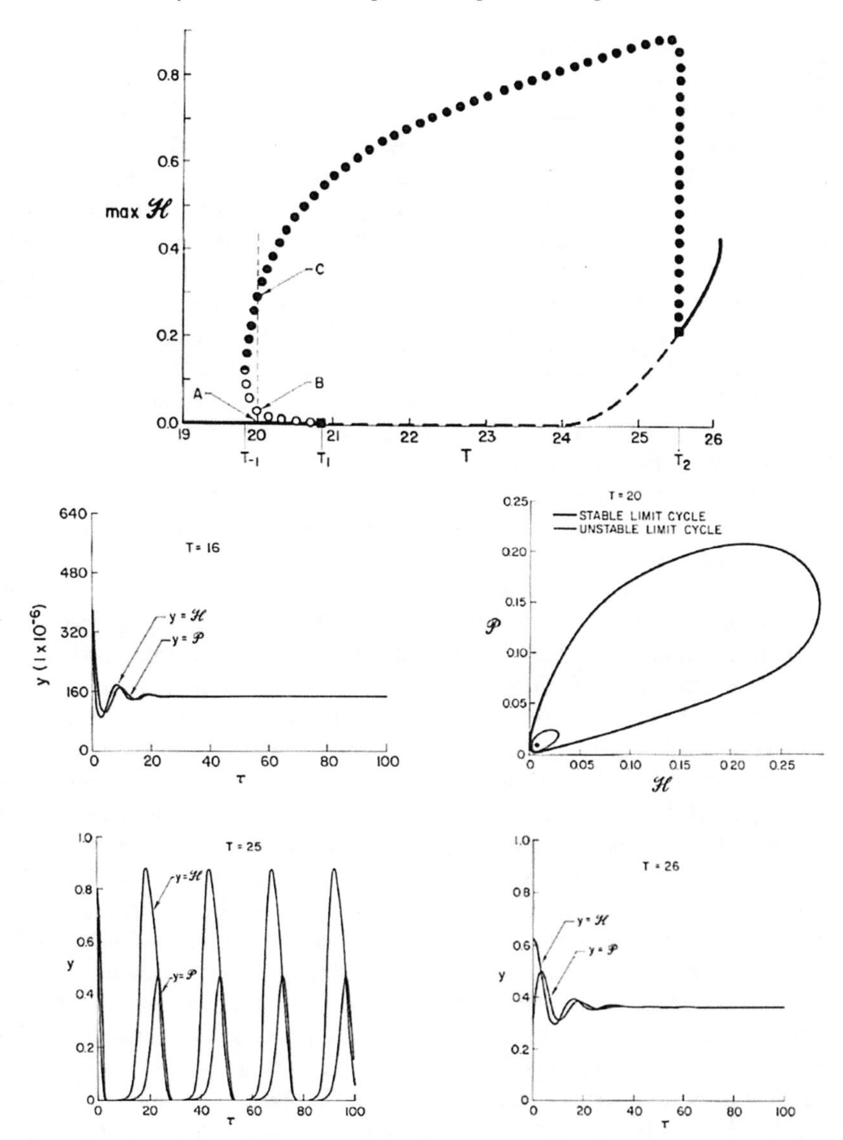

Fig. 3.17 A max $\mathcal{H}$ versus T one-parameter bifurcation diagram for the composite May mite model with $\gamma = 0.15$ and $D = 0.00013$ [21]. The insets, which are $\mathcal{H}$ and $\mathcal{P}$ plots versus τ for $T = 16, 25$, and 26 and a schematic phase-plane plot for $T = 20$, have been chosen at those representative temperatures where the system exhibits qualitatively different behaviors [139].

(repelling) limit cycles trajectories spiral into this semi-stable limit cycle from the outside and away from it on the inside.

Note that the individual stability properties of each branch just described are only local in nature. For any fixed value of temperature, the global stability behavior of the system can be deduced from the bifurcation diagram of Fig. 3.17 by examining the intersection of its solution branches with a vertical line drawn through that T value. Doing this we find that low-population stationary equilibria are the only stable states which can exist for $0 \leq T < 19.86$ and thus can conclude that they are globally stable in that temperature interval. Similarly, we can conclude that high-population stationary equilibria are globally stable for $25.56 < T \leq 27.21$. Further, we can also conclude that globally stable limit cycles occur for $20.89 < T < 25.56$.

As a check on the validity of these predictions, a numerical differential equation solver can be applied to our composite May mite model for various temperatures in the specific intervals just delineated and for a variety of different possible combinations of initial conditions. A partial compilation of these results are presented as insets in Fig. 3.17. From the representative values included in those insets, we observe that there is a stationary globally stable equilibrium at $T = 16$ of $\mathcal{H}_e = 1.457 \times 10^{-4}$ (1 mite per 23 leafs) and at $T = 26$ of $\mathcal{H}_e = 3.627 \times 10^{-1}$ (109 mites per leaf), while at $T = 25$ there is a globally stable limit cycle with the maximum of prey oscillation approaching 90% of carrying capacity or 270 mites per leaf. These results, which are independent of the choice of initial conditions and give rise to only one stable solution for any fixed temperature, corroborate our AUTO predictions in those intervals.

Upon examination of the branches for the interval $19.86 < T < 20.89$ in Fig. 3.17, however, we can conclude that there exist multiple stable states. Drawing a vertical line through $T = 20$, a typical temperature in that interval, we see its three intersection points with those branches correspond to a stable stationary equilibrium point, an unstable limit cycle, and a stable limit cycle, respectively. As a check on this prediction we again apply a numerical differential equation solver to our system for a variety of different possible combinations of initial conditions but now with $T = 20$. The results of those computations are summarized in the $T = 20$ inset of Fig. 3.17. This schematic phase-plane plot confirms our AUTO predictions in that interval: Namely there exists a locally stable stationary equilibrium point, surrounded by an unstable limit cycle, which in turn is itself surrounded by a stable limit cycle. The unstable limit cycle serves as a separatrix dividing the phase plane into two basins of attraction. Trajectories starting inside it terminate as a focus at the equilibrium point while those starting outside it converge to the stable limit cycle. Such a situation for which an equilibrium point is stable to small disturbances but can be unstable to sufficiently large ones is commonly referred to as *metastability* or a *subcritical instability*. Note that, although its occurrence violates our initial conjecture, it does not violate Kolmogorov's Theorem since the community equilibrium point is either globally stable for $T \in [0, T_{-1}) \cup (T_2, T_M)$ or is surrounded by a closed integral curve (the stable limit cycle) that is globally stable from the outside for $T \in (T_{-1}, T_2)$.

Observe that for $T = T_{-1}$ the stable and unstable limit cycles coincide and become a semi-stable limit cycle. Further, note that the community equilibrium point being linearly unstable is a sufficient condition for the existence of a stable limit cycle in this instance. Finally, by considering states possessing amplitudes in the neighborhood of the low-temperature stable portion of the stationary solution branch in the bifurcation diagram of Fig. 3.17 and gradually increasing temperature, we can see that limit cycle behavior will not spontaneously occur until T is greater than T_1 while if we start in the neighborhood of the stable portion of the periodic solution branch in that figure and gradually decrease temperature limit cycle behavior will not spontaneously cease occurring until T is less than T_{-1}. A *hysteretic* effect of this sort is often associated with metastable phenomenon.

So far all the results obtained for our composite May mite model have been with $\gamma = 0.15$ and $D = 0.00013$. In order to examine the sensitivity of our system to other choices for these quantities, we next use AUTO to produce bifurcation diagrams as two parameters are varied. In such two-parameter bifurcation diagrams curves of Hopf bifurcation and limit points are plotted in the plane of the varying parameters. Three diagrams of this sort are presented in Fig. 3.18. Here Fig. 3.18a contains plots in the T-γ plane with $D = 0.00013$; Fig. 3.18b, plots in the T-D plane with $\gamma = 0.15$; and Fig. 3.18c, a plot in the D-γ plane with $T = 16.7$. In Fig. 3.18a, b the solid curves represent Hopf bifurcations, and the dashed ones, limit points. From the latter we can observe that there is no measurable interval of metastability unless $\gamma > 0.09$ in Fig. 3.18a or $D < 0.005$ in Fig. 3.18b. Further since these curves delineate the instability region in their respective spaces we can see that there exists a point (T_c, γ_c) in Fig. 3.18a with $T_c = 14.02$ and $\gamma_c = 0.0672$ such that for $\gamma < \gamma_c$ or $T > T_c$ the community equilibrium point would be globally stable and a value $D_c = 0.0177$ in Fig. 3.18b such that the same thing would be true for $D > D_c$. These results of Fig. 3.18a may be generalized to other choices for D by replacing the critical quantities in the above inequalities by $T_c(D)$ and $\gamma_c(D)$, respectively. Note, in particular, that the intersection points of the dashed line $\gamma = 0.15$ in Fig. 3.18a with its relevant plots reproduce T_{-1}, T_1, and T_2 of Fig. 3.17. Observe that small deviations of γ from that value yield identical qualitative behavior and hence our system is resilient to such parameter variation. In this context, such resiliency is expected for realistic models of biological systems. Finally, for any fixed temperature the region in D-γ space characterized by globally stable limit cycle oscillatory behavior can be represented with a plot of the Hopf bifurcation curves as we have done in Fig. 3.18c when $T = 16.7$. Specifically, the point (D_0, γ_c) in that figure where $D_0 = 0.00254$ and $\gamma_c = 0.0872$ corresponds to $T_c(D_0) = 16.7$ and $\gamma_c(D_0) = \gamma_c$. The actual region of oscillations is slightly larger than that depicted once metastability is taken into account. For instance when $D = 0.00013$, consistent with Fig. 3.18a when $T = 16.7$, the Hopf bifurcations occur at $\gamma_1 = 0.074$ and $\gamma_2 = 0.0966$ while the limit point occurs at $\gamma_{-1} = 0.0980$. Once a particular choice of D and γ has been found that yields a desired period at a given temperature it is possible for AUTO to trace a curve of solutions with this fixed period in the D-γ plane. Such a curve is also shown

(a)

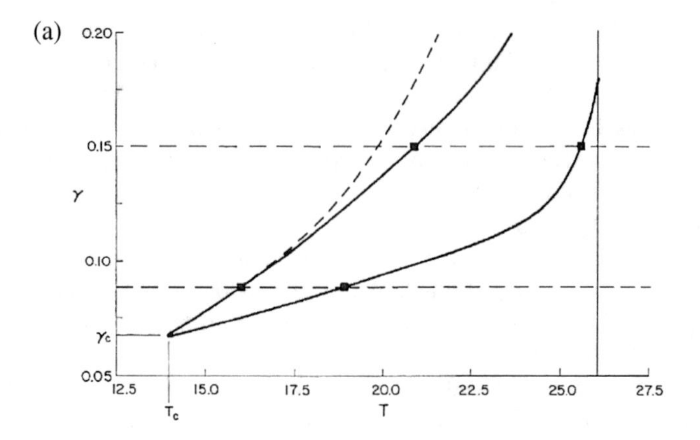

(b)

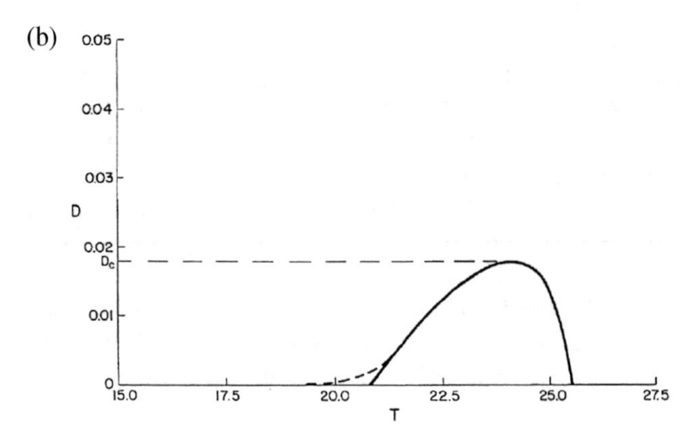

(c)

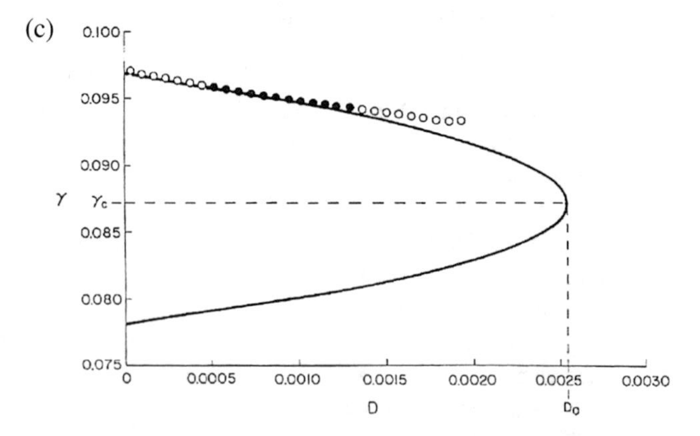

Fig. 3.18 Two-parameter bifurcation diagrams for the composite May mite model for (a) γ versus T for $D = 0.00013$, (b) D versus T for $\gamma = 0.15$, and (c) γ versus D for $T = 16.7$ [21].

in Fig. 3.18c where the locus of solutions of period $\Omega = 18.8$ (corresponding at $T = 16.7$ to 73.75 days) appears as a sequence of circles some of which occur in the region of metastability. The choice of temperature and period for this sequence are related to the laboratory experiments of Huffaker et al. [51] mentioned in Section 3.4 and to be discussed in greater detail below.

We close by recalling that our goal in developing this temperature-dependent composite May mite model was to produce a unified predator–prey system capable of representing both periodic oscillations and outbreak occurrences in populations simultaneously. From Fig. 3.18 we can conclude that increasing γ or decreasing D has a destabilizing influence upon our model which ecologically corresponds to triggering outbreaks through food quality or inducing the paradox of enrichment by carrying capacity, respectively. Specifically γ_c of Fig. 3.18a serves as a threshold for outbreak behavior which can occur provided $\gamma > \gamma_c$ as temperature increases and that occurrence does not depend upon the presence of a region of hysteresis. In particular, for $\gamma = 0.068 > \gamma_c$, an instance in which hysteresis is absent when $D = 0.00013$, with $T = 14.2$ just above T_1 and employing a numerical differential equation solver for the extremely low initial population levels $\mathcal{H}_0 = 10^{-5}$ and $\mathcal{P}_0 = 0.5 \times 10^{-5}$ we find limit cycle behavior having a prey maximum during the first oscillation of 0.44 of carrying capacity and not settling down to its constant very much lower amplitude until the eighth cycle while the period of all these oscillations stays virtually constant. Those are exactly the same aspects exhibited by our system when $\gamma > 0.09$ in the hysteresis region $T_{-1} < T < T_1$ of Fig. 3.18a for very small initial populations – *e.g.*, the situation depicted in the $T = 20$ phase plane inset of Fig. 3.17. In a more general ecological context, we note that the existence of such a threshold is reminiscent of the behavior of the grasshopper *Austroicetes cruciata*, whose outbreaks are triggered by a rise in food quality [16], while the temperature interval $T_1 < T < T_2$ characterized by limit cycle dynamics for $\gamma > \gamma_c$ is analogous to the variation of larch bud moth populations with altitude in Switzerland if one assumes a simple inverse relationship between temperature and altitude. More precisely the larch bud moth exhibits stable oscillations at altitudes between 1700 and 1900 m whereas no such phenomena occur above or below this range [80].

In a similar more general ecological vein, we note that Fig. 3.18b can be used to represent the paradox of enrichment [106] graphically. May [80] defined this term as a tendency for a given population to shift its dynamics from a stable equilibrium point to a stable limit cycle as life gets better for it in the sense that its environmental carrying capacity increases. Taking any constant $T_1 < T < T_2$ and decreasing D along such a line in Fig. 3.18b, we are able to demonstrate the desired result since D is inversely proportional to this carrying capacity. Specifically for large enough $D \cong 1$ the stable stationary equilibrium point is a focus which converges to the equilibrium point after a single oscillation while for very small values of D the oscillations associated with the limit cycle behavior may become sufficiently drastic to drive the prey density low enough so that stochastic effects cause extinction of the system to occur. This was the case with the experiments of Huffaker et al. [51] in which an originally stable periodic oscillatory situation was destabilized by tripling the food supply available to the herbivorous six-spotted spider mite. The interaction in the enriched universe was brought to a halt after only a single cycle because predator overexploitation had reduced the prey population to such an extremely low level that the predator did not stand an appreciable chance of survival. Hence, in the universes considered by Huffaker et al. [51] during their laboratory experiments, this paradox was manifested by the occurrence of periodic oscillatory behavior becoming so drastic upon enrichment that extinction eventually resulted.

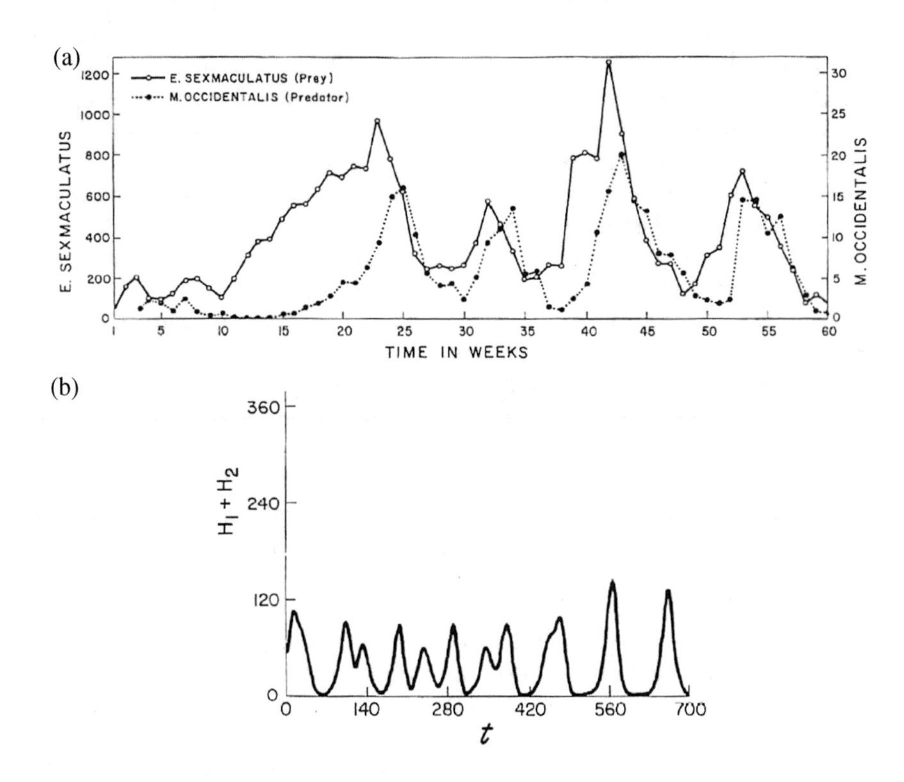

Fig. 3.19 Plots of population versus time at $T = 16.7$: (a) as observed by Huffaker et al. [51] in one of their experimental universes where the densities are measured in number per whole orange equivalents and time in weeks; (b) as predicted by applying the composite May mite model to two independent subpopulation pairs with $\gamma_1 = 0.09$, $D_1 = 0.00013$ and $\gamma_2 = 0.08$, $D_2 = 0.00020$ [21].

In order to simplify our analysis we have assumed from the outset that the environment being considered is spatially homogeneous with mite population densities being distributed uniformly. Such an assumption oversimplifies the ecological situation not only for field environments but also for some laboratory experiments as well including those of Huffaker et al. [51]. They studied the interaction of *M. occidentalis* with its prey, the six-spotted spider mite *Eotetranychus sexmaculatus* Riley, in a carefully controlled laboratory environment which was maintained at a temperature of 26.7°C and consisted of cabinets containing arrays of oranges, replenished at regular intervals, that served as the food source for the prey with various pathways and barriers between the oranges. Huffaker et al. [51] were able to construct systems in which the populations of the two mite species did not converge to an equilibrium value but rather oscillated with a relatively constant period and somewhat variable amplitude as shown in Fig. 3.19a and these oscillations tended to persist in ecological time. Indeed that experiment lasted 60 weeks and was terminated not by predator overexploitation but only when the six-spotted spider mite contracted a viral infection which caused its population density to crash. The limit cycle oscillations characteristic of our model summarized in Fig. 3.17 show this constant period and those dramatic differences between minimum and maximum amplitudes exhibited by these data but not their variation in amplitude.

Variations in amplitude of that sort can be achieved if we relax this assumption of spatial uniformity. The easiest way to introduce spatial heterogeneity into a previously homogeneous model system is to postulate the existence of nonmigrating subpopulations. Since Huffaker et al. [51] inadvertently used oranges of an inferior quality for the prey food source during part of their experiment while only one-seventh of the oranges were replaced in each cabinet per week, there is a very real possibility that the prey mites were divided into subpopulations with different carrying capacities and inherent food qualities. If we assume, for simplicity, the existence of just two independent prey subpopulations with corresponding predator subpopulations such that each subpopulation pair obeys the dynamics of our composite May mite model with $T = 16.7$, we can produce variable amplitudes in the oscillations of the total population densities obtained by summing these subpopulations, as illustrated in Fig. 3.19b. The amplitude fluctuations in the total population densities are due to the different values of carrying capacity and food quality selected for each of the two prey subpopulations, as measured by their (D, γ) pairs chosen appropriately from Fig. 3.18c. In particular, these parameters and the relevant initial densities for those two subpopulation pairs of Fig. 3.19b are given by

$$\gamma_1 = 0.09,\ D_1 = 0.00013 \text{ or } K_1 = 300 \text{ mites per leaf},\ (\mathcal{H}_1)_0 = 0.1370,\ (\mathcal{P}_1)_0 = 0.0517; \tag{3.8.1a}$$

$$\gamma_2 = 0.08,\ D_1 = 0.00020 \text{ or } K_2 = 200 \text{ mites per leaf},\ (\mathcal{H}_2)_0 = 0.0117,\ (\mathcal{P}_2)_0 = 0.0047; \tag{3.8.1b}$$

respectively. In addition,

$$H_1 = K_1\mathcal{H}_1,\ H_2 = K_2\mathcal{H}_2,\ t = \frac{\tau}{r_1(16.7)} = 3.92\tau \text{ days} \tag{3.8.1c}$$

in that figure. Upon comparison of the two parts of Fig. 3.19, it can be seen that our theoretical predictions are in good qualitative agreement with the experimental data. This approach is entirely analogous to the situation of a patchy environment with D and γ varying between the patches in the absence of interpatch migration. A natural extension of such an analysis can be accomplished by introducing dispersal effects into our model through the consideration of either passive diffusion between discrete patches allowing for colonization and invasion or active diffusion in space incorporating the flux mechanisms of mite motility and predator aggregation. In the latter case our model would consist of a governing system of interaction–diffusion partial differential equations. We shall consider phenomena more properly modeled by systems of this sort in later chapters after examining the soap film problem to be introduced in Chapter 4.

Problems

3.1. Consider the cubic equation for $x = x(\varepsilon)$:

$$\varepsilon x^3 - x + 1 = 0, 0 < \varepsilon << 1.$$

(a) Find an asymptotic solution to it of the form

$$x^{(1)}(\varepsilon) \sim x_0 + x_1\varepsilon \text{ as } \varepsilon \to 0.$$

(b) Now let

$$x(\varepsilon) = \frac{\xi(\varepsilon)}{\varepsilon^s} \text{ for } s > 0$$

and determine the value of s such that $\xi(\varepsilon) = O^*(1)$ as $\varepsilon \to 0$. Here $O^*(1)$ means $O(1)$ but not $o(1)$.

(c) With s so determined, find two asymptotic solutions of $\xi(\varepsilon)$ of the form

$$\xi(\varepsilon) \sim \xi_0 + \xi_1\varepsilon^s \text{ as } \varepsilon \to 0 \text{ where } \xi \neq 0$$

and hence deduce expansions for the other two roots given by

$$x^{(2,3)}(\varepsilon) \sim \frac{\xi_0^{(2,3)}}{\varepsilon^s} + \xi_1^{(2,3)} \text{ as } \varepsilon \to 0.$$

(d) Finally, rewrite the cubic equation in the form

$$\varepsilon x^2 - 1 = -\frac{1}{x} = y.$$

Then by graphing y versus x for these two curves represent the three roots of the cubic schematically and identify them with $x^{(1,2,3)}(\varepsilon)$.

3.2. Consider the following differential equation for $y(x;\varepsilon)$:

$$\varepsilon\frac{d^2y}{dx^2} + \frac{dy}{dx} + b_0 y = 0; 0 < x < 1, b_0 > 0, 0 < \varepsilon << 1;$$

$$y(0;\varepsilon) = 0, y(1;\varepsilon) = 1.$$

(a) Given that the boundary layer occurs at $x = 0$ and has thickness ε, find the one-term outer, one-term inner, and one-term uniformly valid additive composite for this singularly perturbed boundary value problem by the Method of Matched Asymptotic Expansions.

(b) Introducing

$$x = 1 - \frac{T}{T_M}, \; y(x;\varepsilon) = \frac{r(T)}{\psi}, \; \varepsilon = \frac{\Delta T}{T_M}, \; b_0 = \rho T_M,$$

where $r \equiv$ the per capita growth rate for arthropods per day, $T =$°C -10°C $\equiv$ temperature measured in °C above a given base temperature of 10°C, $\psi \equiv$ growth rate at the base temperature $T = 0$, $\rho \equiv$ rate of increase to optimum temperature, $T_M \equiv$ the lethal maximum temperature threshold, and $\Delta T \equiv$ width of the high-temperature boundary layer, show that the one-term uniformly valid additive composite of part (a) corresponds to

$$r(T) = \psi[e^{\rho T} - e^{\rho T_M} e^{-(T_M - T)/\Delta T}] \text{ for } T \in [0, T_M].$$

(c) Using the previous results determine the following relationship between the optimum temperature T_0 and the parameters ε, b_0, and T_M:

$$T_0 = T_M \left[1 + \varepsilon \frac{\ln(\varepsilon b_0)}{1 - \varepsilon b_0} \right].$$

(d) Employing the following numerical values relevant to the predaceous mite *M. occidentalis*:

$$\psi = 0.089, \; \rho = 0.055, \; T_M = 27.215, \; \Delta T = 2.070;$$

make a schematic plot of the representation for rate-temperature phenomenon in arthropods identifying the optimum temperature T_0. Note in this context that $\varepsilon b_0 = \rho \Delta T$.

3.3. Consider a two-species competitive interaction governed by the following nondimensional model dynamical system for $x = x(t;\alpha)$ and $y = y(t;\alpha)$:

$$\frac{dx}{dt} = x(1-x) - \alpha xy = F(x,y;\alpha), \; \frac{dy}{dt} = y(1-y) - \alpha xy = G(x,y;\alpha), \; t > 0;$$

with initial conditions

$$x(0;\alpha) = x_0(\alpha), \; y(0;\alpha) = y_0(\alpha).$$

Here $t \equiv$ time and $x, y \equiv$ the population densities of the competing species while the parameter α, a measure of the effect of that symmetric interaction, is assumed to be positive but unequal to unity.

(a) Find the four equilibrium points of this system

$$x_e = x_e(\alpha), y_e = y_e(\alpha)$$

which satisfy

$$F(x_e, y_e; \alpha) = G(x_e, y_e; \alpha) = 0 \text{ for } x_e, y_e \geq 0.$$

(b) Determine the linear stability of each of these equilibrium points by seeking a solution of the governing differential equations of the form

$$x(t;\alpha) = x_e(\alpha) + \varepsilon x_{11} e^{\lambda t} + O(\varepsilon^2)$$

and

$$y(t;\alpha) = y_e(\alpha) + \varepsilon y_{11} e^{\lambda t} + O(\varepsilon^2)$$

where $|\varepsilon| << 1$ and $|x_{11}|^2 + |y_{11}|^2 \neq 0$; neglecting terms of $O(\varepsilon^2)$; cancelling the resultant common exponential factor; evaluating the relevant eigenvalues $\lambda_{1,2} = \lambda_{1,2}(\alpha)$ explicitly; and examining their signs as functions of α.

(c) Finally, draw two schematic phase-plane plots of the trajectories of this system in the first quadrant of the x-y plane for $0 < \alpha < 1$ and $\alpha > 1$, respectively. Label these two cases competitive exclusion or coexistence as appropriate and give an ecological interpretation of those results involving the initial conditions (Hint: Plot the isoclines of the system, identify their intercept with each other as well as with the x- and y-axes, compute the eigenvectors associated with the eigenvalues of the community equilibrium point, and examine the signs of dx/dt and dy/dt in the relevant portions of the first quadrant of the x-y plane).

3.4. Consider the following exploitation model system for $H = H(t)$ and $P = P(t)$:

$$\frac{dH}{dt} = r_1 H\left(1 - \frac{H}{K}\right) - a\frac{HP}{K}, \frac{dP}{dt} = r_2 P\left(1 - \frac{P}{\gamma H}\right).$$

Here t denotes time while H, P and $r_{1,2}$ represent prey and predator densities and growth rates, respectively. Further, γ is a measure of the food quality of the prey for conversion to predator births while a is the per capita consumption rate when the prey density is at its carrying capacity K. This model differs from the May model in that it employs a Holling type I functional response $f_1(H) = aH/K$ rather than the type II response.

(a) Obtain the nondimensional form of this governing system of equations

$$\frac{d\mathcal{H}}{d\tau} = \mathcal{H}(1-\mathcal{H}) - \varphi\mathcal{H}\mathcal{P} = \mathcal{F}(\mathcal{H},\mathcal{P}), \frac{d\mathcal{P}}{d\tau} = \theta\mathcal{P}\left(1 - \frac{\mathcal{P}}{\mathcal{H}}\right) = \mathcal{G}(\mathcal{H},\mathcal{P})$$

by introducing the following dimensionless variables and parameters:

$$\tau = r_1 t, \mathcal{H} = \frac{H}{P}, \mathcal{P} = \frac{P}{\gamma K}, \theta = \frac{r_2}{r_1}, \varphi = \frac{a\gamma}{r_1} = \frac{\alpha}{1-\alpha} \text{ for } 0 < \alpha < 1.$$

(b) Show that the community equilibrium point $(\mathcal{H}_e, \mathcal{P}_e)$ of this system satisfying

$$\mathcal{F}(\mathcal{H}_e, \mathcal{P}_e) = \mathcal{G}(\mathcal{H}_e, \mathcal{P}_e) = 0$$

is given by

$$\mathcal{H}_e = \mathcal{P}_e = 1 - \alpha.$$

(c) Perform a linear perturbation analysis of this equilibrium point by seeking a solution of that system of the form

$$\mathcal{H}(\tau) = \mathcal{H}_e + \varepsilon\mathcal{H}_1(\tau) + O(\varepsilon^2), \mathcal{P}(\tau) = \mathcal{P}_e + \varepsilon\mathcal{P}_1(\tau) + O(\varepsilon^2) \text{ where } |\varepsilon| << 1,$$

and obtain

$$\frac{d\mathcal{H}_1}{d\tau} = (\alpha - 1)\mathcal{H}_1 - \alpha\mathcal{P}_1, \quad \frac{d\mathcal{P}_1}{d\tau} = \theta(\mathcal{H}_1 - \mathcal{P}_1).$$

(d) Letting

$$\begin{bmatrix} \mathcal{H}_1 \\ \mathcal{P}_1 \end{bmatrix}(\tau) = \begin{bmatrix} \mathcal{H}_{11} \\ \mathcal{P}_{11} \end{bmatrix} e^{\sigma\tau} \text{ with } |\mathcal{H}_{11}|^2 + |\mathcal{P}_{11}|^2 \neq 0,$$

deduce the characteristic equation

$$\sigma^2 + (\theta + 1 - \alpha)\sigma + \theta = 0.$$

(e) Represent the loci (i) $\text{Re}(\sigma) < 0$, $\text{Im}(\sigma) \neq 0$ and (ii) $\sigma < 0$ graphically in the θ-α plane for $\theta > 0$ and $0 < \alpha < 1$. Hint: The locus on which the discriminant of the quadratic deduced in part (f) is zero satisfies $\alpha = (\sqrt{\theta} - 1)^2$.

(f) If $\varphi = \sqrt{\theta^2}$ or equivalently $\alpha = \theta^2/(\theta^2 + 1)$ where $\theta = \theta(T)$ is a decreasing function of temperature T discuss the linear stability behavior of the community equilibrium point of the system as T increases over the interval $T \in [0, T_M]$ given that $\theta(0) > 4$ and $\theta(T_M) = 0$.

(g) Finally construct schematic phase portraits in the $\mathcal{H}$-$\mathcal{P}$ plane for representative values of T, assuming that the local phase portrait at the community equilibrium point is qualitatively equivalent to the global phase portrait of the system. Then consider the community equilibrium line $\mathcal{H} = \mathcal{P} = 1 - \alpha$ in that plane for $0 \leq \mathcal{H}$, $\mathcal{P} \leq 1$ and describe the global stability behavior along that line for $T \in [0, T_M]$ in the same relevant subintervals of part (f).

Chapter 4
Canonical Soap Film Problem

In this chapter, the canonical Plateau problem of determining the shape of the soap film formed between two concentric circular wire rings of possibly different radii, as a function of the axial distance between the rings, is considered. Since that shape is a catenoid of revolution, which is a minimal surface determined by a Calculus of Variations approach, the concepts of a catenary curve, surface area integrals, and the initial treatment of the Calculus of Variations are all introduced as pastoral interludes. Then the behavior of the soap film as the distance of separation slowly increases is examined until that surface ruptures and occupies the two concentric circular rings alone. Given that this critical distance requires a determination of the envelope of a one-parameter family of curves, that concept is presented as a pastoral interlude as well. The problem deals with the closely related Calculus of Variations brachistochrone example of determining the shape for the curve which provides the shortest time of descent of a bead sliding along it acted upon by gravitational force alone.

4.1 Soap Films and Minimal Surfaces

When a thin soap film is stretched across a metal frame, it takes the shape of a surface that has minimum area. Assuming constant surface energy per unit area this is a special case of physical phenomena tending to seek a state that renders their Gibbs free energy a minimum and is known as Plateau's problem after the Belgian physicist Joseph Plateau (1801–1883) who conducted extensive experiments on such soap films. We shall consider the canonical Plateau problem of determining the shape of the soap film formed between two concentric circular wire rings of possibly different radii as a function of the axial distance between the rings illustrated in Fig. 4.1. This first entails deducing the generating curve for a volume of revolution of minimum lateral surface area. That is one of the classical minimal surface problems solved by the techniques of the Calculus of Variations and turns out to be a catenary curve which is the shape taken by a chain hanging only under its own weight. Hence our

© Springer International Publishing AG, part of Springer Nature 2017

D. J. Wollkind and B. J. Dichone, *Comprehensive Applied Mathematical Modeling in the Natural and Engineering Sciences*,
https://doi.org/10.1007/978-3-319-73518-4_4

general approach of presenting such developments as much as possible from first principles requires us to introduce as pastoral interludes the concepts of catenaries, surface areas of volumes of revolution, and the Calculus of Variations in Sections 4.2–4.4 before formulating and solving this problem in Section 4.5. Then after another pastoral interlude involving envelopes of one-parameter families of curves presented in Section 4.6 we shall finally be ready to apply these results to our canonical soap film problem in Section 4.7.

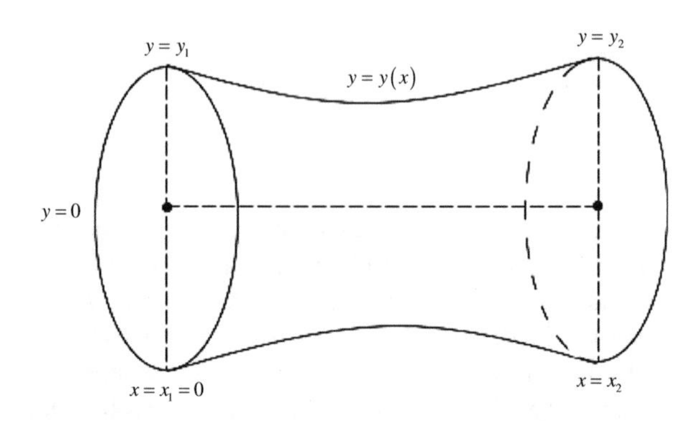

Fig. 4.1 Soap film surface formed between two concentric circular rings of radii y_1 and y_2, respectively. Such a surface can be generated by rotating the curve $y = y(x)$, $x_1 \le x \le x_2$, where $y(x_1) = y_1$ and $y(x_2) = y_2$, about the x-axis $y = 0$. Here for ease of exposition we have taken $x_1 = 0$.

4.2 Pastoral Interlude: Equation Satisfied by Catenaries

Consider a perfectly flexible cable hanging between two points in the x-y plane, satisfying the equation $y = y(x)$, and being acted on only by gravity as depicted in Fig. 4.2. Let $\sigma \equiv$ arc length and define $\boldsymbol{f}(\sigma) \equiv$ force per unit length acting on the cable for any $0 \le \sigma \le s$ where $0 \le s \le \ell$. Here $\boldsymbol{f}(\sigma) = -w\boldsymbol{j}$ where $w \equiv$ weight of the cable per unit length and $\boldsymbol{j} \equiv$ unit vector in the vertical direction. Further let $\boldsymbol{\tau}(s) \equiv$ tension force acting tangentially along the cable at $\sigma = s$. Then, from the balance of forces at equilibrium, we can conclude that

$$\boldsymbol{F}_0 + \int_0^s \boldsymbol{f}(\sigma)\,d\sigma + \boldsymbol{\tau}(s) = \boldsymbol{0}$$

and upon differentiating this vector condition with respect to s obtain

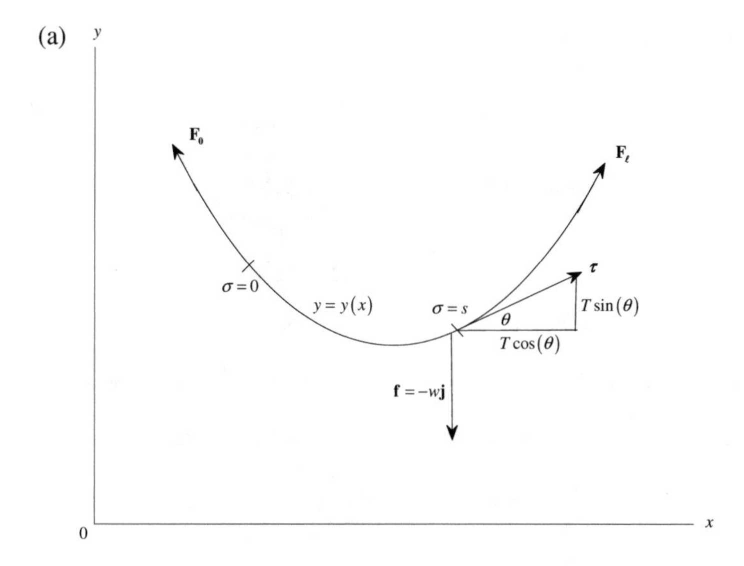

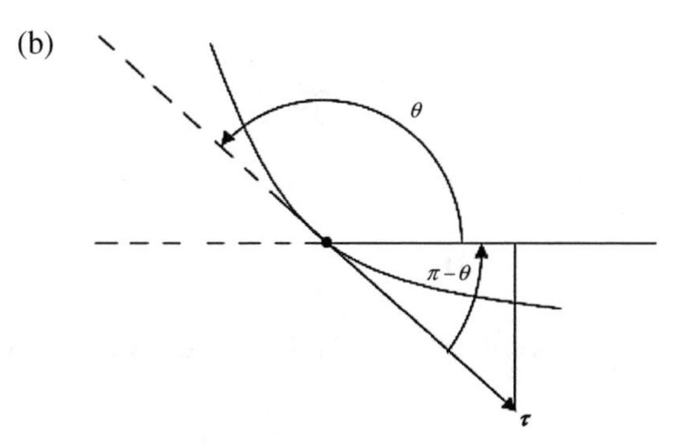

Fig. 4.2 Schematic representation of a catenary curve for the angle of inclination (a) $0 \leq \theta \leq \pi/2$ and (b) $\pi/2 < \theta \leq \pi$.

$$\frac{d\boldsymbol{\tau}(s)}{ds} = -\boldsymbol{f}(s) = w\boldsymbol{j}.$$

We shall first consider Fig. 4.2a in which the angle of inclination θ is restricted to the first quadrant. Hence

$$\boldsymbol{\tau}(s) = T[\cos(\theta)\boldsymbol{i} + \sin(\theta)\boldsymbol{j}] \text{ for } 0 \leq \theta \leq \frac{\pi}{2},$$

where $T = |\boldsymbol{\tau}(s)|$ and $\boldsymbol{i} \equiv$ unit vector in the horizontal direction. Thus from these vector relations we can deduce the two conditions

$$\frac{d}{ds}[T\cos(\theta)] = 0 \text{ and } \frac{d}{ds}[T\sin(\theta)] = w.$$

Now the first condition implies

$$T\cos(\theta) = h \text{ or } T = \frac{h}{\cos(\theta)},$$

which, upon substitution into the second condition, yields

$$\frac{d}{ds}[\tan(\theta)] = \frac{w}{h} = \frac{1}{a}.$$

We next consider Fig. 4.2b in which θ is restricted to the second quadrant. Hence

$$\boldsymbol{\tau}(s) = T[\cos(\pi-\theta)\boldsymbol{i} - \sin(\pi-\theta)\boldsymbol{j}] = -T[\cos(\theta)\boldsymbol{i} + \sin(\theta)\boldsymbol{j}] \text{ for } \pi/2 < \theta \leq \pi.$$

From our vector relations we can deduce the two conditions

$$\frac{d}{ds}[-T\cos(\theta)] = 0 \text{ and } \frac{d}{ds}[-T\sin(\theta)] = w;$$

the first of which implies

$$-T\cos(\theta) = h \text{ or } -T = \frac{h}{\cos(\theta)}.$$

Then, upon substituting this result into the second one, we again obtain that

$$\frac{d}{ds}[\tan(\theta)] = \frac{w}{h} = \frac{1}{a}.$$

Hence we can conclude that this relation is valid for all θ. Recalling the derivatives

$$y'(x) = \tan(\theta) \text{ and } \frac{ds(x)}{dx} = \sqrt{1+[y'(x)]^2},$$

it follows that

$$\frac{d}{ds}[y'(x)] = \frac{\frac{d}{dx}[y'(x)]}{\frac{ds(x)}{dx}} = \frac{\frac{d}{dx}[y'(x)]}{\sqrt{1+[y'(x)]^2}} = \frac{1}{a}$$

or

$$\int^{v=y'(x)} \frac{dv}{\sqrt{1+v^2}} = \frac{1}{a}\int dx.$$

Let

$$e^z = \cosh(z) + \sinh(z) \text{ where } \cosh(-z) = \cosh(z) \text{ and } \sinh(-z) = -\sinh(z).$$

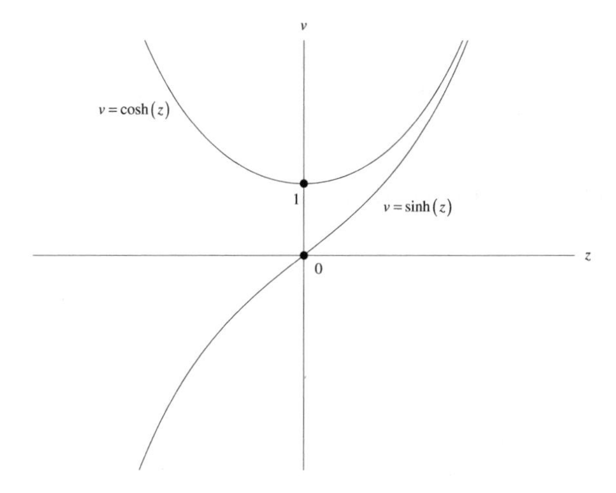

Fig. 4.3 Plots of the hyperbolic cosine and sine functions.

Then (see Fig. 4.3)

$$\cosh(z) = \frac{e^z + e^{-z}}{2}, \ \sinh(z) = \frac{e^z - e^{-z}}{2}.$$

Thus

$$\sinh'(z) = \cosh(z), \ \cosh'(z) = \sinh(z), \ \cosh^2(z) = 1 + \sinh^2(z), \ \cosh(z) \geq 1.$$

Now

$$\int^{v=\sinh(z)} \frac{dv}{\sqrt{1+v^2}} = \int \frac{\cosh(z)dz}{\sqrt{1+\sinh^2(z)}} = \int \frac{\cosh(z)dz}{\cosh(z)} = \int dz = z = \arg\sinh(v).$$

Hence

$$\arg\sinh[y'(x)] = \frac{x}{a} - \alpha$$

or

$$y'(x) = \sinh\left(\frac{x}{a} - \alpha\right) \Rightarrow y(x) = a\cosh\left(\frac{x}{a} - \alpha\right) + \beta.$$

Finally, for the special case of $\beta = 0$, this reduces to

$$y''(x) = \frac{1}{a}\cosh\left(\frac{x}{a} - \alpha\right) = \frac{y(x)}{a^2}$$

and, therefore

$$\frac{\frac{d}{dx}[y'(x)]}{\sqrt{1+[y'(x)]^2}} = \frac{y''(x)}{\sqrt{1+[y'(x)]^2}} = \frac{1}{a} \Rightarrow \frac{y(x)}{\sqrt{1+[y'(x)]^2}} = a.$$

4.3 Pastoral Interlude: Surface Area of Volumes of Revolution

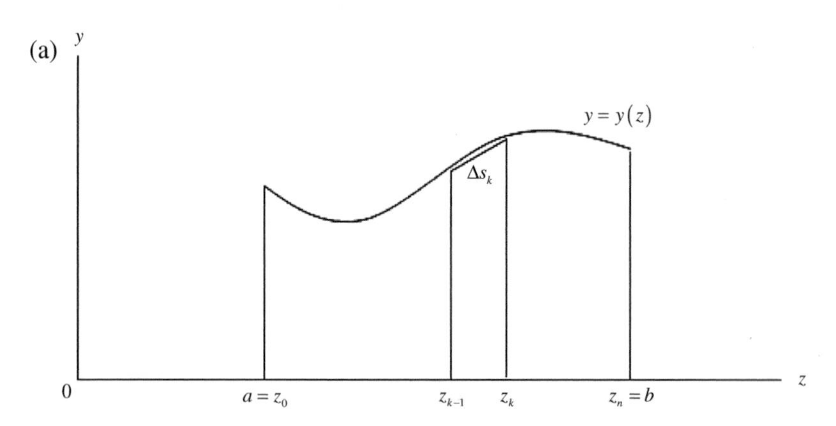

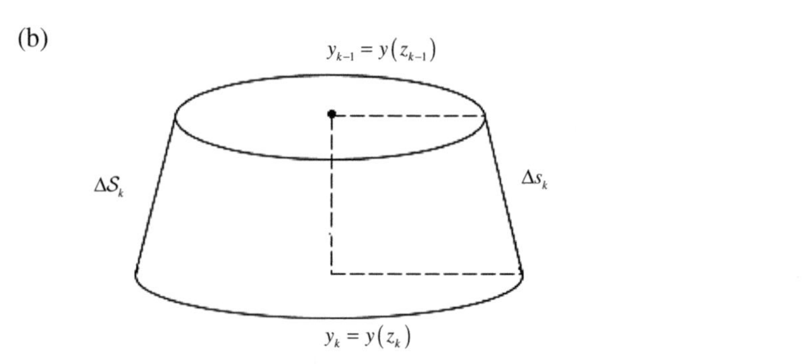

Fig. 4.4 (a) Partition of the z-interval for the volume of revolution generating function $y = y(z)$, $a \leq z \leq b$. (b) Frustum generated by $y = y(z)$ for $z_{k-1} \leq z \leq z_k$.

We now wish to develop an integral formula for the lateral surface area of a volume of revolution generated by the rotation of a rectifiable curve $y = y(z)$, $a \leq z \leq b$, about the z-axis as depicted in Fig. 4.4a. To do so we first need to develop formulae for the lateral surface area of the cone and frustum represented in Figs. 4.5 and 4.6, respectively. The right circular cone in Fig. 4.5a has base radius r and slant height ℓ. Hence the circumference of its base circle is $2\pi r$. Such a cone can be constructed by cutting the appropriate sector from a circle of radius ℓ and folding accordingly as indicated in Fig. 4.5b. The associated arc length of that sector is $2\pi(\ell - r)$. Hence the lateral surface area of the cone $\mathcal{S}_{\ell r}$ can be related to the area of the complete circle $\pi\ell^2$ by the corresponding ratio of their circumferences or

$$\frac{\mathcal{S}_{\ell r}}{\pi\ell^2} = \frac{2\pi r}{2\pi\ell} = \frac{r}{\ell} \Rightarrow \mathcal{S}_{\ell r} = \pi\ell r.$$

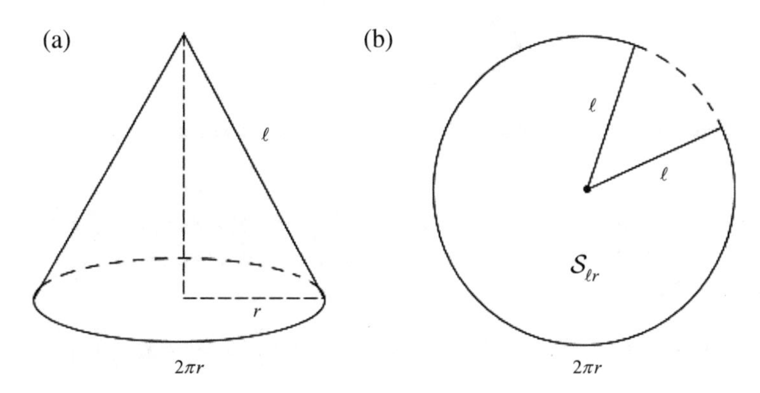

Fig. 4.5 (a) Right circular cone with base radius r and slant height ℓ. (b) Lateral surface area for the cone of part (a).

The right circular frustum, a cone with its top cut off, in Fig. 4.6 has base radius R, top radius $r < R$, and slant height s. Thus the lateral surface area of the frustum $\mathcal{S}_{Rrs}$ can be determined by subtracting the lateral surface area of the top cone $\mathcal{S}_{\ell r}$ from that of the big cone $\mathcal{S}_{LR}$ where $L = \ell + s$. Hence

$$\mathcal{S}_{Rrs} = \mathcal{S}_{LR} - \mathcal{S}_{\ell r} = \pi(\ell + s)R - \pi\ell r = \pi\ell(R - r) + \pi s R.$$

Then by employing the proportionality property for the similar right triangles denoted in Fig. 4.6 with hypotenuse and base of ℓ, r and s, $R - r$, respectively, we can conclude that

$$\frac{\ell}{s} = \frac{r}{R - r} \Rightarrow \ell(R - r) = sr.$$

Finally upon substitution of this relation into our previous result we obtain the required formula

$$\mathcal{S}_{Rrs} = \pi s(R + r).$$

Returning to Fig. 4.4 and our development of a formula for the volume of revolution's lateral surface area, we partition the z-interval $[a, b]$ into the subintervals

$$a = z_0 < z_1 < \cdots < z_{k-1} < z_k < \cdots < z_n = b,$$

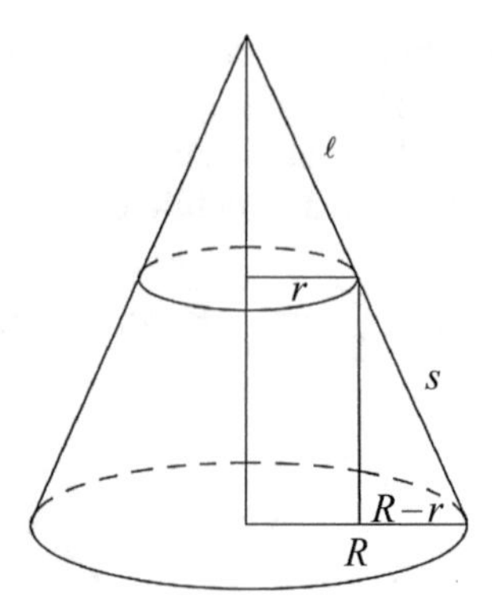

Fig. 4.6 Right circular frustum with base radius R, top radius $r < R$, and slant height s.

rotate the kth trapezoid as depicted in Fig. 4.4a about the z-axis, and generate the frustum of Fig. 4.4b with base radius $y_k = y(z_k)$, top radius $y_{k-1} = y(z_{k-1})$, and slant height Δs_k where

$$(\Delta s_k)^2 = (\Delta z_k)^2 + (\Delta y_k)^2 \text{ for } \Delta z_k = z_k - z_{k-1} \text{ and } \Delta y_k = y_k - y_{k-1}.$$

Making use of the mean value theorem

$$\Delta y_k = y(z_k) - y(z_{k-1}) = y'(c_k)\Delta z_k \text{ where } c_k \in [z_{k-1}, z_k]$$

and substituting this result into our previous relation, we can deduce that

$$\Delta s_k = \sqrt{1 + [y'(c_k)]^2}\Delta z_k.$$

Employing the lateral surface area formula for the frustum of Fig. 4.4b we obtain

$$\Delta \mathcal{S}_k = \mathcal{S}_{y_k y_{k-1} \Delta s_k} = \pi[y(z_k) + y(z_{k-1})]\sqrt{1 + [y'(c_k)]^2}\Delta z_k.$$

Then summing all these lateral surface areas

$$\mathcal{S}_n = \sum_{k=1}^{n} \Delta \mathcal{S}_k,$$

defining the norm of the partition by

$$||\Delta|| \equiv \max_k \Delta z_k \text{ for } k = 1, 2, \ldots, n,$$

and taking the limit of $\mathcal{S}_n$ as $n \to \infty$ with $||\Delta|| \to 0$, we derive the desired Riemann integral formula for the lateral surface area $\mathcal{S}$ of our volume of revolution

$$\mathcal{S} = 2\pi \int_a^b y(z)\sqrt{1 + [y'(z)]^2}\, dz.$$

4.4 Pastoral Interlude: Calculus of Variations

Given that there exists a function $y = y(x) \in C^2[x_1, x_2]$ with $y(x_1) = y_1$ and $y(x_2) = y_2$, which renders the integral functional

$$I(Y) = \int_{x_1}^{x_2} f(x, Y, Y')\, dx$$

an extremum, find the ordinary differential equation satisfied by $y = y(x)$. We first define a one-parameter family of testing or comparison functions

$$Y(x; \varepsilon) = y(x) + \varepsilon s(x)$$

where $s(x) \in C^2[x_1, x_2]$ is an arbitrary function such that $s(x_1) = s(x_2) = 0$. Then, substituting this into our functional, we obtain

$$\mathcal{I}(\varepsilon) = \int_{x_1}^{x_2} f(x, Y, Y')\, dx \text{ where } Y'(x; \varepsilon) = y'(x) + \varepsilon s'(x).$$

Since $\mathcal{I}(\varepsilon)$ has an extremum at $\varepsilon = 0$, it is necessary that $\mathcal{I}'(0) = 0$. Now

$$\mathcal{I}'(\varepsilon) = \int_{x_1}^{x_2} \left(\frac{\partial f}{\partial Y}\frac{\partial Y}{\partial \varepsilon} + \frac{\partial f}{\partial Y'}\frac{\partial Y'}{\partial \varepsilon} \right) dx = \int_{x_1}^{x_2} [f_2 s(x) + f_3 s'(x)]\, dx$$

and

$$\mathcal{I}'(0) = \int_{x_1}^{x_2} [f_2(x, y, y')s(x) + f_3(x, y, y')s'(x)]\, dx.$$

Since, upon integration by parts,

$$\int_{x_1}^{x_2} f_3(x, y, y')s'(x)\, dx = f_3(x, y, y')s(x)|_{x_1}^{x_2} - \int_{x_1}^{x_2} \frac{d}{dx}[f_3(x, y, y')]s(x)\, dx$$

and $s(x_1) = s(x_2) = 0$, we obtain

$$\mathcal{I}'(0) = \int_{x_1}^{x_2} \left\{ f_2(x, y, y') - \frac{d}{dx}[f_3(x, y, y')] \right\} s(x)\, dx = 0.$$

By the arbitrariness of $s(x)$, we can conclude from the fundamental lemma of the Calculus of Variations, *i.e.*,

$$\int_{x_1}^{x_2} g(x, y, y')s(x)\,dx = 0$$

where $g(x, y, y') \in C[0, 1] \Rightarrow g(x, y, y') = 0$ which is proved by contradiction for a specific choice of $s(x)$, that

$$f_2(x, y, y') - \frac{d}{dx}[f_3(x, y, y')] = 0 \text{ or more simply } \frac{\partial f}{\partial y} - \frac{d}{dx}\left(\frac{\partial f}{\partial y'}\right) = 0,$$

which is called the Euler–Lagrange equation.

4.5 Calculus of Variations Application to Minimal Soap Film Surfaces

We are ready to determine the generating curve for the volume of revolution of minimum lateral surface area. Consider the scenario depicted in Fig. 4.1 and let $y = y(x)$ be a rectifiable function defined for $x_1 \le x \le x_2$ such that $y(x_1) = y_1$ and $y(x_2) = y_2$. Then from our result of Section 4.3 the lateral surface area of the volume of revolution in that figure is given by

$$S = 2\pi I(y) \text{ where } I(y) = \int_{x_1}^{x_2} f(x, y, y')\,dx \text{ with } f(x, y, y') = y\sqrt{1 + (y')^2}.$$

Now, applying the Calculus of Variations approach of Section 4.4 to the functional $I(y)$ and realizing any extrema associated with it must be a minimum, we find that the $y(x)$ which minimizes this lateral surface area satisfies the Euler–Lagrange equation

$$\frac{\partial f}{\partial y} - \frac{d}{dx}\left(\frac{\partial f}{\partial y'}\right) = 0 \text{ where } f(x, y, y') = y\sqrt{1 + (y')^2}.$$

Noting that the integrand of the functional is actually independent of x in this instance since $f(x, y, y') = f(y, y')$, we can deduce a first integral of our Euler–Lagrange equation by taking the derivative of the energy-type function $f - y' f_{y'}$ with respect to x as follows (here the subscript denotes a partial derivative):

$$\frac{d}{dx}(f - y' f_{y'}) = f_x + f_y y' + f_{y'} y'' - y'' f_{y'} - y' \frac{d}{dx}(f_{y'}) = f_x + y'\left[f_y - \frac{d}{dx}(f_{y'})\right].$$

Then, since

$$f_y - \frac{d}{dx}(f_{y'}) = f_x = 0,$$

by virtue of the Euler–Lagrange equation and the x-independence of f, respectively, we can conclude that

$$\frac{d}{dx}(f - y'f_{y'}) = 0 \Rightarrow f - y'f_{y'} \equiv a, \text{ a constant.}$$

Particularizing this result to our integrand

$$f(y,y') = y\sqrt{1+(y')^2} \Rightarrow f_{y'}(y,y') = \frac{yy'}{\sqrt{1+(y')^2}},$$

we find that

$$f(y,y') - y'f_{y'}(y,y') = y\sqrt{1+(y')^2} - \frac{y(y')^2}{\sqrt{1+(y')^2}} = \frac{y[1+(y')^2] - y(y')^2}{\sqrt{1+(y')^2}} = \frac{y}{\sqrt{1+(y')^2}},$$

which yields the relation

$$\frac{y(x)}{\sqrt{1+[y'(x)]^2}} = a.$$

Since, as was shown in Section 4.2, this is exactly the equation satisfied by

$$y(x) = a\cosh\left(\frac{x}{a} - \alpha\right),$$

we have deduced that such a catenary curve is the generator which minimizes the lateral surface area of our volume of revolution and hence this minimal surface is a catenoid.

4.6 Pastoral Interlude: Envelope of a One-Parameter Family of Curves

Consider a one-parameter family of curves in the x-y plane that satisfies the relation

$$F(x,y,\alpha) \equiv 0.$$

Taking the partial derivative of that relation with respect to the parameter α yields

$$F_\alpha(x,y,\alpha) \equiv 0.$$

Now, solving that second relation for $\alpha = A(x,y)$ and substituting this back into the first relation, we obtain the envelope function of our one-parameter family of curves defined by

$$E(x,y) = F[x,y,A(x,y)] \equiv 0.$$

We next demonstrate that a member of this family must be tangent to the envelope at any point of intersection with it and hence cannot pass through it. Select a fixed value of α denoted by α_0. If $F(x,y,\alpha_0) \equiv 0$ intersects $E(x,y) \equiv 0$ at the point (x_0,y_0),

$$F(x_0,y_0,\alpha_0) = E(x_0,y_0) = 0 \Rightarrow \alpha_0 = A(x_0,y_0).$$

Representing this $\alpha = \alpha_0$ member of that family by $y = y_1(x)$, we have

$$F[x, y_1(x), \alpha_0] \equiv 0,$$

which, upon differentiation with respect to x, yields

$$F_x[x, y_1(x), \alpha_0] + F_y[x, y_1(x), \alpha_0]y_1'(x) \equiv 0.$$

Evaluating this relation at $x = x_0$ and noting that $y_1(x_0) = y_0$, we find

$$y_1'(x_0) = -\frac{F_x(x_0, y_0, \alpha_0)}{F_y(x_0, y_0, \alpha_0)}.$$

In a similar manner, representing the envelope function $E(x,y) \equiv 0$ by $y = y_2(x)$, we have

$$F[x, y_2, \mathcal{A}(x)] \equiv 0 \text{ where } \mathcal{A}(x) = A[x, y_2(x)],$$

which, upon differentiation with respect to x, yields

$$F_x[x, y_2(x), \mathcal{A}(x)] + F_y[x, y_2(x), \mathcal{A}(x)]y_2'(x) + F_\alpha[x, y_2(x), \mathcal{A}(x)]\mathcal{A}'(x) \equiv 0.$$

Evaluating this relation at $x = x_0$ and noting that $y_2(x_0) = y_0$ while $\mathcal{A}(x_0) = A(x_0, y_0) = \alpha_0$, we obtain

$$F_x(x_0, y_0, \alpha_0) + F_y(x_0, y_0, \alpha_0)y_2'(x_0) + F_\alpha(x_0, y_0, \alpha_0)\mathcal{A}'(x_0) \equiv 0.$$

Since, $F_\alpha(x_0, y_0, \alpha_0) = 0$ by virtue of the way the envelope function was defined, we find that

$$y_2'(x_0) = -\frac{F_x(x_0, y_0, \alpha_0)}{F_y(x_0, y_0, \alpha_0)} = y_1'(x_0).$$

Hence, demonstrating the result. Here, should $F_y(x_0, y_0, \alpha_0) = 0$, there will be vertical tangency at the intersection point.

In order to illustrate this procedure we consider the one-parameter family of circles centered at the point (α, α) with radius $\alpha > 0$ [44] represented in Fig. 4.7. This family satisfies the relation

$$F(x, y, \alpha) = (x - \alpha)^2 + (y - \alpha)^2 - \alpha^2 \equiv 0 \text{ for } \alpha > 0.$$

Then

$$F_\alpha(x, y, \alpha) = 2(\alpha - x) + 2(\alpha - y) - 2\alpha \equiv 0 \Rightarrow \alpha = A(x, y) = x + y.$$

Thus

$$E(x, y) = F[x, y, A(x, y)] = F(x, y, x + y) = y^2 + x^2 - (x + y)^2 = -2xy \equiv 0.$$

Hence the envelope for this family is given by

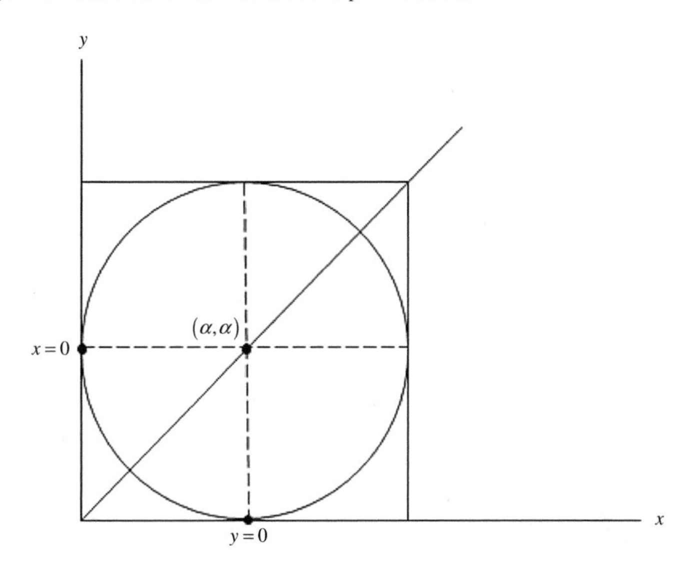

Fig. 4.7 Plot of a member of the one-parameter family of circles $(x-\alpha)^2+(y-\alpha)^2=\alpha^2$ for $\alpha>0$.

$$x = 0 \text{ or } y = 0.$$

Note that every member of this family has an intersection point $(\alpha,0)$ of horizontal tangency with the line $y=0$ since $F_x(\alpha,0,\alpha)=0$ and an intersection point $(0,\alpha)$ of vertical tangency with the line $x=0$ since $F_y(0,\alpha,\alpha)=0$.

4.7 Diagrammatic Results for the Canonical Soap Film Problem

Having developed all the necessary machinery to analyze our canonical soap film problem, we now return to the scenario depicted in Fig. 4.1. We have shown that a catenary of the form

$$y(x) = a\cosh\left(\frac{x}{a}-\alpha\right)$$

provides a minimum for the lateral surface area of the volume of revolution generated by a twice-differentiable curve satisfying the conditions $y(x_1)=y_1$ and $y(x_2)=y_2$. Next we reduce this two-parameter family of catenaries to a one-parameter family by requiring all these curves to emanate from the same fixed point (x_1,y_1) where for ease of exposition x_1 is being taken equal to zero (see Fig. 4.8). That is

$$y(0) = a\cosh(\alpha) = y_1 \Rightarrow a = \frac{y_1}{\cosh(\alpha)}$$

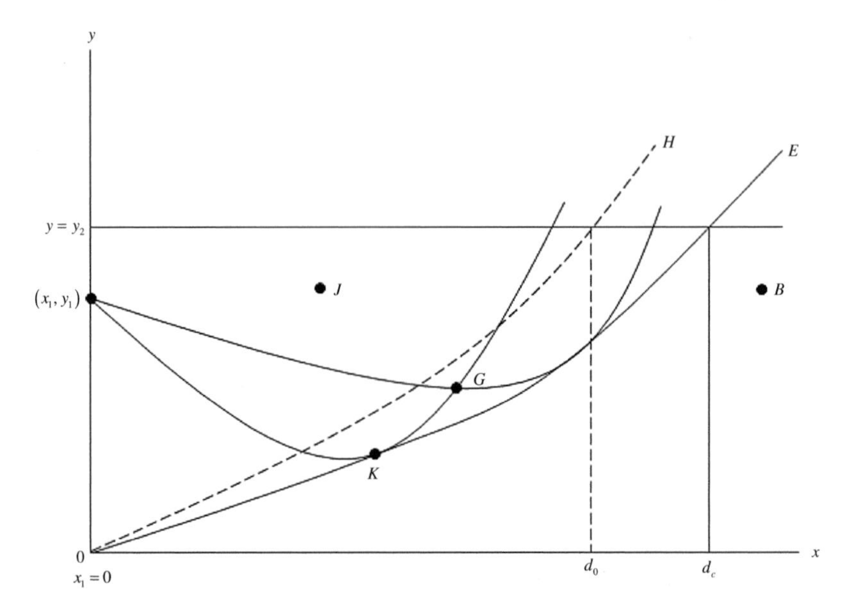

Fig. 4.8 Graphical diagram relevant to the canonical soap film problem.

and our one-parameter family is given by

$$F(x,y,\alpha) = y\cosh(\alpha) - y_1\cosh\left[\cosh(\alpha)\frac{x}{y_1} - \alpha\right] \equiv 0.$$

Thus

$$F_\alpha(x,y,\alpha) = y\sinh(\alpha) - [\sinh(\alpha)x - y_1]\sinh\left[\cosh(\alpha)\frac{x}{y_1} - \alpha\right] \equiv 0 \Rightarrow \alpha = A(x,y)$$

and the envelope function for this family is, as usual, defined by

$$E(x,y) = F[x,y,A(x,y)] \equiv 0.$$

In what follows we shall be following the approach of Weinstock [133] while adding some details omitted from that exposition. We wish to discover which if any of these catenaries passes through the end point (x_2, y_2) and can actually provide the minimum we seek. Several of these curves as well as their envelope function are plotted in Fig. 4.8. After Weinstock [133] we make the following observations about the points in that diagram and summarize them in Table 4.1. As demonstrated in Section 4.6 every one of these catenaries is tangent to the envelope function and hence none of them can pass through a point such as B that lies in the region to the right of that function. It can be shown [8] in this event that the so-called Goldschmidt discontinuous solution (GDS) defined by

$$\begin{cases} x = x_1 = 0, & \text{for } 0 \leq y \leq y_1 \\ y = 0, & \text{for } 0 = x_1 \leq x \leq x_2 \\ x = x_2, & \text{for } 0 \leq y \leq y_2 \end{cases}$$

generates the minimum surface area given by the sum of the areas of the two circles of radii y_1 and y_2, respectively, or

$$\mathcal{A}(y_1, y_2) = \pi(y_1^2 + y_2^2).$$

For any point in the region to the left of the envelope function such as G or J in Fig. 4.8, two catenaries, which are explicitly drawn for G, can fit the required end point condition but only the upper one generates a minimum while the lower one whose point of tangency lies within the interval $x_1 < x < x_2$ does not. Even for a point such as K, which lies on the envelope function and has a unique catenary with this end point, that catenary does not generate a minimum. Again this minimum is generated by the GDS.

In an analogous manner to the way the envelope function was defined we need to develop the locus for which the lateral surface area of the catenoid generated by the upper catenary at a point in the region to the left of the envelope function is equal to that associated with the GDS. That is

$$\begin{aligned} S(y) &= 2\pi \int_0^x y(z)\sqrt{1+[y']^2}\,dz = 2\pi a \int_0^x \cosh^2\left(\frac{z}{a} - \alpha\right) dz \text{ with } a = \frac{y_1}{\cosh(\alpha)} \\ &= \pi a \int_0^x \left[1 + \cosh\left(\frac{2z}{a} - 2\alpha\right)\right] dz = \pi a x + \frac{\pi a^2}{2}\left[\sinh\left(2\frac{x}{a} - 2\alpha\right) + \sinh(2\alpha)\right] \\ &= \mathcal{A}(y_1, y) = \pi(y_1^2 + y^2) \Rightarrow \alpha = C(x, y) \end{aligned}$$

and

$$H(x, y) = F[x, y, C(x, y)] \equiv 0.$$

This locus is also plotted in Fig. 4.8, denoted by H, and lies above and to the left of the envelope function denoted by E in that figure. For points such as J in the region to the left of H the upper catenary generates a catenoid that provides an absolute minimum for the lateral surface area and the GDS provides a relative minimum while for points such as G that lie in the region between H and E the GDS provides the absolute minimum and the upper catenary generates a catenoid that provides a relative minimum for the lateral surface area.

By examining the horizontal line $y = y_2$ in Fig. 4.8 we can finally determine the experimental shape of the soap film formed between the two circular rings as a function of the axial distance d between the two rings which is the x-coordinate on that line. The catenoid generated by the upper catenary provides the absolute minimum of the lateral surface area for $0 < d < d_0$ where $H(d_0, y_2) = 0$ and a relative minimum for $d_0 < d < d_c$ where $E(d_c, y_2) = 0$ while the GDS provides a relative minimum for $0 < d < d_0$ and the absolute minimum for $d \geq d_0$. It can be shown

Table 4.1 Minimum lateral surface area behavior for the reference points in Fig. 4.8

Reference point	Minimum surface area behavior
J	$\begin{cases} \text{upper catenary} \equiv \text{absolute minimum} \\ \text{GDS} \equiv \text{relative minimum} \end{cases}$
G	$\begin{cases} \text{upper catenary} \equiv \text{relative minimum} \\ \text{GDS} \equiv \text{absolute minimum} \end{cases}$
K	GDS $\equiv$ absolute minimum
B	GDS $\equiv$ absolute minimum

that the catenoid generated by the upper catenary is stable to small disturbances for $0 < d < d_c$ with that linear stability being stationary for $0 < d \le d_0$ and oscillatory for $d_0 < d < d_c$ [32]. Hence if one starts with the two rings coincident and begins slowly pulling them apart by small increments, allowing each new configuration to stabilize before continuing this process, the soap film will take the shape of the catenoid generated by the upper catenary for $0 < d < d_c$ with the onset of that configuration occurring from a stationary state for $0 < d \le d_0$ and from an oscillatory one for $d_0 < d < d_c$. Even though the GDS provides the absolute minimum for $d \ge d_0$, the soap film will not rupture and occupy the circular areas of the two rings separately until $d \ge d_c$ since the disturbances to an existing state are small during this procedure.

Problem

4.1. The *brachistochrone* ("shortest time") problem, posed by Johann Bernoulli in 1696, states that a particle of mass m starts at a point $P_1(x_1, y_1)$ with speed V and moves, without friction, under the influence of gravity (assumed to be acting in the negative y-direction) along a curve $y = y(x)$ to a point $P_2(x_2, y_2)$ where $y_1 > y_2$ and $x_2 > x_1$, and requires the curve along which the elapsed time T is a minimum.

(a) If v is the speed and s the distance travelled along the arc at time t, use the relations

$$v = \frac{ds}{dt} \text{ and } \frac{v^2}{2g} + y \equiv \frac{V^2}{2g} + y_1 = y_0$$

to obtain the formulation

$$T = \frac{1}{\sqrt{2g}} \int_{x_1}^{x_2} \frac{\sqrt{1 + [y'(x)]^2}}{\sqrt{y_0 - y(x)}}\, dx \text{ where } y(x_1) = y_1 \text{ and } y(x_2) = y_2.$$

Hint: Note that $ds/dt = s'(x)/t'(x)$ and $T = \int_{x_1}^{x_2} t'(x)\, dx$.

(b) If a minimizing curve $x = x(y)$ exists, deduce by means of the Calculus of Variations, that it satisfies

$$x'(y) = -\frac{\sqrt{y_0 - y}}{\sqrt{c_1 - (y_0 - y)}}$$

where c_1 is a positive constant satisfying $c_1 > y_1 - y_2 + V^2/(2g)$.
Hint: Find an equation satisfied by $y'(x)$ and invert it. Note that the integrand of the integral to which T is proportional does not depend explicitly on the independent variable x.

(c) Letting

$$y_0 - y = 2a\sin^2\left(\frac{\theta}{2}\right) \text{ where } a = \frac{c_1}{2},$$

show that this potential minimizing curve can be defined by parametric equations of the form

$$x = x_0 + a[\theta - \sin(\theta)], \; y = y_0 - a[1 - \cos(\theta)],$$

where x_0 and a are constants to be determined such that P_1 and P_2 lie on the curve.
Hint: $dx/d\theta = x'[y(\theta)]dy/d\theta$.

(d) These extremals are actually *cycloids*. Unlike the family of catenary curves it is known that x_0 and a can be uniquely determined so that P_1 and P_2 are on the same "arch" of one such cycloid and the corresponding arc(P_1P_2) then truly

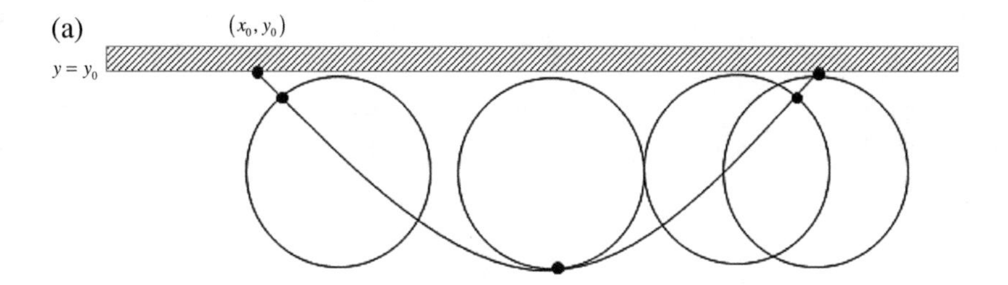

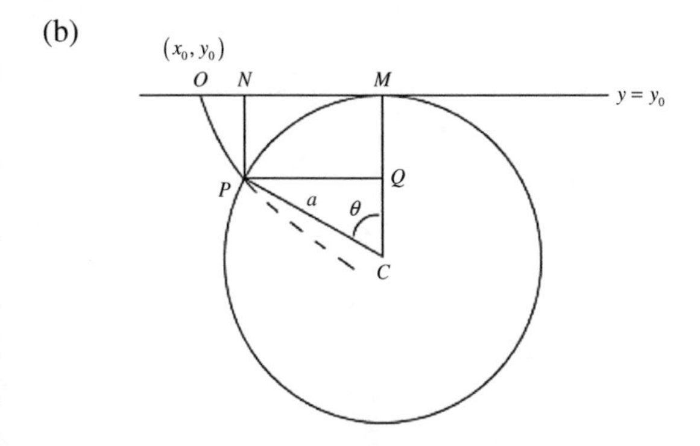

Fig. 4.9 (a) Plot of the cycloid curve. (b) Enlargement of the first circle in part (a) denoting the central angle θ.

minimizes T. Deduce from Fig. 4.9 appearing above that this cycloid is generated by the motion of a fixed point $P(x, y)$ on the circumference of a circle of radius a which starts at point $O(x_0, y_0)$ and rolls without slipping along the underside of the line $y = y_0$. You need only verify this for the central angle θ of the circle in the first quadrant or $0 \leq \theta \leq \pi/2$.
Hint: Since the circle rolls along the underside of the given line without slipping, the $\text{arc}(PM) = a\theta = \overline{OM}$.

Chapter 5
Heat Conduction in a Finite Bar with a Linear Source

This chapter considers heat conduction in a laterally insulated finite bar with a linear source term. That requires a derivation of the heat equation in a nonmoving continua from a conservation of energy balance law, which involves source and flux terms by using both the divergence theorem and the DuBois–Reymond Lemma. Then an equation of state and constitutive relations must be introduced. Finally, for appropriate boundary conditions, the partial differential heat equation is solved by a separation of variables eigenvalue approach and initial conditions satisfied by a Fourier series methodology, which is introduced as a pastoral interlude. This allows for an examination of the long-time behavior of that heat conduction situation and a deduction of a stability criterion. The problems deal with the complex exponential definitions of the trigonometric functions employed in this analysis of heat conduction in a finite bar and with a similar situation involving different boundary conditions.

5.1 Heat Equation in Nonmoving Continua, Divergence Theorem, and DuBois–Reymond Lemma

Particle mechanics deals with the equilibrium and motion of systems of point masses. In continuum mechanics one deals with continuous media in which smoothly varying properties such as density (ρ) and temperature (T) must be assigned at each point (x, y, z) in space instantaneously occupied by a material body under examination at the time (t) in question using the so-called continuum hypothesis. There exist a large number of phenomena to which this hypothesis can be applied and hence may be modeled by the various techniques of continuum mechanics. Such techniques usually associated with postulated balance-type conservation laws produce governing systems of partial differential equations (PDEs). Often consistent with experimental, thermodynamic, or empirical evidence it becomes necessary to adopt particular constitutive relations, equations of state, or some other relationship between the dependent variables. All of these equations along with appropriate boundary (space) and initial (time) conditions constitute the mathematical formulation of the problem.

© Springer International Publishing AG, part of Springer Nature 2017

D. J. Wollkind and B. J. Dichone, *Comprehensive Applied Mathematical Modeling in the Natural and Engineering Sciences*,
https://doi.org/10.1007/978-3-319-73518-4_5

To illustrate this procedure we shall consider a number of problems involving nonmoving continua that give rise to governing reaction–diffusion type PDEs in this and the following three chapters. We begin in this chapter and the next with heat conduction phenomena.

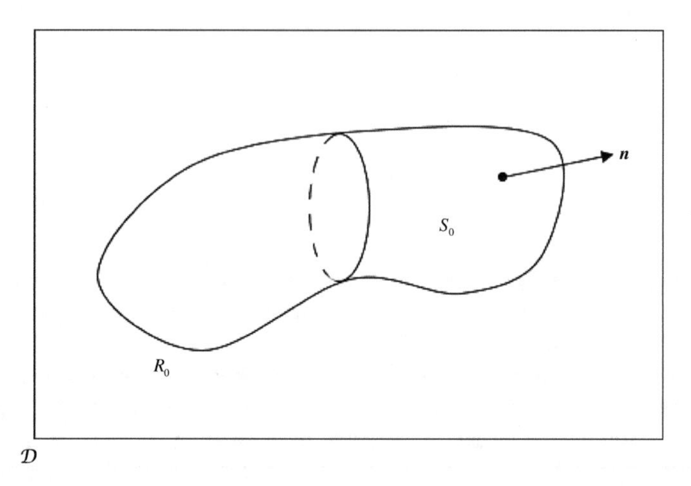

Fig. 5.1 Schematic diagram of an arbitrary fixed material region $R_0 \subset \mathcal{D}$ with bounding surface S_0 and outward-pointing unit normal $\boldsymbol{n}$.

Let $R_0 \subset \mathcal{D}$ be a *fixed arbitrary* material region with bounding surface S_0 possessing outward-pointing unit normal $\boldsymbol{n}$ (see Fig. 5.1). We postulate the following balance-type conservation law for heat energy $Q(t)$ in this region:

$$\frac{dQ(t)}{dt} = \frac{d}{dt}\iiint\limits_{R_0} (\rho e)(\boldsymbol{x},t)d\tau = \iiint\limits_{R_0} r(\boldsymbol{x},t)d\tau - \iint\limits_{S_0} \boldsymbol{J}^{(Q)}(\boldsymbol{x},t)\bullet\boldsymbol{n}\,d\sigma$$

where $\boldsymbol{x} = (x,y,z) \equiv$ position vector and $e \equiv$ specific internal heat energy while r and $\boldsymbol{J}^{(Q)}$ represent the heat source and flux vector in the bulk of R_0 and across the boundary out of S_0, respectively. Here $\boldsymbol{u}\bullet\boldsymbol{v} \equiv u_1v_1 + u_2v_2 + u_3v_3$ is the dot product of $\boldsymbol{u} = (u_1,u_2,u_3)$ and $\boldsymbol{v} = (v_1,v_2,v_3)$. Finally $d\tau \equiv dx\,dy\,dz$ and $d\sigma \equiv$ differential surface element.

Writing this balance law in the form ① = ② − ③ we now convert ① and ③ to volume integrals of the same form as ②:

$$① = \lim_{\Delta t\to 0}\frac{Q(t+\Delta t)-Q(t)}{\Delta t} = \iiint\limits_{R_0}\lim_{\Delta t\to 0}\frac{(\rho e)(\boldsymbol{x},t+\Delta t)-(\rho e)(\boldsymbol{x},t)}{\Delta t}\,d\tau = \iiint\limits_{R_0}\frac{\partial(\rho e)}{\partial t}(\boldsymbol{x},t)\,d\tau$$

where we have employed the definition of the derivative and the fact that R_0 is a fixed region;

$$\textcircled{3} = \iint_{S_0} \boldsymbol{J}^{(Q)}(\boldsymbol{x},t) \bullet \boldsymbol{n}\, d\sigma = \iiint_{R_0} \nabla \bullet \boldsymbol{J}^{(Q}(\boldsymbol{x},t)\, d\tau;$$

where we have employed the *divergence theorem* for a vector $\boldsymbol{w}(x,y,z) = (w_1, w_2, w_3)$

$$\iint_{S_0} \boldsymbol{w} \bullet \boldsymbol{n}\, d\sigma = \iiint_{R_0} \nabla \bullet \boldsymbol{w}\, d\tau \text{ with } \nabla \bullet \boldsymbol{w} \equiv \frac{\partial w_1}{\partial x} + \frac{\partial w_2}{\partial y} + \frac{\partial w_3}{\partial z}.$$

Now substitution of these expressions for $\textcircled{1}$ and $\textcircled{3}$ into our balance law yields

$$\iiint_{R_0} \left[\frac{\partial(\rho e)}{\partial t} + \nabla \bullet \boldsymbol{J}^{(Q)} - r \right] (\boldsymbol{x},t)\, d\tau = 0.$$

We next employ the *DuBois–Reymond Lemma* of continuum mechanics which is philosophically equivalent to the fundamental lemma of the Calculus of Variations introduced in Chapter 4:

If for every *arbitrary* subset $R_0 \subset \mathcal{D}$,

$$\iiint_{R_0} g(\boldsymbol{x},t)\, d\tau = 0 \text{ and } g \in C[\mathcal{D}] \Rightarrow g(\boldsymbol{x},t) \equiv 0 \text{ for all } \boldsymbol{x} \in \mathcal{D}.$$

Then applying this lemma to our previous result and assuming the requisite continuity of the partial derivatives, we obtain the basic governing PDE for heat conduction:

$$\left[\frac{\partial(\rho e)}{\partial t} + \nabla \bullet \boldsymbol{J}^{(Q)} - r \right] (\boldsymbol{x},t) \equiv 0 \text{ for all } \boldsymbol{x} \in \mathcal{D}.$$

5.2 Equation of State, Constitutive Relations, and Boundary and Initial Conditions

In order to complete our formulation for heat conduction we now adopt the following *equation of state and constitutive relations*:

(i) Equation of State: The five state variables of a system of this sort consist of $p \equiv$ pressure and $s \equiv$ specific entropy in addition to ρ, T, and e. In general two of them are considered to be the independent variables and the other three are functions of them determined by prescribed thermodynamic relationships. We shall return to this topic and develop it in more detail in later chapters, but for our present purposes, it is only necessary to adopt the following equation of state:

$$e = c_0 T \text{ where } c_0 \equiv \text{specific heat assumed constant and } T \text{ is measured in K.}$$

(ii) Constitutive Relations: These are relationships based on the material properties of a system. We shall adopt two of them:

(a) Homogeneous Medium or $\rho \equiv \rho_0$;
(b) Newton's Law of Cooling or $\boldsymbol{J}^{(Q)} = -k_0 \nabla T$,

where

$$k_0 \equiv \text{thermal conductivity assumed constant and } \nabla T = \left(\frac{\partial T}{\partial x}, \frac{\partial T}{\partial y}, \frac{\partial T}{\partial z} \right).$$

Let us investigate the heat conduction implication of this latter assumption. For ease of exposition, consider a simplified temperature distribution given by $T = T(x)$. Then $\nabla T = T'(x)\boldsymbol{i}$ where $\boldsymbol{i} = (1, 0, 0)$ and $\boldsymbol{J}^{(Q)} = J_1(x)\boldsymbol{i}$ where $J_1(x) = -k_0 T'(x)$. Assume that for any $x_2 > x_1$, $T(x_2) = T_2 > T_1 = T(x_1)$. Hence $T'(x) > 0$ and $J_1'(x) < 0$ as depicted in Fig. 5.2. Thus heat flows in a direction opposite to the temperature gradient or from hot to cold which is what one would expect for a thermodynamically valid such assumption.

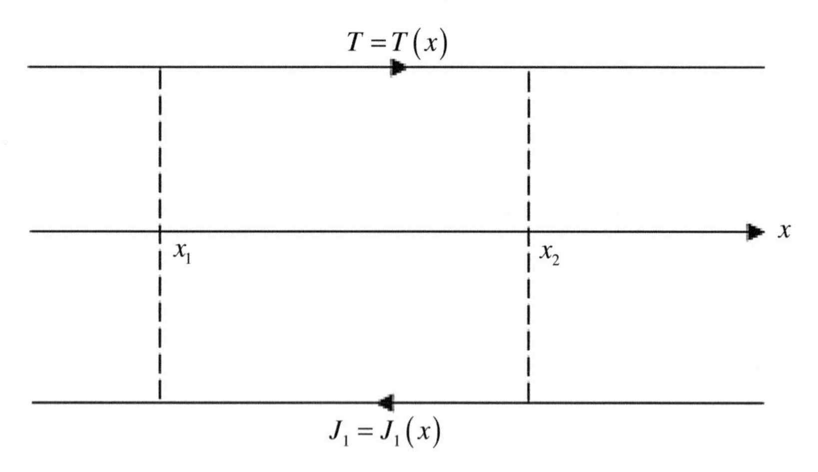

Fig. 5.2 Schematic plot of a temperature field $T = T(x)$ and its flux vector $\boldsymbol{J}^{(Q)}(x) = J_1(x)\boldsymbol{i}$ where $J_1(x) = -k_0 T'(x)$ and $T'(x) > 0$.

Now, adopting these three relationships and substituting them into our basic governing PDE, we obtain the following reaction–diffusion heat equation

$$\frac{\partial T}{\partial t} = \kappa_0 \nabla^2 T + s,$$

where

$$\kappa_0 = \frac{k_0}{\rho_0 c_0} \equiv \text{thermal diffusivity or theormometric conductivity and } s = \frac{r}{\rho_0 c_0},$$

since

$$-\nabla \bullet \boldsymbol{J}^{(Q)} = k_0 \nabla \bullet \nabla T = k_0 \left(\frac{\partial^2 T}{\partial x^2} + \frac{\partial^2 T}{\partial y^2} + \frac{\partial^2 T}{\partial z^2} \right) \equiv k_0 \nabla^2 T.$$

Finally we assume a linear heat source term based upon the deviation from an ambient temperature T_e

$$r = r_0(T - T_e) \Rightarrow s = s_0(T - T_e) \text{ where } s_0 = \frac{r_0}{\rho_0 c_0} > 0.$$

Hence our heat equation for this linear source situation becomes

$$\frac{\partial T}{\partial t} = \kappa_0 \nabla^2 T + s_0(T - T_e).$$

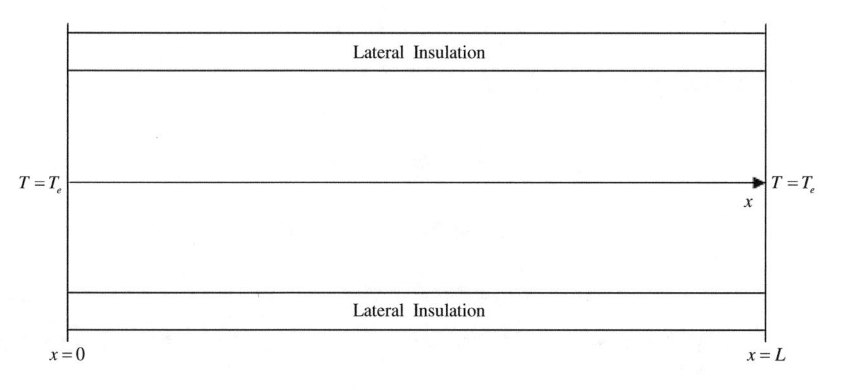

Fig. 5.3 Schematic diagram of the laterally insulated finite bar of length L with heat conduction driven by a linear source.

Further we shall consider heat conduction in the laterally insulated finite bar of length L depicted in Fig. 5.3. Then its temperature distribution $T = T(x, t)$ in the bar is governed by the reaction–diffusion equation

$$\frac{\partial T}{\partial t} = \kappa_0 \frac{\partial^2 T}{\partial x^2} + s_0(T - T_e), 0 < x < L, \ t > 0. \tag{5.2.1a}$$

This PDE is defined on the open interval in space and time. To complete the formulation for our heat conduction finite-bar problem we require *boundary conditions* (BC's) at the end points of the bar $x = 0$, L and an *initial condition* (IC) at $t = 0$. Toward that end we impose the conditions

$$T(0,t) = T(L,t) = T_e \text{ for } t > 0; \tag{5.2.1b}$$

$$T(x,0) = T_e + F(x) \text{ for } 0 < x < L. \tag{5.2.1c}$$

5.3 Separation of Variables Solution

Note that the subsystem of (5.2.1) consisting of the PDE of (5.2.1a) and the BC's of (5.2.1b) admits the ambient temperature solution

$$T(x,t) \equiv T_e. \tag{5.3.1}$$

Since the separation of variables method to be developed in this section requires that subsystem to be homogeneous – *i.e.*, zero must be a solution of it, we first introduce the change of dependent variable

$$u(x,t) = T(x,t) - T_e, \tag{5.3.2}$$

which transforms system (5.2.1) into

$$\frac{\partial u}{\partial t} = \kappa_0 \frac{\partial^2 u}{\partial x^2} + s_0 u, \ 0 < x < L, \ t > 0; \tag{5.3.3a}$$

$$u(0,t) = u(L,t) = 0 \text{ for } t > 0; \tag{5.3.3b}$$

$$u(x,0) = F(x) \text{ for } 0 < x < L. \tag{5.3.3c}$$

To simplify this procedure further we next introduce the additional change of dependent variable

$$u(x,t) = e^{\delta t} w(x,t), \tag{5.3.4a}$$

which transforms the PDE of (5.3.3a) into

$$\frac{\partial w}{\partial t} + \delta w = \kappa_0 \frac{\partial^2 w}{\partial x^2} + s_0 w. \tag{5.3.4b}$$

Observe, from an examination of (5.3.4) that, upon selection of

$$\delta = s_0 \text{ or } u(x,t) = e^{s_0 t} w(x,t), \tag{5.3.5}$$

system (5.3.3) is transformed into

$$\frac{\partial w}{\partial t} = \kappa_0 \frac{\partial^2 w}{\partial x^2}, 0 < x < L, t > 0; \tag{5.3.6a}$$

$$w(0,t) = w(L,t) = 0 \text{ for } t > 0; \tag{5.3.6b}$$

$$w(x,0) = F(x) \text{ for } 0 < x < L. \tag{5.3.6c}$$

We are finally ready to seek a separation of variables solution of subsystem (5.3.6a) and (5.3.6b) of the form

$$w(x,t) = X(x)G(t). \tag{5.3.7}$$

Then, upon substitution of (5.3.7) into that subsystem, we obtain

$$\frac{X''(x)}{X(x)} = \frac{\dot{G}(t)}{\kappa_0 G(t)} \equiv \sigma, 0 < x < L, t > 0; \tag{5.3.8a}$$

$$X(0) = X(L) = 0; \tag{5.3.8b}$$

which, in particular, yields an eigenvalue problem for σ, the separation constant, since $X(x) \equiv 0$ is a solution to (5.3.8) and we wish to find values of σ for which there exist nontrivial $X(x) \not\equiv 0$. Such σ are called eigenvalues and the corresponding nontrivial solutions $X(x)$ eigenfunctions. We shall consider $\sigma \in \mathbb{C}$ and examine the possibilities that $\sigma = 0$ or $\sigma \neq 0$, sequentially.

(i) $\sigma = 0$: $X''(x) \equiv 0 \Rightarrow X(x) = ax + b$ and $X(0) = b = 0 \Rightarrow X(x) = ax$ while then $X(L) = aL = 0 \Rightarrow a = 0 \Rightarrow X(x) \equiv 0 \Rightarrow \sigma = 0$ cannot be an eigenvalue for this problem.

(ii) $\sigma \neq 0$: $X''(x) - \sigma X(x) = 0 \Rightarrow X(x) = c_1 e^{\sqrt{\sigma}x} + c_2 e^{-\sqrt{\sigma}x}$ where $\mathrm{Re}(\sqrt{\sigma}) \geq 0$ and $X(0) = c_1 + c_2 = 0 \Rightarrow X(x) = c_1(e^{\sqrt{\sigma}x} - e^{-\sqrt{\sigma}x})$ while then

$$X(L) = c_1(e^{\sqrt{\sigma}L} - e^{-\sqrt{\sigma}L}) = 0$$

$$\Rightarrow e^{2\sqrt{\sigma}L} = 1 = e^{2n\pi i} \text{ for } i = \sqrt{-1} \text{ and } n = \pm 1, \pm 2, \pm 3, \ldots \text{ since } c_1 \neq 0 \Rightarrow \sqrt{\sigma_n} = \frac{n\pi i}{L}$$

$$\Rightarrow \sigma_n = \frac{-n^2\pi^2}{L^2}, X_n(x) = c_1(e^{n\pi x i/L} - e^{-n\pi x i/L}) = c_n \sin\left(\frac{n\pi x}{L}\right) \text{ where } c_n = 2ic_1.$$

Then

$$\dot{G}_n(t) = \sigma_n \kappa_0 G_n(t) = -n^2\pi^2 \kappa_0 \frac{G_n(t)}{L^2} \Rightarrow G_n(t) = c_0 e^{-n^2\pi^2\kappa_0 t/L^2}.$$

Thus, we have obtained the solutions

$$w_n(x,t) = X_n(t)G_n(t) = c_n e^{-n^2\pi^2\kappa_0 t/L^2} \sin\left(\frac{n\pi x}{L}\right) \text{ for } n = \pm 1, \pm 2, \pm 3, \ldots, \tag{5.3.9}$$

where, with no loss of generality, c_0 has been taken equal to 1. Finally, without any loss of generality, we may restrict the n in (5.3.9) to $n = 1, 2, 3, \ldots$, since the corresponding negative integers only yield linearly dependent solutions. Note that we did not include the possibility of $n = 0$ here given that this was taken into account by the $\sigma = 0$ case already treated above.

5.4 Pastoral Interlude: Fourier Series

The solutions of (5.3.9) satisfy the subsystem (5.3.6a) and (5.3.6b). Ultimately in order to satisfy the IC (5.3.6c) we must employ the concept of Fourier series which is presented in this section. We begin with a statement of the Fourier series Theorem for a function $f = f(x)$:

Let f, f' be piecewise continuous for $x \in (-L, L)$ and extended periodically outside this interval such that $f(x+2L) = f(x)$ for all $x \in \mathbb{R}$ (see Fig. 5.4). Then the series

$$\frac{a_0}{2} + \sum_{n=1}^{\infty}\left[a_n \cos\left(\frac{n\pi x}{L}\right) + b_n \sin\left(\frac{n\pi x}{l}\right)\right]$$

where

$$a_n = \frac{1}{L}\int_{-L}^{L} f(u)\cos\left(\frac{n\pi u}{L}\right) du,\ b_n = \frac{1}{L}\int_{-L}^{L} f(u)\sin\left(\frac{n\pi u}{L}\right) du,$$

converges to

$$\frac{f(x+0)+f(x-0)}{2} \doteq f(x) \text{ with } f(x\pm 0) = \lim_{u\to x^{\pm}} f(u).$$

Here we have used the notation "$\doteq$" to indicate that the series converges to $f(x)$ at any point x where the function is continuous and hence $f(x\pm 0) = f(x)$.

We shall give a heuristic justification of this result by letting $f(x)$ represent the series with general coefficients where it converges and then finding relationships between those coefficients and that function consistent with the Fourier series Theorem. Recall the orthogonality relations

$$\int_{-L}^{L} \cos\left(\frac{n\pi x}{L}\right) dx = \int_{-L}^{L} \sin\left(\frac{n\pi x}{L}\right) dx = \int_{-L}^{L} \sin\left(\frac{n\pi x}{L}\right)\cos\left(\frac{m\pi x}{L}\right) dx = 0$$

while

$$\int_{-L}^{L} \cos\left(\frac{n\pi x}{L}\right)\cos\left(\frac{m\pi x}{L}\right) dx = \int_{-L}^{L} \sin\left(\frac{n\pi x}{L}\right)\sin\left(\frac{m\pi x}{L}\right) dx = L\delta_{mn}$$

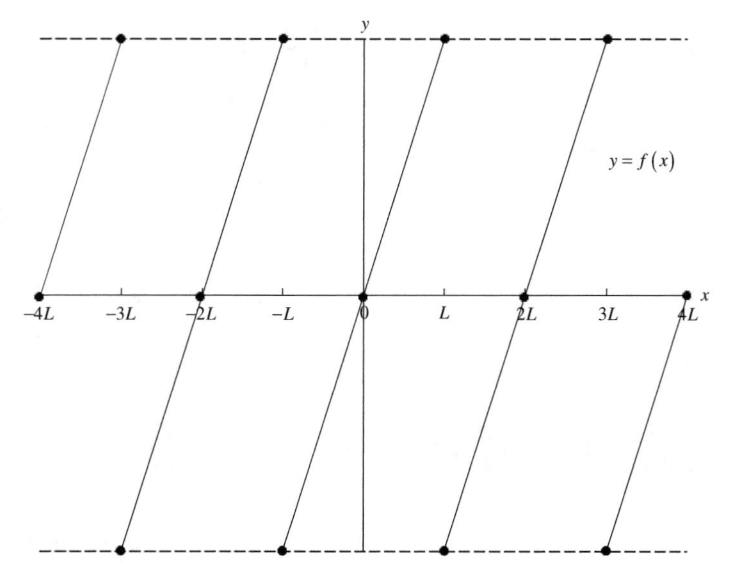

Fig. 5.4 Schematic plot of a square-wave type function $y = f(x)$ defined for $-L < x < L$ and extended periodically outside this interval such that $f(x+L) = f(x)$.

where

$$\delta_{mn} = \begin{cases} 1, & \text{for } m = n \\ 0, & \text{for } m \neq n \end{cases}.$$

Now define

$$f(x) = \frac{a_0}{2} + \sum_{n=1}^{\infty} \left[a_n \cos\left(\frac{n\pi x}{L}\right) + b_n \sin\left(\frac{n\pi x}{L}\right) \right].$$

(i) Integrating this equation term by term on x from $-L$ to L and employing the orthogonality relations, we obtain

$$\int_{-L}^{L} f(x)\,dx = \frac{a_0}{2} \int_{-L}^{L} dx + \sum_{n=1}^{\infty} \left[a_n \int_{-L}^{L} \cos\left(\frac{n\pi x}{L}\right) dx + b_n \int_{-L}^{L} \sin\left(\frac{n\pi x}{L}\right) dx \right] = a_0 L.$$

(ii) Multiplying this equation by $\cos(m\pi x/L)$ and proceeding as in (i), we obtain

$$\int_{-L}^{L} f(x)\cos\left(\frac{m\pi x}{L}\right)dx = \frac{a_0}{2}\int_{-L}^{L}\cos\left(\frac{m\pi x}{L}\right)dx$$
$$= \sum_{n=1}^{\infty}\left[a_n\int_{-L}^{L}\cos\left(\frac{n\pi x}{L}\right)\cos\left(\frac{m\pi x}{L}\right)dx + b_n\int_{-L}^{L}\sin\left(\frac{n\pi x}{L}\right)\cos\left(\frac{m\pi x}{L}\right)dx\right]$$
$$= L\sum_{n=1}^{\infty} a_n\delta_{mn} = La_m.$$

(iii) Multiplying this equation by $\sin(m\pi x/L)$ and proceeding as in (ii), we obtain

$$\int_{-L}^{L} f(x)\sin\left(\frac{m\pi x}{L}\right)dx = \frac{a_0}{2}\int_{-L}^{L}\sin\left(\frac{m\pi x}{L}\right)dx$$
$$= \sum_{n=1}^{\infty}\left[a_n\int_{-L}^{L}\cos\left(\frac{n\pi x}{L}\right)\sin\left(\frac{m\pi x}{L}\right)dx + b_n\int_{-L}^{L}\sin\left(\frac{n\pi x}{L}\right)\sin\left(\frac{m\pi x}{L}\right)dx\right]$$
$$= L\sum_{n=1}^{\infty} b_n\delta_{mn} = Lb_m.$$

We shall next consider the special Fourier series for an even or odd function. Recall that

$$\int_{-L}^{L} g(x)\,dx = \begin{cases} 2\int_0^L g(x)\,dx, & \text{if } g(-x)=g(x) \text{ or } g \text{ is even} \\ 0, & \text{if } g(-x)=-g(x) \text{ or } g \text{ is odd} \end{cases}.$$

(i) Suppose $f(x)$ is even. Then

$$a_nL = 2\int_0^L f(x)\cos\left(\frac{n\pi x}{L}\right)dx,\ b_n = 0 \Rightarrow f(x) = \frac{a_0}{2} + \sum_{n=1}^{\infty} a_n\cos\left(\frac{n\pi x}{L}\right).$$

(ii) Suppose $f(x)$ is odd. Then

$$a_n = 0, b_nL = 2\int_0^L f(x)\sin\left(\frac{n\pi x}{L}\right)dx \Rightarrow f(x) = \sum_{n=1}^{\infty} b_n\sin\left(\frac{n\pi x}{L}\right).$$

Finally we shall introduce a so-called half-range function $F(x)$ restricted to the subinterval $x \in (0,L)$ such that F, F' are piecewise continuous there. Then we can define its odd extension $f(x)$ as depicted in Fig. 5.5 by

$$f(x) = \begin{cases} F(x), & \text{for } x \in (0,L) \\ -F(-x), & \text{for } x \in (-L,0) \end{cases}.$$

Note, as so defined $f(-x) = -f(x)$, and thus

$$f(x) = \sum_{n=1}^{\infty} b_n\sin\left(\frac{n\pi x}{L}\right) \text{ where } b_nL = 2\int_0^L f(x)\sin\left(\frac{n\pi x}{L}\right)dx = 2\int_0^L F(x)\sin\left(\frac{n\pi x}{L}\right)dx$$

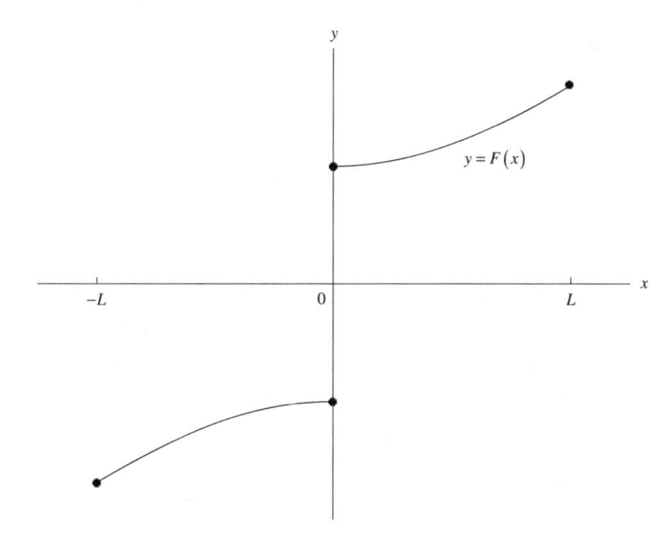

Fig. 5.5 A plot of the odd extension $y = f(x)$ of the function $y = F(x)$ defined for $x \in (0, L)$.

or, for the restricted interval $x \in (0, L)$, we obtain the half-range Fourier series

$$F(x) = \sum_{n=1}^{\infty} b_n \sin\left(\frac{n\pi x}{L}\right) \text{ where } b_n L = 2\int_0^L F(x)\sin\left(\frac{n\pi x}{L}\right) dx.$$

5.5 Fourier Series Application to Heat Conduction in the Finite Bar

Recall from (5.3.9) we have shown that

$$w_n(x,t) = c_n e^{-n^2\pi^2\kappa_0 t/L^2} \sin\left(\frac{n\pi x}{L}\right) \text{ for } n = 1, 2, 3, \ldots,$$

are each individually solutions of (5.3.6a) and (5.3.6b). We now wish to select $w(x,t)$ to solve the IC (5.3.6c)

$$w(x, 0) = F(x) \text{ for } 0 < x < L.$$

So far we have not specified the initial temperature distribution $F(x)$ precisely. Let us now examine our prospective solution for several possibilities all of which satisfy the compatibility conditions that $F(0) = F(L) = 0$. In what follows we shall implicitly assume that this function is measured in K temperature units.

(i) Suppose $F(x) = 3\sin(4\pi x/L)$. We then select

$$w(x,t) = w_4(x,t) \text{ with } c_4 = 3.$$

(ii) Suppose

$$F(x) = \sum_{n=1}^{N} b_n \sin\left(\frac{n\pi x}{L}\right) \text{ for } N \geq 2,$$

where the b_n are prescribed. Since our problem is linear a *finite* sum of solutions of (5.3.6a, 5.3.6b) is also a solution. Thus we then select

$$w(x,t) = \sum_{n=1}^{N} w_n(x,t) \text{ with } c_n = b_n \text{ for } n = 1,2,\ldots,N.$$

(iii) Finally suppose that $F(x)$ is any function such that F, F' are piecewise continuous for $x \in (0,L)$ and satisfy the compatibility conditions $F(0) = F(L) = 0$. Employing our half-range Fourier series result from the last section

$$F(x) = \sum_{n=1}^{\infty} b_n \sin\left(\frac{n\pi x}{L}\right) \text{ where } b_n L = 2\int_0^L F(x)\sin\left(\frac{n\pi x}{L}\right) dx.$$

We shall next assume that a *superposition principle* holds for our linear problem in which an *infinite* sum of solutions is also a solution. Hence we select

$$w(x,t) = \sum_{n=1}^{\infty} w_n(x,t) \text{ with } c_n = b_n.$$

Thus, from (5.3.3) and (5.3.5) we obtain the finite-bar temperature solution

$$T(x,t) = T_e + \sum_{n=1}^{\infty} e^{s_0 t} w_n(x,t) = T_e + \sum_{n=1}^{\infty} b_n e^{(s_0 - n^2\pi^2\kappa_0/L^2)t} \sin\left(\frac{n\pi x}{L}\right).$$

5.6 Long-Time Behavior of Solution

We now wish to examine the long-time behavior of the finite-bar temperature solution just deduced in the previous section. Such an examination is similar in philosophy to the linear stability analysis of solutions to the predator–prey systems of ordinary differential equations treated earlier but differs from that investigation in two respects: Namely it involves a partial differential equation boundary-initial value problem to which we have found an explicit solution and there are no nonlinear terms. These two occurrences are intimately related in that the method of separation of variables used to obtain this solution can only be employed for linear equations. Here, in that context, we are implicitly investigating the stability of the ambient temperature solution $T(x,t) \equiv T_e$ to our original problem by considering $T(x,t) = T_e + u(x,t)$ and then examining the long-time behavior of $u(x,t)$. Toward this end observe from our exact solution that the time dependence of $u(x,t)$ is of the form of modes proportional to

$$e^{\sigma_n t} \text{ where now } \sigma_n = s_0 - n^2\pi^2\kappa_0/L^2 \text{ for } n = 1,2,3,\ldots. \tag{5.6.1}$$

Note that the exponential nature of (5.6.1) is crucially dependent upon the linearity of the equation under investigation. Hence, as before, we can determine the requisite stability criterion by examining the sign of $\text{Re}(\sigma_n)$. Since, in this instance, $\text{Re}(\sigma_n) = \sigma_n$, we have stability provided

$$\sigma_n < 0 \text{ for } n = 1,2,3,\ldots. \tag{5.6.2}$$

Observe from (5.6.1) that

$$\sigma_n > \sigma_{n+1}. \tag{5.6.3}$$

Thus, if

$$\sigma_1 < 0 \Rightarrow 0 > \sigma_1 > \sigma_2 > \sigma_3 > \cdots > \sigma_n > \cdots. \tag{5.6.4}$$

Therefore, our stability criterion becomes

$$\sigma_1 = s_0 - \frac{\pi^2\kappa_0}{L^2} < 0 \text{ or } \frac{s_0}{\kappa_0} = \frac{r_0}{k_0} < \left(\frac{\pi}{L}\right)^2 \text{ or } \beta = \frac{s_0 L^2}{\kappa_0 \pi^2} < 1. \tag{5.6.5}$$

Physically this means that the bar can conduct heat away faster than it is being generated and the ambient temperature solution T_e is stable. In that event $u(x,t) \to 0$ as $t \to \infty$ and represents the transient part of the solution involving the initial temperature distribution $F(x)$ contained in the integrand of b_n. Should this criterion be violated, then

$$\beta \begin{Bmatrix} > \\ = \end{Bmatrix} 1 \Rightarrow u(x,t) \to \begin{cases} \infty, \\ b_1 \sin(\pi x/L) \end{cases} \text{as } t \to \infty \Rightarrow \begin{cases} \text{thermal runaway} \\ \text{marginal stability} \end{cases}.$$

Here $n = 1$ is called "the most dangerous mode" associated with the dominant eigenvalue possessing the largest real part for the ordering of (5.6.4) which always occurs for this sort of ordinary differential equation boundary value problem provided its coefficients are continuous functions (see [53]) and the thermal runaway instability situation corresponds to what occurs in a nuclear reactor when there is overheating causing core meltdown. Observe that for the standard heat conduction problem in a finite bar with $s_0 = 0$ and hence lacking a source term, $\sigma_n = -n^2\pi^2\kappa_0/L^2 < 0$ for $n = 1,2,3,\ldots$. Thus $T(x,t) \equiv T_e$, the ambient temperature solution, is always stable.

Problems

5.1. Define

$$e^w = \sum_{n=0}^{\infty} \frac{w^n}{n!}, \cos(z) = \sum_{k=0}^{\infty} \frac{(-1)^k z^{2k}}{(2k)!}, \sin(z) = \sum_{k=0}^{\infty} \frac{(-1)^k z^{2k+1}}{(2k+1)!}$$

for $w \in \mathbb{C}$ and $z \in \mathbb{R}$.

(a) Now taking $w = iz$ where $i^2 = -1$, show that

$$e^{iz} = \cos(z) + i\sin(z).$$

Hint: You may consider

$$\sum_{n=0}^{\infty} \frac{(iz)^n}{n!} = \sum_{k=0}^{\infty} \frac{i^{2k} z^{2k}}{(2k)!} + \sum_{k=0}^{\infty} \frac{i^{2k+1} z^{2k+1}}{(2k+1)!}.$$

(b) Using part (a) deduce that

$$\cos(z) = \frac{e^{iz} + e^{-iz}}{2} \text{ and } \sin(z) = \frac{e^{iz} - e^{-iz}}{2i}; e^{2\pi n i} = 1 \text{ and } e^{(2n+1)\pi i} = -1,$$

for any integer n.

5.2. Consider the heat conduction problem for $T = T(x,t)$, the temperature distribution in a laterally insulated bar with a linear source that is also insulated at one end:

$$\frac{\partial T}{\partial t} = \kappa_0 \frac{\partial^2 T}{\partial x^2} + s_0(T - T_e), 0 < x < L, t > 0;$$

$$\frac{\partial T(0,t)}{\partial x} = 0, T(L,t) = T_e, t > 0;$$

$$T(x,0) = T_e + F(x), 0 < x < L.$$

(a) Proceeding exactly as in this chapter show that

$$T(x,t) = T_e + \sum_{n=1}^{\infty} a_n e^{(s_0 - [n-1/2]^2 \pi^2 \kappa_0 / L^2)t} \cos\left(\left[n - \frac{1}{2}\right] \frac{\pi x}{L}\right)$$

where

$$a_n = \frac{2}{L} \int_0^L F(x) \cos\left(\left[n - \frac{1}{2}\right] \frac{\pi x}{L}\right) dx.$$

Hint: You may assume that $\cos([n-1/2]\pi x/L)$ behaves exactly as $\cos(n\pi x/L)$ did with respect to orthogonality properties, Fourier-type series, and even extensions [29].

(b) Show that $n = 1$ corresponds to the most dangerous mode for this problem, and hence, deduce that the stability criterion for its ambient temperature solution is given by

$$\frac{s_0}{\kappa_0} < \frac{\pi^2}{4L^2}.$$

(c) In addition to F, F' being piecewise continuous for $0 < x < L$ it must also satisfy this problem's compatibility conditions $F'(0) = F(L) = 0$. Find the a_n's if $F(x)$ is equal to (i) $5\cos(\pi x/[2L])$, (ii) $\cos^3(\pi x/[2L])$, and (iii) $x^2 - L^2$, respectively. Hint: Recall for

$$\text{(ii) } 4\cos^3(z) = 3\cos(z) + \cos(3z)$$

and

$$\text{(iii) } \int q(x)\cos(bx)\,dx = \frac{q(x)}{b}\sin(bx) + \frac{q'(x)}{b^2}\cos(bx) - \frac{q''(x)}{b^3}\sin(bx) + C$$

where $q(x)$ is a quadratic.

Chapter 6
Heat Conduction in a Semi-Infinite Bar

In the last chapter, heat conduction in a laterally insulated *finite* bar generated by a linear bulk source was considered. This chapter considers heat conduction in a similar laterally insulated *semi-infinite* bar driven by a heat reservoir in contact with its end. When this reservoir temperature is independent of time, that problem can be solved by a similarity solution method, this methodology being introduced in a pastoral interlude. That yields a temperature distribution involving the complementary error function, the asymptotic behavior of which is deduced by means of Watson's Lemma, introduced in pastoral interludes. Then this problem is solved by using an approximation solution method and the results compared with the exact similarity solution. Finally, a periodic boundary condition in time is imposed at its end and the separation of variables type solution to that problem is used to examine heat conduction underground in the Earth's crust from which it is deduced that, when it is summer on the surface, it is winter 4.44 meters underground. The problems consider heat conduction in a laterally insulated infinite bar and in an axially insulated planar layer with a one-time introduction of heat at the respective center lines of each configuration, both of which are solved by similarity solution methods; and a real solution method for semi-infinite bar heat conduction in contact with a periodic bath, which is solved by a complex solution method in this chapter.

6.1 Governing Equation and Boundary and Initial Conditions for Impulsive Heat Conduction

We wish to examine heat conduction in a laterally insulted semi-infinite bar driven by a heat reservoir in contact with its $x = 0$ end in the absence of any bulk sources. To determine how diffusive effects develop with time and space, we first consider so-called impulsive heat conduction in such a laterally insulated semi-infinite bar, originally at some ambient temperature T_e, which, from time $t \geq 0$, is placed in contact with a heat reservoir of elevated temperature $T_e + T_1$ as depicted in Fig. 6.1. Heat conduction for this scenario is then governed by the following boundary-initial value

© Springer International Publishing AG, part of Springer Nature 2017

D. J. Wollkind and B. J. Dichone, *Comprehensive Applied Mathematical Modeling in the Natural and Engineering Sciences*,
https://doi.org/10.1007/978-3-319-73518-4_6

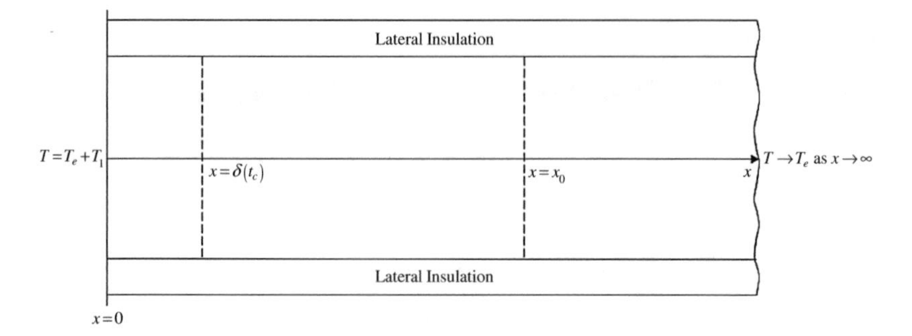

Fig. 6.1 Schematic diagram of impulsive heat conduction in a laterally insulated semi-infinite bar.

partial differential diffusion equation problem for $T = T(x,t;T_e,T_1,\kappa_0) \equiv$ temperature distribution in the bar:

$$\frac{\partial T}{\partial t} = \kappa_0 \frac{\partial^2 T}{\partial x^2}, \; 0 < x < \infty; \; t > 0; \tag{6.1.1a}$$

$$T(0,t;T_e,T_1,\kappa_0) = T_e + T_1, \tag{6.1.1b}$$

$$T(x,t;T_e,T_1,\kappa_0) \to T_e \text{ as } x \to \infty \text{ for } t > 0; \tag{6.1.1c}$$

$$T(x,0;T_e,T_1,\kappa_0) = T_e \text{ for } 0 < x < \infty. \tag{6.1.1d}$$

The exact and approximate solutions for this problem to be developed in this chapter are based upon the methodologies of Segel [114] and Noble [86], respectively. Since the similarity solution method to be employed for determining the exact solution requires a nondimensional dependent variable, we now introduce $u = u(x,t;\kappa_0)$ defined by

$$T(x,t;T_e,T_1,\kappa_0) = T_e + T_1 u(x,t;\kappa_0) \tag{6.1.2}$$

which transforms (6.1.1) into

$$\frac{\partial u}{\partial t} = \kappa_0 \frac{\partial^2 u}{\partial x^2}, \; 0 < x < \infty, \; t > 0; \tag{6.1.3a}$$

$$u(0,t;\kappa_0) = 1, \tag{6.1.3b}$$

$$u(x,t;\kappa_0) \to 0 \text{ as } x \to \infty \text{ for } t > 0; \tag{6.1.3c}$$

$$u(x,0;\kappa_0) = 0 \text{ for } 0 < x < \infty. \tag{6.1.3d}$$

As mentioned above we shall solve this problem exactly by means of the similarity solution method to be introduced in the next section. That method is crucially dependent upon the BC's of (6.1.3b) and (6.1.3c) being independent of t. If they are not, then a problem of this sort must be solved by the Laplace transform method. We shall resolve it exactly, for the sake of completeness, by that more generally applicable method in a section of Chapter 20 devoted to Laplace transform methods of solution. Further, later in this chapter, we offer an approximate solution method as well since, in our opinion, the process required for that solution provides more physical insight into the problem than the exact solution methodology.

6.2 Pastoral Interlude: The Buckingham Pi Theorem and Similarity Solutions of PDE's

Note, as indicated above, that u as defined by (6.1.2), is nondimensional hence its dimension is denoted by $[u] = 1$. Since it is a nondimensional dependent variable and the BC's of (6.1.3b) and (6.1.3c) are independent of time, we can employ a similarity solution method which allows us to assume that this dependent variable is a function of a dimensionless combination of independent variables and parameter(s). In order to determine that quantity we shall use the so-called Buckingham Pi Theorem which is standard operating procedure for deducing all possible such dimensionless combinations of a particular problem. We have deferred until now a determination of the dimension of our parameter κ_0, the thermal diffusivity, which is required for the application of that theorem to this problem. To deduce its dimensionality, we consider the PDE (6.1.3a). Then assigning our independent variables the dimensions

$$[x] = L \text{ and } [t] = \tau, \tag{6.2.1a}$$

where L and τ are length and time scales, respectively, we can conclude from (6.1.3a) that

$$\left[\frac{\partial u}{\partial t}\right] = \frac{[u]}{[t]} = \frac{1}{\tau} = \left[\kappa_0 \frac{\partial^2 u}{\partial x^2}\right] = [\kappa_0]\left[\frac{\partial^2 u}{\partial x^2}\right] = [\kappa_0]\frac{[u]}{[x]^2} = \frac{[\kappa_0]}{L^2} \text{ or } [\kappa_0] = \frac{L^2}{\tau}. \tag{6.2.1b}$$

The Buckingham Pi Theorem entails defining a quantity

$$\Pi = x^{\alpha_1} t^{\alpha_2} \kappa_0^{\alpha_3} \tag{6.2.2a}$$

consisting of the product of the independent variables and parameter of the problem each raised to a power. Then we wish to select these powers so that Π is a dimensionless such combination. Thus, since

$$[\Pi] = [x^{\alpha_1} t^{\alpha_2} \kappa_0^{\alpha_3}] = [x^{\alpha_1}][t^{\alpha_2}][\kappa_0^{\alpha_3}] = [x]^{\alpha_1}[t]^{\alpha_2}[\kappa_0]^{\alpha_3} = L^{\alpha_1}\tau^{\alpha_2}\left(\frac{L^2}{\tau}\right)^{\alpha_3}$$
$$= L^{\alpha_1+2\alpha_3}\tau^{\alpha_2-\alpha_3},$$

$$[\Pi] = 1 \Rightarrow \alpha_1 + 2\alpha_3 = \alpha_2 - \alpha_3 = 0 \text{ or } \alpha_2 = \alpha_3 = -\frac{\alpha_1}{2}. \tag{6.2.2b}$$

Then, (6.2.2a) and (6.2.2b) yield

$$\Pi_{\alpha_1} = \left(\frac{x}{\sqrt{\kappa_0 t}}\right)^{\alpha_1} \tag{6.2.2c}$$

and, taking $\alpha_1 = 1$ with no loss of generality, we obtain

$$\Pi_1 = \frac{x}{\sqrt{\kappa_0 t}}, \tag{6.2.2d}$$

as our nondimensional combination. Finally, in order to simplify the calculations of the next section, Segel [114] selected

$$\eta = \frac{x}{2\sqrt{\kappa_0 t}} \tag{6.2.3a}$$

and let

$$u(x,t;\kappa_0) = F(\eta), \tag{6.2.3b}$$

where F is some twice differentiable function.

6.3 Similarity Solution Method

We first rewrite (6.1.3a) and (6.2.3) in the forms

$$\mathcal{L}[u] = 0 \text{ where } \mathcal{L} \equiv \kappa_0 \frac{\partial^2}{\partial x^2} - \frac{\partial}{\partial t}, \; 0 < x < \infty, \; t > 0 \tag{6.3.1a}$$

and

$$u = F(\eta) \text{ where } \eta = \frac{xt^{-1/2}}{2\sqrt{\kappa_0}}. \tag{6.3.1b}$$

Then computing

$$\frac{\partial \eta}{\partial t} = -\frac{x}{4t\sqrt{\kappa_0 t}}, \; \frac{\partial \eta}{\partial x} = \frac{1}{2\sqrt{\kappa_0 t}}; \tag{6.3.2}$$

and calculating

$$\frac{\partial u}{\partial t} = F'(\eta)\frac{\partial \eta}{\partial t} = -\frac{xF'(\eta)}{4t\sqrt{\kappa_0 t}}, \tag{6.3.3a}$$

$$\frac{\partial^2 u}{\partial x^2} = F''(\eta)\left(\frac{\partial \eta}{\partial x}\right)^2 = \frac{F''(\eta)}{4\kappa_0 t}; \tag{6.3.3b}$$

we find that

$$\mathcal{L}[u] = \frac{F''(\eta) + \frac{x}{\sqrt{\kappa_0 t}}F'(\eta)}{4t} = 0, \; 0 < x < \infty, \; t > 0,$$

which implies

$$F''(\eta) + 2\eta F'(\eta) = 0, \; 0 < \eta < \infty. \tag{6.3.4a}$$

Further, noting that $x = 0$ and $x \to \infty$ for $t > 0$ correspond to $\eta = 0$ and $\eta \to \infty$, respectively, BC's (6.1.3b) and (6.1.3c) imply

$$F(0) = 1, \tag{6.3.4b}$$

$$F(\eta) \to 0 \text{ as } \eta \to \infty. \tag{6.3.4c}$$

Finally, rewriting IC (6.1.3d) in its equivalent limiting form

$$u(x,t;\kappa_0) \to 0 \text{ as } t \to 0^+ \text{ for } 0 < x < \infty, \tag{6.3.5}$$

we see that (6.3.5) is satisfied identically by (6.3.4c).

Since (6.3.4a) is a linear first-order ordinary differential equation in F', it can be solved for

$$F'(\eta) = ce^{-\eta^2}. \tag{6.3.6}$$

Rewriting (6.3.6) in terms of a dummy variable z and integrating it from 0 to η, yields

$$\int_0^{\eta} F'(z)\,dz = F(\eta) - F(0) = F(\eta) - 1 = c\int_0^{\eta} e^{-z^2}\,dz$$

or

$$F(\eta) = 1 + c\int_0^{\eta} e^{-z^2}\,dz.$$

Now, making use of the fact, demonstrated below, that

$$\int_0^{\infty} e^{-z^2}\,dz = \frac{\sqrt{\pi}}{2}$$

and imposing BC (6.3.4c), we finally obtain

$$F(\eta) = 1 - \frac{2}{\sqrt{\pi}}\int_0^{\eta} e^{-z^2}\,dz = 1 - \text{erf}(\eta) \equiv \text{erf}_c(\eta) = \frac{2}{\sqrt{\pi}}\int_{\eta}^{\infty} e^{-z^2}\,dz,$$

where both $\text{erf}(\eta) \equiv$ error function and $\text{erf}_c(\eta) \equiv$ complementary error function are represented by the appropriate areas under the curve $y = 2e^{-z^2}/\sqrt{\pi}$ in Fig. 6.2. Note that this error function which arises naturally in modeling a variety of diffusion phenomena differs from the one formerly employed in conjunction with the normal distribution of probability theory

$$\text{Erf}(\eta) \equiv \frac{1}{\sqrt{2\pi}}\int_{-\infty}^{\eta} e^{-z^2/2}\,dz = \frac{1}{2}\left[1 + \text{erf}\left(\frac{\eta}{\sqrt{2}}\right)\right].$$

Thus, from (6.3.2) and (6.3.3), we can conclude that the solution of our original problem (6.3.1) is given by

$$T(x,t;T_e,T_1,\kappa_0) = T_e + T_1\text{erf}_c(\eta) \text{ where } \eta = \frac{x}{2\sqrt{\kappa_0 t}}. \tag{6.3.7}$$

It only remains for us to demonstrate conclusively that the total area under the curve in Fig. 6.2 has been normalized to 1. We do so by showing that

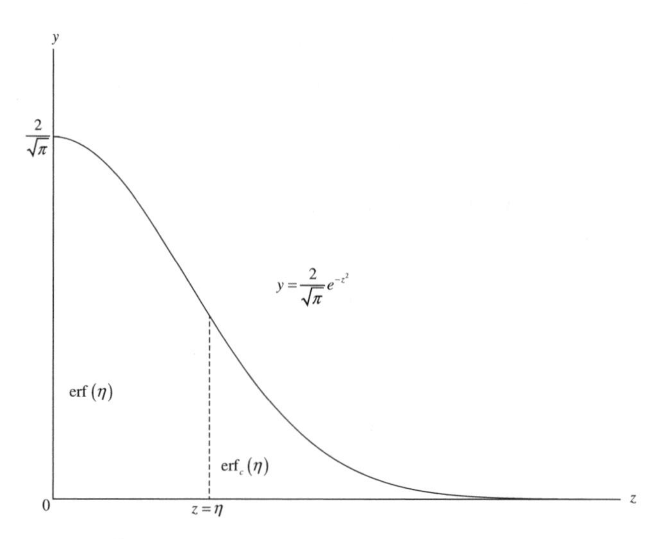

Fig. 6.2 Plot of $y = 2e^{-z^2}/\sqrt{\pi}$ with the erf(η) and the erf$_c$(η) represented by the indicated areas.

$$I = \lim_{R_\infty} I_R = \lim_{R\to\infty} \int_0^R e^{-x^2}\,dx = \int_0^\infty e^{-x^2}\,dx = \frac{\sqrt{\pi}}{2}$$

as follows:

Consider

$$I_R^2 = \left[\int_0^R e^{-x^2}\,dx\right]\left[\int_0^R e^{-y^2}\,dy\right] = \int_0^R \left[\int_0^R e^{-y^2}\,dy\right] e^{-x^2}\,dx$$
$$= \int_0^R \int_0^R e^{-x^2+y^2}\,dx\,dy = \iint_{S_R} e^{-(x^2+y^2)}\,dx\,dy$$

as depicted in Fig. 6.3 where the equivalence of the iterated and double integrals appearing above depends only on the continuity of the integrand. Then from Fig. 6.3 we can observe that

$$\iint_{C_R} e^{-(x^2+y^2)}\,dx\,dy < I_R^2 < \iint_{C_{R\sqrt{2}}} e^{-(x^2+y^2)}\,dx\,dy.$$

Now introducing the change of variables

$$x = r\cos(\theta),\ y = r\sin(\theta),\ dx\,dy = J(r,\theta)\,dr\,d\theta$$

where

$$J(r,\theta) = \begin{vmatrix} x_r & x_\theta \\ y_r & y_\theta \end{vmatrix} = \begin{vmatrix} \cos(\theta) & -r\sin(\theta) \\ \sin(\theta) & r\cos(\theta) \end{vmatrix} = r,$$

$$\iint\limits_{C_R} e^{-(x^2+y^2)}\,dx\,dy = \int_0^R \int_0^{\pi/2} e^{-r^2}\,r\,dr\,d\theta = \frac{\pi}{2}\int_0^R e^{-r^2}\,r\,dr = \frac{\pi}{4}[e^{-r^2}]_R^0 = \frac{\pi}{4}(1-e^{-R^2}).$$

Thus

$$\frac{\pi}{4}(1-e^{-R^2}) < I_R^2 < \frac{\pi}{4}(1-e^{-2R^2}).$$

Noting that

$$\lim_{R\to\infty}\frac{\pi}{4}(1-e^{-R^2}) = \lim_{R\to\infty}\frac{\pi}{4}(1-e^{-2R^2}) = \frac{\pi}{4},$$

and applying the Pinching Theorem yields the desired result

$$\lim_{R\to\infty} I_R^2 = \left[\lim_{R\to\infty} I_R\right]^2 = I^2 = \frac{\pi}{4} \Rightarrow I = \frac{\sqrt{\pi}}{2} > 0.$$

6.4 Pastoral Interlude: Asymptotic Series Revisited

We now consider in more depth a topic introduced in Section 2.4: Namely, asymptotic series. One says, in general, that

$$\sum_{n=1}^{N} a_n\varphi_n(x)$$

is an N-term asymptotic series for $f(x)$ as $x \longrightarrow x_0$ iff

(i) $\{\varphi_n(x)\}_{n=1}^N$ is an asymptotic sequence as $x \longrightarrow x_0$.
(ii) $f(x) = \sum_{n=1}^N a_n\varphi_n(x) + o(\varphi_N(x))$ as $x \longrightarrow x_0$ or $\lim_{x\to x_0}\left\{\left[f(x) - \sum_{n=1}^N a_n\varphi_n(x)\right]/\varphi_N(x)\right\} = 0$.

One then writes

$$f(x) \sim \sum_{n=1}^{N} a_n\varphi_n(x) \text{ as } x \longrightarrow x_0.$$

If $N = 1$ or $f(x) \sim a_1\varphi_1(x)$ as $x \longrightarrow x_0$, we say that f has a *one-term asymptotic representation* while should

$$f(x) \sim \sum_{n=1}^{\infty} a_n\varphi_n(x) \text{ as } x \longrightarrow x_0,$$

we say that f has a *complete asymptotic series*. A function $f(x)$ does not necessarily have a unique asymptotic series however given a specific asymptotic sequence

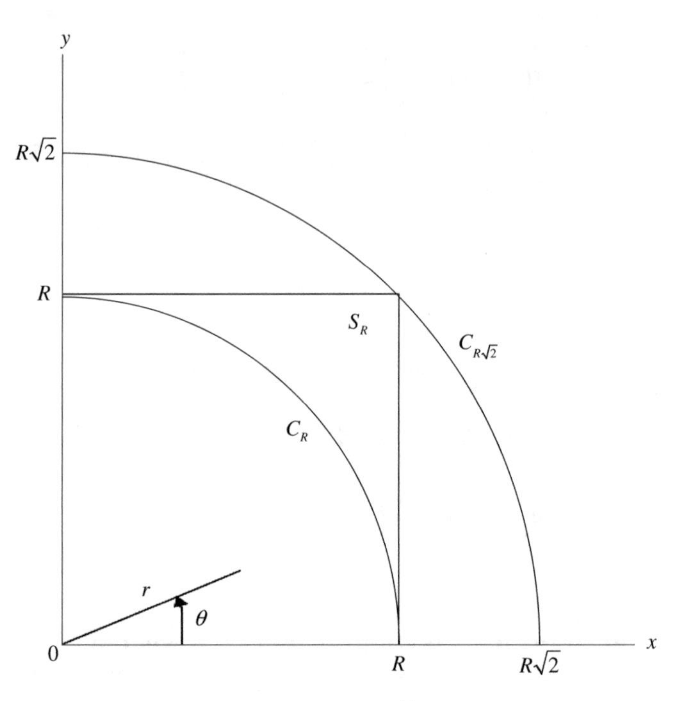

Fig. 6.3 Plots of the quarter circles with radii R and $R\sqrt{2}$ as well as the square with side-length R in the first-quadrant of the x–y plane.

$\{\varphi_n(x)\}_{n=1}^N$, the a_n's are unique and can be determined by

$$a_1 = \lim_{x\to x_0} \frac{f(x)}{\varphi_1(x)}$$

while

$$a_m = \lim_{x\to x_0} \left\{ \left[f(x) - \sum_{n=1}^{m-1} a_n\varphi_n(x) \right] / \varphi_m(x) \right\} \text{ for } m = 2, \ldots, N.$$

We demonstrate this result by the so-called procedure of Transfinite Mathematical Induction (see Section 12.6 on Leibniz's Rule of Differentiation for an example of an application of the regular procedure of Mathematical Induction):

(i) $N = 1$:
Then $f(x) = a_1\varphi_1(x) + o(\varphi_1(x))$ as $x \to x_0$. Therefore

$$\lim_{x\to x_0} \frac{f(x) - a_1\varphi_1(x)}{\varphi_1(x)} = 0 \Rightarrow a_1 = \lim_{x\to x_0} \frac{f(x)}{\varphi(x)}.$$

(ii) $m = N$:
Then

$$\lim_{x\to x_0} \frac{f(x) - \sum_{n=1}^{N} a_n\varphi_n(x)}{\varphi_N(x)} = 0 \text{ and } \sum_{n=1}^{N} a_n\varphi_n(x) = \sum_{n=1}^{N-1} a_n\varphi_n(x) + a_N\varphi_N(x).$$

Hence it follow directly that

$$a_N = \lim_{x\to x_0} \frac{f(x) - \sum_{n=1}^{N-1} a_n\varphi_n(x)}{\varphi_N(x)}.$$

(iii) $m \leq N-1$:
Examine

$$\lim_{x\to x_0} \frac{f(x) - \sum_{n=1}^{N-1} a_n\varphi_n(x)}{\varphi_m(x)} = \lim_{x\to x_0} \left[\left(\frac{f(x) - \sum_{n=1}^{N-1} a_n\varphi_n(x)}{\varphi_N(x)}\right)\left(\frac{\varphi_N(x)}{\varphi_m(x)}\right)\right].$$

Note:

$$\frac{\varphi_N(x)}{\varphi_m(x)} = \prod_{n=m}^{N-1} \frac{\varphi_{n+1}(x)}{\varphi_n(x)} \longrightarrow 0 \text{ as } x \longrightarrow x_0 \text{ given that } \lim_{x\to x_0} \frac{\varphi_{n+1}(x)}{\varphi_n(x)} = 0.$$

Thus

$$\lim_{x\to x_0} \frac{f(x) - \sum_{n=1}^{N-1} a_n\varphi_n(x)}{\varphi_m(x)} = a_N \cdot 0 = 0.$$

Since

$$\sum_{n=1}^{N-1} a_n\varphi_n(x) = \sum_{n=1}^{m-1} a_n\varphi_n(x) + a_m\varphi_m(x) + \sum_{n=m+1}^{N-1} a_n\varphi_n(x)$$

and again

$$\frac{\varphi_n(x)}{\varphi_m(x)} = \prod_{k=m}^{n-1} \frac{\varphi_{k+1}(x)}{\varphi_k(x)} \longrightarrow 0 \text{ as } x \longrightarrow x_0 \text{ for } n \geq m+1,$$

we obtain

$$a_m = \lim_{x\to x_0} \frac{f(x) - \sum_{n=1}^{m-1} a_n\varphi_n(x)}{\phi_m(x)} \text{ for } m = 2, \ldots, N-1,$$

which completes our demonstration.

Finally observe that

$$f(x) = \sum_{n=1}^{m} a_n\varphi_n(x) + O(\varphi_{m+1}(x)) \text{ as } x \longrightarrow x_0 \text{ for } m = 1, 2, \ldots, N-1$$

since

$$a_{m+1} = \lim_{x \to x_0} \frac{f(x) - \sum_{n=1}^{m} a_n \varphi_n(x)}{\varphi_{m+1}(x)} \text{ for } m = 1, 2, \ldots, N-1.$$

Note that in Section 2.4 we considered asymptotic series valid in the limit as $x \longrightarrow 0$. Here we have generalized this limit to $x \longrightarrow x_0$, where $x_0 \in \mathbb{R}$, and that also includes the possibility that $x \longrightarrow \infty$, which is the relevant limit for the asymptotic behavior of the complementary error function to be examined in the next section.

6.5 Pastoral Interlude: The Complementary Error Function and Watson's Lemma

Consider

$$\text{erf}_c(x) = \frac{2}{\sqrt{\pi}} \int_x^{\infty} e^{-z^2} \, dz.$$

In order to examine its asymptotic behavior as $x \longrightarrow \infty$ we wish to employ Watson's Lemma which states that given an

$$I(k) = \int_0^{\infty} f(v) e^{-kv} \, dv$$

where $f(v)$ has the asymptotic series

$$f(v) \sim \sum_{n=0}^{N} a_n v^{\alpha_n} \text{ as } v \longrightarrow 0 \text{ and } -1 < \alpha_0 < \alpha_1 < \cdots < \alpha_N,$$

then

$$I(k) \sim \sum_{n=0}^{N} a_n \int_0^{\infty} v^{\alpha_n} e^{-kv} \, dv \text{ as } k \longrightarrow \infty,$$

or since, on introducing the change of variables $v = t/k$,

$$\int_0^{\infty} v^{\alpha_n} e^{-kv} \, dv = \frac{1}{k^{\alpha_n+1}} \int_0^{\infty} t^{\alpha_n} e^{-t} \, dt = \frac{\Gamma(\alpha_n + 1)}{k^{\alpha_n+1}},$$

$$I(k) \sim \sum_{n=0}^{N} \frac{a_n \Gamma(\alpha_n + 1)}{k^{\alpha_n+1}} \text{ as } k \longrightarrow \infty.$$

Here the gamma function is defined by

$$\Gamma(z) = \int_0^{\infty} t^{z-1} e^{-t} \, dt \text{ for } \text{Re}(z) > 0.$$

Note that $\Gamma(n+1) = n!$ for n a nonnegative integer (see Chapter 19). In order to apply Watson's Lemma to $\text{erf}_c(x)$, we must sequentially make the change of variables

$z = xu^{1/2}$ and $u = v + 1$ in its integral which transforms that function into

$$\text{erf}_c(x) = \frac{x}{\sqrt{\pi}} \int_1^\infty \frac{e^{-x^2 u}}{u^{1/2}}\, du = \frac{xe^{-x^2}}{\sqrt{\pi}} I(x^2) \text{ where } f(v) = (1+v)^{-1/2}.$$

To apply Watson's Lemma to $I(x^2)$ we must find the asymptotic series for its $f(v)$. Since functions of the form $(1+v)^\alpha$ have so-called binomial series which converge for $|v| < 1$ and such convergent series are also asymptotic, we need only employ that series representation with $\alpha = -1/2$. For the sake of completeness and because it is difficult to show the remainder term of the resulting Maclaurin polynomial goes to zero for this function, we shall deduce that representation by the following indirect method. Toward this end we consider the formal series

$$F(v) = \sum_{n=0}^\infty \frac{f^{(n)}(0)}{n!} = 1 + \sum_{n=1}^\infty \frac{\alpha(\alpha-1)\cdots(\alpha-n+1)}{n!} v^n = 1 + \sum_{n=1}^\infty u_n(v).$$

Since

$$\left| \frac{u_{n+1}(v)}{u_n(v)} \right| = \frac{|\alpha - n|}{n+1} |v| \to |v| \text{ as } n \to \infty,$$

this power series converges for $|v| < 1$ by the ratio test. Given that a power series can be differentiated term by term in its interval of convergence

$$\begin{aligned} F'(v) &= \sum_{n=1}^\infty \frac{\alpha(\alpha-1)\cdots(\alpha-n+1)}{n!} n v^{n-1} = \alpha + \sum_{n=2}^\infty \frac{\alpha(\alpha-1)\cdots(\alpha-n+1)}{(n-1)!} v^{n-1} \\ &= \alpha + \sum_{j=1}^\infty \frac{\alpha(\alpha-1)\cdots(\alpha-j+1)(\alpha-j)}{j!} v^j = \alpha + \sum_{n=1}^\infty \frac{\alpha(\alpha-1)\cdots(\alpha-n+1)(\alpha-n)}{n!} v^n. \end{aligned}$$

Thus

$$vF'(v) = \sum_{n=1}^\infty \frac{\alpha(\alpha-1)\cdots(\alpha-n+1)}{n!} n v^n$$

and

$$\begin{aligned} (1+v)F'(v) &= \alpha + \sum_{n=1}^\infty \frac{\alpha(\alpha-1)\cdots(\alpha-n+1)(\alpha-n+n)}{n!} v^n \\ &= \alpha + \sum_{n=1}^\infty \frac{\alpha(\alpha-1)\cdots(\alpha-n+1)}{n!} \alpha v^n = \alpha \left[1 + \sum_{n=1}^\infty \frac{\alpha(\alpha-1)\cdots(\alpha-n+1)}{n!} v^n \right] \\ &= \alpha F(v) \end{aligned}$$

for $|v| < 1$ while $F(0) = 1$. Solving this first-order ordinary differential equation we find that $F(v) = (1+v)^\alpha$ which implies that

$$(1+v)^{\alpha} = 1 + \sum_{n=1}^{\infty} \frac{\alpha(\alpha-1)\cdots(\alpha-n+1)}{n!} v^n \text{ for } |v| < 1.$$

Finally when $\alpha \leq -1$ this series converges for $|v| < 1$; when $-1 < \alpha < 0$, for $-1 < v \leq 1$; and when $\alpha > 0$ and a non-integer, for $|v| \leq 1$ [1].

Now setting $\alpha = -1/2$ we find that

$$\begin{aligned}(1+v)^{-1/2} &= 1 + \sum_{n=1}^{\infty} \frac{(-1/2)(-3/2)\cdots([-2n+1]/2)}{n!} v^n \\ &= 1 + \sum_{n=1}^{\infty} \frac{(-1)^n 1\cdot 3\cdot 3\cdot \cdots \cdot (2n-1)}{2^n n!} v^n = \sum_{n=0}^{\infty} \frac{(-1)^n (2n)!}{(2^n n!)^2} v^n \text{ for } -1 < v \leq 1,\end{aligned}$$

where in the above use has been made of the fact that $1\cdot 3\cdot \cdots \cdot (2n-1) = (2n)!/(2^n n!)$. Thus

$$a_n = \frac{(-1)^n (2n)!}{(2^n n!)^2} \text{ and } \alpha_n = n \Rightarrow \Gamma(\alpha_n + 1) = \Gamma(n+1) = n! \text{ for } n = 0, 1, \ldots.$$

Hence

$$I(x^2) \sim \sum_{n=0}^{\infty} \frac{a_n \Gamma(\alpha_n+1)}{x^{2\alpha_n+2}} = \sum_{n=0}^{\infty} \frac{(-1)^n (2n)! n!}{(2^n n!)^2} \frac{1}{x^{2n+2}} = \sum_{n=0}^{\infty} \frac{(-1)^n (2n)!}{2^{2n} n!} \frac{1}{x^{2n+2}} \text{ as } x \to \infty$$

and

$$\operatorname{erf}_c(x) = \frac{xe^{-x^2}}{\sqrt{\pi}} I(x^2) \sim \frac{e^{-x^2}}{\sqrt{\pi}x} \sum_{n=0}^{\infty} \frac{(-1)^n (2n)!}{n!(2x)^{2n}} = \frac{e^{-x^2}}{\sqrt{\pi}x} \sum_{n=0}^{\infty} w_n(x) \text{ as } x \to \infty.$$

Let us examine

$$\left| \frac{w_{n+1}(x)}{w_n(x)} \right| = \frac{n+1/2}{x^2} \to \begin{cases} 0, & \text{as } x \to \infty \text{ for fixed } n \\ \infty, & \text{as } n \to \infty \text{ for fixed } x \end{cases}.$$

This is a complete asymptotic series as $x \to \infty$ but it diverges for a fixed value of x. The natural question then is how many terms does one take to get the best approximation say for $x = 2$?

Since the terms in the asymptotic series for the complementary error function alternate in sign, we have averaged the pairs appearing in Table 6.1. From this table we can conclude that the best approximation occurs by averaging the values from the 4 and 5 term pair which amazingly yields the actual value of 0.004678 [1] to six significant figures.

Table 6.1 Asymptotic behavior for $\text{erf}_c(x)$ at $x = 2$ where the actual value is 0.004678.

Number of terms $(N+1)$	Asymptotic representation for $\text{erf}_c(2)$	Average
1	0.005167	0.005167
$\begin{cases}2\\3\end{cases}$	$\begin{cases}0.004521\\0.004763\end{cases}$	0.004642
$\begin{cases}4\\5\end{cases}$	$\begin{cases}0.004612\\0.004744\end{cases}$	$0.004678 = \text{erf}_c(2)$
$\begin{cases}6\\7\end{cases}$	$\begin{cases}0.004595\\0.004800\end{cases}$	0.004698

6.6 Spatial and Temporal Behavior of the Exact Solution

Having determined a complete asymptotic series for the complementary error function valid in the limit of large values of its argument, we return to our impulsive heat conduction problem in a laterally insulated semi-infinite bar. We begin this investigation by examining the qualitative and quantitative behavior of $F(\eta) = \text{erf}_c(\eta) = \frac{2}{\sqrt{\pi}}\int_\eta^\infty e^{-z^2}\,dz > 0$. We plot this function versus η in Fig. 6.4. Observe from the BC's (6.3.4) satisfied by $F(\eta)$ that $F(0) = 1$ and $F(\eta) \to 0$ as $\eta \to \infty$ while since $F'(\eta) = 4\eta e^{-\eta^2}/\sqrt{\pi} < 0$ and $F''(\eta) = 4\eta e^{-\eta^2}/\sqrt{\pi} > 0$ for $\eta > 0$, it is positive, decreasing, and concave-upward. We now summarize its quantitative behavior [1] in Table 6.2 that includes for comparison purposes the corresponding values of the function $G(\eta)$, which is also plotted in Fig. 6.4 and defined by

$$G(\eta) = \begin{cases}(1-\eta/\sqrt{3})^2 & \text{for } 0 \le \eta \le \sqrt{3}\\ 0 & \text{for } \eta > \sqrt{3}\end{cases}, \tag{6.6.1}$$

relevant to the approximate solution to be developed in the next section.

We are now ready to determine how diffusive effects develop in both (i) time and (ii) space for our semi-infinite bar (see Fig. 6.1):

(i) Let $x \equiv x_0 > 0$ represent a cross-section of that bar and define a reference time $t = t_0$ by

$$\eta_0 = 0.02 = \frac{x_0}{2\sqrt{\kappa_0 t_0}} \text{ or } t_0 = x_0^2 \times \frac{10^4}{16\kappa_0}.$$

Consider $t > t_0$. Then

$$\eta(x_0, t) = \frac{x_0}{2\sqrt{\kappa_0 t}} < \frac{x_0}{2\sqrt{\kappa_0 t_0}} = \eta_0 = 0.02.$$

Since $F'(\eta) > 0$,

$$F[\eta(x_0, t)] > F(\eta_0) = F(0.02) = 0.977.$$

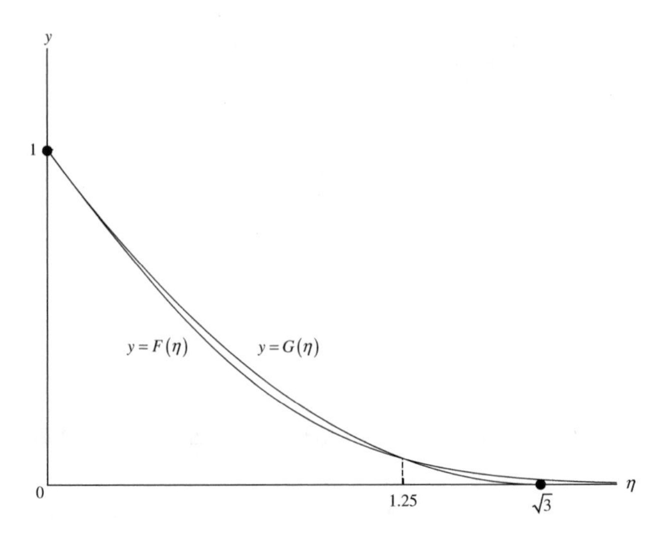

Fig. 6.4 Plots of $F(\eta) = \mathrm{erf}_c(\eta)$ and $G(\eta)$ of (6.6.1) versus $\eta \geq 0$.

Table 6.2 The values of $F(\eta)$ and $G(\eta)$ to three significant figures for representative values of η. Note that the value of $F(2)$ is consistent with the six significant figures one introduced in the previous section while its values at $\eta = 0$ and as $\eta \to \infty$ are consistent with the BC's.

η	$F(\eta)$	$G(\eta)$
0	1.000	1.000
0.02	0.977	0.977
0.20	0.777	0.782
0.50	0.480	0.506
1.00	0.157	0.179
1.25	0.077	0.077
$\sqrt{2}$	0.046	0.034
$\sqrt{3}$	0.014	0.000
2.00	0.005	0.000
∞	0.000	0.000

Hence, suppressing the parameter designations appearing in (6.3.7) for ease of exposition,

$$T(x_0, t) > T_e + T_1 F(0.02) = T_e + 0.977 T_1.$$

Indeed, if $t >> t_0$, then $T(x_0, t) \cong T_e + T_1$. Thus, for a fixed position in the bar, if one waits long enough, the temperature at that position will virtually be $T_e + T_1$, the bath temperature of the reservoir, and that waiting time, as measured by t_0, is directly proportional to x_0^2 and inversely proportional to κ_0. In other words, the greater κ_0 is, the shorter the waiting time will be.

(ii) Let $t \equiv t_c > 0$ represent a critical time period and define a critical length x_c by

$$\eta_c = 2.00 = \frac{x_c}{2\sqrt{\kappa_0 t_c}} \text{ or } x_c = 4\sqrt{\kappa_0 t_c}.$$

Consider $x > x_c$. Then

$$\eta(x, t_c) = \frac{x}{2\sqrt{\kappa_0 t_c}} > \frac{x_c}{2\sqrt{\kappa_0 t_c}} = \eta_c = 2.00.$$

Again, since $F'(\eta) > 0$,

$$F[\eta(x, t_c)] < F(\eta) = F(2.00) = 0.005.$$

Hence,

$$T(x, t_c) < T_e + T_1 F(2.00) = T_e + 0.005 T_1.$$

Indeed, if $x >> x_c$, then $T(x, t_c) \cong T_e$. Thus, for a fixed time period, if one travels out far enough along the bar, the temperature at that time will virtually be T_e, the ambient temperature, and that boundary-layer type distance, as measured by x_c, is proportional to $\sqrt{\kappa_0 t_c}$, the diffusion length. In other words, the greater κ_0 is, the longer the diffusion length will be. This behavior with thermometric conductivity as well as that in (i) is characteristic of all diffusion phenomena with their diffusivity whether it be heat, momentum, population density, or chemical concentration.

6.7 Approximate Solution Method

We begin this section by plotting the exact solution $T(x, t)$ of (6.3.7) versus x for $t = t_1$, t_2, and t_3 where $0 < t_1 < t_2 < t_3$ in Fig. 6.5. Although all of these curves have a horizontal asymptote at $T = T_e$, we indicate in that figure a schematic value of x denoted by $\delta(t)$ for $t = t_1$, t_2, and t_3 where the temperature is equal to a fixed threshold just slightly above the ambient. Motivated by the parabolic appearance of these curves, Noble [86] sought an approximate solution to this problem of the form

$$T(x, t) = T_e + T_1 \begin{cases} [1 - x/\delta(t)]^2, & 0 \le x \le \delta(t) \\ 0, & x > \delta(t) \end{cases},$$

where $\delta(t) \equiv$ thermal boundary-layer thickness such that $\delta(0) = 0$. We now consider the region R of constant cross-sectional area $\mathcal{A}_0$ depicted in Fig. 6.6 and let the heat in that region be defined by

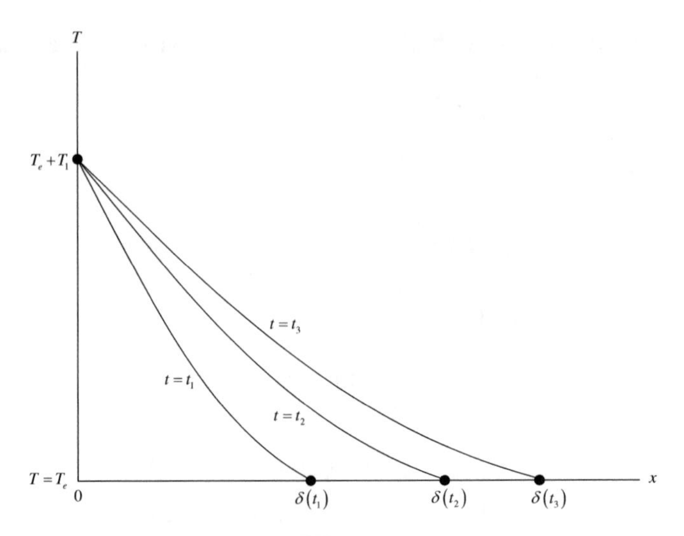

Fig. 6.5 Plots of $T(x, t) = T_e + T_1 \text{erf}_c[x/(2\sqrt{\kappa_0 t})]$ versus x for $t = t_1 < t_2 < t_3$. Here $\delta(t)$ represents $T[\delta(t), t] = T_e + 0.014T_1$.

$$Q(t) = \iiint\limits_R \rho_0 c_0 T(x,t)\, dx\, dy\, dz = Q_e + \iiint\limits_R \rho_0 c_0 [T(x,t) - T_e]\, dx\, dy\, dz$$

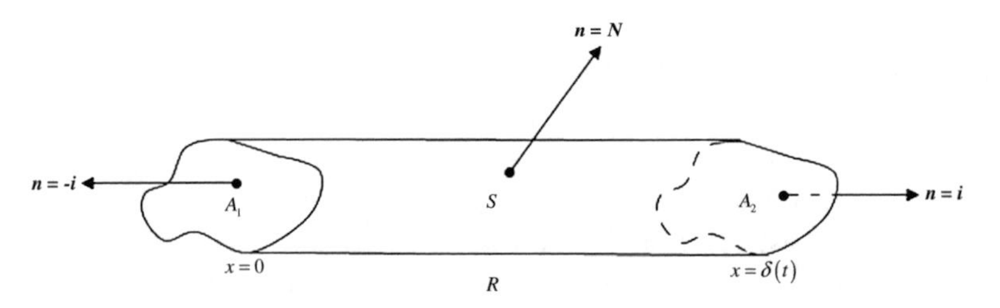

Fig. 6.6 Schematic representation of the region relevant to the approximate solution for the impulsive heat conduction in the semi-infinite laterally insulated bar.

where Q_e represents the heat in the bar associated with the ambient temperature, and employ our balance law

$$\frac{dQ(t)}{dt} = \iiint\limits_R r(x,t)\, d\tau - \iint\limits_S \boldsymbol{J}^{(Q)}(x,t) \bullet \boldsymbol{n}\, d\sigma$$

with $r(x,t) \equiv 0$ and $\boldsymbol{J}^{(Q)}(x,t) = -k_0 \nabla T(x,t)$. Then

$$Q(t) = Q_e + \rho_0 c_0 T_1 \iint_A \left\{ \int_0^{\delta(t)} \left[1 - \frac{x}{\delta(t)} \right]^2 dx \right\} dy\,dz$$

where A is an arbitrary cross section of area $\mathcal{A}_0$. Thus, introducing the change of variables $u = 1 - x/\delta(t)$, we obtain

$$Q(t) = Q_e + \rho_0 c_0 T_1 \mathcal{A}_0 \delta(t) \int_0^1 u^2\,du = Q_e + \frac{\rho_0 c_0 T_1 \mathcal{A}_0 \delta(t)}{3}$$

and hence

$$\textcircled{1} \equiv \frac{dQ(t)}{dt} = \frac{\rho_0 c_0 T_1 \mathcal{A}_0}{3} \frac{d\delta}{dt}(t).$$

We next compute

$$\textcircled{2} \equiv \iint_S \boldsymbol{J}^{(Q)}(x,t) \bullet \boldsymbol{n}\,d\sigma$$

by two different methods, after noting that

$$\boldsymbol{J}^{(Q)}(\boldsymbol{x},t) = J_1(x,t)\boldsymbol{i}$$

where

$$J_1(x,t) = -k_0 \frac{\partial T}{\partial x}(x,t) = \frac{2k_0 T_1}{\delta(t)} \left[1 - \frac{x}{\delta(t)} \right].$$

(i) Direct Method:

$$\textcircled{2} = \iint_{A_1} J_1(0,t)\boldsymbol{i} \bullet (-\boldsymbol{i})\,dy\,dz + \iint_{A_2} J_1[\delta(t),t]\boldsymbol{i} \bullet \boldsymbol{i}\,dy\,dz + \iint_S J_1(x,t)\boldsymbol{i} \bullet \boldsymbol{N}\,d\sigma$$
$$= \mathcal{A}_0 \{ J_1[\delta(t),t] - J_1(0,t) \}$$

since the cross-sectional areas of $A_{1,2}$ are both equal to $\mathcal{A}_0$ while $\boldsymbol{i} \bullet \boldsymbol{i} = 1$ and $\boldsymbol{i} \bullet \boldsymbol{N} = 0$.

(ii) Divergence Theorem Method:

$$\textcircled{2} = \iiint_R \nabla \bullet \boldsymbol{J}^{(Q)}(x,t)\,dx\,dy\,dz = \iint_A \left[\int_{x=0}^{\delta(t)} \frac{\partial J_1}{\partial x}(x,t)\,dx \right] dy\,dz$$
$$= \mathcal{A}_0 \{ J_1[\delta(t),t] - J_1(0,t) \}.$$

Therefore

$$-\textcircled{2} = \mathcal{A}_0 \{ J_1(0,t) - J_1[\delta(t),t] \} = \mathcal{A}_0 J_1(0,t) = \frac{2k_0 T_1 \mathcal{A}_0}{\delta(t)}$$

since $J_1[\delta(t),t] = 0$. Finally our balance law of $\textcircled{1} = -\textcircled{2}$ implies that

$$\delta(t)\frac{d\delta}{dt}(t) = 6\kappa_0.$$

Recalling that $\delta(0) = 0$, this yields the thermal boundary-layer solution of

$$\delta^2(t) = 12\kappa_0 t \text{ or } \delta(t) = 2\sqrt{\kappa_0 t}\sqrt{3}.$$

Substituting this result into our assumed solution, we obtain

$$T(x,t) = T_e + T_1 \begin{cases} [1 - x/(2\sqrt{\kappa_0 t}\sqrt{3})]^2, & 0 \le x \le 2\sqrt{\kappa_0 t}\sqrt{3} \\ 0, & x > 2\sqrt{\kappa_0 t}\sqrt{3} \end{cases}$$

or, again including the parameter designations of (6.3.7),

$$T(x,t;T_e,T_1,\kappa_0) = T_e + T_1 G(\eta) \text{ with } \eta = \frac{x}{2\sqrt{\kappa_0 t}} \tag{6.7.1}$$

where $G(\eta)$ is as defined by (6.6.1). Upon examination of Table 6.2 and Fig. 6.4, we can see that

$$G(1.25) = F(1.25).$$

Although employment of (6.7.1) overpredicts the actual temperature distribution in the bar for $\eta < 1.25$ and underpredicts it for $\eta > 1.25$, nonetheless this relatively simplistic solution involving $G(\eta)$ gives a very good approximation to the exact solution involving $\text{erf}_c(\eta)$. Further its derivation which uses the heat balance law directly and hence is self-contained also in some ways provides a better physical feel for the problem than does the similarity solution method. Finally, in this context, we can deduce that the temperature threshold slightly above the ambient mentioned at the start of the section used in conjunction with the intercept $x = \delta(t)$ corresponds to

$$F(\sqrt{3}) = 0.014 \text{ since } G(\eta) \equiv 0 \text{ for } \eta \ge \sqrt{3}$$

and hence, is given by

$$T_e + 0.014T_1.$$

6.8 Heat Conduction in Contact with a Reservoir of Oscillating Temperature or Why When it is Summer on the Earth's Surface it is Winter 4.44 meters Under Ground in its Crust

In order to model the seasonal variation of temperature underground in the Earth's crust, we consider the prototype heat conduction problem of a laterally insulated semi-infinite bar in contact with a reservoir of oscillating temperature in the absence of any bulk sources as depicted in Fig. 6.7. Heat conduction for this scenario is then governed by the following boundary-value partial differential diffusion equation problem for $T = T(x,t) \equiv$ temperature distribution in the bar:

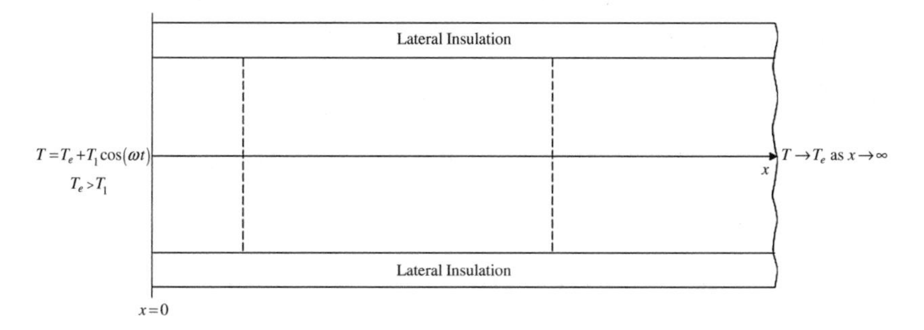

Fig. 6.7 Schematic diagram of heat conduction in a laterally insulated semi-infinite bar driven by a reservoir of oscillating temperature.

$$\frac{\partial T}{\partial t} = \kappa_0 \frac{\partial^2 T}{\partial x^2}, \ 0 < x < \infty; \tag{6.8.1a}$$

$$T(0,t) = T_e + T_1 \cos(\omega t), \tag{6.8.1b}$$

$$T(x,t) \to T_e \text{ as } x \to \infty; \tag{6.8.1c}$$

where $T_e > T_1$ and $P = 2\pi/\omega \equiv$ one year. As in (6.1.2) we now introduce the nondimensional dependent variable $u = u(x,t)$ defined by

$$T(x,t) = T_e + T_1 u(x,t), \tag{6.8.2}$$

which transforms (6.8.1) into

$$\mathcal{L}[u] = 0, \ 0 < x < \infty; \tag{6.8.3a}$$

$$u(0,t,) = \cos(\omega t), \tag{6.8.3b}$$

$$u(x,t) \to 0 \text{ as } x \to \infty; \tag{6.8.3c}$$

where $\mathcal{L}$ is the linear partial differential operator defined in (6.3.1a). Although this problem could be solved by a real variables technique as sketched in Problem 6.3, we shall instead pose and solve the companion complex variables problem for $U = U(x,t)$:

$$\mathcal{L}[U] = 0, \ 0 < x < \infty; \tag{6.8.4a}$$

$$U(0,t) = e^{i\omega t}, \tag{6.8.4b}$$

$$U(x,t) \to 0 \text{ as } x \to \infty; \tag{6.8.4c}$$

where

$$u(x,t) = \text{Re}[U(x,t)]. \tag{6.8.5}$$

The equality of (6.8.5) can be demonstrated as follows: Let

$$U(x,t) = R(x,t) + iI(x,t) \text{ where } R(x,t) = \text{Re}[U(x,t)] \text{ and } I(x,t) = \text{Im}[U(x,t)].$$

Then since $\mathcal{L}$ is a linear operator

$$\mathcal{L}[U] = \mathcal{L}[R] + i\mathcal{L}[I] = 0 = 0 + 0i \Rightarrow \mathcal{L}[R] = \mathcal{L}[I] = 0 \tag{6.8.6a}$$

while

$$\begin{aligned} U(0,t) &= R(0,t) + iI(0,t) = e^{i\omega t} = \cos(\omega t) + i\sin(\omega t) \\ &\Rightarrow R(0,t) = \cos(\omega t), I(0,t) = \sin(\omega t) \end{aligned} \tag{6.8.6b}$$

and

$$\begin{aligned} &U(x,t) = R(x,t) + iI(x,t) \to 0 = 0 + 0i \text{ as } x \to \infty \\ &\Rightarrow R(x,t) \to 0 \text{ as } x \to \infty, I(x,t) \to 0 \text{ as } x \to \infty. \end{aligned} \tag{6.8.6c}$$

Upon comparison of (6.8.4) with (6.8.6), the equality of (6.8.5) is established directly since both satisfy the same boundary-value problem and that problem has a unique solution. Hence to find $u(x,t)$ we merely have to solve (6.8.4) for $U(x,t)$ and take its real part.

We shall solve (6.8.4) by a separation of variables type method. Let

$$U(x,t) = e^{i\omega t}F(x). \tag{6.8.7}$$

Thus

$$\begin{aligned} \mathcal{L}[U] &= e^{i\omega t}\left[F''(x) - \frac{i\omega}{\kappa_0}F(x)\right] \\ &\Rightarrow F''(x) - \frac{i\omega}{\kappa_0}F(x) = 0, 0 < x < \infty; \end{aligned} \tag{6.8.8a}$$

while

$$U(0,t) = e^{i\omega t}F(0) = e^{i\omega t} \Rightarrow F(0) = 1, \tag{6.8.8b}$$

$$U(x,t) = e^{i\omega t}F(x) \to 0 \text{ as } x \to \infty \Rightarrow F(x) \to 0 \text{ as } x \to \infty. \tag{6.8.8c}$$

Seeking a solution of (6.8.8a) of the form

$$F(x) = e^{mx}, \tag{6.8.9}$$

we find that

$$m^2 = \frac{i\omega}{\kappa_0} \text{ or } m_{1,2} = \sqrt{\frac{\omega}{\kappa_0}}(i)^{1/2} \tag{6.8.10a}$$

where the subscripts denote the two roots of $(i)^{1/2}$ which we next develop from first principles. Consider the complex variable $z = x + iy$ where as usual $x, y \in \mathbb{R}$ and $i = \sqrt{-1}$. Introduction of the polar coordinate transformation $x = r\cos(\theta)$ and $y = r\sin(\theta)$ yields the polar form of z given by

$$z = r[\cos(\theta) + i\sin(\theta)] = re^{i\theta} = re^{i(\theta+2\pi k)}$$

where k is an integer. Now, taking the nth root of z, we obtain

$$z^{1/n} = \sqrt[n]{r}e^{i(\theta+2\pi k)/n} = \sqrt[n]{r}\left[\cos\left(\frac{\theta+2\pi k}{n}\right) + i\sin\left(\frac{\theta+2\pi k}{n}\right)\right] \text{ for } k = 0, 1, \ldots, n-1$$

where n is an integer and $\sqrt[n]{r}$ denotes the principal n^{th} root. Note that both $k = 0$ and n yield the same value since the cos and sin functions are periodic with period 2π. Particularizing this result to $n = 2$ and $z = i$, for which $r = 1$ and $\theta = \pi/2$, we find that

$$(i)^{1/2} = \cos\left(\frac{\pi}{4} + k\pi\right) + i\sin\left(\frac{\pi}{4} + k\pi\right) \text{ for } k = 0 \text{ and } 1$$

or

$$k = 0 : (i)^{1/2} = \cos\left(\frac{\pi}{4}\right) + i\sin\left(\frac{\pi}{4}\right) = \frac{1+i}{\sqrt{2}}$$

and

$$k = 1 : (i)^{1/2} = \cos\left(\frac{5\pi}{4}\right) + i\sin\left(\frac{5\pi}{4}\right) = -\frac{1+i}{\sqrt{2}}$$

Then the $m_{1,2}$ of (6.8.10a) are given by

$$m_{1,2} = \pm\alpha(1+i) \text{ where } \alpha = \sqrt{\frac{\omega}{2\kappa_0}}. \tag{6.8.10b}$$

and (6.8.8a) has the general solution

$$F(x) = c_1 e^{\alpha(1+i)x} + c_2 e^{-\alpha(1+i)x}. \tag{6.8.11a}$$

Appling the far-field BC (6.8.8c) implies that

$$c_1 = 0 \text{ or } F(x) = c_2 e^{-\alpha(1+i)x} \tag{6.8.11b}$$

while BC (6.8.8b) yields

$$F(0) = c_2 = 1 \text{ or } F(x) = e^{-\alpha(1+i)x}. \tag{6.8.11c}$$

Finally from (6.8.7) we find that our companion complex variables problem has solution

$$U(x,t) = e^{i\omega t}F(x) = e^{-\alpha x}e^{i(\omega t - \alpha x)}. \tag{6.8.12a}$$

and hence from (6.8.5)

$$u(x,t) = \text{Re}[u(x,t)] = e^{-\alpha x}\cos(\omega t - \alpha x). \tag{6.8.12b}$$

Observe from (6.8.12b) that this solution has the same time dependence as the bath reservoir but is out of phase with it by an angle $\varphi(x) = \alpha x$ and has an attenuating amplitude of $\mathcal{A}(x) = e^{-\alpha x}$. Specifically for

$$x = x_0 \text{ such that } \alpha x_0 = \sqrt{\frac{\omega}{2\kappa_0}} x_0 = \pi, \tag{6.8.13a}$$

$$\cos(\omega t - \alpha x_0) = \cos(\omega t - \pi) = -\cos(\omega t) \text{ and } e^{-\alpha x_0} = e^{-\pi} = 0.043. \tag{6.8.13b}$$

We now apply this result for our prototype problem to the seasonal variation of temperature under ground in the Earth's crust. We consider x to be the depth under ground with $x = 0$ corresponding to the Earth's surface. Further we are interpreting the temperature at the surface given by BC (6.8.1a) with $\omega = 2\pi/P$ for $P \equiv$ one year

$$T(0,t) = T_e + T_1 \cos\left(\frac{2\pi t}{P}\right) \tag{6.8.14a}$$

to represent a very simplistic seasonal diurnal variation where the successive quarter cycles for $t \in [0, P]$ correspond to summer, fall, winter, and spring, respectively. Then from (6.8.13) the temperature under ground at the depth x_0 is

$$T(x_0,t) = T_e - 0.043 T_1 \cos\left(\frac{2\pi t}{P}\right) \tag{6.8.14b}$$

which has a variation opposite in sign from that at the surface but with its amplitude reduced by a factor of 0.043. Hence at this depth the successive quarter cycles being two cycles out of phase with the surface correspond to winter, spring, summer, and fall, respectively. Let us now calculate this depth. From (6.8.13a)

$$x_0 = \pi\sqrt{\kappa_0\left(\frac{2}{\omega}\right)} \tag{6.8.15a}$$

while [69]

$$\kappa_0 \equiv \text{ thermal diffusivity for soil } = 2 \times 10^{-3}\,\text{cm}^2/\text{sec} \tag{6.8.15b}$$

and

$$P = \frac{2\pi}{\omega} \equiv \text{ one year } = 31{,}536{,}000\,\text{sec} \cong \pi \times 10^7\,\text{sec} \Rightarrow \frac{2}{\omega} \cong 10^7\,\text{sec}. \tag{6.8.15c}$$

Thus

$$x_0 \cong \pi\sqrt{(2 \times 10^{-3}\,\text{cm}^2/\text{sec})10^7\,\text{sec}} = \pi\sqrt{2}(100\,\text{cm}) = 4.44\,\text{m} \tag{6.8.15d}$$

and, in this sense, when it is summer on the Earth's surface, it is winter 4.44 m under ground in its crust [69].

Finally, rewriting (6.8.12b) in the form

$$u(x,t) = e^{-\alpha x}\cos(\alpha x - \omega t) \tag{6.8.16a}$$

and noting that

$$\alpha x - \omega t = \alpha(x - ct) \text{ where } c = \frac{\omega}{\alpha} = \sqrt{2\kappa_0\omega}, \tag{6.8.16b}$$

we can view this temperature function as a damped traveling wave with speed

$$c = \sqrt{2(2\times 10^{-3}\text{cm}^2/\text{sec})(2\times 10^{-7}\text{sec})} = 2\sqrt{2}\times 10^{-5}\text{cm/sec} \tag{6.8.16c}$$

upon application of (6.8.15b) and (6.8.15c).

Problems

6.1. Consider an infinite laterally insulated bar of constant cross-sectional area $\mathcal{A}$ as depicted in Fig. 6.8. At time $t = 0$ a supply of heat of amount Q_0 is instantaneously introduced at position $s = 0$ (this is not a constantly replenished source). Heat conduction in the bar is assumed to be dependent only on the s coordinate.

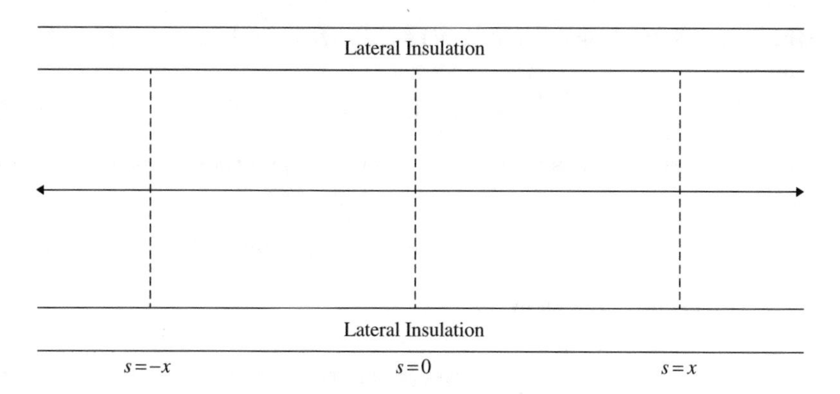

Fig. 6.8 Schematic diagram of heat conduction in the laterally insulated infinite bar.

(a) Consider a finite region, R, of the bar between $s = \pm x$. Show that the amount of heat in this region is given by

$$Q(x,t) = \rho_0 c_0 \mathcal{A} \int_{-x}^{x} T(s,t)\,ds + Q_e, \tag{P1.1}$$

where $\rho_0 \equiv$ density and $c_0 \equiv$ specific heat of the bar, assumed to be constant, while $T(s,t) \equiv$ deviation of temperature in the bar from the ambient T_e and $Q_e \equiv$ heat in the bar associated with that ambient temperature.

(b) For $t > 0$ deduce that the heat conservation law governing this situation reduces to

$$\frac{\partial Q}{\partial t}(x,t) = -\iint_S \boldsymbol{J}^{(Q)}(s,t) \bullet \boldsymbol{n}\, d\sigma, \tag{P1.2}$$

where $S \equiv$ bounding surface of R, $\boldsymbol{J}^{(Q)}(s,t) \equiv$ heat flux vector, and $\boldsymbol{n} \equiv$ unit outward-pointing normal to S, by explaining why there is no source term in (P1.2). Depict S and $\boldsymbol{n}$ for this problem by means of a diagram analogous to Fig. 6.6 for the semi-infinite bar.

(c) Either by evaluating the surface integral in (P1.2) directly or by applying the divergence theorem to it in conjunction with the fundamental theorem of the calculus and then making use of Fick's First Law of Diffusion or, equivalently, Newton's Law of Cooling for this situation

$$\boldsymbol{J}^{(Q)}(s,t) = J_1(s,t)\boldsymbol{i} \text{ with } J_1(s,t) = -k_0 T_1(s,t) \text{ and } T_1(s,t) \equiv \frac{\partial T}{\partial s}(s,t)$$

where $k_0 \equiv$ thermal conductivity and $\boldsymbol{i} \equiv$ unit vector in the s-direction, conclude that

$$\frac{\partial Q}{\partial t}(x,t) = k_0 \mathcal{A}[T_1(x,t) - T_1(-x,t)]. \tag{P1.3}$$

(d) Using the fundamental theorem of the calculus and the chain rule prove Leibniz's rule

$$\frac{d}{dx}\int_{g(x)}^{h(x)} f(s)\, ds = f[h(x)]h'(x) - f[g(x)]g'(x). \tag{P1.4}$$

(e) Using (P1.1) and (P1.4) show that

$$\frac{\partial Q}{\partial x}(x,t) = \rho_0 c_0 \mathcal{A}[T(x,t) + T(-x,t)]. \tag{P1.5}$$

(f) Differentiating (P1.5) with respect to x and comparing this result with (P1.3) show that $Q(x,t)$ satisfies the following equation for $t > 0$ sometimes called Fick's Second Law of Diffusion:

$$\frac{\partial Q}{\partial t} = \kappa_0 \frac{\partial^2 Q}{\partial x^2}, \; 0 < x < \infty, \tag{P1.6}$$

where $\kappa_0 = k_0/(\rho_0 c_0) \equiv$ thermometric conductivity or thermal diffusivity.

(g) From (P1.1) and the stated physics of the problem, respectively, conclude that for $t > 0$:

$$Q(0,t) = Q_e,\; Q(x,t) \to Q_e + Q_0 \text{ as } x \to \infty. \tag{P1.7}$$

(h) Using the idea of nondimensionalization and the concept of similarity solutions conclude that $Q(x,t)$ satisfying (P1.6) and (P1.7) is of the form

$$Q(x,t) = Q_e + Q_0 F(\eta) \text{ where } \eta = \frac{x}{2\sqrt{\kappa_0 t}}. \tag{P1.8}$$

[Note that here, as well as in parts (i) and (j), any result demonstrated previously in this chapter may be employed, rather than having to be reproduced.]

(i) Substituting the similarity solution of (P1.8) into (P1.6) and (P1.7) show that F satisfies the following differential equation and boundary conditions:

$$F''(\eta) + 2\eta F'(\eta) = 0,\ 0 < \eta < \infty;\ F(0) = 0,\ F(\eta) \to 1 \text{ as } \eta \to \infty. \tag{P1.9}$$

(j) Solve (P1.9) for $F(\eta)$ and hence obtain

$$F(\eta) = \operatorname{erf}(\eta) = \frac{2}{\sqrt{\pi}} \int_0^{\eta} e^{-z^2}\,dz. \tag{P1.10}$$

(k) Using the fact that $T(x,t) = T(-x,t)$ [why?] in conjunction with (P1.4), (P1.5), (P1.8), and (P1.10), finally deduce that

$$T(x,t) = \frac{Q_0 e^{-x^2/(4\kappa_0 t)}}{2\mathcal{A}\sqrt{\rho_0 c_0 k_0 \pi t}}. \tag{P1.11}$$

(l) By direct differentiation of (P1.11) show that this $T(x,t)$ satisfies the same diffusion equation as $Q(x,t)$. Namely

$$\frac{\partial T}{\partial t} = \kappa_0 \frac{\partial^2 T}{\partial x^2}. \tag{P1.12}$$

(m) Schematically plot T versus x for different values of t and discuss the implications of these results.

(n) Examine the behavior of T with t for a fixed value of $x \equiv x_0 > 0$ and represent this graphically.

6.2. Consider an axially insulated solid layer of depth ℓ that is infinite in planar extent. At time $t = 0$ a supply of heat of amount Q_0 is instantaneously introduced along the $s = 0$ axis (this is not a constantly replenished source). Heat conduction in the layer is assumed to be dependent only upon s, the radial coordinate.

(a) Consider a finite cylindrical region of the layer between $0 \le s \le r$, $0 \le \theta \le 2\pi$, and $0 \le z \le \ell$ as depicted in Fig. 6.9 where θ and z are the circumferential and axial coordinates, respectively. Show that the amount of heat in this region is given by

$$Q(r,t) = 2\pi\rho_0 c_0 \ell \int_0^r T(s,t)s\,ds + Q_e, \tag{P2.1}$$

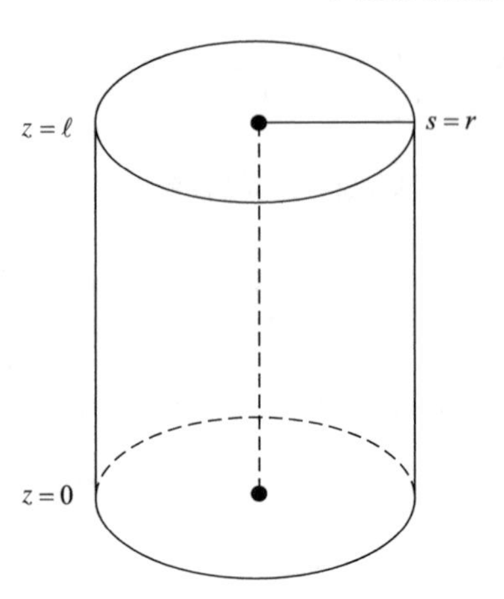

Fig. 6.9 Schematic diagram of the finite cylindrical region of the axially-insulated solid layer of infinite extent.

where $\rho_0 \equiv$ density and $c_0 \equiv$ specific heat of the layer, assumed to be constant, while $T(s,t) \equiv$ deviation of temperature in the layer from the ambient T_e and $Q_e \equiv$ heat in the layer associated with that ambient temperature.

(b) For $t > 0$ deduce the following conservation law:

$$\frac{\partial Q}{\partial t} = -\iint_S \boldsymbol{J}^{(Q)}(s,t) \bullet \boldsymbol{n}\, d\sigma, \tag{P2.2}$$

where $S \equiv$ bounding surface of R, $\boldsymbol{J}^{(Q)}(s,t) \equiv$ heat flux vector, and $\boldsymbol{n} \equiv$ unit outward-pointing normal to S, by explaining why there is no source term in (P2.2). Depict S and $\boldsymbol{n}$ for this problem by means of a figure analogous to Fig. 6.6 for the semi-infinite bar.

(c) Adopting the constitutive relation

$$\boldsymbol{J}^{(Q)}(s,t) = -k_0 T_1(s,t)\boldsymbol{e}_r \text{ with } T_1(s,t) \equiv \frac{\partial T}{\partial s}(s,t),$$

where $k_0 \equiv$ thermal conductivity and $\boldsymbol{e}_r \equiv$ unit vector in the radial direction, conclude by directly evaluating the surface integral in (P2.2) that

$$\frac{\partial Q}{\partial t}(r,t) = 2\pi\ell k_0 r T_1(r,t). \tag{P2.3}$$

(d) Using (P2.1) and the fundamental theorem of the calculus show that

$$\frac{\partial Q}{\partial r}(r,t) = 2\pi\ell\rho_0 c_0 r T(r,t). \tag{P2.4}$$

(e) Differentiating (P2.4) with respect to r and comparing this result with (P2.3) show that $Q(r,t)$ satisfies the following diffusion equation for $t > 0$:

$$\frac{\partial Q}{\partial t} = \kappa_0 \left[\frac{\partial^2 Q}{\partial r^2} - \left(\frac{1}{r}\right)\frac{\partial Q}{\partial r}\right], 0 < r < \infty, \tag{P2.5}$$

where $\kappa_0 = k_0/(\rho_0 c_0) \equiv$ thermometric conductivity.

(f) From (P2.1) and the stated physics of the problem, respectively, conclude that for $t > 0$:

$$Q(0,t) = Q_e,\ Q(r,t) \to Q_e + Q_0 \text{ as } r \to \infty. \tag{P2.6}$$

(g) Using the idea of nondimensionalization and the concept of similarity solutions conclude that $Q(x,t)$ satisfying (P2.5) and (P2.6) is of the form

$$Q(r,t) = Q_e + Q_0 F(\eta) \text{ where } \eta = \frac{r}{2\sqrt{\kappa_0 t}}. \tag{P2.7}$$

[Note that here as well as in part (h) any result demonstrated previously in this chapter may be employed rather than having to be reproduced].

(h) Substituting the similarity solution of (P2.7) into (P2.5) and (P2.6) show that F satisfies the following differential equation and boundary conditions:

$$F''(\eta) + \left(2\eta - \frac{1}{\eta}\right)F'(\eta) = 0,\ 0 < \eta < \infty; \tag{P2.8a}$$

$$F(0) = 0,\ F(\eta) \to 1 \text{ as } \eta \to \infty. \tag{P2.8b}$$

(i) Solve the boundary-value problem (P2.8) by integrating (P2.8a) twice and employing (P2.8b) to obtain

$$F(\eta) = 1 - e^{-\eta^2}. \tag{P2.9}$$

(j) Using (P2.4), (P2.7), and (P2.9), finally deduce that

$$T(r,t) = \left(\frac{Q_0}{\ell}\right)\frac{e^{-r^2/(4\kappa_0 t)}}{4\pi k_0 t}. \tag{P2.10}$$

(k) Now by direct differentiation of (P2.10) demonstrate that this $T(r,t)$ satisfies the usual axisymmetric radially dependent heat equation in cylindrical coordinates (see Chapter 12): Namely,

$$\frac{\partial T}{\partial t} = \kappa_0\left[\frac{\partial^2 T}{\partial r^2} + \left(\frac{1}{r}\right)\frac{\partial T}{\partial r}\right] = \frac{\kappa_0}{r}\frac{\partial}{\partial r}\left(r\frac{\partial T}{\partial r}\right). \tag{P2.11}$$

Observe in this context that unlike for the previous exercise involving Cartesian coordinates the one-dimensional temperature and heat functions no longer satisfy the same diffusion equation.

(ℓ) Extrapolating from this result of (P2.11) that

$$\nabla \bullet \boldsymbol{J}^{(Q)}(s,t) = -\frac{k_0}{s}\frac{\partial}{\partial s}\left(s\frac{\partial T}{\partial s}\right),$$

deduce (P2.3) by applying the divergence theorem to the surface integral in (P2.2).

6.3. The purpose of this problem is to solve (6.8.3) by using a real variables approach. Toward that end we reconsider this problem for $u = u(x,t)$:

$$\frac{\partial u}{\partial t} = \kappa_0 \frac{\partial^2 u}{\partial x^2},\ 0 < x < \infty; \tag{P3.1a}$$

$$u(0,t) = \cos(\omega t),\ u(x,t) \to 0 \text{ as } x \to \infty; \tag{P3.1b}$$

and seek a solution of it of the form

$$u(x,t) = f(x)\cos(\omega t) + g(x)\sin(\omega t). \tag{P3.2}$$

(a) Substituting (P3.2) into (P3.1) and employing the linear independence of $\cos(\omega t)$ and $\sin(\omega t)$, show that f and g satisfy the coupled system

$$f'' = 2\alpha^2 g;\ g'' = -2\alpha^2 f \text{ where } 2\alpha^2 = \frac{\omega}{\kappa_0} \text{ and } \alpha > 0; \tag{P3.3a}$$

$$f(0) = 1, f(x) \to 0 \text{ as } x \to \infty;\ g(0) = 0,\ g(x) \to 0 \text{ as } x \to \infty. \tag{P3.3b}$$

(b) Eliminating g between the differential equations of (P3.3a), show that f satisfies

$$f^{(4)} + 4\alpha^4 f = 0 \text{ where } f^{(4)} \equiv \frac{d^4 f}{dx^4}. \tag{P3.4}$$

(c) Letting

$$f(x) = e^{rx}, \tag{P3.5a}$$

substituting (P3.5a) into (P3.4), obtaining

$$r^4 + 4\alpha^4 = (r^2)^2 + (2\alpha^2)^2 = 0, \tag{P3.5b}$$

and noting that upon completing the square on its linear term (P3.5b) can be factored to yield

$$(r^2)^2 + 2(r^2)(2\alpha^2) + (2\alpha^2)^2 - (2\alpha r)^2 = (r^2 + 2\alpha r + 2\alpha^2)(r^2 - 2\alpha r + 2\alpha^2) = 0, \tag{P3.5c}$$

find that

$$r_{1,2,3,4} = \alpha(\pm 1 \pm i) \qquad \text{(P3.5d)}$$

and hence conclude

$$f(x) = e^{\alpha x}[c_1 \cos(\alpha x) + c_2 \sin(\alpha x)] + e^{-\alpha x}[c_3 \cos(\alpha x) + c_4 \sin(\alpha x)]. \qquad \text{(P3.5e)}$$

Observe that the roots of (P3.5d) could have been determined by computing the four complex roots $(-4\alpha^4)^{1/4} = \sqrt{2}\alpha(-1)^{1/4}$ using the formula for $z^{1/n}$ with $z = -1$, $n = 4$, and $k = 0, 1, 2$, and 3 but here except for the complex roots resulting from the application of the quadratic formula to (P3.5c) the analysis has been restricted to the reals.

(d) Applying the boundary conditions for f in (P3.5b) deduce from (P3.5e) that

$$f(x) = e^{-\alpha x}h(x) \text{ where } h(x) = \cos(\alpha x) + c_4 \sin(\alpha x). \qquad \text{(P3.6a)}$$

Now use (P3.3a) to show from (P3.6a) that

$$g(x) = -e^{-\alpha x}\frac{h'(x)}{\alpha}. \qquad \text{(P3.6b)}$$

(e) Applying the boundary conditions for g in (P3.3b) deduce from (P3.6b) that $h(x) = \cos(\alpha x)$ and hence finally obtain the result, equivalent to (6.8.16a), from (P3.6) and (P3.2)

$$u(x,t) = e^{-\alpha x}[\cos(\alpha x)\cos(\omega t) + \sin(\alpha x)\sin(\omega t)] = e^{-\alpha x}\cos(\alpha x - \omega t). \qquad \text{(P3.7)}$$

Chapter 7
Initiation of Cellular Slime Mold Aggregation Viewed as an Instability

The material to be presented sequentially in the next two chapters is based upon the seminal work of Keller and Segel [59] and Prigogine and Lefever [99] involving two fundamental problems in developmental biological modeling: Namely, cellular slime mold aggregation and the chemical Brusselator diffusive instability system, respectively. In this chapter, the initiation of cellular slime mold aggregation is identified as the onset of a self-organized linear instability of a simplified reaction–diffusion model system defined on an unbounded planar domain for the slime mold amoeba density and the concentration of the extracellular chemical acrasin produced by them to which they are chemotactically attracted. To derive these governing equations, a general balance law involving source and flux terms must be deduced by employing the divergence, Stokes, and Green's theorems, which are the subject of a pastoral interlude and applied to the amoeba and acrasin, as well as to the two other dependent variables included in this process. These are the enzyme acrasinase, a second chemical produced by the amoeba, that degrades the acrasin to a product to which they are not chemotactically attracted, and an intermediate complex formed by the interaction of those two chemicals in a reversible equilibrium reaction. Here, the law of mass action is introduced to determine the proper source terms for those chemical components. Then, by making two additional assumptions valid in the earliest stages of aggregation, that four-component system is reduced to the simplified amoeba–acrasin reaction–diffusion model mentioned above upon elimination of the acrasinase and intermediate complex. The initial conditions for the linear perturbation system are satisfied by means of Fourier integrals, introduced by another pastoral interlude that deduces the relevant formula, and, in so doing, also includes the concept of Laplace transforms. The factors that favor the initiation of such aggregation are predicted by examining the linear instability criterion. The problem considers the equivalent normal-mode linear stability analysis of a slightly more general four-component model system explicitly including the acrasinase and intermediate complex.

© Springer International Publishing AG, part of Springer Nature 2017

D. J. Wollkind and B. J. Dichone, *Comprehensive Applied Mathematical Modeling in the Natural and Engineering Sciences*,
https://doi.org/10.1007/978-3-319-73518-4_7

7.1 Introduction

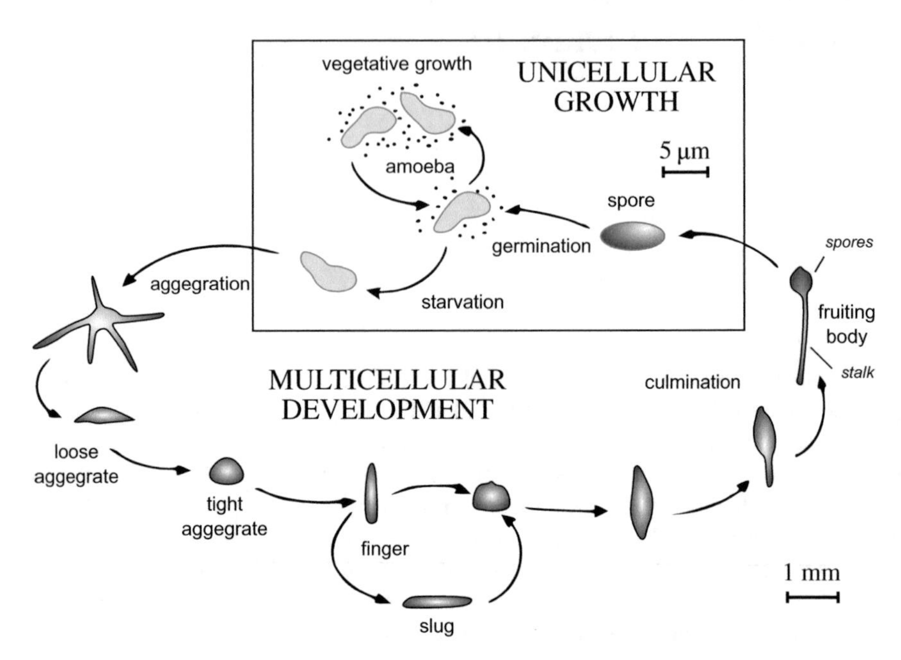

Fig. 7.1 Schematic diagram of the life cycle for the cellular slime mold reproduced from a Wikimedia image under a Creative Commons license.

Cellular slime molds represent perhaps the simplest life form to exhibit the leap between the apparent uniformity of cells produced in the earliest stages of embryonic development during mitosis and the structural complexity of various tissues comprising the full-grown organism. Cellular slime molds are amoeba which have the interesting property of existing and reproducing both as individual cells and as differentiated multicellular aggregates. This transition from uni- to multicellular development leading to differentiation is probably the simplest example of its kind that can be studied in biology. The life cycle of the cellular slime mold depicted in Fig. 7.1 may be described as starting with a spore, which under favorable conditions – *i.e.*, the availability of bacteria, its favorite food supply, germinates into an individual single-celled amoeba. These amoebae are mobile by virtue of pseudo pod formation and swarm toward the bacteria, engulfing and consuming them in the usual way by means of contractual vacuoles. They then reproduce asexually employing the fission process wherein each amoeba divides itself into two individuals. Having exhausted their food supply the amoebae first disperse in a more or less random fashion and distribute themselves uniformly over their environment. They then aggregate into regularly spaced clumps, which later form migrating slugs that erect fruiting bodies consisting of dead stalk cells bearing clusters of spore cells and the cycle repeats itself.

Enough was known about the various processes involved in such aggregation for Keller and Segel [59] to model this phenomenon. That aggregation is mediated by a chemical substance generically called acrasin secreted by them which attracts other amoeba through chemotaxis. They also secrete an enzyme that degrades the acrasin to a product to which the amoeba are not chemotactically attracted called an acrasinase. The identification of the chemical cyclic adenosine monophosphate (cAMP) that acted as an acrasin for the cellular slime mold *Dictyostelium discoidium* caused renewed interest in investigating this aggregation process about forty-five years ago. The same species also produced a phosphodiesterace that converted cAMP to chemotactically inactive AMP and hence acted as an acrasinase. The main purpose of Keller and Segel's [59] analysis was to determine the conditions that favored the onset of aggregation after the individual cellular slime mold amoeba had exhausted its bacterial food supply, divided by asexual fission, and distributed themselves uniformly over their environment, assumed to be flat and of infinite planar extent. In order to derive the governing equations for this situation, we need to deduce a two-dimensional version of the divergence theorem which is accomplished in the next section.

7.2 Pastoral Interlude: Divergence Theorem Revisited, Stokes Theorem, and Green's Theorem

We shall deduce the required two-dimensional version of the divergence theorem by applying the three-dimensional version of it introduced in Chapter 5 to the region R_0 depicted in Fig. 7.2. That is we shall reconsider this divergence theorem

$$\iint_{S_0} \boldsymbol{w} \bullet \boldsymbol{n}\, d\sigma = \iiint_{R_0} \nabla \bullet \boldsymbol{w}\, d\tau \text{ where } \nabla \bullet \boldsymbol{w} \equiv \frac{\partial w_1}{\partial x} + \frac{\partial w_2}{\partial y} + \frac{\partial w_3}{\partial z}$$

with $\boldsymbol{w} = \boldsymbol{v}(x, y, t) = v_1(x, y, t)\boldsymbol{i} + v_2(x, y, t)\boldsymbol{j}$ for $\boldsymbol{i}$, $\boldsymbol{j} \equiv$ unit vectors in the x-, y-directions and $R_0 \equiv$ a pill box of depth D replicated in the z-direction having a constant cross-section possessing bounding closed curve C: $\boldsymbol{r}(s) = x(s)\boldsymbol{i} + y(s)\boldsymbol{j}$ for $s \equiv$ arc length. Hence $S_0 \equiv A_1 \cap A \cap \mathcal{S}$ where A_1, A, and $\mathcal{S}$ represent the top ($z = D$), bottom ($z = 0$), and lateral surface of that pill box having associated outward-pointing unit normals $\boldsymbol{n} \equiv \boldsymbol{k} \equiv$ unit vector in the z-direction, $-\boldsymbol{k}$, and $\boldsymbol{N}$, respectively. Thus

$$\iiint_{R_0} \nabla \bullet \boldsymbol{w}\, d\tau = D \iint_{A} \nabla_2 \bullet \boldsymbol{v}\, dx\, dy \text{ where } \nabla_2 \bullet \boldsymbol{v} \equiv \frac{\partial v_1}{\partial x} + \frac{\partial v_2}{\partial y}$$

while

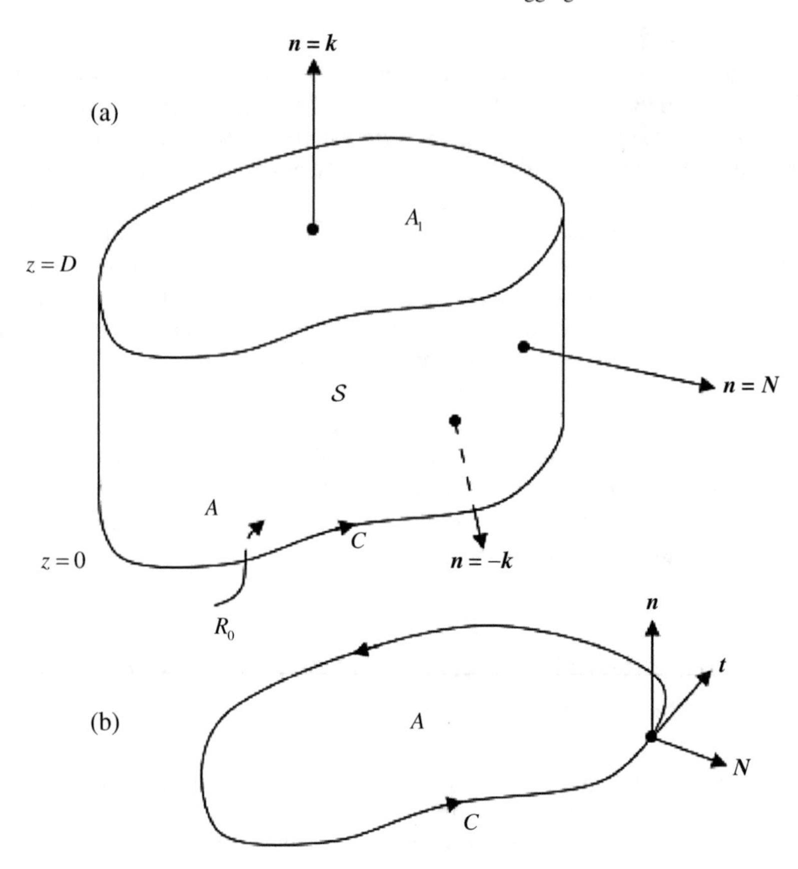

Fig. 7.2 Plots for (a) the derivation of the two-dimensional version of the divergence theorem from the three-dimensional version and (b) both the two-dimensional version of the divergence and Stokes theorems as well as the derivation of the governing equations for the cellular slime mold aggregation model.

$$\iint_{S_0} \boldsymbol{w} \bullet \boldsymbol{n}\, d\sigma = \iint_{A_1} \boldsymbol{v} \bullet \boldsymbol{k}\, dx\, dy + \iint_{A} \boldsymbol{v} \bullet (-\boldsymbol{k})\, dx\, dy + \int_{z=0}^{D} \left(\oint_C \boldsymbol{v} \bullet \boldsymbol{N}\, ds \right) dz$$
$$= D \oint_C \boldsymbol{v} \bullet \boldsymbol{N}\, ds.$$

Equating these two expressions and cancelling the common D factor, we obtain the desired two-dimensional version of the divergence theorem

$$\oint_C \boldsymbol{v} \bullet \boldsymbol{N}\, ds = \iint_A \nabla_2 \bullet \boldsymbol{v}\, dx\, dy \text{ where } \nabla_2 \bullet \boldsymbol{v} \equiv \frac{\partial v_1}{\partial x} + \frac{\partial v_2}{\partial y}.$$

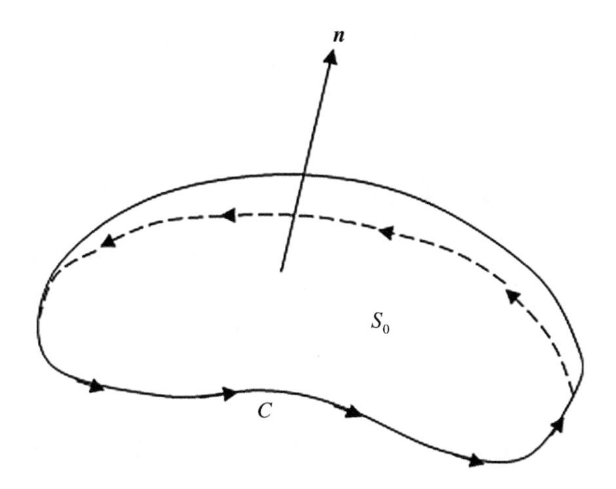

Fig. 7.3 Plot for the three-dimensional version of Stokes theorem.

For the sake of completeness and because it is related both to the material just presented and that to be introduced in later chapters, we now consider Stokes theorem for the surface S_0 depicted in Fig. 7.3 with bounding curve C: $\boldsymbol{r}(s) = x(s)\boldsymbol{i} + y(s)\boldsymbol{j} + z(s)\boldsymbol{k}$ and having outward-pointing unit normal $\boldsymbol{n}$:

$$\iint_{S_0} (\text{curl } \boldsymbol{w}) \bullet \boldsymbol{n}, d\sigma = \oint_C \boldsymbol{w} \bullet \boldsymbol{t}\, ds$$

where

$$\text{curl } \boldsymbol{w} = \left(\frac{\partial w_3}{\partial y} - \frac{\partial w_2}{\partial z}\right)\boldsymbol{i} + \left(\frac{\partial w_1}{\partial z} - \frac{\partial w_3}{\partial x}\right)\boldsymbol{j} + \left(\frac{\partial w_2}{\partial x} - \frac{\partial w_1}{\partial y}\right)\boldsymbol{k}$$

and

$$\boldsymbol{t}(s) = \dot{\boldsymbol{r}}(s) = \dot{x}(s)\boldsymbol{i} + \dot{y}(s)\boldsymbol{j} + \dot{z}(s)\boldsymbol{k},$$

the latter being a unit tangent vector to C since s represents arc length. Thus

$$\oint_C \boldsymbol{w} \bullet \boldsymbol{t}\, ds = \oint_C [w_1\dot{x}(s) + w_2\dot{y}(s) + w_3\dot{z}(s)]\, ds = \oint_C w_1\, dx + w_2\, dy + w_3\, dz.$$

We now deduce its two-dimensional form by applying Stokes theorem to the surface depicted in Fig. 7.2b. That is we take $\boldsymbol{w} = \boldsymbol{v}(x, y, t)$ as employed above, $S_0 = A$, and $\boldsymbol{n} = \boldsymbol{k}$. Then

$$\iint_{S_0} (\text{curl } \boldsymbol{w}) \bullet \boldsymbol{n} \, d\sigma = \iint_A (\text{curl } \boldsymbol{v}) \bullet \boldsymbol{k} \, dx \, dy = \iint_A \left(\frac{\partial v_2}{\partial x} - \frac{\partial v_1}{\partial y} \right) dx \, dy$$

while

$$\oint_C w_1 \, dx + w_2 \, dy + w_3 \, dz = \oint v_1 \, dx + v_2 \, dy.$$

Again equating these two expressions yields the desired two-dimensional version of Stokes theorem

$$\iint_A \left(\frac{\partial v_2}{\partial x} - \frac{\partial v_1}{\partial y} \right) dx \, dy = \oint_C v_1 \, dx + v_2 \, dy.$$

We shall next show that the two-dimensional forms of either of these vector-integral relations can be used to deduce Green's theorem for $[P, Q] = [P, Q](x, y)$ given by

$$\oint_C P \, dx + Q \, dy = \iint_A \left(\frac{\partial Q}{\partial x} - \frac{\partial P}{\partial y} \right) dx \, dy.$$

This follows directly from the two-dimensional version of Stokes theorem with $v_1 = P$, $v_2 = Q$. It can also be demonstrated by employing the two-dimensional version of the divergence theorem as follows: Let

$$\boldsymbol{v} = Q\boldsymbol{i} - P\boldsymbol{j} \Rightarrow v_1 = Q, \, v_2 = -P.$$

Further from Fig. 7.2b note that for $\boldsymbol{t} = \dot{x}(s)\boldsymbol{i} + \dot{y}(s)\boldsymbol{j}$ and $\boldsymbol{n} = \boldsymbol{k}$,

$$\boldsymbol{N} = \boldsymbol{t} \times \boldsymbol{n} = \boldsymbol{t} \times \boldsymbol{k} = \begin{vmatrix} \boldsymbol{i} & \boldsymbol{j} & \boldsymbol{k} \\ \dot{x}(s) & \dot{y}(s) & 0 \\ 0 & 0 & 1 \end{vmatrix} = \dot{y}(s)\boldsymbol{i} - \dot{x}(s)\boldsymbol{j}.$$

Thus

$$\oint_C \boldsymbol{v} \bullet \boldsymbol{N} \, ds = \oint_C [Q\dot{y}(s) + P\dot{x}(s)] \, ds = \oint_C P \, dx + Q \, dy$$

while

$$\iint_A \nabla_2 \bullet \boldsymbol{v} \, dx \, dy = \iint_A \left(\frac{\partial Q}{\partial x} - \frac{\partial P}{\partial y} \right) dx \, dy \text{ since } \frac{\partial v_1}{\partial x} + \frac{\partial v_2}{\partial y} = \frac{\partial Q}{\partial x} - \frac{\partial P}{\partial y}.$$

Hence again demonstrating the result. This also shows that the two-dimensional versions of both the divergence and Stokes theorems are, in some sense, equivalent.

Using the same geometry as in Fig. 7.2b we now consider the special case of Green's theorem with $Q = x$ and $P = -y$. Since then

$$\frac{\partial Q}{\partial x} - \frac{\partial P}{\partial y} = 1 - (-1) = 1 + 1 = 2 \text{ and } \iint_A dx\,dy = \mathcal{A}$$

where $\mathcal{A} \equiv$ enclosed area of A, we have shown that

$$\mathcal{A} = \frac{1}{2} \oint_C x\,dy - y\,dx.$$

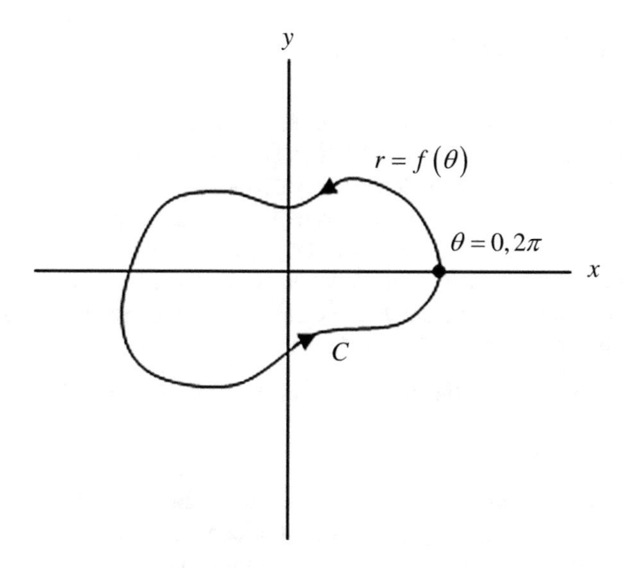

Fig. 7.4 Plot for the Green's theorem in space example.

Let us apply this result to the scenario depicted in Fig. 7.4. Here introducing polar coordinates where $r = f(\theta)$, we have

$$x(\theta) = f(\theta)\cos(\theta),\ y(\theta) = f(\theta)\sin(\theta) \text{ for } 0 \le \theta \le 2\pi \text{ with } f(0) = f(2\pi).$$

Thus

$$\dot{x}(\theta) = \dot{f}(\theta)\cos(\theta) - f(\theta)\sin(\theta) \text{ and } \dot{y}(\theta) = \dot{f}(\theta)\sin(\theta) + f(\theta)\cos(\theta).$$

Since then

$$\oint x\,dy - y\,dx = \int_0^{2\pi} [x(\theta)\dot{y}(\theta) - y(\theta)\dot{x}(\theta)]\,d\theta$$

and

$$x(\theta)\dot{y}(\theta) - y(\theta)\dot{x}(\theta) = f(\theta)\dot{f}(\theta)\cos(\theta)\sin(\theta) + f^2(\theta)\cos^2(\theta)$$
$$-f(\theta)\dot{f}(\theta)\sin(\theta)\cos(\theta) + f^2(\theta)\sin^2(\theta) = f^2(\theta),$$

this yields the well-known result that

$$\mathcal{A} = \frac{1}{2}\int_0^{2\pi} f^2(\theta)\,d\theta.$$

7.3 Formulation of the Problem

We now present a development of the Keller and Segel model [59] for this phenomenon discussed in Section 7.1. Let $(x, y) \equiv$ two-dimensional spatial coordinate system, $t \equiv$ time be the independent variables and $a(x, y, t) \equiv$ concentration of amoebae per unit area, $\rho(x, y, t) \equiv$ concentration of acrasin per unit area, $\eta(x, y, t) \equiv$ concentration of acrasinase per unit area, $c(x, y, t) \equiv$ concentration of the intermediate complex per unit area formed by these latter two chemicals in a reversible reaction (see (3) below), be the dependent variables. Five basic processes are to be described in the governing equations for that model:

(1) Acrasin is produced by the amoebae at the rate of $f(\rho)$ per amoeba.

(2) Acrasinase is produced by the amoebae at the net rate of $g(\rho, \eta)$ per amoeba.

(3) Acrasine and acrasinase react to form an intermediate complex that dissociates into the free enzyme plus a degraded product of concentration P per unit area. This can be represented by the skeletal chemical equation [111]

$$\rho + \eta \underset{k_{-1}}{\overset{k_1}{\rightleftharpoons}} c \overset{k_2}{\longrightarrow} \eta + P,$$

where k_1, k_{-1}, and k_2 are reaction rates assumed to be constant.

(4) Acrasin, acrasinase, and the complex diffuse according to Fick's first law with diffusion coefficients assumed to be constant.

(5) Amoebae concentration changes as a result of an oriented chemotactic motion in the direction of a positive acrasin gradient and a random motion analogous to that of diffusive dispersal.

The Keller and Segel [59] model for cellular slime mold aggregation did not take either the amoebas' reproduction or its interaction with bacteria into account because it was intended for that scenario after they had already undergone asexual fission and exhausted their food supply. This also included the possibility of bacteria being absent from the outset since then germinating individual cells have been observed to proceed directly to aggregation. The acquisition of stickiness was also ignored which is permissible in the early stages of aggregation under investigation when clumping can be considered negligible. Finally any convective or advective effects were also neglected.

The next step in this development is to incorporate processes ① – ⑤ into governing equations for a, ρ, η, and c. To do so, consider an *arbitrary* fixed region A in the x-y plane (with bounding curve C having outward-pointing unit normal $\boldsymbol{N}$ as depicted in Fig. 7.2b) on which the amoebae are located. We shall proceed exactly as with the three-dimensional situation treated in Section 5.1 but restrict our attention to two-dimensions instead. Conservation of amoebae mass requires

$$\frac{d}{dt}\iint_A a(x,y,t)\,dx\,dy = \iint_A Q^{(a)}(x,y,t)\,dx\,dy - \oint_C \boldsymbol{J}^{(a)}(x,y,t)\bullet\boldsymbol{N}\,ds,$$

where $Q^{(a)}(x,y,t) \equiv$ net amoebae mass produced per unit area per unit time in A and $\boldsymbol{J}^{(a)}(x,y,t) \equiv$ flux vector of amoebae mass (with dimension mass per unit length per unit time) across C out of A for $C: \boldsymbol{r}(s) = x(s)\boldsymbol{i} + y(s)\boldsymbol{j}$, $0 \le s \le L \equiv$ perimeter of C, a closed non-self-intersecting curve satisfying $\boldsymbol{r}(0) = \boldsymbol{r}(L)$. Given that A is fixed and on applying the two-dimensional version of the divergence theorem to its line-integral term, this balance law becomes

$$\iint_A \left[\frac{\partial a}{\partial t} - Q^{(a)} + \nabla_2\bullet\boldsymbol{J}^{(a)}\right](x,y,t)\,dx\,dy = 0.$$

Now by virtue of the DuBois-Reymond Lemma since A is also arbitrary and assuming the requisite continuity that equation can only be valid provided its integrand vanishes identically in the plane which yields the governing equation for a:

$$\frac{\partial a}{\partial t} = Q^{(a)} - \nabla_2\bullet\boldsymbol{J}^{(a)}. \tag{7.3.1a}$$

Governing equations for the other dependent variables can be deduced in an analogous manner:

$$\frac{\partial \rho}{\partial t} = Q^{(\rho)} - \nabla_2\bullet\boldsymbol{J}^{(\rho)}, \tag{7.3.1b}$$

$$\frac{\partial \eta}{\partial t} = Q^{(\eta)} - \nabla_2\bullet\boldsymbol{J}^{(\eta)}, \tag{7.3.1c}$$

$$\frac{\partial c}{\partial t} = Q^{(c)} - \nabla_2\bullet\boldsymbol{J}^{(c)}. \tag{7.3.1d}$$

To complete the formulation of this problem it is necessary to determine the source and flux terms in system (7.3.1). In general the source terms may be considered of the form $Q = \lambda - \mu$ where λ and μ correspond to rates of increase and decrease, respectively, for the quantity in question. Keller and Segel [59], as has become standard operating procedure for cellular slime mold aggregation models, took

$$Q^{(a)} \equiv 0. \tag{7.3.2a}$$

Since $\lambda^{(a)} \equiv 0$, given that there is no reproduction in the time frame of interest for their model from when the amoebae are distributed uniformly until they begin to aggregate, this implicitly assumed $\mu^{(a)} \equiv 0$ as well. Thus no amoebae are either being created or destroyed, the latter assumption neglecting mortality effects due to nematode predation, poisoning by bacterial produced toxins, cannibalism, or simply of natural causes related to aging, all of which can occur in their normal soil environment.

The deduction of the other source terms for the chemical quantities requires an understanding of the law of mass action. Consider the generic chemical equation

$$\alpha R_1 + \beta R_2 \underset{r_{-1}}{\overset{r_1}{\rightleftharpoons}} \gamma P_1 + \delta P_2,$$

where $R_{1,2}$ and $P_{1,2}$ represent concentrations of reactants and products, respectively; r_1 and r_{-1}, reaction rates; and α, β, γ, and δ, integer balance coefficients. The law of mass action states that the rate at which P_1 or P_2 is produced (or equivalently R_1 or R_2 is degraded) equals $r_1 R_1^{\alpha} R_2^{\beta}$ while the rate at which R_1 or R_2 is produced (or equivalently P_1 or P_2 is degraded) equals $r_{-1} P_1^{\gamma} P_2^{\delta}$. For the occurrence of chemical equilibrium these two rates must be the same.

Thus the source terms for the chemical quantities have contributions from the equation in ③ described by this law of mass action while those for ρ and η also have contributions reflecting their production by the amoebae in accordance with ① and ②, respectively. Note in this context that the integer balance coefficients relevant to these chemical reactions are either 1 or 0. Hence

$$Q^{(\rho)} = k_{-1}c - k_1\rho\eta + af(\rho), \tag{7.3.2b}$$

$$Q^{(\eta)} = (k_{-1} + k_2)c - k_1\rho\eta + ag(\rho, \eta), \tag{7.3.2c}$$

and

$$Q^{(c)} = k_1\rho\eta - (k_{-1} + k_2)c, \tag{7.3.2d}$$

where the positive and negative terms tend to correspond to $\lambda^{(\rho,\eta,c)}$ and $\mu^{(\rho,\eta,c)}$, respectively, although the possibility exists for $g(\rho, \eta) \leq 0$ due to inactivation caused by amoebae-production of an η-inhibitor during aggregation [102].

It only remains for us to determine the flux terms for the chemical quantities and the amoebae from processes ④ and ⑤, which imply that

$$\boldsymbol{J}^{(\rho)} = -D_\rho \nabla_2 \rho, \tag{7.3.3a}$$

$$\boldsymbol{J}^{(\eta)} = -D_\eta \nabla_2 \eta, \tag{7.3.3b}$$

$$\boldsymbol{J}^{(c)} = -D_c \nabla_2 c, \tag{7.3.3c}$$

where $D_{\rho,\eta,c}$ are positive diffusion constants; and

$$\boldsymbol{J}^{(a)} = \chi(a,\rho)\nabla_2\rho - \nu(a,\rho)\nabla_2 a, \tag{7.3.3d}$$

where the chemotaxis and motility coefficients, χ and ν, are assumed positive, and represent cross- and self-diffusion effects, respectively. Although only their general form will be employed in what follows, we note, for the sake of definiteness, that Kareiva and Odell [57], when formulating a mathematically similar model relevant to the predator–prey interaction of lady bird beetles and aphids, took

$$\chi(a,\rho) = \chi_0 \frac{a}{\rho} \text{ and } \nu(a,\rho) \equiv \nu_0. \tag{7.3.4}$$

Finally, substituting the source and flux terms of (7.3.2) and (7.3.3) into (7.3.1), yields the following system of governing reaction–diffusion partial differential equations for this problem:

$$\frac{\partial a}{\partial t} = \nabla_2 \bullet [\nu(a,\rho)\nabla_2 a] - \nabla_2 \bullet [\chi(a,\rho)\nabla_2 \rho], \tag{7.3.5a}$$

$$\frac{\partial \rho}{\partial t} = k_{-1}c - k_1\rho\eta + af(\rho) + D_\rho \nabla_2^2 \rho \text{ where } \nabla_2^2 \equiv \nabla_2 \bullet \nabla_2 = \frac{\partial^2}{\partial x^2} + \frac{\partial^2}{\partial y^2}, \tag{7.3.5b}$$

$$\frac{\partial \eta}{\partial t} = (k_{-1} + k_2)c - k_1\rho\eta + ag(\rho,\eta) + D_\eta \nabla_2^2 \eta, \tag{7.3.5c}$$

$$\frac{\partial c}{\partial t} = k_1\rho\eta - (k_{-1} + k_2)c + D_c \nabla_2^2 c. \tag{7.3.5d}$$

Here as a simplification it has been assumed that these equations hold in the entire x-y plane for $0 \le x^2 + y^2 < \infty$. Implicit in such an assumption is that the dependent variables satisfy far-field boundary conditions requiring

$$\alpha, \rho, \eta, \text{ and } c \text{ to remain bounded as } x^2 + y^2 \to \infty. \tag{7.3.5e}$$

Finally, given initial conditions at time $t = 0$ for those dependent variables, this system could in principle be solved to obtain a, ρ, η, and c for all later times. Similar to our analysis of predator–prey dynamical systems involving ordinary differential equations in Chapter 3 we shall see that in practice stability methods may often be employed on such systems of PDE's to yield much useful information without having to resort to solving the system exactly. Further as mentioned in Chapter 1 it is sometimes necessary to simplify a model in order to make its analysis more tractable while simultaneously preserving enough salient features so that the resulting predictions obtained from this simplified model are still worthwhile. In this context it is often convenient to reduce a more complex system to its most significant subsystem involving just two dependent variables. A procedure of that sort is demonstrated in the next section.

7.4 Simplified Model of Aggregation

Devoting their main attention to the aggregation process, Keller and Segel's [59] basic point of view was to regard the initiation of aggregation as a manifestation of instability in a uniform distribution of amoebae a and acrasin ρ. Toward that end they employed the simplest reasonable model which allowed them to reduce their four-equation system (7.3.5) to a two-equation subsystem involving those dependent variables. Keller and Segel [59] accomplished this simplification by making two additional assumptions: Namely that the complex c is in chemical equilibrium and the total concentration of the acrasinase enzyme in both its free (η) and bound (c) states is a constant η_0. These assumptions imply that

$$Q^{(c)} = k_1\rho\eta - (k_{-1}+k_2)c \equiv 0, \tag{7.4.1a}$$

$$\eta + c \equiv \eta_0, \tag{7.4.1b}$$

respectively. Those conditions are equivalent to the ones employed by Briggs and Haldane [13] who studied the kinetics of an enzyme reaction satisfying the equation of (3) in the absence of any other sources for the production of the chemicals involved. Specifically they assumed these chemicals to be homogeneously distributed in space and hence considered them to be functions of time alone. In that context η_0 was the initial enzyme concentration while Briggs and Haldane [13] took the corresponding initial complex concentration $c_0 = 0$. Consequently (7.4.1b) represented an enzyme conservation law independent of whether or not the complex was in chemical equilibrium and (7.4.1a) held. To see this most easily we need only add (7.3.5c) and (7.3.5d) obtaining

$$\frac{\partial(\eta+c)}{\partial t} = ag(\rho,\eta) + D_\eta\nabla_2^2\eta + D_c\nabla_2^2 c. \tag{7.4.2}$$

Then, in the absence of nonchemical production or any diffusive effects for their situation, the result follows immediately given the above initial conditions. Further in that situation (7.4.1a) represented a quasi-steady-state assumption for (7.3.5d). They finally employed (7.4.1a, 7.4.1b) to deduce a Michaelis–Menten type production rate for P which has been the standard approach for deriving the latter ever since [111]. It can be observed from (7.4.2) that any process critically dependent upon the amoeba production of η and the diffusion of η or c would preclude the enzyme conservation law (7.4.1b). Nevertheless, since for the preaggregation stage, the amoeba production rate of η is small and nonhomogeneities in η or c are negligible, it is permissible to adopt (7.4.1b) as well as (7.4.1a) when modeling the onset of aggregation. Thus, after Briggs and Haldane [13], solving (7.4.1a, 7.4.1b) simultaneously to obtain

$$c = \eta_0 \frac{\rho}{\rho + K_M} \text{ where } K_M = \frac{k_{-1} + k_2}{k_1} \tag{7.4.3}$$

and employing (7.4.1a) in conjunction with (7.4.3), (7.3.5b) becomes

$$\frac{\partial \rho}{\partial t} = af(\rho) - k(\rho) + D_\rho \nabla_2^2 \rho \text{ where } k(\rho) = k_2 \eta_0 \frac{\rho}{\rho + K_M}. \tag{7.4.4}$$

Here $K_M \equiv$ Michaelis–Menten constant [111]. Now (7.3.5a) and (7.4.4) form the desired simplified two-equation subsystem for a and ρ.

7.5 Linear Stability Analysis of its Uniform State

For ease of exposition we now rewrite the governing equations of the simplified subsystem involving $a = a(x, y, t)$ and $\rho = \rho(x, y, t)$ in the form

$$\frac{\partial a}{\partial t} = [\nabla_2 \nu(a, \rho)] \bullet \nabla_2 a - [\nabla_2 \chi(a, \rho)] \bullet \nabla_2 \rho + \nu(a, \rho) \nabla_2^2 a - \chi(a, \rho) \nabla_2^2 \rho, \tag{7.5.1a}$$

$$\frac{\partial \rho}{\partial t} = G(a, \rho) + D_\rho \nabla_2^2 \rho \text{ where } G(a, \rho) = af(\rho) - k(\rho)$$
$$\text{with } k(\rho) = k_2 \eta_0 \frac{\rho}{\rho + K_M} \text{ and } K_M = \frac{k_{-1} + k_2}{k_1}; \tag{7.5.1b}$$

$$a, \rho \text{ remain bounded as } x^2 + y^2 \to \infty. \tag{7.5.1c}$$

Since a uniform state for a and ρ prior to aggregation seems to be consistent with observation, we seek a community equilibrium solution of system (7.5.1) of the form

$$a(x, y, t) \equiv a_e > 0, \ \rho(x, y, t) \equiv \rho_e > 0, \tag{7.5.2a}$$

$$\text{such that } G(a_e, \rho_e) = 0 \Rightarrow a_e f(\rho_e) = k(\rho_e). \tag{7.5.2b}$$

Condition (7.5.2b) means that the production rate of acrasin by the amoebae is balanced exactly by the rate at which the chemical reaction is degrading acrasin.

In what follows we shall examine the linear stability of this equilibrium state by considering solutions of system (7.5.1) of the form

$$a(x, y, t) = a_e[1 + \varepsilon a_1(x, y, t) + O(\varepsilon^2)], \ \rho(x, y, t) = \rho_e[1 + \varepsilon \rho_1(x, y, t) + O(\varepsilon^2)]; \tag{7.5.3}$$

where $|\varepsilon| << 1$. Substituting (7.5.3) into the basic equations of (7.5.1), using Taylor's series for a function of two variables, neglecting terms of $O(\varepsilon^2)$, and cancelling the resulting common ε factor, we obtain the following linear homogeneous system in the perturbation quantities a_1 and ρ_1 where as in Chapter 3 use has been made of

$$G(a,\rho) = G(a_e,\rho_e) + G_1(a_e,\rho_e)(a-a_e) + G_2(a_e,\rho_e)(\rho-\rho_e) + O(\varepsilon^2)$$
$$= G_e + \varepsilon[a_e(G_1)_e a_1 + \rho_e(G_2)_e\rho_1] + O(\varepsilon^2) = \varepsilon[a_e(G_1)_e a_1 + \rho_e(G_2)_e\rho_1] + O(\varepsilon^2)$$

with analogous expansions for $v(a,\rho)$ and $\chi(a,\rho)$:

$$a_e\frac{\partial a_1}{\partial t} = v_e a_e \nabla_2^2 a_1 - \chi_e\rho_e\nabla_2^2\rho_1, \tag{7.5.4a}$$

$$\rho_e\frac{\partial \rho_1}{\partial t} = a_e(G_1)_e a_1 + \rho_e(G_2)_e\rho_1 + D_\rho\rho_e\nabla_2^2\rho_1, \tag{7.5.4b}$$

$$a_1,\rho_1 \text{ remain bounded as } x^2+y^2 \to \infty, \tag{7.5.4c}$$

where $v_e = v(a_e,\rho_e)$, $\chi_e = \chi(a_e,\rho_e)$, $(G_1)_e = f(\rho_e)$, and $(G_2)_e = a_e f'(\rho_e) - k'(\rho_e)$; (7.5.4d)

or

$$\frac{\partial a_1}{\partial t} = v_e\nabla_2^2 a_1 - \chi_e\left(\frac{\rho_e}{a_e}\right)\nabla_2^2\rho_1, \tag{7.5.5a}$$

$$\frac{\partial \rho_1}{\partial t} = \left(\frac{a_e}{\rho_e}\right)f(\rho_e)a_1 + [a_e f'(\rho_e) - k'(\rho_e) + D_\rho\nabla_2^2]\rho_1, \tag{7.5.5b}$$

$$a_1,\rho_1 \text{ remain bounded as } x^2+y^2 \to \infty. \tag{7.5.5c}$$

Note, in this context, that since $\nabla_2(v,\chi,a,\rho) = \boldsymbol{O}(\varepsilon)$ then $\nabla_2 v \bullet \nabla_2 a$, $\nabla_2\chi \bullet \nabla_2\rho = O(\varepsilon^2)$ while

$$k'(\rho_e) = k_2\eta_0\frac{K_M}{(\rho_e+K_M)^2} > 0. \tag{7.5.6}$$

We now assume a normal-mode type solution for these perturbation quantities of the form

$$[a_1,\rho_1](x,y,t) = [a_{11},\rho_{11}]\exp(i\{q_1x+q_2y\}+\sigma t) \tag{7.5.7}$$

where $|a_{11}|^2 + |\rho_{11}|^2 \neq 0$, $q_{1,2} \in \mathbb{R}$ to satisfy (7.5.5c), and $\sigma \in \mathbb{C}$. Although this solution may seem to be somewhat special at first, we shall show in the next two sections that (7.5.7) actually represents an arbitrary component of a Fourier decomposable disturbance and hence is quite general. Since all such components must remain stable for stability to occur and only one need become unstable to guarantee instability, by examining the long-time behavior of such a mode we can determine the stability of the uniform state to the fairly wide class of Fourier decomposable disturbances. Substituting (7.5.7) into (7.5.5a, 7.5.5b), noting that

$$\nabla_2^2[a_1,\rho_1] = -q^2[a_1,\rho_1] \text{ where } q^2 = q_1^2+q_2^2 \geq 0, \tag{7.5.8}$$

and cancelling the exponential common factor, we obtain the following linear homogeneous equations for a_{11} and ρ_{11}:

$$(\sigma + v_e q^2)a_{11} - \chi_e\left(\frac{\rho_e}{a_e}\right)q^2\rho_{11} = 0, \tag{7.5.9a}$$

$$-\left(\frac{a_e}{\rho_e}\right)f(\rho_e)a_{11} + (\sigma - F)\rho_{11} = 0 \text{ where } F = a_e f'(\rho_e) - k'(\rho_e) - D_\rho q^2. \quad (7.5.9b)$$

A system of this sort admits a nontrivial solution only if its determinant vanishes, yielding the quadratic satisfied by σ

$$\begin{vmatrix} \sigma + \nu_e q^2 & -\chi_e\left(\frac{\rho_e}{a_e}\right)q^2 \\ -\left(\frac{a_e}{\rho_e}\right)f(\rho_e) & \sigma - F \end{vmatrix} = \sigma^2 + (\nu_e q^2 - F)\sigma - [\chi_e f(\rho_e) + \nu_e F]q^2 = 0. \quad (7.5.10a)$$

Calculating the discriminant of this quadratic

$$\mathcal{D} = (\nu_e q^2 - F)^2 + 4[\chi_e f(\rho_e) + \nu_e F]q^2 = (\nu_e q^2 + F)^2 + 4\chi_e f(\rho)q^2 = 4\omega^2 \geq 0, \quad (7.5.10b)$$

we can conclude that $\sigma \in \mathbb{R}$.

Recalling from Chapter 3 that a quadratic of this sort in σ has roots such that $\text{Re}(\sigma) = \sigma < 0$ if and only if both its coefficients are positive, we can deduce the following stability criteria

$$F < \nu_e q^2 \text{ and } F < -\chi_e \frac{f(\rho_e)}{\nu_e} \quad (7.5.11)$$

which reduces to the single stability criterion

$$F = a_e f'(\rho_e) - k'(\rho_e) - D_\rho q^2 < -\chi_e \frac{f(\rho_e)}{\nu_e}. \quad (7.5.12)$$

Hence we have instability with $\sigma > 0$ provided this inequality is reversed or

$$\chi_e \frac{f(\rho_e)}{\nu_e} + a_e f'(\rho_e) > k'(\rho_e) + D_\rho q^2. \quad (7.5.13)$$

We next introduce a threshold condition on the critical level of the relative acrasin gradient necessary to trigger this aggregative instability. That is, aggregation will only occur provided

$$\left|\frac{\nabla_2 \rho}{\rho}\right| > \alpha > 0. \quad (7.5.14)$$

Now, given (7.5.3) and (7.5.7),

$$|\rho| \cong \rho_e \text{ and } \nabla_2\rho \cong \varepsilon\rho_e\nabla_2\rho_1 = \varepsilon\rho_e i(q_1, q_2)\rho_1 \Rightarrow |\nabla_2\rho| \cong \rho_e|\varepsilon q\rho_{11}|e^{\sigma t}. \quad (7.5.15)$$

Therefore let us define α by requiring

$$\left|\frac{\nabla_2\rho}{\rho}\right| = \frac{|\nabla_2\rho|}{|\rho|} \cong |\varepsilon q\rho_{11}|e^{\sigma t} > |\varepsilon q\rho_{11}| > \alpha \quad (7.5.16a)$$

which implies that for aggregation to occur q^2 and its associated wavelength λ must satisfy

$$q^2 > \frac{\alpha^2}{|\varepsilon\rho_{11}|^2} \text{ or } \lambda = \frac{2\pi}{|q|} < \frac{2\pi|\varepsilon\rho_{11}|}{\alpha}. \tag{7.5.16b}$$

We note in our formulation it has been implicitly assumed that the average distance between individual amoebae is relatively small when compared with the length scale characteristic of the territory over which they are distributed so that a continuum mechanical model may be introduced. Further if λ should also be relatively small when compared with this scale it is permissible to assume infinite planar extent since then any prospective boundary conditions for a finite region no longer have much influence on spatial patterns [43].

Thus combining the inequalities of (7.5.13) and (7.5.16b) we deduce the linear instability criterion

$$\chi_e \frac{f(\rho_e)}{v_e} + a_e f'(\rho_e) > k'(\rho_e) + D_\rho \frac{\alpha^2}{|\varepsilon\rho_{11}|^2}. \tag{7.5.17}$$

We shall provide a mechanistic interpretation of that criterion with respect to cellular slime mold aggregation in the last section of this chapter after the concept of Fourier integrals has been developed and then employed to satisfy arbitrary initial conditions for the perturbation quantities, sequentially, in the next two sections. The main purpose of this interlude is to demonstrate the general utility of a normal-mode analysis of the perturbation system as mentioned earlier.

7.6 Pastoral Interlude: Fourier Integrals and Laplace Transforms

Consider a real-valued function $f(x)$ of a real variable such that f and f' are piecewise continuous on any finite interval and

$$\int_{-\infty}^{\infty} |f(x)|^2 \, dx \text{ converges.}$$

Then the Fourier integral theorem states that

$$\frac{f(x+0)+f(x-0)}{2} \doteq f(x) = \int_{-\infty}^{\infty} C(q)e^{iqx} \, dq \text{ where } C(q) = \frac{1}{2\pi}\int_{-\infty}^{\infty} f(x)e^{-iqx} \, dx.$$

We shall demonstrate the plausibility of this result directly as follows. Let

$$f(x) = \int_{-\infty}^{\infty} C(k)e^{ikx} \, dk$$

and assume that

$$\int_{-\ell}^{\ell} f(x)e^{-iqx}\,dx = \int_{-\infty}^{\infty} C(k)\left[\int_{-\ell}^{\ell} e^{i(k-q)x}\,dx\right] dk.$$

This is a demonstration rather than a proof partially because we have interchanged the order of integration in the above equation without formal justification. Noting that

$$\int_{-\ell}^{\ell} e^{i(k-q)x}\,dx = \begin{cases} 2\ell & \text{for } k = q \\ \left.\frac{e^{i(k-q)x}}{i(k-q)}\right|_{x=-\ell}^{\ell} & \text{for } k \neq q \end{cases} = \frac{2\sin([k-q]\ell)}{k-q}$$

since

$$\left.\frac{e^{i(k-q)x}}{i(k-q)}\right|_{x=-\ell}^{\ell} = \frac{e^{i(k-q)\ell} - e^{-i(k-q)\ell}}{i(k-q)} = \frac{2\sin([k-q]\ell)}{k-q} \quad \text{for } k \neq q$$

while

$$\lim_{k \to q} \frac{2\sin([k-q]\ell)}{k-q} = 2\ell,$$

and substituting this result into that equation, we obtain

$$\int_{-\ell}^{\ell} f(x)e^{-iqx}\,dx = 2\int_{-\infty}^{\infty} C(k)\frac{2\sin([k-q]\ell)}{k-q}\,dk.$$

Now, making the change of variables $t = (k-q)\ell$ or $k = q + t/\ell$ with $dk = dt/\ell$, yields

$$\int_{-\ell}^{\ell} f(x)e^{-iqx}\,dx = 2\int_{-\infty}^{\infty} C\left(q + \frac{t}{\ell}\right)\frac{\sin(t)}{t}\,dt.$$

Finally assuming that

$$\lim_{\ell \to \infty}\int_{-\ell}^{\ell} f(x)e^{-iqx}\,dx = \int_{-\infty}^{\infty} f(x)e^{-iqx}\,dx = 2C(q)\int_{-\infty}^{\infty}\frac{\sin(t)}{t}\,dt = 2\pi C(q),$$

we can deduce the desired Fourier integral transform relation

$$C(q) = \frac{1}{2\pi}\int_{-\infty}^{\infty} f(x)e^{-iqx}\,dx$$

since, as will be derived below,

$$\int_{-\infty}^{\infty}\frac{\sin(t)}{t}\,dz = \pi.$$

The second reason this is a demonstration rather than a proof is that we have taken the limit of an improper integral by taking the limit of its integrand without formal justification.

It only remains for us to evaluate this last improper integral which is implicitly meant in the sense of a Cauchy Principal Value. Hence

$$\int_{-\infty}^{\infty} \frac{\sin(t)}{t} dt = 2\int_{0}^{\infty} \frac{\sin(t)}{t} dt.$$

Thus we shall evaluate

$$\int_{0}^{\infty} \frac{\sin(t)}{t} dt$$

in order to derive that result. Note although the integrand is not defined at its lower limit this is a removable discontinuity since the limit exists. We shall use a Laplace transform technique in order to perform that evaluation.

Define the Laplace transform of a real function $f(t)$ where $s \in \mathbb{C}$ by

$$\mathcal{L}\{f(t)\} = \int_{0}^{\infty} e^{-st} f(t)\, dt = F(s) \text{ for } \mathrm{Re}(s) > \alpha.$$

The easiest way to understand Laplace transforms is to compute a few examples. First let

$$f(t) = e^{at} \text{ for } a \in \mathbb{C}.$$

Then

$$\mathcal{L}\{e^{at}\} = \int_{0}^{\infty} e^{-(s-a)t}\, dt = -\left.\frac{e^{-(s-a)t}}{s-a}\right|_{t=0}^{\infty} = \frac{1}{s-a} \text{ for } \mathrm{Re}(s) > \mathrm{Re}(a) = \alpha$$

since

$$\lim_{t\to\infty} e^{-(s-a)t} = 0 \text{ for } \mathrm{Re}(s) > \mathrm{Re}(a).$$

Now let

$$f(t) = \sin(\omega t) = \frac{e^{i\omega t} - e^{-i\omega t}}{2i}.$$

Then since $\mathcal{L}$ is a linear operator

$$\mathcal{L}\{\sin(\omega t)\} = \frac{\mathcal{L}\{e^{i\omega t}\} - \mathcal{L}\{e^{-i\omega t}\}}{2i} = \frac{1}{2i}\left(\frac{1}{s-i\omega} - \frac{1}{s+i\omega}\right) = \frac{\omega}{s^2+\omega^2}.$$

Here we are restricting our Laplace transform exposition precisely to those concepts needed to evaluate the integral in question. This topic will be presented more completely in Chapter 20.

We next develop the useful identity involving $f(t)$ and its Laplace transform $F(s)$:

$$\int_{0}^{\infty} F(s)\, ds = \int_{0}^{\infty}\int_{0}^{\infty} e^{-st} f(t)\, dt\, ds = \int_{0}^{\infty} f(t) \int_{0}^{\infty} e^{-st}\, ds\, dt$$
$$= \int_{0}^{\infty} f(t)\left[\left.\frac{e^{-st}}{t}\right|_{s\to\infty}^{s=0}\right] dt = \int_{0}^{\infty} \frac{f(t)}{t} dt$$

where improper integration has again been interchanged with no formal justification. Thus, employing this identity with $f(t) = \sin(t)$ for which $F(s) = 1/(s^2+1)$, we obtain

$$\int_0^\infty \frac{f(t)}{t}\,dt = \int_0^\infty \frac{\sin(t)}{t}\,dt = \int_0^\infty F(s)\,ds = \int_0^\infty \frac{ds}{s^2+1}$$
$$= \lim_{s\to\infty} \arctan(s) - \arctan(0) = \frac{\pi}{2},$$

and hence, finally demonstrate the desired result.

7.7 Satisfaction of Initial Conditions

We are now ready to consider initial conditions for our perturbation system (7.5.4). One would normally assign an initial condition to each of its two dependent variables a_1 and ρ_1. Since they are related, however, it is possible with no loss of generality to assign initial conditions to a_1 and $\partial a_1/\partial t$, instead. Toward that end we impose the following initial conditions for this system:

$$a_1(x,y,0) = h_1(x,y), \quad \frac{\partial a_1(x,y,0)}{\partial t} = h_2(x,y).$$

Analogous to our approach with infinite series in Chapter 5, we next assume a superposition principle for integrals related to our normal-mode solution (7.5.7) and represent

$$a_1(x,y,t) = \int_{-\infty}^{\infty}\left[\int_{-\infty}^{\infty}(C_1 e^{\sigma^+ t} + C_2 e^{\sigma^- t})e^{iq_2 y}\,dq_2\right]e^{iq_1 x}\,dq_1$$

where $\sigma^\pm$ are the roots of (7.5.10) given by $\sigma^\pm = -b \pm \omega$ for $2b = v_e q^2 - F$. Even though these roots are real rather than complex it is often advantageous to use a similar approach and rewrite

$$C_1 e^{\sigma^+ t} + C_2 e^{\sigma^- t} = e^{-bt}(C_1 e^{\omega t} + C_2 e^{-\omega t}) = e^{-bt}\{c_1 \cosh(\omega t) + c_2 \sinh(\omega t)\}$$

where $c_1 = C_1 + C_2$ and $c_2 = C_1 - C_2$. Hence

$$a_1(x,y,t) = \int_{-\infty}^{\infty}\left[\int_{-\infty}^{\infty} e^{-bt}\{c_1 \cosh(\omega t) + c_2 \sinh(\omega t)\}e^{iq_2 y}\,dq_2\right]e^{iq_1 x}\,dq_1.$$

Thus

$$h_1(x,y) = \int_{-\infty}^{\infty} \left[\int_{-\infty}^{\infty} c_1 e^{iq_2 y} \, dq_2 \right] e^{iq_1 x} \, dq_1,$$
$$h_2(x,y) = \int_{-\infty}^{\infty} \left[\int_{-\infty}^{\infty} (\omega c_2 - bc_1) e^{iq_2 y} \, dq_2 \right] e^{iq_1 x} \, dq_1.$$

Applying our Fourier transform result from the previous section to these two double integrals, we find the bracketed terms in each satisfy

$$\int_{\infty}^{\infty} [c_1, \omega c_2 - bc_1] e^{iq_2 y} \, dq_2 = \frac{1}{2\pi} \int_{-\infty}^{\infty} [h_1(x,y), h_2(x,y)] e^{-iq_1 x} \, dx.$$

Finally, applying it again to the resulting integrals, we obtain

$$[c_1, \omega c_2 - bc_1] = \frac{1}{4\pi^2} \int_{-\infty}^{\infty} \int_{-\infty}^{\infty} [h_1(x,y), h_2(x,y)] e^{-i(q_1 x + q_2 y)} \, dx \, dy,$$

completing the process and allowing us to synthesize arbitrary initial conditions.

7.8 Mechanistic Interpretation of the Linear Instability Aggregative Criterion

Returning to the linear instability criterion (7.5.17), reproduced below for convenience

$$\chi_e \frac{f(\rho_e)}{v_e} + a_{ef}'(\rho_e) > k'(\rho_e) + D_\rho \frac{\alpha^2}{|\varepsilon \rho_{11}|^2}, \tag{7.8.1}$$

we can observe that any factor which tends to increase its left-hand side or decrease its right-hand one is a condition favoring the onset of instability or, in this case, the initiation of cellular slime mold aggregation. Thus, upon examining the form of the second term on the right-hand side of (7.8.1), we note that, since $|\varepsilon| << 1$, aggregation is unlikely to occur unless its relative gradient critical threshold value of α is correspondingly very low as well. This is consistent with the suggestion that, given the long-range nature of the chemotactic attraction of amoebae for acrasin, they are able to respond to quite small gradients in the acrasin level [59]. Hence, we now make the further assumption that

$$\alpha = \alpha_0 |\varepsilon| \text{ where } \alpha_0 = O(1) > 0 \text{ as } \varepsilon \to 0. \tag{7.8.2}$$

Upon substitution of (7.8.2) into (7.8.1), we obtain the final linear instability criterion

$$\chi_e \frac{f(\rho_e)}{v_e} + a_{ef}'(\rho_e) > k'(\rho_e) + D_\rho \frac{\alpha_0^2}{|\rho_{11}|^2}, \tag{7.8.3}$$

which is independent of the perturbation expansion parameter ε. Let us investigate its terms relative to the factors mentioned above favoring aggregation in conjunction with a mechanism for such aggregative motion. Consider a uniform distribution of amoebae and acrasin with the amoebae at rest. Introduce a perturbation of acrasin at some point in the plane. Consistent with the factors favoring aggregation assume that $|\rho_{11}|$, a measure of the size of this disturbance, is relatively large while D_ρ, α_0, and $k'(\rho_e)$ are small. Since $|\rho_{11}|$ is large an imbalance of acrasin will be generated locally. Since D_ρ is small it will not have much tendency to diffuse. Further, since $k'(\rho_e)$ is small, the rate of degradation will not be increased much by this addition in acrasin level. All these factors will tend to preserve that acrasin imbalance. Finally since α_0 is small, the threshold triggering amoebae motion will be exceeded and they will tend to move toward the region where the local imbalance of acrasin is located. Again consistent with those factors favoring aggregation assume that $\chi_e, f(\rho_e)$, and $a_e f'(\rho_e)$ are relatively large while v_e is small. Since χ_e is large the amoebae are strongly attracted to an acrasin imbalance and rapidly move toward such a region. Once there they have little tendency to disperse since v_e is small. Further since $f(\rho_e)$ is large these amoebae cause an even greater level of acrasin imbalance. Since $a_e f'(\rho_e)$ is large as well this increase will cause an even greater production of acrasin by the amoebae which will in turn attract other amoebae and that cycle will continue, ultimately resulting in the onset of an aggregative instability for the whole system.

Problem

7.1. The purpose of this problem is to perform a linear stability analysis of the equilibrium solution for a slightly more general model than the simplified version of the one introduced by Keller and Segel [59] to describe the initiation of cellular slime mold aggregation. Toward that end consider the governing system of equations (7.3.5) under the *single* simplifying assumption that the complex is in chemical equilibrium. In other words, unlike the simplified version of this model treated by Keller and Segel [59], it is not assumed that an enzyme conservation law for the acrasinase holds as well. All the nomenclature to be employed here is the same as that used in this chapter.

(a) Assume the complex to be in chemical equilibrium and hence let

$$k_1 \rho \eta = (k_{-1} + k_2)c.$$

Show that under this simplifying assumption the governing system of equations (7.3.5) becomes

$$\frac{\partial a}{\partial t} = \nabla_2 \bullet [v(a,\rho)\nabla_2 a] - \nabla_2 \bullet [\chi(a,\rho)\nabla_2 \rho],$$

$$\frac{\partial \rho}{\partial t} = af(\rho) - k_2 c + D_\rho \nabla_2^2 \rho \text{ where } \nabla_2^2 \equiv \nabla_2 \bullet \nabla_2 \frac{\partial^2}{\partial x^2} + \frac{\partial^2}{\partial y^2},$$

$$\frac{\partial \eta}{\partial t} = ag(\rho, \eta) + D_\eta \nabla_2^2 \eta,$$

$$\frac{\partial c}{\partial t} = D_c \nabla_2^2 c;$$

a, ρ, η and c to remain bounded as $x^2 + y^2 \to \infty$.

(b) Find a community equilibrium solution of this modified system of the form

$$a(x,y,t) \equiv a_e > 0,\ \rho(x,y,t) \equiv \rho_e > 0,\ \eta(x,y,t) \equiv \eta_e > 0,\ c(x,y,t) \equiv c_e > 0,$$

by determining the relationships between these constants and then explain the physical significance of those results.

(c) Perform a linear perturbation analysis of this equilibrium state by seeking a solution of the basic governing equations of the form

$$a(x,y,t) = a_e[1 + \varepsilon a_1(x,y,t) + O(\varepsilon^2)],\ \rho(x,y,t) = \rho_e[1 + \varepsilon \rho_1(x,y,t) + O(\varepsilon^2)];$$

$$\eta(x,y,t) = \eta_e[1 + \varepsilon \eta_1(x,y,t) + O(\varepsilon^2)],\ c(x,y,t) = c_e[1 + \varepsilon c_1(x,y,t) + O(\varepsilon^2)];$$

where $|\varepsilon| << 1$. Substituting that solution into the modified system, neglecting all terms of $O(\varepsilon^2)$, cancelling the resulting common ε factor, and dividing each equation by its relevant equilibrium-state constant, derive a linear homogeneous partial differential system for the perturbation quantities.

(d) Assume a normal-mode type solution for these perturbation quantities of the form

$$[a_1, \rho_1, \eta_1, c_1](x,y,t) = [a_{11}, \rho_{11}, \eta_{11}, c_{11}] \exp(i\{q_1 x + q_2 y\} + \sigma t)$$

where $|a_{11}|^2 + |\rho_{11}|^2 + |\eta_{11}|^2 + |c_{11}|^2 \neq 0, q_{1,2} \in \mathbb{R}$ to satisfy the far-field conditions with $q_1^2 + q_2^2 = q^2 \geq 0$, and $\sigma \in \mathbb{C}$. Upon substituting this normal-mode solution into the perturbation system and cancelling the common exponential factor, obtain a set of four linear homogeneous algebraic equations in the perturbation constants.

(e) Since, to ensure nontrivial solutions for this linear homogeneous system, it is required that the determinant of its coefficients must vanish, deduce a fourth-order polynomial satisfied by σ.

(f) This fourth-order polynomial in σ will be a product of two linear and one quadratic factor. Show that its roots are real and develop linear *stability* criteria by analyzing their sign. Discuss the significance of these criteria with respect to the initiation of cellular slime mold aggregation recalling that this is equivalent to the onset of a linear *instability* for the model system under examination.

Chapter 8
Chemical Turing Patterns and Diffusive Instabilities

The study of diffusive effects in nonmoving continua is completed in this chapter by examining the so-called Turing instabilities, characteristic of certain chemical reactions. The Brusselator activator–inhibitor reaction–diffusion model is derived using the law of mass action for its source terms and Fick's first law for its flux terms and conditions deduced by a normal-mode linear stability analysis for the development of chemical Turing instabilities over a parameter range for which the dynamical system in the absence of diffusion would exhibit a stable homogeneous distribution. The effect the introduction of an immobilizer would have on such diffusive instabilities is also examined. The limitations of linear stability predictions of this sort are discussed, and the results of a nonlinear stability analysis, which will be treated in detail in later chapters, are sketched for the Brusselator. In the problems, similar normal-mode linear stability analyses of the Schnackenberg simplification of the Brusselator and a simplified version of the so-called CDIMA (*C*hlorine *D*ioxide *I*odine *M*alonic *A*cid) chemical reaction–diffusion system are considered.

8.1 Brusselator Reaction–Diffusion Activator–Inhibitor Model System

More than sixty years ago, Turing [121] proposed the chemical basis of morphogenesis in a landmark paper with that title. In particular, he postulated the existence of chemical morphogens which formed the basis of embryo-morphogenesis through the development of prepatterns. Specifically, he investigated the possibility of an instability occurring in purely dissipative systems involving chemical reactions far from equilibrium and the transport process of diffusion but no hydrodynamic motion. When restricted to two chemical species, an activator and an inhibitor, the existence of such instabilities requires an autocatalytic or positive feedback reaction for the activator and a diffusive advantage for the inhibitor as necessary conditions. Then an initially homogeneous state which would be stable in the absence of diffusion can be destabilized resulting in a re-equilibrated inhomogeneous symmetry breaking

© Springer International Publishing AG, part of Springer Nature 2017

D. J. Wollkind and B. J. Dichone, *Comprehensive Applied Mathematical Modeling in the Natural and Engineering Sciences*,
https://doi.org/10.1007/978-3-319-73518-4_8

pattern. That concept along with the one of positional information has made Turing theory a fundamental paradigm for explaining developmental biological processes ranging from embryology to limb formation and coat patterning [85]. A variety of artificial reaction schemes have been proposed that exhibit this sort of chemical instability behavior in conjunction with diffusion, dating from the well-known "Brusselator" due to Prigogine and Lefever [99] and its variants such as the Schnackenberg [110] model. Given the historical importance of that seminal model for describing chemical Turing patterns, we have chosen to concentrate our presentation of diffusive instabilities in this chapter on it. We begin by considering the skeletal Brusselator chemical reaction

$$A \xrightarrow{k_1} X,\ X + B \xrightarrow{k_2} Y + D,\ 2X + Y \xrightarrow{k_3} 3X,\ X \xrightarrow{k_4} E;$$

which is chemically unrealistic because of its third trimolecular step. Here A, B and X, Y represent the concentrations of the reactant pool species, assumed to be constant, and the intermediate species, which are our dynamical variables, assumed to be functions of $\boldsymbol{s} = (s_1, s_2) \equiv$ position and $\tau \equiv$ time, respectively, while $k_{1,2,3,4} \equiv$ reaction-rate constants. Using the same methodology as employed in the last chapter, we can deduce by the law of mass action and Fick's first law that X, Y satisfy the reaction–diffusion system with $D_{1,2} \equiv$ constant diffusivities

$$\frac{\partial X}{\partial \tau} = Q^{(X)} - \nabla_2 \bullet \boldsymbol{J}^{(X)},\ \frac{\partial Y}{\partial \tau} = Q^{(Y)} - \nabla_2 \bullet \boldsymbol{J}^{(Y)},\ \text{where } \nabla_2 = \left(\frac{\partial}{\partial s_1}, \frac{\partial}{\partial s_2}\right)$$

and

$$Q^{(X)} = k_1 A - k_2 BX + k_3 X^2 Y - k_4 X,\ Q^{(Y)} = k_2 BX - k_3 X^2 Y;\ \boldsymbol{J}^{(X,Y)} = -D_{1,2}\nabla_2(X, Y);$$

or

$$\frac{\partial X}{\partial \tau} = k_1 A - k_2 BX + k_3 X^2 Y - k_4 X + D_1 \nabla_2^2 X, \tag{8.1.1a}$$

$$\frac{\partial Y}{\partial \tau} = k_2 BX - k_3 X^2 Y + D_2 \nabla_2^2 Y; \tag{8.1.1b}$$

where $\nabla_2^2 \equiv \nabla_2 \bullet \nabla_2 = \partial^2/\partial s_1^2 + \partial^2/\partial s_2^2$ and

$$X, Y \text{ remain bounded as } s_1^2 + s_2^2 \to \infty. \tag{8.1.1c}$$

8.2 Community Equilibrium Point and its Linear Stability Analysis

There exists a community equilibrium solution to (8.1.1)

$$X(\boldsymbol{s}, \tau) \equiv X_e > 0,\ Y(\boldsymbol{s}, \tau) \equiv Y_e > 0, \tag{8.2.1a}$$

such that

$$Q^{(X)}(X_e, Y_e) = Q^{(Y)}(X_e, Y_e) = 0 \tag{8.2.1b}$$

where

$$X_e = \frac{k_1 A}{k_4}, Y_e = \frac{k_2 B}{k_3 X_e} = \frac{k_2 k_4 B}{k_1 k_3 A}. \tag{8.2.1c}$$

We now introduce the nondimensional variables and parameters

$$\boldsymbol{r} = \frac{\boldsymbol{s}}{(D_2/k_4)^{1/2}}, t = k_4 \tau, x(\boldsymbol{r}, t) = \frac{X(\boldsymbol{s}, \tau)}{X_e}, y(\boldsymbol{r}, t) = \frac{Y(\boldsymbol{s}, \tau)}{Y_e}, \tag{8.2.2a}$$

$$\alpha = \frac{k_2 B}{k_4}, \beta = \frac{k_3 X_e^2}{k_4} = \frac{k_1^2 k_3 A^2}{k_4^3}, \mu = \frac{D_1}{D_2}, \tag{8.2.2b}$$

which transforms (8.1.1) into

$$\frac{\partial x}{\partial t} = F(x, y; \alpha) + \mu \nabla^2 x, \tag{8.2.3a}$$

$$\frac{\partial y}{\partial t} = \beta G(x, y) + \nabla^2 y, \tag{8.2.3b}$$

where

$$F(x, y; \alpha) = 1 - (\alpha + 1)x + \alpha x^2 y, G(x, y) = x - x^2 y, \nabla^2 = \frac{\partial^2}{\partial r_1^2} + \frac{\partial^2}{\partial r_2^2}; \tag{8.2.3c}$$

$$x, y \text{ remain bounded as } r_1^2 + r_2^2 \to \infty. \tag{8.2.3d}$$

Here x and y are the activator and inhibitor, respectively, of our reaction–diffusion system. Note that, given (8.2.2a), the community equilibrium solution (8.2.1) corresponds to $x_e = y_e \equiv 1$ in our nondimensional variables which is consistent with the fact that $F(1, 1; \alpha) = G(1, 1) = 0$. It is the linear stability of this equilibrium state with which we are concerned in what follows. Hence we seek a perturbation solution of (8.2.3) of the form

$$x(\boldsymbol{r}, t) = 1 + \varepsilon_1 x_1(\boldsymbol{r}, t) + O(\varepsilon_1^2), y(\boldsymbol{r}, t) = 1 + \varepsilon_1 y_1(\boldsymbol{r}, t) + O(\varepsilon_1^2), |\varepsilon_1| << 1. \tag{8.2.4}$$

Substituting (8.2.4) into (8.2.3), neglecting terms of $O(\varepsilon_1^2)$, and cancelling the resulting ε_1 common factor, we obtain in the usual way the following linear homogeneous perturbation system

$$\frac{\partial x_1}{\partial t} = F_1(1, 1; \alpha) x_1 + F_2(1, 1; \alpha) y_1 + \mu \nabla^2 x_1, \tag{8.2.5a}$$

$$\frac{\partial y_1}{\partial t} = \beta [G_1(1, 1) x_1 + G_2(1, 1) y_1] + \nabla^2 y_1, \tag{8.2.5b}$$

$$x_1, y_1 \text{ remain bounded as } r_1^2 + r_2^2 \to \infty, \tag{8.2.5c}$$

where

$$F_1(1,1;\alpha) = \alpha - 1, F_2(1,1;\alpha) = \alpha, G_1(1,1) = G_2(1,1) = -1; \tag{8.2.5d}$$

or

$$\frac{\partial x_1}{\partial t} = (\alpha - 1)x_1 + \alpha y_1 + \mu \nabla^2 x_1, \tag{8.2.6a}$$

$$\frac{\partial y_1}{\partial t} = -\beta x_1 - \beta y_1 + \nabla^2 y_1, \tag{8.2.6b}$$

$$x_1, y_1 \text{ remain bounded as } r_1^2 + r_2^2 \to \infty. \tag{8.2.6c}$$

Recalling from Chapter 7 that a normal-mode solution of the form

$$\exp(i\{q_1 r_1 + q_2 r_2\} + \sigma t)$$

when assumed for a linear homogeneous perturbation system involving ∇^2 alone, only yields results dependent upon $q^2 = q_1^2 + q_2^2 \geq 0$ for $q_{1,2} \in \mathbb{R}$, we can without loss of generality let $\nabla^2 = \partial^2/\partial r^2$ and assume instead a normal-mode solution of (8.2.6) of the form

$$[x_1, y_1](r,t) = [x_{11}, y_{11}]e^{\sigma t}\cos(qr) \tag{8.2.7}$$

where $|x_{11}|^2 + |y_{11}|^2 \neq 0$; $q \geq 0$ to satisfy x_1, y_1 remain bounded as $r \to \pm\infty$; and $\sigma \in \mathbb{C}$. Substituting (8.2.7) into (8.2.6a, 8.2.6b), noting that

$$\nabla^2[x_1, y_1] = -q^2[x_1, y_1], \tag{8.2.8}$$

and cancelling the $e^{\sigma t}\cos(qr)$ common factor, we obtain the following linear homogeneous system of equations for x_{11} and y_{11}

$$(\sigma + 1 - \alpha + \mu q^2)x_{11} - \alpha y_{11} = 0, \tag{8.2.9a}$$

$$\beta x_{11} + (\sigma + \beta + q^2)y_{11} = 0. \tag{8.2.9b}$$

Again requiring the vanishing of the determinant of its coefficients to guarantee a nontrivial solution yields the following quadratic secular equation that σ satisfies

$$\begin{vmatrix} \sigma + 1 - \alpha + \mu q^2 & -\alpha \\ \beta & \sigma + \beta + q^2 \end{vmatrix} = 0$$

or

$$\sigma^2 + [(1+\mu)q^2 + \beta + 1 - \alpha]\sigma + \mu q^4 + (\mu\beta + 1 - \alpha)q^2 + \beta = 0. \tag{8.2.10a}$$

Since a Turing instability must occur from a homogeneous equilibrium state that is linearly stable in the absence of diffusion, we examine the stability behavior of (8.2.10a) for $q^2 = 0$, the latter condition being equivalent to eliminating diffusive effects from our system. Then (8.2.10a) reduces to

$$\sigma^2 + (\beta + 1 - \alpha)\sigma + \beta = 0.$$

Recalling that there is stability for a quadratic of this sort such that $\mathrm{Re}(\sigma) < 0$ provided its coefficients are positive, we restrict our parameter range to

$$\beta > \alpha - 1 \tag{8.2.10b}$$

and note that this only imposes a substantive requirement when $\alpha > 1$ since $\beta > 0$. Should $\beta < \alpha - 1$, $\mathrm{Re}(\sigma) > 0$ and $\mathrm{Im}(\sigma) \neq 0$. In the next section, we shall examine the occurrence of Turing instabilities by including diffusive effects and hence considering (8.2.10) for $q^2 > 0$.

8.3 Diffusive Instabilities and Chemical Turing Pattern Formation

We first observe that for $q^2 > 0$ and under the condition of (8.2.10b), the coefficient of the linear term in (8.2.10a) is identically positive. Thus if any instability is to occur it must take place by the constant term in (8.2.10a) becoming negative and that onset would then be from a stationary state. Let us examine this instability criterion

$$\beta < \beta_0(q^2) = \frac{q^2(\alpha - 1 - \mu q^2)}{\mu q^2 + 1}, \tag{8.3.1}$$

in some detail. Note that unless

$$\alpha > 1, \tag{8.3.2}$$

it cannot be satisfied at all since for $\alpha - 1 \leq 0$, $\beta_0(q^2) < 0$. Further this necessary condition for instability implies $F_1(1, 1; \alpha) = \alpha - 1 > 0$ from (8.2.5d) and hence is equivalent to requiring an autocatalytic or positive feedback reaction at equilibrium for the activator x species. Then combining (8.2.10b) and (8.3.2), we require that

$$\beta > \alpha - 1 > 0 \tag{8.3.3}$$

for the occurrence of a chemical Turing diffusive instability. We next wish to plot the marginal stability curve $\beta = \beta_0(q^2)$ associated with that instability criterion. We note that it has two zeroes at $q^2 = 0$ and $q^2 = q_0^2 = (\alpha - 1)/\mu$; a linear asymptote given by $\beta = -q^2 + \mu/\alpha$ as $q^2 \to \infty$; and a maximum at the point (q_c^2, β_c) as depicted in Fig. 8.1 where $d\beta_0(q_c^2)/dq^2 = 0$. Therefore

$$(\alpha - 1 - 2\mu q_c^2)(\mu q_c^2 + 1) = \mu[(\alpha - 1)q_c^2 - \mu q_c^4] \tag{8.3.4a}$$

$$\Rightarrow \mu^2 q_c^4 + 2\mu q_c^2 + 1 = (\mu q_c^2 + 1)^2 = \alpha \Rightarrow q_c^2 = \frac{\alpha^{1/2} - 1}{\mu} \tag{8.3.4b}$$

$$\Rightarrow \beta_c = \beta_0(q_c^2) = \frac{(\alpha-1)q_c^2 - \mu q_c^4}{\mu q_c^2 + 1} = \frac{\alpha - 1 - 2\mu q_c^2}{\mu} = \frac{(\alpha^{1/2}-1)^2}{\mu}, \tag{8.3.4c}$$

where use has been made of (8.3.4a, 8.3.4b) in (8.3.4c). Thus for fixed values of α and μ that marginal stability curve separates the linearly stable region in the first quadrant of the q^2-β plane where $\beta > \beta_0(q^2)$ from the unstable region where $0 < \beta < \beta_0(q^2)$. The marginal stability curve, the linearly stable region, and the unstable region can be characterized by $\sigma_0 = 0$, $\mathrm{Re}(\sigma_0) < 0$, and $\sigma_0 > 0$, respectively, where σ_0 represents that root of (8.2.10a) having the largest real part. Hence, when $\beta > \beta_c$, there exists no squared wavenumbers q^2 corresponding to growing disturbances, while when $0 < \beta < \beta_c$ there exists a band of them, centered about q_c^2, corresponding to such disturbances with associated wavelength $\lambda = 2\pi/|q|$ (see Fig. 8.1). Combining this result with (8.3.3) and (8.3.4c), we can finally deduce the following Turing chemical diffusive instability criterion

$$0 < \alpha - 1 < \beta < \frac{(\alpha^{1/2}-1)^2}{\mu}. \tag{8.3.5}$$

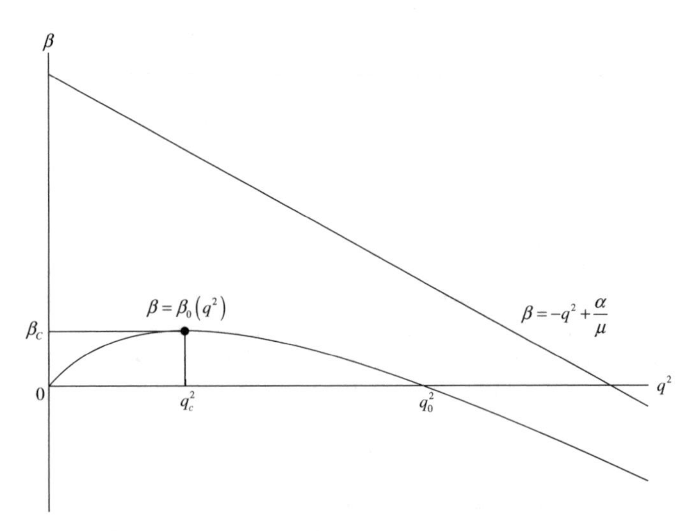

Fig. 8.1 Plot of $\beta = \beta_0(q^2)$ of (8.3.1) in the q^2-β plane for $\alpha > 1$.

We represent the instability region of (8.3.5) for a fixed value of μ in the α-β plane of Fig. 8.2 by plotting both the Hopf ($\beta = \alpha - 1$) and Turing ($\beta = \beta_c$) instability boundary curves for this region provided they actually intersect at a point (α_0, β_0) where $\alpha_0 > 1$ as depicted in that figure. The existence of such an intersection point serves as a necessary condition for the occurrence of the chemical Turing diffusive instability under investigation. To determine that condition, we shall find explicit formulae for the coordinates of this point as follows: Consider

$$\beta_0 = \alpha_0 - 1 = \beta_c(\alpha_0) = \frac{(\alpha_0^{1/2} - 1)^2}{\mu} \text{ where } \alpha_0 > 1. \tag{8.3.6}$$

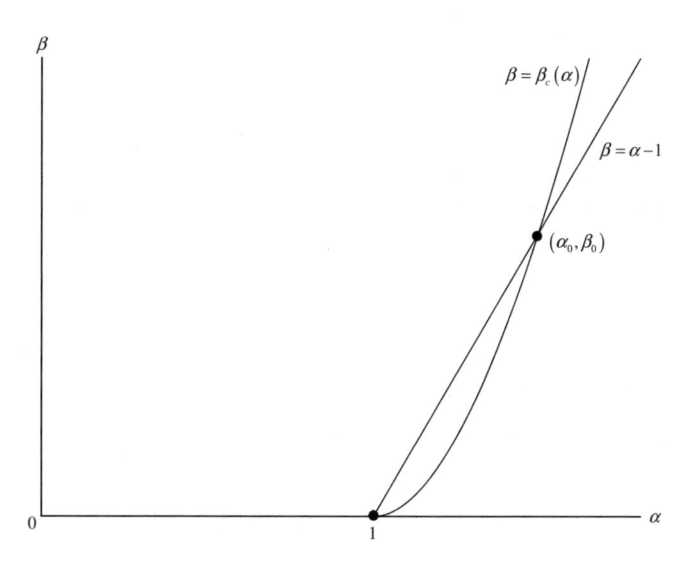

Fig. 8.2 Plots of the Turing boundary $\beta = \beta_c$ of (8.3.4c) and the Hopf boundary $\beta = \alpha - 1$ in the α-β plane for $\alpha > 1$ and $0 < \mu < 1$.

Note that $\alpha_0 = 1$ is a zero for both of these functions and thus they clearly have a geometric intersection point at (1,0), although this value does not satisfy the requirement that $\alpha_0 > 1$. To derive the formulae in question let

$$z_0 = \alpha_0^{1/2} > 1. \tag{8.3.7a}$$

Then (8.3.6) becomes

$$z_0^2 - 1 = (z_0 + 1)(z_0 - 1) = \frac{(z_0 - 1)^2}{\mu}, \tag{8.3.7b}$$

which, upon cancellation of the common factor $z_0 - 1 > 0$ and multiplication by μ, yields

$$\mu(z_0 + 1) = z_0 - 1 \Rightarrow z_0 = \frac{1 + \mu}{1 - \mu} > 1 \text{ for } \mu \neq 1. \tag{8.3.7c}$$

Observe from the inequality of (8.3.7c) that a necessary condition for its satisfaction is

$$0 < \mu = \frac{D_1}{D_2} < 1 \Rightarrow D_2 > D_1. \tag{8.3.8}$$

Hence a necessary condition for the occurrence of a Turing diffusive instability is that the inhibitor y species have a diffusive advantage over the activator x species. Under this condition, we can deduce from (8.3.6) and (8.3.7) that the intersection point has coordinates

$$\alpha_0 = \frac{(1+\mu)^2}{(1-\mu)^2}, \beta_0 = \frac{4\mu}{(1-\mu)^2}. \tag{8.3.9}$$

Although (8.3.8) is a necessary condition for that occurrence, the Turing instability criterion of (8.3.5) may actually require μ to be considerably less than one. We demonstrate this assertion most easily by noting that, consistent with (8.3.8), condition (8.3.5) implies

$$0 < \mu < \mu_c(\alpha) = \frac{(\alpha^{1/2}-1)^2}{\alpha - 1} = \frac{\alpha^{1/2}-1}{\alpha^{1/2}+1} \text{ for } \alpha > 1 \tag{8.3.10a}$$

and, particularizing (8.3.10a) to $\alpha = 2$, obtain

$$0 < \mu < \mu_c = 0.17 \Rightarrow c = \frac{1}{\mu} = \frac{D_2}{D_1} > \frac{1}{\mu_c} = c_{\min} = 5.83. \tag{8.3.10b}$$

Hence the inhibitor must have a significant diffusivity advantage to generate Turing patterns.

8.4 Extension to the Brusselator/Immobilizer Model System

Although theoretical investigations of the variety of artificial model systems mentioned in Section 8.1 yielded precisely the sort of structures predicted by Turing as just demonstrated for the Brusselator, experimental confirmation of such behavior proved to be far more elusive. The need for the activator species to diffuse significantly less rapidly than the inhibitor posed a major obstacle for designing an experiment which exhibited chemical Turing instability patterns since in aqueous media nearly all simple molecules and ions have diffusion coefficients within a factor of two of $1.5 \times 10^{-5}\text{cm}^2/\text{sec}$. Finally, Castets et al. [15] and Quyang and Swinney [89] managed to overcome this difficulty by conducting their experiments involving the iodide–chlorite chemical reaction–diffusion activator–inhibitor system in a gel reactor with a starch indicator which besides preventing convection resulted in a marked reduction of the effective iodide–chlorite diffusion coefficient ratio. Turing structures consisting of parallel stripes, rhombi, and hexagonal arrays of spots or net-like structures appeared as the system's control parameters of reservoir pool species concentrations consisting of chlorine dioxide, iodine, and malonic acid were tuned. The mechanism suggested for this reduction by Lengyel and Epstein [66] was that the starch indicator reversibly formed an immobile complex with the activator iodide species and the iodine rapidly enough to allow, in essence, a circumvention of the differential diffusivity requirement. Thus Turing instabilities could be generated over a parameter range where ordinarily the system would have exhibited oscillatory behavior in the absence of starch. We next quantify this reduction mechanism proposed by Lengyel and Epstein [66] and apply it to the Brusselator model system. In doing so, we shall follow the approach of Stephenson and Wollkind [117]. This can be accomplished by the introduction of an indicator such as starch which reversibly forms an immobile complex with the activator species making it possible to produce Turing patterns in a system that would otherwise not exhibit them. Consider the Brusselator reaction to which has been appended the immobile activator complex formation reversible reaction

$$X + S \underset{k_r}{\overset{k_f}{\rightleftharpoons}} X^*,$$

where $X^* \equiv$ concentration of the immobile activator complex assumed to be a third dynamical variable, $S \equiv$ concentration of the Turing pattern indicator assumed to be constant, and $k_{f,r} \equiv$ reaction rates associated with the formation of the activator complex assumed to be constant. Then by the law of mass action and the fact that the complex is immobile, we deduce

$$\frac{\partial X^*}{\partial \tau} = k_f SX - k_r X^* \tag{8.4.1a}$$

and incorporate that equation into the basic system of (8.1.1) while adding the same source terms but with the sign reversed to the right-hand side of (8.1.1a) to obtain

$$\frac{\partial X}{\partial \tau} = k_1 A - k_2 BX + k_3 X^2 Y - k_4 X - \frac{\partial X^*}{\partial \tau} + D_1 \nabla_2^2 X \tag{8.4.1b}$$

and retaining (8.1.1b).

Employing the nondimensional variables and parameters of (8.2.2) and, in addition, introducing

$$x^*(\boldsymbol{r},t) = \frac{X^*(\boldsymbol{s},\tau)}{X_e^*} \text{ where } X_e^* = KX_e \text{ with } K = \frac{k_f S}{k_r}, \; \varepsilon = \frac{k_4}{k_r}, \tag{8.4.2}$$

we obtain the dimensionless form of the Brusselator/Immobilizer model

$$\frac{\partial x}{\partial t} = F(x,y;\alpha) - K\frac{\partial x^*}{\partial t} + \mu\nabla^2 x, \tag{8.4.3a}$$

$$\frac{\partial y}{\partial t} = \beta G(x,y) + \nabla^2 y, \tag{8.4.3b}$$

$$\varepsilon\frac{\partial x^*}{\partial t} = x - x^*; \tag{8.4.3c}$$

$$x, y, x^* \text{ remain bounded as } r_1^2 + r_2^2 \to \infty; \tag{8.4.3d}$$

where (8.4.3c, 8.4.3a) are the transformed versions of (8.4.1a, 8.4.1b), respectively, while (8.4.3b) is a reproduction of (8.2.3b) which has, of course, been unchanged. Here $F(x,y;\alpha)$, $G(x,y)$, and ∇^2 are as defined in (8.2.3c) and (8.4.3d) represent the far-field condition relevant to this system. That system admits the equilibrium point $x_e = y_e = x_e^* = 1$ and then performing a linear normal-mode stability analysis of it by considering solutions of the form of (8.2.4) and (8.2.7) with, in addition,

$$x^*(r,t) = 1 + \varepsilon_1 x_{11}^* e^{\sigma t} \cos(qr) + O(\varepsilon_1^2), \tag{8.4.4}$$

where $|x_{11}|^2 + |y_{11}|^2 + |x_{11}^*|^2 \neq 0$, yields the cubic secular equation

$$\varepsilon\sigma^3 + [1 + K - \varepsilon\mathrm{tr}(M)]\sigma^2 + [\varepsilon\det(M) - \mathrm{tr}(M)\varepsilon - Km_{22}]\sigma + \det(M) = 0. \tag{8.4.5a}$$

Here $\mathrm{tr}(M) = m_{11} + m_{22}$, $\det(M) = m_{11}m_{22} - m_{12}m_{21}$, and m_{jk} for j, $k = 1,2$ are the components of the 2×2 matrix M such that $\det(M - \sigma I) = 0$, where $I \equiv 2 \times 2$ identity matrix, is equivalent to the quadratic secular equation (8.2.10). In this context, note that (8.4.5) reduces to (8.2.10) for $\varepsilon = K = 0$. Using the tools of singular perturbation theory as applied to cubic equations (see Problem 3.1), it follows that (8.4.5a) possesses roots with the asymptotic behavior

$$\sigma_{1,2} = O(1) \text{ and } \sigma_3 \sim -\frac{1+K}{\varepsilon} \text{ as } \varepsilon \to 0. \tag{8.4.5b}$$

We shall now, after Lengyel and Epstein [66], assume that

$$0 < \varepsilon = \frac{k_4}{k_r} << 1. \tag{8.4.6a}$$

Hence with no loss of generality for the diffusive instabilities under examination, we may consider (8.4.3c) in the limit as $\varepsilon \to 0$ since the neglected root σ_3 from (8.4.5b) is highly stabilizing under this assumption. Then in that limit (8.4.3c) implies the chemical quasi-equilibrium condition

$$x^* = x \tag{8.4.6b}$$

and thus (8.4.3a) becomes

$$(1+K)\frac{\partial x}{\partial t} = F(x,y;\alpha) + \mu\nabla^2 x. \tag{8.4.6c}$$

Next, we rescale the Brusselator/Immobilizer system of (8.4.6c) and (8.4.3b) by replacing the independent time and position variables with

$$t' = \frac{t}{1+K}, \boldsymbol{r}' = \frac{\boldsymbol{r}}{(1+K)^{1/2}}, \tag{8.4.7a}$$

which also requires the redesignation of the dependent chemical species variables

$$x'(\boldsymbol{r}',t') = x(\boldsymbol{r},t), y'(\boldsymbol{r}',t') = y(\boldsymbol{r},t). \tag{8.4.7b}$$

Then since

$$\frac{\partial}{\partial t} = \left(\frac{1}{1+K}\right)\frac{\partial}{\partial t'}, \ \nabla^2 = \left(\frac{1}{1+K}\right)\nabla'^2, \tag{8.4.7c}$$

we would obtain a system of the same form as (8.2.3a, 8.2.3b)

$$\frac{\partial x}{\partial t} = F(x,y;\alpha) + \mu_K \nabla^2 x, \tag{8.4.8a}$$

$$\frac{\partial y}{\partial t} = \beta_K G(x,y) + \nabla^2 y, \tag{8.4.8b}$$

but now

$$\mu_K = \frac{\mu}{1+K}, \ \beta_K = (1+K)\beta, \tag{8.4.8c}$$

where, for ease of exposition, the primes have been dropped from the new variables. From (8.4.8) we see that the introduction of an immobilizer reduces the effective diffusion activator–inhibitor coefficient ratio. Thus we can deduce directly from (8.3.5) upon the replacement of β and μ with β_K and μ_K, respectively, the linear chemical Turing diffusive instability criterion for this system

$$0 < \alpha - 1 < \beta_K < \frac{(\alpha^{1/2}-1)^2}{\mu_K}. \tag{8.4.9a}$$

Upon employment of (8.4.8c), this yields

$$0 < \alpha - 1 < (1+K)\beta < \frac{(1+K)(\alpha^{1/2}-1)^2}{\mu} \tag{8.4.9b}$$

which implies

$$0 < \beta_1(\alpha;K) = \frac{\alpha-1}{1+K} < \beta < \frac{(\alpha^{1/2}-1)^2}{\mu} = \beta_2(\alpha;\mu). \tag{8.4.9c}$$

From (8.3.9) we can conclude that the intersection point (α_0, β_0) of these two bounding curves, such that $\beta_1(\alpha_0;K) = \beta_2(\alpha_0;\mu) = \beta_0$, has coordinates

$$\alpha_0 = \frac{(1+\mu_K)^2}{(1-\mu_K)^2} = \frac{(1+K+\mu)^2}{(1+K-\mu)^2} = \alpha_0(K,\mu) \tag{8.4.10a}$$

and

$$\beta_0 = \frac{4\mu_K}{(1-\mu_K)^2(1+K)} = \frac{4\mu}{(1+K-\mu)^2}. \tag{8.4.10b}$$

The instability region of (8.4.9c) is represented in the α-β plane of Fig. 8.3 for $\mu = 1$ and $K = 15$. As demonstrated by this figure for the special case of $\mu = 1$, in conjunction with (8.4.10), it is now possible to produce Turing instabilities even when the ratio of diffusivities

$$\mu = \frac{D_1}{D_2} \geq 1 \tag{8.4.11a}$$

provided K, the complexation strength, satisfies

$$K > \mu - 1. \tag{8.4.11b}$$

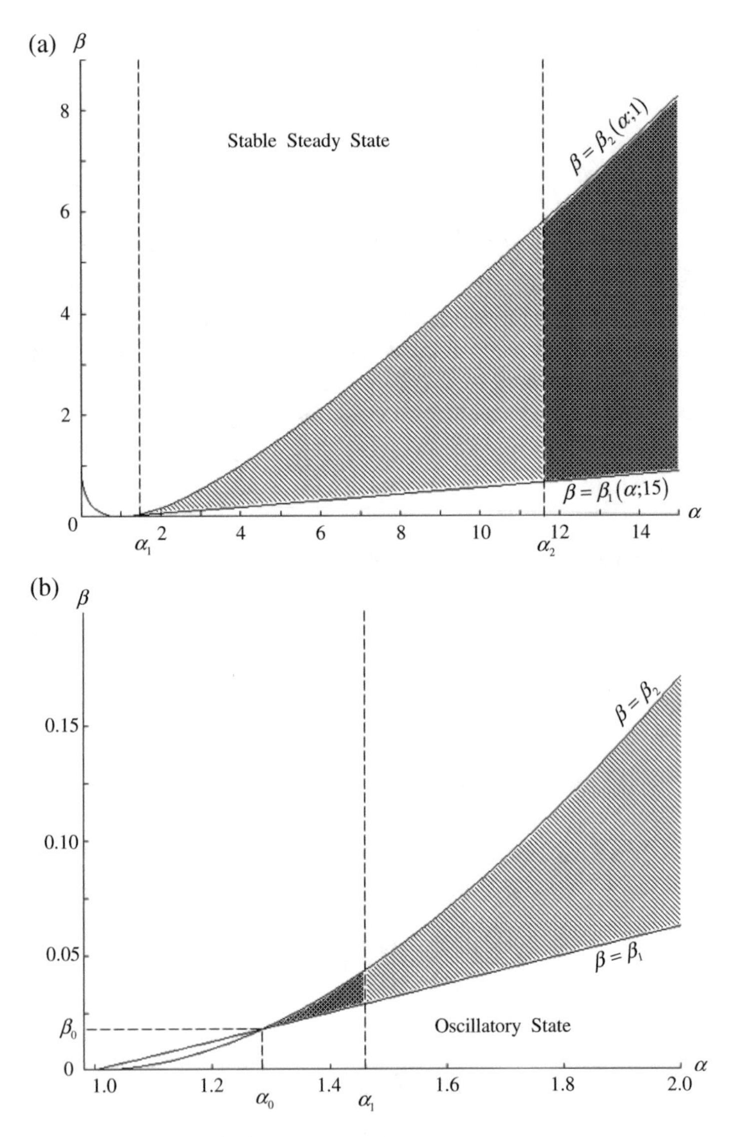

Fig. 8.3 Stability diagram in the α-β plane for the Brusselator/Immobilizer model system when $\mu = 1$ and $K = 15$.

Note that when $K = 0$ this situation reverts to the Brusselator in the absence of an immobilizer, the instability region for which is represented in Fig. 8.2 and requires $\mu < \mu_c(\alpha) < 1$. Consider any point (α_p, β_p) lying below the Hopf boundary line $\beta = \beta_1(\alpha; K) = (\alpha - 1)/(1 + K)$ with $K = 0$ in that figure or $\beta = \alpha - 1$, where there is oscillatory behavior. This implies that

$$\beta_p < \beta_1(\alpha_p; 0) \text{ for } \alpha_p > 1. \tag{8.4.12a}$$

Since the slope of the Hopf boundary line decreases with increasing K, there exists a $K_c > 0$ such that

$$\beta_p = \beta_1(\alpha_p; K_c). \tag{8.4.12b}$$

Then

$$\beta_p > \beta_1(\alpha_p; K_p) \text{ for any } K_p > K_c. \tag{8.4.12c}$$

If

$$\alpha_p > \alpha_0(K_p, \mu), \tag{8.4.13a}$$

as well, the point will lie in the Turing instability region, and if not, since $\alpha_0(K, \mu)$ of (8.4.10a) is also a decreasing function of K, there exists a $K_0 \geq K_p$ such that

$$\alpha_p = \alpha_0(K_0, \mu). \tag{8.4.13b}$$

Then

$$\alpha_p > \alpha_0(K, \mu) \text{ for any } K > K_0, \tag{8.4.13c}$$

and again the point will lie in the Turing instability region. In all of these cases, we have demonstrated that the introduction of an immobilizer can produce Turing instabilities under conditions for which there would be none without it provided its complexation strength is sufficiently large. An analogous argument may be used to conclude that increasing K in Fig. 8.3 can also cause Turing instabilities to appear where oscillatory behavior would occur for $K = 15$.

Given the importance of the role played by the use of a gel medium in the presence of an immobilizer on this analysis, there is some merit in examining just how that affects the wavelength of the resulting Turing patterns. To do so we first consider the dependence of that Turing pattern's wavelength on K. Since from (8.4.7a)

$$q_c r = q_c[(1+K)^{1/2} r'] = [(1+K)^{1/2} q_c] r' = q_K r' \Rightarrow q_K = (1+K)^{1/2} q_c, \tag{8.4.14a}$$

we can conclude that the corresponding dimensional wavelength, given the scaling introduced in (8.2.2a) and (8.4.7a), satisfies

$$\lambda_K^* = \left(\frac{D_2}{k_0}\right)^{1/2} \left(\frac{2\pi}{q_K}\right) \text{ with } k_0 = \frac{k_4}{1+K} \Rightarrow \lambda_K^* = \left(\frac{D_2}{k_4}\right)^{1/2} \left(\frac{2\pi}{q_c}\right) = \lambda_c^* \tag{8.4.14b}$$

and hence is independent of K. We note that in the experiments of Ouyang and Swinney [89] and Lengyel and Epstein [66], involving the iodide–chlorite chemical reaction in a gel medium with starch as the Turing pattern indicator, λ_K^* was indeed independent of K. It remains to decide what values should be assigned for the diffusion coefficients themselves. Initially, Ouyang and Swinney [89] and Lengyel and Epstein [66] assumed that the gel did not affect the diffusivity of the reactant species and simply took $D_{1,2} = D_{x,y} \equiv$ aqueous diffusion coefficients of X, Y. Such an assumption in conjunction with (8.4.14a) implies that the wavelength of the patterns would be unaltered as well by the presence of the gel medium when compared with an aqueous solution situation. Since the employment of the resulting dimensional wavelength formula

$$\lambda_K^* = \left(\frac{D_y}{k_4}\right)^{1/2} \lambda_c = \lambda_c^* \tag{8.4.15}$$

tended to over predict the actual observed wavelength for these gel media situations, Ouyang et al. [90] designed a set of experiments to test this hypothesis definitively. They found that the presence of the gel reduced both diffusion coefficients and this reduction was exactly the same for each one. That is, for polyacrylamide gels, Stephenson and Wollkind [117] took

$$D_{1,2} = \chi^2 D_{x,y} \text{ for } \chi = 0.40. \tag{8.4.16a}$$

The constitutive relation of (8.4.16a) reflects the fact that a fully hydrolyzed saturated gel will result in an ionic diffusion coefficient which has been uniformly reduced from its aqueous solution value, the amount of that reduction being dependent on the characteristic pore diameter of the gel itself as observed by Ouyang et al. [90]. Note that the adoption of (8.4.16a) preserves the standard diffusivity ratio relation $\mu = D_x/D_y$ but reduces the wavelength of (8.4.15) by the factor χ

$$\lambda_K^* = \chi \lambda_c^*. \tag{8.4.16b}$$

We note that when Pearson [93] assigned the parameters of his activator–inhibitor/immobilizer model system typical values while taking $\chi = 1$ in (8.4.16b), he predicted a $\lambda_c^* \cong 0.40$ mm instead of the observed experimental value of $\lambda_K^* \cong 0.17$ mm observed by DeKepper et al. [27], an over prediction which would be adjusted correctly to $\lambda_K^* \cong 0.16$ mm upon adoption of the χ of (8.4.16a).

In this context, since the experimental patterns under investigation typically had a gel disk diameter to characteristic wavelength ratio of 150, it seems reasonable as a first approximation to have considered the Brusselator/Immobilizer equations on an unbounded spatial domain given that, under these circumstances, Graham et al. [43] showed the actual boundaries do not significantly influence the patterns.

We close this section by noting that, just as with the dynamical systems discussed in Chapter 3, the main fruit of linear theory involving partial differential equation models is the determination of the critical conditions for the onset of instability to initially infinitesimal disturbances ($\beta < \beta_c$). For models of this type, these critical conditions also include the wavenumber squared (q_c^2) of those disturbances most likely to grow first.

There are three major deficiencies of linear stability theory in this instance:
(1) It does not determine the long-time behavior of such growing disturbances.

(2) It is unable to resolve which of the possible $q_{1,2}$ pairs such that $q_1^2 + q_2^2 = q_c^2$ are actually selected and hence it cannot predict pattern formation in the plane.

(3) It does not ascertain the effect of initially finite-amplitude disturbances and thus cannot predict global behavior.

In order to alleviate these deficiencies, the nonlinear terms of the perturbation systems must be taken into account. We give a brief overview of nonlinear stability theory applied to the Brusselator/Immobilizer model in the next section and defer a more detailed treatment of such methods until after the basic equations for fluid mechanics have been developed and applied to a variety of models for convective flow phenomena.

8.5 Nonlinear Stability Theory: An Overview

Stephenson and Wollkind [117] examined the stability of its community equilibrium point to one-dimensional perturbations by considering solutions of (8.4.8) of the form

$$\begin{aligned} x(r,t) \sim 1 + A(t)\cos(q_K r) + A^2(t)[x_{20} + x_{22}\cos(2q_K r)] \\ + A^3(t)[x_{31}\cos(q_K r) + x_{33}\cos(3q_K r)] \end{aligned} \tag{8.5.1a}$$

with an analogous expansion for $y(r,t)$ where the amplitude function $A(t)$ satisfied

$$\frac{dA}{dt} \sim \sigma_0 A - a_1 A^3. \tag{8.5.1b}$$

Here q_K was the critical wavenumber of linear stability theory while σ_0 and a_1 denoted the growth rate associated with that most dangerous mode and the corresponding Landau constant. A method of this sort, although incorporating the nonlinearities of the relevant model system, basically pivots a perturbation procedure about the critical point of linear stability theory.

Substituting this solution into (8.4.8) and expanding F, G in a Mclaurin series about $A = 0$, they obtained a series of problems, each of which was proportional to a term appearing explicitly in the expansion of (8.5.1a). The $A(t)\cos(q_K r)$ problem was equivalent to the linear one with $q \equiv q_K$ where, as indicated by (8.4.14a) and (8.3.4b),

$$q_K^2 = (\beta + 1)q_c^2 = \frac{(\beta + 1)(\alpha^{1/2} - 1)}{\mu} \tag{8.5.1c}$$

and a diffusive instability ($\sigma_0 > 0$) occurred for β satisfying (8.4.9c). The two $O(A^2)$ problems could be solved in a straightforward manner, while the $A^3(t)\cos(q_K r)$ problem yielded the Fredholm-type solvability condition

$$a_1 = a_1(\alpha; \mu, K) \tag{8.5.1d}$$

in the limit as $\beta \to \beta_2$. The function a_1 of (8.5.1d) is plotted versus α in Fig. 8.4 for the μ and K values of Fig. 8.3. From this figure, we can see there exist two zeroes $\alpha_{1,2}$ of a_1 such that

$$a_1 < 0 \text{ for } 1 < \alpha < \alpha_1 \text{ or } \alpha > \alpha_2, \; a_1 > 0 \text{ for } \alpha_1 < \alpha < \alpha_2; \tag{8.5.2a}$$

where

$$\alpha_1 = 1.46, \alpha_2 = 11.60. \tag{8.5.2b}$$

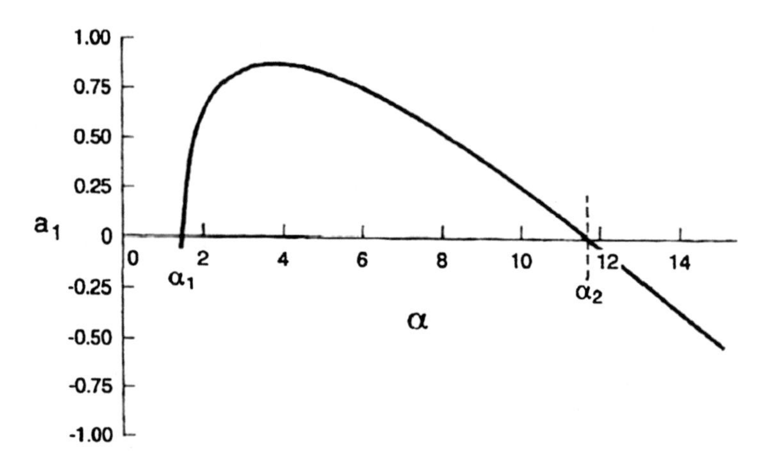

Fig. 8.4 Plot of a_1 of (8.5.1d) versus α with $\beta = \beta_2$ when $\mu = 1$ and $K = 15$ [117].

Given these conditions, the amplitude function $A(t)$ of (8.4.1b) undergoes a standard supercritical pitchfork bifurcation [127] at the Turing boundary $\beta = \beta_2$ when $\alpha_1 < \alpha < \alpha_2$, from which we may conclude that:

1. For $\beta > \beta_2$ and $\alpha_1 < \alpha < \alpha_2$, the undisturbed state $A = 0$ is stable yielding a uniform homogeneous pattern $x(r,t) \to x_e = 1$ as $t \to \infty$.
2. For $\beta_1 < \beta < \beta_2$ and $\alpha_1 < \alpha < \alpha_2$, $A = A_e = \sqrt{\sigma_0/a_1}$ is stable, yielding a one-dimensional periodic Turing pattern consisting of stationary parallel stripes

$$x(r,t) \longrightarrow x_e(r) \sim 1 + A_e \cos\left(\frac{2\pi r}{\lambda_K}\right) \text{ as } t \longrightarrow \infty \tag{8.5.3a}$$

of characteristic wavelength

$$\lambda_K = \frac{2\pi}{q_K}. \tag{8.5.3b}$$

The system undergoes a Hopf bifurcation at the boundary $\beta = \beta_1$ and hence there is oscillatory behavior for $0 < \beta < \beta_1$ while when $\alpha_a = 1 < \alpha < \alpha_1$ or $\alpha > \alpha_2$, the bifurcation is subcritical (the implication of such subcriticality is the subject of the nonlinear stability analysis of that model equation introduced in Chapter 16). These one-dimensional nonlinear stability results are summarized by means of the indicated regions in Fig. 8.3, where the vertical dotted lines denote the loci $\alpha = \alpha_{1,2}$ of Fig. 8.4; the stripes, the Turing patterns; and the shading, the subcritical behavior.

One-dimensional analyses of the sort outlined in this section can only predict stripes while Ouyang and Swinney [89] also observed hexagonal and rhombic Turing patterns. In order to investigate the possibility of occurrence of such planar two-dimensional structures, it is necessary to let $\nabla^2 = \partial^2/\partial r_1^2 + \partial^2/\partial r_2^2$ and consider an extension of the one-dimensional analysis which in particular replaces the $x-1$ perturbation expansion of (8.5.1) by [143]

$$\begin{aligned} &A_1(t)\cos[q_K r_1 + \varphi_1(t)] + A_2(t)\cos[q_K(r_1 - \sqrt{3}r_2)/2 - \varphi_2(t)] \\ &\quad + A_3(t)\cos[q_K(r_1 + \sqrt{3}r_2)/2 - \varphi_3(t)] \end{aligned} \tag{8.5.4a}$$

to lowest order where, for $(i,j,k) \equiv$ even permutation of $(1,2,3)$,

$$\frac{dA_i}{dt} \sim \sigma_0 A_i - 4a_0 A_j A_k \cos(\varphi_i + \varphi_j + \varphi_k) - A_i[a_1 A_i^2 + 2a_2(A_j^2 + A_k^2)], \tag{8.5.4b}$$

$$A_i \frac{d\varphi_i}{dt} \sim 4a_0 A_j A_k \sin(\varphi_i + \varphi_j + \varphi_k); \tag{8.5.4c}$$

and by

$$A(t)\cos(q_K r_1) + B(t)\cos(q_K r_3) \text{ for } r_3 = r_1\cos(\psi) + r_2\sin(\psi) \tag{8.5.5a}$$

where

$$\frac{dA}{dt} \sim \sigma_0 A - A(a_1 A^2 + b_1 B^2),\ \frac{dB}{dt} \sim \sigma_0 B - B(a_1 B^2 + b_1 A^2); \tag{8.5.5b}$$

when examining potential hexagonal and rhombic patterns, respectively, with corresponding expansions for $y-1$ analogous to them. While mentioned here for the sake of completeness, investigations of that sort are the primary subjects of Chapters 17 and 18 which are concerned with nonlinear optical hexagonal and vegetative rhombic pattern formation, respectively. The advantage of such an approach over strictly numerical procedures is that it allows one to deduce quantitative relationships between system parameters and stable patterns which are valuable for comparison with experimental or observational evidence and difficult to accomplish using simulation alone.

Problems

8.1. The Schnackenberg simplification of the Brusselator replaces its second reaction with $B \xrightarrow{k_2} Y$. Consider the nondimensionalized Schnackenberg chemical reaction–diffusion model system given by

$$\frac{\partial x}{\partial t} = 1 - \frac{\alpha}{2} - x + \alpha \frac{x^2 y}{2} + \mu \frac{\partial^2 x}{\partial r^2}, \frac{\partial y}{\partial t} = \beta(1 - x^2 y) + \frac{\partial^2 y}{\partial r^2} \quad \text{(P1.1a)}$$

where x and y represent the concentrations of the activator and inhibitor species, respectively, which are dynamical variables of space r and time t, while the quantities

$$\alpha = \frac{2k_2 B}{k_1 A + k_2 B}, \beta = \frac{k_3(k_1 A + k_2 B)^2}{k_4^3}, \mu = \frac{D_x}{D_y} \quad \text{(P1.1b)}$$

are parameters since the pool species concentrations (A, B), reaction rates (k_1, k_2, k_3, k_4), and diffusion coefficients (D_x, D_y) have been assumed to remain constant.

(a) Show that system (P1.1) possesses a single community equilibrium point

$$x(r,t) \equiv x_e, y(r,t) \equiv y_e \text{ satisfying } x_e = y_e = 1. \quad \text{(P1.2)}$$

(b) Seeking a normal-mode solution of system (P1.1) of the form

$$x(r,t) = 1 + \varepsilon_1 x_{11} \cos(qr) e^{\sigma t} + O(\varepsilon_1^2), y(t,y) = 1 + \varepsilon_1 y_1 1 \cos(qr) e^{\sigma t} + O(\varepsilon_1^2) \quad \text{(P1.3a)}$$

where

$$|\varepsilon_1| << 1, |x_{11}|^2 + |y_{11}|^2 \neq 0, q \geq 0, \quad \text{(P1.3b)}$$

demonstrate that σ satisfies the same quadratic secular equation as the Brusselator

$$\sigma^2 + [(1+\mu)q^2 + \beta + 1 - \alpha]\sigma + \mu q^4 + (\mu\beta + 1 - \alpha)q^2 + \beta = 0, \quad \text{(P1.3c)}$$

and hence conclude that there exists a Turing-type diffusive instability provided

$$0 < \alpha - 1 < \beta < \frac{(\sqrt{\alpha} - 1)^2}{\mu} \text{ for } 0 < \mu < 1. \quad \text{(P1.3d)}$$

(c) Particularizing the result of (P1.3d) to $\alpha = 2$ obtains the instability criteria

$$1 < \beta < \frac{\mu_c}{\mu} \text{ and } 0 < \mu < \mu_c = \frac{1}{3 + 2\sqrt{2}}, \quad \text{(P1.4)}$$

represent this instability region of (P1.5) graphically in the μ-β plane, and discuss the physical significance of $c_{\min} = 1/\mu_c$ with reference to the parameter c defined by

$$c = \frac{1}{\mu} = \frac{D_y}{D_x}. \tag{P1.5}$$

8.2. Consider the simplified chemical reaction–diffusion model system

$$\frac{\partial X}{\partial \tau} = k_1 - k_2 X - \frac{4k_3 Y}{X} + D_x \frac{\partial^2 X}{\partial s^2}, \tag{P2.1a}$$

$$\frac{\partial Y}{\partial \tau} = k_2 X - \frac{k_3 Y}{X} + D_y \frac{\partial^2 Y}{\partial s^2}, \tag{P2.1b}$$

where X and Y represent iodide and chlorite concentrations, respectively, which are functions of space s and time τ, while the reaction rates (k_1, k_2, k_3) and diffusion coefficients (D_x, D_y) have been assumed to remain constant.

(a) Show that system (P2.1) possesses a community equilibrium point of the form

$$X(s,\tau) \equiv X_e,\ Y(s,\tau) \equiv Y_e \text{ such that } Y_e = \frac{k_2 X_e^2}{k_3} \tag{P2.2}$$

by explicitly finding $X_e > 0$.

(b) Introducing the nondimensional variables and parameters

$$r = \frac{s}{(D_y/k_2)^{1/2}},\ t = k_2\tau,\ x = \frac{X}{X_e},\ y = \frac{Y}{Y_e}; \tag{P2.3a}$$

$$\beta = \frac{5k_3}{3k_1},\ \mu = \frac{D_x}{D_y}; \tag{P2.3b}$$

transform system (P2.1) into

$$\frac{\partial x}{\partial t} = F(x,y) + \mu \frac{\partial^2 x}{\partial r^2}, \tag{P2.4a}$$

$$\frac{\partial y}{\partial t} = \beta G(x,y) + \frac{\partial^2 y}{\partial r^2}, \tag{P2.4b}$$

where

$$F(x,y) = 5 - x - \frac{4y}{x},\ G(x,y) = 3\left(x - \frac{y}{x}\right); \tag{P2.4c}$$

and show that $F(1,1) = G(1,1) = 0$.

(c) Seeking a linear perturbation solution of system (P2.4) of the form

$$x(r,t) = 1 + \varepsilon_1 x_1(r,t) + O(\varepsilon_1^2),\ y(r,t) = 1 + \varepsilon_1 y_1(r,t) + O(\varepsilon_1^2) \text{ for } |\varepsilon_1| << 1, \tag{P2.5}$$

deduce that

$$\frac{\partial x_1}{\partial t} = F_1(1,1)x_1 + F_2(1,1)y_1 + \mu \frac{\partial^2 x_1}{\partial r^2}, \tag{P2.6a}$$

$$\frac{\partial y_1}{\partial t} = \beta[G_1(1,1)x_1 + G_2(1,1)y_1] + \frac{\partial^2 y_1}{\partial r^2}, \tag{P2.6b}$$

where

$$F_1(1,1) = 3,\ F_2(1,1) = -4,\ G_1(1,1) = 6,\ G_2(1,1) = -3. \tag{P2.6c}$$

(d) Then assuming a normal-mode resolution for the perturbation quantities in system (P2.6) by letting

$$[x_1, y_1](r,t) = [x_{11}, y_{11}]\cos(qr)e^{\sigma t} \text{ where } |x_{11}|^2 + |y_{11}|^2 \neq 0 \text{ and } q \geq 0, \tag{P2.7}$$

obtain the following quadratic secular equation satisfied by σ

$$\sigma^2 + [(1+\mu)q^2 + 3(\beta - 1)]\sigma + \mu q^4 + 3(\beta\mu - 1)q^2 + 15\beta = 0. \tag{P2.8}$$

(e) Show that the community equilibrium point is stable to homogeneous pertubations ($q^2 = 0$) provided

$$\beta > 1 \tag{P2.9}$$

and that there then exists a diffusive instability when $q^2 > 0$ for

$$\beta < \beta_0(q^2) = \frac{3q^2 - \mu q^4}{3(\mu q^2 + 5)} \leq \beta_c = \beta_0(q_c^2) \tag{P2.10a}$$

where

$$q_c^2 = \frac{2\sqrt{10} - 5}{\mu} \tag{P2.10b}$$

and

$$\beta_c = \frac{13 - 4\sqrt{10}}{3\mu} = \frac{3}{\mu(13 + 4\sqrt{10})}. \tag{P2.10c}$$

(f) Deduce, from (P2.9) and (P2.10), the following criteria for the occurrence of Turing diffusive instabilities

$$1 < \beta < \frac{\mu_c}{\mu} \text{ and } 0 < \mu < \mu_c = \frac{3}{13 + 4\sqrt{10}}. \tag{P2.11}$$

(g) Defining the parameter

$$c = \frac{1}{\mu} = \frac{D_y}{D_x}, \tag{P2.12a}$$

demonstrate that (P2.11) is equivalent to

$$1 < \beta < \frac{c}{c_{\min}} \text{ and } c > c_{\min} = \frac{1}{\mu_c} = \frac{13 + 4\sqrt{10}}{3}, \tag{P2.12b}$$

represent the region of (P2.12b) graphically in the c-β plane, and discuss the physical significance of $c_{\min}$.

Part II
Chapters 9–14

Chapter 9
Governing Equations of Fluid Mechanics

So far, all the models for the phenomena developed in previous chapters have been concerned with nonmoving continua involving either rigid materials or those for which convective effects could be considered negligible when compared to diffusive ones. To model moving continua in general and fluid mechanical phenomena, in particular, it is necessary to deduce the appropriate governing equations and boundary conditions. Toward that end, the former are considered in this chapter and the latter, in the next one. The basic equations of continuum mechanics for moving continua are derived from first principles employing the continuum hypothesis, substantial derivative, and material and spatial coordinates. After the continuity equation is deduced from conservation of mass using the Reynolds Transport Theorem, a general balance law is developed and applied to momentum (linear and angular) and energy for nonpolar continua. Then equations of state relevant to ideal and adiabatic gases and constitutive relations relevant to Newtonian fluid flow are introduced involving thermodynamics and Cartesian tensor notation, respectively. In addition, the continuity equation in Cartesian coordinates is deduced by a fixed volume big box method as well and then that equation is converted to cylindrical coordinates by direct transformation. The problems examine various aspects of these concepts including incompressibility, stream functions, a big box derivation of the continuity equation in cylindrical coordinates, a transformation of the continuity equation in Cartesian coordinates to spherical coordinates, the constitutive relations for Newtonian fluids, the governing equation for chemical species conservation, the balance of angular momentum for polar continua, the terms in the energy equations for ideal gases, and the Clausius–Duhem inequality.

9.1 Continuum Hypothesis, Substantial Derivative, and Reynolds Transport Theorem

Continuum mechanics for moving continua deals with continuous media in which smoothly varying material properties such as density ρ, a scalar quantity, and veloc-

© Springer International Publishing AG, part of Springer Nature 2017

D. J. Wollkind and B. J. Dichone, *Comprehensive Applied Mathematical Modeling in the Natural and Engineering Sciences*,
https://doi.org/10.1007/978-3-319-73518-4_9

ity $\boldsymbol{v}$, a vector one, must have values assigned to them at those points in space instantaneously contained in a volume under examination at a time of interest. In actuality moving fluid media such as gases and liquids consist of molecules separated by vacuous regions much larger in extent than the one occupied by those molecules. The continuum hypothesis regards physical quantities such as mass and momentum associated with the matter contained in a given small volume as being spread uniformly over that volume instead of being concentrated in the molecular fraction of it.

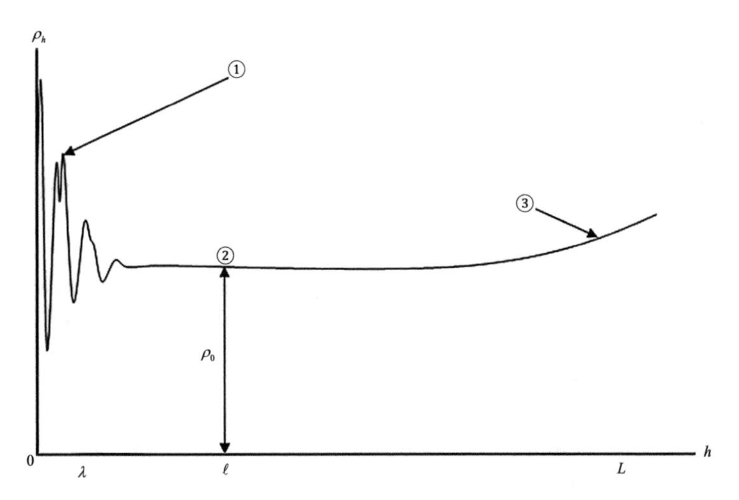

Fig. 9.1 Plot of ρ_h, the density of water in a cube of volume h^3 centered about a point (x_0, y_0, z_0) at time t_0, versus h. Here (1), (2), and (3) designate microscopic, homogeneous, and macroscopic intervals, respectively, with corresponding scales $\lambda << \ell << L$ while ρ_0 represents the continuum density $\rho(x_0, y_0, z_0, t_0)$.

In order to examine the validity of this continuum hypothesis as well as to determine the proper value to assign for a property such as density and reconcile that approach with a purely molecular one, consider the density of water in a cubic volume of dimension h centered about the point (x_0, y_0, z_0) at time t_0

$$\rho_h(x_0, y_0, z_0, t_0) = \frac{M_h(x_0, y_0, z_0, t_0)}{h^3} \tag{9.1.1a}$$

where

$$M_h(x_0, y_0, z_0, t_0) \equiv \text{ mass of the molecules contained in that cube.} \tag{9.1.1b}$$

This function is plotted versus h in Fig. 9.1. Observe from that figure there exist three intervals over which this density exhibits qualitatively different behavior: (1) A microscopic interval such that $h = O(\lambda)$, for $\lambda \equiv$ the mean free path of molecules, where the density undergoes dramatic variations due to molecular fluctuations; (2) a homogeneous interval such that $h = O(\ell)$, for $\ell \equiv$ a scale of local observation, where the density is virtually equal to the constant value of ρ_0; and (3) a macroscopic interval such that $h = O(L)$, for $L \equiv$ a nonhomogeneous scale, where the density undergoes slight variations associated with its spatial distribution. These scales have the typical values of [6, 69]

$$\lambda = 10^{-7} \text{ cm}, \ell = 10^{-3} \text{ cm}, L = 10^2 \text{ cm}. \tag{9.1.1c}$$

Let us determine the number of molecules specifically in a cube of either dimension $h = \lambda$ or ℓ. Recall that a mole or gram-molecular mass of a substance contains $N_0 \equiv$ Avagadro's Number $= 6.02 \times 10^{23}$ of molecules. Further recall that for the most commonly occurring isotopes of hydrogen $\equiv H$ and oxygen $\equiv O$ with gram-molecular mass of 1 gm and 16 gm, respectively, water $\equiv H_2O$ has a gram-molecular mass of 18 gm. Finally, recall that a cubic centimeter cm^3 of water at standard conditions of temperature and pressure has a mass of 1 gm. Now, since a cube of water of dimension $h \equiv h_0$ cm has a volume of $h^3 \equiv h_0^3$ cm^3, its mass is h_0^3 gm at standard conditions. Thus, if y denotes the number of molecules in a cube with that dimension,

$$\frac{y}{N_0} = \frac{h_0^3 \text{ gm}}{18 \text{ gm}} = \frac{h_0^3}{18} \tag{9.1.2a}$$

or

$$y = \frac{h_0^3 N_0}{18} = \left(\frac{6.02}{18}\right) h_0^3 \times 10^{23} \text{ molecules.} \tag{9.1.2b}$$

Particularizing this result to $h_0 = \lambda$ and ℓ or $h_0 = 10^{-7}$ and 10^{-3}, respectively, we find that

$$y_\lambda = \frac{602}{18} \text{ molecules}, y_\ell = \left(\frac{6.02}{18}\right) \times 10^{14} \text{ molecules,} \tag{9.1.3a}$$

or

$$y_\lambda = 33.4 \text{ molecules}, y_\ell = 3.34 \times 10^{13} \text{ molecules.} \tag{9.1.3b}$$

Given that there are only 33 or 34 molecules in a cube with a typical dimension in the microscopic interval the number of molecules in that cube will fluctuate from one observation to the next and hence the density will vary in an irregular way with the size of its volume as indicated in Fig. 9.1. In the homogeneous interval, where the dimension is large with respect to the microscopic scale but small with respect to the macroscopic one, the number of molecules contained in the cubic volume is still enormous enough so that fluctuations caused by slight variations in its volume size have no effect on the observed local average density as also indicated in Fig. 9.1. Then the proper continuum value to assign for the density function ρ at the point

(x_0, y_0, z_0) and time t_0 is the local average

$$\rho(x_0, y_0, z_0, t_0) = \rho_0. \tag{9.1.4}$$

Similar, but slightly more complicated considerations, because it is a vector quantity, allow one to identify the continuum value for velocity $\boldsymbol{v}$ with the average velocity of the molecules in a small but not too small cube centered about the point in question at the time of interest.

Clearly, in order for the continuum hypothesis to be valid the scales in question must satisfy the relation

$$\lambda << \ell << L \tag{9.1.5}$$

as do those of (9.1.1c). There exist a large number of phenomena for which this continuum hypothesis is valid and hence can be modeled by the various techniques of continuum mechanics as discussed for nonmoving continua in Section 5.1.

As a continuum moves it is natural to specify the position of each point in it as a function of time. To identify these points, we specify their initial positions $\boldsymbol{A}$ at some time which, without loss of generality, can be taken as $t = 0$. Then if $\boldsymbol{X}(\boldsymbol{A}, t)$ denotes the position of those points for some later time $t > 0$, the motion of the continuum may be described by (see Fig. 9.2).

$$\boldsymbol{x} = \boldsymbol{X}(\boldsymbol{A}, t), \boldsymbol{X}(\boldsymbol{A}, 0) = \boldsymbol{A}. \tag{9.1.6}$$

For a fixed value of $\boldsymbol{A}$, a point that satisfies this relation is called a *particle* (not to be confused with a molecule) and (9.1.6), its *particle path*. Further

$$\frac{\partial \boldsymbol{X}(\boldsymbol{A}, t)}{\partial t} = \boldsymbol{V}(\boldsymbol{A}, t) \equiv \text{ velocity}. \tag{9.1.7}$$

We shall assume that the Jacobian of the transformation of (9.1.6)

$$J(A_1, A_2, A_3, t) = \frac{\partial(x_1, x_2, x_3)}{\partial(A_1, A_2, A_3)} = \begin{vmatrix} \frac{\partial x_1}{\partial A_1} & \frac{\partial x_1}{\partial A_2} & \frac{\partial x_1}{\partial A_3} \\ \frac{\partial x_2}{\partial A_1} & \frac{\partial x_2}{\partial A_2} & \frac{\partial x_2}{\partial A_3} \\ \frac{\partial x_3}{\partial A_1} & \frac{\partial x_3}{\partial A_2} & \frac{\partial x_3}{\partial A_3} \end{vmatrix} \neq 0$$

and hence that transformation may be inverted to obtain

$$\boldsymbol{A} = \boldsymbol{a}(\boldsymbol{x}, t). \tag{9.1.8}$$

Thus

$$\boldsymbol{X}[\boldsymbol{a}(\boldsymbol{x}, t), t] = \boldsymbol{x},\ \boldsymbol{a}[\boldsymbol{X}(\boldsymbol{A}, t), t] = \boldsymbol{A}, \text{ and } \boldsymbol{V}[\boldsymbol{a}(\boldsymbol{x}, t), t] \equiv \boldsymbol{v}(\boldsymbol{x}, t). \tag{9.1.9}$$

Here we have two different coordinate descriptions of any continuum property such as density: Namely, an *Eulerian* or spatial one, involving x_1, x_2, x_3, and t; and a

Lagrangian or material one, involving A_1, A_2, A_3, and t. Then denoting the density in Lagrangian and Eulerian variables by

$$\delta = \delta(A_1, A_2, A_3, t) = \delta(\boldsymbol{A}, t) \text{ and } \rho = \rho(x_1, x_2, x_3, t) = \rho(\boldsymbol{x}, t), \tag{9.1.10a}$$

respectively, we have point-particle interchangeability such that

$$\delta[\boldsymbol{a}(\boldsymbol{x}, t), t] = \rho(\boldsymbol{x}, t) \text{ and } \rho[\boldsymbol{X}(\boldsymbol{A}, t), t] = \delta(\boldsymbol{A}, t). \tag{9.1.10b}$$

Note that we are using both $\boldsymbol{x} = (x_1, x_2, x_3)$ and $\boldsymbol{x} = (x, y, z)$ to represent position, whichever seems more convenient at the time. We shall also be using $\boldsymbol{v} = \boldsymbol{v}(x_1, x_2, x_3, t) = (v_1, v_2, v_3)$ and $\boldsymbol{v} = \boldsymbol{v}(x, y, z, t) = (u, v, w)$ interchangeably as well.

We next introduce the concept of a substantial or material derivative of density

$$\frac{\partial \delta(\boldsymbol{A}, t)}{\partial t} = \nabla \rho|_{\boldsymbol{x}=\boldsymbol{X}(\boldsymbol{A},t)} \bullet \frac{\partial \boldsymbol{X}(\boldsymbol{A}, t)}{\partial t} + \left.\frac{\partial \rho}{\partial t}\right|_{\boldsymbol{x}=\boldsymbol{X}(\boldsymbol{A},t)}, \tag{9.1.11a}$$

obtained from (9.1.10a) by employment of the chain rule for functions of several variables. Then, upon substitution of (9.1.7), (9.1.11a) becomes

$$\frac{\partial \delta(\boldsymbol{A}, t)}{\partial t} = \nabla \rho|_{\boldsymbol{x}=\boldsymbol{X}(\boldsymbol{A},t)} \bullet \boldsymbol{V}(\boldsymbol{A}, t) + \left.\frac{\partial \rho}{\partial t}\right|_{\boldsymbol{x}=\boldsymbol{X}(\boldsymbol{A},t)}. \tag{9.1.11b}$$

Finally, representing this substantial derivative in Eulerian variables and making use of (9.1.9), yields

$$\left.\frac{\partial \delta(\boldsymbol{A}, t)}{\partial t}\right|_{\boldsymbol{A}=\boldsymbol{a}(\boldsymbol{x},t)} = \left(\nabla \rho \bullet \boldsymbol{v} + \frac{\partial \rho}{\partial t}\right)(\boldsymbol{x}, t) \equiv \frac{D\rho(\boldsymbol{x}, t)}{Dt}. \tag{9.1.11c}$$

Observe that this substantial derivative reduces to the usual partial time derivative in the absence of convection for which $\boldsymbol{v} \equiv \boldsymbol{0}$ but is inherently nonlinear due to its convection term otherwise.

We shall now show that the substantial derivative of the Jacobian

$$\frac{\partial J(\boldsymbol{A}, t)}{\partial t} = J(\boldsymbol{A}, t)\mathrm{div}\boldsymbol{V}(\boldsymbol{A}, t) \text{ where } \mathrm{div}\boldsymbol{V}(\boldsymbol{A}, t) \equiv (\nabla \bullet \boldsymbol{v})[\boldsymbol{X}(\boldsymbol{A}, t), t] \tag{9.1.12}$$

which for ease of exposition will be developed for a two-dimensional situation and extrapolated to three-dimensions. Hence consider

$$J(A_1, A_2, t) = \begin{vmatrix} \frac{\partial X_1}{\partial A_1} & \frac{\partial X_1}{\partial A_2} \\ \frac{\partial X_2}{\partial A_1} & \frac{\partial X_2}{\partial A_2} \end{vmatrix} (A_1, A_2, t) = \begin{vmatrix} a & b \\ c & d \end{vmatrix} (A_1, A_2, t)$$

Note that

$$\frac{\partial}{\partial t}\begin{vmatrix} a & b \\ c & d \end{vmatrix} = \begin{vmatrix} \frac{\partial a}{\partial t} & \frac{\partial b}{\partial t} \\ c & d \end{vmatrix} + \begin{vmatrix} a & b \\ \frac{\partial c}{\partial t} & \frac{\partial d}{\partial t} \end{vmatrix} = \textcircled{1} + \textcircled{2} \text{ and } \frac{\partial}{\partial t}\left(\frac{\partial X_i}{\partial A_j}\right) = \frac{\partial}{\partial A_j}\left(\frac{\partial X_i}{\partial t}\right) = \frac{\partial V_i}{\partial A_j}$$

where in the above we are assuming the requisite continuity to allow for equivalence of the mixed partial derivatives. Then

$$\textcircled{1} = \begin{vmatrix} \frac{\partial V_1}{\partial A_1} & \frac{\partial V_1}{\partial A_2} \\ c & d \end{vmatrix} = \begin{vmatrix} \alpha a + \beta c & \alpha b + \beta d \\ c & d \end{vmatrix} = \alpha \begin{vmatrix} a & b \\ c & d \end{vmatrix} + \beta \begin{vmatrix} c & d \\ c & d \end{vmatrix} = \alpha \begin{vmatrix} a & b \\ c & d \end{vmatrix}$$

for

$$\boldsymbol{V}(\boldsymbol{A},t) = \boldsymbol{v}[\boldsymbol{X}(\boldsymbol{A},t),t],\ \alpha = \left.\frac{\partial v_1}{\partial x_1}\right|_{\boldsymbol{x}=\boldsymbol{X}(\boldsymbol{A},t)},\ \text{and } \beta = \left.\frac{\partial v_1}{\partial x_2}\right|_{\boldsymbol{x}=\boldsymbol{X}(\boldsymbol{A},t)}.$$

Thus

$$\textcircled{1} = J(A_1,A_2,t)\left.\frac{\partial v_1}{\partial x_1}\right|_{\boldsymbol{x}=\boldsymbol{X}(\boldsymbol{A},t)}. \tag{9.1.13a}$$

In the same manner,

$$\textcircled{2} = \begin{vmatrix} a & b \\ \frac{\partial V_2}{\partial A_1} & \frac{\partial V_2}{\partial A_2} \end{vmatrix} = \begin{vmatrix} a & b \\ \gamma a + \delta c & \gamma b + \delta d \end{vmatrix} = \gamma \begin{vmatrix} a & b \\ a & b \end{vmatrix} + \delta \begin{vmatrix} a & b \\ c & d \end{vmatrix} = \delta \begin{vmatrix} a & b \\ c & d \end{vmatrix}$$

for

$$\gamma = \left.\frac{\partial v_2}{\partial x_1}\right|_{\boldsymbol{x}=\boldsymbol{X}(\boldsymbol{A},t)} \text{ and } \delta = \left.\frac{\partial v_2}{\partial x_2}\right|_{\boldsymbol{x}=\boldsymbol{X}(\boldsymbol{A},t)}.$$

Thus

$$\textcircled{2} = J(A_1,A_2,t)\left.\frac{\partial v_2}{\partial x_2}\right|_{\boldsymbol{x}=\boldsymbol{X}(\boldsymbol{A},t)}. \tag{9.1.13b}$$

Hence, employing (9.1.13), we can deduce that

$$\frac{\partial J(A_1,A_2,t)}{\partial t} = \textcircled{1} + \textcircled{2} = J(A_1,A_2,t)\mathrm{div}_2\boldsymbol{V}(A_1,A_2,t) \tag{9.1.14a}$$

where

$$\mathrm{div}_2\boldsymbol{V}(A_1,A_2,t) \equiv \left(\frac{\partial v_1}{\partial x_1} + \frac{\partial v_2}{\partial x_2}\right)[\boldsymbol{X}(A_1,A_2,t)] \equiv (\nabla_2 \bullet \boldsymbol{v})[\boldsymbol{X}(A_1,A_2,t),t]. \tag{9.1.14b}$$

Extrapolating this development to three-dimensions we would obtain in a similar manner that

$$\frac{\partial J(\boldsymbol{A},t)}{\partial t} = \textcircled{1} + \textcircled{2} + \textcircled{3} \text{ where } \textcircled{3} = J(A,t)\left.\frac{\partial v_3}{\partial x_3}\right|_{\boldsymbol{x}=\boldsymbol{X}(\boldsymbol{A},t)}; \tag{9.1.15}$$

hence demonstrating the desired result that

$$\frac{\partial J(\boldsymbol{X},t)}{\partial t} = J(\boldsymbol{A},t)\mathrm{div}\boldsymbol{V}(\boldsymbol{A},t) \tag{9.1.16a}$$

where

$$\operatorname{div} \boldsymbol{V}(\boldsymbol{A}, t) \equiv \left(\frac{\partial v_1}{\partial x_1} + \frac{\partial v_2}{\partial x_2} + \frac{\partial v_3}{\partial x_3}\right)[\boldsymbol{X}(\boldsymbol{A}, t), t] = (\nabla \bullet \boldsymbol{v})[\boldsymbol{X}(\boldsymbol{A}, t), t]. \tag{9.1.16b}$$

Defining the Jacobian of the transformation in Eulerian variables by

$$J[\boldsymbol{a}(\boldsymbol{x}, t), t] = j(\boldsymbol{x}, t), \tag{9.1.17a}$$

we would obtain the equivalent relationship in those variables given by

$$\left.\frac{\partial J(\boldsymbol{A}, t)}{\partial t}\right|_{\boldsymbol{A}=\boldsymbol{a}(\boldsymbol{x}, t)} = \frac{Dj(\boldsymbol{x}, t)}{Dt} = \left(\frac{\partial j}{\partial t} + \nabla j \bullet \boldsymbol{v}\right)(\boldsymbol{x}, t) = j(\boldsymbol{x}, t)(\nabla \bullet \boldsymbol{v})(\boldsymbol{x}, t). \tag{9.1.17b}$$

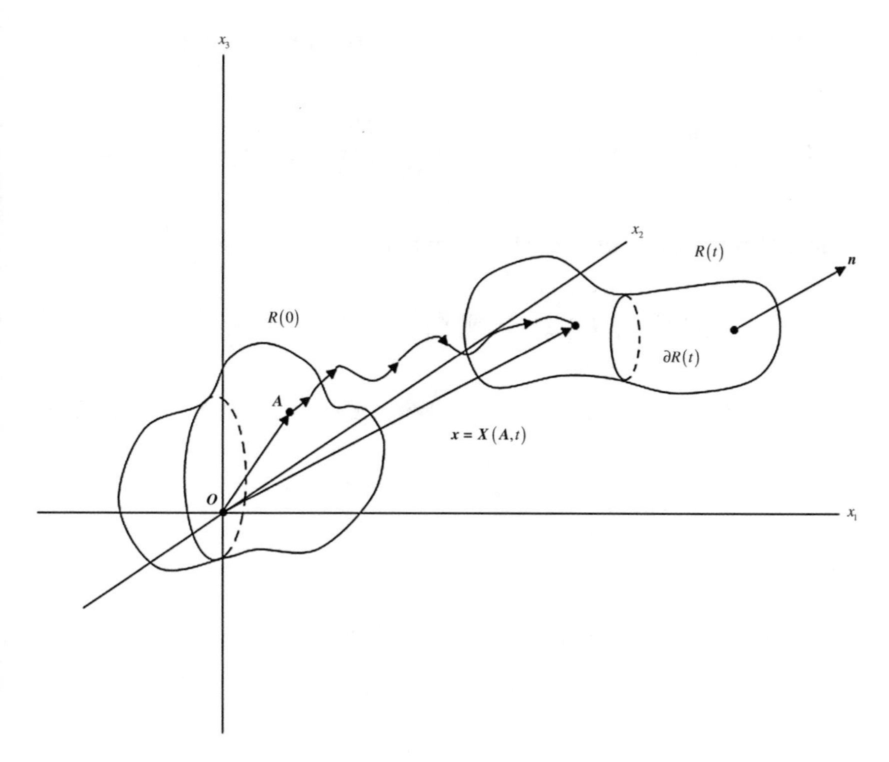

Fig. 9.2 Schematic diagram of a moving material region $R(t)$ where $\boldsymbol{x} = \boldsymbol{X}(\boldsymbol{A}, t)$ is a plot of an associated particle path.

As a final result in this section, we shall now deduce the *Reynolds Transport Theorem* for the material region $R(t)$, depicted in Fig. 9.2, given by

$$\frac{d}{dt} \iiint\limits_{R(t)} f(\boldsymbol{x},t)\, dx_1\, dx_2\, dx_3 = \iiint\limits_{R(t)} \left(\frac{Df}{Dt} + f\nabla \bullet \boldsymbol{v} \right) (\boldsymbol{x},t)\, dx_1\, dx_2\, dx_3 .$$

where $f(x,t)$ is any scalar function and the particles in $R(t)$ are moving according to (9.1.6).

Consider

$$\mathcal{I}(t) = \iiint\limits_{R(t)} f(\boldsymbol{x},t)\, dx_1\, dx_2\, dx_3$$

and make the change of variables $\boldsymbol{x} = \boldsymbol{X}(\boldsymbol{A},t)$, $\boldsymbol{X}(\boldsymbol{A},0) = \boldsymbol{A} \in R(0)$. Then

$$\mathcal{I}(t) = \iiint\limits_{R(0)} (FJ)(\boldsymbol{A},t)\, dA_1\, dA_2\, dA_3$$

where $F(\boldsymbol{A},t) = f[\boldsymbol{X}(\boldsymbol{A},t),t]$ and $dx_1\, dx_2\, dx_3 = J(\boldsymbol{A},t)\, dA_1\, dA_2\, dA_3$. Since $R(0)$ is independent of t

$$\frac{d\mathcal{I}}{dt}(t) = \iiint\limits_{R(0)} \frac{\partial}{\partial t}(FJ)(\boldsymbol{A},t)\, dA_1\, dA_2\, dA_3 = \iiint\limits_{R(0)} \left(\frac{\partial F}{\partial t} J + F \frac{\partial J}{\partial t} \right) (\boldsymbol{A},t)\, dA_1\, dA_2\, dA_3 .$$

Upon employment of (9.1.16a) this becomes

$$\frac{d\mathcal{I}}{dt}(t) = \iiint\limits_{R(0)} \left(\frac{\partial F}{\partial t} + F \mathrm{div} \boldsymbol{V} \right) (\boldsymbol{A},t) J(\boldsymbol{A},t)\, dA_1\, dA_2\, dA_3 .$$

Now, making the inverse change of variables $\boldsymbol{A} = \boldsymbol{a}(\boldsymbol{x},t)$, we deduce the desired result

$$\frac{d\mathcal{I}}{dt}(t) = \iiint\limits_{R(t)} \left(\frac{Df}{Dt} + f\nabla \bullet \boldsymbol{v} \right) (\boldsymbol{x},t)\, dx_1\, dx_2\, dx_3$$

since

$$F[\boldsymbol{a}(\boldsymbol{x},t),t] = f(\boldsymbol{x},t),\ \frac{\partial F}{\partial t}[\boldsymbol{a}(\boldsymbol{x},t),t] = \frac{Df}{Dt}(\boldsymbol{x},t),\ \mathrm{div}\boldsymbol{V}[\boldsymbol{a}(\boldsymbol{x},t),t] = (\nabla \bullet \boldsymbol{v})(\boldsymbol{x},t),$$
$$\text{and } J(\boldsymbol{A},t)\, dA_1\, dA_2\, dA_3 = dx_1\, dx_2\, dx_3 .$$

9.2 Conservation of Mass and Continuity Equation

Let $R(t)$ be a material region with bounding surface $\partial R(t)$ possessing unit outward-pointing normal $\boldsymbol{n}$ (see Fig. 9.2). Here $R(t) \subset \mathcal{D}$ is an *arbitrary* subset of a moving continuum $\mathcal{D}$.

1. The first version of the principle of *conservation mass* states that the mass $M(t)$ contained in $R(t)$ will not change with time or defining

$$M(t) = \iiint\limits_{R(t)} \rho(\boldsymbol{x}, t)\, dx_1\, dx_2\, dx_3,$$

this implies that $M(t) \equiv M(0)$ or equivalently

$$\frac{d}{dt} M(t) = \frac{d}{dt} \iiint\limits_{R(t)} \rho(\boldsymbol{x}, t)\, dx_1\, dx_2\, dx_3 = 0.$$

Now applying the Reynolds Transport Theorem to this integral with $f = \rho$, we obtain

$$\iiint\limits_{R(t)} \left(\frac{D\rho}{Dt} + \rho \nabla \bullet \boldsymbol{v} \right) (\boldsymbol{x}, t)\, dx_1\, dx_2\, dx_3 = 0.$$

Assuming the requisite continuity of its integrand and invoking the DuBois–Reymond Lemma due to the arbitrariness of $R(t)$ this yields the *continuity equation*

$$\left(\frac{D\rho}{Dt} + \rho \nabla \bullet \boldsymbol{v} \right) (\boldsymbol{x}, t) = 0 \text{ for all } \boldsymbol{x} \in \mathcal{D}, \tag{9.2.1a}$$

or, equivalently, upon substitution of (9.1.11c),

$$\left(\frac{\partial \rho}{\partial t} + \nabla \rho \bullet \boldsymbol{v} \right) + \rho \nabla \bullet \boldsymbol{v} = \frac{\partial \rho}{\partial t} + (\nabla \rho \bullet \boldsymbol{v} + \rho \nabla \bullet \boldsymbol{v}) = \frac{\partial \rho}{\partial t} + \nabla \bullet (\rho \boldsymbol{v}) = 0. \tag{9.2.1b}$$

Written in terms of its components, we can represent the continuity equation in either of the following two forms:

$$\frac{\partial \rho}{\partial t} + u \frac{\partial \rho}{\partial x} + v \frac{\partial \rho}{\partial y} + w \frac{\partial \rho}{\partial z} + \rho \left(\frac{\partial u}{\partial x} + \frac{\partial v}{\partial y} + \frac{\partial w}{\partial z} \right) = 0, \tag{9.2.2a}$$

$$\frac{\partial \rho}{\partial t} + \frac{\partial (\rho u)}{\partial x} + \frac{\partial (\rho v)}{\partial y} + \frac{\partial (\rho w)}{\partial z} = 0, \tag{9.2.2b}$$

the latter of which implies that

$$\iiint\limits_{R(t)} \frac{\partial \rho}{\partial t}\, dx_1\, dx_2\, dx_3 = - \iiint\limits_{R(t)} \nabla \bullet (\rho \boldsymbol{v})\, dx_1\, dx_2\, dx_3 = - \iint\limits_{\partial R(t)} \rho \boldsymbol{v} \bullet \boldsymbol{n}\, d\sigma.$$

Let R_0 be a *fixed volume* in space with bounding surface S_0 possessing unit outward-pointing normal $\boldsymbol{n}$ (see Fig. 5.1). Then particularizing our previous result to $R(t) \equiv R_0$ and $\partial R(t) \equiv S_0$ we obtain

$$\iiint_{R_0} \frac{\partial \rho}{\partial t} dx_1\, dx_2\, dx_3 = \iint_{S_0} \rho \boldsymbol{v} \bullet (-\boldsymbol{n})\, d\sigma$$

or, equivalently,

$$\frac{d}{dt} \iiint_{R_0} \rho\, dx_1\, dx_2\, dx_3 = \frac{d}{dt} M_0(t) = \iint_{S_0} \rho v_n\, d\sigma$$

since R_0 is independent of t where $M_0(t) \equiv$ mass in R_0 and $v_n = \boldsymbol{v} \bullet (-\boldsymbol{n}) \equiv$ normal component of velocity across the boundary S_0 into that volume. Here the dimension of both of these terms is mass per unit time and that of the integrand in the surface integral, mass per unit time per unit area which is a flux. Motivated by the form of this relation we now formulate the following alternate statement for conservation of mass:

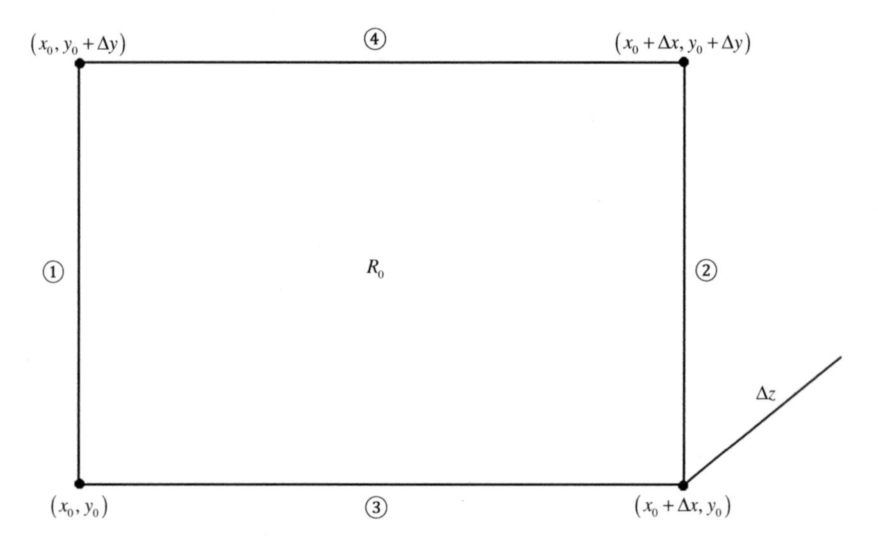

Fig. 9.3 Schematic diagram of a rectangular big box R_0 in Cartesian coordinates where the circled numbers designate its relevant surfaces for a two-dimensional situation.

2. The second version of the principle of conservation of mass states that the time-rate of change of mass $M_0(t)$ in R_0 is equal to the net mass flux across its boundary S_0 into that fixed volume. We shall now apply this version of the conservation of mass to the big box depicted in Fig. 9.3 in order to rederive the continuity equation. For ease of exposition we shall restrict our analysis to a two-dimensional situation and again extrapolate to three-dimensions. That is, we shall consider the density ρ and velocity vector $\boldsymbol{v} = u\boldsymbol{i} + v\boldsymbol{j}$ where $[\rho, u, v] = [\rho, u, v](x, y, t)$. Then as in Section 5.1

$$M_0(t) = \iiint_{R_0} \rho(x,y,t)\,dx\,dy\,dz = \Delta z \int_{x_0}^{x_0+\Delta x} \int_{y_0}^{y_0+\Delta y} \rho(x,y,t)\,dx\,dy$$

and

$$\frac{d}{dt}M_0(t) = \lim_{\Delta t \to 0} \frac{M_0(t+\Delta t) - M_0(t)}{\Delta t} = \Delta z \int_{x_0}^{x_0+\Delta x} \int_{y_0}^{y_0+\Delta y} \frac{\partial \rho}{\partial t}(x,y,t)\,dx\,dy.$$

Further, as per our second version of the conservation of mass, let

$$\frac{dM_0(t)}{dt} = F_{①} + F_{②} + F_{③} + F_{④} \tag{9.2.3}$$

where the right-hand side of (9.2.3) represents the net flux through the sides of the box (note there is no flow into or out of the box from its front or back since it has been assumed that $w = 0$).

We now compute the fluxes $F_{①}$ and $F_{②}$:

①: $x = x_0$, $y_0 \le y \le y_0 + \Delta y$:

In order to determine the proper expression for $F_{①}$ we first consider the special case of constant dependent variables,

$$\rho \equiv \rho_0,\ u \equiv u_0,\ v \equiv v_0, \tag{9.2.4}$$

and calculate the mass flux passing through this surface into the box during the time interval Δt (see Fig. 9.4a, b)

$$F_{①} = \frac{\Delta m_{①}}{\Delta t} = \rho_0(u_0 \Delta t)\frac{\Delta y \Delta z}{\Delta t} = \rho_0 u_0 \Delta y \Delta z. \tag{9.2.5}$$

Generalizing this flux to the nonconstant variable situation we can deduce that

$$F_{①} = \Delta z \int_{y_0}^{y_0+\Delta y} (\rho u)(x_0, y, t)\,dy$$

which reduces to (9.2.5) for the special case of (9.2.4).

② $x = x_0 + \Delta x$, $y_0 \le y \le y_0 + \Delta y$:

The mass flux at this surface is of the same form as $F_{①}$ but with a minus sign since the flow passes *out of* the box. Hence it is given by

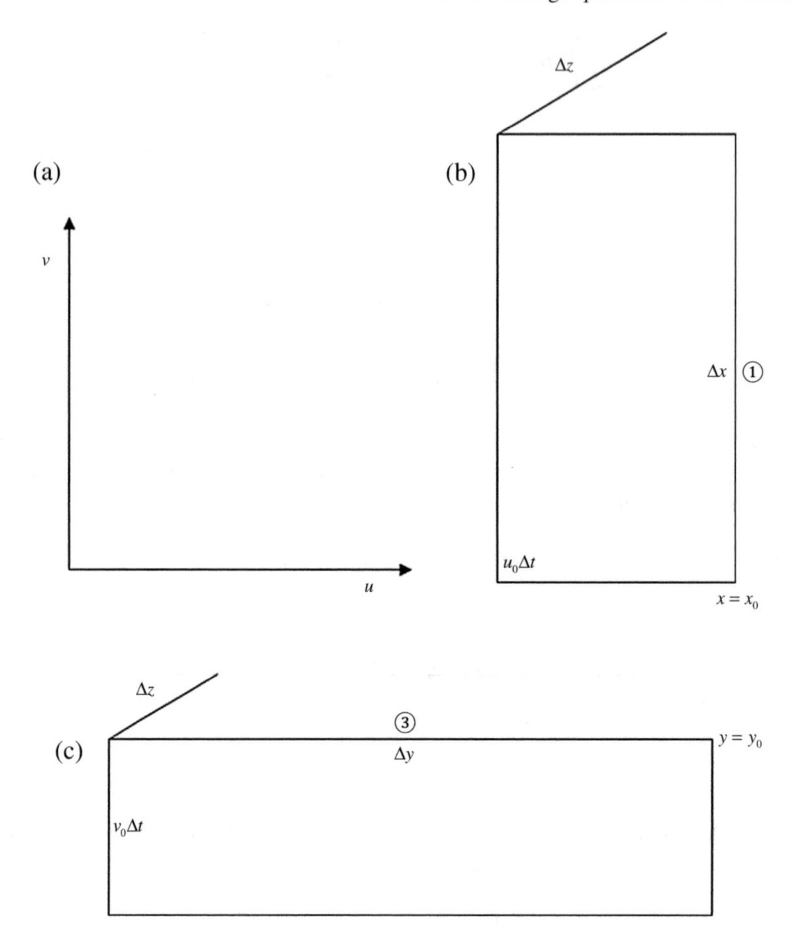

Fig. 9.4 Schematic diagrams associated with the two-dimensional flow situation of Fig. 9.3 representing (a) velocity components and fluxes under constant conditions through surfaces (b) ①, and (c) ③.

$$F_{②} = -\Delta z \int_{y_0}^{y_0+\Delta y} (\rho u)(x_0 + \Delta x, y, t)\, dy.$$

Then adding $F_{①}$ and $F_{②}$ together we obtain

$$
\begin{aligned}
F_{\textcircled{1}} + F_{\textcircled{2}} &= \Delta z \int_{y_0}^{y_0+\Delta y} [(\rho u)(x_0, y, t) - (\rho u)(x_0 + \Delta x, y, t)]\, dy \\
&= -\Delta z \int_{y_0}^{y_0+\Delta y} \int_{x_0}^{x_0+\Delta x} \frac{\partial}{\partial x}(\rho u)(x, y, t)\, dx\, dy,
\end{aligned}
$$

upon employment of the fundamental theorem of the calculus.

We next compute the fluxes $F_{\textcircled{3}}$ and $F_{\textcircled{4}}$:

$\textcircled{3}$ $y = y_0,\ x_0 \le x \le x + \Delta x$:

Again assuming the constant variable situation of (9.2.4), we calculate the mass flux passing through this surface into the box during the time interval Δt obtaining (see Fig. 9.4a, c)

$$
F_{\textcircled{3}} = \frac{\Delta m_{\textcircled{3}}}{\Delta t} = \rho_0 (v_0 \Delta t) \frac{\Delta x \Delta z}{\Delta t} = \rho_0 v_0 \Delta x \Delta z. \tag{9.2.6}
$$

Generalizing this flux to the nonconstant dependent variable situation we can deduce that

$$
F_{\textcircled{3}} = \Delta z \int_{x_0}^{x_0+\Delta x} (\rho v)(x, y_0, t)\, dx
$$

which reduces to (9.2.6) for the special case of (9.2.4).

$\textcircled{4}$ $y = y_0 + \Delta y,\ x_0 \le x \le x_0 + \Delta x$:

The mass flux at this surface is of the same form as $F_{\textcircled{3}}$ but with a minus sign since the flow passes *out of* the box. Hence it is given by

$$
F_{\textcircled{4}} = -\Delta z \int_{x_0}^{x_0+\Delta x} (\rho v)(x, y_0 + \Delta y, t)\, dx.
$$

Then adding $F_{\textcircled{3}}$ and $F_{\textcircled{4}}$ together we obtain

$$
\begin{aligned}
F_{\textcircled{3}} + F_{\textcircled{4}} &= \Delta z \int_{x_0}^{x_0+\Delta x} [(\rho v)(x, y_0, t) - (\rho v)(x, y_0 + \Delta y, t)]\, dx \\
&= -\Delta z \int_{y_0}^{y_0+\Delta y} \int_{x_0}^{x_0+\Delta x} \frac{\partial}{\partial y}(\rho v)(x, y, t)\, dx\, dy,
\end{aligned}
$$

upon employment of the fundamental theorem of the calculus.
Now combining all these results and using (9.2.3) yields

$$\int_{y_0}^{y_0+\Delta y}\int_{x_0}^{x_0+\Delta x}\left[\frac{\partial\rho}{\partial t}+\frac{\partial(\rho u)}{\partial x}+\frac{\partial(\rho v)}{\partial y}\right](x,y,t)\,dx\,dy=0$$

upon cancellation of $\Delta z > 0$. Further employing the integral mean value theorem and cancelling $\Delta x, \Delta y > 0$ we obtain

$$\left[\frac{\partial\rho}{\partial t}+\frac{\partial(\rho u)}{\partial x}+\frac{\partial(\rho v)}{\partial y}\right](x^*,y^*,t)=0 \tag{9.2.7a}$$

where

$$x_0 < x^* < x_0+\Delta x,\; y_0 < y^* < y_0+\Delta y. \tag{9.2.7b}$$

Finally, taking the limit of (9.2.7) as $\Delta x, \Delta y \to 0$, we rederive the two-dimensional version of the continuity equation in the form (9.2.2b)

$$\left[\frac{\partial\rho}{\partial t}+\frac{\partial(\rho u)}{\partial x}+\frac{\partial(\rho v)}{\partial y}\right](x_0,y_0,t)=0 \tag{9.2.8}$$

since (x_0, y_0) may be considered any arbitrary point in the plane.
Extrapolating this method to three-dimensions by employing a big box with $z_0 \le z \le z_0+\Delta z$, considering a nonzero w, and assuming all dependent variables to be functions of (x,y,z,t), the net flux into and out of that box through its front ⑤ and back ⑥, respectively, would yield in a similar manner to that employed for two-dimensions

$$F_{⑤}+F_{⑥}=-\int_{x_0}^{x_0+\Delta x}\int_{y_0}^{y_0+\Delta y}\int_{z_0}^{z_0+\Delta z}\frac{\partial}{\partial z}(\rho w)(x,y,z,t)\,dx\,dy\,dz.$$

Then extending (9.2.3) to three-dimensions by letting

$$\frac{dM_0(t)}{dt}=F_{①}+F_{②}+F_{③}+F_{④}+F_{⑤}+F_{⑥} \tag{9.2.9}$$

and proceeding exactly as for the two-dimensional case we would rederive the continuity equation in the form (9.2.2b)

$$\left[\frac{\partial\rho}{\partial t}+\frac{\partial(\rho u)}{\partial x}+\frac{\partial(\rho v)}{\partial y}+\frac{\partial(\rho w)}{\partial z}\right](x_0,y_0,z_0,t)=0 \tag{9.2.10}$$

since (x_0, y_0, z_0) may be considered any arbitrary point in the plane.

We complete this section by converting the two-dimensional version of the continuity equation to polar coordinates and hence its three-dimensional version to cylindrical coordinates while leaving for Problem 9.4 the conversion of that equation to spherical coordinates.

Recapitulating, the two-dimensional continuity equation is given by

$$\left[\frac{\partial\rho}{\partial t}+\frac{\partial}{\partial x}(\rho u)+\frac{\partial}{\partial y}(\rho v)\right](x,y,t)=0.$$

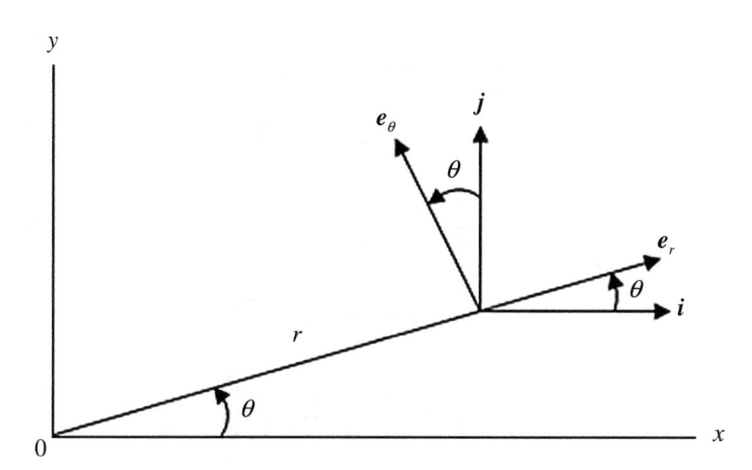

Fig. 9.5 Representations of the two-dimensional basis vectors for both Cartesian and polar coordinates.

We now introduce the polar coordinate transformation (see Fig. 9.5)

$$x=r\cos(\theta),\ y=r\sin(\theta),\ \rho(x,y,t)=R(r,\theta,t); \tag{9.2.11a}$$

$$\boldsymbol{v}(x,y,t)=u(x,y,t)\boldsymbol{i}+v(x,y,t)\boldsymbol{j}=\boldsymbol{V}(r,\theta,t)=U(r,\theta,t)\boldsymbol{e}_r+V(r,\theta,t)\boldsymbol{e}_\theta; \tag{9.2.11b}$$

$$\boldsymbol{e}_r=(\boldsymbol{i}\bullet\boldsymbol{e}_r)\boldsymbol{i}+(\boldsymbol{j}\bullet\boldsymbol{e}_r)\boldsymbol{j}=\cos(\theta)\boldsymbol{i}+\cos\left(\frac{\pi}{2}-\theta\right)\boldsymbol{j}=\cos(\theta)\boldsymbol{i}+\sin(\theta)\boldsymbol{j}, \tag{9.2.11c}$$

$$\boldsymbol{e}_\theta=(\boldsymbol{i}\bullet\boldsymbol{e}_\theta)\boldsymbol{i}+(\boldsymbol{j}\bullet\boldsymbol{e}_\theta)\boldsymbol{j}=\cos\left(\frac{\pi}{2}+\theta\right)\boldsymbol{i}+\cos(\theta)\boldsymbol{j}=-\sin(\theta)\boldsymbol{i}+\cos(\theta)\boldsymbol{j}. \tag{9.2.11d}$$

Then, taking the inner product of (9.2.11b) with $\boldsymbol{e}_r$ and $\boldsymbol{e}_\theta$ sequentially, we obtain

$$U=u(\boldsymbol{i}\bullet\boldsymbol{e}_r)+v(\boldsymbol{j}\bullet\boldsymbol{e}_r)=u\cos(\theta)+v\sin(\theta); \tag{9.2.12a}$$

$$V=V(\boldsymbol{i}\bullet\boldsymbol{e}_\theta)+v(\boldsymbol{j}\bullet\boldsymbol{e}_\theta)=-u\sin(\theta)+v\cos(\theta). \tag{9.2.12b}$$

Further

$$\frac{\partial}{\partial x}=\frac{\partial r}{\partial x}\frac{\partial}{\partial r}+\frac{\partial\theta}{\partial x}\frac{\partial}{\partial\theta},\ \frac{\partial}{\partial y}=\frac{\partial r}{\partial y}\frac{\partial}{\partial r}+\frac{\partial\theta}{\partial y}\frac{\partial}{\partial\theta}.$$

We next compute the relevant partial derivatives contained in the above expression directly by inverting the polar coordinate transformation of (9.2.11a) which yields

$$r = \sqrt{x^2 + y^2},\ \theta = \arctan\left(\frac{y}{x}\right). \tag{9.2.13}$$

Thus

$$\frac{\partial r}{\partial x} = \frac{x}{\sqrt{x^2+y^2}} = \cos(\theta), \tag{9.2.14a}$$

$$\frac{\partial r}{\partial y} = \frac{y}{\sqrt{x^2+y^2}} = \sin(\theta); \tag{9.2.14b}$$

$$\frac{\partial \theta}{\partial x} = -\frac{y/x^2}{1+y^2/x^2} = \frac{-y}{x^2+y^2} = -\frac{\sin(\theta)}{r}, \tag{9.2.14c}$$

$$\frac{\partial \theta}{\partial y} = \frac{1/x}{1+y^2/x^2} = \frac{x}{x^2+y^2} = \frac{\cos(\theta)}{r}. \tag{9.2.14d}$$

Hence

$$\frac{\partial}{\partial x} = \cos(\theta)\frac{\partial}{\partial r} - \left(\frac{\sin(\theta)}{r}\right)\frac{\partial}{\partial \theta}, \tag{9.2.15a}$$

$$\frac{\partial}{\partial y} = \sin(\theta)\frac{\partial}{\partial r} + \left(\frac{\cos(\theta)}{r}\right)\frac{\partial}{\partial \theta}. \tag{9.2.15b}$$

Finally,

$$\frac{\partial \rho}{\partial t} = \frac{\partial R}{\partial t}; \tag{9.2.16a}$$

$$\begin{aligned}
\frac{\partial(\rho u)}{\partial x} + \frac{\partial(\rho v)}{\partial y} &= \left\{\cos(\theta)\frac{\partial}{\partial r} - \left(\frac{\sin(\theta)}{r}\right)\frac{\partial}{\partial \theta}\right\}(Ru) \\
&\quad + \left\{\sin(\theta)\frac{\partial}{\partial r} + \left(\frac{\cos(\theta)}{r}\right)\frac{\partial}{\partial \theta}\right\}(Rv) \\
&= \frac{\partial}{\partial r}\{R[u\cos(\theta) + v\sin(\theta)]\} + \frac{1}{r}\frac{\partial}{\partial \theta}\{R[-u\sin(\theta) + v\cos(\theta)]\} \\
&\quad + \frac{R[u\cos(\theta) + v\sin(\theta)]}{r} \\
&= \frac{\partial(RU)}{\partial r} + \frac{1}{r}\frac{\partial(RV)}{\partial \theta} + \frac{RU}{r} = \frac{1}{r}\frac{\partial(rRU)}{\partial r} + \frac{1}{r}\frac{\partial(RV)}{\partial \theta}.
\end{aligned} \tag{9.2.16b}$$

Therefore from (9.2.16) the continuity equation (9.2.10) in polar coordinates is given by

$$\frac{\partial R}{\partial t} + \frac{1}{r}\frac{\partial(rRU)}{\partial r} + \frac{1}{r}\frac{\partial(RV)}{\partial \theta} = 0. \tag{9.2.17}$$

If we were to introduce the cylindrical coordinate transformation

$$x = r\cos(\theta),\ y = r\sin(\theta),\ z = z;\ \rho(x, y, z, t) = R(r, \theta, z, t); \tag{9.2.18a}$$

$$\boldsymbol{v}(x,y,z,t) = u(x,y,z,t)\boldsymbol{i} + v(x,y,z,t)\boldsymbol{j} + w(x,y,z,t)\boldsymbol{k}$$
$$= \boldsymbol{V}(r,\theta,z,t) = U(r,\theta,z,t)\boldsymbol{e}_r + V(r,\theta,z,t)\boldsymbol{e}_\theta + W(r,\theta,z,t)\boldsymbol{k}, \qquad (9.2.18\text{b})$$

it is clear that

$$w(x,y,z,t) = W(r,\theta,z,t) \qquad (9.2.18\text{c})$$

and thus

$$\frac{\partial(\rho w)}{\partial z} = \frac{\partial(RW)}{\partial z}. \qquad (9.2.19)$$

From (9.2.16) and (9.2.19) we can conclude that the continuity equation (9.2.2b) in cylindrical coordinates is given by

$$\frac{\partial R}{\partial t} + \frac{1}{r}\frac{\partial(rRU)}{\partial r} + \frac{1}{r}\frac{\partial(RV)}{\partial \theta} + \frac{\partial(RW)}{\partial z} = 0. \qquad (9.2.20)$$

Since it is sometimes difficult in actual practice to use the approaches employed above to deduce (9.2.12) and (9.2.13), the transformed system's velocity components in terms of those of the Cartesian coordinate system and the inverse transformation used to calculate the relevant partial derivatives of (9.2.14) directly, we close this section with alternative indirect methods to obtain those quantities included in (9.2.12) and (9.2.14). These methods to be developed below are particularly convenient for the conversion of the continuity equation to spherical coordinates and their employment in this capacity are suggested in Problem 9.4 concerned with that specific transformation.

Toward this end we consider a "particle path"-type description of the medium given by

$$\boldsymbol{x}(t) = x(t)\boldsymbol{i} + y(t)\boldsymbol{j} \qquad (9.2.21\text{a})$$

and

$$\dot{\boldsymbol{x}}(t) = \boldsymbol{v}(t) = u(t)\boldsymbol{i} + v(t)\boldsymbol{j} \qquad (9.2.21\text{b})$$

where

$$\dot{x}(t) = u(t) \qquad (9.2.21\text{c})$$

and

$$\dot{y}(t) = v(t). \qquad (9.2.21\text{d})$$

Then for polar coordinates

$$\boldsymbol{x}(t) = r(t)\boldsymbol{e}_r(t) \qquad (9.2.22\text{a})$$

and

$$\boldsymbol{v}(t) = \boldsymbol{V}(t) = U(t)\boldsymbol{e}_r(t) + V(t)\boldsymbol{e}_\theta(t) \qquad (9.2.22\text{b})$$

where

$$\boldsymbol{e}_r(t) = \cos[\theta(t)]\boldsymbol{i} + \sin[\theta(t)]\boldsymbol{j} \qquad (9.2.22\text{c})$$

and

$$\boldsymbol{e}_\theta(t) = -\sin[\theta(t)]\boldsymbol{i} + \cos[\theta(t)]\boldsymbol{j}. \qquad (9.2.22\text{d})$$

Now, upon differentiating (9.2.22a) and employing (9.2.22c, 9.2.22d),

$$\dot{\boldsymbol{x}}(t) = \dot{r}(t)\boldsymbol{e}_r(t) + r(t)\dot{\boldsymbol{e}}_r(t) = \dot{r}(t)\boldsymbol{e}_r(t) + r(t)\dot{\theta}(t)\boldsymbol{e}_\theta(t) = \boldsymbol{V}(t), \tag{9.2.23a}$$

which implies, from (9.2.22b), that

$$U(t) = \dot{r}(t) \tag{9.2.23b}$$

and

$$V(t) = r(t)\dot{\theta}(t). \tag{9.2.23c}$$

Since, from (9.2.21a) and (9.2.22a),

$$x(t) = r(t)\cos[\theta(t)] \tag{9.2.24a}$$

and

$$y(t) = r(t)\sin[\theta(t)], \tag{9.2.24b}$$

differentiation of (9.2.24) and employment of (9.2.21c, d) and (9.2.23b, c), yields

$$\dot{x}(t) = u(t) = \dot{r}(t)\cos[\theta(t)] - r(t)\dot{\theta}(t)\sin[\theta(t)] = U(t)\cos[\theta(t)] - V(t)\sin[\theta(t)] \tag{9.2.25a}$$

and

$$\dot{y}(t) = v(t) = \dot{r}(t)\sin[\theta(t)] + r(t)\dot{\theta}(t)\cos[\theta(t)] = U(t)\sin[\theta(t)] - V(t)\cos[\theta(t)]. \tag{9.2.25b}$$

Finally (9.2.25) can be inverted to obtain (9.2.12). This is the method suggested to obtain the relationships between velocity components in the spherical coordinate continuity equation transformation Problem 9.4 where the identifications analogous to (9.2.23b, c) are provided and do not have to be derived.

Note that our approach of converting the continuity equation into cylindrical coordinates differs from developments which treat the spatial derivative terms separately and substitute (9.2.25) directly into the continuity equation as a first step in this transformation rather than treating them together and employing (9.2.12) only in the last step of that process. The former approach requires much more tedious calculation than does the latter one which is the reason we have chosen to employ it instead. This approach is also suggested for Problem 9.4.

In order to obtain the required partial derivatives contained in (9.2.14) without having to invert the transformation and computing them directly, we now introduce the following indirect approach employing implicit partial differentiation. Consider the polar coordinate transformation of (9.2.11a)

$$x = r\cos(\theta), y = r\sin(\theta) \tag{9.2.26a}$$

which implicitly defines

$$r = r(x, y), \theta = \theta(x, y). \tag{9.2.26b}$$

Now assume that x and y are independent variables such that

$$\frac{\partial x}{\partial y} = \frac{\partial y}{\partial x} = 0,\ \frac{\partial x}{\partial x} = \frac{\partial y}{\partial y} = 1; \tag{9.2.27a}$$

take the partial derivatives of (9.2.26) with respect to them, sequentially, obtaining

$$\begin{aligned} 1 &= \cos(\theta)\frac{\partial r}{\partial x} - r\sin(\theta)\frac{\partial \theta}{\partial x}, \\ 0 &= \sin(\theta)\frac{\partial r}{\partial x} + r\cos(\theta)\frac{\partial \theta}{\partial x}; \end{aligned} \tag{9.2.27b}$$

and

$$\begin{aligned} 1 &= \cos(\theta)\frac{\partial r}{\partial y} - r\sin(\theta)\frac{\partial \theta}{\partial y}, \\ 0 &= \sin(\theta)\frac{\partial r}{\partial y} + r\cos(\theta)\frac{\partial \theta}{\partial y}; \end{aligned} \tag{9.2.27c}$$

Solving each of these systems (9.2.27b, 9.2.27c) simultaneously yields

$$\frac{\partial r}{\partial x} = \cos(\theta),\ \frac{\partial \theta}{\partial x} = -\frac{\sin(\theta)}{r}; \tag{9.2.28a}$$

and

$$\frac{\partial r}{\partial y}\sin(\theta),\ \frac{\partial \theta}{\partial y} = \frac{\cos(\theta)}{r}; \tag{9.2.28b}$$

which rederives the partial derivatives contained in (9.2.14) without having to resort to finding explicit representations for the functions of (9.2.26b).

Sometimes it is more convenient to find these partial derivatives by "mixing the modes" as it were and calculating some of them directly by inverting the transformation for those variables while calculating the rest of them by employing implicit partial differentiation as demonstrated above. This is the case for the continuity equation transformation Problem 9.4 involving the spherical coordinates (r, θ, φ) representing distance and the azimuthal and planar sweep angles, respectively. Here the derivatives of r and φ may be calculated directly while then those for θ can be determined most easily by implicit differentiation of the z-coordinate relation.

9.3 Balance of Linear and Angular Momentum and Conservation of Energy

We postulate the general balance law

$$\frac{d}{dt}\iiint\limits_{R(t)} (\rho F)(\boldsymbol{x},t)\,d\tau = \iiint\limits_{R(t)} (\rho Q)(\boldsymbol{x},t)\,d\tau - \iint\limits_{\partial R(t)} j(\boldsymbol{x},\boldsymbol{n},t)\,d\sigma$$

where $R(t)$ is the same material region as employed for conservation of mass, $d\tau \equiv dx_1\,dx_2\,dx_3$, and F is the quantity per unit mass the balance of which concerns us while Q and j are related to its source in the bulk and flux across the boundary of that region, respectively. Now rewriting this balance law in the form ① = ② − ③, we first convert ① to a volume integral of the same form as ② by employing the Reynolds Transport Theorem with $f = \rho F$ and obtain

$$① = \iiint\limits_{R(t)} \left[\frac{D}{Dt}(\rho F) + \rho F \nabla \bullet \boldsymbol{v}\right] d\tau = \iiint\limits_{R(t)} \left[\rho\frac{DF}{Dt} + F\left(\frac{D\rho}{Dt} + \rho\nabla \bullet \boldsymbol{v}\right)\right] d\tau$$

$$= \iiint\limits_{R(t)} \rho\frac{DF}{Dt}\,d\tau.$$

Thus, this balance law becomes

$$\iint\limits_{\partial R(t)} j(\boldsymbol{x},\boldsymbol{n},t)\,d\sigma = \iiint\limits_{R(t)} \rho\left(Q - \frac{DF}{Dt}\right)(\boldsymbol{x},t)\,d\tau.$$

Now, letting $R(t) = R_L(t) \equiv$ a family of similar star-shaped regions centered about the point $\boldsymbol{x}_0$ such that its volume $\mathcal{V}[R_L(t)] = \lambda_R L^3$ and its surface area $\mathcal{S}[R_L(t)] = \mu_R L^2$ where λ_R and μ_R are constants characteristic of the shape of this family – *e.g.*, for a family of spheres of radius L, $\lambda_R = 4\pi/3$ and $\mu_R = 4\pi$; we can deduce the *principle of local flux equilibrium* as follows

$$\iint\limits_{\partial R_L(t)} j(\boldsymbol{x},\boldsymbol{n},t)\,d\sigma = \left[\rho\left(Q - \frac{DF}{Dt}\right)\right](\boldsymbol{x}^*,t)\lambda_R L^3 \Rightarrow \lim_{L\to 0}\frac{1}{L^2}\iint\limits_{\partial R_L(t)} j(\boldsymbol{x},\boldsymbol{n},t)\,d\sigma = 0.$$

Applying this principle sequentially to (see [69])

(i) $R_L(t) \equiv$ a family of flakes implies that $j(\boldsymbol{x},-\boldsymbol{n},t) = -j(\boldsymbol{x},\boldsymbol{n},t)$.
(ii) $R_L(t) \equiv$ a family of tetrahedrons implies that $j(\boldsymbol{x},\boldsymbol{n},t) = \boldsymbol{J}(\boldsymbol{x},t) \bullet \boldsymbol{n}$ where $\boldsymbol{J}(\boldsymbol{x},t) \equiv$ the flux vector with components $j(\boldsymbol{x},\boldsymbol{e}_i,t)$ for $\boldsymbol{e}_i \equiv$ unit vector in the x_i-direction; $i = 1,2,3$.

Next using this relation and employing the divergence theorem, we convert ③ to

$$③ = \iiint\limits_{R(t)} (\nabla \bullet \boldsymbol{J})(\boldsymbol{x},t)\,d\tau.$$

Then substituting both these results for ① and ③ in the balance law we obtain

$$\iiint_{R(t)} \left[\rho\left(\frac{DF}{Dt} - Q\right) + \nabla \bullet \boldsymbol{J}\right](\boldsymbol{x}, t) = 0.$$

Assuming the requisite continuity of its integrand and invoking the DuBois–Reymond Lemma due to the arbitrariness of $R(t)$ this yields the *general balance equation*

$$\rho\frac{DF}{Dt} = \rho Q - \nabla \bullet \boldsymbol{J} \text{ for all } \boldsymbol{x} \in \mathcal{D}.$$

We now use this law to deduce the governing partial differential equations for the balance of linear and angular momentum and the conservation of energy as follows:

(a) Balance of Linear Momentum: Since this is a vector relation whereas our general balance law was for a scalar quantity we shall use the designation "$\sim$" to denote the relevant identifications. Then $F \sim \boldsymbol{v}$, $Q \sim \boldsymbol{f}$, $-j \sim \boldsymbol{t}$ where $\boldsymbol{f} \equiv$ body force vector per unit mass and $\boldsymbol{t} \equiv$ stress vector while $-\boldsymbol{J} \sim \boldsymbol{T} \equiv$ stress tensor with components $T_{ij} = t_j(\boldsymbol{x}, \boldsymbol{e}_i, t)$ for $i, j = 1, 2, 3$. Thus the governing so-called Cauchy vector equation for balance of linear momentum is given by

$$\rho\frac{D\boldsymbol{v}}{Dt} = \rho\boldsymbol{f} + \nabla \bullet \boldsymbol{T} \tag{9.3.1a}$$

where

$$(\nabla \bullet \boldsymbol{T})_j = \partial_i T_{ij} \text{ for } \partial_i \equiv \frac{\partial}{\partial x_i} \text{ and } \partial_i T_{ij} \equiv \sum_{i=1}^{3} \partial_i T_{ij}$$

or, in component form,

$$\rho\frac{Dv_j}{Dt} = \rho f_j + \partial_i T_{ij} \text{ for } j = 1, 2, 3. \tag{9.3.1b}$$

(b) Balance of Angular Momentum: Then, assuming that we are dealing with a nonpolar substance having no internal moments and giving rise to neither body torques nor surface stress couples, $F \sim \boldsymbol{r} \times \boldsymbol{v}$, $Q \sim \boldsymbol{r} \times \boldsymbol{f}$, $-j \sim \boldsymbol{r} \times \boldsymbol{t}$ where $\boldsymbol{r} \equiv$ position vector with components x_j for $j = 1, 2, 3$ while $-\boldsymbol{J} \sim \boldsymbol{r} \times \boldsymbol{T}$. Thus the governing vector equation for balance of angular momentum in this event is given by

$$\rho\frac{D(\boldsymbol{r} \times \boldsymbol{v})}{Dt} = \rho\boldsymbol{r} \times \boldsymbol{f} + \nabla \bullet (\boldsymbol{r} \times \boldsymbol{T})$$

where

$$[\nabla \bullet (\boldsymbol{r} \times \boldsymbol{T})]_i = \partial_p(\epsilon_{ijk} x_j T_{pk}) = \epsilon_{ijk}\partial_p(x_j T_{pk}) \text{ for } i = 1, 2, 3$$

with

$$\epsilon_{ijk} = \begin{cases} 1, & \text{if indices are in cyclic order} \\ 0, & \text{if any two of them are equal} \\ -1, & \text{if they are in anticyclic order} \end{cases}.$$

Now since

$$\frac{D(\boldsymbol{r} \times \boldsymbol{v})}{Dt} = \frac{D\boldsymbol{r}}{Dt} \times \boldsymbol{v} + \boldsymbol{r} \times \frac{D\boldsymbol{v}}{dt} = \boldsymbol{v} \times \boldsymbol{v} + \boldsymbol{r} \times \frac{D\boldsymbol{v}}{dt} = \boldsymbol{r} \times \frac{D\boldsymbol{v}}{dt}$$

and

$$\begin{aligned} \epsilon_{ijk}\partial_p(x_j T_{pk}) &= \epsilon_{ijk}[(\partial_p x_j)T_{pk} + x_j \partial_p T_{pk}] = \epsilon_{ijk}(\delta_{pj} T_{pk} + x_j \partial_p T_{pk}) \\ &= \epsilon_{ijk}(T_{jk} + x_j \partial_p T_{pk}), \end{aligned}$$

this equation becomes in component form, upon making use of (9.3.1)

$$\epsilon_{ijk} T_{jk} = \epsilon_{ijk} x_j \left[\rho \frac{Dv_k}{Dt} - \rho f_k - \partial_p T_{pk} \right] = 0 \text{ for } i = 1,2,3. \tag{9.3.2a}$$

Equation (9.3.2a) implies that

$$\begin{aligned} \epsilon_{1jk} T_{jk} &= \epsilon_{123} T_{23} + \epsilon_{132} T_{32} = T_{23} - T_{32} = 0 \text{ or } T_{23} = T_{32}; \\ \epsilon_{2jk} T_{jk} &= \epsilon_{231} T_{12} + \epsilon_{213} T_{13} = T_{31} - T_{13} = 0 \text{ or } T_{31} = T_{13}; \\ \epsilon_{3jk} T_{jk} &= \epsilon_{312} T_{12} + \epsilon_{321} T_{21} = T_{12} - T_{21} = 0 \text{ or } T_{12} = T_{21}; \\ &\text{or } T_{ij} = T_{ji};\ i,j = 1,2,3. \end{aligned} \tag{9.3.2b}$$

Hence for a nonpolar substance the balance of angular momentum merely implies that the stress tensor is symmetric. Finally combining (9.3.1) and (9.3.2) we have the following balance of momentum equations for such a substance given by

$$\rho \frac{Dv_i}{Dt} = \rho f_i + \partial_j T_{ij} \text{ for } i = 1,2,3. \tag{9.3.3}$$

(c) Conservation of Energy: For a nonpolar substance $F = e + (\boldsymbol{v} \bullet \boldsymbol{v})/2$ which represents the sum of the internal plus the kinetic energies, $Q = r/\rho + \boldsymbol{f} \bullet \boldsymbol{v}$ which represents the sum of the energy source plus the rate of work done by the body forces, and $j = h - \boldsymbol{t} \bullet \boldsymbol{v}$ which represents the energy flux plus the rate of work done by the surface stresses while $\boldsymbol{J} = -\boldsymbol{T} \bullet \boldsymbol{v} + \boldsymbol{q}$ where $\boldsymbol{q} \equiv$ energy flux vector. This leads to the following energy equation

$$\rho \frac{D(e + \boldsymbol{v} \bullet \boldsymbol{v}/2)}{Dt} = r + \rho \boldsymbol{f} \bullet \boldsymbol{v} - \nabla \bullet \boldsymbol{q} + \nabla \bullet (\boldsymbol{T} \bullet \boldsymbol{v})$$

where

$$\nabla \bullet (\boldsymbol{T} \bullet \boldsymbol{v}) = \partial_j (T_{ij} v_i) = v_i \partial_j T_{ij} + T_{ij} \partial_j v_i.$$

Further

$$\frac{D(e+\boldsymbol{v}\bullet\boldsymbol{v}/2)}{Dt} = \frac{De}{Dt} + \frac{D\boldsymbol{v}}{Dt}\bullet\boldsymbol{v}$$

and

$$\nabla\bullet(\boldsymbol{T}\bullet\boldsymbol{v}) = \boldsymbol{v}\bullet(\nabla\bullet\boldsymbol{T}) + T_{ij}\varepsilon_{ij}$$

where

$$\boldsymbol{v}\bullet(\nabla\bullet\boldsymbol{T}) = v_i\partial_j T_{ij} \text{ and } \partial_j v_i = \varepsilon_{ij} + \Omega_{ij}$$

with

$$\varepsilon_{ij} = \varepsilon_{ji} = \frac{\partial_j v_i + \partial_i v_j}{2} \text{ and } \Omega_{ij} = -\Omega_{ji} = \frac{\partial_j v_i - \partial_i v_j}{2} \tag{9.3.4a}$$

since

$$T_{ij}\partial_j v_i = T_{ij}(\varepsilon_{ij} + \Omega_{ij}) = T_{ij}\varepsilon_{ij}$$

given that

$$T_{ij}\Omega_{ij} = T_{ji}\Omega_{ji} = -T_{ij}\Omega_{ij} \Rightarrow T_{ij}\Omega_{ij} = 0.$$

Thus substituting these results in the conservation of energy equation we obtain

$$\rho\frac{De}{Dt} + \left(\rho\frac{D\boldsymbol{v}}{Dt} - \rho\boldsymbol{f} - \nabla\bullet T\right)\bullet\boldsymbol{v} = r - \nabla\bullet\boldsymbol{q} + T_{ij}\varepsilon_{ij}$$

which, upon employment of (9.3.3), yields that equation in its final form

$$\rho\frac{De}{Dt} = r - \nabla\bullet\boldsymbol{q} + T_{ij}\varepsilon_{ij}. \tag{9.3.4b}$$

These continuity, momentum, and energy equations of (9.2.1), (9.3.3), and (9.3.4), respectively, constitute the governing partial differential equations for phenomena involving nonpolar moving continua. As indicated in Section 5.1 it is still necessary to adopt constitutive relations and equations of state in order to complete the mathematical formulation of the fluid mechanical situations we wish to model. Toward that end we shall develop such relationships between our dependent variables in the next two sections.

9.4 Constitutive Relation

We begin this section by developing a rationale for a constitutive relationship between the stress tensor with components T_{ij} and the rate of strain tensor with components ε_{ij}, both defined in the last section. First of all we give a definition of Cartesian tensors of various orders. Consider the Cartesian coordinate system $OX_1X_2X_3$ and a rotated such system $OX_1'X_2'X_3'$ as depicted in Fig. 9.6. Let a point P have coordinates (x_1, x_2, x_3) and (x_1', x_2', x_3'), respectively, in these two coordinate systems. Then

$$x_j' = \ell_{ij}x_i \text{ for } j = 1,2,3 \text{ or } x_i = \ell_{ij}x_j' \text{ for } i = 1,2,3 \tag{9.4.1}$$

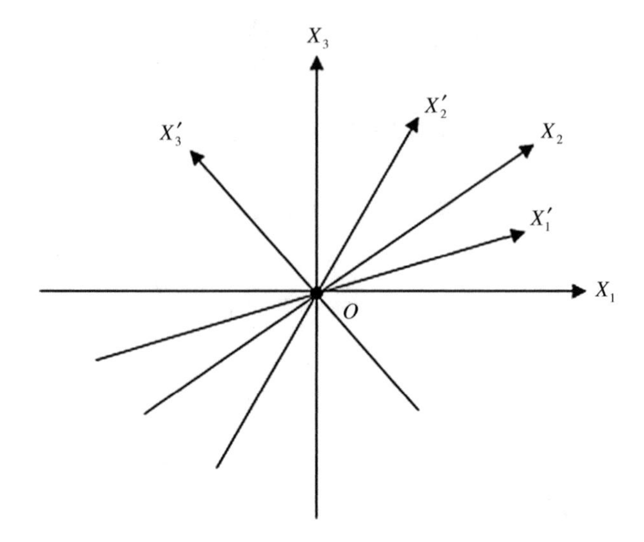

Fig. 9.6 Representation of both an right-handed orthogonal Cartesian coordinate system $OX_1X_2X_3$ and a rotated such system $OX_1'X_2'X_3'$.

where $\ell_{ij} = \boldsymbol{e}_i \bullet \boldsymbol{e}_j'$ and $\boldsymbol{e}_j' \equiv$ unit vector in the x_j'-direction.

A Cartesian tensor of order 0 or a scalar is a quantity with $3^0 = 1$ component $\varphi(x_1, x_2, x_3)$ in OX_i and 1 component $\varphi'(x_1', x_2', x_3')$ in OX_j' related by

$$\varphi = \varphi'. \tag{9.4.2}$$

A Cartesian tensor of order 1 or a vector is a quantity with $3^1 = 3$ components $v_i(x_1, x_2, x_3)$ in OX_i' and 3 components $v_j'(x_1', x_2', x_3')$ in OX_j' related by

$$v_j' = \ell_{ij} v_i \text{ for } j = 1, 2, 3. \tag{9.4.3}$$

Note in this context that (9.4.1) satisfies (9.4.3) and hence represents the usual position vector. As an interlude let us show that the gradient of a scalar is a vector using this definition. Define a quantity with 3 components $v_i = \partial\varphi/\partial x_i$ in OX_i and 3 components $v_j' = \partial\varphi'/\partial x_j'$ in OX_j'. Then

$$v_j' = \frac{\partial \varphi'}{\partial x_j'} = \frac{\partial \varphi}{\partial x_j'} = \left(\frac{\partial \varphi}{\partial x_i}\right)\left(\frac{\partial x_i}{\partial x_j'}\right) = \ell_{ij} v_i \text{ for } j = 1, 2, 3$$

and hence, the gradient is a vector.

A Cartesian tensor of order 2 is a quantity with $3^2 = 9$ components T_{ij} in OX_i and 9 components T'_{pq} in OX'_j related by

$$T'_{pq} = \ell_{ip}\ell_{jq}T_{ij}. \tag{9.4.4}$$

Let us show that the stress "matrix" considered in the previous section

$$T_{ij} = T_{ji} = t_i(\boldsymbol{x}, \boldsymbol{e}_j, t)$$

is a tensor in this sense. Define $T'_{pq} = t'_p(\boldsymbol{x}', \boldsymbol{e}'_q, t)$. Then, since $\boldsymbol{t} \equiv$ stress vector and $\boldsymbol{e}'_q = \ell_{jq}\boldsymbol{e}_j$,

$$T'_{pq} = t'_p(\boldsymbol{x}', \boldsymbol{e}'_q, t) = \ell_{ip}t_i(\boldsymbol{x}, \ell_{jq}\boldsymbol{e}_j, t) = \ell_{ip}\ell_{jq}t_i(\boldsymbol{x}, \boldsymbol{e}_j, t) = \ell_{ip}\ell_{jq}T_{ij},$$

demonstrating the result. If we define $\varepsilon'_{pq} = (\partial'_q v'_p + \partial'_p v'_q)/2$, this is a tensor as well.

Generalizing this definition a Cartesian tensor of order n is a quantity with 3^n components $T_{i_1 i_2 \ldots i_n}$ in OX_i and 3^n components $T'_{j_1 j_2 \cdots j_n}$ in OX'_j related by

$$T'_{j_1 j_2 \cdots j_n} = \ell_{i_1 j_1}\ell_{i_2 j_2}\ldots\ell_{i_n j_n}T_{i_1 i_2 \ldots i_n} \text{ for } j_1, j_2, \ldots, j_n = 1,2,3. \tag{9.4.5}$$

Returning to the stress tensor let us examine the behavior of various prototype stress matrices. Assume a normal stress of the form

$$\boldsymbol{T} = \begin{pmatrix} \tau & 0 & 0 \\ 0 & 0 & 0 \\ 0 & 0 & 0 \end{pmatrix}$$

and recall that $t_j = T_{ij}n_i$ for $j = 1,2,3$. We shall examine how this affects the two-dimensional element aligned along the x_1-axis depicted in Fig. 9.7a by computing the stress on its four sides as designated in that figure from ① through ④.

①: $\boldsymbol{n} = \boldsymbol{e}_1 \Rightarrow t_j = T_{ij}n_i = T_{1j}$ for $j = 1,2,3 \Rightarrow \boldsymbol{t} = \tau\boldsymbol{e}_1$;

②: $\boldsymbol{n} = -\boldsymbol{e}_1 \Rightarrow t_j = T_{ij}n_i = -T_{1j}$ for $j = 1,2,3 \Rightarrow \boldsymbol{t} = -\tau\boldsymbol{e}_1$;

③: $\boldsymbol{n} = \boldsymbol{e}_2 \Rightarrow t_j = T_{ij}n_i = T_{2j}$ for $j = 1,2,3 \Rightarrow \boldsymbol{t} = \boldsymbol{0}$;

②: $\boldsymbol{n} = -\boldsymbol{e}_2 \Rightarrow t_j = T_{ij}n_i = -T_{2j}$ for $j = 1,2,3 \Rightarrow \boldsymbol{t} = -\boldsymbol{0}$;

Hence the result is an extension of the element aligned along the x_1-axis if $\tau > 0$ as pictured in Fig. 9.7a and a compression if $\tau < 0$ where the stress vectors act with strength τ in the direction of the normal to those sides on which they are nonzero. Similarly the general normal stress matrix

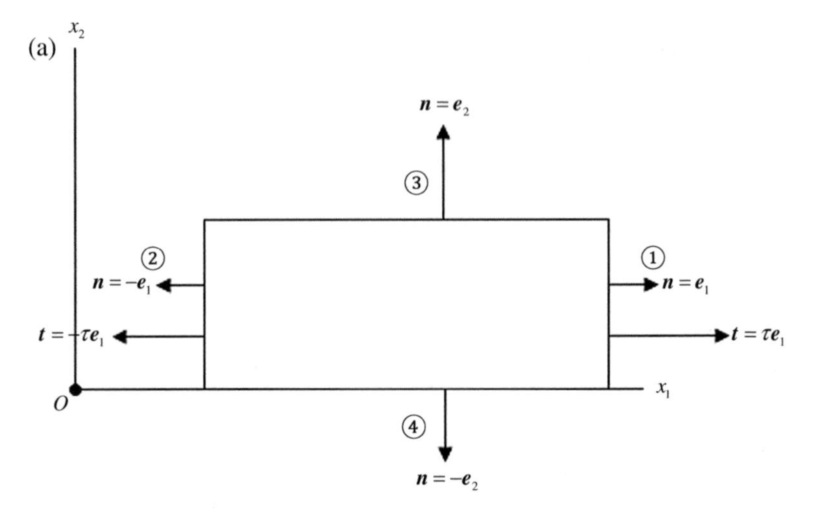

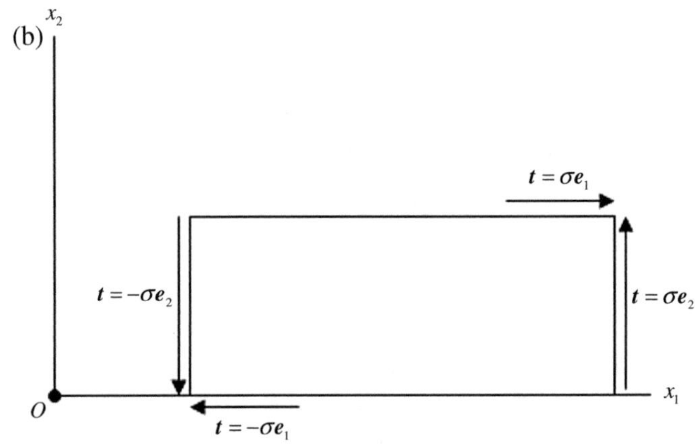

Fig. 9.7 Schematic diagrams of the effect (a) normal and (b) shear stresses have on an element aligned along the x_1-axis in the x_1-x_2 plane.

$$
\boldsymbol{T} = \begin{pmatrix} \tau_1 & 0 & 0 \\ 0 & \tau_2 & 0 \\ 0 & 0 & \tau_2 \end{pmatrix},
$$

gives rise to an extension or compression of an element aligned along the x_i-axis depending upon whether $\tau_i > 0$ or $\tau_i < 0$, respectively.

Assume a shear stress of the form

$$T = \begin{pmatrix} 0 & \sigma & 0 \\ \sigma & 0 & 0 \\ 0 & 0 & 0 \end{pmatrix}.$$

Now, proceeding in the same manner as for the normal stress case, we obtain:

$$\textcircled{1}: \boldsymbol{t} = \sigma \boldsymbol{e}_2; \textcircled{2}: \boldsymbol{t} = -\sigma \boldsymbol{e}_2; \textcircled{3}: \boldsymbol{t} = \sigma \boldsymbol{e}_1; \textcircled{4}: \boldsymbol{t} = -\sigma \boldsymbol{e}_1;.$$

Hence the result is a shear stress couple of strength σ acting to deform the right angles in the x_1-x_2 plane that originally occur at the corners of the element aligned along the x_1-axis as pictured in Fig. 9.7b. Similarly the general shear stress matrix

$$T = \begin{pmatrix} 0 & \sigma_{12} & \sigma_{13} \\ \sigma_{12} & 0 & \sigma_{23} \\ \sigma_{13} & \sigma_{23} & 0 \end{pmatrix}$$

is such that each shear couple of strength σ_{ij} causes an angle of the element in the x_i-x_j plane originally at $\pi/2$ to be deformed.

Since the stress relation $\boldsymbol{t} = \boldsymbol{T} \bullet \boldsymbol{n}$ is linear the effect of the most general stress matrix

$$T = \begin{pmatrix} \tau_1 & \sigma_{12} & \sigma_{13} \\ \sigma_{12} & \tau_2 & \sigma_{23} \\ \sigma_{13} & \sigma_{23} & \tau_3 \end{pmatrix}$$

can be determined by merely superposing those of the general normal and shear stress matrices just deduced.

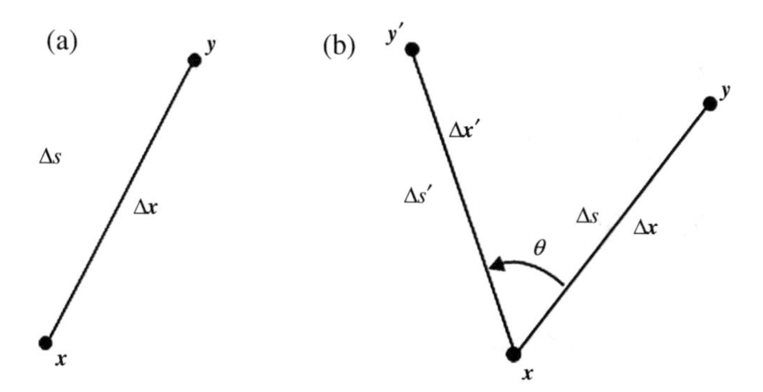

Fig. 9.8 Representations of (a) a vector $\Delta\boldsymbol{x}$ of length Δs joining $\boldsymbol{x}$ to nearby point $\boldsymbol{y}$ and (b) $\Delta\boldsymbol{x}$ plus a second vector $\Delta\boldsymbol{x}'$ with length $\Delta s'$ joining $\boldsymbol{x}$ to another nearby point $\boldsymbol{y}' \neq \boldsymbol{y}$ which makes an angle θ with the first one.

Next we examine the effect of the rate of strain tensor on the elements depicted in Fig. 9.7. To do so we consider Fig. 9.8. In Fig. 9.8a we depict a vector $\Delta\boldsymbol{x}$ joining the point $\boldsymbol{x}$ to a nearby point $\boldsymbol{y}$ such that

$$\Delta\boldsymbol{x} = \boldsymbol{y} - \boldsymbol{x} \text{ or } \Delta x_i = y_i - x_i \text{ for } i = 1,2,3 \text{ with } |\Delta\boldsymbol{x}| = \Delta s \tag{9.4.6a}$$

and

$$\frac{D(\Delta x_i)}{Dt} = \frac{Dy_i}{Dt} - \frac{Dx_i}{Dt} = v_i(\boldsymbol{y},t) - v_i(\boldsymbol{x},t) \cong \partial_j v_i(\boldsymbol{x},t)\Delta x_j. \tag{9.4.6b}$$

Thus

$$\Delta\boldsymbol{x} \bullet \Delta\boldsymbol{x} = \Delta x_i \Delta x_i = (\Delta s)^2. \tag{9.4.7}$$

Then

$$\begin{aligned}\frac{D(\Delta x_i \Delta x_i)}{Dt} &= \frac{D(\Delta x_i)}{Dt}\Delta x_i + \Delta x_i \frac{D(\Delta x_i)}{Dt} \\ &\cong \partial_j v_i(\boldsymbol{x},t)\Delta x_j \Delta x_i + \Delta x_i \partial_j v_i(\boldsymbol{x},t)\Delta x_j = \partial_j v_i(\boldsymbol{x},t)\Delta x_j \Delta x_i + \Delta x_j \partial_i v_j(\boldsymbol{x},t)\Delta x_i \\ &= 2\varepsilon_{ij}\Delta x_i \Delta x_j \text{ and } \frac{D(\Delta s)^2}{Dt} = 2\Delta s \frac{D(\Delta s)}{Dt}.\end{aligned} \tag{9.4.8}$$

Hence, we can conclude that

$$\varepsilon_{ij}\Delta x_i \Delta x_j \cong \Delta s \frac{D(\Delta s)}{Dt}. \tag{9.4.9a}$$

Finally, selecting $\Delta\boldsymbol{x} = \Delta s \boldsymbol{e}_1$, we obtain

$$\varepsilon_{11}(\Delta x_1)^2 = \varepsilon_{11}(\Delta s)^2 \cong \Delta s \frac{D(\Delta s)}{Dt} \Rightarrow \varepsilon_{11} \cong \frac{1}{\Delta s}\frac{D(\Delta s)}{Dt}. \tag{9.4.9b}$$

In Fig. 9.8b we depict both $\Delta\boldsymbol{x}$ and another vector $\Delta\boldsymbol{x}'$ joining $\boldsymbol{x}$ to a second nearby point $\boldsymbol{y}' \neq \boldsymbol{y}$ such that

$$\Delta\boldsymbol{x}' = \boldsymbol{y}' - \boldsymbol{x} \text{ or } \Delta x_i' = y_i' - x_i \text{ for } i = 1,2,3 \text{ with } |\Delta\boldsymbol{x}'| = \Delta s' \tag{9.4.10a}$$

and

$$\frac{D(\Delta x_i')}{Dt} = \frac{Dy_i'}{Dt} - \frac{Dx_i}{Dt} = v_i(\boldsymbol{y}',t) - v_i(\boldsymbol{x},t) \cong \partial_j v_i(\boldsymbol{x},t)\Delta x_j'. \tag{9.4.10b}$$

Thus

$$\Delta\boldsymbol{x} \bullet \Delta\boldsymbol{x}' = \Delta x_i \Delta x_i' = \Delta s \Delta s' \cos(\theta) \tag{9.4.11}$$

where θ is the angle between $\Delta\boldsymbol{x}$ and $\Delta\boldsymbol{x}'$. Then taking D/Dt of (9.4.11), employing (9.4.6) and (9.4.10), and proceeding in the same manner as was used to deduce (9.4.9) while selecting

$$\Delta\boldsymbol{x} = \Delta s \boldsymbol{e}_1, \ \Delta\boldsymbol{x}' = \Delta s' \boldsymbol{e}_2, \text{ and } \theta = \frac{\pi}{2}, \tag{9.4.12}$$

we would obtain that

$$\varepsilon_{12} \cong -\frac{1}{2}\frac{D\theta}{Dt}, \tag{9.4.13}$$

the details of which are left to Problem 9.8.

It is now clear that the effects produced by the normal stress due to τ_1 and the shear stress due to the σ_{12}-couple are consistent with the formulae for ε_{11} of (9.4.9) and for ε_{12} of (9.4.13), respectively. Hence we shall assume that there is a constitutive relation between $\boldsymbol{T}$ and $\boldsymbol{\varepsilon}$ of the form

$$\boldsymbol{T} = \boldsymbol{f}(\boldsymbol{\varepsilon}; p, T) \text{ or } T_{ij} = f_{ij}(\boldsymbol{\varepsilon}; p, T) \text{ for } i, j = 1, 2, 3, \tag{9.4.14}$$

where $p \equiv$ pressure and $T \equiv$ temperature in K are scalar quantities, such that

(1) $\boldsymbol{f}$ is *autonomous* in $\boldsymbol{x}$ and t; (2) $\boldsymbol{f}$ is *continuous* in its arguments; (3) If $\boldsymbol{T}' = \boldsymbol{f}'(\boldsymbol{\varepsilon}'; p, T)$, then $\boldsymbol{f} \equiv \boldsymbol{f}'$; (4) $\boldsymbol{f}(\boldsymbol{0}; p, T) = -p\boldsymbol{I}$ where $\boldsymbol{0}$ and $\boldsymbol{I}$ are the zero and identity tensors, respectively; (5) $\boldsymbol{f}$ is *linear* in $\boldsymbol{\varepsilon}$.

A continuum satisfying properties (1)–(5) is said to be a *Newtonian fluid*, and in what follows we shall be dealing with such fluids. Then properties (1)–(5) imply that

$$T_{ij} = -p\delta_{ij} + c_{ijk\ell}\varepsilon_{k\ell} \tag{9.4.15a}$$

where

$$\delta_{ij} = \begin{cases} 1, & \text{if } i = j \\ 0, & \text{if } i \neq j \end{cases} \text{ and } c_{ijk\ell} = \lambda\delta_{ij}\delta_{k\ell} + \mu(\delta_{ik}\delta_{j\ell} + \delta_{i\ell}\delta_{jk}) + \gamma(\delta_{ik}\delta_{j\ell} - \delta_{i\ell}\delta_{jk}). \tag{9.4.15b}$$

Here $c_{ijk\ell}$ represents the fourth-order isotropic tensor that is the same in all coordinate systems while λ, μ, and γ are scalar functions of p and T assumed to be slowly varying [114].

In addition, by virtue of (9.3.2b), we require

$$T_{ij} = -p\delta_{ij} + c_{ijk\ell}\varepsilon_{k\ell} = T_{ji} = -p\delta_{ji} + c_{jik\ell}\varepsilon_{k\ell} = -p\delta_{ij} + c_{ijk\ell}\varepsilon_{k\ell} \Rightarrow c_{ijk\ell}\varepsilon_{k\ell} = c_{jik\ell}\varepsilon_{k\ell}$$

or, upon employing (9.4.15b)

$$\begin{aligned}&[\lambda\delta_{ij}\delta_{k\ell} + \mu(\delta_{ik}\delta_{j\ell} + \delta_{i\ell}\delta_{jk}) + \gamma(\delta_{ik}\delta_{j\ell} - \delta_{i\ell}\delta_{jk})]\varepsilon_{k\ell} \\ &= [\lambda\delta_{ji}\delta_{k\ell} + \mu(\delta_{jk}\delta_{i\ell} + \delta_{j\ell}\delta_{ik}) + \gamma(\delta_{jk}\delta_{i\ell} - \delta_{j\ell}\delta_{ik})]\varepsilon_{k\ell}\end{aligned}$$

which implies that

$$\gamma(\delta_{ik}\delta_{j\ell} - \delta_{i\ell}\delta_{jk})\varepsilon_{k\ell} = \gamma(\varepsilon_{ij} - \varepsilon_{ji}) = 0 \tag{9.4.16a}$$

and is satisfied, provided either

$$\gamma = 0 \text{ or } \varepsilon_{ij} = \varepsilon_{ji}. \tag{9.4.16b}$$

Since the rate of strain tensor $\boldsymbol{\varepsilon}$ as defined by (9.3.4a) is symmetric, this is satisfied identically. Thus, (9.4.15) in conjunction with (9.4.16) yields

$$T_{ij} = -p\delta_{ij} + [\lambda\delta_{ij}\delta_{k\ell} + \mu(\delta_{ik}\delta_{j\ell} + \delta_{i\ell}\delta_{jk})]\varepsilon_{k\ell} = (-p + \lambda\varepsilon_{kk})\delta_{ij} + 2\mu\varepsilon_{ij} \tag{9.4.17a}$$

or, upon substitution of the explicit form for ε_{ij} from (9.3.4a),

$$T_{ij} = (-p + \lambda\partial_k v_k)\delta_{ij} + \mu(\partial_j v_i + \partial_i v_j). \tag{9.4.17b}$$

We now calculate $\partial_j T_{ij}$ and $T_{ij}\varepsilon_{ij}$ from (9.4.17) and substitute them into our momentum and energy equations (9.3.3) and (9.3.4), respectively. Thus, assuming λ and μ to be so slowly varying functions of p and T that we may treat them as constants for our modeling purposes,

$$\partial_j T_{ij} = (-\partial_j p + \lambda\partial_j\partial_k v_k)\delta_{ij} + \mu(\partial_j\partial_j v_i + \partial_j\partial_i v_j) = -\partial_i p + (\lambda+\mu)\partial_i\partial_j v_j + \mu\partial_j\partial_j v_i, \tag{9.4.18}$$

$$T_{ij}\varepsilon_{ij} = [(-p + \lambda\varepsilon_{kk})\delta_{ij} + 2\mu\varepsilon_{ij}]\varepsilon_{ij} = -p\varepsilon_{jj} + \lambda\varepsilon_{kk}\varepsilon_{jj} + 2\mu\varepsilon_{ij}\varepsilon_{ij} = -p\theta + \Phi \tag{9.4.19a}$$

where

$$\theta = \varepsilon_{jj} = \partial_j v_j = \nabla \bullet \boldsymbol{v} \text{ and } \Phi = \lambda\theta^2 + 2\mu\varepsilon_{ij}\varepsilon_{ij} \equiv \text{ viscous dissipation.} \tag{9.4.19b}$$

Substituting (9.4.18) and (9.4.19) into (9.3.3) and (9.3.4), respectively, we obtain

$$\rho\frac{Dv_i}{Dt} = -\partial_i p + \rho f_i + (\lambda+\mu)\partial_i\partial_j v_j + \mu\partial_j\partial_j v_i \text{ for } i = 1, 2, 3; \tag{9.4.20}$$

$$\rho\frac{De}{Dt} = -p\theta + \Phi + r - \nabla \bullet \boldsymbol{q}, \tag{9.4.21a}$$

or, upon adoption of Newton's Law of Cooling $\boldsymbol{q} = -k\nabla T$ as a constitutive relation for $\boldsymbol{q}$,

$$\rho\frac{De}{Dt} = -p\theta + \Phi + r + \nabla \bullet (k\nabla T). \tag{9.4.21b}$$

9.5 Equations of State

Upon examination of our governing system of five equations (9.2.1), (9.4.20), and (9.4.21), we can see that they involve the seven dependent variables ρ, $v_1 = u$, $v_2 = v$, $v_3 = w$, p, e, and T (in this context $\boldsymbol{f}$, the body force vector per unit mass, and r, the energy source term per unit mass per unit time, are imposed quantities). Hence in general we need two equations of state to complete the formulation of our system. In this instance we have the five state variables $\rho = 1/V$ where $V \equiv$ specific volume, p, e, T, and $s \equiv$ specific entropy, any two of which are independent while the other three are related to them by so-called equations of state. There are certain situations in which a particular additional constitutive relation or a specific equation of state allows us to uncouple the four equations of flow (9.2.1) and (9.4.20) from the energy equation (9.4.21). Before examining the more general case let us consider two instances, one of each type, that allow for such uncoupling.

(i) *Incompressibility*: $D\rho/Dt = 0$. This is a constitutive relation that in conjunction with the continuity equation (9.2.1) implies a solenoidal flow satisfying $\nabla \bullet \boldsymbol{v} = 0$. Hence, along with (9.4.20) that in vector form now reduces to

$$\rho \frac{D\boldsymbol{v}}{Dt} = -\nabla p + \rho \boldsymbol{f} + \mu \nabla^2 \boldsymbol{v}, \tag{9.5.1}$$

we have five equations in the five variables ρ, u, v, w, and p. A special case of incompressibility is a homogeneous fluid for which $\rho \equiv \rho_0$. Then the incompressibility equation is satisfied identically, and our system reduces still further to the four Navier–Stokes equations

$$\nabla \bullet \boldsymbol{v} = 0, \ \frac{D\boldsymbol{v}}{Dt} = -\frac{\nabla p}{\rho_0} + \boldsymbol{f} + \nu \nabla^2 \boldsymbol{v} \text{ where } \nu = \frac{\mu}{\rho_0} \equiv \text{kinematic viscosity} \tag{9.5.2}$$

in the four variables u, v, w, and p. In the instance of a two-dimensional incompressible flow for which $\boldsymbol{v}(x,y,t) = u(x,y,t)\boldsymbol{i} + v(x,y,t)\boldsymbol{j}$ the solenoidal condition becomes

$$\nabla_2 \bullet \boldsymbol{v} = \frac{\partial u}{\partial x} + \frac{\partial v}{\partial y} = 0. \tag{9.5.3a}$$

This implies the existence of a stream function $\psi = \psi(x,y,t)$ such that

$$u = \frac{\partial \psi}{\partial y} \text{ and } v = -\frac{\partial \psi}{\partial x} \tag{9.5.3b}$$

which satisfies (9.5.3a) identically assuming the requisite continuity.

(ii) *Barotropic*: $p = \mathcal{P}(\rho)$: This is an equation of state that reduces (9.4.20) in vector form to

$$\rho \frac{D\boldsymbol{v}}{Dt} = -\nabla \mathcal{P}(\rho) + \rho \boldsymbol{f} + (\lambda + \mu)\nabla(\nabla \bullet \boldsymbol{v}) + \mu \nabla^2 \boldsymbol{v} \tag{9.5.4}$$

which along with the continuity equation (9.2.1) gives us four equations in the four variables ρ, u, v, and w. We shall offer an example of such a barotropic relationship at the conclusion of this section by developing the so-called adiabatic equation of state after returning to the more general case discussed above.

Toward this end and assuming for the sake of definiteness that ρ and T are the independent state variables, we wish to complete the formulation of our general problem involving (9.2.1), (9.4.20), and (9.4.21) by finding, in particular, equations of state of the form

$$p = P(\rho, T) \text{ and } \frac{De}{Dt} = c_0 \frac{DT}{Dt}. \tag{9.5.5}$$

To do so we first must introduce some thermodynamic concepts. Consider

$$e(\boldsymbol{x}, t) = E(s, V) \tag{9.5.6}$$

where for this development s and V are assumed to be the independent state variables. Then

$$dE = \left(\frac{\partial E}{\partial s}\right)_V ds + \left(\frac{\partial E}{\partial V}\right)_s dV \tag{9.5.7a}$$

and we define T and p such that

$$T = \left(\frac{\partial E}{\partial s}\right)_V, p = -\left(\frac{\partial E}{\partial V}\right)_s. \tag{9.5.7b}$$

Thus

$$dE = T\,ds - p\,dV. \tag{9.5.7c}$$

From these, we can deduce Gibbs relation which is the continuum analog of the first law of thermodynamics

$$\frac{De}{Dt} = T\frac{Ds}{Dt} - p\frac{D(1/\rho)}{Dt} = T\frac{Ds}{Dt} + \frac{p}{\rho^2}\frac{D\rho}{Dt} \tag{9.5.8}$$

and assuming the requisite continuity of the second partial derivatives one of Maxwell's relations

$$\left[\frac{\partial(\partial E/\partial s)_v}{\partial V}\right]_s = \left[\frac{\partial(\partial E/\partial V)_s}{\partial s}\right]_V \text{ or } \left(\frac{\partial T}{\partial V}\right)_s = -\left(\frac{\partial p}{\partial s}\right)_V. \tag{9.5.9}$$

Further, defining another energy measure *enthalpy* by

$$H = E + pV \text{ or } h = e + \frac{p}{\rho}, \tag{9.5.10}$$

then

$$dH = dE + p\,dV + V\,dp = T\,ds + V\,dp = \left(\frac{\partial H}{\partial s}\right)_p dV + \left(\frac{\partial H}{\partial p}\right)_s dp \tag{9.5.11a}$$

which implies that

$$T = \left(\frac{\partial H}{\partial s}\right)_p, V = \left(\frac{\partial H}{\partial p}\right)_s; \tag{9.5.11b}$$

and hence, we can deduce another of Maxwell's relations

$$\left(\frac{\partial T}{\partial p}\right)_s = \left(\frac{\partial V}{\partial s}\right)_p; \tag{9.5.11c}$$

while from (9.5.11a) and (9.5.8)

$$\frac{Dh}{Dt} = T\frac{Ds}{Dt} + \frac{1}{\rho}\frac{Dp}{Dt} = \frac{De}{Dt} - \left(\frac{p}{\rho^2}\right)\frac{D\rho}{Dt} + \frac{1}{\rho}\frac{Dp}{Dt} \tag{9.5.12a}$$

and thus, upon employment of (9.4.21) and (9.2.1), this yields the alternate energy equation for enthalpy

$$\rho\frac{Dh}{Dt} = \rho\frac{De}{Dt} - \frac{p}{\rho}\frac{D\rho}{Dt} + \frac{Dp}{Dt} = \frac{Dp}{Dt} + \Phi + r + \nabla \bullet (k\nabla T). \tag{9.5.12b}$$

We now consider the ideal gas equation of state

$$pV_n = n\mathcal{R}T$$

where $n \equiv$ number of moles, $V_n \equiv$ volume of gas occupied by n moles, and $\mathcal{R} \equiv$ universal gas constant = 8.314 joules/(mole K). Then representing this in the form of (9.5.5) we obtain

$$p = \left(\frac{nM}{V_n}\right)\left(\frac{\mathcal{R}}{M}\right)T = \rho RT, \tag{9.5.13}$$

where $M \equiv$ gram-molecular mass of the gas, as the desired equation of state. For such an equation of state it can be deduced that (see Problem 9.9)

$$\frac{De}{Dt} = C_V\frac{DT}{Dt} \tag{9.5.14a}$$

and

$$\frac{Dh}{Dt} = C_p\frac{DT}{Dt} \tag{9.5.14b}$$

where $C_{V,p} \equiv T(\partial s/\partial T)_{V,p}$ are specific heats at constant volume and pressure, respectively. Therefore our energy equations (9.4.21) and (9.5.12b) become

$$\rho C_V\frac{DT}{Dt} = -p\theta + \Phi + r + \nabla \bullet (k\nabla T) \tag{9.5.15a}$$

and

$$\rho C_p\frac{DT}{Dt} = \frac{Dp}{Dt} + \Phi + r + \nabla \bullet (k\nabla T). \tag{9.5.15b}$$

We complete this introduction of thermodynamic concepts with the Clausius–Duhem inequality, which is the continuum analog of the second law of thermodynamics and is deduced from the balance-type law (see Problem 9.11)

$$\frac{d}{dt}\iiint\limits_{R(t)} \rho s\, d\tau \geq \iiint\limits_{R(t)} T^{-1} r\, d\tau - \iint\limits_{\partial R(t)} T^{-1}\boldsymbol{q}\bullet\boldsymbol{n}\, d\sigma$$

which implies that

$$\rho\frac{Ds}{Dt} \geq T^{-1}r - \nabla\bullet(T^{-1}\boldsymbol{q}). \tag{9.5.16}$$

It only remains to derive the adiabatic equation of state from that for an ideal gas under *isentropic* conditions ($s \equiv s_0$). To do so we first convert the equation of state (9.5.13) for pressure $p = \rho RT$ in which ρ and T are considered to be the independent variables to one in which the independent variables are ρ and s instead. From (9.5.8), (9.5.14a), and (9.5.13) we can deduce that

$$dE = C_V dT = T\,ds + p\frac{d\rho}{\rho^2} = T\,ds + RT\frac{d\rho}{\rho}, \tag{9.5.17}$$

which upon division by T becomes

$$C_V\frac{dT}{T} = ds + R\frac{d\rho}{\rho}. \tag{9.5.18}$$

Then, taking the differential of (9.5.13)

$$dp = R(\rho\, dT + T\, d\rho), \tag{9.5.19}$$

and dividing by p yields

$$\frac{dp}{p} = R\frac{\rho\, dT + T\, d\rho}{\rho RT} = \frac{dT}{T} + \frac{d\rho}{\rho} \tag{9.5.20a}$$

or

$$\frac{dT}{T} = \frac{dp}{p} - \frac{d\rho}{\rho}. \tag{9.5.20b}$$

Now, upon substitution of (9.5.20b) into (9.5.18) and collection of terms, we obtain

$$C_V\frac{dp}{p} = ds + (C_V + R)\frac{d\rho}{\rho} = ds + C_p\frac{d\rho}{\rho}, \tag{9.5.21}$$

where use has been made of the fact that $C_p = C_V + R$ as demonstrated in Problem 9.9 in the exercises. Next, dividing (9.5.21) by C_V

$$\frac{dp}{p} = \frac{ds}{C_V} + \left(\frac{C_p}{C_V}\right)\frac{d\rho}{\rho} = \frac{ds}{C_V} + \gamma\frac{d\rho}{\rho} \tag{9.5.22}$$

and integrating this result, it follows that

$$\ln\left(\frac{p}{p_0}\right) = \frac{s - s_0}{C_V} + \ln\left(\frac{\rho}{\rho_0}\right)^{\gamma}, \tag{9.5.23a}$$

where $p = p_0$ corresponds to $s = s_0$ and $\rho = \rho_0$, or equivalently

$$p = p_0 e^{(s-s_0)/C_V} \left(\frac{\rho}{\rho_0}\right)^{\gamma}, \tag{9.5.23b}$$

which is the desired equation of state. Finally, assuming the isentropic situation of constant entropy $s \equiv s_0$, this reduces to the adiabatic equation of state

$$p = p_0 \left(\frac{\rho}{\rho_0}\right)^{\gamma} = \mathcal{P}(\rho). \tag{9.5.24}$$

which is clearly barotropic as well.

Problems

9.1. Consider a two-dimensional steady flow with density $\rho = \rho(x,y)$ and velocity $\boldsymbol{v}(x,y) = u(x,y)\boldsymbol{i} + v(x,y)\boldsymbol{j}$.

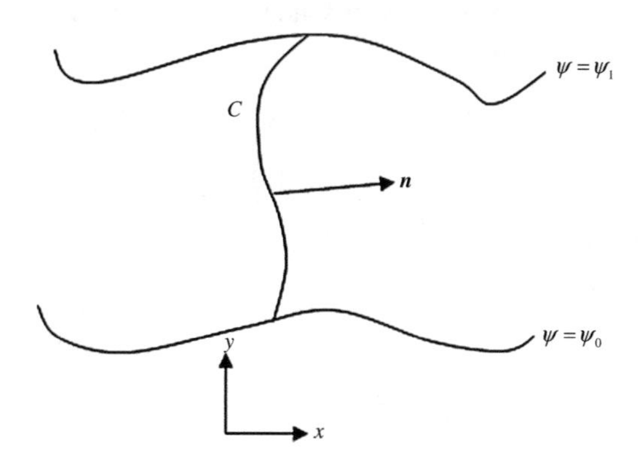

Fig. 9.9 A plot in a two-dimensional situation of a curve C intersecting two streamlines in the x-y plane relevant to Problem 9.1.

(a) Show that the continuity equation then implies that there exists a function $\psi = \psi(x,y)$ such that $\rho u = \partial\psi/\partial y$ and $\rho v = -\partial\psi/\partial x$.
(b) ψ is called a *stream function*. Justify this name by showing that along a *streamline* $x = x(\sigma)$ and $y = y(\sigma)$, defined by $dx/d\sigma = u$ and $dy/d\sigma = v$, ψ is constant. Hint: In order to deduce this simply define $\Psi(\sigma) = \psi[x(\sigma), y(\sigma)]$ and demonstrate that $d\Psi(\sigma)/d\sigma \equiv 0$.
(c) Let C be a curve joining a pair of streamlines, denoted by $\psi \equiv \psi_0$ and $\psi \equiv \psi_1$ in Fig. 9.9. This curve is described parametrically by $C = \{[x(s), y(s)], s_0 \leq s \leq s_1\}$

where s_0 and s_1 denote the values of s at which C intersects $\psi \equiv \psi_0$ and $\psi \equiv \psi_1$, respectively. Show that

$$\int_{s_0}^{s_1} (\rho \boldsymbol{v})[x(s), y(s)] \bullet \boldsymbol{n}\, ds = \psi_1 - \psi_0$$

where $\boldsymbol{n} = y'(s)\boldsymbol{i} - x'(s)\boldsymbol{j}$ is the normal to C.

9.2. (a) Show that the material or Lagrangian form of the continuity equation takes the form

$$(\delta J)(\boldsymbol{A}, t) = \delta(\boldsymbol{A}, 0)$$

where $\delta = \delta(\boldsymbol{A}, t)$ is the density and $J = J(\boldsymbol{A}, t)$ the Jacobian in material or Lagrangian variables.

(b) Use the result of (a) to show that

$$\frac{d}{dt} \iiint\limits_{R(t)} (\rho g)(\boldsymbol{x}, t)\, d\tau = \iiint\limits_{R(t)} \left(\rho \frac{Dg}{Dt}\right)(\boldsymbol{x}, t)\, d\tau$$

where $\rho = \rho(\boldsymbol{x}, t)$ is the density in spatial or Eulerian variables and $R(t)$ is a moving material region.

9.3. Derive the two-dimensional continuity equation (9.2.17) in polar coordinates

$$\frac{\partial R}{\partial t} + \frac{1}{r}\frac{\partial}{\partial r}(rRU) + \frac{1}{r}\frac{\partial}{\partial \theta}(RV) = 0$$

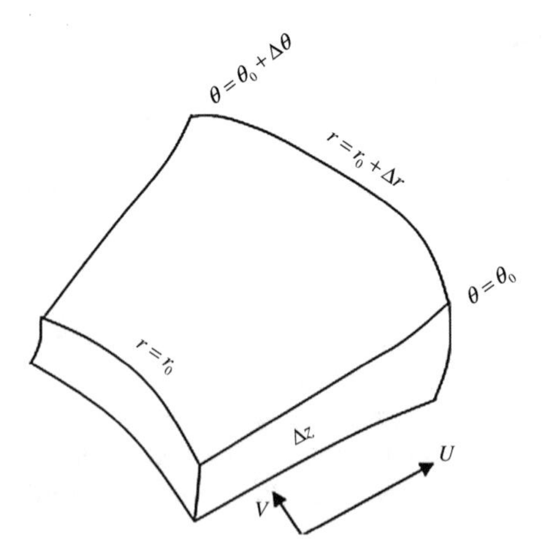

Fig. 9.10 Schematic diagram of the big box volume element in polar coordinates relevant to Problem 9.3.

by direct application of the big box argument to the volume element pictured in Fig. 9.10. Here $R(r,\theta,t)$ is the density while $U(r,\theta,t)$ and $V(r,\theta,t)$ are the velocity components in the direction of increasing r and θ, respectively.

9.4. Consider a fluid of density $\rho(x,y,z,t)$ and a flow situation with velocity field $\boldsymbol{v}(x,y,z,t) = u(x,y,z,t)\boldsymbol{i} + v(x,y,z,t)\boldsymbol{j} + w(x,y,z,t)\boldsymbol{k}$. Show that the continuity equation in spherical coordinates (see Fig. 9.11) takes the form

$$\frac{\partial R}{\partial t} + \frac{1}{r^2}\frac{\partial}{\partial r}(r^2 UR) + \frac{1}{r\sin(\theta)}\left[\frac{\partial}{\partial \theta}(VR\sin(\theta) + \frac{\partial}{\partial \phi}(WR)\right] = 0$$

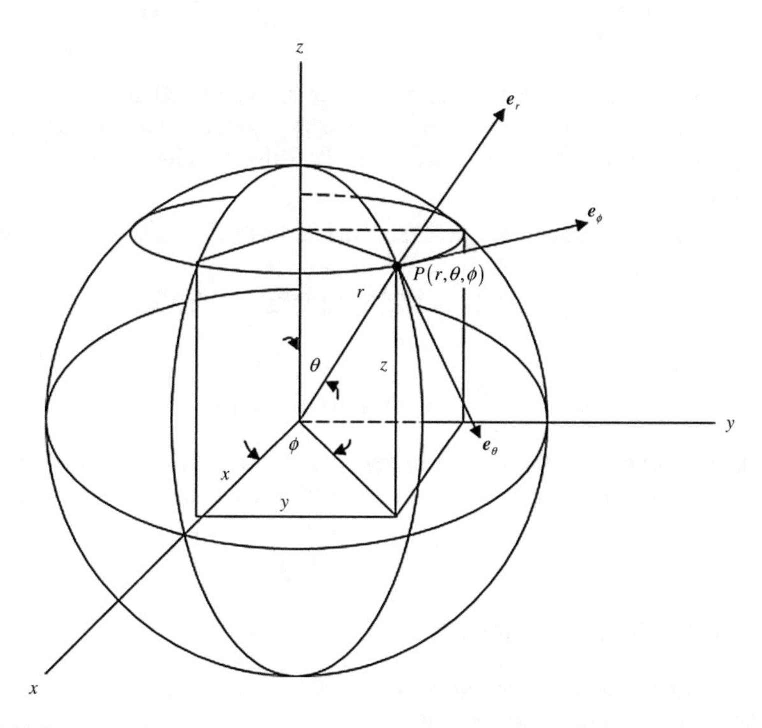

Fig. 9.11 Representation of the spherical coordinate system (r, θ, ϕ) and its basis vectors relevant to Problem 9.4. Note that here θ and ϕ represent the azimuthal and planar angles, respectively, which is a *right-handed* system since $\boldsymbol{e}_r \times \boldsymbol{e}_\theta = \boldsymbol{e}_\phi$, unlike many spherical coordinate representations that switch these angular designations and hence are implicitly *left-handed*.

where

$$x = r\sin(\theta)\cos(\phi),\ y = r\sin(\theta)\sin(\phi),\ z = r\cos(\theta),\ \rho(x,y,z,t) = R(r,\theta,\phi,t),$$

and

$$\boldsymbol{v}(x,y,z,t) = U(r,\theta,\phi,t)\boldsymbol{e}_r + V(r,\theta,\phi,t)\boldsymbol{e}_\theta + W(r,\theta,\phi,t)\boldsymbol{e}_\phi.$$

Hint: $U\sin(\theta) + V\cos(\theta) = u\cos(\phi) + v\sin(\phi)$ and if one considers the particle path

$$\boldsymbol{x}(t) = r(t)\sin[\theta(t)]\cos[\phi(t)]\boldsymbol{i} + r(t)\sin[\theta(t)]\sin[\phi(t)]\boldsymbol{j} + r(t)\cos[\theta(t)]\boldsymbol{k}$$

then $\dot{r} = U$, $r\dot{\theta} = V$, and $r\sin(\theta)\dot{\phi} = W$.

9.5. Consider the postulated conservation law for a chemical in an aqueous solution

$$\frac{d}{dt}\iiint\limits_{R(t)} c\,dx\,dy\,dz = \iiint\limits_{R(t)} r^{(c)}\,dx\,dy\,dz + \iint\limits_{\partial R(T)} \mathcal{D}(c)\frac{\partial c}{\partial n}\,d\sigma$$

where $c(x,y,z,t)$ is the concentration of this chemical with reaction term $r^{(c)}(x,y,z,t)$ and diffusion coefficient $\mathcal{D}(c)$ while $R(t)$ is an arbitrary material region moving with velocity $\boldsymbol{v}(x,y,z,t)$ and possessing boundary $\partial R(t)$ with unit outward-pointing normal $\boldsymbol{n} = (n_1,n_2,n_3)$.

(a) Given that $\partial c/\partial n$ represents the directional derivative in the $\boldsymbol{n}$-direction defined by

$$\lim_{\lambda\to 0}\frac{c(x+\lambda n_1, y+\lambda n_2, z+\lambda n_3, t) - c(x,y,z,t)}{\lambda}$$

show that

$$\frac{\partial c}{\partial n}(x,y,z,t) = \nabla c(x,y,z,t)\bullet\boldsymbol{n}.$$

(b) Using the Reynolds transport and divergence theorems and the DuBois–Reymond Lemma, conclude that c satisfies the advection–reaction–diffusion equation given by

$$\frac{\partial c}{\partial t} = r^{(c)} - \nabla\bullet\boldsymbol{J}^{(c)}$$

where the effective flux vector $\boldsymbol{J}^{(c)} = -\mathcal{D}(c)\nabla c + c\boldsymbol{v}$.

9.6. Consider the balance law for angular momentum in a polar fluid such as a polymer involving large chain molecules

$$\frac{d}{dt}\iiint\limits_{R(t)} (\boldsymbol{r}\times\rho\boldsymbol{v} + \rho\boldsymbol{\ell})(\boldsymbol{x},t)\,d\tau = \iiint\limits_{R(t)} (\boldsymbol{r}\times\rho\boldsymbol{f} + \rho\boldsymbol{m})(\boldsymbol{x},t)\,d\tau + \iint\limits_{\partial R(t)} (\boldsymbol{r}\times\boldsymbol{t}+\boldsymbol{c})(\boldsymbol{x},\boldsymbol{n},t)\,d\sigma$$

where

$$\boldsymbol{\ell} \equiv \text{internal angular momentum vector per unit mass} = (\ell_1,\ell_2,\ell_3),$$

$$\boldsymbol{m} \equiv \text{body torque vector per unit mass} = (m_1,m_2,m_3),$$

$$\boldsymbol{c} \equiv \text{surface stress couple per unit area} = (c_1,c_2,c_3).$$

(a) Show that this yields the governing partial differential equations for angular momentum balance

$$\rho \frac{D\ell_i}{Dt} = \rho m_i + \partial_j C_{ij} + \epsilon_{ijk} T_{jk} \text{ for } i = 1, 2, 3,$$

where

$$c_i(\boldsymbol{x}, \boldsymbol{n}, t) = c_i(\boldsymbol{x}, n_j \boldsymbol{e}_j, t) = c_i(x, \boldsymbol{e}_j, t) n_j = C_{ij} n_j$$

and

$$C_{ij} \equiv c_i(x, \boldsymbol{e}_j, t).$$

(b) Representing

$$T_{jk} = S_{jk} + A_{jk}$$

in terms of its symmetric and antisymmetric parts such that

$$S_{kj} = S_{jk} \text{ and } A_{kj} = -A_{jk},$$

show that

$$\epsilon_{ijk} T_{jk} = \epsilon_{ijk} A_{jk}.$$

(c) Now in a nonpolar fluid for which $\boldsymbol{l} = \boldsymbol{m} = \boldsymbol{c} = \boldsymbol{0}$, show that these equations reduce to

$$\epsilon_{ijk} A_{jk} = 0 \text{ for } i = 1, 2, 3 \Rightarrow A_{jk} = 0 \text{ for } j, k = 1, 2, 3 \Rightarrow T_{jk} = S_{jk} \text{ for } j, k = 1, 2, 3;$$

which is equivalent to the result of (9.3.2b).

9.7. Show that the rate of strain or deformation tensor $\boldsymbol{\varepsilon}$ defined in component form by (9.3.4a)

$$\varepsilon_{ij} = \frac{\partial_j v_i + \partial_i v_j}{2} \text{ for } i, j = 1, 2, 3$$

is identically equal to zero

$$\boldsymbol{\varepsilon} \equiv \boldsymbol{0} \Rightarrow \varepsilon_{ij} = 0 \text{ for } i, j = 1, 2, 3$$

if and only if there is rigid body motion – *i.e.*,

$$\boldsymbol{v}(\boldsymbol{x}, t) = \boldsymbol{\omega}^{(0)}(t) \times \boldsymbol{r} + \boldsymbol{v}^{(0)} \Rightarrow v_i(x, t) = \epsilon_{ijk} \omega_j^{(0)}(t) x_k + v_i^{(0)}(t), \text{ for } i = 1, 2, 3$$

or the velocity field can be represented by a translation plus a rotation. Demonstrate the if part by direct partial differentiation in conjunction with the definition of ε_{ij} and the only if part by direct partial integration. Hint: For the only if part it is preferable to proceed by examining the ε_{ij}'s in the following order:

$$\varepsilon_{11}, \varepsilon_{12}, \varepsilon_{13}, \varepsilon_{22}, \varepsilon_{33}, \varepsilon_{23};$$

obtaining general expressions for the velocity components v_i from $\varepsilon_{1i} = 0$ for $i = 1, 2, 3$, and noting that $a(x_2, x_3) x_1 + b(x_2, x_3) \equiv 0 \Rightarrow a(x_2, x_3) = b(x_2, x_3) \equiv 0$ while

$f(x_2,x_3)$ is linear in its arguments provided all its second partial derivatives are identically equal to zero or, assuming the requisite continuity, $f_{22} = f_{33} = f_{23} \equiv 0$ where here the subscripts denote partial differentiation with respect to that numbered spatial variable.

9.8. Consider the configuration of Fig. 9.8b and the dot product relation of (9.4.11)

$$\Delta\boldsymbol{x} \bullet \Delta\boldsymbol{x}' = \Delta x_i \Delta x_i' = \Delta s \Delta s' \cos(\theta).$$

(a) Taking D/Dt of this relation, employing (9.4.6b) and (9.4.10b), and proceeding in an analogous manner to the one used to deduce (9.4.9a), show that

$$2\varepsilon_{ij}\Delta x_i \Delta x_j' \cong \left[\Delta s' \frac{D(\Delta s)}{Dt} + \Delta s \frac{D(\Delta s')}{Dt}\right] \cos(\theta) - \Delta s \Delta s' \sin(\theta) \frac{D\theta}{Dt}.$$

(b) By employing the selection of (9.4.12) $\Delta\boldsymbol{x} = \Delta s \boldsymbol{e}_1$, $\Delta\boldsymbol{x}' = \Delta s' \boldsymbol{e}_2$, and $\theta = \pi/2$, in the result of (a) deduce (9.4.13)

$$\varepsilon_{12} \cong -\frac{1}{2}\frac{D\theta}{Dt},$$

and hence, demonstrate that ε_{12} is approximately equal to one-half the instantaneous rate of decrease at time t of the right angle between two elements originally aligned along the x_1- and x_2-axes, respectively.

9.9. Recall Gibb's relation (9.5.8) for e

$$\frac{De}{Dt} = T\frac{Ds}{Dt} + \frac{p}{\rho^2}\frac{D\rho}{Dt} \tag{P9.1a}$$

and the corresponding relationship (9.5.12a) for $h = e + p/\rho$

$$\frac{Dh}{Dt} = T\frac{Ds}{Dt} + \frac{1}{\rho}\frac{Dp}{Dt}. \tag{P9.1b}$$

(a) Given A, Helmholtz free energy, and G, Gibb's free energy, defined by

$$A = E - Ts \text{ and } G = A + pV, \tag{P9.2}$$

use (9.5.7c)

$$dE = T\,ds - p\,dV \tag{P9.3}$$

to show that

$$dA = -p\,dV - s\,dT \text{ and } dG = V\,dp - s\,dT. \tag{P9.4}$$

(b) Making use of the total differential relationships implicit to (P9.4), deduce the Maxwell's relations

$$\left(\frac{\partial s}{\partial V}\right)_T = \left(\frac{\partial p}{\partial T}\right)_V \text{ and } \left(\frac{\partial s}{\partial p}\right)_T = -\left(\frac{\partial V}{\partial T}\right)_p. \tag{P9.5}$$

(c) From the definition of specific heat at either constant volume or pressure it can be concluded

$$\left(\frac{\partial s}{\partial T}\right)_{V,p} = \frac{C_{V,p}}{T}. \tag{P9.6}$$

Use (P9.5) and (P9.6) in conjunction with the total differential of ds to show that

$$ds = C_V \frac{dT}{T} + \left(\frac{\partial p}{\partial T}\right)_V dV = C_p \frac{dT}{T} - \left(\frac{\partial V}{\partial T}\right)_p dp. \tag{P9.7}$$

(d) For an ideal gas which satisfies the equation of state $p = \rho RT = RT/V$, show that

$$\left(\frac{\partial p}{\partial T}\right)_V = \frac{p}{T} \text{ and } \left(\frac{\partial V}{\partial T}\right)_p = \frac{1}{\rho T}. \tag{P9.8}$$

(e) Use the results of (P9.7) and (P9.8) to obtain that

$$T\frac{Ds}{Dt} = C_V \frac{DT}{Dt} - \frac{p}{\rho^2}\frac{D\rho}{Dt} = C_p \frac{DT}{Dt} - \frac{1}{\rho}\frac{Dp}{Dt} \tag{P9.9}$$

and hence, finally conclude from (P9.1) that

$$\frac{De}{Dt} = C_V \frac{DT}{Dt} \text{ and } \frac{Dh}{Dt} = C_p \frac{DT}{Dt}. \tag{P9.10}$$

(f) Since for an ideal gas $h = e + p/\rho = e + RT$, show that (P9.10) implies

$$C_p = C_V + R. \tag{P9.11}$$

9.10. Show that the viscous dissipation defined in (9.4.19b)

$$\Phi = \lambda\theta^2 + 2\mu\varepsilon_{ij}\varepsilon_{ij} \text{ where } \theta = \varepsilon_{jj} = \varepsilon_{11} + \varepsilon_{22} + \varepsilon_{33}$$

can be written as

$$\Phi = \left(\lambda + \frac{2\mu}{3}\right)\theta^2 + 4\mu(\varepsilon_{12}^2 + \varepsilon_{23}^2 + \varepsilon_{31}^2) + \left(\frac{2\mu}{3}\right)[(\varepsilon_{11} - \varepsilon_{22})^2 + (\varepsilon_{22} + \varepsilon_{33})^2 + (\varepsilon_{33} + \varepsilon_{11})^2]$$

and hence, $\Phi \geq 0$ provided the bulk viscosity $\lambda + 2\mu/3 \geq 0$ and the *shear viscosity* $\mu > 0$ which are always true. Hint: Note that

$$\theta^2 = \varepsilon_{11}^2 + \varepsilon_{22}2 + \varepsilon_{33}^2 + 2(\varepsilon_{11}\varepsilon_{22} + \varepsilon_{22}\varepsilon_{33} + \varepsilon_{33} + \varepsilon_{11}),$$

$$\varepsilon_{ij}\varepsilon_{ij} = \varepsilon_{11}^2 + \varepsilon_{22}^2 + \varepsilon_{33}^2 + 2(\varepsilon_{12}^2 + \varepsilon_{23}^2 + \varepsilon_{31}^2).$$

9.11. (a) From the postulated balance-type law for an arbitrary material region $R(t)$ with bounding surface $\partial R(t)$ possessing outward-pointing unit normal $\boldsymbol{n}$

$$\frac{d}{dt}\iiint\limits_{R(t)} \rho s\, d\tau \geq \iiint\limits_{R(t)} T^{-1} r\, d\tau - \iint\limits_{\partial R(t)} T^{-1}\boldsymbol{q} \bullet \boldsymbol{n}\, d\sigma,$$

deduce the Clausius–Duhem inequality (9.5.16)

$$\rho\frac{Ds}{Dt} \geq T^{-1}r - \nabla \bullet (T^{-1}\boldsymbol{q}), \quad \text{(P11.1)}$$

which is the continuum analog of the second law of thermodynamics, by applying the Reynolds transport and divergence theorems and a generalization of the DuBois–Reymond Lemma.

(b) Given the energy equation (9.4.21a) $\rho De/Dt = -p\theta + \Phi + r - \nabla \bullet \boldsymbol{q}$, Gibb's relation (9.5.8) $De/Dt = TDs/Dt + (p/\rho^2)D\rho/Dt$, and the continuity equation (9.2.1) written in the form $D\rho/Dt + \rho\theta = 0$, show that

$$\rho\frac{Ds}{Dt} = T^{-1}\Phi + T^{-1}r - T^{-1}\nabla \bullet \boldsymbol{q}. \quad \text{(P11.2)}$$

(c) Substituting (P11.2) into (P11.1), deduce the requirement that

$$\Phi - T^{-1}\nabla T \bullet \boldsymbol{q} \geq 0 \quad \text{(P11.3)}$$

which is satisfied provided

$$\Phi \geq 0 \text{ and } \nabla T \bullet \boldsymbol{q} \leq 0 \quad \text{(P11.4)}$$

and discuss the physical significance of these sufficient conditions.

Chapter 10
Boundary Conditions for Fluid Mechanics

With this chapter, we complete the formulation of fluid mechanical model systems by considering the requisite boundary conditions for the governing partial differential equations developed in the last one. The no-penetration and no-slip fluid mechanical boundary conditions at rigid surfaces are catalogued for various configurations and the kinematic boundary condition at material surfaces is deduced from the relative normal speed of such a moving interface. The concept of a surface of discontinuity is introduced and careful application of the balance laws to that surface is shown to yield jump-type boundary conditions satisfied by the dependent variables across them. Then these are applied to the two-dimensional propagation of a shock front through an inviscid fluid to deduce the Rankine Hugoniot jump conditions for that situation. The problems deal with those specific jump-type conditions to be imposed for a variety of other continuum processes containing such surfaces when modeled in this manner, which involves its curvature and surface tension.

10.1 No-Penetration and No-Slip or Adherence Boundary Conditions

In this section we restrict our attention to a boundary surface that does not change with time and hence satisfies the equation $f(x_1, x_2, x_3) = 0$, as depicted in Fig. 10.1, and has unit outward-pointing normal $\boldsymbol{n}$ and tangent vectors $\boldsymbol{\tau}_{1,2}$ associated with it. In this context we shall pose two types of boundary conditions at such rigid surfaces: Namely, the no-penetration and the no-slip or adherence boundary conditions, involving the normal and tangential components of the fluid velocity, respectively, as follows:

$$\text{No-Penetration Boundary Condition: } \boldsymbol{v} \bullet \boldsymbol{n} = 0 \text{ at } f(x_1, x_2, x_3) = 0, \qquad (10.1.1)$$

which requires that the normal component of the fluid velocity must equal zero at this surface;

© Springer International Publishing AG, part of Springer Nature 2017

D. J. Wollkind and B. J. Dichone, *Comprehensive Applied Mathematical Modeling in the Natural and Engineering Sciences*,
https://doi.org/10.1007/978-3-319-73518-4_10

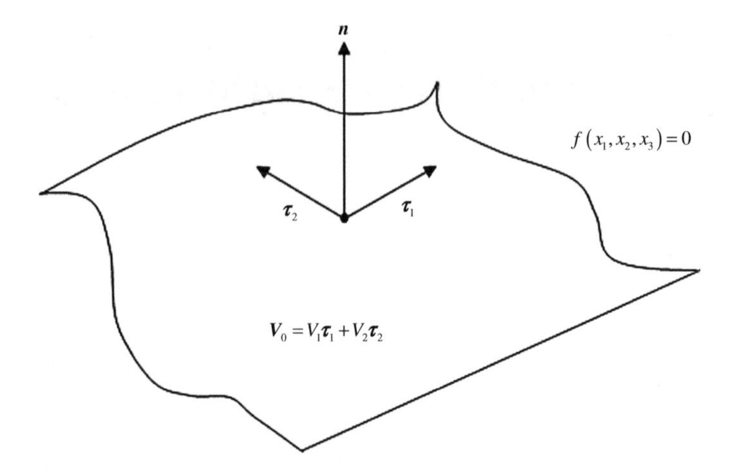

Fig. 10.1 Schematic diagram of a time-independent boundary surface $f(x_1,x_2,x_3) = 0$ with unit upward-pointing normal $\boldsymbol{n}$ and tangent vectors $\boldsymbol{\tau}_{1,2}$. Here $\boldsymbol{V}_0$ represents the experimentally imposed tangential velocity at that boundary.

No-Slip or Adherence Boundary Conditions: $\boldsymbol{v} \bullet \boldsymbol{\tau}_{1,2} = V_{1,2}$ at $f(x_1,x_2,x_3) = 0$,
$$(10.1.2)$$
where $\boldsymbol{V}_0 = V_1\boldsymbol{\tau}_1 + V_2\boldsymbol{\tau}_2$ represents the constant experimentally imposed tangential surface velocity , which requires that the tangential components of the fluid velocity must equal these imposed velocity components at this surface.

As a demonstration of how these boundary conditions are applied in actual practice, we consider the configurations depicted in Fig. 10.2:

Fig. 10.2a represents the situation of a fluid confined between two rigid parallel surfaces located at $z = 0$ and d in Cartesian coordinates where the velocity is given by $\boldsymbol{v} = u\boldsymbol{i} + v\boldsymbol{j} + w\boldsymbol{k}$, the upper surface is moving at a constant speed U_0 in the positive x-direction, and the lower surface is at rest. Then $\boldsymbol{n} = \mp\boldsymbol{k}$ at $z = 0$ and d, respectively, while $\boldsymbol{\tau}_1 = \boldsymbol{i}$ and $\boldsymbol{\tau}_2 = \boldsymbol{j}$. Thus upon application of the boundary conditions (10.1.1) and (10.1.2) at these surfaces we obtain

$$\boldsymbol{v} \bullet (-\boldsymbol{k}) = 0 \Rightarrow w = 0 \text{ and } \boldsymbol{v} \bullet \boldsymbol{i} = u = 0, \boldsymbol{v} \bullet \boldsymbol{j} = v = 0 \text{ at } z = 0; \tag{10.1.3a}$$

$$\boldsymbol{v} \bullet \boldsymbol{k} = w = 0 \text{ and } \boldsymbol{v} \bullet \boldsymbol{i} = u = U_0, \boldsymbol{v} \bullet \boldsymbol{j} = v = 0 \text{ at } z = d. \tag{10.1.3b}$$

Fig. 10.2b represents the situation of a fluid exterior to a rigid surface located at $r = a$ in cylindrical coordinates where the velocity is given by $\boldsymbol{v} = v_r\boldsymbol{e}_r + v_\theta\boldsymbol{e}_\theta + v_z\boldsymbol{k}$ and that cylinder is rotating counterclockwise with a constant angular speed Ω_0. Then $\boldsymbol{n} = -\boldsymbol{e}_r$ at $r = a$ while $\boldsymbol{\tau}_1 = \boldsymbol{e}_\theta$ and $\boldsymbol{\tau}_2 = \boldsymbol{k}$. Thus upon application of the boundary conditions (10.1.1) and (10.1.2) at this surface we obtain

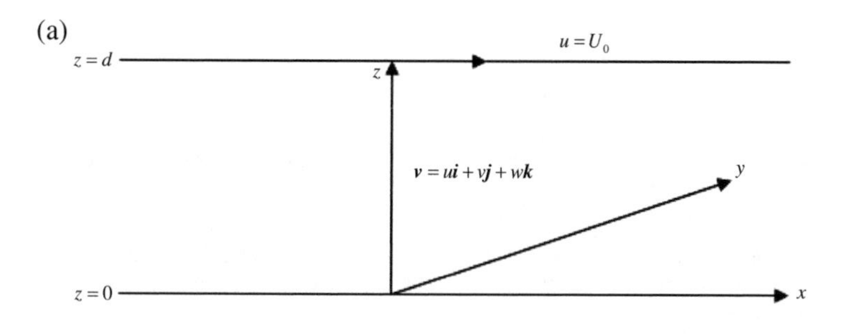

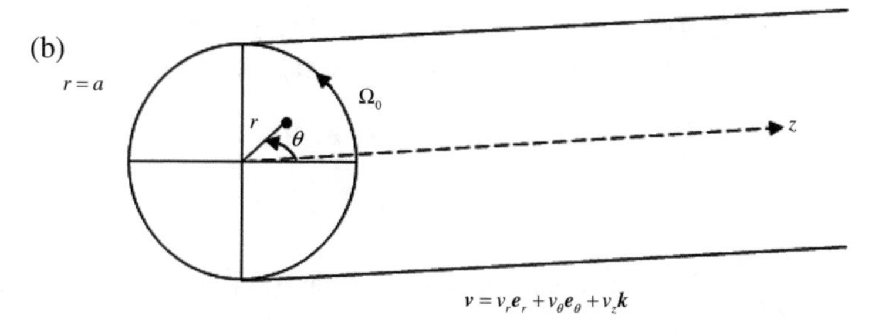

Fig. 10.2 Schematic diagrams of a fluid (a) confined between two rigid parallel surfaces located at $z = 0, d$ in Cartesian coordinates with velocity $\boldsymbol{v} = u\boldsymbol{i} + v\boldsymbol{j} + w\boldsymbol{k}$ where the upper surface is moving at speed U_0 in the positive x-direction while the lower one is at rest and (b) outside a cylinder of radius $r = a$ in cylindrical coordinates with velocity $\boldsymbol{v} = v_r\boldsymbol{e}_r + v_\theta\boldsymbol{e}_\theta + v_z\boldsymbol{k}$ where the cylinder is rotating at angular speed Ω_0 in the counterclockwise direction.

$$\boldsymbol{v} \bullet (-\boldsymbol{e}_r) = 0 \Rightarrow v_r = 0 \text{ and } \boldsymbol{v} \bullet \boldsymbol{e}_\theta = v_\theta = a\Omega_0,\ \boldsymbol{v} \bullet \boldsymbol{k} = v_z = 0 \text{ at } r = a, \qquad (10.1.4)$$

where, in the above, use has been made of the fact that $v_\theta = |\boldsymbol{\omega} \times \boldsymbol{a}|$ for $\boldsymbol{a} = a\boldsymbol{e}_r$ and $\boldsymbol{\omega} \equiv \Omega_0\boldsymbol{k}$.

10.2 Relative Normal Speed and Kinematic Boundary Condition

In this section and the next we shall deduce the proper boundary conditions to assign at moving surfaces separating immiscible fluids or phases at or near equilibrium. We begin with material interfaces such as the one depicted in Fig. 10.3 between a gas and a liquid satisfying the equation $f(x, y, z, t) = z - \zeta(x, y, t) = 0$. Here we shall be concentrating upon the dependence of the unit normal $\boldsymbol{n}$ to that interface and its normal speed w_n on this function f, or equivalently, on ζ. Toward that end let

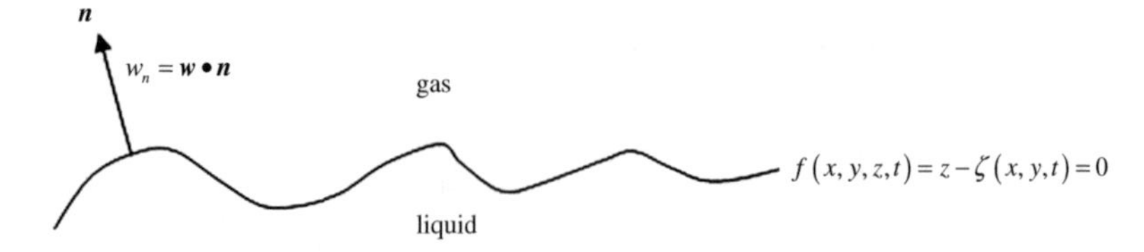

Fig. 10.3 Schematic diagram of a moving material surface $z = \zeta(x, y, t)$ separating a gas from a liquid where the normal speed w_n of that surface in the direction of the unit normal $\boldsymbol{n}$ pointing into the gas is indicated.

time $t = t_0$ and consider a curve C lying totally within this surface. If it is described parametrically in terms of its arc length by $C = \{[x(s), y(s), z(s)], 0 \leq s \leq L\}$, then

$$F(s) = f[x(s), y(s), z(s), t_0] \equiv 0 \text{ for } 0 \leq s \leq L. \tag{10.2.1}$$

Since, now

$$\frac{dF(s)}{ds} = \nabla f[x(s), y(s), z(s), t_0] \bullet \boldsymbol{u}(s) \equiv 0 \text{ for } 0 \leq s \leq L \tag{10.2.2a}$$

where

$$\boldsymbol{u}(s) = \dot{x}(s)\boldsymbol{i} + \dot{y}(s)\boldsymbol{j} + \dot{z}(s)\boldsymbol{k} \equiv \text{ unit tangent vector to } C, \tag{10.2.2b}$$

the gradient vector ∇f is orthogonal to this tangent vector $\boldsymbol{u}$ by virtue of (10.2.2) and hence normal to the surface $f = 0$ or equivalently to $z = \zeta$. Therefore the unit normal vector $\boldsymbol{n}$ to this interface pointing into the gas is given by

$$\boldsymbol{n} = \frac{\nabla f(x, y, z, t)}{|\nabla f(x, y, z, t)|} \text{ at } f(x, y, z, t) = 0 \text{ or } z = \zeta(x, y, t) \tag{10.2.3a}$$

$$\Rightarrow \boldsymbol{n} = \frac{-\zeta_x(x, y, t)\boldsymbol{i} - \zeta_y(x, y, t)\boldsymbol{j} + \boldsymbol{k}}{\sqrt{1 + \zeta_x^2(x, y, t) + \zeta_y^2(x, y, t)}}. \tag{10.2.3b}$$

We next wish to deduce an expression for the normal speed w_n of the interface. Let us describe the points on this interface by $\boldsymbol{x} = \boldsymbol{y}(\boldsymbol{Y}, t)$ such that

$$F(\boldsymbol{Y}, t) = f[\boldsymbol{y}(\boldsymbol{Y}, t), t] \equiv 0 \text{ where } \boldsymbol{y}(\boldsymbol{Y}, 0) = \boldsymbol{Y} \text{ and } \boldsymbol{w} \equiv \frac{\partial \boldsymbol{y}}{\partial t}. \tag{10.2.4}$$

Then

$$\frac{\partial F}{\partial t} = \nabla f \bullet \boldsymbol{w} + \frac{\partial f}{\partial t} \equiv 0 \Rightarrow w_n = \boldsymbol{w} \bullet \boldsymbol{n} = \nabla f \bullet \frac{\boldsymbol{w}}{|\nabla f|} = -\frac{\partial f / \partial t}{|\nabla f|}. \tag{10.2.5}$$

For a material surface the normal velocity v_n of a fluid particle adjacent to the interface

$$v_n = \boldsymbol{v} \bullet \boldsymbol{n} = \nabla f \bullet \frac{\boldsymbol{v}}{|\nabla f|} = w_n = -\frac{\partial f/\partial t}{|\nabla f|} \Rightarrow \frac{Df}{Dt} = 0 \text{ at } f = 0, \qquad (10.2.6a)$$

which is called the kinematic boundary condition and implies that the relative normal speed of the interface $\equiv s_n = w_n - v_n = 0$. For $f = z - \zeta$, (10.2.6a) takes the form (see Problem 10.5)

$$w = \zeta_t + u\zeta_x + v\zeta_y \text{ at } z = \zeta. \qquad (10.2.6b)$$

10.3 Jump Conditions at Surfaces of Discontinuity for Mass, Momentum, and Energy

Some phenomena when modeled by the various techniques of continuum mechanics give rise to inherent surfaces at which the dependent variables exhibit discontinuities. These are usually referred to as surfaces of discontinuity and *careful* application of the postulated conservation or balance laws to them produce jump-type boundary conditions satisfied by the dependent variables across those surfaces. Often consistent with experimental, thermodynamic, or empirical evidence it becomes necessary to adopt constitutive relations, equations of state, or some additional constraints on the dependent variables both at these surfaces and in the bulk of the sub-regions separated by them (see Chapter 9). Physically such surfaces of discontinuity can be true interfaces separating immiscible fluids (see Fig. 10.3) or phases at/near equilibrium (solid–liquid; vapor–liquid) as well as convenient approximations for thin regions over which there are dramatic changes of properties (shocks/detonations, fog boundaries, flame fronts). The mathematical formulation describing all of these phenomena is an example of a Stefan Problem having the fundamental feature that its diffusion equations are to be satisfied in a region the boundaries of which must be determined.

Consider a material region $R(t)$ which is separated by an arbitrary moving surface Σ into sub-regions $R^{\pm}(t)$ having boundaries $S^{\pm}$ possessing outward-pointing unit normals designated by $\boldsymbol{N}^{\pm}$, respectively, and a curve of intersection C with Σ as depicted in Figs. 10.4 and 10.5. Dependent variables including density ρ and velocity $\boldsymbol{v}$ may be discontinuous across Σ. This discontinuity surface has velocity $\boldsymbol{w}$ and unit normal $\boldsymbol{n}_{\Sigma}$ pointing into R^+. If $a \in \Sigma$, we define for any quantity $E(\boldsymbol{x}, t)$,

$$E^{\pm}(\boldsymbol{a}, t) = \lim_{\boldsymbol{x}\to\boldsymbol{a}}[\,E(\boldsymbol{x}, t)|_{\boldsymbol{x}\in R^{\pm}}] \text{ and } [\![E(\boldsymbol{a}, t)]\!] = E^{+}(\boldsymbol{a}, t) - E^{-}(\boldsymbol{a}, t).$$

Then we postulate the following balance-type conservation law for the quantity ρF:

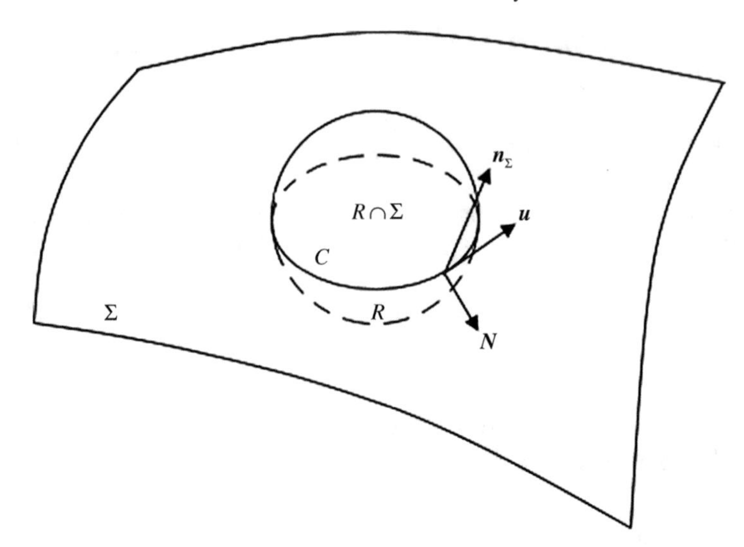

Fig. 10.4 Schematic representation of a material region R traversed by a moving discontinuity surface Σ possessing unit normal $\boldsymbol{n}_\Sigma$ depicting that section with bounding curve C having unit tangent vector $\boldsymbol{u}$ and binormal vector $\boldsymbol{N} = \boldsymbol{u} \times \boldsymbol{n}_\Sigma$.

$$\frac{d}{dt}\iiint\limits_R \rho F\, d\tau = \iiint\limits_R \rho Q\, d\tau - \iint\limits_{\partial R} j d\sigma + \iint\limits_{R\cap\Sigma} g_\Sigma\, d\sigma + \oint\limits_C g_C\, ds$$

where $g_{\Sigma,C}$ are related to sources for that quantity within $R\cap\Sigma$ or along C, respectively, and state without proof a generalized Reynolds Transport Theorem for a context-free (not necessarily material) region [69]

$$A(t) = \{\boldsymbol{x}|\boldsymbol{x} = \boldsymbol{y}(\boldsymbol{Y}, t) \text{ for every } \boldsymbol{Y} = \boldsymbol{y}(\boldsymbol{Y}, 0) \in A(0)\},$$

$$\frac{d}{dt}\iiint\limits_{A(t)} \rho F\, d\tau = \iiint\limits_{A(t)} \frac{\partial(\rho F)}{\partial t}\, d\tau + \iint\limits_{\partial A(t)} \rho F \boldsymbol{u} \bullet \boldsymbol{n}\, d\sigma$$

where $\boldsymbol{n} \equiv$ unit outward-pointing normal to $\partial A(t)$ and $\boldsymbol{u} \equiv \partial \mathbf{y}/\partial t$.

We now apply this theorem to the sub-regions $R^\pm(t)$. That is let

(1) & (2): $A(t) = R^\pm(t)$:

$$\frac{d}{dt}\iiint\limits_{R^\pm(t)} \rho F\, d\tau = \iiint\limits_{R^\pm(t)} \frac{\partial(\rho F)}{\partial t}\, d\tau + \iint\limits_{\partial R^\pm(t)} \rho F \boldsymbol{u}^\pm \bullet \boldsymbol{n}^\pm\, d\sigma$$

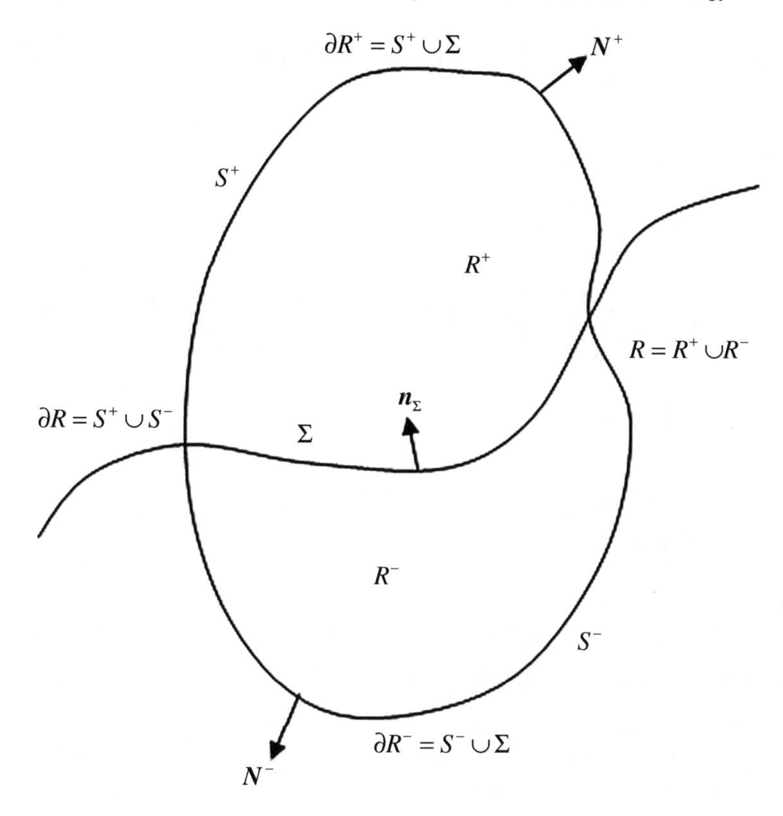

Fig. 10.5 Schematic representation of the sub-regions $R^{\pm}$ into which Σ divides the region R of Fig. 10.4.

where $R^{\pm}(t) = S^{\pm} \cup \Sigma$, $\boldsymbol{u}^{\pm} = \begin{cases} \boldsymbol{v}, & \boldsymbol{x} \in S^{\pm} \\ \boldsymbol{w}, & \boldsymbol{a} \in \Sigma \end{cases}$, and $\boldsymbol{n}^{\pm} = \begin{cases} \boldsymbol{N}^{\pm}, & \boldsymbol{x} \in S^{\pm} \\ \mp \boldsymbol{n}_\Sigma, & \boldsymbol{a} \in \Sigma \end{cases}$. Then

$$\iint\limits_{\partial R^{\pm}(t)} \rho F \boldsymbol{u}^{\pm} \bullet \boldsymbol{n}^{\pm} \, d\sigma = \iint\limits_{S^{\pm}} \rho F \boldsymbol{v} \bullet \boldsymbol{N}^{\pm} \, d\sigma + \iint\limits_{\Sigma} \rho^{\pm} F^{\pm} \boldsymbol{w} \bullet (\mp \boldsymbol{n}_\Sigma) \, d\sigma.$$

Define $s_n^{\pm} = (\boldsymbol{w} - \boldsymbol{v}^{\pm}) \bullet \boldsymbol{n}_\Sigma$ or $\boldsymbol{w} \bullet \boldsymbol{n}_\Sigma = \boldsymbol{v}^{\pm} \bullet \boldsymbol{n}_\Sigma + s_n^{\pm}$. Hence

$$\iint\limits_{\Sigma} \rho^{\pm} F^{\pm} \boldsymbol{w} \bullet (\mp \boldsymbol{n}_\Sigma) \, d\sigma = \iint\limits_{\Sigma} \rho^{\pm} F^{\pm} \boldsymbol{v}^{\pm} \bullet (\mp \boldsymbol{n}_\Sigma) \, d\sigma \mp \iint\limits_{\Sigma} \rho^{\pm} F^{\pm} s_n^{\pm} \, d\sigma.$$

Thus

$$\iint_{S^\pm} \rho F \boldsymbol{v} \bullet \boldsymbol{N}^\pm \, d\sigma + \iint_{\Sigma} \rho^\pm F^\pm \boldsymbol{v}^\pm (\mp \boldsymbol{n}_\Sigma) \, d\sigma = \iint_{\partial R^\pm(t)} \rho F \boldsymbol{v} \bullet \boldsymbol{n}^\pm \, d\sigma = \iiint_{R^\pm(t)} \nabla \bullet (\rho F \boldsymbol{v}) \, d\tau.$$

Therefore

$$\frac{d}{dt} \iiint_{R^\pm(t)} \rho F \, d\tau = \iiint_{R^\pm(t)} \left[\frac{\partial(\rho F)}{\partial t} + \nabla \bullet (\rho F \boldsymbol{v}) \right] d\tau \mp \iint_{\Sigma} \rho^\pm F^\pm s_n^\pm \, d\sigma$$

$$= \iiint_{R^\pm(t)} \rho \frac{DF}{Dt} \, d\tau \mp \iint_{\Sigma} \rho^\pm F^\pm s_n^\pm \, d\sigma \text{ since } \frac{\partial \rho}{\partial t} + \nabla \bullet (\rho \boldsymbol{v}) = 0.$$

③ $R(t) = R^+(t) \cup R^-(t)$:

$$\frac{d}{dt} \iiint_{R(t)} \rho F \, d\tau = ① + ② = \iiint_{R(t)} \rho \frac{DF}{Dt} \, d\tau - \iint_{\Sigma} [\![\rho F s_n]\!] \, d\sigma$$

where

$$[\![\rho F s_n]\!] = \rho^+ F^+ s_n^+ - \rho^- F^- s_n^-. \tag{10.3.1a}$$

Substituting this result in our balance law we obtain

$$\iiint_{R(t)} \rho \left(\frac{DF}{DT} - Q \right) d\tau = \iint_{R \cap \Sigma} \{ [\![\rho F s_n]\!] + g_\Sigma \} \, d\sigma - \iint_{\partial R} j \, d\sigma + \oint_C g_C \, ds$$

where

$$\iint_{\partial R} j \, d\sigma = \iint_{S^+} j(\boldsymbol{x}, \boldsymbol{N}^+, t) \, d\sigma + \iint_{S^-} j(\boldsymbol{x}, \boldsymbol{N}^-, t) \, d\sigma$$

and

$$\oint_C g_C(\boldsymbol{x}, \boldsymbol{N}, t) \, ds = \iint_{R \cap \Sigma} G_\Sigma(\boldsymbol{x}, \boldsymbol{n}_\Sigma, t) \, d\sigma,$$

the latter symbolically representing a surface divergence theorem [12] with $\boldsymbol{N} \equiv$ binormal vector as depicted in Fig. 10.4. We shall now collapse the region R to one flanking Σ. Then (see Figs. 10.4 and 10.5)

$$\iiint_{R(t)} \left(\rho \frac{DF}{Dt} - Q \right) d\tau \to 0; \, R \cap \Sigma, \, S^\pm \to \Sigma; \text{ and } \boldsymbol{N}^\pm \to \pm \boldsymbol{n}_\Sigma;$$

in that limit, assuming the requisite continuity of the integrand for $\boldsymbol{x} \in R^\pm$; hence our balance law reduces to

$$\iint_{\Sigma} \{ [\![\rho F s_n]\!] + g_\Sigma + G_\Sigma - [\![j]\!] \} \, d\sigma = 0$$

and, finally applying a DuBois-Reymond Lemma based on the arbitrariness of Σ, we can deduce the jump-type boundary condition satisfied by our surface of discontinuity

$$[\![\rho F s_n]\!] + g = [\![j]\!] \text{ for any } \boldsymbol{a} \in \Sigma \tag{10.3.1b}$$

where

$$g = g_\Sigma + G_\Sigma \tag{10.3.1c}$$

and

$$[\![j]\!] = [\![j(\boldsymbol{a}, \boldsymbol{n}_\Sigma, t)]\!] = j^+(\boldsymbol{a}, \boldsymbol{n}_\Sigma, t) - j^-(\boldsymbol{a}, \boldsymbol{n}_\Sigma, t). \tag{10.3.1d}$$

We next deduce the jump conditions at a surface of discontinuity for ① mass, ② momentum, and ③ energy from (10.3.1) by identifying the appropriate components in the balance laws relevant to each of these quantities in the same manner as employed in the last chapter.

① *Mass*: Since for mass

$$\frac{d}{dt}\iiint_R \rho \, d\tau = 0,$$

we make the identifications $F = 1$ and $Q = j = g = 0$. Thus from (10.3.1) we can deduce that the jump condition for mass at a surface of discontinuity is given by

$$[\![\rho s_n]\!] = 0 \text{ for any } \boldsymbol{a} \in \Sigma. \tag{10.3.2}$$

② *Momentum*: Since for *linear* momentum

$$\frac{d}{dt}\iiint_R \rho \boldsymbol{v} \, d\tau = \iiint_R \rho \boldsymbol{f} \, d\tau + \iint_{\partial R} \boldsymbol{t} \, d\sigma + \oint_C \boldsymbol{\ell} \, ds$$

where $\boldsymbol{\ell} \equiv$ surface tension vector and assuming the constitutive relation $\boldsymbol{\ell} = \gamma \boldsymbol{N}$ while from the surface divergence theorem (see Problem 10.2)

$$\oint_C \gamma \boldsymbol{N} \, ds = \iint_{R \cap \Sigma} \boldsymbol{L} \, d\sigma$$

where

$$\boldsymbol{L} = \gamma \kappa \boldsymbol{n}_\Sigma + \nabla_{(\Sigma)} \gamma \tag{10.3.3a}$$

for $\kappa \equiv$ curvature of Σ and $\nabla_{(\Sigma)}\gamma \equiv$ surface gradient vector satisfying

$$\kappa = \nabla_2 \bullet \left[\frac{\nabla_2 \zeta}{|\nabla f|}\right] \text{ and } \nabla_{(\Sigma)} = \boldsymbol{\tau}_1 \frac{\partial}{\partial \tau_1} + \boldsymbol{\tau}_2 \frac{\partial}{\partial \tau_2} \tag{10.3.3b}$$

with $\boldsymbol{\tau}_{1,2} \equiv$ unit orthogonal tangent vectors to that Σ given by $f(x, y, z, t) = z - \zeta(x, y, t) = 0$, we make the identifications $F \sim \boldsymbol{v}$, $Q \sim \boldsymbol{f}$, $j \sim -\boldsymbol{t}$, $g_\Sigma \sim \boldsymbol{0}$, and $G_\Sigma \sim \boldsymbol{L}$.

Thus from (10.3.1) we can deduce that the jump condition for linear momentum at a surface of discontinuity is given by

$$[\![\rho \boldsymbol{v} s_n + \boldsymbol{t}]\!] + \boldsymbol{L} = \boldsymbol{0} \text{ for any } \boldsymbol{a} \in \Sigma \tag{10.3.3c}$$

We shall show below that (10.3.3) satisfies the jump condition for angular momentum as well.

(3) *Energy*: Since for energy

$$\frac{d}{dt} \iiint_R \rho\left(e + \frac{\boldsymbol{v} \bullet \boldsymbol{v}}{2}\right) d\tau = \iiint_R \rho(\boldsymbol{f} \bullet \boldsymbol{v} + r)\, d\tau + \iint_{\partial R} (\boldsymbol{t} \bullet \boldsymbol{v} - h)\, d\sigma + \iint_{R \cap \Sigma} H\, d\sigma + \oint_C \boldsymbol{\ell} \bullet \boldsymbol{w}\, ds$$

where $H \equiv$ energy of transformation while [12]

$$\oint_C \gamma \boldsymbol{N} \bullet \boldsymbol{w}\, ds = \iint_{R \cap \Sigma} \boldsymbol{L} \bullet \boldsymbol{w}\, d\sigma,$$

then $F = e + (\boldsymbol{v} \bullet \boldsymbol{v})/2$, $Q = \boldsymbol{f} \bullet \boldsymbol{v} + r$, $j = h - \boldsymbol{t} \bullet \boldsymbol{v}$, $g_\Sigma = H$, and $G_\Sigma = \boldsymbol{L} \bullet \boldsymbol{w}$. Thus from (10.3.1) we can deduce that the jump condition for energy at a surface of discontinuity is given by

$$\left[\!\!\left[\rho\left(e + \frac{\boldsymbol{v} \bullet \boldsymbol{v}}{2}\right) s_n + \boldsymbol{t} \bullet \boldsymbol{v} - h\right]\!\!\right] + H + \boldsymbol{L} \bullet \boldsymbol{w} = 0 \text{ for any } \boldsymbol{a} \in \Sigma. \tag{10.3.4}$$

We have deferred until now an examination of the jump condition for angular momentum at a surface of discontinuity in a nonpolar continuum. Since for angular momentum in such a continuum

$$\frac{d}{dt} \iiint_R \rho(\boldsymbol{r} \times \boldsymbol{v})\, d\tau = \iiint_R \rho(\boldsymbol{r} \times \boldsymbol{f})\, d\tau + \iint_{\partial R} \boldsymbol{r} \times \boldsymbol{t}\, d\sigma + \oint_C \boldsymbol{r} \times \boldsymbol{\ell}\, ds$$

while (see Problem 10.3)

$$\oint_C \boldsymbol{r} \times \gamma \boldsymbol{N}\, ds = \iint_{R \cap \Sigma} \boldsymbol{r} \times \boldsymbol{L}\, d\sigma,$$

we make the identifications $F \sim \boldsymbol{r} \times \boldsymbol{v}$, $Q \sim \boldsymbol{r} \times \boldsymbol{f}$, $j \sim -\boldsymbol{r} \times \boldsymbol{t}$, $g_\Sigma \sim \boldsymbol{0}$, and $G_\Sigma \sim \boldsymbol{r} \times \boldsymbol{L}$. Thus from (10.3.1) we can deduce that the jump condition for angular momentum at a surface of discontinuity is given by

$$[\![\rho(\boldsymbol{r} \times \boldsymbol{v}) s_n + \boldsymbol{r} \times \boldsymbol{t}]\!] + \boldsymbol{r} \times \boldsymbol{L} = \boldsymbol{r} \times \{[\![\rho \boldsymbol{v} s_n + \boldsymbol{t}]\!] + \boldsymbol{L}\} = \boldsymbol{0} \text{ for any } \boldsymbol{a} \in \Sigma. \tag{10.3.5}$$

As indicated above this jump condition for angular momentum is satisfied identically by (10.3.3), the jump condition for linear momentum, since

$$\boldsymbol{r} \times \{[\![\rho \boldsymbol{v} s_n + \boldsymbol{t}]\!] + \boldsymbol{L}\} = \boldsymbol{r} \times \boldsymbol{0} = \boldsymbol{0}.$$

Note that then the satisfaction of (10.3.5) does not add any additional jump conditions for a surface of discontinuity in a nonpolar continuum and hence we have chosen to categorize our development for deducing (10.3.3) under the heading of momentum rather than linear momentum in this section.

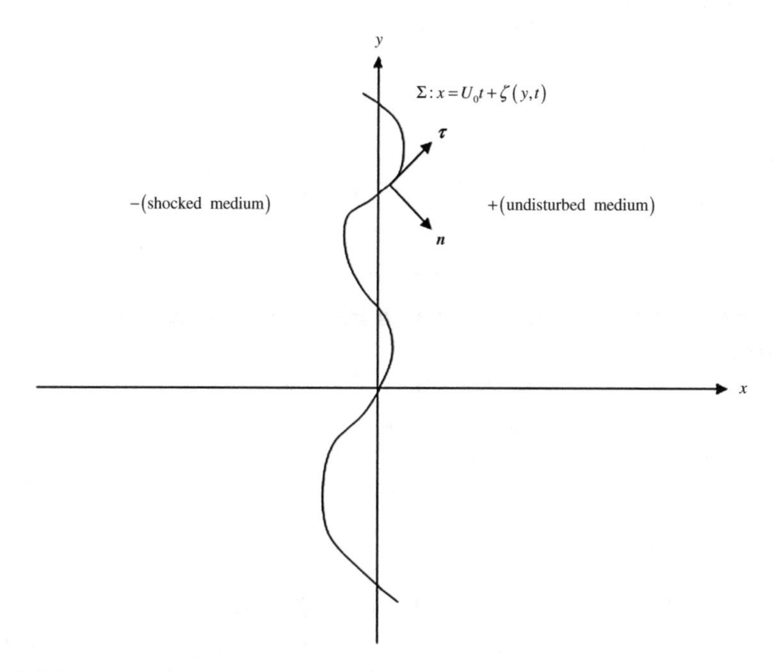

Fig. 10.6 Schematic representation of the propagation of a two-dimensional plane parallel shock front satisfying the equation $x = U_0 t + \zeta(y, t)$ where the y-axis corresponds to $x = U_0 t$.

We close with an examination of these three jump conditions (10.3.2)–(10.3.4) for the propagation of a plane parallel shock front through an inviscid fluid – *i.e.*, $\lambda = \mu = 0$ – in the absence of body forces and with no heat flow. We shall analyze the two-dimensional x-y situation depicted in Fig. 10.6 and consider this shock front to be a discontinuity surface possessing no source terms which fluid actually crosses – *i.e.*, the relevant relative normal speeds of that interface are negative or

$$s_n^{\pm} = (\boldsymbol{w} - \boldsymbol{v}^{\pm}) \bullet \boldsymbol{n} = w_n - v_n^{\pm} < 0. \tag{10.3.6}$$

Here the "–" and "+" phases represent the shocked and undisturbed media, respectively; the shock front Σ is propagating into the "+" phase; and, for ease of exposition, we have denoted the unit normal $\boldsymbol{n}_\Sigma$ to this front pointing into that phase by $\boldsymbol{n}$.

We shall now examine (10.3.2)–(10.3.4) for (1) mass, (2) momentum, and (3) energy, sequentially, under these conditions.

(1) *Mass*: Then

$$[\![\rho s_n]\!] = 0 \Rightarrow \rho^+ s_n^+ = \rho^- s_n^-. \tag{10.3.7}$$

Note that if

$$\rho^+ = \rho^- \Rightarrow s_n^+ = s_n^- \Rightarrow v_n^+ = v_n^-$$

while if

$$\rho^+ \neq \rho^- \Rightarrow s_n^+ \neq s_n^- \Rightarrow v_n^+ \neq v_n^-.$$

(2) *Momentum*: Then

$$[\![\rho \boldsymbol{v} s_n + \boldsymbol{t}]\!] + \boldsymbol{L} = \boldsymbol{0}$$

with

$$\boldsymbol{L} = \boldsymbol{0}, \boldsymbol{t} = -p\boldsymbol{n}, \text{ and } \boldsymbol{v} = v_n\boldsymbol{n} + v_\tau\boldsymbol{\tau}, \tag{10.3.8}$$

where $\boldsymbol{\tau} \equiv$ unit tangent vector to Σ in the increasing y-direction such that $\boldsymbol{n} \bullet \boldsymbol{\tau} = 0$, becomes

$$[\![\rho(v_n\boldsymbol{n} + v_\tau\boldsymbol{\tau})s_n - p\boldsymbol{n}]\!] = 0. \tag{10.3.9a}$$

Now, from (10.3.6), we can conclude that

$$v_n = w_n - s_n. \tag{10.3.9b}$$

Then

$$\begin{aligned} \rho(v_n\boldsymbol{n} + v_\tau\boldsymbol{\tau})s_n - p\boldsymbol{n} &= [\rho s_n(w_n - s_n) - p]\boldsymbol{n} + \rho s_n v_\tau \boldsymbol{\tau} \\ &= [\rho s_n w_n - (\rho s_n^2 + p)]\boldsymbol{n} + \rho s_n v_\tau \boldsymbol{\tau}. \end{aligned} \tag{10.3.9c}$$

Hence

$$[\rho^+ s_n^+ w_n - (\rho^+ s_n^{+2} + p^+)]\boldsymbol{n} + \rho^+ s_n^+ v_\tau^+ \boldsymbol{\tau} = [\rho^- s_n^- w_n - (\rho^- s_n^{-2} + p^-)]\boldsymbol{n} + \rho^- s_n^- v_\tau^- \boldsymbol{\tau} \tag{10.3.9d}$$

which, equating the normal and tangential components of (10.3.9d), implies that

$$\rho^+ s_n^+ w_n - (\rho^+ s_n^{+2} + p^2) = \rho^- s_n^- w_n - (\rho^- s_n^{-2} + p^-) \tag{10.3.10a}$$

and

$$\rho^+ s_n^+ v_\tau^+ = \rho^- s_n^- v_\tau^-. \tag{10.3.10b}$$

Finally, employing (10.3.7), these momentum jump conditions of (10.3.10) reduce to

$$\rho^+ s_n^{+2} + p^+ = \rho^- s_n^{-2} + p^- \text{ or } [\![\rho s_n^2 + p]\!] = 0 \tag{10.3.11a}$$

and

$$v_\tau^+ = v_\tau^- \text{ or } [\![v_\tau]\!] = 0. \tag{10.3.11b}$$

(3) *Energy*: Then

$$\left[\!\!\left[\rho\left(e + \frac{\boldsymbol{v} \bullet \boldsymbol{v}}{2}\right) s_n + \boldsymbol{t} \bullet \boldsymbol{v} - h\right]\!\!\right] + H + \boldsymbol{L} \bullet \boldsymbol{w} = 0$$

with (10.3.8) and $h = H = 0$ becomes

$$\left[\!\!\left[\rho\left(e + \frac{\boldsymbol{v} \bullet \boldsymbol{v}}{2}\right) s_n + \boldsymbol{t} \bullet \boldsymbol{v}\right]\!\!\right] = \left[\!\!\left[\rho s_n\left(e + \frac{v_n^2 + v_\tau^2}{2}\right) - p v_n\right]\!\!\right] = 0. \tag{10.3.12a}$$

Now making use of (10.3.9b)

$$\begin{aligned}
\rho s_n\left(e + \frac{v_n^2 + v_\tau^2}{2}\right) - p v_n &= \rho s_n\left[e + \frac{(w_n - s_n)^2}{2} + \frac{v_\tau^2}{2}\right] + p(s_n - w_n) \\
&= \rho s_n\left(e + \frac{w_n^2 - 2w_n s_n + s_n^2 + v_\tau^2}{2}\right) + p(s_n - w_n) \\
&= \rho s_n\left(e + \frac{p}{\rho} + \frac{w_n^2 + v_\tau^2 + s_n^2}{2}\right) - w_n(\rho s_n^2 + p).
\end{aligned} \tag{10.3.12b}$$

Hence

$$\begin{aligned}
&\rho^+ s_n^+\left(e^+ + \frac{p^+}{\rho^+} + \frac{w_n^2 + v_\tau^{+2} + s_n^{+2}}{2}\right) - w_n(\rho^+ s_n^{+2} + p^+) \\
&= \rho^- s_n^-\left(e^- + \frac{p^-}{\rho^-} + \frac{w_n^2 + v_\tau^{-2} + s_n^{-2}}{2}\right) - w_n(\rho^- s_n^{-2} + p^-).
\end{aligned} \tag{10.3.12c}$$

Finally employing (10.3.7) and (10.3.11) this energy jump condition of (10.3.12c) reduces to

$$e^+ + \frac{p^+}{\rho^+} + \frac{s_n^{+2}}{2} = e^- + \frac{p^-}{\rho^-} + \frac{s_n^{-2}}{2} \text{ or } \left[\!\!\left[\eta + \frac{s_n^2}{2}\right]\!\!\right] = 0 \tag{10.3.13a}$$

where

$$\eta = e + \frac{p}{\rho} = e + pV \equiv \text{ enthalpy.} \tag{10.3.13b}$$

Here we are employing η to denote enthalpy rather than the h used heretofore both to avoid confusion with our energy flux term of (10.3.4) and for historical accuracy since this was its notation originally. That the Greek capital letter for eta is H helps explain the present day usage of h.

Observe that should Σ be a material surface rather than a shock front then (10.3.7) would be satisfied identically without imposing any constraints on $\rho^{\pm}$ while (10.3.11a) and (10.3.13) would reduce to

$$[\![p]\!] = 0 \tag{10.3.14a}$$

and

$$[\![\eta]\!] = 0 \tag{10.3.14b}$$

since in this event

$$s_n^{\pm} = 0 \Rightarrow v_n^+ = v_n^- \text{ or } [\![v_n]\!] = 0. \tag{10.3.14c}$$

Further, note that (10.3.14c) in conjunction with (10.3.11b) would then imply that

$$\boldsymbol{v}^+ = \boldsymbol{v}^- \text{ or } [\![\boldsymbol{v}]\!] = \boldsymbol{0}. \tag{10.3.14d}$$

In other words, for two-dimensional inviscid fluids with no heat flow and interfacial sources, their pressure, enthalpy, and velocity would exhibit no discontinuities at such a material surface although density could. This is the case for interfaces separating two immiscible fluids such as a gas and a liquid.

We next convert (10.3.11a) and (10.3.13) to the forms usually employed for the Rankine Hugoniot jump conditions of plane parallel shock dynamics. Proceeding sequentially let us first consider (10.3.11a) in conjunction with (10.3.7) which yields

$$\begin{aligned} p^+ - p^- &= (p^- s_n^-) s_n^- - (p^+ s_n^+) s_n^+ = (\rho^+ s_n^+) s_n^- - (\rho^- s_n^-) s_n^+ \\ &= \left(\frac{1}{\rho^-} - \frac{1}{\rho^+}\right)(\rho^+ s_n^+)(\rho^- s_n^-) = (V^- - V^+)(\rho^{\pm} s_n^{\pm})^2 \end{aligned}$$

$$\Rightarrow s_n^{\pm 2} = V^{\pm 2} \frac{p^+ - p^-}{V^- - V^+} \Rightarrow s_n^{\pm} = -V^{\pm} \sqrt{\frac{p^+ - p^-}{V^- - V^+}} \tag{10.3.15}$$

since $s_n^{\pm} < 0$ by virtue of (10.3.6). Then from (10.3.6) and (10.3.15) we can deduce that

$$s_n^+ - s_n^- = v_n^- - v_n^+ = (V^- - V^+) \sqrt{\frac{p^+ - p^-}{V^- - V^+}}$$

$$\Rightarrow (v_n^+ - v_n^-)^2 = (V^- - V^+)^2 \frac{p^+ - p^-}{V^- - V^+} = (V^- - V^+)(p^+ - p^-). \tag{10.3.16}$$

Finally considering (10.3.13) in conjunction with (10.3.15) we obtain

$$
\begin{aligned}
e^+ - e^- &= \frac{s_n^{-2} - s_n^{+2}}{2} + V^- p^- - V^+ p^+ \\
&= \frac{[(V^-)^2 - (V^+)^2](p^+ - p^-)}{2(V^- - V^+)} + V^- p^- - V^+ p^+ \\
&= \frac{(V^- - V^+)(p^+ - p^-)}{2} + V^- p^- - V^+ p^+ \\
&= \frac{V^- p^+ + V^+ p^+ - V^- p^- - V^+ p^-}{2} + V^- p^- - V^+ p^+
\end{aligned}
$$

or

$$e^+ - e^- = \frac{V^- p^+ + V^- p^- - V^+ p^+ - V^+ p^-}{2} = \frac{(V^- - V^+)(p^+ + p^-)}{2}. \qquad (10.3.17)$$

Consider a two-dimensional planar shock front Σ as indicated in Fig. 10.6 satisfying the equation $f(x, y, t) = x - U_0 t - \zeta(y, t) = 0$ where

$$\lim_{L \to \infty} \frac{1}{2L} \int_{-L}^{L} \zeta(y, t)\, dy = 0.$$

Thus the mean position of the front at time t will be at $x = U_0 t$ which is why the term planar has been used to describe it. We conclude, as in Section 10.2, with a determination of the dependence of $\boldsymbol{\tau}$, $\boldsymbol{n}$, and w_n upon this deviation $\zeta(y, t)$ from the planar shock front.

$\boldsymbol{\tau}$: We can represent Σ parametrically by

$$\boldsymbol{r}(y; t) = [U_0 t + \zeta(y, t)]\boldsymbol{i} + y\boldsymbol{j} \text{ with } -\infty < y < \infty.$$

Hence

$$\boldsymbol{\tau} = \frac{\frac{d\boldsymbol{r}}{dy}(y; t)}{\left|\frac{d\boldsymbol{r}}{dy}(y; t)\right|} \Rightarrow \boldsymbol{\tau} = \frac{\zeta_y(y, t)\boldsymbol{i} + \boldsymbol{j}}{[1 + \zeta_y^2(y, t)]^{1/2}}.$$

$\boldsymbol{n}$: From (10.2.3)

$$\boldsymbol{n} = \frac{\nabla_2 f(x, y, t)}{|\nabla_2 f(x, y, t)|} \Rightarrow \boldsymbol{n} = \frac{\boldsymbol{i} - \zeta_y(y, t)\boldsymbol{j}}{[1 + \zeta_y^2(y, t)]^{1/2}}.$$

Note that $\boldsymbol{n} \bullet \boldsymbol{n} = \boldsymbol{\tau} \bullet \boldsymbol{\tau} = 1$ and $\boldsymbol{n} \bullet \boldsymbol{\tau} = 0$ as required. In this context, for $\boldsymbol{v}^\pm = u^\pm \boldsymbol{i} + v^\pm \boldsymbol{j}$, $v_n^\pm = \boldsymbol{v}^\pm \bullet \boldsymbol{n} = (u^\pm - v^\pm \zeta_y)/(1 + \zeta_y^2)^{1/2}$ and $v_\tau^\pm = \boldsymbol{v}^\pm \bullet \boldsymbol{\tau} = (u^\pm \zeta_y + v^\pm)/(1 + \zeta_y^2)^{1/2}$.

w_n: From (10.2.5)

$$
\begin{aligned}
w_n &= \frac{-\partial f(x, y, t)/\partial t}{|\nabla_2 f(x, y, t)|} \\
\Rightarrow w_n &= \frac{U_0 + \zeta_t(y, t)}{[1 + \zeta_y^2(y, t)]^{1/2}}.
\end{aligned}
$$

Note that this reduces to $w_n = U_0$ for the planar shock front with $\zeta(y,t) \equiv 0$ as expected. Observe that the replacement of (10.3.11a) and (10.3.13) with the Rankine Hugoniot jump conditions of (10.3.16) and (10.3.17) eliminates w_n from those conditions and it then appears exclusively in (10.3.7). Finally note that only w_n can be determined in this manner and that the tangential component of this speed w_τ must be imposed as a constitutive relation most often taken to satisfy $w_\tau = v_\tau^+ = v_\tau^-$, by virtue of (10.3.11b), which is an interfacial no-slip type condition at the shock front.

Problems

10.1. The curvature of an interface satisfying the equation

$$f(x,y,z,t) = z - \zeta(x,y,t) = 0$$

was given in (10.3.3a) by $\kappa = \nabla_2 \bullet [\nabla_2\zeta/|\nabla f|]$. The object of this problem is to derive the formula

$$\kappa = \frac{\zeta_{xx}(1+\zeta_y^2) - 2\zeta_x\zeta_y\zeta_{xy} + \zeta_{yy}(1+\zeta_x^2)}{(1+\zeta_x^2+\zeta_y^2)^{3/2}}.$$

(a) First, by the definition of the relevant gradient vectors, demonstrate that

$$\nabla_2\zeta = \zeta_x\boldsymbol{i} + \zeta_y\boldsymbol{j};\ \nabla f = -\zeta_x\boldsymbol{i} - \zeta_y\boldsymbol{j} + \boldsymbol{k};\ \frac{\nabla_2\zeta}{|\nabla f|} = \frac{\zeta_x\boldsymbol{i} + \zeta_y\boldsymbol{j}}{(1+\zeta_x^2+\zeta_y^2)^{1/2}}.$$

(b) Second, by the definition of divergence, conclude that

$$\nabla_2 \bullet \frac{\nabla_2\zeta}{|\nabla f|} = \frac{\partial}{\partial x}[\zeta_x(1+\zeta_x^2+\zeta_y^2)^{-1/2}] + \frac{\partial}{\partial y}[\zeta_y(1+\zeta_x^2+\zeta_y^2)^{-1/2}].$$

(c) Now, by direct partial differentiation, show that

$$\frac{\partial}{\partial x}[\zeta_x(1+\zeta_x^2+\zeta_y^2)^{-1/2}] = \frac{\zeta_{xx}(1+\zeta_y^2) - \zeta_x\zeta_y\zeta_{yx}}{(1+\zeta_x^2+\zeta_y^2)^{3/2}}.$$

(d) Next, by interchanging the roles of x and y in (c), deduce that

$$\frac{\partial}{\partial y}[\zeta_y(1+\zeta_y^2+\zeta_x^2)^{-1/2}] = \frac{\zeta_{yy}(1+\zeta_x^2) - \zeta_y\zeta_x\zeta_{xy}}{(1+\zeta_x^2+\zeta_y^2)^{3/2}}.$$

(e) Finally, by adding (c) and (d), obtain the desired result after assuming the requisite continuity of the second partial derivatives of ζ which implies that

$$\zeta_{yx} = \zeta_{xy}.$$

10.2. The object of this problem is to derive the surface divergence theorem of (10.3.3) in the form

$$\oint_C \gamma \boldsymbol{N}\, ds = \iint_\Sigma [\gamma\kappa\boldsymbol{n} + \nabla_{(\Sigma)}\gamma]\, d\sigma$$

for the steady-state two-dimensional case by employing Stokes Theorem. Hence consider Fig. 10.4 particularized to that Σ which is the section of the surface $z = \zeta(x)$ bounded by the curve C, parametrically represented by $\boldsymbol{r}(s) = x(s)\boldsymbol{i} + y(s)\boldsymbol{j} + z(s)\boldsymbol{k}$ where s is arc length, while $\gamma = \gamma(x,z)$. Thus $\boldsymbol{N} = \boldsymbol{u}\times\boldsymbol{n}$ where $\boldsymbol{u}(s) = \dot{\boldsymbol{r}}(s) \equiv$ unit tangent vector to C such that $\dot{\boldsymbol{r}}(s) = \dot{x}(s)\boldsymbol{i} + \dot{y}(s)\boldsymbol{j} + \dot{z}(s)\boldsymbol{k}$ and $\boldsymbol{n} = n_1(x)\boldsymbol{i} + n_2(x)\boldsymbol{j} + n_3(x)\boldsymbol{k} \equiv$ unit upward-pointing normal vector to Σ with $n_3(x) = \{1 + [\zeta'(x)]^2\}^{-1/2}$, $n_2(x) \equiv 0$, $n_1(x) = -\zeta'(x)n_3(x)$. Finally,

$$\kappa = \frac{\zeta''(x)}{\{1 + [\zeta'(x)]^2\}^{3/2}} \text{ and } \nabla_{(\Sigma)}\gamma = \frac{\partial\gamma}{\partial\tau}\boldsymbol{\tau} \text{ with } \frac{\partial\gamma}{\partial\tau} = \nabla_2\gamma \bullet \boldsymbol{\tau}$$

where $\nabla_2\gamma = \gamma_x\boldsymbol{i} + \gamma_z\boldsymbol{k}$ and $\boldsymbol{\tau} = n_3(x)\boldsymbol{i} - n_1(x)\boldsymbol{k} \equiv$ unit tangent vector to Σ.

(a) From the definition of the cross product and noting $\dot{x}_i(s)\, ds = dx_i$ $(i = 1,2,3)$ show that

$$\oint_C \gamma \boldsymbol{N}\, ds = \boldsymbol{i}\oint_C \gamma n_3\, dy + \boldsymbol{j}\oint_C \gamma(n_1\, dz - n_3\, dx) + \boldsymbol{k}\oint_C (-\gamma n_1)\, dy.$$

(b) Now applying Stokes Theorem

$$\oint_C A_1\, dx + A_2\, dy + A_3\, dz$$

$$= \iint_\Sigma \left[\left(\frac{\partial A_3}{\partial y} - \frac{\partial A_2}{\partial z}\right)n_1 + \left(\frac{\partial A_1}{\partial z} - \frac{\partial A_3}{\partial x}\right)n_2 + \left(\frac{\partial A_2}{\partial x} - \frac{\partial A_1}{\partial y}\right)n_3\right] d\sigma$$

in the form

$$\oint_C [A_1(x,z)\, dx + A_2(x,z)\, dy + A_3(x,z)\, dz] = \iint_\Sigma [A_{2x}(x,z)n_3(x) - A_{2z}(x,z)n_1(x)]\, d\sigma$$

to each of the line integrals on the right-hand side of part (a) deduce the desired result. Hint: Note that $n_3'(x) = -\kappa\zeta'(x)$, $n_1'(x) = -\kappa$, and $\nabla_2\gamma \bullet \boldsymbol{\tau} = \gamma_x n_3(x) - \gamma_z n_1(x)$.

10.3. Demonstrate the validity of

$$\oint_C \boldsymbol{r}\times\gamma\boldsymbol{N}\, ds = \iint_{R\cap\Sigma} \boldsymbol{r}\times[\gamma\kappa\boldsymbol{n} + \nabla_{(\Sigma)}\gamma]\, d\sigma$$

by making use of the facts that [12]

$$\oint_C \boldsymbol{N} \times \boldsymbol{a}\, ds = \iint_{R\cap\Sigma} [\nabla_{(\Sigma)} \times \boldsymbol{a} + \kappa \boldsymbol{n} \times \boldsymbol{a}]\, d\sigma$$

for any arbitrary vector $\boldsymbol{a}$ where $\nabla_{(\Sigma)} \times \boldsymbol{a}$ is defined in an analogous manner to $\nabla \times \boldsymbol{a} \equiv \text{curl}\, \boldsymbol{a}$;

$$\nabla_{(\Sigma)} \times (\alpha \boldsymbol{a}) = [\nabla_{(\Sigma)}\alpha] \times \boldsymbol{a} + \alpha \nabla_{(\Sigma)} \times \boldsymbol{a} \text{ for any arbitrary scalar } \alpha;$$

and

$$\nabla_{(\Sigma)} \times \boldsymbol{r} = \boldsymbol{0} \text{ for the position vector } \boldsymbol{r}.$$

10.4. Consider a phase transformation such that $\gamma \equiv$ surface free energy is constant. In the absence of convection – *i.e.*, $\boldsymbol{v} = \boldsymbol{0}$ in both phases—show that the jump conditions for mass, momentum, and energy given by (10.3.2)–(10.3.4) yield

$$\rho^+ = \rho^- = \rho_0,\ p^+ - p^- = \gamma\kappa,$$

$$k^- \frac{\partial T^-}{\partial n} - k^+ \frac{\partial T^+}{\partial n} = w_n[L + \rho_0(c^+T^+ - c^-T^-) + (\Delta\gamma)\kappa],$$

once we adopt the constitutive relations for a Newtonian fluid and take $e = cT$, $h = -k\partial T/\partial n$, and $H = m\lambda - \gamma_0 \kappa w_n$ where $m = \rho^+ s_n^+ = \rho^- s_n^- = \rho_0 w_n$, $\lambda \equiv$ latent heat of transformation per unit mass, $L = \rho_0 \lambda \equiv$ latent heat of transformation per unit volume, $\gamma_0 \equiv$ surface energy, and $\Delta\gamma = \gamma - \gamma_0$.

10.5. For a surface of discontinuity, which can be described by the interfacial equation of Problem 10.1

$$f(x,y,z,t) = z - \zeta(x,y,t) = 0,$$

note that

$$\boldsymbol{n} = (-\zeta_x, -\zeta_y, 1)(1 + \zeta_x^2 + \zeta_y^2)^{-1/2} \text{ and } \boldsymbol{\tau}_1 = (1, 0, \zeta_x)(1 + \zeta_x^2)^{-1/2}$$

consistent with Problem 10.2, while

$$\boldsymbol{\tau}_2 = \boldsymbol{n} \times \boldsymbol{\tau}_1 = (-\zeta_x\zeta_y, 1 + \zeta_x^2, \zeta_y)(1 + \zeta_x^2 + \zeta_y^2)^{-1/2}(1 + \zeta_x^2)^{-1/2}.$$

Let this surface be a material one – *i.e.*, $s_n^{\pm} = 0$ – such that $\boldsymbol{v}^+ = \boldsymbol{0}$. Designating $\boldsymbol{v}^-$ by $\boldsymbol{v}$ deduce from the kinematic boundary condition of (10.2.6) and the momentum jump conditions of (10.3.3) that then

$$w = \zeta_t + u\zeta_x + v\zeta_y$$

and

$$T_{ij}^- n_j - T_{ij}^+ n_j = \gamma\kappa n_i + [\nabla_{(\Sigma)}\gamma]_i \text{ for } i = 1, 2, 3 \text{ at } \zeta(x,y,t),$$

where

$$T_{ij}^{+} = -p_A\delta_{ij} \text{ and } T_{ij}^{-} = -p\delta_{ij} + \mu_0(\partial_j v_i + \partial_i v_j).$$

Here $\gamma \equiv$ interfacial surface tension coefficient, κ is as given in Problem 10.1, and as usual

$$\nabla_{(\Sigma)}\gamma = \frac{\partial\gamma}{\partial\tau_1}\boldsymbol{\tau}_1 + \frac{\partial\gamma}{\partial\tau_2}\boldsymbol{\tau}_2 \text{ where } \frac{\partial\gamma}{\partial\tau_{1,2}} = \nabla\gamma \bullet \boldsymbol{\tau}_{1,2}.$$

10.6. Consider a stationary planar material surface $z = \zeta(x, y, t) \equiv 0$ separating the two phases of Problem 10.5. Assume the lower phase to be of depth d, has thermal conductivity diffusivity κ_0, and is subject to an adverse temperature gradient $-\beta_0 < 0$.

(a) Show that for this situation the kinematic boundary and momentum jump conditions become

$$w = 0, p - p_A = 2\mu_0\frac{\partial w}{\partial z}, \mu_0\left(\frac{\partial w}{\partial x} + \frac{\partial u}{\partial z}\right) = \frac{\partial\gamma}{\partial x}, \mu_0\left(\frac{\partial w}{\partial y} + \frac{\partial v}{\partial z}\right) = \frac{\partial\gamma}{\partial y} \text{ at } z = 0.$$

(b) Conclude that the kinematic boundary and *tangential* components of the momentum jump conditions of (a) are equivalent to

$$w = 0, \ \mu_0\frac{\partial u}{\partial z} = \frac{\partial\gamma}{\partial x}, \ \mu_0\frac{\partial v}{\partial z} = \frac{\partial\gamma}{\partial y} \text{ at } z = 0.$$

(c) Let $\rho^- \equiv \rho_0$. Using the continuity equation in the "–" phase which now takes the form

$$\frac{\partial u}{\partial x} + \frac{\partial v}{\partial y} + \frac{\partial w}{\partial z} = 0 \text{ for } z < 0$$

and assuming the requisite continuity at $z = 0$ of the second partial derivatives of the velocity components with respect to their spatial variables deduce from the momentum jump conditions of (b) that

$$-\mu_0\frac{\partial^2 w}{\partial z^2} = \nabla_2^2\gamma \text{ at } z = 0$$

where $\nabla_2^2 \equiv \nabla_2 \bullet \nabla_2 = \partial^2/\partial x^2 + \partial/\partial y^2$.

(d) Consider surface tension to vary linearly with temperature such that

$$\gamma = \gamma_0 - \gamma_1 T$$

where $\gamma_1 \equiv$ surface entropy change and introduce the nondimensional variables

$$(x^*, y^*, z^*) = \frac{(x, y, z)}{d}, t^* = \frac{\kappa_0 t}{d^2}, w^* = \frac{dw}{\kappa_0}, T^* = \frac{T}{\beta_0 d}.$$

Show that (c), upon dropping the *'s for ease of exposition, then transforms into

$$\frac{\partial^2 w}{\partial z^2} = \mathcal{M}\nabla_2^2 T \text{ at } z = 0$$

where $\mathcal{M} = \gamma_1\beta_0 d^2/(\rho_0\nu_0\kappa_0) \equiv$ Marangoni number with $\nu_0 = \mu_0/\rho_0 \equiv$ kinematic viscosity.

Chapter 11
Subsonic Sound Waves Viewed as a Linear Perturbation in an Inviscid Fluid

We have developed the governing equations of motion and boundary conditions for fluid mechanics in the previous two chapters. At the end of the last chapter, we deduced jump conditions at a plane shock front for a two-dimensional inviscid fluid. In each of the next two chapters, we shall investigate a particular phenomenon also involving an inviscid fluid before turning our attention to the effect of viscosity on such modeling endeavors. The object of this chapter is to use linear perturbation theory to derive the equation for one-dimensional subsonic sound waves and then solve that equation exactly. Subsonic sound waves are treated as a linear perturbation to an initially quiescent homogeneous state of a one-dimensional inviscid compressible barotropic fluid of infinite extent. This results in a wave equation satisfied by the density condensation function. The concept of characteristic coordinates relevant to first-order quasi-linear and second-order constant coefficient linear partial differential equations is introduced in a pastoral interlude. Then that method is used to obtain D'Alembert's solution to the sound wave equation, and the physical interpretation of that solution as a traveling wave propagating, either to the left or right, is discussed. The problems consider two examples giving rise to models involving a first-order linear partial differential equation that are solved by the method of characteristics and a parallel flow situation of a one-dimensional homogeneous inviscid fluid layer the linear normal-mode perturbation analysis of which produces a governing Orr-Sommerfeld ordinary differential equation of motion that is then investigated.

11.1 Governing Equations of Motion

In our development, we shall view sound waves as an infinitesimal disturbance to a homogeneous static state of a compressible one-dimensional ($v = w = \partial/\partial y = \partial/\partial z \equiv 0$) inviscid ($\lambda = \mu = 0$) barotropic [$p = \mathcal{P}(\rho)$] fluid in the absence of body forces ($\boldsymbol{f} = f_1 \boldsymbol{i} \equiv \boldsymbol{0}$). Hence this undisturbed state is that of no motion $u \equiv 0$ and constant density $\rho \equiv \rho_0$ in an infinite unbounded spatial x-domain. To a good approximation, such sound waves can be thought of as small variations of pressure accompanying

© Springer International Publishing AG, part of Springer Nature 2017

D. J. Wollkind and B. J. Dichone, *Comprehensive Applied Mathematical Modeling in the Natural and Engineering Sciences*,
https://doi.org/10.1007/978-3-319-73518-4_11

small deviations of density and velocity from this undisturbed state.
Since, under these conditions,

$$\boldsymbol{v}(x,t) = u(x,t)\boldsymbol{i} \text{ and } \rho = \rho(x,t),$$

our governing equations of motion for this situation from (9.2.1) and (9.5.4) are given by:

$$\frac{\partial \rho}{\partial t} + \nabla\rho \bullet \boldsymbol{v} + \rho\nabla \bullet \boldsymbol{v} = \frac{\partial \rho}{\partial t} + u\frac{\partial \rho}{\partial x} + \rho\frac{\partial u}{\partial x} = 0; \tag{11.1.1a}$$

$$\rho\frac{D\boldsymbol{v}}{Dt} = \left(\rho\frac{Du}{Dt}\right)\boldsymbol{i} = \rho\left(\frac{\partial u}{\partial t} + u\frac{\partial u}{\partial x}\right)\boldsymbol{i}$$
$$= -\nabla\mathcal{P}(\rho) + \rho\boldsymbol{f} + (\lambda+\mu)\nabla(\nabla\bullet\boldsymbol{v}) + \mu\nabla^2\boldsymbol{v} = \left[-\frac{\partial\mathcal{P}(\rho)}{\partial x}\right]\boldsymbol{i}$$

or

$$\rho\left(\frac{\partial u}{\partial t} + u\frac{\partial u}{\partial x}\right) + \frac{\partial\mathcal{P}(\rho)}{\partial x} = 0; \tag{11.1.1b}$$

for

$$-\infty < x < \infty \text{ and } t > 0; \tag{11.1.1c}$$

with *far-field* boundary conditions:

$$u(x,t),\ \rho(x,t) \text{ remain bounded as } x^2 \to \infty \text{ for } t > 0; \tag{11.1.1d}$$

and initial conditions:

$$\rho(x,0) = \rho_0[1+\varepsilon f(x)],\ \frac{\partial\rho(x,0)}{\partial t} = \rho_0[1+\varepsilon h'(x)],\ |\varepsilon| << 1;\ \text{for } -\infty < x < \infty; \tag{11.1.1e}$$

where a physical interpretation of our small perturbation parameter ε will be offered in Section 11.5.

11.2 Linear Perturbation Analysis of its Homogeneous Static Solution

The homogeneous static state represented by

$$\rho(x,t) \equiv \rho_0 \text{ and } u(x,t) \equiv 0$$

is an exact solution to our governing equations and boundary conditions of (11.1.1). It is a linear perturbation analysis of that undisturbed state with which we are concerned in this section. Hence we seek a solution of (11.1.1) of the form

$$\rho(x,t) = \rho_0[1 + \varepsilon s(x,t) + O(\varepsilon^2)] \text{ and } u(x,t) = 0 + \varepsilon u_1(x,t) + O(\varepsilon^2),$$

where $s(x,t) \equiv$ *condensation*. Upon substitution of this solution into (11.1.1); noting that

$$\frac{\partial \rho}{\partial t} = \varepsilon\rho_0\frac{\partial s}{\partial t} + O(\varepsilon^2),\ \frac{\partial \rho}{\partial x} = \varepsilon\rho_0\frac{\partial s}{\partial x} + O(\varepsilon^2),\ \frac{\partial u}{\partial x} = \varepsilon\frac{\partial u_1}{\partial x} + O(\varepsilon^2),$$

$$u\frac{\partial \rho}{\partial x} = O(\varepsilon^2), \rho\frac{\partial u}{\partial x} = \varepsilon\rho_0\frac{\partial u_1}{\partial x} + O(\varepsilon^2)$$

$$\Rightarrow \frac{\partial \rho}{\partial t} + u\frac{\partial \rho}{\partial x} + \rho\frac{\partial u}{\partial x} = \varepsilon\rho_0\left(\frac{\partial s}{\partial t} + \frac{\partial u_1}{\partial x}\right) + O(\varepsilon^2) = 0;$$

while, in addition

$$\frac{\partial u}{\partial t} = \varepsilon\frac{\partial u_1}{\partial t} + O(\varepsilon^2),\ \rho\frac{\partial u}{\partial t} = \varepsilon\rho_0\frac{\partial u_1}{\partial t} + O(\varepsilon^2),\ \rho u\frac{\partial u}{\partial x} = O(\varepsilon^2),$$

$$\mathcal{P}(\rho) = \mathcal{P}(\rho_0) + \mathcal{P}'(\rho_0)(\rho - \rho_0) + O([\rho - \rho_0]^2) = \mathcal{P}(\rho_0) + \varepsilon\mathcal{P}'(\rho_0)\rho_0 s + O(\varepsilon^2),$$

$$\frac{\partial \mathcal{P}(\rho)}{\partial x} = \varepsilon\rho_0 c_0^2\frac{\partial s}{\partial x} + O(\varepsilon^2) \text{ where } c_0^2 \equiv \mathcal{P}'(\rho_0) > 0$$

$$\Rightarrow \rho\left(\frac{\partial u}{\partial t} + u\frac{\partial u}{\partial x}\right) + \frac{\partial \mathcal{P}(\rho)}{\partial x} = \varepsilon\rho_0\left(\frac{\partial u_1}{\partial t} + c_0^2\frac{\partial s}{\partial x}\right) + O(\varepsilon^2) = 0;$$

neglecting terms of $O(\varepsilon^2)$; and cancelling the resulting common $\varepsilon\rho_0$ factor; we obtain

$$\frac{\partial s}{\partial t} + \frac{\partial u_1}{\partial x} = 0 \tag{11.2.1a}$$

and

$$\frac{\partial u_1}{\partial t} + c_0^2\frac{\partial s}{\partial x} = 0; -\infty < x < \infty, t > 0; \tag{11.2.1b}$$

with far-field boundary conditions:

$$s(x,t), u_1(x,t) \text{ remain bounded as } x^2 \to \infty \text{ for } t > 0; \tag{11.2.1c}$$

and initial conditions:

$$s(x,0) = f(x), \frac{\partial s(x,0)}{\partial t} = h'(x) \text{ for } -\infty < x < \infty. \tag{11.2.1d}$$

Then, eliminating u_1 from (11.2.1a, 11.2.1b) by taking $\partial(11.2.1a)/\partial t$ and $\partial(11.2.1b)/\partial x$ yields the *wave equation* in the condensation s

$$\frac{\partial^2 s}{\partial t^2} = -\frac{\partial^2 u_1}{\partial t \partial x} = -\frac{\partial^2 u_1}{\partial x \partial t} = c_0^2\frac{\partial^2 s}{\partial x^2}; -\infty < x < \infty, t > 0, \tag{11.2.1e}$$

upon assuming the requisite continuity of its second partial derivatives so that $u_{1xt} = u_{1tx}$.

11.3 Pastoral Interlude: Characteristic Coordinates

(i) First-Order Quasi-Linear Partial Differential Equations:

Here we consider the general quasi-linear first-order partial differential equation for $u = u(x, y)$:

$$a(x, y, u)u_x + b(x, y, u)u_y = c(x, y, u) \tag{11.3.1a}$$

where u is prescribed to be $f(t)$ along the curve $\boldsymbol{r}(t) = g(t)\boldsymbol{i} + h(t)\boldsymbol{j}$ or

$$u[g(t), h(t)] = f(t). \tag{11.3.1b}$$

Introducing the characteristic coordinates $x = x(s, t)$ and $y = y(s, t)$ which satisfy

$$\frac{\partial x}{\partial s} = a \tag{11.3.2a}$$

and

$$\frac{\partial y}{\partial s} = b \tag{11.3.2b}$$

such that

$$x(0, t) = g(t) \tag{11.3.2c}$$

and

$$y(0, t) = h(t), \tag{11.3.2d}$$

and defining $U(s, t) = u[x(s, t), y(s, t)]$,

$$\frac{\partial U}{\partial s} = u_x\frac{\partial x}{\partial s} + u_y\frac{\partial y}{\partial s} = au_x + bu_y = c \tag{11.3.2e}$$

while

$$U(0, t) = u[x(0, t), y(0, t)] = u[g(t), h(t)] = f(t). \tag{11.3.2f}$$

Solving system (11.3.2) for $x = x(s, t)$, $y = y(s, t)$, and $U = U(s, t)$, we can invert the transformation $x = x(s, t)$ and $y = y(s, t)$ to obtain $s = s(x, y)$ and $t = t(x, y)$ provided its Jacobian is nonzero and finally deduce the solution to the original problem (11.3.1) given by

$$u(x, y) = U[s(x, y), t(x, y)]. \tag{11.3.3}$$

In order to illustrate this procedure, we solve a *linear* example of (11.3.1a) for $u = u(x, y)$ where

$$a_0u_x + u_y = m_0u,\ y > 0;\ u(x, 0) = f(x). \tag{11.3.4}$$

If we let $u \equiv$ density, $a_0 \equiv$ constant velocity, $x \equiv$ position, $y \equiv$ time, and $m_0u \equiv$ mass source, this represents the one-dimensional continuity equation with a mass source term m given by $\partial\rho/\partial t + \nabla\rho \bullet \boldsymbol{v} + \rho\nabla \bullet \boldsymbol{v} = m$, upon making the identifications $\rho = u$, $t = y$, $\boldsymbol{v} = a_0\boldsymbol{i}$, and $m = m_0\rho$. Introducing the characteristic coordinates of (11.3.1) with $g(t) = t$ and $h(t) = 0$, we find

$$\frac{\partial x}{\partial s} = a_0, \ \frac{\partial y}{\partial s} = 1, \ \frac{\partial U}{\partial s} = m_0 U; \ x(0,t) = t, \ y(0,t) = 0, \ U(0,t) = f(t);$$

$$\Rightarrow x(s,t) = a_0 s + t, \ y(s,t) = s, \ U(s,t) = f(t)e^{m_0 s};$$

$$\Rightarrow s(x,y) = y, \ t(x,y) = x - a_0 y, \ u(x,y) = U(y, x - a_0 y) = f(x - a_0 y)e^{m_0 y}. \quad (11.3.5)$$

(ii) Second-Order Constant Coefficient Linear Partial Differential Equations:

Here we consider the general constant coefficient second-order linear partial differential equation for $u = u(x,y)$

$$Au_{xx} + Bu_{xy} + Cu_{yy} + H = 0 \text{ where } H = au_x + bu_y + cu + d. \quad (11.3.6)$$

We categorize this equation as being hyperbolic, parabolic, or elliptic depending upon whether $B^2 - 4AC$ is positive, zero, or negative, respectively. In this subsection we shall be concentrating on the hyperbolic case of $B^2 - 4AC > 0$. We now introduce the characteristic coordinates

$$\xi = \xi(x,y), \ \eta = \eta(x,y), \ U(\xi,\eta) = u(x,y)$$

which, since $\partial/\partial x = \xi_x\partial/\partial\xi + \eta_x\partial/\partial\eta$, $\partial/\partial y = \xi_y\partial/\partial\xi + \eta_y\partial/\partial\eta$, transforms (11.3.6) into

$$\mathcal{A}U_{\xi\xi} + \mathcal{B}U_{\xi\eta} + \mathfrak{C}U_{\eta\eta} + \mathcal{H} = 0 \quad (11.3.7)$$

where $\mathcal{A} = A\xi_x^2 + B\xi_x\xi_y + C\xi_y^2$, $\mathcal{B} = 2A\xi_x\eta_x + B(\xi_x\eta_y + \xi_y\eta_x) + 2C\xi_y\eta_y$, $\mathfrak{C} = A\eta_x^2 + B\eta_x\eta_y + C\eta_y^2$, while $\mathcal{H} = (a\xi_x + b\xi_y)U_\xi + (a\eta_x + b\eta_y)U_\eta + cU + d$. Observe that since

$$\mathcal{B}^2 - 4\mathcal{A}\mathfrak{C} = (\xi_x\eta_y - \xi_y\eta_x)^2(B^2 - 4AC),$$

and $\xi_x\eta_y - \xi_y\eta_x \neq 0$, being the Jacobian of the transformation, the sign of $\mathcal{B}^2 - 4\mathcal{A}\mathfrak{C}$ is invariant under this transformation. We wish to select characteristic coordinates for the hyperbolic case so that $\mathcal{A} = \mathfrak{C} = 0$ or

$$A\xi_x^2 + B\xi_x\xi_y + C\xi_y^2 = A\eta_x^2 + B\eta_x\eta_y + C\eta_y^2 = 0$$

and our transformed equation becomes $\mathcal{B}U_{\xi\eta} + \mathcal{H} = 0$. Note this implies that

$$A + B\left(\frac{\xi_y}{\xi_x}\right) + C\left(\frac{\xi_y}{\xi_x}\right)^2 = A + B\left(\frac{\eta_y}{\eta_x}\right) + C\left(\frac{\eta_y}{\eta_x}\right)^2 = 0.$$

Observe that these ratios satisfy the same quadratic equation but to guarantee that the Jacobian of the transformation is nonzero, we, in particular, take

$$\left(\frac{\xi_y}{\xi_x}\right) = \frac{-B+(B^2-4AC)^{1/2}}{2C} = -a_0, \quad \left(\frac{\eta_y}{\eta_x}\right) = \frac{-B-(B^2-4AC)^{1/2}}{2C} = -b_0;$$

with $\xi(x,0) = \eta(x,0) = x$. Then, applying the solution (11.3.5) of our example of subsection (i) with $m_0 = 0$ and $f(x) = x$, we obtain the characteristic coordinates

$$\xi(x,y) = x - a_0 y \text{ and } \eta(x,y) = x - b_0 y. \tag{11.3.8}$$

11.4 D'Alembert's Method of Solution of its Wave Equation Formulation

Upon comparing our wave equation (11.2.1e) to the canonical second-order linear constant coefficient partial differential equation (11.3.7) with $u = s$ and $y = t$, we can make the identifications that $A = c_0^2$, $B = 0$, and $C = -1$. Then

$$a_0 = -b_0 = c_0 \tag{11.4.1a}$$

in (11.3.8). Employing D'Alembert's method of solution, we introduce the characteristic coordinate change of variables of (11.3.8)

$$\xi = x - c_0 t, \ \eta = x + c_0 t, \ S(\xi,\eta) = s(x,t) \tag{11.4.1b}$$

into (11.2.1e). Since $\xi_x = \eta_x = 1$, $\eta_t = -\xi_t = c_0$;

$$\frac{\partial}{\partial x} = \xi_x \frac{\partial}{\partial \xi} + \eta_x \frac{\partial}{\partial \eta} = \frac{\partial}{\partial \xi} + \frac{\partial}{\partial \eta}, \tag{11.4.2a}$$

$$\frac{\partial}{\partial t} = \xi_t \frac{\partial}{\partial \xi} + \eta_t \frac{\partial}{\partial \eta} = c_0\left(\frac{\partial}{\partial \eta} - \frac{\partial}{\partial \xi}\right). \tag{11.4.2b}$$

Thus

$$\left(c_0^2 \frac{\partial^2}{\partial x^2} - \frac{\partial^2}{\partial t^2}\right) s(x,t) = c_0^2 \left[\left(\frac{\partial}{\partial \xi} + \frac{\partial}{\partial \eta}\right)^2 - \left(\frac{\partial}{\partial \eta} - \frac{\partial}{\partial \xi}\right)^2\right] S(\xi,\eta)$$
$$= 4c_0^2 \frac{\partial^2 S(\xi,\eta)}{\partial \xi \partial \eta} = 0 \Rightarrow \left(\frac{\partial}{\partial \xi}\right)\left[\frac{\partial S(\xi,\eta)}{\partial \eta}\right] = 0. \tag{11.4.3a}$$

Hence

$$\frac{\partial S(\xi,\eta)}{\partial \eta} = M'(\eta) \Rightarrow S(\xi,\eta) = M(\eta) + N(\xi) \Rightarrow s(x,t) = M(x+c_0 t) + N(x - c_0 t) \tag{11.4.3b}$$

where the functions M and N are determined from the initial conditions (11.2.1d) as follows:

$$s(x,0) = M(x) + N(x) = f(x), \tag{11.4.4a}$$

$$\frac{\partial s(x,0)}{\partial t} = c_0[M'(x) - N'(x)] = h'(x) \Rightarrow M(x) - N(x) = \frac{h(x)}{c_0} + k. \tag{11.4.4b}$$

Solving (11.4.3a, 11.4.3b) simultaneously, we obtain

$$M(x) = \frac{1}{2}\left[f(x) + \frac{h(x)}{c_0} + k\right], N(x) = \frac{1}{2}\left[f(x) - \frac{h(x)}{c_0} - k\right] \tag{11.4.4c}$$

which upon substitution into (11.4.3b) yields

$$\begin{aligned} s(x,t) &= M(x + c_0 t) + N(x - c_0 t) \\ &= \frac{1}{2}\left[f(x + c_0 t) + \frac{h(x + c_0 t)}{c_0} + k\right] + \frac{1}{2}\left[f(x - c_0 t) - \frac{h(x - c_0 t)}{c_0} - k\right] \end{aligned}$$

or

$$s(x,t) = \frac{1}{2}[f(x + c_0 t) + f(x - c_0 t)] + \frac{1}{2c_0}[h(x + c_0 t) - h(x - c_0 t)]. \tag{11.4.5}$$

11.5 Physical Interpretation of that Solution

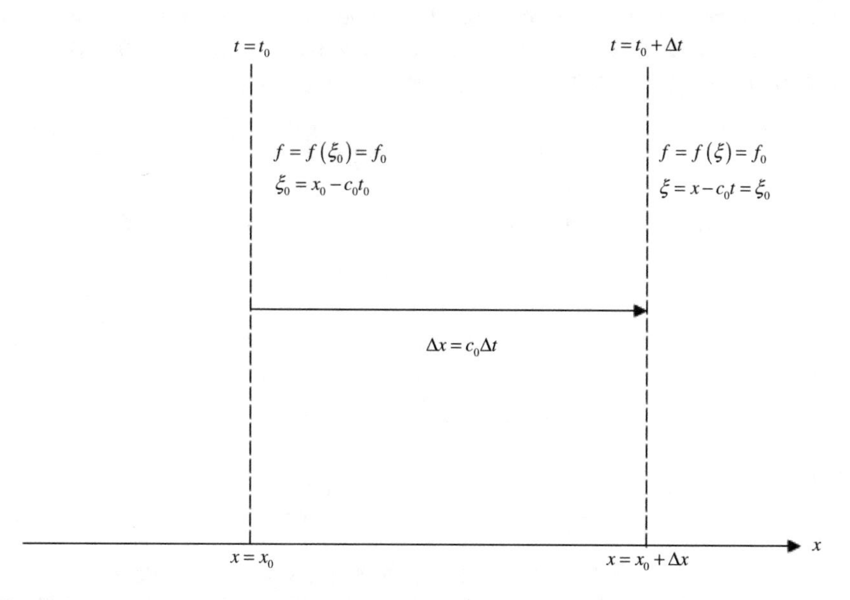

Fig. 11.1 A schematic diagram of wave propagation for the $f(x - c_0 t)$ term in (11.4.5).

In order to make a physical interpretation of the solution of (11.4.5), we first examine the behavior of its $f(x - c_0t)$ term. At some time t_0 and position $x_0, f = f(\xi_0) = f_0$ where $\xi_0 = x_0 - c_0t_0$. For some later time $t_0 + \Delta t$, we wish to find a position $x_0 + \Delta x$ such that $f(\xi) = f_0$ or

$$\xi = x - c_0(t_0 + \Delta t) = \xi_0 = x_0 - c_0t_0 \Rightarrow x = x_0 + c_0\Delta t \Rightarrow \Delta x = c_0\Delta t, \tag{11.5.1}$$

as depicted in Fig. 11.1. Hence this term represents a wave front traveling to the right with speed c_0 while a term of the form $f(x + c_0t)$ represents a wave traveling to the left with that speed c_0. Let us consider this wave speed c_0 in more detail under the assumption that our barotropic fluid satisfies the adiabatic equation of state (9.5.24)

$$p = \mathcal{P}(\rho) = p_0\left(\frac{\rho}{\rho_0}\right)^{\gamma}. \tag{11.5.2}$$

Then since

$$\mathcal{P}'(\rho) = \left(\frac{\gamma p_0}{\rho_0}\right)\left(\frac{\rho}{\rho_0}\right)^{\gamma-1}, \tag{11.5.3a}$$

$$c_0^2 \equiv \mathcal{P}'(\rho_0) = \frac{\gamma p_0}{\rho_0} > 0 \Rightarrow c_0 = \left(\frac{\gamma p_0}{\rho_0}\right)^{1/2}, \tag{11.5.3b}$$

which represents the speed of sound in an adiabatic fluid [6]. It remains only to discuss the subsonic nature of our sound wave propagation. Let position, time, and velocity have scale factors L, τ, and V, respectively. Then the wave equation (11.2.1e) implies that (here the bracket notation is being used to indicate size)

$$[c_0]^2 = \frac{L^2}{\tau^2} \Rightarrow [c_0] = \frac{L}{\tau}, \tag{11.5.4}$$

while our small perturbation analysis of (11.1.1b) requires, in particular, that

$$\left[u\frac{\partial u}{\partial x}\right] << \left[\frac{\partial u}{\partial t}\right]$$

which implies that

$$\frac{V^2}{L} << \frac{V}{\tau} \Rightarrow V << \frac{L}{\tau} = [c_0] \tag{11.5.5a}$$

or, equivalently,

$$M_a \equiv \text{ Mach number } = \frac{V}{c_0} << 1. \tag{11.5.5b}$$

If $M_a = 1$ or $M_a > 1$ then the flow is said to be trans- or supersonic, respectively, the latter resulting in the sort of shock waves treated in the last chapter. Given that our nondimensional perturbation parameter $|\varepsilon| << 1$ is a measure of the size of the disturbance velocity, it seems appropriate at this time to offer the physical interpretation that ε can be related to M_a by

$$|\varepsilon| = M_a, \tag{11.5.6}$$

for the sake of definiteness.

Problems

11.1. Consider a material layer with kinematic boundary condition at its surface $z = \zeta(x, \tau)$

$$w = \zeta_\tau + u\zeta_x$$

where, since this fluid layer is homogeneous, the components of its velocity field are solenoidal

$$u_x + w_z = 0.$$

(a) Show that, for $u = -x$ and $w = z$, the solenoidal condition is satisfied identically and this solution, which represents layer shortening in the x-direction and spreading in the z-direction, converts the kinematic boundary condition to

$$-x\zeta_x + \zeta_\tau = \zeta.$$

(b) Letting time $\tau = y$, obtain the first-order linear partial differential equation for $\zeta = \zeta(x, y)$:

$$-x\zeta_x + \zeta_y = \zeta \text{ for } y > 0;$$

and, assuming the initial condition $\zeta(x, 0) = A_0 \cos(q_0 x)$, use the method of characteristic coordinates to find its solution

$$\zeta(x, y) = A_0 e^y \cos(q_0 e^y x).$$

11.2. Consider the age-structure-type continuity equation for a population density $c = c(x, y)$ with $x \equiv$ age and $y \equiv$ time given by

$$c_y + uc_x = -\mu_0 c \text{ for } 0 < x \le D_0$$

where $u \equiv$ rate of aging, $\mu_0 \equiv$ per capita mortality rate, and $D_0 \equiv$ maximum lifetime. Since for most populations it may be assumed that each year they age a year or $u = 1$, this age-structure continuity equation is standardly taken to be

$$c_x + c_y = -\mu_0 c \text{ for } 0 < x \le D_0; \text{ with } c(0, y) = b(y) \equiv \text{ birthrate.}$$

(a) Solve this problem by the method of characteristic coordinates to find that

$$c(x, y) = b(y - x)e^{-\mu_0 x}.$$

(b) Then conclude that (a) corresponds to a total population density at time y of

$$\mathfrak{C}(y) = \int_0^{D_0} c(x,y)\,dx = e^{-\mu_0 y}\int_{y-D_0}^{y} e^{\mu_0\sigma} b(\sigma)\,d\sigma.$$

11.3. The purpose of this problem is to use the technique of linear perturbation theory to determine a necessary condition for the instability of a parallel flow situation of an inviscid fluid in a channel. It concerns the two-dimensional ($v^* = \partial/\partial y^* \equiv 0$) motion of an inviscid ($\lambda = \mu = 0$) fluid of uniform density ($\rho \equiv \rho_0$) confined between two infinite rigid plates located at $z^* = 0$ and d. The governing equations of motion with no-penetration and far-field boundary conditions for this situation are as follows:

$$\rho_0 \frac{D\boldsymbol{v}^*}{Dt^*} = \rho\left(\frac{\partial \boldsymbol{v}^*}{\partial t^*} + \nabla^*\boldsymbol{v}^* \bullet \boldsymbol{v}^*\right) = -\nabla^*(p^* + \rho_0 g z^*),\ \nabla^* \bullet \boldsymbol{v}^* = 0; \qquad \text{(P3.1a)}$$

$$w^* = 0 \text{ at } z^* = 0, d;\ u^*, w^*, p^* \text{ remain bounded as } {x^*}^2 + {z^*}^2 \to \infty. \qquad \text{(P3.1b)}$$

Here $(x^*, z^*) \equiv$ position, $t^* \equiv$ time, $\boldsymbol{v}^* = (u^*, w^*) \equiv$ velocity, $p^* \equiv$ pressure, $\boldsymbol{f}^* = (0, -g) \equiv$ body force per unit mass due to gravity. Note that, in the absence of viscosity, the no-slip or adherence boundary condition must be dropped and only the no-penetration condition, retained.

(a) Introducing the following nondimensional variables:

$$(x,z) = \frac{(x^*, z^*)}{d},\ t = \frac{t^*}{(d/g)^{1/2}},\ \boldsymbol{v} = (u,w) = \frac{(u^*, w^*)}{(dg)^{1/2}},\ p = \frac{p^*}{\rho_0 d g};$$

show that this transforms (P3.1a, P3.1b) into

$$\frac{D\boldsymbol{v}}{Dt} = \frac{\partial \boldsymbol{v}}{\partial t} + \nabla\boldsymbol{v} \bullet \boldsymbol{v} = -\nabla(p + z),\ \nabla \bullet \boldsymbol{v} = 0; \qquad \text{(P3.2a)}$$

$$w = 0 \text{ at } z = 0, 1;\ u, w, p \text{ remain bounded as } x^2 + z^2 \to \infty. \qquad \text{(P3.2b)}$$

(b) Show that (P3.2a, P3.2b) has a steady-state parallel flow solution given by

$$u = H(z),\ w \equiv 0,\ p = -z + p_A \text{ with } p_A \equiv \text{ constant atmospheric pressure;} \qquad \text{(P3.3)}$$

where $H(z) \equiv$ arbitrary real-valued second-continuously differentiable function of z.

(c) Assume a solution of (P3.2a, P3.2b) of the form

$$u = H(z) + \varepsilon u_1 + O(\varepsilon^2),\ w = 0 + \varepsilon w_1 + O(\varepsilon^2),\ p = -z + p_0 + \varepsilon p_1 + O(\varepsilon^2) \text{ for } |\varepsilon| << 1. \qquad \text{(P3.4)}$$

Substituting (P3.4) into (P3.2a, P3.2b), neglecting terms of $O(\varepsilon^2)$, and cancelling the resultant common ε factor, determine a set of linear homogeneous partial differential equations and boundary conditions that govern the infinitesimal perturbations to the exact solution given by (P3.3).

(d) Assume a normal-mode solution to the perturbation system of (c) of the form

$$[u_1, w_1, p_1](x, z, t) = [U, W, P](z)\exp(i\alpha\{x - ct\}),\ \alpha \in \mathbb{R}. \tag{P3.5}$$

Substituting (P3.5) into that perturbation system, cancelling the common exponential factor, and eliminating $U(z)$ and $P(z)$, show that this problem can be reduced to the following second-order inviscid Orr-Sommerfeld ordinary differential equation for $W(z)$:

$$(H - c)(W'' - \alpha^2 W) - H''W = 0;\ W(0) = W(1) = 0. \tag{P3.6}$$

(e) Let

$$c = c_r + ic_i \text{ where } c_{r,i} \in \mathbb{R},$$

and divide (P3.6) by $H - c$, assuming that $c_i \neq 0$. Then multiply this equation by

$\overline{W}$, the *complex conjugate* of W=Re(W)+iIm(W) defined by $\overline{W}$ = Re(W)−iIm(W),

integrate with respect to z from 0 to 1, and perform the appropriate integration by parts. Finally by taking the imaginary part of the resulting expression deduce that

$$c_i \int_0^1 H''|W|^2|H - c|^{-2}\,dz = 0.$$

Hint: Note that

$$|W|^2 = W\overline{W} = [\text{Re}(W)]^2 + [\text{Im}(W)]^2 \geq 0,$$

$$\frac{1}{H - c} = \frac{\overline{H - c}}{(H - c)(\overline{H - c})} = (H - \overline{c})|H - c|^{-2},$$

and

$$\overline{c} = c_r - ic_i$$

while

$$\frac{dW}{dz} = \frac{d}{dz}\text{Re}(W) + i\frac{d}{dz}\text{Im}(W)$$

and

$$\frac{d}{dz}\overline{W} = \frac{\overline{dW}}{dz}.$$

(f) Conclude from the imaginary part integral relation of (e) that H must have at least one inflection point in $0 < z < 1$ if $c_i \neq 0$ and comment on the significance of this result with respect to a necessary condition for linear instability.

Hint: Recall that an inflection point z_0 of H satisfies

$$H''(z_0) = 0 \text{ such that } H'' \text{ changes sign at } z = z_0$$

and note that

$$|\exp(i\alpha\{x - ct\})| = \exp(\text{Re}\{-i\alpha ct\}) = \exp(\alpha c_i t).$$

Chapter 12
Potential Flow Past a Circular Cylinder of a Homogeneous Inviscid Fluid

In this chapter, the steady-state two-dimensional potential flow of an inviscid fluid past a circular cylinder is considered. The resulting Laplace's equation for the velocity potential is converted to cylindrical coordinates by the Calculus of Variations method of transformation of coordinates, introduced in a pastoral interlude, and that equation is then solved by a separation of variables technique. Then integration of the pressure, determined by Bernoulli's relation, about the cylinder yields D'Alembert's paradox for a two-dimensional situation that the drag on the cylinder is zero. That physically unrealistic result is a consequence of the assumption that the small viscosity coefficient can be neglected. The problems fill in some details involving vector identities employing the alternating tensor introduced in Chap. 9, examine the properties of the orthogonal Hermite polynomials similar in behavior to the Legendre polynomials discussed below, and consider the corresponding companion situation of three-dimensional potential flow past a sphere. This requires that the resulting Laplace's equation for the velocity potential be converted to spherical coordinates by the Calculus of Variations transformation method. Then the separation of variables technique of solution gives rise to Legendre polynomials, the properties of which have been deduced by means of the two pastoral interludes that conclude the chapter. Integration of the pressure about the sphere yields D'Alembert's paradox for a three-dimensional situation, that the force on the sphere is zero. Although the modeling of sound waves using inviscid fluid equations in the last chapter yielded physically reasonable results, this demonstrates that an inviscid fluid model, dropping the no-slip boundary condition out of necessity, cannot be employed to deduce the proper fluid flow past bodies.

12.1 Governing Equations of Motion

For a homogeneous ($\rho \equiv \rho_0$) inviscid ($\mu = 0$) fluid, we deduced from (9.5.2) the equations of motion

© Springer International Publishing AG, part of Springer Nature 2017

D. J. Wollkind and B. J. Dichone, *Comprehensive Applied Mathematical Modeling in the Natural and Engineering Sciences*,
https://doi.org/10.1007/978-3-319-73518-4_12

$$\nabla \bullet \boldsymbol{v} = 0, \tag{12.1.1a}$$

$$\frac{D\boldsymbol{v}}{Dt} = \frac{\partial \boldsymbol{v}}{\partial t} + (\boldsymbol{v} \bullet \nabla)\boldsymbol{v} = -\frac{\nabla p'}{\rho_0} + \boldsymbol{f}, \tag{12.1.1b}$$

where $\boldsymbol{v} = u\boldsymbol{i} + v\boldsymbol{j} + w\boldsymbol{k} \equiv$ velocity, $p' \equiv$ pressure, and $\boldsymbol{f} \equiv$ body force. If $\boldsymbol{f}$ is conservative and hence derivable from a scalar potential such that $\boldsymbol{f} = -\nabla\Omega$, then (12.1.1b) can be written as

$$\frac{\partial \boldsymbol{v}}{\partial t} + (\boldsymbol{v} \bullet \nabla)\boldsymbol{v} = -\frac{\nabla p}{\rho_0}, \tag{12.1.1c}$$

where $p = p' + \rho_0\Omega \equiv$ reduced pressure. Note, for a gravitational body force with orientation $\boldsymbol{f} = -g\boldsymbol{j}$, this implies that $\Omega = gy - p_A/\rho_0$. Observing that (see Problem 12.1)

$$(\boldsymbol{v} \bullet \nabla)\boldsymbol{v} = \frac{1}{2}\nabla(\boldsymbol{v} \bullet \boldsymbol{v}) - \boldsymbol{v} \times \boldsymbol{\omega} \tag{12.1.2a}$$

where $(\boldsymbol{v} \bullet \nabla)\boldsymbol{v} \equiv (v_j\partial_j)\boldsymbol{v}$ and $\boldsymbol{\omega} = \text{curl}(\boldsymbol{v}) \equiv$ vorticity, (12.1.1c) becomes

$$\frac{\partial \boldsymbol{v}}{\partial t} = -\nabla\left(\frac{p}{\rho_0} + \frac{\boldsymbol{v} \bullet \boldsymbol{v}}{2}\right) + \boldsymbol{v} \times \boldsymbol{\omega}. \tag{12.1.2b}$$

For a steady-state ($\partial/\partial t \equiv 0$) irrotational ($\boldsymbol{\omega} \equiv \boldsymbol{0}$) situation, (12.1.2b) reduces to

$$\nabla\left(\frac{p}{\rho_0} + \frac{\boldsymbol{v} \bullet \boldsymbol{v}}{2}\right) = \boldsymbol{0} \Rightarrow p + \frac{\rho_0|\boldsymbol{v}|^2}{2} \equiv \text{constant (the Bernoulli relation)}. \tag{12.1.3}$$

Now taking the curl of (12.1.2b), assuming the continuity of the requisite second partial derivatives, and noting that

$$\text{curl}(\nabla\varphi) \equiv \begin{vmatrix} \boldsymbol{i} & \boldsymbol{j} & \boldsymbol{k} \\ \partial_x & \partial_y & \partial_z \\ \varphi_x & \varphi_y & \varphi_z \end{vmatrix} = (\varphi_{zy} - \varphi_{yz})\boldsymbol{i} + (\varphi_{xz} - \varphi_{zx})\boldsymbol{j} + (\varphi_{yx} - \varphi_{xy})\boldsymbol{k} = \boldsymbol{0}, \tag{12.1.4}$$

we obtain

$$\frac{\partial \boldsymbol{\omega}}{\partial t} = \text{curl}(\boldsymbol{v} \times \boldsymbol{\omega}) \tag{12.1.5a}$$

which, upon observing that (see Problem 12.1)

$$\text{curl}(\boldsymbol{v} \times \boldsymbol{\omega}) = (\boldsymbol{\omega} \bullet \nabla)\boldsymbol{v} - (\boldsymbol{v} \bullet \nabla)\boldsymbol{\omega} + (\nabla \bullet \boldsymbol{\omega})\boldsymbol{v} - (\nabla \bullet \boldsymbol{v})\boldsymbol{\omega}, \tag{12.1.5b}$$

while employing (12.1.1a) and

$$\nabla \bullet \boldsymbol{\omega} = \nabla \bullet \text{curl}(\boldsymbol{v}) = (w_y - v_z)_x + (u_z - w_x)_y + (v_x - u_y)_z = 0, \tag{12.1.5c}$$

implies

$$\frac{\partial \boldsymbol{\omega}}{\partial t} + (\boldsymbol{v} \bullet \nabla)\boldsymbol{\omega} = \frac{D\boldsymbol{\omega}}{Dt} = (\boldsymbol{\omega} \bullet \nabla)\boldsymbol{v}. \tag{12.1.5d}$$

Defining vorticity in Lagrangian variables by $\mathbf{\Omega}(\mathbf{A},t) = \boldsymbol{\omega}[\mathbf{X}(\mathbf{A},t),t]$, note that (12.1.5d) is equivalent to

$$\frac{\partial \Omega_i(\mathbf{A},t)}{\partial t} = \Omega_k(\mathbf{A},t)\left(\frac{\partial v_i}{\partial x_k}\right)\bigg|_{\mathbf{x}=\mathbf{X}(\mathbf{A},t)} \quad \text{for } i = 1,2,3. \tag{12.1.6a}$$

Further then

$$\Omega_i(\mathbf{A},t) = \Omega_j(\mathbf{A},0)\left[\frac{\partial X_i(\mathbf{A},t)}{\partial A_j}\right] \quad \text{for } i = 1,2,3; \tag{12.1.6b}$$

which will be demonstrated as follows: Let

$$\Omega_i(\mathbf{A},t) = c_j(\mathbf{A},t)\left[\frac{\partial X_i(\mathbf{A},t)}{\partial A_j}\right] \quad \text{for } i = 1,2,3 \tag{12.1.6c}$$

which is permissible since the Jacobian of the transformation $J(\mathbf{A},t) \neq 0$. Thus

$$\frac{\partial \Omega_i}{\partial t} = \left(\frac{\partial c_j}{\partial t}\right)\left(\frac{\partial X_i}{\partial A_j}\right) + c_j\left(\frac{\partial V_i}{\partial A_j}\right) \tag{12.1.7a}$$

while from (12.1.6b) and the chain rule

$$\frac{\partial \Omega_i}{\partial t} = \Omega_k\left(\frac{\partial v_i}{\partial x_k}\right)\bigg|_{\mathbf{x}=\mathbf{X}(\mathbf{A},t)} = c_j\left(\frac{\partial X_k}{\partial A_j}\right)\left(\frac{\partial v_i}{\partial x_k}\right)\bigg|_{\mathbf{x}=\mathbf{X}(\mathbf{A},t)} = c_j\left(\frac{\partial V_i}{\partial A_j}\right). \tag{12.1.7b}$$

Upon comparing (12.1.7a, 12.1.7b) we can conclude that

$$\left(\frac{\partial c_j}{\partial t}\right)\left(\frac{\partial X_i}{\partial A_j}\right) = 0 \text{ for } i = 1,2,3 \Rightarrow \frac{\partial c_j}{\partial t} = 0 \text{ or } c_j(\mathbf{A},t) = c_j(\mathbf{A},0) \text{ for } j = 1,2,3. \tag{12.1.8a}$$

Now, from (12.1.7a), for $i = 1,2$, or 3,

$$\Omega_i(\mathbf{A},0) = c_j(\mathbf{A},0)\left[\frac{\partial X_i(\mathbf{A},0)}{\partial A_j}\right] = c_j(\mathbf{A},t)\frac{\partial A_i}{\partial A_j} = c_j(\mathbf{A},0)\delta_{ij} = c_i(\mathbf{A},0), \tag{12.1.8b}$$

which demonstrates the result. Note, from (12.1.6) that should the initial vorticity

$$\Omega_i(\mathbf{A},0) = 0 \Rightarrow \Omega_i(\mathbf{A},t) = 0 \text{ for } i = 1,2,3 \Rightarrow \boldsymbol{\omega} = \text{curl}(\mathbf{v}) \equiv \mathbf{0}. \tag{12.1.9}$$

Since the result of (12.1.4) is actually an "if and only if" one, we can finally conclude that for such an irrotational flow there must exist a velocity potential φ satisfying

$$\mathbf{v} = \nabla\varphi \Rightarrow u = \varphi_x,\ v = \varphi_y,\ w = \varphi_z \Rightarrow \nabla \bullet \mathbf{v} = \varphi_{xx} + \varphi_{yy} + \varphi_{zz} \equiv \nabla^2\varphi = 0, \tag{12.1.10a}$$

while the Bernoulli relation (12.1.3) in this event takes the form

$$p + \rho_0\frac{\varphi_x^2 + \varphi_y^2 + \varphi_z^2}{2} \equiv \text{ constant.} \tag{12.1.10b}$$

Having deduced the governing equation of motion (12.1.10) for generic potential flow, we now consider the special case of two-dimensional uniform stream flow past a circular cylinder of radius a of an inviscid homogeneous fluid acted upon by conservative body forces as depicted in Fig. 12.1. Here

$$\boldsymbol{v}(x,y,t) = u(x,y,t)\boldsymbol{i} + v(x,y,t)\boldsymbol{j} \tag{12.1.11a}$$

where

$$\boldsymbol{v}(x,y,t) \sim U_0\boldsymbol{i} \text{ as } x^2+y^2 \to \infty. \tag{12.1.11b}$$

Note that the vorticity for the two-dimensional case is given by

$$\boldsymbol{\omega} = \text{curl}(\boldsymbol{v}) = \omega_3\boldsymbol{k} \text{ with } \omega_3 = v_x - u_y = \Omega_3(A_1,A_2,t). \tag{12.1.12}$$

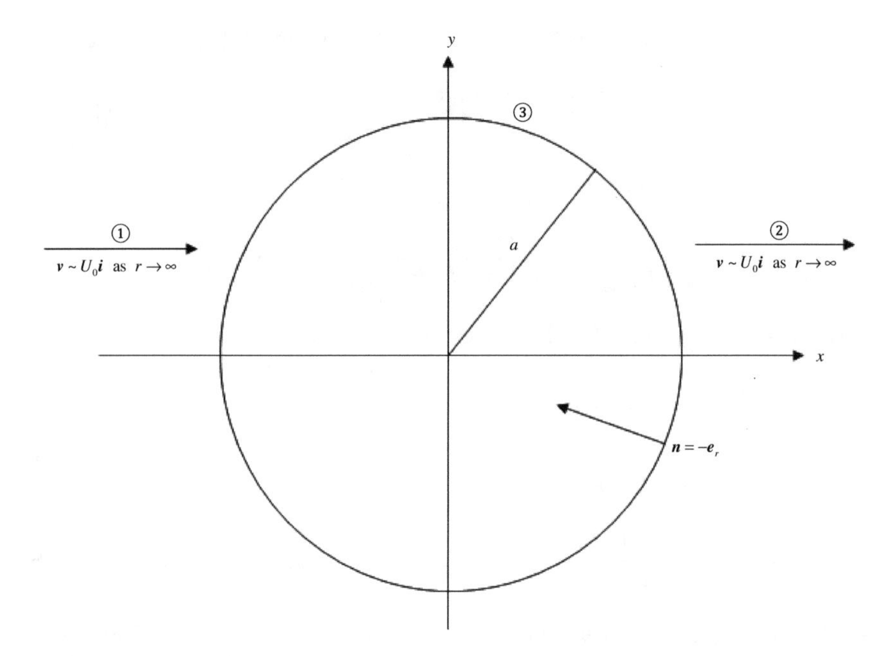

Fig. 12.1 A schematic diagram of two-dimensional inviscid potential flow past a circular cylindrical body of radius $r = a$. Here the circled numerals represent the following three regions: ① ≡ downstream, ② ≡ upstream, and ③ ≡ near the cylindrical body, respectively.

In Fig. 12.1 there are three designated regions: Namely, ① ≡ downstream, ② ≡ up stream, and ③ ≡ near the cylinder. Observe that in the free-stream flow of ① and ② by virtue of (12.1.1b) $u = U_0$ and $v = 0$ and hence $\omega_3 = \Omega_3 = 0$. Since material particles that originally start at ① either end up at ② or ③ and those that start at ③ either remain there or move to ② while these particles retain their initial vorticity by virtue of (12.1.6b) this implies that $\omega_3 = \Omega_3 = 0$ at ③ as well. Thus by virtue of (12.1.12) we can conclude that $\boldsymbol{\omega} \equiv \boldsymbol{0}$, the flow is irrotational, and hence there exists a velocity potential $\varphi = \varphi(x, y, t)$ such that

$$\boldsymbol{v} = \nabla_2\varphi \Rightarrow u = \varphi_x, v = \varphi_y \Rightarrow \nabla_2 \bullet \boldsymbol{v} = \nabla_2^2\varphi = \varphi_{xx} + \varphi_{yy} = 0 \text{ for } a^2 < x^2 + y^2 < \infty. \tag{12.1.13a}$$

As in Problem 11.3 of the last chapter we now impose the no-penetration condition

$$\boldsymbol{v} \bullet \boldsymbol{n} = \nabla_2\varphi \bullet \boldsymbol{n} = \frac{\partial\varphi}{\partial n} = 0 \text{ for } \boldsymbol{n} = -\boldsymbol{e}_r \text{ at } x^2 + y^2 = a^2 \tag{12.1.13b}$$

and the far-field or free-stream condition of (12.1.11b)

$$u = \varphi_x \sim U_0, v = \varphi_y \sim 0 \Rightarrow \varphi(x, y, t) \sim U_0 x \text{ as } x^2 + y^2 \to \infty. \tag{12.1.13c}$$

as our boundary conditions. Finally we note that (12.1.13c) implies this exterior Dirichlet problem is steady-state as well and therefore conclude

$$\varphi = \varphi(x, y). \tag{12.1.13d}$$

Due to the circular domain for this problem it is natural to reformulate it in polar coordinates by means of the transformation of (9.2.11). Although it is possible to convert the Laplace equation of (12.1.13a) to polar coordinates by means of direct substitution, we have in the Calculus of Variations an indirect method for doing so that significantly reduces the amount of labor involved. Toward that end we now revisit the Calculus of Variations to develop that method.

12.2 Pastoral Interlude: Calculus of Variations Method of Change of Variables

As a preliminary step, we must first extend our Calculus of Variations approach to functionals involving more independent variables and hence wish to determine a function w that extremalizes

$$I(W) = \iint_{\mathcal{D}} f(x, y, W, W_x, W_y)\,dx\,dy$$

among all sufficiently smooth functions W prescribed on $\partial\mathcal{D}$. Consider admissible functions of the form $W(x, y; \varepsilon) = w(x, y) + \varepsilon s(x, y)$ where $s(x, y) = 0$ for $(x, y) \in \partial\mathcal{D}$. Our usual argument requires that $\mathcal{I}'(0) = 0$ where

$$\mathcal{I}(\varepsilon) = \iint_{\mathcal{D}} f(x, y, w + \varepsilon s, w_x + \varepsilon s_x, w_y + \varepsilon s_y)\,dx\,dy$$

which implies that

$$\iint_{\mathcal{D}} (f_3 s + f_4 s_x + f_5 s_y)\, dx\, dy = 0.$$

To accomplish the necessary transferral of derivatives on s to derivatives involving f_4 and f_5, we define a vector function $\boldsymbol{v} = f_4 \boldsymbol{i} + f_5 \boldsymbol{j}$. Using $\boldsymbol{v}$ the terms of interest appearing in the above expression can be represented by

$$\iint_{\mathcal{D}} (f_4 s_x + f_5 s_y)\, dx\, dy = \iint_{\mathcal{D}} (\nabla s \bullet \boldsymbol{v})\, dx\, dy.$$

Since

$$\nabla \bullet (s\boldsymbol{v}) = \sum_{i=1}^{2} \frac{\partial}{\partial x_i}(s v_i) = \sum_{i=1}^{2} \frac{\partial s}{\partial x_i} v_i + s \sum_{i=1}^{2} \frac{\partial v_i}{\partial x_i} = \nabla s \bullet \boldsymbol{v} + s \nabla \bullet \boldsymbol{v},$$

where $x_1 = x$ and $x_2 = y$, using Green's Theorem in space

$$\iint_{\mathcal{D}} \nabla \bullet (s\boldsymbol{v})\, dx\, dy = \iint_{\mathcal{D}} (\nabla s \bullet \boldsymbol{v}\, dx\, dy) + \iint_{\mathcal{D}} (s \nabla \bullet \boldsymbol{v})\, dx\, dy = \oint_{\partial\mathcal{D}} (s\boldsymbol{v}) \bullet \boldsymbol{N}\, ds$$

and employing the fact that $s = 0$ on $\partial\mathcal{D}$, yields

$$\iint_{\mathcal{D}} (\nabla s \bullet \boldsymbol{v})\, dx\, dy = -\iint_{\mathcal{D}} (s \nabla \bullet \boldsymbol{v})\, dx\, dy = -\iint_{\mathcal{D}} \left(\frac{\partial f_4}{\partial x} + \frac{\partial f_5}{\partial y} \right) s\, dx\, dy.$$

With all this our original condition becomes

$$\iint_{\mathcal{D}} \left(f_3 - \frac{\partial f_4}{\partial x} - \frac{\partial f_5}{\partial y} \right) s\, dx\, dy = 0.$$

Then the fundamental lemma of the Calculus of Variations implies that

$$\frac{\partial f}{\partial w} - \frac{\partial}{\partial x}\left(\frac{\partial f}{\partial w_x} \right) - \frac{\partial}{\partial y}\left(\frac{\partial f}{\partial w_y} \right) = 0 \text{ or, more generally, } \frac{\partial f}{\partial w} - \sum_{i=1}^{n} \frac{\partial}{\partial x_i}\left(\frac{\partial f}{\partial w_{x_i}} \right) = 0.$$

We are now ready to develop a Calculus of Variations method of change of independent variables. Consider our Euler–Lagrange equation for $f = f(x_1, x_2, x_3, \varphi, \varphi_{x_1}, \varphi_{x_2}, \varphi_{x_3})$

$$\frac{\partial f}{\partial \varphi} - \sum_{i=1}^{3} \frac{\partial}{\partial x_i}\left(\frac{\partial f}{\partial \varphi_{x_i}} \right) = 0 \text{ where } \varphi = \varphi(x_1, x_2, x_3).$$

In general, it is a second-order partial differential equation. Here (x_1, x_2, x_3) represents a Cartesian coordinate system. There are occasions when the geometrical configuration of a particular problem makes it more convenient to introduce a curvilinear coordinate system such as a cylindrical or spherical one. Although it is possible to transform the Euler–Lagrange equation from Cartesian coordinates to some other coordinate system by direct substitution this often involves a tremendous amount of very tedious calculation. We have, however, in the techniques of the Calculus of Variations, a means for significantly reducing the amount of labor involved to affect this transformation. The advantage of this indirect method is that it only involves the transformation of first-order partial derivatives whereas the method of direct substitution involves the transformation of second-order partial derivatives, a much more complicated procedure.

Let

$$x_i = x_i(r_1, r_2, r_3) \text{ for } i = 1, 2, 3$$

be the equations of transformation from our Cartesian coordinate system to a general system of coordinates (r_1, r_2, r_3) where we are assuming both systems are *right-handed*. Then the triple integral

$$I = \iiint_{\mathcal{D}} f(x_1, x_2, x_3, \varphi, \varphi_{x_1}, \varphi_{x_2}, \varphi_{x_3})\, dx_1\, dx_2\, dx_3$$

becomes under this transformation

$$I = \iiint_{\mathcal{R}} F(r_1, r_2, r_3, \Phi, \Phi_{r_1}, \Phi_{r_2}, \Phi_{r_3}) J(r_1, r_2, r_3)\, dr_1\, dr_2\, dr_3$$

where $\mathcal{R}$ is the region $\mathcal{D}$ described in (r_1, r_2, r_3) variables,

$$\Phi(r_1, r_2, r_3) = \varphi(x_1, x_2, x_3),\ F(r_1, r_2, r_3, \Phi, \Phi_{r_1}, \Phi_{r_2}, \Phi_{r_3}) = f(x_1, x_2, x_3, \varphi, \varphi_{x_1}, \varphi_{x_2}, \varphi_{x_3}),$$

and the Jacobian of the transformation

$$J(r_1, r_2, r_3) = \frac{\partial(x_1, x_2, x_3)}{\partial(r_1, r_2, r_3)}.$$

Then, forming the auxiliary ε one-parameter family of functions for each case

$$\varphi + \varepsilon s \text{ and } \Phi + \varepsilon S \text{ where } s(x_1, x_2, x_3) = S(r_1, r_2, r_3),$$

we find, in the usual way that

$$\begin{aligned} I'(0) &= \iiint_{\mathcal{D}} \left[\frac{\partial f}{\partial \varphi} - \sum_{i=1}^{3} \frac{\partial}{\partial x_i} \left(\frac{\partial f}{\partial \varphi_{x_i}} \right) \right] s\, dx_1\, dx_2\, dx_3 \\ &= \iiint_{\mathcal{R}} \left[\frac{\partial (FJ)}{\partial \Phi} - \sum_{i=1}^{3} \frac{\partial}{\partial r_i} \left(\frac{\partial [FJ]}{\partial \Phi_{r_i}} \right) \right] S\, dr_1\, dr_2\, dr_3 \end{aligned}$$

and transforming the first integral in the above equation it follows that

$$\iiint\limits_{\mathcal{R}} \left\{ J \left[\frac{\partial f}{\partial \varphi} - \sum_{i=1}^{3} \frac{\partial}{\partial x_i} \left(\frac{\partial f}{\partial \varphi_{x_i}} \right) \right]_{x_i = x_i(r_1, r_2, r_3)} - \left[\frac{\partial (FJ)}{\partial \Phi} - \sum_{i=1}^{3} \frac{\partial}{\partial r_i} \left(\frac{\partial [FJ]}{\partial \Phi_{r_i}} \right) \right] \right\} S\, dr_1\, dr_2\, dr_3 = 0,$$

which, upon application of the fundamental lemma of the Calculus of Variations, yields the final result

$$\left[\frac{\partial f}{\partial \varphi} - \sum_{i=1}^{3} \frac{\partial}{\partial x_i} \left(\frac{\partial f}{\partial \varphi_{x_i}} \right) \right]_{x_i = x_i(r_1, r_2, r_3)} = J^{-1} \left[\frac{\partial (FJ)}{\partial \Phi} - \sum_{i=1}^{3} \frac{\partial}{\partial r_i} \left(\frac{\partial [FJ]}{\partial \Phi_{r_i}} \right) \right].$$

For an illustration of the use of this technique, we employ

$$f = \varphi_x^2 = \varphi_y^2 + \varphi_z^2 \text{ with the identification that } x_1 = x,\ x_2 = y,\ x_3 = z$$

and consider the cylindrical coordinate transformation

$$x = r_1 \cos(r_2),\ y = r \sin(r_2),\ z = r_3$$

where, as usual we associate r_1 and r_2 with r and θ, respectively, the polar coordinates of a point in the x-y plane.

Then

$$\varphi_x = \Phi_r \frac{\partial r}{\partial x} + \Phi_\theta \frac{\partial \theta}{\partial x} + \Phi_z \frac{\partial z}{\partial x},$$
$$\varphi_y = \Phi_r \frac{\partial r}{\partial y} + \Phi_\theta \frac{\partial \theta}{\partial y} + \Phi_z \frac{\partial z}{\partial y},$$
$$\varphi_z = \Phi_r \frac{\partial r}{\partial z} + \Phi_\theta \frac{\partial \theta}{\partial z} + \Phi_z \frac{\partial z}{\partial z}.$$

Since

$$x = r\cos(\theta),\ y = r\sin(\theta),\ \text{and } z = z;$$

thus

$$\frac{\partial r}{\partial x} = \cos(\theta),\ \frac{\partial r}{\partial y} = \sin(\theta),\ \frac{\partial r}{\partial z} = 0;$$
$$\frac{\partial \theta}{\partial x} = -\frac{\sin(\theta)}{r},\ \frac{\partial \theta}{\partial y} = \frac{\cos(\theta)}{r},\ \frac{\partial \theta}{\partial z} = 0;$$
$$\frac{\partial z}{\partial x} = \frac{\partial z}{\partial y} = 0,\ \frac{\partial z}{\partial z} = 1.$$

Then

$$\varphi_x = \Phi_r \cos(\theta) - \frac{\sin(\theta)}{r} \Phi_\theta,\ \varphi_y = \Phi_r \sin(\theta) + \frac{\cos(\theta)}{r} \Phi_\theta,\ \varphi_z = \Phi_z.$$

Hence

$$f = \varphi_x^2 + \varphi_y^2 + \varphi_z^2 = \Phi_r^2 + \frac{1}{r^2}\Phi_\theta^2 + \Phi_z^2 = F.$$

Further, since

$$\frac{\partial x}{\partial r} = \cos(\theta), \ \frac{\partial x}{\partial \theta} = -r\sin(\theta), \ \frac{\partial x}{\partial z} = 0;$$

$$\frac{\partial y}{\partial r} = \sin(\theta), \ \frac{\partial y}{\partial \theta} = r\cos(\theta), \ \frac{\partial y}{\partial z} = 0;$$

while

$$\frac{\partial z}{\partial r} = 0, \ \frac{\partial z}{\partial \theta} = 0, \ \frac{\partial z}{\partial z} = 1;$$

$$J = \begin{vmatrix} \cos(\theta) & -r\sin(\theta) & 0 \\ \sin(\theta) & r\cos(\theta) & 0 \\ 0 & 0 & 1 \end{vmatrix} = r,$$

$$FJ = r\Phi_r^2 + \frac{1}{r}\Phi_\theta^2 + r\Phi_z^2.$$

Finally,

$$\frac{\partial f}{\partial \varphi} - \sum_{i=1}^{3} \frac{\partial}{\partial x_i}\left(\frac{\partial f}{\partial \varphi_{x_i}}\right) = -2\left(\sum_{i=1}^{3} \varphi_{x_i x_i}\right) = -2\nabla^2\varphi$$

and

$$J^{-1}\left[\frac{\partial(FJ)}{\partial \Phi} - \sum_{i=1}^{3} \frac{\partial}{\partial r_i}\left(\frac{\partial[FJ]}{\partial \Phi_{r_i}}\right)\right] = -\frac{2}{r}\left[\frac{\partial}{\partial r}(r\Phi_r) + \frac{1}{r}\Phi_{\theta\theta} + r\Phi_{zz}\right].$$

Therefore,

$$\Delta\varphi \equiv \nabla^2\varphi$$

in cylindrical coordinates transforms into

$$\Delta\Phi \equiv \nabla^2\Phi = \frac{1}{r}\frac{\partial}{\partial r}(r\Phi_r) + \frac{1}{r^2}\Phi_{\theta\theta} + \Phi_{zz}$$

or, equivalently,

$$\nabla^2\Phi = \Phi_{rr} + \frac{1}{r}\Phi_r + \frac{1}{r^2}\Phi_{\theta\theta} + \Phi_{zz}.$$

Note for the circular polar coordinate case where $\Phi = \Phi(r, \theta)$ this reduces to

$$\nabla_2^2\Phi = \Phi_{rr} + \frac{1}{r}\Phi_r + \frac{1}{r^2}\Phi_{\theta\theta}.$$

12.3 Governing Laplace's Equation in Circular Polar Coordinates

We now wish to convert our governing Laplace's equation and boundary conditions of (12.1.13) into circular polar coordinates by introducing the change of variables

$$x = r\cos(\theta),\ y = r\sin(\theta),\ \varphi(x,y) = \Phi(r,\theta). \tag{12.3.1a}$$

From our results of the last section, Laplace's equation (12.1.13a) is then transformed into

$$\nabla_2^2\Phi = \Phi_{rr} + \frac{1}{r}\Phi_r + \frac{1}{r^2}\Phi_{\theta\theta} = 0,\ a < r < \infty. \tag{12.3.1b}$$

Next, observing that (12.1.13b) is equivalent to

$$\nabla_2\varphi \bullet \boldsymbol{e}_r = 0 \text{ at } r = a \text{ with } \boldsymbol{e}_r = \cos(\theta)\boldsymbol{i} + \sin(\theta)\boldsymbol{j},$$

from (9.2.11c), this no-penetration boundary condition at $r = a$ transforms into

$$\varphi_x\cos(\theta) + \varphi\sin(\theta) = \varphi_x\frac{\partial x}{\partial r} + \varphi_y\frac{\partial y}{\partial r} = \Phi_r(a,\theta) = 0; \tag{12.3.1c}$$

while the far-field boundary condition (12.1.13c) becomes

$$\Phi(r,\theta) \sim U_0 r\cos(\theta) \text{ as } r \to \infty. \tag{12.3.1d}$$

Whereas Cartesian coordinates have a one-to-one correspondence between its ordered pairs and points in the plane, polar coordinates do not, in that, for instance, (r_0,θ_0) and $(r_0,\theta_0+2\pi)$ both correspond to the same point. This being the case it is necessary to impose periodicity conditions in θ on the polar coordinate representation for the velocity. Toward that end we examine the polar coordinate velocity components in more detail. Let

$$\boldsymbol{v}(x,y) = u\boldsymbol{i} + v\boldsymbol{j} = \varphi_x\boldsymbol{i} + \varphi_y\boldsymbol{j} = U\boldsymbol{e}_r + V\boldsymbol{e}_\theta = \boldsymbol{V}(r,\theta).$$

Employing

$$\varphi_x = \Phi_r\cos(\theta) - \frac{\sin(\theta)}{r}\Phi_\theta \text{ and } \varphi_y = \Phi_r\sin(\theta) + \frac{\cos(\theta)}{r}\Phi_\theta$$

from the previous section as well as (9.2.11c) and (9.2.11d) or $\boldsymbol{e}_\theta = -\sin(\theta)\boldsymbol{i} + \cos(\theta)\boldsymbol{j}$,

$$\varphi_x\boldsymbol{i} + \varphi_y\boldsymbol{j} = \Phi_r\boldsymbol{e}_r + \frac{1}{r}\Phi_\theta\boldsymbol{e}_\theta \Rightarrow U = \Phi_r,\ V = \frac{1}{r}\Phi_\theta.$$

Hence imposing the periodicity condition $\boldsymbol{V}(r,\theta) = \boldsymbol{V}(r,\theta+2\pi)$ implies that

$$\Phi_r(r,\theta) = \Phi_r(r,\theta+2\pi) \text{ and } \Phi_\theta(r,\theta) = \Phi_\theta(r,\theta+2\pi). \tag{12.3.1e}$$

12.4 Separation of Variables Solution

We now seek a separation of variables solution of Laplace's equation (12.3.1b) subject to the periodicity conditions (12.3.1e). Let $\Phi(r,\theta) = R(r)F(\theta)$. Then (12.3.1b, 12.3.1e) imply that

$$\frac{r^2R''(r)+rR'(r)}{R(r)} = -\frac{F''(\theta)}{F(\theta)}; \tag{12.4.1}$$

$$R'(r)F(\theta) = R'(r)F(\theta+2\pi) \tag{12.4.2a}$$

and

$$F'(\theta) = F'(\theta+2\pi); \tag{12.4.2b}$$

since $R(r) \not\equiv 0$. There now exist two cases which must be considered:

(i) $R'(r) \equiv 0$: In that event (12.4.2a) is satisfied identically while (12.4.1) and (12.4.2b) reduce to

$$F''(\theta) \equiv 0 \tag{12.4.3a}$$

and

$$F'(\theta) = F'(\theta+2\pi). \tag{12.4.3b}$$

Then (12.4.3a) implies

$$F^{(0)}(\theta) = c_1^{(0)}\theta + c_2^{(0)} \tag{12.4.4a}$$

which satisfies (12.4.3b) automatically. Since $R'(r) \equiv 0$,

$$R^{(0)}(r) \equiv c_3^{(0)}. \tag{12.4.4b}$$

Thus, from (12.4.4a, 12.4.4b)

$$\Phi^{(0)}(r,\theta) = R^{(0)}(r)F^{(0)}(\theta) = A\theta + \alpha. \tag{12.4.4c}$$

(ii) $R'(r) \not\equiv 0$: In this event (12.4.1) and (12.4.2a, 12.4.2b) imply

$$\frac{r^2R''+rR'(r)}{R(r)} = -\frac{F''(\theta)}{F(\theta)} = \lambda \equiv \text{ separation constant;} \tag{12.4.5a}$$

$$F(\theta) = F(\theta+2\pi), \tag{12.4.5b}$$

$$F'(\theta) = F'(\theta+2\pi). \tag{12.4.5c}$$

This case has two subcases which must be considered:

(a) First let $\lambda = 0$. Then (12.4.5a) implies that

$$F''(\theta) \equiv 0 \tag{12.4.6a}$$

and

$$r^2R''(r)+rR'(r) = 0. \tag{12.4.6b}$$

Now, again (12.4.6a) implies $F_0(\theta) = c_{1_0}\theta + c_{2_0}$, while (12.4.5b) yields

$$F_0(\theta) = c_{1_0}\theta + c_{2_0} = F_0(\theta+2\pi) = c_{1_0}(\theta+2\pi) + c_{2_0} \Rightarrow c_{1_0} = 0 \Rightarrow F_0(\theta) \equiv c_{2_0} \tag{12.4.7a}$$

which satisfies (12.4.5c) automatically. Since (12.4.6b) is an Euler equation, we seek a solution of the form

$$R(r) = r^m \Rightarrow m(m-1) + m = m^2 = 0 \Rightarrow m_{1,2} = 0 \Rightarrow R_0(r) = c_{3_0} + c_{4_0} \ln\left(\frac{r}{a}\right). \tag{12.4.7b}$$

Thus, from (12.4.7a, 12.4.7b),

$$\Phi_0(r,\theta) = R_0(r)F_0(\theta) = B\ln\left(\frac{r}{a}\right) + \beta. \tag{12.4.7c}$$

(b) Next, let $\lambda = \mu^2 \neq 0$ where $\mathrm{Re}(\mu) > 0$. Then (12.4.5a) implies

$$F''(\theta) + \mu^2 F(\theta) = 0 \tag{12.4.8a}$$

and

$$r^2R''(r) + rR'(r) - \mu^2 R(r) = 0. \tag{12.4.8b}$$

Now (12.4.8a) implies

$$F(\theta) = c_1\cos(\mu\theta) + c_2\sin(\mu\theta) \Rightarrow F'(\theta) = \mu[-c_1\sin(\mu\theta) + c_2\cos(\mu\theta)],$$

where $\cos(z)$ and $\sin(z)$ are as defined in Problem 5.1, while (12.4.5b, 12.4.5c) yield

$$c_1[\cos(\mu\{\theta+2\pi\}) - \cos(\mu\theta)] + c_2[\sin(\mu\{\theta+2\pi\}) - \sin(\mu\theta)] = 0,$$

$$c_1[\sin(\mu\{\theta+2\pi\}) - \sin(\mu\theta)] + c_2[\cos(\mu\{\theta+2\pi\}) - \cos(\mu\theta)] = 0.$$

Since this is an eigenvalue problem for μ, the nontriviality condition $|c_1|^2 + |c_2|^2 \neq 0$ implies

$$\begin{gathered}[\cos(\mu\{\theta+2\pi\}) - \cos(\mu\theta)]^2 + [\sin(\mu\{\theta+2\pi\}) - \sin(\mu\theta)]^2 = 0 \\ \Rightarrow 2 - 2[\cos(\mu\{\theta+2\pi\})\cos(\mu\theta) + \sin(\mu\{\theta+2\pi\})\sin(\mu\theta)] = 0 \\ \Rightarrow \cos(\mu\theta + 2\pi\mu - \mu\theta) = \cos(2\pi\mu) = 1 \Rightarrow \mu = n \\ \Rightarrow F_n(\theta) = c_{1_n}\cos(n\theta) + c_{2_n}\sin(n\theta)\end{gathered} \tag{12.4.9a}$$

for $n = 1, 2, 3, \ldots$. Then (12.4.8b) implies that $r^2R''(r) + rR'(r) - n^2R(r) = 0$. Let

$$R(r) = r^m \Rightarrow m^2 - n^2 = 0 \Rightarrow m_{1,2} = \pm n \Rightarrow R_n(r) = c_{3_n}r^n + c_{4_n}r^{-n}. \tag{12.4.9b}$$

Thus, from (12.4.9, 12.4.9b),

$$\Phi_n(r,\theta) = R_n(r)F_n(\theta) = (a_n r^n + b_n r^{-n})\cos(n\theta) + (c_n r^n + d_n r^{-n})\sin(n\theta) \tag{12.4.9c}$$

for $n = 1, 2, 3, \ldots$. Finally, employing a superposition principle, we obtain the general solution

$$\Phi(r,\theta) = \Phi^{(0)}(r,\theta) + \Phi_0(r,\theta) + \sum_{n=1}^{\infty} \Phi_n(r,\theta)$$

$$= A\theta + B\ln\left(\tfrac{r}{a}\right) + C\sum_{n=1}^{\infty}[(a_n r^n + b_n r^{-n})\cos(n\theta) + (c_n r^n + d_n r^{-n})\sin(n\theta)]$$

where $C = \alpha + \beta$.

12.5 D'Alembert's Paradox

We now wish to determine the proper values to assign the constants contained in the general separation of variables solution determined in the last section

$$\Phi(r,\theta) = A\theta + B\ln\left(\frac{r}{a}\right) + \sum_{n=1}^{\infty}[(a_n r^n + b_n r^{-n})\cos(n\theta) + (c_n r^n + d_n r^{-n})\sin(n\theta)].$$

Note that, without loss of generality, we have taken $C = 0$ since velocity potentials can only be determined up to an arbitrary additive constant. In order to determine the rest of the constants contained in that solution, we next impose the boundary conditions (12.3.1c, 12.3.1d). We first consider the far-field or free-stream boundary condition of (12.3.1d) in the form

$$\lim_{r\to\infty}\frac{1}{r}\Phi(r,\theta) = U_0\cos(\theta).$$

Since r^{-n}, $\ln(r/a)/r \to 0$ as $r \to \infty$; this implies $a_1 = U_0$, $c_1 = 0$, $a_n = c_n = 0$ for $n \geq 2$. Thus

$$\Phi(r,\theta) = A\theta + B\ln\left(\frac{r}{a}\right) + U_0 r\cos(\theta) + \sum_{n=1}^{\infty}[b_n\cos(\theta) + d_n\sin(\theta)]r^{-n}.$$

Hence

$$\Phi_r(r,\theta) = \frac{B}{r} + U_0\cos(\theta) - \sum_{n=1}^{\infty} n[b_n\cos(\theta) + d_n\sin(\theta)]r^{-n-1}.$$

Then, imposing the no-penetration boundary condition (12.3.1c), we obtain

$$\Phi_r(a,\theta) = \frac{B}{a} + \left(U_0 - \frac{b_1}{a^2}\right) - \frac{d_1}{a^2}\sin(\theta) - \sum_{n=2}^{\infty} n[b_n\cos(\theta) + d_n\sin(\theta)]a^{-n-1} = 0,$$

which implies $B = 0, b_1 = U_0 a^2, d_1 = 0, b_n = d_n = 0$ for $n \geq 2$.

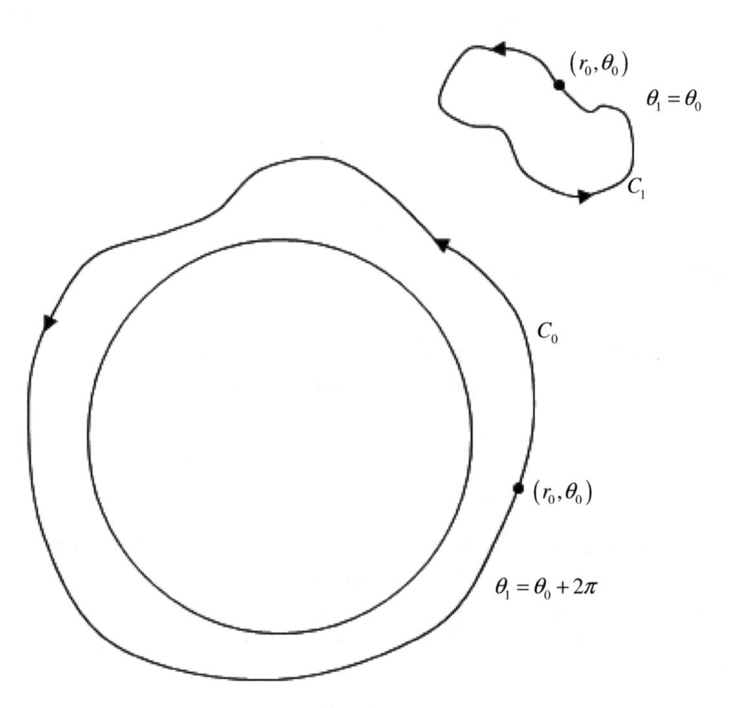

Fig. 12.2 A schematic diagram of the two types of curves that have qualitatively different circulation behavior for the potential flow situation depicted in Fig. 12.1: Namely, C_0 which encircles the cylinder and C_1 which does not. Then, $K(C_0) = 2\pi A = K_0$ and $K(C_1) = 0$.

Finally

$$\Phi(r,\theta) = A\theta + U_0\left(r + \frac{a^2}{r}\right)\cos(\theta). \tag{12.5.1a}$$

This is a one-parameter family of solutions where that parameter A is undetermined. Before imposing an extra condition to determine A uniquely let us examine how this extra term $\Phi^{(0)}(r,\theta) = A\theta$ contributes to the velocity field. Given that

$$U^{(0)} = \Phi_r^{(0)}(r,\theta) = 0,\ V^{(0)} = \frac{\Phi_\theta^{(0)}(r,\theta)}{r} = \frac{A}{r} \Rightarrow \mathbf{V}^{(0)} = \left(\frac{A}{r}\right)\boldsymbol{e}_\theta,$$

this flow is purely circumferential and does not affect the free-stream condition since it disappears in the limit as $r \to \infty$. In 1869, Lord Kelvin defined the concept of circulation to characterize such flows as follows (see Fig. 12.2):

$$K(C) = \oint_C \boldsymbol{v} \bullet \boldsymbol{dr} = \oint_C \varphi_x \, dx + \varphi_y \, dy = \oint_C d\varphi = \Phi(r_0, \theta_1) - \Phi(r_0, \theta_0)$$

where C is a closed curve in the plane traversed once in the counterclockwise direction starting at the point in polar coordinates $r = r_0$, $\theta = \theta_0$ and terminating at the point $r = r_0, \theta = \theta_1$. Note from Fig. 12.2 that for a curve such as C_1 which does not encircle the cylinder θ_1 can be taken as θ_0 or $\theta_1 = \theta_0$ and hence $K(C_1) = 0$ while for a curve such as C_0 which encircles the cylinder θ_1 must be taken as $\theta_0 + 2\pi$ or $\theta_1 = \theta_0 + 2\pi$ and hence, for our velocity potential function (12.5.1a)

$$K(C_0) = A(\theta_0 + 2\pi) - A\theta_0 = 2\pi A = K_0 \Rightarrow A = \frac{K_0}{2\pi} \tag{12.5.1b}$$

since the $\cos(\theta)$ is periodic with period 2π. Here K_0 represents the circulation of the flow defined as the value of $K(C)$ for a curve C encircling the cylinder and traversed once in the counterclockwise direction. Hence by imposing the circulation of the flow the velocity potential can be determined uniquely.

We now wish to determine the force on the cylinder due to this velocity potential function (12.5.1). Recalling that for an inviscid fluid the stress vector which has the dimension of a force per unit area is given by $\boldsymbol{t} = -p\boldsymbol{n}$, the total force on the cylinder per unit length in the z-direction

$$\boldsymbol{F} = \oint_{r=a} \boldsymbol{t} ds = -\int_0^{2\pi} P(a,\theta)[\cos(\theta)\boldsymbol{i} + \sin(\theta)\boldsymbol{j}]a \, d\theta$$

where $p(x, y) = P(r, \theta)$, $\boldsymbol{n} = \boldsymbol{e}_r$, $ds = a\,d\theta$, and the line integral is in the clockwise direction. Next, recalling from Bernoulli's relation (12.1.11) and the Calculus of Variations development of Section 12.2, that

$$p + \rho_0 \frac{\phi_x^2 + \phi_y^2}{2} = P + \rho_0 \frac{\Phi_r^2 + \Phi_\theta^2/r^2}{2} = P_\infty + \rho_0 \frac{U_0^2}{2} \text{ where } P_\infty = \lim_{r \to \infty} P(r, \theta);$$

we can deduce

$$P(a,\theta) = P_\infty + \rho_0 \frac{U_0^2}{2} - \rho_0 \frac{\Phi_\theta^2(a,\theta)}{2a^2} \text{ since } \Phi_r(a,\theta) = 0. \tag{12.5.2a}$$

Now, from (12.5.1)

$$\Phi_\theta(r,\theta) = A - U_0 \left(r + \frac{a^2}{r} \right) \sin(\theta) \Rightarrow \Phi_\theta(a,\theta) = A - 2aU_0 \sin(\theta)$$

$$\Rightarrow \Phi_\theta^2(a,\theta) = A^2 - 4AaU_0\sin(\theta) + 4a^2U_0^2\sin^2(\theta).$$

Substitution of this expression for $\Phi_\theta^2(a,\theta)$ into (12.5.2a) yields

$$P(a,\theta) = P_\infty + \rho_0\frac{U_0^2}{2} - \rho_0\frac{A^2}{2a^2} + 2\rho_0\frac{AU_0\sin(\theta)}{a} - 2\rho_0U_0^2\sin^2(\theta). \qquad (12.5.2b)$$

Employing (12.5.2b) in our integral relation for $\boldsymbol{F}$ and the orthonormality conditions

$$\int_0^{2\pi}\cos(\theta)\,d\theta = \int_0^{2\pi}\sin(\theta)\,d\theta = \int_0^{2\pi}\sin(\theta)\cos(\theta)\,d\theta = 0,$$

$$\int_0^{2\pi}\sin^2(\theta)\cos(\theta)\,d\theta = \int_0^{2\pi}\sin^3(\theta)\,d\theta = 0,\ \int_0^{2\pi}\sin^2(\theta)\,d\theta = \pi;$$

we obtain

$$\boldsymbol{F} = -(2\pi A)\rho_0U_0\boldsymbol{j} = -K_0\rho_0U_0\boldsymbol{j}. \qquad (12.5.3a)$$

Since, in general, the components of this force on a two-dimensional body are defined to be $\boldsymbol{F} = -F_1\boldsymbol{i} + F_2\boldsymbol{j}$ where $F_1 \equiv$ drag and $F_2 \equiv$ lift, (12.5.3a) implies that

$$F_1 = 0 \text{ and } F_2 = -K_0\rho_0U_0. \qquad (12.5.3b)$$

In other words our inviscid fluid analysis predicts that there is no drag on the cylindrical body and a lift on it proportional to the circulation. Indeed this result can be extrapolated to inviscid potential flow past any two-dimensional body. In three dimensions, as demonstrated for the special case of potential flow past a sphere by Problem 12.3, $\boldsymbol{F} = \boldsymbol{0}$. This is referred to as D'Alembert's Paradox and loosely interpreted means that one could stand upright in a wind tunnel without having to exert any force to keep from being blown away. Obviously such a result is physically impossible and follows from the necessity of dropping the no-slip or adherence boundary condition at the body due to our inviscid fluid assumption. Thus although an inviscid fluid model yielded reasonable predictions for the propagation of subsonic sound waves through a medium possessing no boundary treated in the last chapter, it leaves much to be desired when examining flow past bodies. Hence in the next chapter we shall begin considering viscous fluid flow models in an effort to alleviate that deficiency. We close this chapter with two pastoral interludes dealing with Leibniz's Rule of Differentiation and Legendre polynomials, respectively: The first topic required for the development of the second. Any function satisfying Laplace's equation is said to be harmonic. In this context observe that such a potential equation is *elliptic* in the sense defined in Section 11.3 while the heat equation treated in depth earlier and the wave equation investigated in the last chapter were parabolic and hyperbolic, respectively, in this sense. Knowledge of Legendre polynomials is essential for solving Laplace's equation involving spherical harmonics such as Problem 12.3 and these interludes are being offered in the spirit of making our presentations as self-contained as at all possible.

12.6 Pastoral Interlude: Leibniz's Rule of Differentiation

Recall the binomial expansion for

$$(a+b)^n = \sum_{k=0}^{n} \binom{n}{k} a^{n-k} b^k$$

where n is a nonnegative integer and

$$\binom{n}{k} = \frac{n!}{k!(n-k)!} = \frac{n(n-1)\cdots(n-k+1)}{k!} \text{ with } n! \equiv n(n-1)\cdots 2\cdot 1 \text{ and } 0! \equiv 1.$$

We shall consider an extrapolation of this concept called Leibniz's Rule of Differentiation: Namely

$$(fg)^{(N)} = \sum_{k=0}^{N} \binom{N}{k} f^{(N-k)} g^{(k)} \text{ where } (-)^{(k)} \equiv \frac{d^k(-)}{dx^k} \text{ and } f^{(0)} \equiv f,$$

which will be demonstrated by *Mathematical Induction* as follows.

(a) First, we show that the statement is true for $N = 1$. From the product rule for differentiation

$$(fg)^{(1)} = f^{(1)} g^{(0)} + f^{(0)} g^{(1)}$$

while

$$\sum_{k=0}^{1} \binom{1}{k} f^{(1-k)} g^{(k)} = \binom{1}{0} f^{(1)} g^{(0)} + \binom{1}{1} f^{(0)} g^{(1)} = f^{(1)} g^{(0)} + f^{(0)} g^{(1)}$$

since

$$\binom{n}{0} = \binom{n}{n} = 1,$$

hence demonstrating the result.

(b) Next, we assume that the statement is true for some integer $N = p$

$$(fg)^{(p)} = \sum_{k=0}^{p} \binom{p}{k} f^{(p-k)} g^{(k)}.$$

(c) Finally, we test its truth for $N = p + 1$. Here

$$(fg)^{(p+1)} = [(fg)^{(p)}]^{(1)} = \sum_{k=0}^{p} \binom{p}{k} [f^{(p-k)} g^{(k)}]^{(1)}$$
$$= \sum_{k=0}^{p} \binom{p}{k} f^{(p+1-k)} g^{(k)} + \sum_{k=0}^{p} \binom{p}{k} f^{(p-k)} g^{(k+1)}$$
$$= f^{(p+1)} g^{(k)} + \sum_{k=1}^{p} \binom{p}{k} f^{(p+1-k)} g^{(k)} + \sum_{k=0}^{p-1} \binom{p}{k} f^{(p-k)} g^{(k+1)} + f^{(0)} g^{(p+1)}.$$

Now, changing the index in the second sum by introducing $j = k + 1$, we obtain

$$\sum_{k=0}^{p-1} \binom{p}{k} f^{(p-k)} g^{(k+1)} = \sum_{j=1}^{p} \binom{p}{j-1} f^{(p+1-j)} g^{(j)} = \sum_{k=1}^{p} \binom{p}{k-1} f^{(p+1-k)} g^{(k)}.$$

Substituting this back into our previous result yields

$$(fg)^{(p+1)} = f^{(p+1)} g^{(k)} + \sum_{k=1}^{p} \left[\binom{p}{k} + \binom{p}{k-1} \right] f^{(p+1-k)} g^{(k)} + f^{(0)} g^{(p+1)}.$$

It only remains to show that

$$\binom{p}{k} + \binom{p}{k-1} = \binom{p+1}{k} \text{ which we do as follows:}$$

$$\binom{p}{k} + \binom{p}{k-1} = \frac{p!}{k!(p-k)!} + \frac{p!}{(k-1)!(p+1-k)!}$$
$$= \frac{p!}{(k-1)!(p-k)!} \left(\frac{1}{k} + \frac{1}{p+1-k} \right)$$
$$= \frac{p!}{(k-1)!(p-k)!} \frac{p+1}{k(p+1-k)} = \frac{(p+1)!}{k!(p+1-k)!} = \binom{p+1}{k};$$

since then

$$(fg)^{(p+1)} = f^{(p+1)} g^{(k)} + \sum_{k=1}^{p} \binom{p+1}{k} f^{(p+1-k)} g^{(k)} + f^{(0)} g^{(p+1)}$$
$$= \sum_{k=0}^{p+1} \binom{p+1}{k} f^{(p+1-k)} g^{(k)},$$

Hence demonstrating its truth for $N = p + 1$. We can now conclude from the *Principle of Mathematical Induction* that Leibniz's Rule of Differentiation is true for all natural numbers N.

12.7 Pastoral Interlude: Legendre Polynomials

Recall $x = x_0$ is called an ordinary point of the ordinary differential equation for $y = y(x)$

$$y'' + p(x)y' + q(x)y = 0$$

provided $p(x)$ and $q(x)$ have Taylor series representations such that

$$p(x) = \sum_{k=0}^{\infty} p_k(x-x_0)^k \text{ and } q(x) = \sum_{k=0}^{\infty} q_k(x-x_0)^k \text{ for } |x-x_0| < R.$$

Then one seeks a power series solution of that equation of the form

$$y(x) = \sum_{k=0}^{\infty} c_k(x-x_0)^k.$$

Consider the Legendre ordinary differential equation for $y = y(x)$

$$L[y] = (1-x^2)y'' - 2xy' + n(n+1)y = 0 \text{ for } |x| < 1.$$

Since $x = 0$ is an ordinary point for this equation, we shall seek a solution of it of the form

$$y(x) = \sum_{k=0}^{\infty} c_k x^k.$$

Then

$$y'(x) = \sum_{k=0}^{\infty} kc_k x^{k-1} \text{ and } y''(x) = \sum_{k=0}^{\infty} k(k-1)c_k x^{k-2} = \sum_{k=2}^{\infty} k(k-1)c_k x^{k-2}.$$

Thus

$$L[y] = \sum_{k=2}^{\infty} k(k-1)c_k x^{k-2} + \sum_{k=0}^{\infty} [-k(k-1) - 2k + n(n+1)]c_k x^k = 0.$$

Given that

$$\sum_{k=2}^{\infty} k(k-1)c_k x^{k-2} = \sum_{j=0}^{\infty} (j+2)(j+1)c_{j+2}x^j = \sum_{k=0}^{\infty} (k+2)(k+1)c_{k+2}x^k$$

and

$$\begin{aligned} n(n+1) - 2k - k(k-1) &= n(n+1) - k(k-1) \\ &= n(n+k+1) - k(n+k+1) = (n+k+1)(n-k), \end{aligned}$$

$$L[y] = \sum_{k=0}^{\infty} [(k+2)(k+1)c_{k+2} + (n+k+1)(n-k)c_k]x^k = 0$$
$$\Rightarrow (k+2)(k+1)c_{k+2} + (n+k+1)(n-k)c_k = 0 \text{ for } k \geq 0$$
$$\Rightarrow c_{2m} = a_{2m}c_0 \text{ and } c_{2m+1} = a_{2m+1}c_1 \text{ for } m \geq 1.$$

Thus

$$y_1(x) = c_0\left[1 + \sum_{m=1}^{\infty} a_{2m}x^{2m}\right],\ y_2(x) = c_1\left[x + \sum_{m=1}^{\infty} a_{2m+1}x^{2m+1}\right],$$

and $y(x) = y_1(x) + y_2(x)$. Note that for n unequal to a nonnegative integer both these series diverge as $|x| \to 1$, while for n equal to a nonnegative integer we have the following result:

(i) $n = 2p$ with $p \geq 0$: $a_{2p+2j} = 0$ for $j \geq 1 \Rightarrow$

$$y_1(x) = c_0 \text{ for } p = 0,\ y_1(x) = c_0\left[1 + \sum_{m=1}^{p} a_{2m}x^{2m}\right] \text{ for } p \geq 1,$$

and $y_2(x)$ diverges as $|x| \to 1$.

(ii) $n = 2p + 1$ with $p \geq 0$: $a_{2p+1+2j} = 0$ for $j \geq 1 \Rightarrow$

$$y_2(x) = c_1 x \text{ for } p = 0,\ y_2(x) = c_1\left[x + \sum_{m=1}^{p} a_{2m+1}x^{2m+1}\right] \text{ for } p \geq 1,$$

and $y_1(x)$ diverges as $|x| \to 1$. Observe that these polynomials of degree $n \geq 0$ are the only bounded solutions of the Legendre equation as $|x| \to 1$. The Legendre polynomial of order n denoted by $P_n(x)$ satisfies $L[y] = 0$ for a nonnegative integer n such that $P_n(1) = 1$. In order to deduce a convenient series representation for $P_n(x)$, it is necessary to employ Rodrigues Formula

$$P_n(x) = \frac{[(x^2-1)^n]^{(n)}}{2^n n!}$$

which will be demonstrated as follows: Define $u_n(x) = (x^2 - 1)^n$. Then

$$u_n^{(1)}(x) = 2nx(x^2-1)^{n-1} \Rightarrow (x^2-1)u_n^{(1)}(x) = 2nxu_n(x).$$

Recall Leibniz's Rule

$$(fg)^{(N)} = \sum_{k=0}^{N} \binom{N}{k} f^{(N-k)}g^{(k)}$$

and differentiate both sides of this equality $(n + 1)$-times with respect to x, obtaining

$$
\begin{aligned}
[(x^2-1)u_n^{(1)}]^{(n+1)} &= \sum_{k=0}^{n+1} \binom{n+1}{k} u_n^{(n+2-k)}(x^2-1)^{(k)} \\
&= \sum_{k=0}^{2} \binom{n+1}{k} u_n^{(n+2-k)}(x^2-1)^{(k)} \\
&= u_n^{(n+2)}(x^2-1) + (n+1)u_n^{(n+1)}(2x) + \frac{(n+1)n}{2} u_n^{(n)}(2) \\
&= (x^2-1)u_n^{(n+2)} + 2(n+1)xu_n^{(n+1)} + n(n+1)u_n^{(n)} = 2n[xu_n]^{(n+1)} \\
&= 2n\left[\sum_{k=0}^{n+1} \binom{n+1}{k} u_n^{(n+1-k)}(x)^{(k)}\right] = 2n\left[\sum_{k=0}^{1} \binom{n+1}{k} u_n^{(n+1-k)}(x)^{(k)}\right] \\
&= 2nxu_n^{(n+1)} + 2n(n+1)u_n^{(n)}
\end{aligned}
$$

$$
\Rightarrow L[u_n^{(n)}] = (1-x^2)u_n^{(n+2)} - 2xu_n^{(n+1)} + n(n+1)u_n^{(n)} = 0.
$$

Since $\varphi_n(x) = u_n^{(n)}(x) = [(x^2-1)^n]^{(n)}$ is a polynomial in x of order n, we can conclude that $P_n(x) = \alpha_n \varphi_n(x)$ where α_n is such that $P_n(1) = 1$. Now

$$
\begin{aligned}
\varphi_n(x) &= [(x-1)^n(x+1)^n]^{(n)} = \sum_{k=0}^{n} \binom{n}{k} [(x-1)^n]^{(n-k)}[(x+1)^n]^{(k)} \\
&= [(x-1)^n]^{(n)}(x+1)^n + o(x-1) = n!(x+1)^n + o(x-1).
\end{aligned}
$$

Thus $\varphi_n(1) = n!2^n$ which implies that $\alpha_n = 1/(2^n n!)$ and completes the demonstration. Finally

$$
\begin{aligned}
P_n(x) &= \frac{[(x^2-1)^n]^{(n)}}{2^n n!} = \frac{[x^{2n} + O(x^{2n-2})]^{(n)}}{2^n n!} \\
&= \frac{[2n(2n-1)\cdot \cdots \cdot (n+1)x^n + O(x^{n-2})]}{2^n n!}.
\end{aligned}
$$

Therefore

$$
c_n = \frac{2n(2n-1)\cdot \cdots \cdot (n+1)}{2^n n!} = \frac{(2n)!}{2^n (n!)^2} \text{ and } c_{n-1} = 0.
$$

Thus

$$
P_n(x) = \sum_{k=0}^{n} c_k x^k = \sum_{m=0}^{p} c_{n-2m} x^{n-2m} \text{ where } n-2p = \begin{cases} 0, & \text{if } n \text{ is even} \\ 1, & \text{if } n \text{ is odd} \end{cases}
$$

since the recursion relation

$$
(k+2)(k+1)c_{k+2} + (n+k+1)(n-k)c_k = 0 \text{ for } k \geq 0,
$$

in conjunction with $c_{n-1} = 0$, implies that

$$c_{n-(2m+1)} = 0 \text{ for } m = 0, \ldots, q \text{ where } n-(2q+1) = \begin{cases} 1, & \text{if } n \text{ is even} \\ 0, & \text{if } n \text{ is odd} \end{cases}.$$

We shall now determine the c_{n-2m} inductively by employing the recursion relation in conjunction with the c_n deduced from Rodrigues Formula as follows:

$$c_k = -\frac{(k+2)(k+1)}{(n-k)(n+k+1)} c_{k+2} \text{ for } k \leq n-2.$$

Consider:
$k = n-2$ or $m = 1$:

$$\begin{aligned} c_{n-2} &= -\frac{n(n-1)}{2(2n-1)} c_n = -\frac{n(n-1)(2n)!}{2(2n-1)2^n(n!)^2} \\ &= -\frac{n(n-1)(2n)(2n-1)(2n-2)!}{2(2n-1)2^n n(n-1)!n(n-1)(n-2)!} \\ &= \frac{(-1)^1(2n-2)!}{2^n 1!(n-1)!(n-2)!}; \end{aligned}$$

$k = n-4$ or $m = 2$:

$$\begin{aligned} c_{n-4} &= -\frac{(n-2)(n-3)}{4(2n-3)} c_{n-2} = \frac{(-1)^2(n-2)(n-3)(2n-2)!}{4(2n-3)2^n 1!(n-1)!(n-2)!} \\ &= \frac{(-1)^2(n-2)(n-3)(2n-2)(2n-3)(2n-4)!}{4(2n-3)2^n 1!(n-1)(n-2)!(n-2)(n-3)(n-4)!} \\ &= \frac{(-1)^2(2n-4)!}{2^n 2!(n-2)!(n-4)!}; \end{aligned}$$

$k = n-6$ or $m = 3$:

$$\begin{aligned} c_{n-6} &= -\frac{(n-4)(n-5)}{6(2n-5)} c_{n-4} = \frac{(-1)^3(n-4)(n-5)(2n-4)!}{6(2n-5)2^n 2!(n-2)!(n-4)!} \\ &= \frac{(-1)^3(n-4)(n-5)(2n-4)(2n-5)(2n-6)!}{6(2n-5)2^n 2!(n-2)(n-3)!(n-4)(n-5)(n-6)!} \\ &= \frac{(-1)^3(2n-6)!}{2^n 3!(n-3)!(n-6)!} \end{aligned}$$

Hence:

$$k = n-2m \text{ for } m = 1, 2, 3, \ldots, p = \begin{cases} n/2, & \text{if } n \text{ is even} \\ (n-1)/2, & \text{if } n \text{ is odd} \end{cases}:$$

$$c_{n-2m} = \frac{(-1)^m(2n-2m)!}{2^n m!(n-m)!(n-2m)!} \text{ such that } n-2p = \begin{cases} 0, & \text{if } n \text{ is even} \\ 1, & \text{if } n \text{ is odd} \end{cases},$$

which could be proven conclusively by Mathematical Induction. Note that this expression reduces to the c_n deduced from Rodrigues Formula for $m = 0$. Therefore, we have derived the desired series representation

$$P_n(x) = \frac{1}{2^n} \sum_{m=0}^{p} \frac{(-1)^m (2n-2m)!}{m!(n-m)!(n-2m)!} x^{n-2m} \text{ for } n = 0, 1, 2, \ldots.$$

There exists a generating function for these polynomials given by

$$g(x,z) = (1 - 2xz + z^2)^{-1/2} = \sum_{n=0}^{\infty} P_n(x) z^n$$

which is demonstrated as follows: Consider the infinite binomial series deduced in Section 6.5

$$(1+v)^{-1/2} = \sum_{j=0}^{\infty} (-1)^j \frac{(2j)!}{(2^j j!)^2} v^j \text{ with } v = -2xz + z^2 = -(2x - z)z.$$

Then

$$g(x,z) = \sum_{j=0}^{\infty} \frac{(2j)!}{2^{2j}(j!)^2} (2x - z)^j z^j.$$

Consider the binomial expansion

$$(a+b)^j = \sum_{m=0}^{j} \binom{j}{m} a^{j-m} b^m \text{ with } a = 2x \text{ and } b = -z.$$

Then

$$(2x - z)^j = \sum_{m=0}^{j} \binom{j}{m} (2x)^{j-m} (-z)^m = j! \sum_{m=0}^{j} \frac{(-1)^m 2^{j-m} x^{j-m}}{m!(j-m)!} z^m \text{ with } j \ge m.$$

Thus

$$g(x,z) = \sum_{j=0}^{\infty} \sum_{m=0}^{j} \frac{(-1)^m (2j)! x^{j-m}}{2^{j+m} j! m! (j-m)!} z^{j+m}.$$

Now, reordering this double sum by introducing

$$n = j + m \ge 0 \text{ or } j = n - m \Rightarrow j - m = n - 2m \ge 0 \Rightarrow m = 0, \ldots, p;$$

we obtain

$$g(x,z) = \sum_{n=0}^{\infty} \left[\sum_{m=0}^{p} \frac{(-1)^m (2n-2m)! x^{n-2m}}{2^n m!(n-m)!(n-2m)!} \right] z^n = \sum_{n=0}^{\infty} P_n(x) z^n,$$

thus completing the demonstration.

Finally, we tabulate the first six Legendre polynomials:

$$P_0(x) = 1,\ P_1(x) = x;$$
$$P_2(x) = \frac{1}{2}(3x^2 - 1),\ P_3(x) = \frac{1}{2}(5x^3 - 3x)$$
$$P_4(x) = \frac{1}{8}(35x^4 - 30x^2 + 3),\ P_5(x) = \frac{1}{8}(63x^5 - 70x^3 + 15).$$

Note in this context that, as expected,

$$P_0(1) = P_1(1) = 1;$$
$$P_2(1) = \frac{1}{2}(3 - 1) = 1,\ P_3(1) = \frac{1}{2}(5 - 3) = 1;$$
$$P_4(1) = \frac{1}{8}(35 - 30 + 3) = 1,\ P_5(1) = \frac{1}{8}(63 - 70 + 15) = 1.$$

We close with an additional three properties of Legendre polynomials the development of which depends crucially upon the knowledge of this generating function

$$g(x,z) = (1 - 2xz + z^2)^{-1/2} = \sum_{n=0}^{\infty} P_n(x)z^n.$$

(i) A Recursion Relation Involving Legendre polynomials

Computing the partial derivative of $g(x,z)$ with respect to z, we find that

$$\frac{\partial g(x,z)}{\partial z} = (x - z)(1 - 2xz + z^2)^{-3/2} = \sum_{n=0}^{\infty} nP_n(x)z^{n-1}.$$

Then

$$(1 - 2xz + z^2)\frac{\partial g(x,z)}{\partial z} = (x - z)g(x,z).$$

Substituting the series representations for the generating function into this result yields

$$(1 - 2xz + z^2)\sum_{n=0}^{\infty} nP_n(x)z^{n-1} = \sum_{n=1}^{\infty} nP_n(x)z^{n-1} - \sum_{n=0}^{\infty} 2xnP_n(x)z^n + \sum_{n=0}^{\infty} nP_n(x)z^{n+1}$$
$$= (x - z)\sum_{n=0}^{\infty} P_n(x)z^n = \sum_{n=0}^{\infty} xP_n(x)z^n - \sum_{n=0}^{\infty} P_n(x)z^{n+1}.$$

Thus

$$\begin{aligned}
\sum_{n=1}^{\infty} nP_n(x)z^{n-1} &= \sum_{j=0}^{\infty}(j+1)P_{j+1}z^j = \sum_{n=0}^{\infty}(n+1)P_{n+1}z^n \\
&= P_1(x) + \sum_{n=1}^{\infty}(n+1)P_{n+1}z^n \\
&= \sum_{n=0}^{\infty}(2n+1)xP_n(x)z^n - \sum_{n=0}^{\infty}(n+1)P_n(x)z^{n+1} \\
&= \sum_{n=0}^{\infty}(2n+1)xP_n(x)z^n - \sum_{j=1}^{\infty} jP_{j-1}(x)z^j \\
&= \sum_{n=0}^{\infty}(2n+1)xP_n(x)z^n - \sum_{n=1}^{\infty} nP_{n-1}(x)z^n \\
&= xP_0(x) + \sum_{n=1}^{\infty}[(2n+1)xP_n(x) - nP_{n-1}(x)]z^n.
\end{aligned}$$

Hence

$$P_1(x) - xP_0(x) + \sum_{n=1}^{\infty}[(n+1)P_{n+1}(x) - (2n+1)xP_n(x) + nP_{n-1}(x)]z^n = 0$$

$$\Rightarrow P_1(x) = xP_0(x),\ (n+1)P_{n+1}(x) = (2n+1)xP_n(x) - nP_{n-1}(x) \text{ for } n \geq 1.$$

Now, since

$$g(x,0) = 1 = P_0(x),$$

which implies that

$$P_1(x) = xP_0(x) = x;$$

for $n = 1$:

$$2P_2(x) = 3xP_1(x) - P_0(x) = 3x^2 - 1 \Rightarrow P_2(x) = \frac{1}{2}(3x^2 - 1);$$

for $n = 2$:

$$\begin{aligned}
3P_3(x) &= 5xP_2(x) - 2P_1(x) = \frac{1}{2}(15x^3 - 5x) - 2x \\
&= \frac{1}{2}(15x^3 - 5x - 4x) = \frac{1}{2}(15x^3 - 9x) \Rightarrow P_3(x) = \frac{1}{2}(5x^3 - 3x);
\end{aligned}$$

for $n = 3$:

$$4P_4(x) = 7xP_3(x) - 3P_2(x) = \frac{7}{2}(5x^4 - 3x^2) - \frac{3}{2}(3x^2 - 1)$$
$$= \frac{1}{2}(35x^4 - 21x^2 - 9x^2 + 3)$$
$$= \frac{1}{2}(35x^4 - 30x^2 + 3) \Rightarrow P_4(x) = \frac{1}{8}(35x^4 - 30x^2 + 3);$$

for $n = 4$:

$$5P_5(x) = 9xP_4(x) - 4P_3(x) = \frac{9}{8}(35x^5 - 30x^3 + 3x) - \frac{16}{8}(5x^3 - 3x)$$
$$= \frac{1}{8}(9 \cdot 35x^5 - 9 \cdot 30x^3 + 9 \cdot 3x - 16 \cdot 5x^3 + 16 \cdot 3x)$$
$$= \frac{1}{8}(63 \cdot 5x^5 - 54 \cdot 5x^3 - 16 \cdot 5x^3 + 25 \cdot 3x)$$
$$= \frac{1}{8}(63 \cdot 5x^5 - 70 \cdot 5x^3 + 15 \cdot 5x) \Rightarrow P_5(x) = \frac{1}{8}(63x^5 - 70x^3 + 15x);$$

in agreement with our earlier tabulation.

(ii) Integral Representation of Legendre polynomials

We shall now deduce an integral representation for $P_n(x)$ valid when $x > 1$. We proceed by first demonstrating that

$$\int_0^{\pi} \frac{d\theta}{a - b\cos(\theta)} = \frac{1}{2}\int_{-\pi}^{\pi} \frac{d\theta}{a - b\cos(\theta)} = \frac{\pi}{(a^2 - b^2)^{1/2}} \text{ for } a > |b|.$$

We shall evaluate this integral in two different ways: One involving a trigonometric substitution procedure and the other, a complex variables residue theorem approach.

(a) This substitution is employed to evaluate integrals with integrands that are rational functions of $\cos(\theta)$ and $\sin(\theta)$. Let $t = \tan(\theta/2) = \sin(\theta/2)/\cos(\theta/2)$. Then

$$\sin(\theta) = 2\sin\left(\frac{\theta}{2}\right)\cos\left(\frac{\theta}{2}\right) = 2t\cos^2\left(\frac{\theta}{2}\right) = \frac{2t}{\sec^2\left(\frac{\theta}{2}\right)} = \frac{2t}{1+t^2},$$

$$\cos(\theta) = 1 - 2\sin^2\left(\frac{\theta}{2}\right) = 1 - 2t^2\cos^2\left(\frac{\theta}{2}\right) = 1 - \frac{2t^2}{1+t^2} = \frac{1-t^2}{1+t^2}$$

and

$$dt = \sec^2\left(\frac{\theta}{2}\right)\frac{d\theta}{2} \text{ or } d\theta = \frac{2\,dt}{1+t^2}.$$

Thus

$$\frac{1}{2}\int_{-\pi}^{\pi}\frac{d\theta}{a-b\cos(\theta)}=\int_{-\infty}^{\infty}\frac{dt}{a(1+t^2)+b(t^2-1)}=\int_{-\infty}^{\infty}\frac{dt}{a-b+(a+b)t^2}$$
$$=\frac{1}{a-b}\int_{-\infty}^{\infty}\frac{dt}{1+(a+b)t^2/(a-b)}.$$

Let

$$u=\sqrt{\frac{a+b}{a-b}}t\Rightarrow dt=\sqrt{\frac{a-b}{a+b}}du.$$

Finally

$$\frac{1}{2}\int_{-\pi}^{\pi}\frac{d\theta}{a-b\cos(\theta)}=\frac{1}{a-b}\sqrt{\frac{a-b}{a+b}}\int_{-\infty}^{\infty}\frac{du}{1+u^2}$$
$$=\frac{1}{(a^2-b^2)^{1/2}}\arctan(u)\Big|_{-\infty}^{\infty}=\frac{\pi}{(a^2-b^2)^{1/2}}.$$

(b) We shall deduce this result by employing the residue theorem. Toward that end for the sake of completeness we provide a sketch of the derivation of this theorem. If a complex function $f(z)$ with $z=x+iy$ is analytic on and within a closed curve C, it has a Taylor series about any point $z=z_0$ in that domain

$$f(z)=\sum_{n=0}^{\infty}a_n(z-z_0)^n\Rightarrow\oint_C f(z)\,dz=\oint_{|z-z_0|=1}f(z)\,dz=0,$$

while if $z=z_0$ is a single pole of order k for that function, it has a Laurent expansion

$$f(z)=\sum_{n=-k}^{\infty}a_n(z-z_0)^n\Rightarrow\oint_C f(z)\,dz=\oint_{|z-z_0|=1}f(z)\,dz=2\pi i a_{-1}.$$

Both of these results follow from

$$\overset{z-z_0=e^{i\theta},\,dz=ie^{i\theta}\,d\theta}{\oint_{|z-z_0|=1}}(z-z_0)^n\,dz=i\int_{-\pi}^{\pi}e^{i(n+1)\theta}\,d\theta=\begin{cases}2\pi i, & \text{for } n=-1\\ 0, & \text{for } n\neq-1\end{cases}.$$

Further for a kth-order pole this residue

$$a_{-1}^{(k)}=\frac{1}{(k-1)!}\lim_{z\to z_0}\frac{d^{k-1}}{dz^{k-1}}[(z-z_0)^k f(z)].$$

Let us examine that residue formula for a first-order pole or $k=1$ which for ease of exposition will be denoted by

$$a_{-1} = \lim_{z \to z_0} [(z - z_0)f(z)].$$

If $f(z)$ is of the form

$$f(z) = \frac{h(z)}{g(z)} \text{ such that } h(z_0) \neq 0 \Rightarrow g(z_0) = 0 \text{ while } g'(z_0) \neq 0,$$

then

$$a_{-1} = \lim_{z \to z_0} \left[\frac{(z - z_0)h(z)}{g(z)} \right] = \lim_{z \to z_0} \left[\frac{z - z_0}{g(z) - g(z_0)} h(z) \right] = \frac{h(z_0)}{g'(z_0)}.$$

Note that these results can be generalized to functions possessing more than one pole in a particular domain by multiplying $2\pi i$ times the sum of all its residues.

Returning to

$$\frac{1}{2} \int_{-\pi}^{\pi} \frac{d\theta}{a - b\cos(\theta)} \text{ with } \cos(\theta) = \frac{e^{i\theta} + e^{-i\theta}}{2},$$

we make the substitution

$$z = e^{i\theta} \Rightarrow dz = ie^{i\theta}\, d\theta \text{ or } d\theta = \frac{dz}{iz} \text{ and } \cos(\theta) = \frac{z + 1/z}{2}.$$

Thus

$$\frac{1}{2} \int_{-\pi}^{\pi} \frac{d\theta}{a - b\cos(\theta} = \frac{1}{2} \oint_{|z|=1} \frac{dz}{iz[a - b(z + 1/z)/2]} = \frac{1}{2i} \oint_{|z|=1} \frac{2\,dz}{-bz^2 + 2az - b}.$$

Here

$$h(z) \equiv 2, g(z) = -bz^2 + 2az - b,$$

and

$$g(z) = 0 \Rightarrow bz^2 - 2az - b = 0 \Rightarrow z = z_{1,2} = \frac{a \pm (a^2 - b^2)^{1/2}}{b}.$$

Since

$$z_2 = \frac{a - (a^2 - b^2)^{1/2}}{b} = \frac{a - (a^2 - b^2)^{1/2}}{b} \frac{a + (a^2 - b^2)^{1/2}}{a + (a^2 - b^2)^{1/2}} = \frac{b}{a + (a^2 - b^2)^{1/2}},$$

$$|z_2| < 1 \text{ for } a > |b| \text{ while } |z_1| > 1 \text{ because } |z_1 z_2| = 1.$$

Hence

$$z_0 = z_2 \text{ and } g'(z_0) = 2(a - bz_2) = 2(a^2 - b^2)^{1/2}.$$

Finally

$$\frac{1}{2}\int_{-\pi}^{\pi}\frac{d\theta}{a-b\cos(\theta)}=\frac{1}{2i}\oint_{|z|=1}\frac{2\,dz}{-bz^2+2az-b}=\frac{2\pi i}{2i}\frac{h(z_2)}{g'(z_2)}$$

$$=\pi\frac{2}{2(a^2-b^2)^{1/2}}=\frac{\pi}{(a^2-b^2)^{1/2}},$$

which completes our demonstration. Observe that these two methods taken together illustrate the fact it is often convenient to evaluate real integrals by employing complex variables techniques.

In order to deduce an integral representation for the Legendre polynomials we next consider

$$\frac{1}{\pi}\int_0^{\pi}\frac{d\theta}{1-h[\mu+(\mu^2-1)^{1/2}\cos(\theta)]}\text{ for }\mu>1.$$

Observe that this integral is of form of the one just evaluated

$$\frac{1}{\pi}\int_0^{\pi}\frac{d\theta}{a-b\cos(\theta)}=(a^2-b^2)^{-1/2}$$

upon making the identifications

$$a=1-h\mu\text{ and }b=h(\mu^2-1)^{1/2},$$

which in order to guarantee that

$$a>|b|\text{ or }a^2-b^2=(1-h\mu)^2-h^2(\mu^2-1)=1-2\mu h+h^2>0,$$

we require $0<h<\mu-(\mu^2-1)^{1/2}=1/[\mu+(\mu^2-1)^{1/2}]<1/\mu$. Thus

$$\frac{1}{\pi}\int_0^{\pi}\frac{d\theta}{1-h[\mu+(\mu^2-1)^{1/2}\cos(\theta)]}=(1-2\mu h+h^2)^{-1/2}=g(\mu,h)=\sum_{\nu=0}^{\infty}P_\nu(\mu)h^\nu.$$

Since for the infinite geometric series

$$\sum_{\nu=0}^{\infty}[\mu+(\mu^2-1)^{1/2}\cos(\theta)]^\nu h^\nu=\frac{1}{1-h[\mu+(\mu^2-1)^{1/2}\cos(\theta)]}$$

with ratio

$$0<h[\mu+(\mu^2-1)^{1/2}\cos(\theta)]<[\mu+(\mu^2-1)^{1/2}\cos(\theta)]/[\mu+(\mu^2-1)^{1/2}]\leq 1,$$

$$\frac{1}{\pi}\int_0^\pi \frac{d\theta}{1-h[\mu+(\mu^2-1)^{1/2}\cos(\theta)]}$$
$$=\sum_{\nu=0}^{\infty}\left\{\frac{1}{\pi}\int_0^\pi [\mu+(\mu^2-1)^{1/2}\cos(\theta)]^\nu\,d\theta\right\}h^\nu$$

where the requisite uniform convergence has been assumed. Then we can conclude that

$$P_\nu(\mu)=\frac{1}{\pi}\int_0^\pi [\mu+(\mu^2-1)^{1/2}\cos(\theta)]^\nu\,d\theta$$

yielding the desired integral representation. Although this derivation was for $\mu > 1$, observe that the result can be extrapolated to $|\mu| < 1$ by means of analytic continuation.

(iii) Orthogonality of Legendre polynomials

To deduce the orthogonality of $P_n(x)$, we rewrite its Legendre differential equation

$$L[P_n]=(1-x^2)P_n''-2xP_n'+n(n+1)P_n=0 \text{ for } |x|\le 1 \text{ where } n=0,1,2,\ldots;$$

as an eigenvalue problem of the following *Sturm–Liouville* canonical form:

$$\mathcal{L}[y_n]=\frac{-1}{r(x)}[p(x)y_n']'+q(x)y_n=\lambda_n y_n \text{ for } a<x<b \text{ where } n=0,1,2,\ldots;$$

by making the identifications

$$y_n(x)=P_n(x),\ r(x)\equiv 1,\ p(x)=1-x^2,\ q(x)\equiv 0,\ \lambda_n=n(n+1),\ a=-1, \text{ and } b=1.$$

We now show that, if $\mathcal{L}[y_n]=\lambda_n y_n$ and $\mathcal{L}[y_m]=\lambda_m y_m$ where $\lambda_n\neq\lambda_m$ for $n\neq m$, then

$$\int_a^b r(x)y_n(x)y_m(x)\,dx=0 \text{ for } n\neq m \text{ provided } [p(x)(y_ny_m'-y_my_n')]\big|_a^b=0.$$

Consider $r(x)\{\mathcal{L}[y_n]y_m-\mathcal{L}[y_m]y_n\}=(\lambda_n-\lambda_m)r(x)y_n(x)y_m(x)$ and examine

$$\begin{aligned}r(x)\{\mathcal{L}[y_n]y_m-\mathcal{L}[y_m]y_n\}&=[p(x)y_m']'y_n-[p(x)y_n']'y_m\\&=p(x)(y_m''y_n-y_n''y_m)+p'(x)(y_m'y_n-y_n'y_m)\\&=[p(x)(y_m'y_n-y_n'y_m)]'\\&=(\lambda_n-\lambda_m)r(x)y_n(x)y_m(x)\end{aligned}$$

since $(y_m'y_n-y_n'y_m)'=y_m''y_n+y_m'y_n'-y_n'y_m'-y_n''y_m=y_m''y_n-y_n''y_m$. Thus, integrating

$$(\lambda_n - \lambda_m)\int_a^b r(x)y_n(x)y_m(x)\,dx = [p(x)(y_n y_m' - y_m y_n')]\Big|_a^b = 0$$

$$\Rightarrow \int_a^b r(x)y_n(x)y_m(x)\,dx = 0 \text{ given that } \lambda_n \neq \lambda_m \text{ for } n \neq m.$$

Hence, observing for our problem that $p(\pm 1) = 0$ while $r(x) \equiv 1$, we can conclude

$$\int_{-1}^{1} P_n(x)P_m(x)\,dx = 0 \text{ for } n(n+1) \neq m(m+1) \text{ or } n \neq m.$$

It only remains to show that

$$\int_{-1}^{1} P_n^2(x)\,dx = \frac{2}{2n+1} \text{ for } n = 0, 1, 2, \ldots;$$

which we shall accomplish by the following indirect technique. Again consider

$$(1 - 2xz + z^2)^{-1/2} = \sum_{n=0}^{\infty} P_n(x)z^n.$$

Then

$$\text{①} = \int_{-1}^{1} \frac{dx}{1 - 2xz + z^2} = \int_{-1}^{1} (1 - 2zx + z^2)^{-1/2}(1 - 2xz + z^2)^{-1/2}\,dx$$
$$= \int_{-1}^{1} \left[\sum_{n=0}^{\infty} P_n(x)z^n\right]\left[\sum_{n=0}^{\infty} P_n(x)z^n\right] dx = \text{②}.$$

Now

$$\text{①} = -\frac{1}{2z}\ln(|1 - 2xz + z^2|)\Big|_{x=-1}^{1} = -\frac{1}{2z}[\ln(|1 - 2z + z^2|) - \ln(|1 + 2z + z^2|)]$$
$$= \frac{1}{2z}[\ln(1+z)^2 - \ln(1-z)^2] = \frac{1}{z}[\ln(1+z) - \ln(1-z)] \text{ for } |z| < 1.$$

Recall

$$\frac{1}{1+z} = \sum_{n=0}^{\infty} (-1)^n z^n \text{ and } \frac{1}{1-z} = \sum_{n=0}^{\infty} z^n \text{ for } |z| < 1$$

$$\Rightarrow \ln(z+1) = \sum_{n=0}^{\infty} \frac{(-1)^n}{n+1} z^{n+1} \text{ and } -\ln(1-z) = \sum_{n=0}^{\infty} \frac{1}{n+1} z^{n+1}.$$

Thus

$$\text{①} = \sum_{n=0}^{\infty} \frac{(-1)^n + 1}{n+1} z^n = \sum_{m=0}^{\infty} \frac{2}{2m+1} z^{2m} \text{ since } (-1)^n + 1 = \begin{cases} 2, & \text{for } n = 2m, \\ 0, & \text{for } n = 2m+1 \end{cases};$$

while

$$\textcircled{2} = \int_{-1}^{1} \sum_{n=0}^{\infty} \left[\sum_{k=0}^{n} P_{n-k}(x) z^{n-k} \cdot P_k(x) z^k \right] dx = \int_{-1}^{1} \sum_{n=0}^{\infty} \left[\sum_{k=0}^{n} P_{n-k}(x) P_k(x) \right] z^n \, dx$$

since by the *Cauchy Product*

$$\left[\sum_{n=0}^{\infty} a_n \right] \left[\sum_{n=0}^{\infty} b_n \right] = \sum_{n=0}^{\infty} c_n \text{ where } c_n = \sum_{k=0}^{n} a_{n-k} b_k$$

Thus, assuming the requisite uniform convergence,

$$\textcircled{2} = \sum_{n=0}^{\infty} \left[\sum_{k=0}^{n} \int_{-1}^{1} P_{n-k}(x) P_k(x) \, dx \right] z^n = \sum_{m=0}^{\infty} \left[\int_{-1}^{1} P_m^2(x) \, dx \right] z^{2m}$$

since, by the orthogonality of the Legendre polynomials,

$$\int_{-1}^{1} P_{n-k}(x) P_k(x) \, dx = \int_{-1}^{1} P_m^2(x) \, dx \text{ is only nonzero for } n - k = k = m \Rightarrow n = 2m.$$

Finally, given the equality $\textcircled{1} = \textcircled{2}$, we conclude that

$$\int_{-1}^{1} P_m^2(x) \, dx = \frac{2}{2m+1} \text{ for } m = 0, 1, 2, \ldots.$$

Therefore, we have demonstrated the orthonormality-type property,

$$\int_{-1}^{1} P_n(x) P_m(x) \, dx = \frac{2}{2n+1} \delta_{mn}.$$

Further, given any f and f' that are piecewise continuous for $|x| \le 1$, we can represent

$$f(x) = \sum_{n=0}^{\infty} \alpha_n P_n(x) \text{ where } \alpha_n = \left(n + \frac{1}{2} \right) \int_{-1}^{1} f(x) P_n(x) \, dx,$$

which implies that the Legendre polynomials may be considered a complete set of functions in this sense. The plausibility of that assertion may be demonstrated as follows:

Let

$$f(x) = \sum_{n=0}^{\infty} \alpha_n P_n(x).$$

Then, for any fixed $m = 0, 1, 2, \ldots,$

$$\int_{-1}^{1} f(x)P_m(x)\,dx = \int_{-1}^{1}\left[\sum_{n=0}^{\infty}\alpha_n P_n(x)\right]P_m(x)\,dx$$
$$= \sum_{n=0}^{\infty}\alpha_n\int_{-1}^{1}P_n(x)P_m(x)\,dx = \sum_{n=0}^{\infty}\alpha_n\frac{2}{2n+1}\delta_{mn} = \frac{2\alpha_m}{2m+1}$$

$$\Rightarrow \alpha_m = \left(m+\frac{1}{2}\right)\int_{-1}^{1} f(x)P_m(x)\,dx.$$

For the special case of $f(x) \equiv 0$, this implies that

$$\sum_{n=0}^{\infty}\alpha_n P_n(x) \equiv 0 \Rightarrow \alpha_n = 0 \text{ or } \sum_{n=0}^{\infty} a_n P_n(x) = \sum_{n=0}^{\infty} b_n P_n(x) \Rightarrow a_n = b_n$$

which is the form usually employed when satisfying boundary conditions for problems exhibiting spherical symmetry such as Problem 12.3.
Lastly, consider *Parseval's Identity*

$$\int_{-1}^{1} f^2(x)\,dx = 2\sum_{n=0}^{n}\frac{\alpha_n^2}{2n+1},$$

the plausibility of which may again be demonstrated as follows:

$$\int_{-1}^{1} f^2(x)\,dx = \int_{-1}^{1} f(x)\left[\sum_{n=0}^{\infty}\alpha_n P_n(x)\right]dx$$
$$= \sum_{n=0}^{\infty}\alpha_n\int_{-1}^{1} f(x)P_n(x)\,dx = 2\sum_{n=0}^{\infty}\frac{\alpha_n^2}{2n+1}.$$

There also exist other sets of orthogonal polynomials which have generating functions, and it is the purpose of Problem 12.2 to demonstrate that fact for the so-called Hermite Polynomials by using some of the methods employed in this section.

Problems

12.1. If $(A)_{ij} = a_{ij}$,

$$\det(A) \equiv \begin{vmatrix} a_{11} & a_{12} & a_{13} \\ a_{21} & a_{22} & a_{23} \\ a_{31} & a_{32} & a_{33} \end{vmatrix} = \epsilon_{ijk}a_{1i}a_{2j}a_{3k} = \epsilon_{ijk}a_{i1}a_{2j}a_{3k}.$$

Then from the behavior of the determinant of a matrix upon row or column interchange and the definition of the alternator, it follows that

$$\begin{vmatrix} a_{ip} & a_{iq} & a_{ir} \\ a_{jp} & a_{jq} & a_{jr} \\ a_{kp} & a_{kq} & a_{kr} \end{vmatrix} = \epsilon_{ijk}\epsilon_{pqr}\det(A).$$

(a) Now taking $A = I$ where $(I)_{ij} = \delta_{ij}$ and setting $r = k$, which implicitly sums on this index, deduce that

$$\epsilon_{ijk}\epsilon_{pqk} = \delta_{ip}\delta_{jq} - \delta_{iq}\delta_{jp}.$$

Hint: Note that $\delta_{kk} = 3$ and $\det(I) = 1$.

(b) Observing that

$$(\boldsymbol{a}\times\boldsymbol{b})_i \equiv \epsilon_{ijk}a_jb_k \text{ and } [\text{curl}(\boldsymbol{v})]_i \equiv \epsilon_{ijk}\partial_j v_k \text{ for } i = 1,2,3;$$

use the result of (a) to derive (12.1.2a), (12.1.5b), and

$$\text{curl}(\text{curl}\boldsymbol{v}) = \nabla(\nabla\bullet\boldsymbol{v}) - \nabla^2\boldsymbol{v}.$$

(c) Observing that

$$\boldsymbol{a}\bullet\boldsymbol{b} \equiv a_ib_i = |\boldsymbol{a}||\boldsymbol{b}|\cos(\theta);$$

use the definitions of the cross product and the determinant as well as the result of (a) to derive

$$\boldsymbol{a}\times(\boldsymbol{b}\times\boldsymbol{c}) = (\boldsymbol{a}\bullet\boldsymbol{c})\boldsymbol{b} - (\boldsymbol{a}\bullet\boldsymbol{b})\boldsymbol{c} \text{ and } \boldsymbol{a}\bullet(\boldsymbol{b}\times\boldsymbol{c}) = \begin{vmatrix} a_1 & a_2 & a_3 \\ b_1 & b_2 & b_3 \\ c_1 & c_2 & c_3 \end{vmatrix}.$$

(d) Use the second result of (c) to deduce that

$$\boldsymbol{a}\bullet(\boldsymbol{a}\times\boldsymbol{b}) = \boldsymbol{b}\bullet(\boldsymbol{a}\times\boldsymbol{b}) = \boldsymbol{0}.$$

(e) Use the definitions of the cross and dot products as well as the result of (a) to deduce that

$$|\boldsymbol{a}\times\boldsymbol{b}|^2 = |\boldsymbol{a}|^2|\boldsymbol{b}|^2 - (\boldsymbol{a}\bullet\boldsymbol{b})^2 = |\boldsymbol{a}|^2|\boldsymbol{b}|^2[1-\cos^2(\theta)] = |\boldsymbol{a}|^2|\boldsymbol{b}|^2\sin^2(\theta),$$

where $|\boldsymbol{a}|^2 \equiv \boldsymbol{a}\bullet\boldsymbol{a}$ and $0 \le \theta \le \pi$, and hence conclude that

$$|\boldsymbol{a}\times\boldsymbol{b}| = |\boldsymbol{a}||\boldsymbol{b}|\sin(\theta).$$

12.2. Another orthogonal polynomial of note $H_n(x)$, the Hermite polynomial of integer order $n \ge 0$, arises in conjunction with the solution to the harmonic oscillator problem of quantum mechanics; satisfies the ordinary differential equation

$$H_n''(x) - 2xH_n'(x) + 2nH_n(x) = 0, -\infty < x < \infty;$$

and has series representation

$$H_n(x) = n! \sum_{m=0}^{p} \frac{(-1)^m (2x)^{n-2m}}{m!(n-2m)!} \text{ where } p = \begin{cases} n/2, & \text{if } n \text{ is even} \\ (n-1)/2, & \text{if } n \text{ is odd} \end{cases}.$$

It is the purpose of this problem to show that

$$\int_{-\infty}^{\infty} e^{-x^2} H_n(x) H_m(x)\, dx = 2^n n! \sqrt{\pi} \delta_{mn}$$

by identifying a generating function for these polynomials satisfying

$$g(x,z) = e^{2zx-z^2} = \sum_{n=0}^{\infty} \frac{1}{n!} H_n(x) z^n.$$

(a) Using the same technique employed to determine an integrating factor for a first-order linear ordinary differential equation, *i.e.*, given

$$y'(x) + P(x)y(x) = Q(x) \Rightarrow \mu(x) \equiv \text{ integrating factor } = e^{\int P(x)\, dx}, -$$

multiply the Hermite differential equation by $e^{-\int 2x\, dx} = e^{-x^2}$, and hence deduce its Sturm–Liouville form

$$\mathcal{L}[H_n] = -\frac{1}{e^{-x^2}}[e^{-x^2} H_n'(x)]' = 2nH_n(x), -\infty < x < \infty.$$

Now making the canonical Sturm–Liouville identifications

$$y_n(x) = H_n(x), r(x) = p(x) = e^{-x^2}, q(x) \equiv 0, \lambda_n = 2n; a \to -\infty \text{ and } b \to \infty;$$

conclude that

$$\int_{-\infty}^{\infty} e^{-x^2} H_n(x) H_m(x)\, dx = 0 \text{ for } n \neq m$$

since $x^N e^{-x^2} \to 0$ as $x \to \pm\infty$ for N, a nonnegative integer.

(b) Next show that

$$e^{2zx-z^2} = \sum_{j=0}^{\infty} \sum_{m=0}^{j} \frac{(-1)^m (2x)^{j-m}}{m!(j-m)!} z^{j+m}$$

by both of the techniques sketched below:

(i) Use the infinite series definition for the exponential

$$e^w = \sum_{j=0}^{\infty} \frac{w^j}{j!},$$

write

$$2zx - z^2 = z(2x - z),$$

and then employ the binomial expansion on the second factor

$$(2x-z)^j = \sum_{m=0}^{j} \binom{j}{m} (2x)^{j-m}(-z)^m \text{ where } \binom{j}{m} = \frac{j!}{m!(j-m)!}.$$

(ii) Recall that

$$e^{2zx-z^2} = e^{2zx} \cdot e^{-z^2},$$

use the infinite series definition for each of these exponentials, and then employ the Cauchy Product

$$\left[\sum_{j=0}^{\infty} a_j\right]\left[\sum_{j=0}^{\infty} b_j\right] = \sum_{j=0}^{\infty}\left[\sum_{m=0}^{j} a_{j-m}b_m\right].$$

(c) Reorder this double sum employing the same technique used to identify the generating function for the Legendre polynomials by introducing

$$n = j + m \geq 0 \text{ or } j = n - m \Rightarrow j - m = n - 2m \geq 0 \Rightarrow m = 0, \ldots, p;$$

and hence produce the desired result

$$e^{2zx-z^2} = \sum_{n=0}^{\infty} \frac{1}{n!} H_n(x) z^n \text{ since } H_n(x) = n! \sum_{m=0}^{p} \frac{(-1)^m (2x)^{n-2m}}{m!(n-2m)!}.$$

(d) Next use this generating function to deduce the normalization constant

$$\int_{-\infty}^{\infty} e^{-x^2} H_n^2(x)\, dx = 2^n n! \sqrt{\pi},$$

by first demonstrating that

$$\textcircled{1} = \int_{-\infty}^{\infty} e^{4xz-2z^2-x^2}\, dx = \int_{-\infty}^{\infty} e^{-x^2} g^2(x,z)\, dx = \textcircled{2}$$

and assuming the requisite uniform convergence that

$$\textcircled{2} = \sum_{m=0}^{\infty}\sum_{n=0}^{m}\left[\frac{1}{(m-n)!n!}\int_{-\infty}^{\infty} e^{-x^2} H_{m-n}(x) H_n(x)\, dx\right] z^m.$$

Then, completing the square to obtain

$$-x^2 + 4xz = -(x^2 - 4xz) = -(x^2 - 4xz + 4z^2) + 4z^2 = -(x-2z)^2 + 4z^2$$

and making the change of variables $u = x - 2z$, show that

$$\textcircled{1} = e^{2z^2}\int_{-\infty}^{\infty} e^{-u^2}\,du = \sqrt{\pi}\sum_{n=0}^{\infty}\frac{2^n z^{2n}}{n!}$$

and employing the orthogonality property of the Hermite polynomials, that

$$\textcircled{2} = \sum_{n=0}^{\infty}\frac{z^{2n}}{(n!)^2}\int_{-\infty}^{\infty} e^{-x^2}H_n^2(x)\,dx.$$

Finally, comparing these two series and equating their z^{2n}-coefficients, deduce that

$$\int_{-\infty}^{\infty} e^{-x^2}H_n^2(x)\,dx = 2^n n!\sqrt{\pi}$$

and hence conclude in conjunction with part (a) the desired result

$$\int_{-\infty}^{\infty} e^{-x^2}H_n(x)H_m(x)\,dx = 2^n n!\sqrt{\pi}\delta_{mn}.$$

12.3. Consider the potential flow ($\boldsymbol{v} = \nabla\varphi = \boldsymbol{V} = \nabla\Phi$) of an inviscid homogeneous fluid of constant density R_0 past a spherical body of radius $\rho = a$. Orient the axes so that far from the sphere as $\rho \to \infty$ there is uniform flow of constant velocity $-U_0\boldsymbol{k}$ and constant pressure P_∞. Use the spherical coordinate system depicted in Fig. 12.3 such that

$$x = r\cos(\theta) \text{ and } y = r\sin(\theta) \text{ with } r = \rho\sin(\phi),\ z = \rho\cos(\phi).$$

(a) Employing the *Calculus of Variations* technique of transformation of coordinates deduce that the Laplacian in spherical coordinates is given by

$$\nabla^2\Phi = \frac{1}{\rho^2}\left[\frac{\partial}{\partial\rho}(\rho^2\Phi_\rho) + \frac{1}{\sin(\phi)}\frac{\partial}{\partial\phi}\{\sin(\phi)\Phi_\phi\} + \frac{1}{\sin^2(\phi)}\Phi_{\theta\theta}\right].$$

Hint:

$$f = \varphi_x^2 = \varphi_y^2 + \varphi_z^2 = \Phi_\rho^2 + \frac{1}{\rho^2}\Phi_\phi^2 + \frac{1}{\rho^2\sin^2(\phi)}\Phi_\theta^2 = F,\ J = \begin{vmatrix} x_\rho & x_\phi & x_\theta \\ y_\rho & y_\phi & y_\theta \\ z_\rho & z_\phi & z_\theta \end{vmatrix} = \rho^2\sin(\phi).$$

(b) Looking for an axisymmetric ($\partial/\partial\theta \equiv 0$) steady-state ($\partial/\partial t \equiv 0$) flow, show that the problem reduces to the following equation and set of boundary conditions for $\Phi = \Phi(\rho,\phi)$:

$$\sin(\phi)\frac{\partial}{\partial\rho}(\rho^2\Phi_\rho) + \frac{\partial}{\partial\phi}[\sin(\phi)\Phi_\phi] = 0,\ a < \rho < \infty,\ 0 \le \phi \le \pi;$$

$$\Phi_\rho(a,\phi) = 0,\ \Phi(\rho,\phi) \sim -U_0\rho\cos(\phi) \text{ as } \rho \to \infty.$$

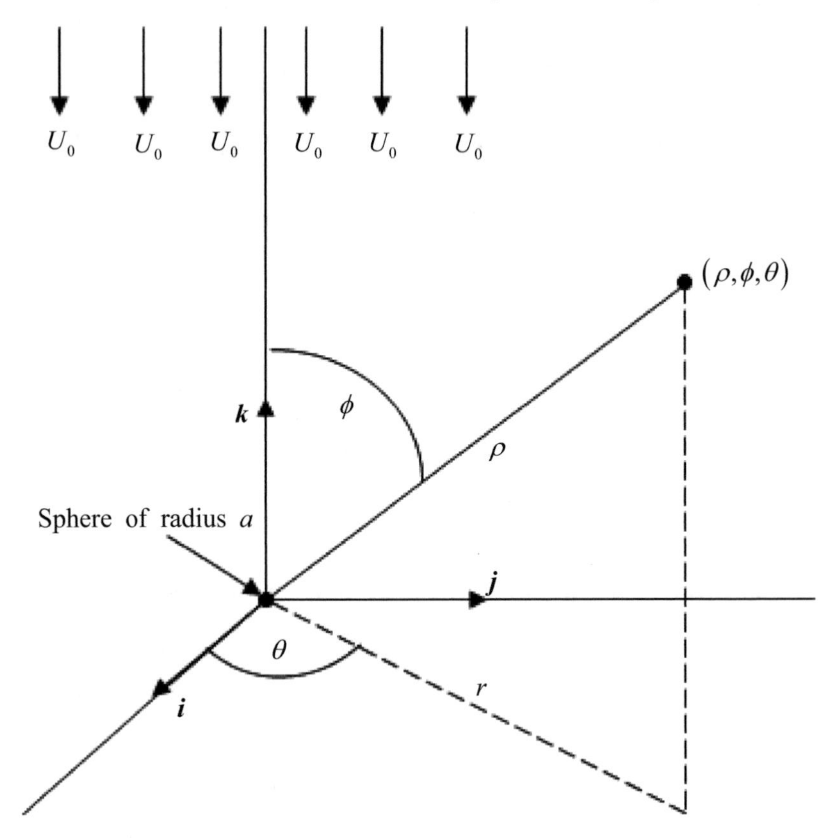

Fig. 12.3 A schematic diagram of the three-dimensional inviscid potential flow past a spherical body of radius a treated in Problem 12.3. Note that the spherical coordinate system employed in this problem is right-handed only for the (ρ, ϕ, θ)-ordering of its independent variables (see Fig. 9.11). Here $r = \rho \sin(\phi)$ represents the radial distance in the x-y plane.

(c) Deduce that the most general solution to this equation which results upon using a separation of variables approach by considering $\Phi(\rho, \phi) = G(\rho)H(\phi)$ is of the form

$$\Phi(\rho, \phi) = \sum_{n=0}^{\infty} (a_n \rho^n + b_n \rho^{-n-1}) P_n[\cos(\phi)].$$

Hint: This approach yields the following equation for $H = H(\phi)$:

$$-\frac{[\sin(\phi)H']'}{\sin(\phi)H} = \lambda, 0 \le \phi \le \pi; \text{ where } \lambda \equiv \text{ separation constant;}$$

which, upon introducing $s = \cos(\phi)$ and $u(s) = H(\phi)$, transforms into the *Legendre equation*

$$-[(1-s^2)u']' = \lambda u, \ -1 \le s \le 1;$$

and the solutions to that equation only remain bounded for $s = \pm 1$ ($\phi = 0$ or π) provided $\lambda = \lambda_n = n(n+1)$ for $n = 0, 1, 2, \ldots$. Hence $u(s) = P_n(s) \Rightarrow H(\phi) = P_n[\cos(\phi)]$ for $n \ge 0$. Then $G = G(\rho)$ satisfies an Euler equation and letting $G(\rho) = \rho^m \Rightarrow m = n$ or $m = -n-1$.

(d) Now determine the constant a_n's and b_n's contained in this general solution so that it satisfies the free-stream and no-penetration boundary conditions, respectively, and thus obtain

$$\Phi(\rho, \phi) = -U_0\left(\rho + \frac{a^3}{2\rho^2}\right)\cos(\phi).$$

Hint: Recall

$$P_0[\cos(\phi)] = 1, P_1[\cos(\phi)] = \cos(\phi),$$

and the $\{P_n[\cos(\phi)]\}_{n=0}^{\infty}$ are a complete set of functions.

(e) Finally, conclude that this flow exerts no net force $\boldsymbol{F}$ on the sphere by showing

$$\boldsymbol{F} = -\iint\limits_{\rho=a} P(a, \phi)\boldsymbol{n}\, d\sigma = \boldsymbol{0}$$

where

$$\iint\limits_{\rho=a} (-)\, d\sigma = \int_{\theta=0}^{2\pi}\int_{\phi=0}^{\pi} (-)a^2\sin(\phi)\, d\theta\, d\phi$$

and

$$\boldsymbol{n} = -\boldsymbol{e}_\rho = -\sin(\phi)[\cos(\theta)\boldsymbol{i} + \sin(\theta)\boldsymbol{j}] - \cos(\phi)\boldsymbol{k}$$

while from Bernoulli's relation for pressure in these coordinates

$$P(\rho, \phi) = P_\infty + \frac{R_0 U_0^2}{2} - \frac{R_0}{2}\left[\Phi_\rho^2(\rho, \phi) + \frac{1}{\rho^2}\Phi_\phi^2(\rho, \phi)\right]$$

we can determine that

$$P(a, \phi) = P_\infty + \frac{R_0 U_0^2}{2} - \frac{R_0}{2a^2}\Phi_\phi^2(a, \phi)$$

since by the no-penetration boundary condition

$$\Phi_\rho(a, \phi) = 0.$$

This result for the special case of a sphere is consistent with the fact that $\boldsymbol{F} = \boldsymbol{0}$ for inviscid potential flow past *any* three-dimensional body.

Chapter 13
Viscous Fluid Flows

Since including viscous effects in the fluid equations plays a fundamental role for the prototype problems of flow past bodies and natural convection to be treated in the next two chapters, respectively, this chapter considers viscosity, in some detail, as a prelude to those investigations. First, after discussing the behavior of the viscosity coefficients and restricting our attention to homogeneous fluids, the resulting Navier–Stokes governing equations of motion are presented in component form for both Cartesian and cylindrical coordinate systems. Then steady-state shear-driven plane Couette and pressure gradient-driven plane Poiseuille parallel flows, which satisfy the Cartesian coordinate form of the Navier–Stokes equations, are examined. Next, regular Couette flow in the annulus between two rotating concentric cylinders and regular Poiseuille flow driven by an axial pressure gradient in a pipe, which satisfy the cylindrical coordinate form of the Navier–Stokes equations, are examined. Finally, Couette flow under the small-gap approximation is analyzed by regular perturbation theory, and the results are compared to lowest order with plane Couette flow. These basic Couette and Poiseuille flows have been chosen for presentation since they are representative of the two major properties of viscosity, respectively: Namely, the adherence of viscous fluids to rigid boundaries and their internal resistance to motion. There are four problems: The first considering the superposition of plane Couette and Poiseuille flows, the second examining the consequences of performing a normal mode linear stability analysis of the regular Couette flow solution to the inviscid Navier–Stokes equations, the third adding an axial and a circumferential pressure gradient to viscous Couette flow, and the fourth performing a small-gap approximation on that flow employing regular perturbation theory.

13.1 Navier–Stokes Equations in Cartesian and Cylindrical Coordinates

Recall from (9.4.17b) that the components of the stress tensor for a Newtonian fluid take the form

© Springer International Publishing AG, part of Springer Nature 2017

D. J. Wollkind and B. J. Dichone, *Comprehensive Applied Mathematical Modeling in the Natural and Engineering Sciences*,
https://doi.org/10.1007/978-3-319-73518-4_13

$$T_{ij} = (-p + \lambda\varepsilon_{kk})\delta_{ij} + 2\mu\varepsilon_{ij} \text{ where } 2\varepsilon_{ij} = \partial_j v_i + \partial_i v_j;\ i,j = 1,2,3. \tag{13.1.1}$$

For an inviscid fluid $\lambda = \mu = 0$; hence $T_{ij} = -p\delta_{ij}$. Thus $T_{ii} = -3p$ while for $\lambda, \mu \neq 0$:

$$T_{ii} = 3(-p + \lambda\theta) + 2\mu\theta = -3p + (3\lambda + 2\mu)\theta \text{ where } \theta = \varepsilon_{ii} = \partial_i v_i = \nabla \bullet \boldsymbol{v}. \tag{13.1.2}$$

Should we still wish to require that $T_{ii} = -3p$ then (13.1.2) implies $\lambda + 2\mu/3 \equiv$ bulk viscosity $= 0$, which is called Stokes assumption. Under this assumption (9.4.20) becomes

$$\rho\frac{D\boldsymbol{v}}{Dt} = -\nabla p + \rho\boldsymbol{f} + \mu\left(\frac{\nabla\theta}{3} + \nabla^2\boldsymbol{v}\right). \tag{13.1.3}$$

Since each term in a given equation must be of equal dimension from (13.1.3) we can deduce that

$$\left[\rho\frac{D\boldsymbol{v}}{Dt}\right] = [\rho]\left[\frac{D\boldsymbol{v}}{Dt}\right] = [\mu\nabla^2\boldsymbol{v}] = [\mu][\nabla^2\boldsymbol{v}]. \tag{13.1.4a}$$

Now, denoting the dimension of length, time, and mass by L, τ, and M, respectively, then

$$[\rho] = \frac{M}{L^3},\ \left[\frac{D\boldsymbol{v}}{Dt}\right] = \frac{[\boldsymbol{v}]}{\tau} = \frac{(L/\tau)}{\tau} = \frac{L}{\tau^2},\ [\nabla^2\boldsymbol{v}] = \frac{[\boldsymbol{v}]}{L^2} = \frac{1}{L\tau}. \tag{13.1.4b}$$

Thus (13.1.4) implies $[\mu] = [\rho][D\boldsymbol{v}/Dt]/[\nabla^2\boldsymbol{v}] = (M/L^3)(L/\tau^2)(L\tau) = M/(L\tau)$. We now tabulate the bulk viscosity for some typical gases [109]:

Gas	$\lambda + 2\mu/3$ (gm/[cm sec])
Argon	0
Helium	0
Air	0.57
Hydrogen	31 2/3
Carbon Dioxide	999

Note $\lambda + 2\mu/3 \geq 0$ as assumed earlier and Stokes assumption is valid for the Noble gases.

For a homogeneous fluid acted on only by the gravitational body force $\boldsymbol{f} = -g\boldsymbol{k}$ we can deduce from (9.5.2) the Navier–Stokes equations

$$\nabla \bullet \boldsymbol{v} = 0,\ \frac{D\boldsymbol{v}}{Dt} = -\frac{\nabla p'}{\rho_0} - g\boldsymbol{k} + \nu\nabla^2\boldsymbol{v} \text{ where } \nu = \frac{\mu}{\rho_0}. \tag{13.1.5a}$$

Here $\boldsymbol{v} = (u, v, w) \equiv$ velocity and $p' \equiv$ pressure. As with our inviscid equations of (12.1.1), we now rewrite (13.1.5a) in the form

$$\nabla \bullet \boldsymbol{v} = 0, \frac{D\boldsymbol{v}}{Dt} = -\frac{\nabla p}{\rho_0} + \nu \nabla^2 \boldsymbol{v} \tag{13.1.5b}$$

where $p = p' + \rho_0 g z - p_A \equiv$ reduced pressure. These are the Navier–Stokes equations in Cartesian coordinates (x, y, z) which in component form are given by

$$\frac{\partial u}{\partial x} + \frac{\partial v}{\partial y} + \frac{\partial w}{\partial z} = 0; \tag{13.1.6a}$$

$$\frac{Du}{Dt} = -\left(\frac{1}{\rho_0}\right)\frac{\partial p}{\partial x} + \nu \nabla^2 u, \tag{13.1.6b}$$

$$\frac{Dv}{Dt} = -\left(\frac{1}{\rho_0}\right)\frac{\partial p}{\partial y} + \nu \nabla^2 v, \tag{13.1.6c}$$

$$\frac{Dw}{Dt} = -\left(\frac{1}{\rho_0}\right)\frac{\partial p}{\partial z} + \nu \nabla^2 w, \tag{13.1.6d}$$

where

$$\frac{D}{Dt} = u\frac{\partial}{\partial x} + v\frac{\partial}{\partial y} + w\frac{\partial}{\partial z}, \tag{13.1.6e}$$

$$\nabla^2 = \frac{\partial^2}{\partial x^2} + \frac{\partial^2}{\partial y^2} + \frac{\partial}{\partial z^2}. \tag{13.1.6f}$$

Upon the introduction of cylindrical coordinates (r, θ, z) with corresponding velocity components (v_r, v_θ, v_z) these equations transform into [52]

$$\left(\frac{1}{r}\right)\frac{\partial (r v_r)}{\partial r} + \left(\frac{1}{r}\right)\frac{\partial v_\theta}{\partial \theta} + \frac{\partial v_z}{\partial z} = 0; \tag{13.1.7a}$$

$$\frac{Dv_r}{Dt} - \frac{v_\theta^2}{r} = -\left(\frac{1}{\rho_0}\right)\frac{\partial p}{\partial r} + \nu\left[\nabla^2 v_r - \frac{v_r}{r^2} - \left(\frac{2}{r^2}\right)\frac{\partial v_\theta}{\partial \theta}\right], \tag{13.1.7b}$$

$$\frac{Dv_\theta}{Dt} + \frac{v_\theta v_r}{r} = -\left(\frac{1}{\rho_0 r}\right)\frac{\partial p}{\partial r} + \nu\left[\nabla^2 v_r - \frac{v_\theta}{r^2} + \left(\frac{2}{r^2}\right)\frac{\partial v_r}{\partial \theta}\right], \tag{13.1.7c}$$

$$\frac{Dv_z}{Dt} = -\left(\frac{1}{\rho_0}\right)\frac{\partial p}{\partial z} + \nu \nabla^2 v_z, \tag{13.1.7d}$$

where

$$\frac{D}{Dt} = \frac{\partial}{\partial t} + v_r\frac{\partial}{\partial r} + \left(\frac{v_\theta}{r}\right)\frac{\partial}{\partial \theta} + v_z\frac{\partial}{\partial z}, \tag{13.1.7e}$$

$$\nabla^2 = \frac{\partial^2}{\partial r^2} + \left(\frac{1}{r}\right)\frac{\partial}{\partial r} + \left(\frac{1}{r^2}\right)\frac{\partial^2}{\partial \theta^2} + \frac{\partial^2}{\partial z^2}. \tag{13.1.7f}$$

Let us determine the dimension of the kinematic viscosity ν:

$$[\nu] = \left[\frac{\mu}{\rho_0}\right] = \frac{[\mu]}{[\rho_0]} = \frac{M}{L\tau}\frac{L^3}{M} = \frac{L^2}{\tau},$$

which as its dimension suggests represents diffusion of momentum. We now tabulate the behavior of this kinematic viscosity with temperature for water and air [109].

	ν (cm^2/sec)	
Temperature (°C)	water	air
20	1.004×10^{-2}	0.150
60	0.475×10^{-2}	0.188
80	0.366×10^{-2}	0.209
100	0.295×10^{-2}	0.230
200	———	0.346

Observe that ν decreases with increasing temperature for water (a liquid) but increases for air (a gas). Hence representing

$$\nu = \nu(T) = \nu_0 + \nu_1(T - T_0)$$

for a reference temperature T_0 where

$$\nu_0 = \nu(T_0) \text{ and } \nu_1 = \nu'(T_0),$$

we note that

$$\nu_1 < 0 \text{ for most liquids}$$

while

$$\nu_1 > 0 \text{ for most gases.}$$

Liquid sulfur is the exception to this rule in that

$$\nu_1 < 0 \text{ for } 120°\text{C} < T_0 < 159°\text{C}$$

while

$$\nu_1 > 0 \text{ for } 159°\text{C} < T_0 < 180°\text{C}.$$

Here 120°C represents the melting point of sulfur and 159°C, the transition point between its crystalline and amorphous states. This difference in behavior of kinematic viscosity between liquids and gases with temperature is often associated with the fact that during natural convection the flow is typically upward in the center of cells for liquids but downward for gases.

13.2 Plane Couette and Poiseuille Flows

In this section we shall examine two canonical exact solutions to the Cartesian coordinate Navier–Stokes equations (13.1.6) which satisfy relevant boundary conditions: Namely, plane Couette and Poiseuille flows, respectively.

(i) Plane Couette Flow

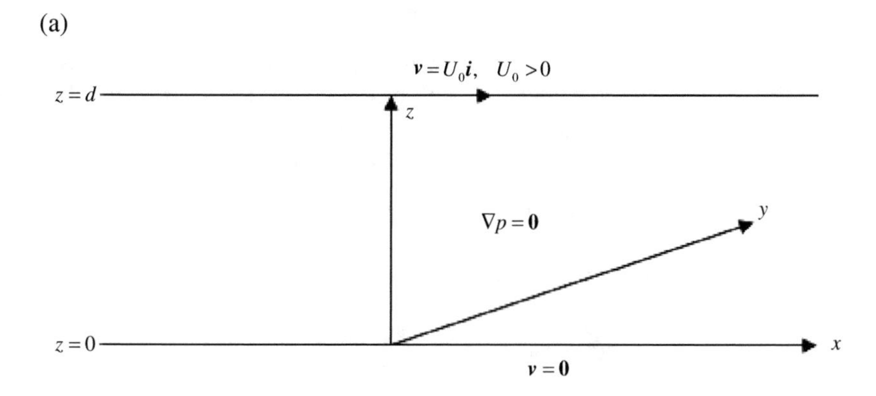

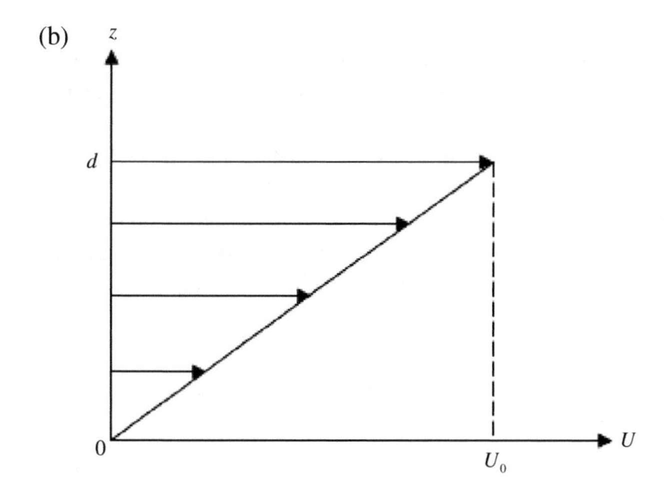

Fig. 13.1 (a) Schematic representation of plane Couette flow and (b) a plot of its velocity profile (13.2.5).

Consider a viscous homogeneous fluid confined between two horizontal plates of infinite planar extent located at $z = 0$ and d, respectively. The lower plate is at rest, while the upper one is moving at a speed U_0 in the positive x-direction (see Fig. 13.1a). Thus the relevant no-penetration, no-slip, and far-field boundary conditions for this situation are given by

$$w = 0 \text{ at } z = 0, d; \tag{13.2.1a}$$

$$u = v = w = 0 \text{ at } z = 0 \text{ and } u = U_0,\ v = 0 \text{ at } z = d; \tag{13.2.1b}$$

$$u, v, w, p \text{ remain bounded as } x^2 + y^2 \to \infty; \tag{13.2.1c}$$

respectively. Note that, since we are now including viscosity, it is possible to impose the no-slip boundary conditions. We shall assume that the flow is steady-state ($\partial/\partial t \equiv 0$), parallel ($v = w \equiv 0$), and stratified in z. Hence we seek a solution of (13.1.6) and boundary conditions (13.2.1) of the form

$$u = U(z),\ v = w \equiv 0,\ p = p(x, y, z). \tag{13.2.2}$$

Then (13.1.6a) is satisfied identically, while (13.1.6b, c, d) yield

$$\frac{\partial p}{\partial x} = \rho_0 \nu U'',\ \frac{\partial p}{\partial y} = \frac{\partial p}{\partial z} = 0 \Rightarrow p = \mathcal{P}(x)$$

$$\Rightarrow U''(z) = \left(\frac{1}{\mu}\right)\mathcal{P}'(x) \text{ since } \rho_0 \nu = \mu. \tag{13.2.3}$$

Given that two equal functions of different variables must be a function of their common variable(s), we can conclude $\mathcal{P}'(x) \equiv -C_0$, where C_0 is constant. Wishing to isolate the effect of the shear caused by the moving upper plate from that of an imposed pressure gradient, we next assume that $C_0 = 0 \Rightarrow \nabla p \equiv \mathbf{0}$. Observe that then the no-penetration and far-field boundary conditions are satisfied identically by (13.2.2) as are the no-slip boundary conditions for v. Therefore (13.2.3) and the no-slip boundary conditions of (13.2.1b) involving u reduce to the following second-order ordinary differential equation for $U(z)$:

$$U''(z) = 0,\ 0 < z < d;\ U(0) = 0,\ U(d) = U_0. \tag{13.2.4}$$

Solving (13.2.4) for $U(z)$ we find the linear velocity profile plotted in Fig. 13.1b

$$U(z) = \frac{U_0 z}{d}. \tag{13.2.5}$$

(ii) Plane Poiseuille Flow

Consider the steady-state parallel stratified flow of a viscous homogenous fluid confined between two horizontal motionless plates of infinite planar extent located at $z = \pm d$ and forced by a pressure gradient of strength C_0 in the negative x-direction (see Fig. 13.2a). The relevant no-slip, no-penetration, and far-field boundary conditions for this situation are

$$u = v = w = 0 \text{ at } z = \pm d;\ u, v, w, p \text{ remain bounded as } x^2 + y^2 \to \infty. \tag{13.2.6}$$

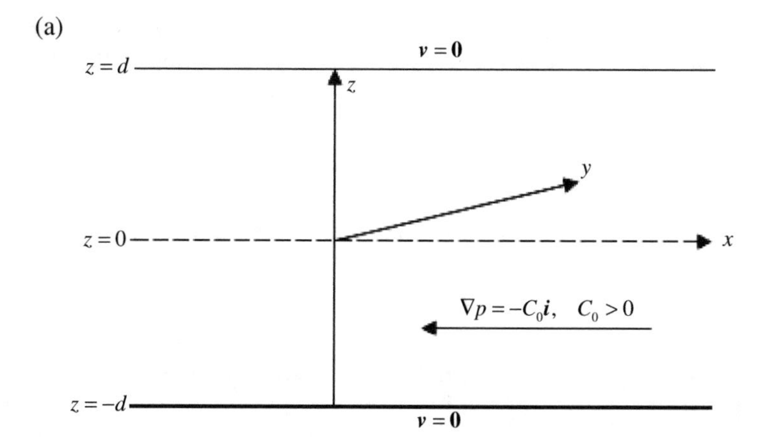

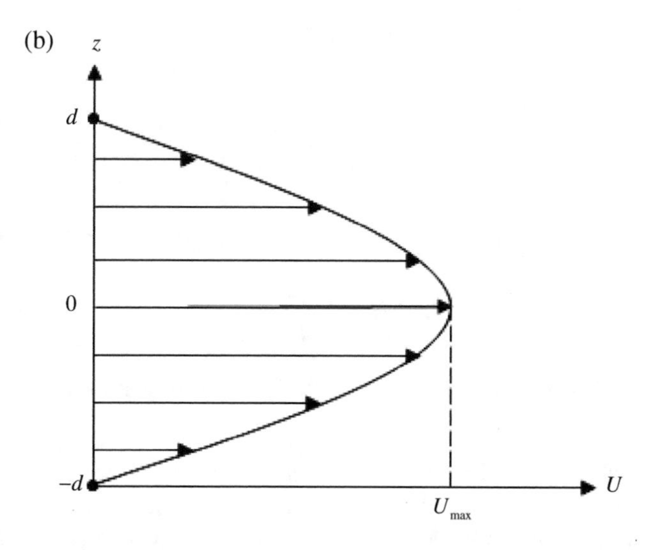

Fig. 13.2 (a) Schematic representation of plane Poiseuille flow and (b) a plot of its velocity profile (13.2.9e) where $U_{\max} = C_0 d^2/(2\mu)$.

Hence we seek a solution of (13.1.6) and boundary conditions (13.2.6) of the form

$$u = U(z), v = w \equiv 0, \nabla p = -C_0 \boldsymbol{i}. \tag{13.2.7}$$

Then (13.1.1a, b, c) and all the boundary conditions for v and w as well as the far-field conditions for u and p are satisfied identically, while (13.1.6b) and the no-slip conditions involving u yield

$$U''(z) = -\frac{C_0}{u}, \ -d < z < d; \ U(\pm d) = 0. \tag{13.2.8}$$

Integrating the differential equation of (13.2.8) twice, we obtain

$$U(z) = -\frac{C_0 z^2}{2\mu} + Az + B \tag{13.2.9a}$$

which upon application of the boundary conditions yields the two equations for A and B

$$Ad + B = \frac{C_0 d^2}{2\mu} \tag{13.2.9b}$$

$$-Ad + B = \frac{C_0 d^2}{2\mu}. \tag{13.2.9c}$$

Solving these simultaneously we find

$$A = 0, \ B = \frac{C_0 d^2}{2\mu}; \tag{13.2.9d}$$

and hence deduce the parabolic velocity profile for $U(z)$ plotted in Fig. 13.2b

$$U(z) = \frac{C_0(d^2 - z^2)}{2\mu} \tag{13.2.9e}$$

which has its maximum value $U_{\max} = C_0 d^2/(2\mu)$ occurring at the centerline of $z = 0$.

Having obtained the plane Couette and Poiseuille solutions (13.2.5) and (13.2.9), respectively, we wish to examine the effect that viscosity has on these flows in conjunction with U_0 for the former and C_0 for the latter. Of particular interest in this regard are (a) the shear stress $\boldsymbol{t}$ on the upper plate at $z = d$ for plane Couette flow and (b) the mass of fluid M that crosses the fixed plane $x = x_0$ per unit time Δt per unit length L in the y-direction for plane Poiseuille flow. We shall close this section by treating each of those effects sequentially in what follows:

(a) We wish to compute the stress vector on the upper plate during plane Couette flow. Note that this vector then has components

$$t_j = T_{ij}\big|_{z=d} n_i \text{ for } j = x, y, z \tag{13.2.10a}$$

where

$$T_{ij} = \mu(\partial_j v_i + \partial_i v_j) \text{ with } \boldsymbol{n} = \boldsymbol{k} \text{ or } n_x = n_y = 0, n_z = 1. \tag{13.2.10b}$$

Thus

$$t_j = T_{zj}\big|_{z=d} \text{ for } j = x, y, z. \tag{13.2.10c}$$

Since, $u = U(z) = U_0 z/d$ while $v = w \equiv 0$,

$$T_{zx} = \mu\left(\frac{\partial u}{\partial z} + \frac{\partial w}{\partial x}\right) = \mu\frac{U_0}{d}, T_{zy} = T_{zz} = 0. \tag{13.2.10d}$$

Hence, we can conclude that

$$\boldsymbol{t} = \mu\frac{U_0}{d}\boldsymbol{i}. \tag{13.2.10e}$$

Note that this stress is *directly* proportional to the product of the *shear viscosity* μ times the speed U_0 of the upper plate and inversely proportional to the channel depth. This is consistent with viscosity tending to cause the moving plate to drag the fluid adhering to it.

(b) We wish to compute the mass of fluid M that crosses the fixed plane $x = x_0$ per unit time Δt per unit length L in the y-direction during plane Poiseuille flow or

$$\begin{aligned}
\frac{M}{L\Delta t} &= \int_{-d}^{d} \rho_0 U(z)\,dz = \frac{C_0\rho_0}{2\mu}\int_{-d}^{d}(d^2 - z^2)\,dz \\
&= \frac{C_0}{\nu}\int_0^d (d^2 - z^2)\,dz = \frac{C_0}{\nu}\left(d^2 z - \frac{z^3}{3}\right)\Bigg|_{z=0}^{d} \\
&= \frac{C_0}{\nu}\left(d^3 - \frac{d^3}{3}\right) = \frac{2C_0 d^3}{3\nu}.
\end{aligned}$$

Note that this quantity is directly proportional to the imposed pressure gradient C_0 times d^3 and *inversely* proportional to the *kinematic viscosity* ν. Thus, the greater the kinematic viscosity, the less mass that can be forced through the channel by the imposed pressure gradient. This is consistent with viscous fluids tending to provide an internal resistance to motion caused by the effect of their kinematic viscosity. Hence these two flows highlight both the principal effects viscosity can have on a fluid, respectively.

13.3 Couette and Poiseuille Flows

In this section we shall examine two canonical exact solutions to the cylindrical coordinate Navier–Stokes equations (13.1.7) which satisfy relevant boundary conditions: Namely, regular Couette and Poiseuille flows, respectively.

(i) Couette Flow

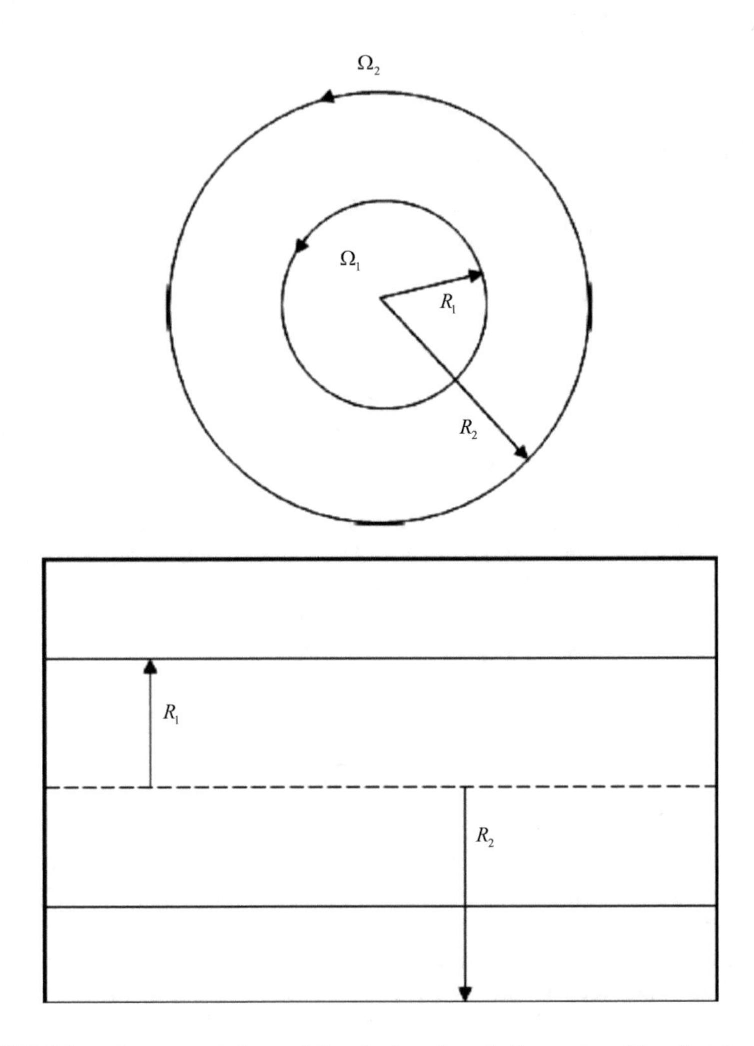

Fig. 13.3 Schematic representations of Couette flow from both an upper (above) and a lateral (below) cross-sectional perspective.

Consider a viscous homogeneous fluid confined in the annulus between two infinitely long concentric circular cylinders of radii R_1 and R_2, which are rotating at speeds Ω_1 and Ω_2 where $R_2 > R_1$ and $\Omega_1 > 0$ as depicted in Fig. 13.3. The relevant no-penetration, no-slip, and far-field boundary conditions for this situation are given by

$$v_r = 0 \text{ at } r = R_{1,2}; \tag{13.3.1a}$$

$$v_\theta = \Omega_1 R_1 \text{ at } r = R_1,\ v_\theta = \Omega_2 R_2 \text{ at } r = R_2, \text{ and } v_z = 0 \text{ at } r = R_{1,2}; \tag{13.3.1b}$$

$$v_r, v_\theta, v_z, p \text{ remain bounded as } z^2 \to \infty; \tag{13.3.1c}$$

respectively. We shall assume that the flow is steady-state ($\partial/\partial t \equiv 0$), axisymmetric ($\partial/\partial\theta = 0$), circumferential ($v_r = v_z \equiv 0$), and independent of the coaxial coordinate ($\partial/\partial z \equiv 0$). Hence we seek a solution of (13.1.7) and boundary conditions (13.3.1) of the form

$$v_\theta = V(r),\ v_r = v_z \equiv 0, p = p(r,\theta,z). \tag{13.3.2}$$

Then (13.1.6a) is satisfied identically, while (13.1.7b, c, d) imply

$$\frac{\partial p}{\partial r} = \frac{\rho_0 V^2}{r},\ \frac{\partial p}{\partial \theta} = \mu\left(rV'' + V' - \frac{V}{r}\right),\ \frac{\partial p}{\partial z} = 0$$

$$\Rightarrow p = \mathcal{P}(r,\theta) = \rho_0 \int \frac{V^2(r)}{r}\,dr + f(\theta),\ \mu\left[rV''(r) + V'(r) - \frac{V(r)}{r}\right] = f'(\theta). \tag{13.3.3}$$

Further, we shall assume there is no stirring and require that $f'(\theta) \equiv 0$ which then implies $f(\theta) \equiv$ a constant, that without loss of generality can be taken to be equal to zero. Thus (13.3.3) becomes

$$p = \mathcal{P}(r) = \rho_0 \int \frac{V^2(r)}{r}\,dr \tag{13.3.4}$$

and

$$r^2 V''(r) + rV'(r) - V(r) = 0,\ R_1 < r < R_2. \tag{13.3.5a}$$

Hence boundary conditions (13.3.1a, 13.3.1c) are satisfied identically as is (13.3.1b) for v_z while that no-slip condition involving v_θ yields

$$V(R_1) = \Omega_1 R_1,\ V(R_2) = \Omega_2 R_2. \tag{13.3.5b}$$

Since (13.3.5a) is an Euler equation for $V(r)$, we seek a solution of it of the form

$$V(r) = r^m \Rightarrow m(m-1) - m = m^2 = 1 \Rightarrow m = m_{1,2} = \pm 1$$

$$\Rightarrow V(r) = Ar + \frac{B}{r}. \tag{13.3.6a}$$

Upon substitution of (13.3.6a) into (13.3.5b), we obtain the following two equations for A and B:

$$R_1^2 A + B = \Omega_1 R_1^2$$

and

$$R_2^2 A + B = \Omega_2 R_2^2.$$

Solving these equations simultaneously, we find that

$$A = \frac{\Omega_2 R_2^2 - \Omega_1 R_1^2}{R_2^2 - R_1^2} \tag{13.3.6b}$$

and

$$B = \frac{(\Omega_1 - \Omega_2) R_1^2 R_2^2}{R_2^2 - R_1^2}; \tag{13.3.6c}$$

thus completing our determination of Couette flow. For the application of this solution to certain experimental situations, it is sometimes convenient to define its rotational velocity in terms of the radial coordinate r by

$$\Omega(r) \equiv \frac{V(r)}{r} = A + \frac{B}{r^2}, \tag{13.3.7a}$$

where the no-slip condition now yields

$$\Omega(R_1) = \Omega_1, \Omega(R_2) = \Omega_2. \tag{13.3.7b}$$

Observe that, should these imposed rotations be equivalent or

$$\Omega_1 = \Omega_2 = \Omega_0, \tag{13.3.8a}$$

then

$$A = \Omega_0 \text{ and } B = 0 \tag{13.3.8b}$$

which implies that

$$\Omega(r) \equiv \Omega_0 \text{ or } V(r) = \Omega_0 r \tag{13.3.8c}$$

and is consistent with rigid body motion for a uniform rotational situation. A reduction of this sort is reminiscent of the correspondence principle in quantum mechanics.

(ii) Poiseuille Flow

Consider the steady-state parallel flow, which depends only on its radial coordinate and thus is axisymmetric and independent of its coaxial coordinate, of a viscous homogeneous fluid in an infinitely long motionless pipe of radius R, driven by an adverse axial pressure gradient of strength D_0 as depicted in Fig. 13.4a. The relevant no-penetration, no-slip, and far-field boundary conditions for this situation are given by

$$v_r = v_\theta = v_z = 0 \text{ at } r = R; v_r, v_\theta, v_z, p \text{ remain bounded as } z \to \infty. \tag{13.3.9}$$

Hence we seek a solution of (13.1.7) and boundary conditions (13.3.9) of the form

$$v_r = v_\theta \equiv 0, v_z = W(r), \nabla p = -D_0 \boldsymbol{k}. \tag{13.3.10}$$

(a)

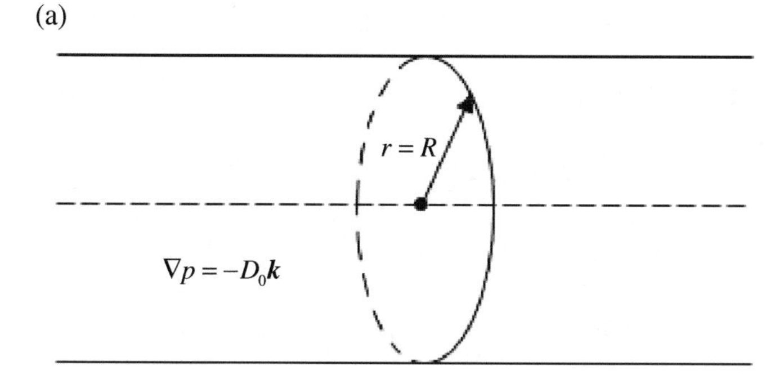

(b)

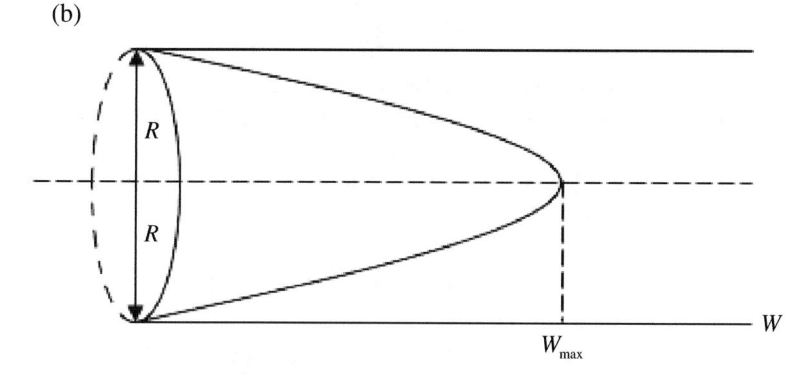

Fig. 13.4 (a) Schematic representation of Poiseuille flow and (b) a plot of its velocity profile (13.3.16c) where $W_{\max} = D_0R^2/(4\mu)$.

Then (13.1.7a, b, c) and all the boundary conditions involving v_r and v_θ as well as the far-field condition for v_z and p are satisfied identically while (13.1.7d) and the no-penetration condition yield

$$W''(r) + \frac{W'(r)}{r} = -\frac{D_0}{\mu}, \quad 0 < r < R; \tag{13.3.11a}$$

$$W(r) \text{ remains bounded as } r \longrightarrow 0, \; W(R) = 0; \tag{13.3.11b}$$

where the boundedness property of $W(r)$ as $r \longrightarrow 0$ is being imposed as a mathematical restriction due to the singularity exhibited by (13.3.11a) at $r = 0$ (there exists no physical boundary condition at $r = 0$, the centerline of the pipe).

We shall solve (13.3.11a) in two different ways as follows:

(a) First-Order Linear Equation Approach

Multiplying (13.3.11a) by r, we obtain the first-order linear equation in $W'(r)$:

$$r[W'(r)]' + [W'(r)] = [rW'(r)]' = -\frac{D_0 r}{\mu}; \tag{13.3.12}$$

Then, successively integrating (13.3.12) twice, we find its solution

$$rW'(r) = -\frac{D_0 r^2}{2\mu} + c_2 \text{ or } W'(r) = -\frac{D_0 r}{2\mu} + \frac{c_2}{r}$$

$$\Rightarrow W(r) = -\frac{D_0 r^2}{4\mu} + c_2 \ln\left(\frac{r}{R}\right) + c_1 \tag{13.3.13}$$

involving the integration constants c_1 and c_2.

(b) Second-Order Nonhomogeneous Euler Equation Approach

Multiplying (13.3.11a) by r^2, we obtain the second-order nonhomogeneous Euler equation:

$$L[W] = r^2 W''(r) + rW'(r) = -\frac{D_0 r^2}{\mu}. \tag{13.3.14}$$

Given the linearity of $L[W]$, we shall seek a solution of (13.3.14) of the form

$$W(r) = W_c(r) + W_p(r) \text{ where } L[W_c] = 0 \text{ and } L[W_p] = -\frac{D_0 r^2}{\mu}. \tag{13.3.15a}$$

Then letting

$$W_c(r) = r^m \Rightarrow m(m-1) + m = m^2 = 0 \Rightarrow m = m_{1,2} = 0$$

$$W_c(r) = c_1 + c_2 \ln\left(\frac{r}{R}\right); \tag{13.3.15b}$$

and

$$W_p(r) = \mathcal{A}r^2 \Rightarrow 2\mathcal{A} + 2\mathcal{A} = 4\mathcal{A} = -\frac{D_0}{\mu} \Rightarrow \mathcal{A} = -\frac{D_0}{4\mu}$$

$$\Rightarrow W_p(r) = -\frac{D_0 r^2}{4\mu}; \tag{13.3.15c}$$

we again obtain

$$W(r) = W_c(r) + W_p(r) = c_1 + c_2 \ln\left(\frac{r}{R}\right) - \frac{D_0 r^2}{4\mu}; \tag{13.3.15d}$$

in agreement with (13.3.13).

Now imposing the boundary conditions of (13.3.11b), we find that

$$W(r) \text{ remain bounded as } r \to 0 \Rightarrow c_2 = 0 \tag{13.3.16a}$$

while then

$$W(R) = c_1 - \frac{D_0 R^2}{4\mu} = 0 \Rightarrow c_1 = \frac{D_0 R^2}{4\mu}; \tag{13.3.16b}$$

and hence deduce the paraboloidal velocity profile for $W(r)$ plotted in Fig. 13.4b

$$W(r) = \frac{D_0(R^2 - r^2)}{4\mu}, \tag{13.3.16c}$$

which has its maximum value of $W_{\max} = D_0 R^2/(4\mu)$ occurring at the centerline of $r = 0$.

We wish to compare these regular Couette and Poiseuille flows with the plane Couette and Poiseuille flows, respectively, deduced in the previous section. Although it would be possible to make the comparison between the Poiseuille flows at this point, we shall defer that comparison until the next section in which the comparison between the Couette flows is included. To make this comparison it is first necessary that we examine the small-gap approximation of regular Couette flow to be developed in that section. For this development we shall assume that the gap between the rotating cylinders $d = R_2 - R_1 << R_1$, introduce the small parameter $0 < \varepsilon = d/R_1 << 1$, and perform a regular perturbation expansion on both the Couette flow solution and the differential equation satisfied by it.

13.4 Small-Gap Regular Perturbation Expansion of Couette Flow

As outlined at the end of the last section, we now assume that $d = R_2 - R_1 << R_1$ and define

$$x = \frac{r - R_1}{d} \text{ or } r = R_1(1 + \varepsilon x); \text{ with } 0 < \varepsilon = \frac{d}{R_1} << 1, R_2 = R_1(1+\varepsilon), 0 \le x \le 1. \tag{13.4.1}$$

We shall develop the small-gap approximation of Couette flow in two different ways:

(i) First, we examine the regular perturbation expansion of the exact solution for Couette flow (13.3.6) under the transformation introduced in (13.4.1): Namely,

$$V(R_1[1+\varepsilon x]) = AR_1(1+\varepsilon x) + \frac{B}{R_1[1+\varepsilon x]} = V_0(x) + \varepsilon V_1(x) + O(\varepsilon^2) \text{ as } \varepsilon \to 0; \tag{13.4.2}$$

to deduce its lowest order small-gap approximation $V_0(x)$. Then, since $0 \le \varepsilon x << 1$,

$$V(R_1[1+\varepsilon x]) = AR_1(1+\varepsilon x) + \left(\frac{B}{R_1}\right)[1-\varepsilon x+\varepsilon^2x^2+O(\varepsilon^2)]$$
$$= AR_1 + \frac{B}{R_1} + \left(AR_1 - \frac{B}{R_1}\right)\varepsilon x + O\left(\frac{B\varepsilon^2}{R_1}\right) \text{ as } \varepsilon \to 0. \quad (13.4.3)$$

To find $V_0(x)$ we need only compute each of the terms appearing on the right-hand side of (13.4.3) by employing the values for A and B of (13.3.6b, 13.3.6c), respectively, in conjunction with (13.4.1):

$$AR_1 + \frac{B}{R_1} = \frac{\Omega_2R_2^2R_1 - \Omega_1R_1^3 + (\Omega_1-\Omega_2)R_1R_2^2}{R_2^2-R_1^2}$$
$$= \frac{\Omega R_1(R_2^2-R_1^2)}{R_2^2-R_1^2} = \Omega_1R_1 \text{ as per (13.3.5b);}$$

$$\left(AR_1 - \frac{B}{R_1}\right)\varepsilon x = \frac{\Omega_2R_2^2R_1 - \Omega_1R_1^3 + (\Omega_2-\Omega_1)R_1R_2^2}{R_2^2-R_1^2}\varepsilon x$$
$$= \frac{2\Omega_2R_2^2R_1 - \Omega_1R_1(R_1^2+R_2^2)}{R_2^2-R_1^2}\varepsilon x$$
$$= \frac{2\Omega_2R_1^3(1+\varepsilon)^2 - \Omega_1R_1^3[1+(1+\varepsilon)^2]}{R_1^2[(1+\varepsilon)^2-1]}\varepsilon x$$
$$= \frac{2\Omega_2R_1(1+\varepsilon)^2 - \Omega_1R_1[1+(1+\varepsilon)^2]}{2\varepsilon+\varepsilon^2}\varepsilon x$$
$$= \frac{2\Omega_2R_1(1+\varepsilon)^2 - \Omega_1R_1[1+(1+\varepsilon)^2]}{2+\varepsilon}x$$
$$= (\Omega_2R_1 - \Omega_1R_1)x + O(\varepsilon)$$
$$= -\Omega_1R_1(1-\mu_0)x + O(\varepsilon) \text{ where } \mu_0 = \frac{\Omega_2}{\Omega_1};$$

$$\frac{B\varepsilon^2}{R_1} = \frac{(\Omega_1-\Omega_2)R_1R_2^2\varepsilon^2}{R_2^2-R_1^2} = \frac{(\Omega_1-\Omega_2)R_1(1+\varepsilon)^2\varepsilon}{2+\varepsilon}$$
$$= \frac{\Omega_1R_1(1-\mu_0)}{2}\varepsilon + O(\varepsilon^2) = O(\varepsilon).$$

Hence, upon substitution of the results of these computations into (13.4.3), we have deduced the following small-gap approximation of Couette flow by an examination of its exact solution:

$$V(r) = V_0(x) + O(\varepsilon) \text{ as } \varepsilon \to 0 \tag{13.4.4a}$$

where

$$V_0(x) = \Omega_1 R_1 [1 - (1 - \mu_0)x]. \tag{13.4.4b}$$

(ii) Next, we introduce the small-gap change of variables directly into the differential equation (13.3.5a) and boundary conditions (13.3.5b) satisfied by $V(r)$ of (13.3.6). Toward that end let

$$r = R_1(1 + \varepsilon x) \text{ or } x = \frac{r - R_1}{\varepsilon R_1}, \; v(x;\varepsilon) = \frac{V(r)}{\Omega_1 R_1}. \tag{13.4.5a}$$

Since then, by the chain rule of differentiation $d/dr = (d/dx)(dx/dr) = (d/dx)/(\varepsilon R_1)$,

$$V' = \frac{\Omega_1 R_1 v'}{\varepsilon R_1} \text{ and } V'' = \frac{\Omega_1 R_1 v''}{(\varepsilon R_1)^2}. \tag{13.4.5b}$$

Thus, $r^2 V'' + rV' - V = 0$, $R_1 < r < R_2$; transforms into

$$R_1^2(1+\varepsilon x)^2 \frac{\Omega_1 R_1 v''}{(\varepsilon R_1)^2} + R_1(1+\varepsilon x)\frac{\Omega_1 R_1 v'}{\varepsilon R_1} - \Omega_1 R_1 v = 0$$

or, upon cancellation of the relevant Ω_1 and R_1 factors and multiplication by ε^2,

$$L[v] = (1+\varepsilon x)^2 v'' + \varepsilon(1+\varepsilon x)v' - \varepsilon^2 v = 0, \; 0 < \varepsilon << 1; 0 < x < 1; \tag{13.4.6a}$$

with boundary conditions

$$v(0;\varepsilon) = \frac{V(R_1)}{\Omega_1 R_1} = \frac{\Omega_1 R_1}{\Omega_1 R_1} = 1, \tag{13.4.6b}$$

$$v(1;\varepsilon) = \frac{V(R_2)}{\Omega_1 R_1} = \frac{\Omega_2 R_2}{\Omega_1 R_1} = \frac{\Omega_2 R_1(1+\varepsilon)}{\Omega_1 R_1} = \mu_0(1+\varepsilon). \tag{13.4.6c}$$

This is a regular perturbation problem for $v = v(x;\varepsilon)$. Therefore, selecting the asymptotic sequence $\varphi_n(\varepsilon) = \varepsilon^n$ and number of terms $N + 1 = 2$, we seek a solution of it of the form

$$v(x;\varepsilon) = v_0(x) + \varepsilon v_1(x) + O(\varepsilon^2) \text{ as } \varepsilon \to 0. \tag{13.4.7a}$$

Hence

$$\begin{aligned} L[v] &= (1 + 2\varepsilon x + \varepsilon^2)[v_0''(x) + \varepsilon v_1''(x) + O(\varepsilon^2)] \\ &+ (\varepsilon + \varepsilon^2 x)[v_0'(x) + \varepsilon v_1'(x) + O(\varepsilon^2)] \\ &- \varepsilon^2[v_0(x) + \varepsilon v_1(x) + O(\varepsilon^2)] \\ &= v_0''(x) + \varepsilon[v_1''(x) + 2xv_0''(x) + v_0'(x)] + O(\varepsilon^2) = 0, \end{aligned}$$

$$v(0;\varepsilon) = v_0(0) + \varepsilon v_1(x) + O(\varepsilon^2) = 1,$$

and

$$v(1;\varepsilon) = v_0(1) + \varepsilon v_1(1) + O(\varepsilon^2) = \mu_0(1+\varepsilon);$$

yield the following problems for $v_0(x)$ and $v_1(x)$, respectively:

$$O(1): v_0''(x) = 0,\ v_0(0) = 1,\ v_0(1) = \mu_0; \tag{13.4.7b}$$

$$O(\varepsilon): v_1''(x) = -v_0'(x),\ v_1(0) = 0,\ v_1(1) = \mu_0. \tag{13.4.7c}$$

Solving these problems sequentially, we find that

$$v_0''(x) = 0 \Rightarrow v_0(x) = ax + b;\ v_0(0) = b = 1,\ v_0(1) = a + 1 = \mu_0$$

$$v_0(x) = 1 + (\mu_0 - 1)x = 1 - (1 - \mu_0)x; \tag{13.4.8a}$$

and

$$v_1''(x) = -v_0'(x) = 1 - \mu_0 \Rightarrow v_1(x) = \frac{(1-\mu_0)x^2}{2} + c_1 x + c_2;$$

$$v_1(0) = c_2 = 0,\ v_1(1) = \frac{1-\mu_0}{2} = c_1 = \mu_0 \Rightarrow v_1(x) = x\frac{(1-\mu_0)x + 3\mu_0 - 1}{2}. \tag{13.4.8b}$$

Thus

$$v(x;\varepsilon) = 1 - (1-\mu_0)x + \varepsilon x\frac{(1-\mu_0)x + 3\mu_0 - 1}{2} + O(\varepsilon^2) \text{ as } \varepsilon \to 0. \tag{13.4.8c}$$

Observe that, while (13.4.8a) is consistent with our previous result of (13.4.4) since

$$V_0(x) = \Omega_1 R_1 v_0(x).$$

it required much less work to determine. Indeed in this context we could have deduced problem (13.4.7b) for it merely by defining

$$v_0(x) = \lim_{\varepsilon \to 0} v(x;\varepsilon)$$

and taking the limit of (13.4.6) as $\varepsilon \to 0$. Correspondingly, $V_1(x)$ of (13.4.2) would be related to (13.4.8b) by

$$V_1(x) = \Omega_1 R_1 v_1(x)$$

but we did not determine it in the previous subsection due to the length of the calculations involved. Even when it is possible to obtain an exact closed-form representation for the solution of a differential equations problem (as in the present case), this example illustrates the fact that method (ii) for finding its asymptotic behavior as a parameter becomes small tends to be superior to method (i) in that it reduces the required labor.

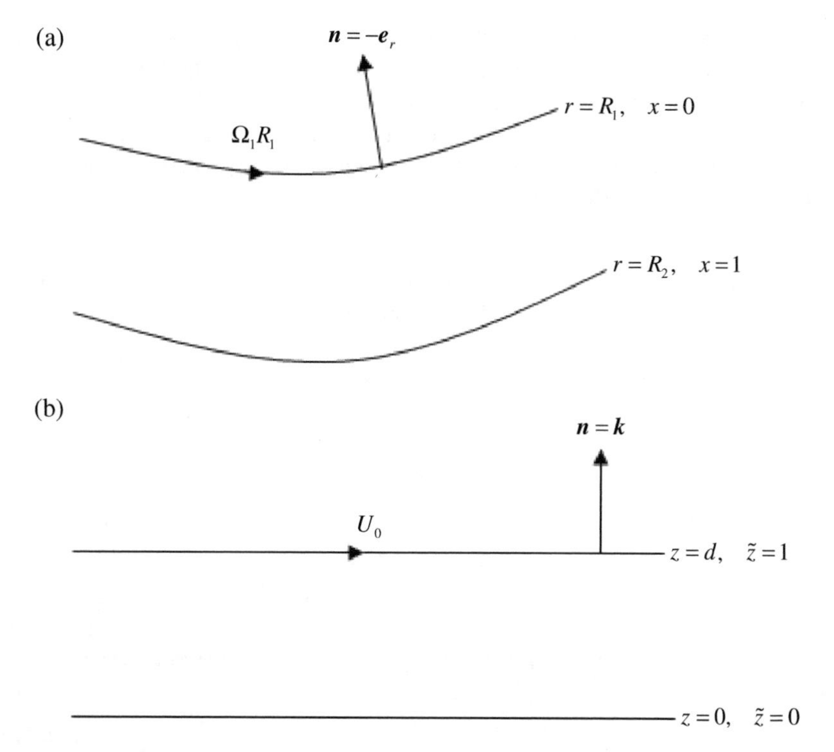

Fig. 13.5 Schematic plots demonstrating the isomorphism between (a) small-gap Couette flow with $\Omega_2 = 0$ given by $V_0(x) = \Omega_1 R_1(1-x)$ where $x = (r - R_1)/d$ and (b) plane Couette flow given by $U(z) = U_0\tilde{z}$ where $\tilde{z} = z/d$. Here d represents the thickness of the layer for both flows.

We shall now compare this small-gap approximation to Couette flow with plane Couette flow. To do so we assume that the outer cylinder is at rest and hence set $\Omega_2 = 0$ which implies $\mu_0 = 0$ as well. Then Couette flow (see Fig. 13.5a) and plane Couette flow (see Fig. 13.5b) satisfy

$$V_0(x) = \Omega_1 R_1(1-x) \text{ and } U(z) = U_0\tilde{z} \text{ where } \tilde{z} = \frac{z}{d}, \tag{13.4.9a}$$

respectively. Hence, if one makes the associations (see Fig. 13.5)

$$U_0 \sim \Omega_1 R_1, \tilde{z} \sim 1 - x \Rightarrow U(z) \sim V_0(x); \tag{13.4.9b}$$

then the small-gap approximation to Couette flow can be considered isomorphic to plane Couette flow with the former's rotating inner cylinder corresponding to the latter's moving upper plate.

Again, having obtained Poiseuille flow in the last subsection and the small-gap approximation to Couette flow in this one, we wish to examine the role that viscosity plays on the shear stress $\boldsymbol{t}$ acting on the inner rotating cylinder for the latter case and the mass M per unit time Δt passing through a fixed circular cross section of radius R in the pipe for the former. We shall close this chapter by treating each of those effects sequentially in what follows:

(a) Toward that end we note that here the components of the stress vector on the inner rotating cylinder during Couette flow are given by

$$t_j = T_{ij}\big|_{r=R_1} n_i \text{ for } j = r, \theta, z \text{ with } \boldsymbol{n} = -\boldsymbol{e}_r \text{ or } n_r = -1, n_\theta = n_z = 0. \quad (13.4.10a)$$

Thus

$$t_j = -T_{rj}\big|_{r=R_1} \text{ for } j = r, \theta, z \quad (13.4.10b)$$

where [52]

$$T_{rr} = 2\mu\frac{\partial v_r}{\partial r},\ T_{r\theta} = \mu\left[\frac{\partial v_r/\partial\theta}{r} + \frac{\partial v_\theta}{\partial r} - \frac{v_\theta}{r}\right],\ T_{rz} = \mu\left(\frac{\partial v_r}{\partial z} + \frac{\partial v_z}{\partial r}\right). \quad (13.4.10c)$$

Since $v_\theta = Ar + B/r$ with $B = (\Omega_1 - \Omega_2)R_1^2R_2^2/(R_2^2 - R_1^2)$ while $v_r = v_z \equiv 0$,

$$T_{rr} = T_{rz} = 0;\ T_{r\theta} = \mu\left(A - \frac{B}{r^2} - A - \frac{B}{r^2}\right) = -2\mu\frac{B}{r^2}. \quad (13.4.10d)$$

Then

$$-T_{r\theta}\big|_{r=R_1} = 2\mu\frac{B}{R_1^2} = 2\mu\frac{(\Omega_1 - \Omega_2)R_2^2}{R_2^2 - R_1^2} \quad (13.4.10e)$$

and hence

$$\boldsymbol{t} = \left[2\mu\frac{(\Omega_1 - \Omega_2)R_2^2}{R_2^2 - R_1^2}\right]\boldsymbol{e}_\theta. \quad (13.4.10f)$$

Let us examine this behavior in the small-gap approximation. To do so we shall take the preferred limit of (13.4.10f) as $R_2 \to R_1$ but keep $R_2 - R_1 = d > 0$. That is considering

$$R_2^2 - R_1^2 = (R_2 + R_1)(R_2 - R_1) = (R_2 + R_1)d, \quad (13.4.11a)$$

(13.4.10f) becomes

$$\boldsymbol{t} = \left(\frac{2\mu}{d}\right)\left[\frac{(\Omega_1 - \Omega_2)R_2^2}{R_2 + R_1}\right]\boldsymbol{e}_\theta \to \left(\frac{\mu\Omega_1 R_1}{d}\right)(1 - \mu_0)\boldsymbol{e}_\theta \text{ as } R_2 \to R_1, \quad (13.4.11b)$$

which for $\Omega_2 = 0$ or $\mu_0 = 0$ reduces to (13.2.10e) for plane Couette flow

$$t \to \frac{\mu\Omega_1 R_1 \boldsymbol{e}_\theta}{d} \text{ as } R_2 \to R_1 \qquad (13.4.11c)$$

upon introducing the isomorphic association of (13.4.9b) in conjunction with $\boldsymbol{i} \sim \boldsymbol{e}_\theta$. It only remains to compare the pressure gradient for the small-gap approximation to Couette flow with that for plane Couette flow. Recall that for Couette flow

$$\mathcal{P}'(r) = \frac{\rho_0 V^2(r)}{r} \qquad (13.4.12a)$$

and define the corresponding nondimensional pressure in the small-gap variables of (13.4.5) by

$$p(x;\varepsilon) = \frac{\mathcal{P}(r)}{\rho_0\Omega_1^2 R_1^2} \text{ where } r = R_1 + xd \text{ and } \varepsilon = \frac{d}{R_1} \qquad (13.4.12b)$$

Then

$$\mathcal{P}'(r) = \frac{\rho_0\Omega_1^2 R_1^2 p'(x;\varepsilon)}{d} = \frac{\rho_0 V^2(r)}{r} = \frac{\rho_0\Omega_1^2 R_1^2 v^2(x;\varepsilon)}{R_1 + xd}$$

$$\Rightarrow p'(x;\varepsilon) = \frac{v^2(x;\varepsilon)d}{R_1 + xd} = \frac{\varepsilon v^2(x;\varepsilon)}{1 + \varepsilon x} = O(\varepsilon) \text{ as } \varepsilon \to 0. \qquad (13.4.12c)$$

Hence defining the pressure of the small-gap approximation to Couette flow by

$$p_0(x) = \lim_{\varepsilon \to 0} p(x;\varepsilon)$$

and taking the limit of (13.4.12c) as $\varepsilon \to 0$, we obtain $p_0'(x) = 0$ consistent with plane Couette flow.

(b) We wish to compute the mass M of fluid per unit time Δt that passes through a fixed circular cross section of radius R in the pipe during Poiseuille flow or

$$\frac{M}{\Delta t} = \int_0^{2\pi}\int_0^R \rho_0 W(r) r\,dr\,d\theta = \frac{2\pi\rho_0 D_0}{4\mu}\int_0^R (R^2 r - r^3)\,dr = \frac{\pi D_0}{2\nu}\left(\frac{R^2 r^2}{2} - \frac{r^4}{4}\right)\Bigg|_{r=0}^{R}$$

$$\Rightarrow \frac{M}{\Delta t} = \frac{\pi D_0 R^4}{8\nu}. \qquad (13.4.13a)$$

Recall that the mass of fluid M that crosses the fixed plane $x = x_0$ per unit time Δt per unit length L in the y-direction during plane Poiseuille flow was given by

$$\frac{M}{L\Delta t} = \frac{2C_0 d^3}{3\nu}. \qquad (13.4.13b)$$

In order to compare these two quantities we determine their associated fluxes $M/(A\Delta t)$ where the relevant cross-sectional area $A = \pi R^2$ or $2Ld$ for regular or plane Poiseuille flows (see Fig. 13.6). Thus (13.4.13a) or (13.4.13b) yields the fluxes

$$\frac{M}{A\Delta t} = \frac{D_0 R^2}{8\nu} \qquad (13.4.14a)$$

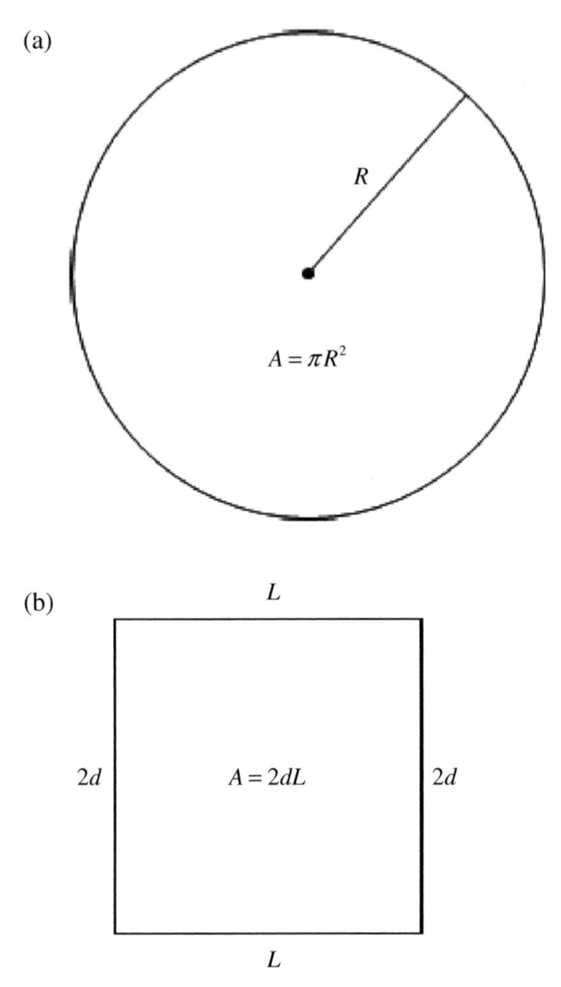

Fig. 13.6 Schematic representations of the cross-sectional areas relevant to mass flux during (a) regular and (b) plane Poiseuille flows.

or

$$\frac{M}{A\Delta t} = \frac{C_0 d^2}{3\nu}. \tag{13.4.14b}$$

Since D_0 and R correspond to C_0 and d, respectively, we see that the flux for plane Poiseuille flow exceeds that for regular Poiseuille flow. The reason for this result is that a no-slip condition occurs along the whole circumference $r = R$ of the circular cross section in the pipe but only on the sides of length L of the perimeter of the rectangular one in the channel located at the plates $z = \pm d$. The fluid can slip along the other two sides of length d of that rectangular perimeter.

Problems

13.1. In deriving plane Couette flow, it was assumed that $\mathcal{P}'(x) \equiv 0$. Keep the same boundary conditions but now set $\mathcal{P}'(x) \equiv -C_0$, where $C_0 > 0$, which implies that $\nabla p = -C_0\boldsymbol{i}$. Show that the resulting flow is a superposition of plane Couette and Poiseuille flows. Sketch typical velocity profiles for the three parameter ranges that produce qualitatively different results.

13.2. This problem outlines the beginnings of an investigation into the stability of Couette flow.

(a) Show that the Couette flow solution

$$v_\theta = V(r) = Ar + \frac{B}{r}, p = \mathcal{P}(r) = \rho_0 \int \frac{V^2(r)}{r}\, dr, v_r = v_z \equiv 0,$$

where $A = (\Omega_2 R_2^2 - \Omega_1 R_1^2)/(R_2^2 - R_1^2)$ and $B = (\Omega_1 - \Omega_2)R_1^2 R_2^2/(R_2^2 - R_1^2)$; satisfies the inviscid equations of motion so that viscosity enters only indirectly by means of the no-slip or adherence boundary conditions. As a first attempt at examining its stability, Couette flow will be investigated in this inviscid limit: Namely, by setting $v = 0$ in (13.1.7) and only retaining the no-penetration and far-field boundary conditions of (13.3.1).

(b) Performing a linear stability analysis on this exact solution by seeking a normal mode solution of these basic equations and boundary conditions of the form

$$v_r = \varepsilon X(r)\exp(ikz + \sigma t) + O(\varepsilon^2), v_\theta = V(r) + \varepsilon Y(r)\exp(ikz + \sigma t) + O(\varepsilon^2),$$

$$v_z = \varepsilon Z(r)\exp(ikz + \sigma t) + O(\varepsilon^2), p = \mathcal{P}(r) + \varepsilon\Pi(r)\exp(ikz + \sigma t) + O(\varepsilon^2),$$

where $|\varepsilon| << 1$, $k \in \mathbb{R}$, and the disturbances are axisymmetric, a system of linear homogeneous ordinary differential equations in the radial stratification functions is obtained upon the neglect of terms of $O(\varepsilon^2)$ and cancellation of the resulting common factor of ε times the exponential. Elimination of Y, Z, and Π from these equations by using the same methodology as employed for Problem 11.3 yields the following boundary-value problem satisfied by $X = X(r)$:

$$[r^{-1}(rX)']' - k^2[\sigma^{-2}\Phi(r) + 1]X = 0, R_1 < r < R_2 \text{ where } \Phi(r) = r^{-3}[r^2V^2(r)]';$$

$$X(R_1) = X(R_2) = 0.$$

Upon multiplying this equation by $r\bar{X}$, where $\bar{X}$ denotes the complex conjugate of X, integrating the result on r from R_1 to R_2, and performing the appropriate integration by parts, demonstrate that this flow is neutrally stable ($\sigma^2 < 0$) provided $\Phi(r) > 0$.

(c) Show that $\Phi(r) > 0$ as long as $\mu_0 > \eta_0^2$ where $\mu_0 = \Omega_2/\Omega_1$ and $\eta_0 = R_1/R_2$. Explain the physical significance of this criterion in terms of the angular momentum given by $r^2\Omega(r)$.

13.3. Consider the flow of a viscous homogeneous fluid confined in the annulus between two infinitely long concentric circular cylinders of radii R_1 and $R_2 > R_1$ rotating with constant angular speeds Ω_1 and Ω_2, respectively. Suppose that in addition there is a constant circumferential component of the pressure gradient and a similar such coaxial component so that $\partial p/\partial\theta = -C_0$ and $\partial p/\partial z = -D_0$ where C_0, $D_0 > 0$. Seeking a solution of (13.1.7) and boundary conditions (13.3.1) of the form

$$v_r \equiv 0,\ v_\theta = V(r),\ v_z = W(r),\ p = \mathcal{P}(r,\theta,z)$$

such that

$$\nabla p = \left[\frac{\rho_0 V^2(r)}{r}\right]\boldsymbol{e}_r - C_0\boldsymbol{e}_\theta - D_0\boldsymbol{k};$$

determine the exact solution of this resulting more general Couette flow.

Hint: If an Euler equation for $V = V(r)$ is of the form $L[V] = c_0 r^m$ and $L[r^m] = 0$ where m is not a double root, one seeks a $V_p(x)$ where $L[V_p] = c_0 r^m$ of the form $V_p(x) = \mathcal{A} r^m \ln(r/R_1)$.

13.4. If in Problem 13.3 the gap width between the cylinders $d = R_2 - R_1 << R_1$ determine the lowest order small-gap approximation to this flow by introducing appropriate nondimensional variables into the original differential equations and boundary conditions for $V(r)$ and $W(r)$ and then solving for them by use of regular perturbation theory. Nondimensionalize $V(r)$ and $W(r)$ by employing the scale factors $V_m = d^2C_0/(12\mu R_1)$ and $W_m = d^2D_0/(12\mu)$, where V_m and W_m are the mean velocity to lowest order of $V(r)$ across the channel in the absence of rotation and the mean velocity to lowest order of $W(r)$ across the channel, respectively, and hence are of $O(1)$ as $\varepsilon = d/R_1 \to 0$ while the same thing may be assumed for $\Omega_{1,2}R_1/V_m$.

That is, introduce the change of variables in those differential equations and boundary conditions

$$v(x;\varepsilon) = \frac{V(r)}{V_m},\ w(x;\varepsilon) = \frac{W(r)}{W_m} \text{ where } r = R_1(1+\varepsilon x),$$

and seek the lowest order regular perturbation solution for $v_0(x)$ and $w_0(x)$ such that

$$v(x;\varepsilon) = v_0(x) + O(\varepsilon) \text{ and } w(x;\varepsilon) = w_0(x) + O(\varepsilon) \text{ as } \varepsilon \to 0;$$

or, more simply,

$$v_0(x) = \lim_{\varepsilon\to 0} v(x;\varepsilon) \text{ and } w_0(x) = \lim_{\varepsilon\to 0} w(x;\varepsilon).$$

Do not attempt either to expand the exact solutions determined in Problem 13.3 to find $v_0(x)$ and $w_0(x)$ (since this requires an enormous amount of labor) nor to calculate the $O(\varepsilon)$ contributions $v_1(x)$ and $w_1(x)$ to those regular perturbation expansions (since that is not being requested for this problem), both of which were done for the small-gap approximation of Couette flow in Section 13.4.

Chapter 14
Blasius Flow Past a Flat Plate

In this chapter, after a pastoral interlude revisiting singular perturbation theory by determining higher-order terms in the Method of Matched Asymptotic Expansions and comparing these results with those obtained by the method of multiple scales for the constant coefficient second-order ordinary differential equation treated in Chapter 3, steady-state Blasius boundary-layer flow of a viscous fluid streaming uniformly past a flat plate, which represents the prototype boundary-layer problem for viscous fluid mechanics, is analyzed. This is accomplished by applying singular perturbation theory techniques to the governing partial differential equation satisfied by the relevant stream function governing that flow. The solution of the resulting boundary-layer equation to lowest order requires a similarity solution argument. When the drag on the plate is calculated for that lowest order solution it is nonzero, resolving D'Alembert's paradox in this instance. In order to make an interpretation of these results, the second-order term in the outer free-stream solution is calculated and compared with that deduced for inviscid flow past an effective body consisting of the flat plate plus the boundary layer. The problems involve two extensions: The first extending the singular perturbation techniques developed in the pastoral interlude for a constant coefficient example to a variable coefficient one and the second extending the Blasius flow methodology employed in the chapter for a uniform stream flow situation to a variable flow one.

14.1 Pastoral Interlude: Singular Perturbation Theory Revisited

In order to examine Blasius flow past a flat plate, it is necessary for us to extend the development of singular perturbation theory contained in Section 3.2. Recall that in this section, we began an investigation of the singularly perturbed boundary value problem for $y = y(x; \varepsilon)$:

$$\varepsilon \frac{d^2y}{dx^2} + 2\frac{dy}{dx} + y = 0, \; 0 < x < 1, \; 0 < \varepsilon << 1; \tag{14.1.1a}$$

© Springer International Publishing AG, part of Springer Nature 2017

D. J. Wollkind and B. J. Dichone, *Comprehensive Applied Mathematical Modeling in the Natural and Engineering Sciences*,
https://doi.org/10.1007/978-3-319-73518-4_14

$$y(0;\varepsilon) = 0, y(1;\varepsilon) = 1; \tag{14.1.1b}$$

and deduced its asymptotic behavior to $O(1)$ in the limit as $\varepsilon \to 0$ by means of the Method of Matched Asymptotic Expansions. We need to extend that analysis to the next higher-order term or to $O(\varepsilon)$ as $\varepsilon \to 0$. Toward that end, we reconsider Fig. 3.6 and proceed as follows:

(i) The outer solution: For $x \in I$, which is an interval flanking $x = 1$, we assumed that

$$y, \frac{dy}{dx}, \frac{d^2y}{dx^2} = O^*(1) \text{ as } \varepsilon \to 0.$$

We now seek a solution of (14.1.1) of the form

$$y(x;\varepsilon) = y_0(x) + \varepsilon y_1(x) + O(\varepsilon^2) \text{ as } \varepsilon \to 0 \text{ for } x \in I \tag{14.1.2}$$

where $d^n y_k(x)/dx^n = O(1)$ as $\varepsilon \to 0$ for $k = 0, 1$ and $n = 0, 1, 2$. Substituting (14.1.2) into (14.1.1a) and that boundary condition of (14.1.1b) at $x = 1$, we deduce the following linear first-order differential equation and boundary condition for $y_0(x)$ and $y_1(x)$:

$$O(1): 2y_0'(x) + y_0(x) = 0, y_0(1) = 1; \tag{14.1.3a}$$

$$O(\varepsilon): 2y_1'(x) + y_1(x) = -y_0''(x), y_1(1) = 0; \tag{14.1.3b}$$

respectively. Not surprisingly, as denoted in Section 3.2

$$y_0(x) = \lim_{\varepsilon \to 0} y(x;\varepsilon),$$

and thus, we can conclude from (3.2.23) that

$$y_0(x) = e^{(1-x)/2}; \tag{14.1.4a}$$

while then, from (14.1.3b) and (14.1.4a), $y_1(x)$ satisfies

$$2y_1'(x) + y_1(x) = \frac{-e^{1/2}e^{-x/2}}{4}, y_1(1) = 0 \Rightarrow y_1(x) = \frac{e^{1/2}(1-x)e^{-x/2}}{8}. \tag{14.1.4b}$$

Hence

$$y(x;\varepsilon) = e^{1/2}e^{-x/2}\left[1 + \frac{\varepsilon(1-x)}{8} + O(\varepsilon^2)\right] \text{ as } \varepsilon \to 0 \text{ for } x \in I. \tag{14.1.4c}$$

(ii) The inner or boundary-layer solution: For $x \in I'$ or $x = O^*(\varepsilon)$ as $\varepsilon \to 0$, we introduced the inner variables $\xi = x/\varepsilon$ and $Y(\xi;\varepsilon) = y(x;\varepsilon)$ which transformed (14.1.1a) and the boundary condition of (14.1.1b) at $x = 0$ into the boundary-layer equation

$$\frac{d^2Y}{d\xi^2} + 2\frac{dY}{d\xi} + \varepsilon Y = 0, \xi > 0; Y(0;\varepsilon) = 0; \tag{14.1.5}$$

where Y, $dY/d\xi$, $d^2Y/d\xi^2 = O^*(1)$ for $\xi = O^*(1)$ as $\varepsilon \to 0$. We now seek a solution of (14.1.5) of the form

$$Y(\xi;\varepsilon) = Y_0(\xi) + \varepsilon Y_1(\xi) + O(\varepsilon^2) \text{ for } \xi = O^*(1) \text{ as } \varepsilon \to 0 \tag{14.1.6}$$

where $d^nY_k(\xi)/d\xi^n = O^*(1)$ as $\varepsilon \to 0$ for $n = 0, 1, 2$ and $k = 0, 1$. Substituting (14.1.6) into (14.1.5), we deduce the following second-order linear equation and boundary condition for $Y_0(\xi)$ and $Y_1(\xi)$:

$$O(1): Y_0''(\xi) + 2Y_0'(\xi) = 0,\ Y_0(0) = 0; \tag{14.1.7a}$$

$$Y_1''(\xi) = 2Y_1'(\xi) = -Y_0(\xi),\ Y_1(0) = 0; \tag{14.1.7b}$$

respectively. Again, not surprisingly, as denoted in Section 3.2

$$Y_0(\xi) = \lim_{\varepsilon \to 0} Y(x;\varepsilon),$$

and thus we can conclude from (3.2.27) that

$$Y_0(\xi) = C_0(1 - e^{-2\xi}).$$

Although the determination of C_0 could be deferred until a solution for $Y_1(\xi)$ had been obtained, it is more convenient to impose the intermediate limit matching condition at each order sequentially. Therefore, at this time, employing the one-term matching rule of Section 3.2

$$y_0(0) = \lim_{\xi \to \infty} Y_0(\xi)$$

which results from such a procedure, we obtain that $e^{1/2} = C_0$. Hence

$$Y_0(\xi) = e^{1/2}(1 - e^{-2\xi}) \text{ for } \xi = x/\varepsilon, \tag{14.1.8}$$

in agreement with (3.2.28) while then, from (14.1.7b) and (14.1.8), $Y_1(\xi)$ satisfies

$$L[Y_1] = Y_1''(\xi) + 2Y_1'(\xi) = e^{1/2}(e^{-2\xi} - 1),\ Y_1(0) = 0. \tag{14.1.9a}$$

We shall solve (14.1.9a) by the method of undetermined coefficients. Let

$$Y_1(\xi) = Y_{1_c}(\xi) + Y_{1_p}(\xi) \tag{14.1.9b}$$

where

$$L[Y_{1_c}] = 0 \text{ and } L[Y_{1_p}] = e^{1/2}(e^{-2\xi} - 1). \tag{14.1.9c}$$

Then, since

$$Y_{1_c}(\xi) = c_1 + c_2e^{-2\xi}, \tag{14.1.10a}$$

we can deduce that $Y_{1_p}(\xi)$ is of the form

$$Y_{1_p}(\xi) = \xi(\mathcal{A} + \mathcal{B}e^{-2\xi}).$$

Thus

$$L[Y_{1_p}] = [4\mathcal{B}(\xi - 1)]e^{-2\xi} + 2[\mathcal{A} + \mathcal{B}(1 - 2\xi)e^{-2\xi}] = 2\mathcal{A} - 2\mathcal{B}e^{-2\xi} = e^{1/2}(e^{-2\xi} - 1),$$

which implies that

$$\mathcal{A} = \mathcal{B} = -\frac{e^{1/2}}{2}$$

and therefore

$$Y_{1_p}(\xi) = \frac{-e^{1/2}\xi(1 + e^{-2\xi})}{2}. \tag{14.1.10b}$$

From (14.1.9b) and (14.1.10a, 14.1.10b) we can now conclude that

$$Y_1(\xi) = Y_{1_c}(\xi) + Y_{1_p}(\xi) = c_1 + c_2e^{-2\xi} - \frac{e^{1/2}\xi(1 + e^{-2\xi})}{2}. \tag{14.1.10c}$$

Finally, imposing the boundary condition

$$Y_1(0) = c_1 + c_2 = 0 \Rightarrow c_2 = -c_1, \tag{14.1.10d}$$

yields

$$Y_1(\xi) = c_1(1 - e^{-2\xi}) - \frac{e^{1/2}\xi(1 + e^{-2\xi})}{2} \text{ for } \xi = \frac{x}{\varepsilon}. \tag{14.1.10e}$$

Hence

$$Y(\xi;\varepsilon) = e^{1/2}(1 - e^{-2\xi}) + \varepsilon\left[c_1(1 - e^{-2\xi}) - \frac{e^{1/2}\xi(1 + e^{-2\xi})}{2}\right] + O(\varepsilon^2) \tag{14.1.11a}$$

for

$$\xi = \frac{x}{\varepsilon} = O^*(1) \text{ as } \varepsilon \to 0 \text{ or } x \in I'. \tag{14.1.11b}$$

(i)–(ii) Combining (14.1.4c) and (14.1.11), we have obtained the following asymptotic behavior for $y(x;\varepsilon)$ as $\varepsilon \to 0$:

$$y(x;\varepsilon) \sim \begin{cases} e^{1/2}e^{-x/2}[1 + \varepsilon(1 - x)/8], & \text{for } x \in I \\ e^{1/2}[1 - x/2 - (1 + x/2)e^{-2x/\varepsilon}] + \varepsilon c_1(1 - e^{-2x/\varepsilon}), & \text{for } x \in I' \end{cases}. \tag{14.1.12}$$

(iii) Intermediate limit technique of matching: Our two-term outer solution $y_0(x) + \varepsilon y_1(x)$ is asymptotically valid for $x = O^*(1)$ in I while our two-term inner solution $Y_0(x/\varepsilon) + \varepsilon Y_1(x/\varepsilon)$ is asymptotically valid for $x = O^*(\varepsilon)$ in I' as $\varepsilon \to 0$. Let $x \in I''$ such that $\gamma = x/\theta(\varepsilon) = O^*(1)$ as $\varepsilon \to 0$ where $\theta(\varepsilon) > 0$ (see Fig. 3.6) and

$$\lim_{\varepsilon\to 0}\theta(\varepsilon) = \lim_{\varepsilon\to 0}\frac{\varepsilon}{\theta(\varepsilon)} = 0.$$

Since such $x \in I''$ are intermediate between $x \in I$ and $x \in I'$ we can assume that both the two-term outer and inner solutions are valid for $x \in I''$ as $\varepsilon \to 0$ and hence equal in this limit. Thus for the outer solution

$$\begin{aligned} y(\gamma\theta;\varepsilon) &= y_0(\gamma\theta) + \varepsilon y_1(\gamma\theta) + O(\varepsilon^2) \text{ as } \varepsilon \to 0 \\ &= y_0(0) + y_0'(0)\gamma\theta + \varepsilon y_1(0) + O(\theta^2) + O(\varepsilon\theta) + O(\varepsilon^2) \text{ as } \varepsilon \to 0. \end{aligned}$$

We now assume that $\theta^2 = o(\varepsilon)$ or $\varepsilon < \theta(\varepsilon) < \varepsilon^{1/2}$ as $\varepsilon \to 0$. Then,

$$y(\gamma\theta;\varepsilon) \sim y_0(0) + y_0'(0)\gamma\theta + \varepsilon y_1(0) \text{ as } \varepsilon \to 0, \tag{14.1.13a}$$

or, for our problem

$$y(\gamma\theta;\varepsilon) \sim e^{1/2}\left(1 - \frac{\gamma\theta}{2}\right) + \varepsilon\frac{e^{1/2}}{8} \text{ as } \varepsilon \to 0, \tag{14.1.13b}$$

while, for the inner solution

$$\begin{aligned} Y\left(\frac{\gamma\theta}{\varepsilon};\varepsilon\right) &= Y_0\left(\frac{\gamma\theta}{\varepsilon}\right) + \varepsilon Y_1\left(\frac{\gamma\theta}{\varepsilon}\right) + O(\varepsilon^2) \text{ as } \varepsilon \to 0 \\ &= e^{1/2}\left[1 - \frac{\gamma\theta}{2} - \left(1 + \frac{\gamma\theta}{2}\right)e^{-2\gamma\theta/\varepsilon}\right] \\ &\quad + \varepsilon c_1(1 - e^{-2\gamma\theta/\varepsilon}) + O(\varepsilon^2) \text{ as } \varepsilon \to 0 \\ &\sim e^{1/2}\left(1 - \frac{\gamma\theta}{2}\right) + \varepsilon c_1 + \text{TST as } \varepsilon \to 0, \end{aligned} \tag{14.1.14a}$$

where TST denotes transcendentally small terms involving $e^{-2\gamma\theta/\varepsilon}$ since

$$e^{-2\gamma\theta/\varepsilon} \to 0 \text{ as } \varepsilon \to 0 \text{ given that } \gamma = O^*(1) \text{ and } \frac{\theta(\varepsilon)}{\varepsilon} \to \infty \text{ in this limit.} \tag{14.1.14b}$$

Now, matching these two expansions of (14.1.13) and (14.1.14), by equating their asymptotic behavior as $\varepsilon \to 0$ for $x \in I'$, we can determine that

$$c_1 = \frac{e^{1/2}}{8}; \tag{14.1.15}$$

which, upon substitution of (14.1.15) into (14.1.12), yields the asymptotic behavior for $y(x;\varepsilon)$ as $\varepsilon \to 0$:

$$y(x;\varepsilon) \sim \begin{cases} e^{1/2}e^{-x/2}[1 + \varepsilon(1-x)/8], & \text{for } x \in I \\ e^{1/2}[1 - x/2 - (1 + x/2)e^{-2x/\varepsilon}] + \varepsilon e^{1/2}(1 - e^{-2x/\varepsilon})/8, & \text{for } x \in I' \end{cases}. \tag{14.1.16}$$

As in Section 3.2, it would be convenient to offer an asymptotic representation for $y(x;\varepsilon)$ that was uniformly valid for all $x \in [0,1]$ as $\varepsilon \to 0$. Toward that end we shall again construct the uniformly valid additive composite obtained by adding the inner expansion to the outer and subtracting their common part. That is

$$y_u^{(1)}\left(x,\frac{x}{\varepsilon};\varepsilon\right)=y_0(x)+\varepsilon y_1(x)+Y_0\left(\frac{x}{\varepsilon}\right)+\varepsilon Y_1\left(\frac{x}{\varepsilon}\right)-[y_0(0)+y_0'(0)x+\varepsilon y_1(0)]. \tag{14.1.17a}$$

From (14.1.16) and (14.1.13), this yields

$$\begin{aligned}y_u^{(1)}\left(x,\frac{x}{\varepsilon};\varepsilon\right)&=e^{1/2}e^{-x/2}\left[1+\frac{\varepsilon(1-x)}{8}\right]+e^{1/2}\left[1-\frac{x}{2}-\left(1+\frac{x}{2}\right)e^{-2x/\varepsilon}\right]\\&\quad+\varepsilon e^{1/2}\left(\frac{1-e^{-2x/\varepsilon}}{8}\right)-e^{1/2}\left(1-\frac{x}{2}+\frac{\varepsilon}{8}\right)\\&=y_{u_0}^{(1)}\left(x,\frac{x}{\varepsilon}\right)+\varepsilon y_{u_1}^{(1)}\left(x,\frac{x}{\varepsilon}\right)\end{aligned} \tag{14.1.17b}$$

where

$$y_{u_0}^{(1)}\left(x,\frac{x}{\varepsilon}\right)=e^{1/2}\left[e^{-x/2}-\left(1+\frac{x}{2}\right)e^{-2x/\varepsilon}\right] \tag{14.1.17c}$$

and

$$y_{u_1}^{(1)}\left(x,\frac{x}{\varepsilon}\right)=\frac{e^{1/2}[(1-x)e^{-x/2}-e^{-2x/\varepsilon}]}{8}; \tag{14.1.17d}$$

as our two-term $O(\varepsilon)$ uniformly valid additive composite.

Recall from (3.2.29) that our one-term $O(1)$ uniformly valid additive composite was given by

$$y_u^{(0)}\left(x,\frac{x}{\varepsilon}\right)=y_0(x)+Y_0\left(\frac{x}{\varepsilon}\right)-y_0(0)=e^{1/2}(e^{-x/2}-e^{-2x/\varepsilon})=y_{u_0}^{(0)}\left(x,\frac{x}{\varepsilon}\right) \tag{14.1.18a}$$

Observe, upon comparison of (14.1.17c) with (14.1.18a), that

$$y_{u_0}^{(0)}\left(x,\frac{x}{\varepsilon}\right)\neq y_{u_0}^{(1)}\left(x,\frac{x}{\varepsilon}\right) \tag{14.1.18b}$$

due to the presence of the extra term proportional to $x/2$ in the latter.

In order to see exactly what is transpiring here, we shall next try to find a uniformly valid solution to our singularly perturbed boundary value problem (14.1.1) from the outset by seeking a multiple scales expansion of $y(x;\varepsilon)$ of the form

$$y(x;\varepsilon)=f(x,\xi;\varepsilon)=f_0(x,\xi)+\varepsilon f_1(x,\xi)+O(\varepsilon^2)\text{ as }\varepsilon\to 0 \tag{14.1.19}$$

where $\xi=x/\varepsilon$ and to ensure uniformity of this expansion it is required that $|f_1/f_0|$, $|f_{1_\xi}/f_{0_\xi}|$ remain bounded for all $0\le x\le 1$ and $\xi\ge 0$ – *e.g.*, as $\xi\to\infty$.

Then, noting that

$$\frac{df_k(x,\xi)}{dx} = f_{k_x}(x,\xi) + f_{k_\xi}(x,\xi)\frac{d\xi}{dx} = f_{k_x}(x,\xi) + \frac{f_{k_\xi}(x,\xi)}{\varepsilon} \quad (14.1.20a)$$

and

$$\frac{d^2f_k(x,\xi)}{dx^2} = f_{k_{xx}}(x,\xi) + \frac{2f_{k_{x\xi}}(x,\xi)}{\varepsilon} + \frac{f_{k_{\xi\xi}}(x,\xi)}{\varepsilon^2} \text{ for } k = 0 \text{ and } 1; \quad (14.1.20b)$$

multiplying (14.1.1a) by ε to obtain

$$L[y] = \varepsilon^2\frac{d^2y}{dx^2} + 2\varepsilon\frac{dy}{dx} + \varepsilon y = 0; \quad (14.1.21)$$

substituting (14.1.19) into (14.1.21); and employing (14.1.20); yields

$$\begin{aligned} L[f] &= f_{0_{\xi\xi}} + 2\varepsilon f_{0_{x\xi}} + \varepsilon^2 f_{0_{xx}} + \varepsilon f_{1_{\xi\xi}} + 2\varepsilon^2 f_{1_{x\xi}} + \varepsilon^3 f_{1_{xx}} + O(\varepsilon^4) \\ &\quad + 2[f_{0_\xi} + \varepsilon f_{0_x} + \varepsilon f_{1_\xi} + \varepsilon^2 f_{1_x} + O(\varepsilon^3)] + \varepsilon f_0 + \varepsilon^2 f_1 + O(\varepsilon^3) \\ &= f_{0_{\xi\xi}} + 2f_{0_\xi} + \varepsilon(2f_{0_{x\xi}} + f_{1_{\xi\xi}} + 2f_{0_x} + 2f_{1_\xi} + f_0) + O(\varepsilon^2) = 0. \end{aligned} \quad (14.1.22)$$

Next, observing that $x = 0$ corresponds to $\xi = 0$ and $x = 1$, to $\xi = 1/\varepsilon \to \infty$ as $\varepsilon \to 0$, the boundary conditions of (14.1.1b) become

$$y(0;\varepsilon) = f(0,0;\varepsilon) = f_0(0,0) + \varepsilon f_1(0,0) + O(\varepsilon^2) = 0 \quad (14.1.23a)$$

and

$$y(1;\varepsilon) = \lim_{\xi\to\infty} f(1,\xi;\varepsilon) = \lim_{\xi\to\infty} f_1(1,\xi) + \varepsilon \lim_{\xi\to\infty} f_1(1,\xi) + O(\varepsilon^2) = 1, \quad (14.1.23b)$$

respectively. Now, from (14.1.22) and (14.1.23), we obtain the following $O(1)$ and $O(\varepsilon)$ problems for $f_0(x,\xi)$ and $f_1(x,\xi)$ as sufficient conditions (note that, unlike all our previous developments, they are not both necessary and sufficient conditions because $\xi = x/\varepsilon$ depends on ε as well):

$$O(1): f_{0_{\xi\xi}} + 2f_{0_\xi} = 0; f_0(0,0) = 0, \lim_{\xi\to\infty} f_0(1,\xi) = 1;$$

$$O(\varepsilon): f_{1_{\xi\xi}} + 2f_{1_\xi} = -2f_{0_{x\xi}} - 2f_{0_x} - f_0; f_1(0,0) = \lim_{\xi\to\infty} f_1(1,\xi) = 0.$$

Solving the differential equation of the $O(1)$ problem for $f_0(x,\xi)$ we find that

$$f_0(x,\xi) = c_0(x) + d_0(x)e^{-2\xi}, \quad (14.1.24)$$

where the functional dependence on x of $c_0(x)$ and $d_0(x)$ is to be determined from the $O(\varepsilon)$ problem for $f_1(x,\xi)$ to ensure that the uniformity requirements are satisfied.

Thus

$$f_{0_x}(x,\xi) = c_0'(x) + d_0'(x)e^{-2\xi} \text{ and } f_{0_{x\xi}}(x,\xi) = -2d_0'(x)e^{-2\xi}.$$

Hence

$$-2f_{0_{x\xi}}(x,\xi)-2f_{0_x}(x,\xi)-f_0(x,\xi)=[2d_0'(x)-d_0(x)]e^{-2\xi}-[2c_0'(x)+c_0(x)]. \quad (14.1.25)$$

Therefore, substituting (14.1.25) into the nonhomogeneous terms in the differential equation of the $O(\varepsilon)$ problem for $f_1(x,\xi)$, we obtain

$$f_{1_{\xi\xi}}(x,\xi)+2f_{1_\xi}(x,\xi)=[2d_0'(x)-d_0(x)]e^{-2\xi}-[2c_0'(x)+c_0(x)]. \quad (14.1.26a)$$

To guarantee that $|f_1/f_0|$ and $|f_{1_\xi}/f_{0_\xi}|$ remain bounded in the limit as $\xi\rightarrow\infty$, we require that

$$2c_0'(x)+c_0(x)=0 \text{ and } 2d_0'(x)-d_0(x)=0, \quad (14.1.26b)$$

since should these bracketed coefficients in (14.1.26a) not be zero they would give rise to terms proportional to ξ and $\xi e^{-2\xi}$, respectively, in $f_1(x,\xi)$ which would violate these uniformity conditions. Now, solving (14.1.26b) for $c_0(x)$ and $d_0(x)$, we obtain

$$c_0(x)=C_0e^{-x/2} \text{ and } d_0(x)=D_0e^{x/2}. \quad (14.1.26c)$$

Substituting (14.1.26c) into (14.1.24) yields

$$f_0(x,\xi)=C_0e^{-x/2}+D_0e^{x/2}e^{-2\xi}, \quad (14.1.27a)$$

where the constants C_0 and D_0 are to be determined from the boundary conditions of the $O(1)$ problem for $f_0(x,\xi)$. That is

$$f_0(0,0)=C_0+D_0=0\Rightarrow D_0=-C_0\Rightarrow f_0(x,\xi)=C_0(e^{-x/2}-e^{x/2}e^{-2\xi}) \quad (14.1.27b)$$

and

$$\lim_{\xi\rightarrow\infty}f_0(1,\xi)=C_0\lim_{\xi\rightarrow\infty}(e^{-1/2}-e^{1/2}e^{-2\xi})=C_0e^{-1/2}=1\Rightarrow C_0=e^{1/2}$$

$$\Rightarrow f_0(x,\xi)=e^{1/2}(e^{-x/2}-e^{x/2}e^{-2\xi}) \text{ where } \xi=\frac{x}{\varepsilon}. \quad (14.1.27c)$$

Finally, we have deduced the asymptotic behavior of $y(x;\varepsilon)$ uniformly valid for all $0\leq x\leq 1$

$$y(x;\varepsilon)\sim f_0\left(x,\frac{x}{\varepsilon}\right)=e^{1/2}(e^{-x/2}-e^{x/2}e^{-2x/\varepsilon}) \text{ as } \varepsilon\rightarrow 0, \quad (14.1.28)$$

which differs from both (14.1.17c) and (14.1.18a) due to the presence of its $e^{x/2}$ factor.

If we were to construct a uniformly valid additive composite to $O(\varepsilon^n)$ by the Method of Matched Asymptotic Expansions in the usual way by adding the inner expansion to the outer determined to that order and subtracting their common part

$$y_u^{(n)}\left(x, \frac{x}{\varepsilon}; \varepsilon\right) = \sum_{k=0}^{n}\left[y_k(x) + Y_k\left(\frac{x}{\varepsilon}\right) - \sum_{j=0}^{n-k}\frac{y_k^{(j)}(0)}{j!}x^j\right]\varepsilon^k$$

and then rearranging it to form

$$y_u^{(n)}\left(x, \frac{x}{\varepsilon}; \varepsilon\right) = \sum_{k=0}^{n} y_{u_k}^{(n)}\left(x, \frac{x}{\varepsilon}\right)\varepsilon^k,$$

we would find that [135]

$$y_{u_0}^{(n)}\left(x, \frac{x}{\varepsilon}\right) = e^{1/2}\left[e^{-x/2} - \sum_{j=0}^{n}\frac{(x/2)^j}{j!}e^{-2x/\varepsilon}\right],$$

consistent with (14.1.18a) and (14.1.17c) for $n = 0$ and 1, respectively. Hence, since

$$e^{x/2} = \sum_{j=0}^{\infty}\frac{(x/2)^j}{j!}$$

we can deduce that the uniformly valid asymptotic representations determined upon employing the Methods of Multiple Scales and Matched Asymptotic Expansions are related by

$$f_0\left(x, \frac{x}{\varepsilon}\right) = \lim_{n\to\infty} y_{u_0}^{(n)}\left(x, \frac{x}{\varepsilon}\right).$$

Further, for an $(n+1)$-term multiple scales expansion of the form

$$y(x;\varepsilon) = f\left(x, \frac{x}{\varepsilon}; \varepsilon\right) = \sum_{k=0}^{n} f_k\left(x, \frac{x}{\varepsilon}\right)\varepsilon^k + O(\varepsilon^{n+1}) \text{ as } \varepsilon \to 0,$$

we would find, in general, that [135]

$$f_k\left(x, \frac{x}{\varepsilon}\right) = \lim_{n\to\infty} y_{u_k}^{(n)}\left(x, \frac{x}{\varepsilon}\right).$$

In this context note that [135]

$$f_1\left(x, \frac{x}{\varepsilon}\right) = \frac{e^{1/2}[(1-x)e^{-x/2} - (1+x)e^{x/2}e^{-2x/\varepsilon}]}{8}, \tag{14.1.29}$$

which bears the same partial sum relationship to (14.1.17d) as (14.1.18a) did to (14.1.27c).

We have deferred until now a discussion of why the boundedness conditions of the absolute values of the ratios involving successive terms in the multiple scales expansion ensure its uniformity. Van Dyke [125] perhaps explains it best when he states that if the $(n+1)$-term multiple scales expansion is to be uniformly valid then the ratios f_{k+1}/f_k must remain of order unity as $\varepsilon \to 0$, since this implies that each such approximation shall be no more singular than its predecessor or vanish no more slowly as $\varepsilon \to 0$ for arbitrary values of its independent variables. He also adds that the same thing must be true of all its derivatives.

We finally observe that in order to employ the method of multiple scales which involves the boundary-layer variable $\xi = x/\delta(\varepsilon)$ the boundary-layer thickness $\delta(\varepsilon)$ must be determined *a priori* as in Section 3.2 since the primary purpose of this method is to convert the inner solution determined by the Method of Matched Asymptotic Expansions into a uniformly valid approximation as occurred naturally in (3.2.21c) while the necessity of examining the problem for the next higher term in the expansion to determine any previous order precisely should also be borne in mind.

It only remains for us to re-examine our procedure for developing the uniformly valid expansion $y_{u0}^{(0)}(x, x/\varepsilon)$ of (3.2.17e) from the exact solution (3.2.17c) of (14.1.1). Recall that this exact solution was given by

$$y(x;\varepsilon) = \frac{e^{m^{(1)}(\varepsilon)x} - e^{m^{(2)}(\varepsilon)x}}{e^{m^{(1)}(\varepsilon)} - e^{m^{(2)}(\varepsilon)}} \text{ where } m^{(1,2)}(\varepsilon) = \frac{-1 \pm (1-\varepsilon)^{1/2}}{\varepsilon}.$$

In that development of Section 3.2 we retained only the lead term of the asymptotic series for each of these roots – *i.e.*, $m^{(1)}(\varepsilon) \sim -1/2$, $m^{(2)}(\varepsilon) \sim -2/\varepsilon$ as $\varepsilon \to 0$. Let us now retain those terms for each of these roots through $O(1)$ instead. That is replace the asymptotic representation for $m^{(2)}(\varepsilon)$ by $m^{(2)}(\varepsilon) \sim -2/\varepsilon + 1/2$ as $\varepsilon \to 0$. Then

$$y(x;\varepsilon) \sim \frac{e^{-x/2} - e^{x/2}e^{2x/\varepsilon}}{e^{-1/2} - e^{1/2}e^{-2/\varepsilon}} \sim e^{1/2}(e^{-x/2} - e^{x/2}e^{-2x/\varepsilon}) = f_0\left(x, \frac{x}{\varepsilon}\right) \text{ as } \varepsilon \to 0,$$

resolving the issue.

Finally for the sake of completeness we note that the singularly perturbed boundary value problem (3.3.1) yields the uniformly valid asymptotic representation of this type given by

$$f_0\left(x, \frac{x}{\varepsilon}\right) = e^{b_0}(e^{-b_0 x} - e^{b_0 x}e^{-x/\varepsilon}), \tag{14.1.30a}$$

which would result in the following refinement of our temperature-rate relation of (3.3.4a)

$$R(T) = \psi[e^{\rho T} - e^{\rho T_M - (T_M - T)/\Delta_1 T}] \text{ where } \frac{1}{\Delta_1 T} = \frac{1}{\Delta T} - \rho. \tag{14.1.30b}$$

Since this is of the same form as (3.3.4a) with the only change being the replacement of ΔT by $\Delta_1 T$ the mite parameter identification of (3.3.6) would be unaltered upon its employment instead.

14.2 Governing Equations of Motion, Vorticity, and the Stream Function

In order to resolve D'Alembert's paradox which predicted zero drag on a body for inviscid potential flow, we shall now use singular perturbation theory to examine Blasius flow past an infinite flat plate of a constant property homogeneous ($\rho \equiv \rho_0$) viscous fluid of relatively low kinematic viscosity (ν_0). Toward that end we consider such a fluid flowing past a semi-infinite flat plate located at $y^* = 0$, $x^* > 0$. We shall restrict our attention to a two-dimensional situation ($w^* = \partial/\partial z^* \equiv 0$) and a velocity profile satisfying the uniform parallel free-stream condition $\boldsymbol{v}^* = u^*\boldsymbol{i} + v^*\boldsymbol{j} \sim U_0\boldsymbol{i}$ as $y^* \to \infty$, $x^* > 0$ (see Fig. 14.1a).

Our governing Navier–Stokes equations and boundary conditions for this situation are as follows:

$$\nabla^* \bullet \boldsymbol{v}^* = 0, \tag{14.2.1a}$$

$$\frac{D^*\boldsymbol{v}^*}{D^*t^*} = \frac{\partial \boldsymbol{v}^*}{\partial t^*} + (\boldsymbol{v}^* \bullet \nabla^*)\boldsymbol{v}^* = -\rho_0^{-1}\nabla^* p^* + \nu_0 \nabla^{*2}\boldsymbol{v}^*; \tag{14.2.1b}$$

where $t^* \equiv$ time, $p^* \equiv$ reduced pressure, $\nabla^* \equiv (\partial/\partial x^*, \partial/\partial y^*, \partial/\partial z^*)$, and $\nabla^{*2} \equiv \nabla^* \bullet \nabla^*$; with

$$u^* = v^* = 0 \text{ at } y^* = 0; \tag{14.2.1c}$$

$$u^* \sim U_0,\ v^* \sim 0 \text{ as } y^* \to \infty; \text{ for } x^* > 0. \tag{14.2.1d}$$

Defining a length scale by selecting a reference point $x^* = \ell$ on the flat plate, we introduce the following nondimensional variables and parameter (see Fig. 14.1b)

$$(x, y) = \frac{(x^*, y^*)}{\ell},\ t = \frac{U_0 t^*}{\ell},\ (u, v) = \frac{(u^*, v^*)}{U_0},\ p = \frac{p^*}{\rho_0 U_0^2},\ R = \frac{U_0 \ell}{\nu_0} \equiv \text{Reynolds number};$$

which transforms the governing equations and boundary conditions of (14.2.1) into

$$\nabla \bullet \boldsymbol{v} = 0 \tag{14.2.2a}$$

$$\frac{D\boldsymbol{v}}{Dt} = \frac{\partial \boldsymbol{v}}{\partial t} + (\boldsymbol{v} \bullet \nabla)\boldsymbol{v} = -\nabla p + R^{-1}\nabla^2 \boldsymbol{v}; \tag{14.2.2b}$$

where $\nabla \equiv (\partial/\partial x, \partial/\partial y, \partial/\partial z)$ and $\nabla^2 \equiv \nabla \bullet \nabla$ with

$$u = v = 0 \text{ at } y = 0; \tag{14.2.2c}$$

$$u \sim 1,\ v \sim 0 \text{ as } y \to \infty; \text{ for } x > 0. \tag{14.2.2d}$$

Recall from (12.1.2a) and part (b) of Problem 12.1 that

$$(\boldsymbol{v} \bullet \nabla)\boldsymbol{v} = \frac{1}{2}\nabla(\boldsymbol{v} \bullet \boldsymbol{v}) - \boldsymbol{v} \times \boldsymbol{\omega} \text{ where } \boldsymbol{\omega} = \text{curl}(\boldsymbol{v}), \tag{14.2.3a}$$

$$\nabla^2 \boldsymbol{v} = \nabla(\nabla \bullet \boldsymbol{v}) - \text{curl}(\text{curl}\boldsymbol{v}) = -\text{curl}(\boldsymbol{\omega}); \tag{14.2.3b}$$

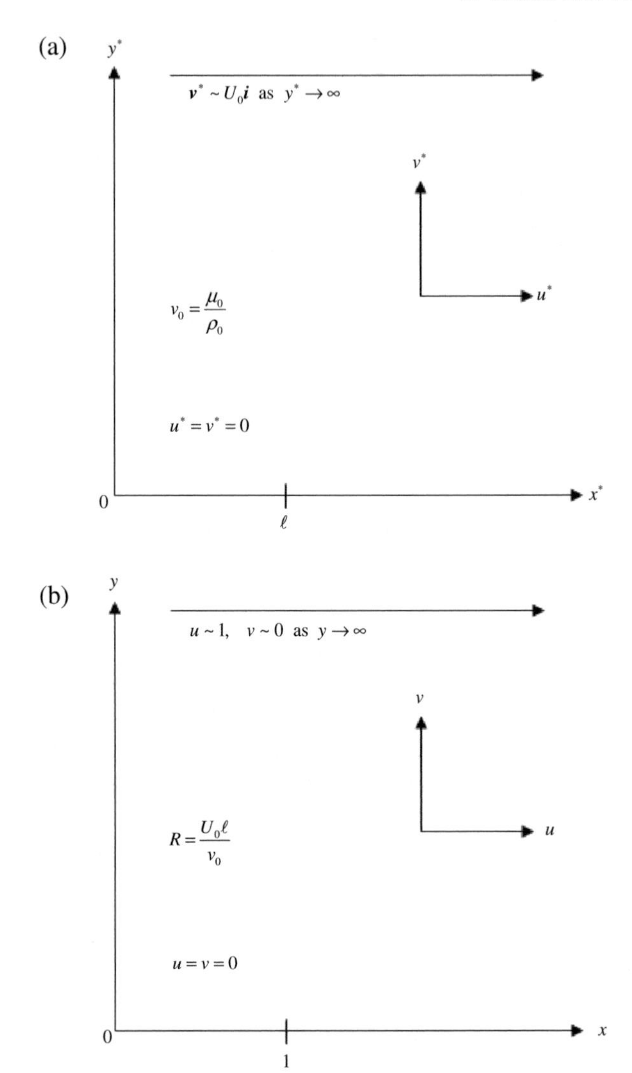

Fig. 14.1 Schematic representation of the Blasius flow scenario past an infinite flat plate located at $y^* = 0$ and $y = 0$ in (a) dimensional and (b) nondimensional variables, respectively.

the latter following by virtue of (14.2.2a). Substituting (14.2.3) into (14.2.2b) yields

$$\frac{\partial \boldsymbol{v}}{\partial t} = -\nabla\left(p + \frac{\boldsymbol{v} \bullet \boldsymbol{v}}{2}\right) + \boldsymbol{v} \times \boldsymbol{\omega} - R^{-1}\mathrm{curl}(\boldsymbol{\omega}). \tag{14.2.3c}$$

Note that for a steady-state ($\partial/\partial t \equiv 0$) irrotational ($\boldsymbol{\omega} \equiv \mathbf{0}$) situation, (14.2.3c) reduces to

$$\nabla\left(p + \frac{\boldsymbol{v} \bullet \boldsymbol{v}}{2}\right) = \mathbf{0} \Rightarrow p + \frac{|\boldsymbol{v}|^2}{2} \equiv \text{ constant.} \tag{14.2.3d}$$

Hence, Bernoulli's relation also holds for a steady-state viscous fluid provided it is irrotational.

Now taking the curl of (14.2.3c), assuming the continuity of the requisite second partial derivatives, and noting that R is a constant, we obtain

$$\frac{\partial \boldsymbol{\omega}}{\partial t} = \mathrm{curl}(\boldsymbol{v} \times \boldsymbol{\omega}) - R^{-1}\mathrm{curl}(\mathrm{curl}\boldsymbol{\omega}), \tag{14.2.4}$$

since $\mathrm{curl}(\nabla\varphi) = \mathbf{0}$. Recall from (12.1.5b) and (14.2.3b) with $\boldsymbol{v} = \boldsymbol{\omega}$ that

$$\mathrm{curl}(\boldsymbol{v} \times \boldsymbol{\omega}) = (\boldsymbol{\omega} \bullet \nabla)\boldsymbol{v} - (\boldsymbol{v} \bullet \nabla)\boldsymbol{\omega} + (\nabla \bullet \boldsymbol{\omega})\boldsymbol{v} - (\nabla \bullet \boldsymbol{v})\boldsymbol{\omega} = (\boldsymbol{\omega} \bullet \nabla)\boldsymbol{v} - (\boldsymbol{v} \bullet \nabla)\boldsymbol{\omega}, \tag{14.2.5a}$$

$$-\mathrm{curl}(\mathrm{curl}\boldsymbol{\omega}) = \nabla^2\boldsymbol{\omega} - \nabla(\nabla \bullet \boldsymbol{\omega}) = \nabla^2\boldsymbol{\omega}; \tag{14.2.5b}$$

by virtue of (14.2.2a) and (12.1.5c). Then substituting (14.2.5) into (14.2.4) yields

$$\frac{D\boldsymbol{\omega}}{Dt} = \frac{\partial \boldsymbol{\omega}}{\partial t} + (\boldsymbol{v} \bullet \nabla)\boldsymbol{\omega} = (\boldsymbol{\omega} \bullet \nabla)\boldsymbol{v} + R^{-1}\nabla^2\boldsymbol{\omega}, \tag{14.2.6}$$

which is the viscous form of (12.1.5d). For our two-dimensional situation

$$\boldsymbol{\omega} = \omega_3\boldsymbol{k} \text{ where } \omega_3 = \frac{\partial v}{\partial x} - \frac{\partial u}{\partial y} \tag{14.2.7a}$$

$$\Rightarrow (\boldsymbol{\omega} \bullet \nabla)\boldsymbol{v} = \left(\omega_3\frac{\partial}{\partial z}\right)\boldsymbol{v} = \mathbf{0}. \tag{14.2.7b}$$

Thus, substituting (14.2.7) into (14.2.6) and assuming a steady-state situation ($\partial/\partial t \equiv 0$) as well due to the uniformity of the free-stream boundary condition (14.2.2d), we obtain

$$u\frac{\partial \omega_3}{\partial x} + v\frac{\partial \omega_3}{\partial y} = R^{-1}\nabla_2^2\omega_3 \tag{14.2.8a}$$

$$\text{where, for } \nabla_2 \equiv \left(\frac{\partial}{\partial x}, \frac{\partial}{\partial y}\right), \nabla_2^2 \equiv \nabla_2 \bullet \nabla_2 = \frac{\partial^2}{\partial x^2} + \frac{\partial^2}{\partial y^2}; \tag{14.2.8b}$$

while our other governing equation (14.2.2a) takes the form

$$\nabla_2 \bullet \boldsymbol{v} = \frac{\partial u}{\partial x} + \frac{\partial v}{\partial y} = 0. \tag{14.2.8c}$$

Finally, as in (9.5.3), let us introduce a stream function

$$\psi = \psi(x,y) \in C^2[\{(x,y) : x > 0, y \geq 0\}] \text{ such that } u = \frac{\partial \psi}{\partial y} \text{ and } v = -\frac{\partial \psi}{\partial x}, \quad (14.2.9)$$

which satisfies (14.2.8c) identically while $\omega_3 = -\nabla_2^2 \psi$. Hence our governing equation (14.2.8) becomes

$$\left(\psi_y \frac{\partial}{\partial x} - \psi_x \frac{\partial}{\partial y} - R^{-1}\nabla_2^2\right)\nabla_2^2 \psi = 0; \; 0 < y < \infty \text{ for } x > 0; \quad (14.2.10a)$$

while our no-slip, no-penetration, and far-field boundary conditions (14.2.2c, 14.2.2d) become:

$$u(x,0) = \psi_y(x,0) = 0, \; v(x,0) = -\psi_x(x,0) = 0 \Rightarrow \psi(x,0) = 0 \text{ for } x > 0; \quad (14.2.10b)$$

$$u(x,y) = \psi_y(x,y) \sim 1, \; v(x,y) = -\psi_x(x,y) \sim 0 \text{ as } y \to \infty \text{ for } x > 0$$

$$\Rightarrow \psi(x,y) \sim y \text{ as } y \to \infty \text{ for } x > 0. \quad (14.2.10c)$$

14.3 Free-Stream and Boundary-Layer Solutions

We have deduced the following boundary value problem (14.2.10) satisfied by the stream function $\psi = \psi(x,y;\varepsilon)$ for Blasius flow when $x > 0$:

$$\left(\psi_y \frac{\partial}{\partial x} - \psi_x \frac{\partial}{\partial y}\right)\nabla^2 \psi = \varepsilon \nabla^4 \psi; \; 0 < \varepsilon = R^{-1} << 1; \; 0 < y < \infty; \quad (14.3.1a)$$

$$\psi_y(x,0;\varepsilon) = 0 \quad (14.3.1b)$$

and

$$\psi(x,0;\varepsilon) = 0; \quad (14.3.1c)$$

$$\psi(x,y;\varepsilon) \sim y \text{ as } y \to \infty. \quad (14.3.1d)$$

Here for ease of exposition we have denoted the two-dimensional Laplacian ∇_2^2 of (14.2.8b) by ∇^2 and in that context $\nabla^4 \equiv (\nabla^2)^2 = (\partial^2/\partial x^2 + \partial^2/\partial y^2)^2 = \partial^4/\partial x^4 + 2\partial^4/\partial x^2 \partial y^2 + \partial^4/\partial y^4$. This is a singular perturbation problem for $\psi(x,y;\varepsilon)$ as $\varepsilon = \nu_0/(U_0 \ell) \to 0$ since in that limit its highest order partial derivatives disappear. Hence in this section we shall examine its behavior by employing the Method of Matched Asymptotic Expansions.

(i) The outer or free-stream solution: For y in the free stream, we shall assume that ψ and all its derivatives are of $O(1)$ as $\varepsilon \to 0$. Then let

$\psi(x,y;\varepsilon) = \psi_0(x,y) + O(\varphi_1(\varepsilon))$ as $\varepsilon \to 0$ for y in the free stream where

$$\lim_{\varepsilon \to 0} \varphi_1(\varepsilon) = 0.$$

Thus

$$\psi_0(x,y) = \lim_{\varepsilon \to 0} \psi(x,y;\varepsilon)$$

satisfies

$$\left(\psi_{0_y}\frac{\partial}{\partial x} - \psi_{0_x}\frac{\partial}{\partial y}\right)\nabla^2\psi_0 = 0,\ 0 < y < \infty. \tag{14.3.2a}$$

Since the limit as $\varepsilon \to 0$ is essentially an inviscid one, we must drop the no-slip boundary condition (14.3.1b) while retaining the no-penetration condition (14.3.1c). Thus the boundary conditions are

$$\psi_0(x,0) = 0, \tag{14.3.2b}$$

$$\psi_0(x,y) \sim y \text{ as } y \to \infty. \tag{14.3.2c}$$

Given that (14.3.2a) is equivalent to the Jacobian

$$\frac{\partial(\nabla^2\psi_0, \psi_0)}{\partial(x,y)} = 0,$$

we deduce by the Implicit Function Theorem that $\nabla^2\psi_0 = -\omega(\psi_0)$. Noting that (14.3.2c) implies $\nabla^2\psi_0(x,y) \equiv 0$ as $y \to \infty$, we conclude that $\omega(\psi_0) \equiv 0$ everywhere in the free stream. Hence

$$\nabla^2\psi_0 = 0,\ 0 < y < \infty. \tag{14.3.3}$$

Finally (14.3.3) in conjunction with boundary conditions (14.3.2b, 14.3.2c) yields the unique solution

$$\psi_0(x,y) = y. \tag{14.3.4}$$

(ii) The inner or boundary-layer solution: We know that the outer or free-stream solution cannot be valid near the flat-plate body since the no-slip condition was dropped and use of that solution to calculate drag on the body results in D'Alembert's paradox. Given that u goes from 0 to 1 as y increases from 0 at the plate to its value in the free stream, there is a great change in $\partial u/\partial y$ as $\varepsilon \to 0$. Hence we must rescale y while retaining our original scale for x. We must also rescale ψ since for nonlinear and/or nonhomogeneous problems it is often necessary to rescale the dependent variable as well as the independent one. As a prelude to rescaling (14.3.1a) we now rewrite it in the form

$$\left(\psi_y\frac{\partial}{\partial x} - \psi_x\frac{\partial}{\partial y}\right)\left(\frac{\partial^2}{\partial x^2} + \frac{\partial^2}{\partial y^2}\right)\psi = \varepsilon\left(\frac{\partial^2}{\partial x^2} + \frac{\partial^2}{\partial y^2}\right)^2\psi. \tag{14.3.5}$$

We wish to select $s, p > 0$ such that if we define $\eta = y/\varepsilon^p$ and $\varphi = \psi/\varepsilon^s$ then φ and all its partial derivatives with respect to η are $O^*(1)$ as $\varepsilon \to 0$ near the plate.

This change of variables transforms the partial differential equation (14.3.5) upon multiplication by ε^{4p-s-1} into

$$\left(\frac{\partial^2}{\partial\eta^2}+\varepsilon^{2p}\frac{\partial^2}{\partial x^2}\right)^2\varphi$$
$$=\varepsilon^{3p+s-1}\left(\varphi_\eta\frac{\partial}{\partial x}-\varphi_x\frac{\partial}{\partial\eta}\right)\varphi_{xx}+\varepsilon^{p+s-1}\left(\varphi_\eta\frac{\partial}{\partial x}-\varphi_x\frac{\partial}{\partial\eta}\right)\varphi_{\eta\eta}. \tag{14.3.6}$$

We wish to select $s,p>0$ such that, in particular, $\partial^4\varphi/\partial\eta^4=O^*(1)$ as $\varepsilon\to 0$ near the plate. Thus as in equation (3.2.6) we have the following two cases to consider:

(a) $3p+s-1=0\Rightarrow p+s-1=-2p<0$ and hence must be rejected;
(b) $p+s-1=0\Rightarrow 3p+s-1=2p>0$ and hence is the correct relation.

Therefore we can conclude that $p+s=1$ and thus

$$\varphi_0(x,\eta)=\lim_{\varepsilon\to 0}\varphi(x,\eta;\varepsilon)$$

satisfies

$$\left(\frac{\partial^2}{\partial\eta^2}-\varphi_{0_\eta}\frac{\partial}{\partial x}+\varphi_{0_x}\frac{\partial}{\partial\eta}\right)\varphi_{0_{\eta\eta}}=0. \tag{14.3.7a}$$

Further since $\varphi_\eta(x,\eta;\varepsilon)=\varepsilon^{p-s}\psi_y(x,y;\varepsilon)=O^*(1)$ as $\varepsilon\to 0$ near the body we can conclude that $s=p=1/2$. Hence

$$\eta=\frac{y}{\varepsilon^{1/2}},\ \varphi(x,\eta;\varepsilon)=\frac{\psi(x,y;\varepsilon)}{\varepsilon^{1/2}},\ \text{and } \varphi_\eta(x,\eta;\varepsilon)=\psi_y(x,y;\varepsilon).$$

Finally boundary conditions (14.3.1b) at $y=0$ imply that

$$\varphi_{0_\eta}(x,0)=\varphi_0(x,0)=0. \tag{14.3.7b}$$

(iii) Intermediate limit technique of matching: Let $\gamma=y/\theta(\varepsilon)=O^*(1)$ as $\varepsilon\to 0$ for y in an intermediate region where $\theta(\varepsilon)>0$ and

$$\lim_{\varepsilon\to 0}\theta(\varepsilon)=\lim_{\varepsilon\to 0}\frac{\varepsilon^{1/2}}{\theta(\varepsilon)}=0.$$

Then matching on the horizontal velocity component to lowest order across the boundary layer

$$\lim_{\varepsilon\to 0}[\psi_{0_y}(x,\gamma\theta)]=\lim_{\varepsilon\to 0}\left[\varphi_{0_\eta}\left(x,\frac{\gamma\theta}{\varepsilon^{1/2}}\right)\right],$$

which implies that

$$\varphi_{0_\eta}(x,\eta)\sim\psi_{0_y}(x,0)\equiv 1 \text{ as } \eta\to\infty. \tag{14.3.7c}$$

We now shall solve (14.3.7) for $\varphi_0(x,\eta)$ by first observing that (14.3.7a) is equivalent to

$$\frac{\partial}{\partial\eta}(\varphi_{0_{\eta\eta\eta}} + \varphi_{0_x}\varphi_{0_{\eta\eta}} - \varphi_{0_\eta}\varphi_{0_{x\eta}}) = 0$$

since upon expansion the left-hand side of the above equation becomes

$$\begin{aligned}\varphi_{0_{\eta\eta\eta\eta}} + \varphi_{0_{x\eta}}\varphi_{0_{\eta\eta}} + \varphi_{0_x}\varphi_{0_{\eta\eta\eta}} - \varphi_{0_{\eta\eta}}\varphi_{0_{x\eta}} - \varphi_{0_\eta}\varphi_{0_{x\eta\eta}}\\ = \varphi_{0_{\eta\eta\eta\eta}} + \varphi_{0_x}\varphi_{0_{\eta\eta\eta}} - \varphi_{0_\eta}\varphi_{0_{x\eta\eta}}.\end{aligned}$$

Hence we can conclude that

$$\varphi_{0_{\eta\eta\eta}} + \varphi_{0_x}\varphi_{0_{\eta\eta}} - \varphi_{0_\eta}\varphi_{0_{x\eta}} = g(x). \tag{14.3.8a}$$

Since (14.3.7c) implies that

$$\varphi_{0_{\eta\eta}}(x,\eta),\ \varphi_{0_{\eta\eta\eta}}(x,\eta) \sim 0 \text{ while } \varphi_{0_{\eta x}}(x,\eta) \sim \psi_{0_{yx}}(x,0) \text{ as } \eta \to \infty,$$

we can deduce the following formula for $g(x)$ from this behavior as $\eta \to \infty$:

$$g(x) = -\psi_{0_y}(x,0)\psi_{0_{yx}}(x,0) \equiv 0. \tag{14.3.8b}$$

Thus denoting $\varphi_0(x,\eta)$ by $\varphi(x,\eta)$ for ease of exposition the problem for the lowest order term in the boundary-layer solution reduces to

$$\varphi_{\eta\eta\eta} + \varphi_x\varphi_{\eta\eta} - \varphi_\eta\varphi_{x\eta} = 0;\ 0 < \eta < \infty; \tag{14.3.9a}$$

$$\varphi(x,0) = \varphi_\eta(x,0) = 0; \tag{14.3.9b}$$

$$\varphi_\eta(x,\eta) \sim 1 \text{ as } \eta \to \infty. \tag{14.3.9c}$$

Let us first consider the transformation

$$x' = \frac{x}{c^2},\ \eta' = \frac{\eta}{c},\ \varphi' = \frac{\varphi}{c} \text{ for } c > 0.$$

Then (14.3.9) becomes for $\varphi' = \varphi'(x',\eta')$:

$$\varphi'_{\eta'\eta'\eta'} + \varphi'_{x'}\varphi'_{\eta'\eta'} - \varphi'_{\eta'}\varphi'_{x'\eta'} = 0;\ 0 < \eta' < \infty; \tag{14.3.10a}$$

$$\varphi'(x',0) = \varphi'_{\eta'}(x',0) = 0; \tag{14.3.10b}$$

$$\varphi'_{\eta'}(x',\eta') \sim 1 \text{ as } \eta' \to \infty. \tag{14.3.10c}$$

Since the transformed system (14.3.10) is identical in form to the original system (14.3.9), this tells us that our original system has a *similarity solution* such that $\varphi/\sqrt{x}$ must be a function of $\eta/\sqrt{x}$ which will convert the partial differential equation (14.3.9a) into an ordinary differential equation. Hence we assume that for $x > 0$

$$\varphi(x,\eta) = \sqrt{2x}f(\alpha) \text{ where } \alpha = \frac{\eta}{\sqrt{2x}}. \tag{14.3.11}$$

Note that

$$\frac{\partial\alpha}{\partial\eta}=\frac{1}{\sqrt{2x}} \text{ and } \frac{\partial\alpha}{\partial x}=-\frac{\eta}{2x\sqrt{2x}};$$

$$\varphi_\eta=f'(\alpha),\ \varphi_{\eta\eta}=\frac{f''(\alpha)}{\sqrt{2x}},\ \varphi_{\eta\eta\eta}=\frac{f'''(\alpha)}{2x},$$

$$\varphi_x=\frac{f(\alpha)}{\sqrt{2x}}-\eta\frac{f'(\alpha)}{2x}=\frac{f(\alpha)-\alpha f'(\alpha)}{\sqrt{2x}},$$

$$\varphi_{x\eta}=\frac{f'(\alpha)-f'(\alpha)-\alpha f''(\alpha)}{2x}=-\alpha\frac{f''(\alpha)}{2x}.$$

Therefore (14.3.11) transforms (14.3.9a) into

$$\begin{aligned}\varphi_{\eta\eta\eta}+\varphi_x\varphi_{\eta\eta}-\varphi_\eta\varphi_{x\eta}&=\frac{f'''(\alpha)+f''(\alpha)f(\alpha)-\alpha f''(\alpha)f'(\alpha)+\alpha f'(\alpha)f''(\alpha)}{2x}\\&=\frac{f'''(\alpha)+f''(\alpha)f(\alpha)}{2x}=0.\end{aligned}$$

Thus (14.3.9) under this transformation becomes for $x>0$

$$f'''(\alpha)+f''(\alpha)f(\alpha)=0,\ 0<\alpha<\infty; \tag{14.3.12a}$$

$$f(0)=f'(0)=0; \tag{14.3.12b}$$

$$f'(\alpha)\sim 1 \text{ as } \alpha\to\infty. \tag{14.3.12c}$$

Next consider

$$f(\alpha)=aF(A) \text{ where } A=a\alpha \text{ for } a>0. \tag{14.3.13}$$

Then

$$f'(\alpha)=a^2F'(A), f''(\alpha)=a^3F''(A), f'''(\alpha)=a^4F'''(A).$$

Therefore (14.3.13) transforms (14.3.12a) into

$$f'''(\alpha)+f''(\alpha)f(\alpha)=a^4[F'''(A)+F''(A)F(A)]=0.$$

Thus (14.3.12) under this transformation becomes

$$F'''(A)+F''(A)F(A)=0,\ 0<A<\infty; \tag{14.3.14a}$$

$$F(0)=F'(0)=0; \tag{14.3.14b}$$

$$a^2F'(A)\sim 1 \text{ as } A\to\infty. \tag{14.3.14c}$$

We now select $F''(0)=1$ and find a power series solution for

$$F(A)=\sum_{n=0}^{\infty}\frac{F^{(n)}(0)}{n!}A^n$$

where by virtue of Leibniz's Rule of Differentiation of a product in conjunction with (14.3.14a)

$$F^{(N+3)}(0) = -\sum_{k=0}^{N} \binom{N}{k} F^{(N+2-k)}(0)F^{(k)}(0) \text{ for } N \geq 0.$$

Computing these derivatives we find that [1]

$$F(A) = \frac{A^2}{2!} - \frac{A^5}{5!} + \frac{11A^8}{8!} - \frac{375A^{11}}{11!} + \frac{27,897A^{14}}{14!} - \cdots \tag{14.3.15}$$

Hence

$$a^2 \lim_{A\to\infty} F'(A) = 1 \Rightarrow a = [F'(\infty)]^{-1/2}. \tag{14.3.16}$$

From this result we obtain in particular

$$f''(0) = a^3F''(0) = a^3 = [F'(\infty)]^{-3/2} = 0.4696, \tag{14.3.17a}$$

from which it can be deduced that

$$f(\alpha) \sim \alpha - \beta_1 \text{ as } \alpha \to \infty \text{ where } \beta_1 = 1.2168. \tag{14.3.17b}$$

Both of these values included in (14.3.17) will play a significant role for the parameters of the boundary-layer to be developed in the next section.

14.4 Parameters of the Boundary Layer

To synthesize our results from the previous section, we have found for Blasius flow past a flat plate that

$$\psi(x,y;\varepsilon) \sim \begin{cases} \psi_0(x,y) = y, & \text{for } y \text{ in the free-stream} \\ \varepsilon^{1/2}\varphi_0(x,\eta) = \sqrt{2x\varepsilon}f(y/\sqrt{2x\varepsilon}), & \text{for } y \text{ in the boundary layer} \end{cases} \text{ as } \varepsilon \to 0; \tag{14.4.1a}$$

where

$$f(0) = f'(0) = 0, f''(0) = 0.4696, f(\alpha) \sim \alpha - \beta_1 \text{ as } \alpha \to \infty \text{ with } \beta_1 = 1.2168. \tag{14.4.1b}$$

We now wish to calculate the following parameters of the boundary layer for this flow:

(i) Boundary-layer thickness: Recall that $u = \partial\psi/\partial y$, denote $\psi_{0_y}(x,y)$ by $\mathcal{U}(x,y)$, and for $0 \leq y \leq \delta$ in the boundary layer examine (see Fig. 14.2)

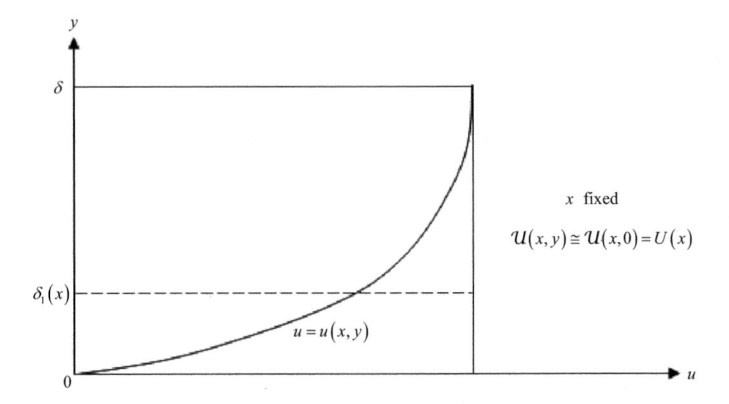

Fig. 14.2 Plots relevant to the definition of boundary-layer thickness $\delta_1(x)$.

$$\begin{aligned}\int_0^{\delta}[\mathcal{U}(x,y)-u(x,y)]\,dy &\cong \int_0^{\delta}[\mathcal{U}(x,0)-u(x,y)]\\ &\cong U(x)\int_0^{\delta}\left[1-\frac{u(x,y)}{U(x)}\right]dy\\ &\cong U(x)\int_0^{\infty}\left[1-\frac{u(x,y)}{U(x)}\right]dy\\ &= U(x)\delta_1(x)\end{aligned}$$

where

$$U(x)=\mathcal{U}(x,0)\text{ and }\delta_1(x)=\int_0^{\infty}\left[1-\frac{u(x,y)}{U(x)}\right]dy\equiv\text{ boundary-layer thickness.}$$

For Blasius flow

$$U(x)\equiv 1\text{ and }u(x,y)=f'(\alpha)\text{ where }\alpha=\frac{y}{\sqrt{2x\varepsilon}}=\sqrt{\frac{R}{2x}}y.$$

Therefore

$$\delta_1(x) = \int_0^\infty [1 - f'(\alpha)]\,dy = \left(\frac{2x}{R}\right)^{1/2} \int_0^\infty [1 - f'(\alpha)]\,d\alpha$$
$$= \left(\frac{2x}{R}\right)^{1/2} [\alpha - f(\alpha)]_0^\infty = \left(\frac{2x}{R}\right)^{1/2} \beta_1$$
$$= 1.2168\sqrt{2}\left(\frac{x}{R}\right)^{1/2} = 1.7208\left(\frac{x}{R}\right)^{1/2},$$

which is the parabolic curve plotted in Fig. 14.3.

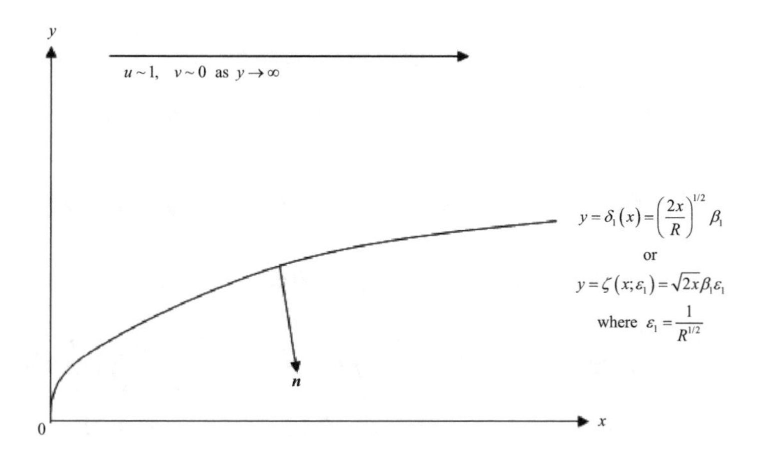

Fig. 14.3 Schematic representation of both the boundary layer $0 \le y \le \delta_1(x)$ to $O(1/R^{1/2})$ for Blasius flow past an infinite plate and inviscid free-stream flow past the corresponding parabolic body with boundary $y = \zeta(x;\varepsilon_1)$.

(ii) Drag: Consider the dimensional component of the stress tensor

$$T^*_{x^*y^*} = \mu_0\left(\frac{\partial u^*}{\partial y^*} + \frac{\partial v^*}{\partial x^*}\right)$$

and introduce its corresponding nondimensionalized form

$$T_{xy} = \frac{T^*_{x^*y^*}}{\rho_0 U_0^2} = \frac{\mu_0}{\rho_0 U_0^2}\frac{U_0}{\ell}\left(\frac{\partial u}{\partial y} + \frac{\partial v}{\partial x}\right) = \frac{\nu_0}{U_0\ell}\left(\frac{\partial u}{\partial y} + \frac{\partial v}{\partial x}\right) = \frac{1}{R}\left(\frac{\partial u}{\partial y} + \frac{\partial v}{\partial x}\right).$$

Then

$$T_{xy}\big|_{y=0} = \frac{1}{R}\left[\frac{\partial u}{\partial y}(x,0) + \frac{\partial v}{\partial x}(x,0)\right] = \frac{1}{R}\frac{\partial u}{\partial y}(x,0)$$

given that $\partial v(x,0)/\partial x \equiv 0$ by virtue of $v(x,0) \equiv 0$. Since

$$\frac{\partial u}{\partial y}(x,y) = \frac{\partial f'(\alpha)}{\partial y} = \left(\frac{R}{2x}\right)^{1/2} f''(\alpha),$$

$$C_D \equiv \text{coefficient of drag} = T_{xy}\big|_{y=0} = \frac{1}{R}\left(\frac{R}{2x}\right)^{1/2} f''(0) = \frac{f''(0)}{\sqrt{2Rx}}.$$

Thus

$$D = \int_0^1 C_D\,dx = \frac{f''(0)}{\sqrt{R}}\int_0^1 \frac{dx}{\sqrt{2x}} = \frac{f''(0)}{\sqrt{R}}\sqrt{2x}\Big|_0^1 = \frac{f''(0)\sqrt{2}}{\sqrt{R}} = \frac{0.4696\sqrt{2}}{\sqrt{R}} = \frac{0.6641}{\sqrt{R}}.$$

Hence the stress vector on the plate has components

$$t_i = T_{ij}\big|_{y=0} n_j \text{ with } \boldsymbol{n} = -\boldsymbol{j} \text{ or } n_x = 0,\ n_y = -1 \Rightarrow t_i = -T_{iy}\big|_{y=0} \text{ for } i = x, y \Rightarrow t_x = -C_D.$$

Therefore the total force on the plate from $0 < x^* \leq \ell$ per unit length in the z-direction

$$\boldsymbol{F} = \int_0^1 \boldsymbol{t}\,dx = \int_0^1 (t_x\boldsymbol{i} + t_y\boldsymbol{j})\,dx = -F_1\boldsymbol{i} + F_2\boldsymbol{j}$$

$$\Rightarrow F_1 \equiv \text{drag} = -\int_0^1 t_x\,dx = \int_0^1 C_D\,dx = D = \frac{0.6641}{\sqrt{R}} = \frac{0.6641}{\sqrt{U_0\ell}}\sqrt{\nu_0} > 0;$$

thus resolving D'Alembert's paradox of Section 12.5 no matter how small ν_0 might be.

14.5 Physical Interpretation

In order to motivate the physical interpretation to be presented in this section, we now determine the next term in the free-stream solution by considering

$$\psi(x,y;\varepsilon) = \psi_0(x,y) + \psi_1(x,y)\varphi_1(\varepsilon) + o(\varphi_1(\varepsilon)) \text{ as } \varepsilon \to 0 \text{ for } y \text{ in the free stream.} \tag{14.5.1a}$$

That is we represent (14.4.1a) in the form

$$\psi(x,y;\varepsilon) = \begin{cases} y + \psi_1(x,y)\varphi_1(\varepsilon) + o(\varphi_1(\varepsilon)), & \text{for } y \text{ in the free stream} \\ \sqrt{2x\varepsilon}f(y/\sqrt{2x\varepsilon}) + o(\varepsilon^{1/2}), & \text{for } y \text{ in the boundary layer} \end{cases} \text{ as } \varepsilon \to 0;$$

where

$$f(\alpha) \sim \alpha - \beta_1 \text{ as } \alpha \to \infty.$$

For y in the intermediate region such that

$$\gamma = \frac{y}{\theta(\varepsilon)} = O^*(1) \text{ where } \lim_{\varepsilon\to 0}\theta(\varepsilon) = \lim_{\varepsilon\to 0}\frac{\varepsilon^{1/2}}{\theta(\varepsilon)} = 0 \text{ as } \varepsilon \to 0,$$

these asymptotic representations are both valid and hence equal in this limit. Thus

$$\begin{aligned}\gamma\theta + \psi_1(x,\gamma\theta)\varphi_1(\varepsilon) + o(\varphi_1(\varepsilon)) &= \gamma\theta + \psi_1(x,0)\varphi_1(\varepsilon) + O(\theta\varphi_1(\varepsilon)) + o(\varphi_1(\varepsilon)) \text{ as } \varepsilon \to 0\\ &= \sqrt{2x\varepsilon} f\left(\frac{\gamma\theta}{\sqrt{2x\varepsilon}}\right) + o(\varepsilon^{1/2})\\ &\sim \sqrt{2x\varepsilon}\left(\frac{\gamma\theta}{\sqrt{2x\varepsilon}} - \beta_1\right) + o(\varepsilon^{1/2})\\ &= \gamma\theta - \sqrt{2x}\beta_1\varepsilon^{1/2} + o(\varepsilon^{1/2}) \text{ as } \varepsilon \to 0\end{aligned}$$

which implies that

$$\varphi_1(\varepsilon) = \varepsilon^{1/2} \text{ and } \psi_1(x,0) = -\sqrt{2x}\beta_1. \tag{14.5.1b}$$

Therefore we seek a free-stream solution of the form

$$\psi(x,y;\varepsilon) = \psi_0(x,y) + \psi_1(x,y)\varepsilon^{1/2} + o(\varepsilon^{1/2}) \text{ as } \varepsilon \to 0 \text{ for } y \text{ in the stream;} \tag{14.5.2a}$$

where

$$\psi_0(x,y) = y \text{ and } \psi_1(x,0) = -\sqrt{2x}\beta_1. \tag{14.5.2b}$$

Note that the nonhomogeneous boundary condition for ψ_1 at $y = 0$ is not consistent with the no-penetration boundary condition (14.3.1c) from which we deduced that $\psi_0(x,0) = 0$.

Substituting (14.5.2) into (14.3.1a, 14.3.1d), we obtain the following equation and free-stream boundary condition satisfied by $\psi_1(x,y)$:

$$\begin{aligned}O(\varepsilon^{1/2}) : &\left(\psi_{0_y}\frac{\partial}{\partial x} - \psi_{0_x}\frac{\partial}{\partial y}\right)\nabla^2\psi_1 + \left(\psi_{1_y}\frac{\partial}{\partial x} - \psi_{1_x}\frac{\partial}{\partial y}\right)\nabla^2\psi_0\\ &= \left(\psi_{0_y}\frac{\partial}{\partial x} - \psi_{0_x}\frac{\partial}{\partial y}\right)\nabla^2\psi_1 = \frac{\partial(\psi_0, \nabla^2\psi_1)}{\partial(x,y)} = 0\end{aligned}$$

since $\nabla^2\psi_0 = 0$ which implies that

$$\nabla^2\psi_1 = -\omega_1(\psi_0) \tag{14.5.3a}$$

and

$$\psi_1(x,y) = o(y) \text{ as } y \to \infty$$

which implies that

$$\psi_1(x,y) \sim y^a \text{ as } y \to \infty \text{ where } 0 < a < 1, \tag{14.5.3b}$$

respectively. Far-field condition (14.5.3b) implies that $\nabla^2\psi_1 = 0$ as $y \to \infty$ from which we can conclude by analytic continuation that $\omega_1(\psi_0) \equiv 0$ in the free stream and hence obtain the following problem for $\psi_1(x,y)$:

$$\nabla^2\psi_1 = 0, 0 < y < \infty; \psi_1(x,0) = -\sqrt{2x}\beta_1, \psi_1(x,y) = o(y) \text{ as } y \to \infty. \qquad (14.5.4a)$$

This problem has solution

$$\psi_1(x,y) = -\beta_1\sqrt{2}\mathrm{Re}\left(2\sqrt{x+iy}\right). \qquad (14.5.4b)$$

which is demonstrated as follows:
Let $z = x + iy$. Then $\sqrt{x+iy} = \sqrt{z}$ is analytic for $x > 0$, $y \geq 0$ which implies that for $\sqrt{z} = \mathcal{u}(x,y) + i\mathcal{v}(x,y)$, $\mathcal{u}$ and $\mathcal{v} \in \mathbb{R}$ satisfy the Cauchy–Riemann conditions $\mathcal{u}_x = \mathcal{v}_y$ and $\mathcal{u}_y = -\mathcal{v}_x$. Thus $\mathcal{u}_{xx} = \mathcal{v}_{yx} = \mathcal{v}_{xy} = -\mathcal{u}_{yy}$ which demonstrates that $\nabla^2\psi_1 = 0$.

Let $z = re^{i\theta}$. Then $|z| = r = \sqrt{x^2+y^2} \sim |y|$ as $y \to \infty$. Thus $|\sqrt{z}| \sim |y|^{1/2}$ as $y \to \infty$ or in particular (14.5.3b) holds for $a = 1/2$ which demonstrates that $\psi_1(x,y) = o(y)$ as $y \to \infty$.

Finally, $\psi_1(x,0) = -\beta_1\sqrt{2}\mathrm{Re}(\sqrt{x}) = -\beta_1\sqrt{2x}$, completing the demonstration. It remains for us to explain the apparent inconsistency of this boundary condition with $\psi_0(x,0) = 0$ and in the process offer a physical interpretation of Blasius flow past an infinite flat plate.

Consider a two-dimensional uniform stream flow of an inviscid homogeneous fluid past a parabolic body located at (see Fig. 14.3)

$$y = \zeta(x;\varepsilon_1) = \sqrt{2x}\beta_1\varepsilon_1 \text{ where } 0 < \varepsilon_1 << 1. \qquad (14.5.5a)$$

In what follows we shall treat ε_1 as a small positive arbitrary parameter although it will ultimately be identified with $\varepsilon^{1/2} = 1/R^{1/2}$ and $\zeta(x;\varepsilon_1)$ with $\delta_1(x)$ the boundary-layer thickness. This sort of fluid is both incompressible and irrotational as well as steady-state. Thus

$$\frac{\partial u}{\partial x} + \frac{\partial v}{\partial y} = \frac{\partial v}{\partial x} - \frac{\partial u}{\partial y} = 0$$

and hence there exists a stream function $\psi = \psi(x,y;\varepsilon_1)$ such that

$$u = \frac{\partial \psi}{\partial y} \text{ and } v = -\frac{\partial \psi}{\partial x}$$

$$\Rightarrow \nabla^2\psi = 0; x > 0, \zeta(x;\varepsilon_1) < y < \infty. \qquad (14.5.5b)$$

Further, we have the usual far-field condition

$$\psi(x,y;\varepsilon_1) \sim y \text{ as } y \to \infty \qquad (14.5.5c)$$

while the no-penetration condition at the boundary $f(x,y;\varepsilon_1) = \zeta(x;\varepsilon_1) - y = 0$ yields

$$\boldsymbol{v} \bullet \boldsymbol{n} = 0$$

where

$$\boldsymbol{v} = u\boldsymbol{i} + v\boldsymbol{j} = \psi_y \boldsymbol{i} - \psi_x \boldsymbol{j}$$

and

$$\boldsymbol{n} = \frac{\nabla f}{|\nabla f|} = \frac{\zeta'(x, y; \varepsilon)\boldsymbol{i} - \boldsymbol{j}}{\sqrt{1 + [\zeta'(x, y; \varepsilon)]^2}}$$

or

$$\psi_x[x, \zeta(x; \varepsilon_1); \varepsilon_1] + \psi_y[x, \zeta(x; \varepsilon_1); \varepsilon_1]\zeta'(x, y, \varepsilon_1) = \frac{d}{dx}\psi[x, \zeta(x; \varepsilon_1); \varepsilon_1] = 0$$

$$\Rightarrow \psi[x, \zeta(x; \varepsilon_1); \varepsilon_1] = 0, \qquad (14.5.5d)$$

with no loss of generality, which implies (14.5.5a) is a streamline for the flow and completes the formulation of the problem.

Let

$$\psi(x, y; \varepsilon_1) = \psi_0(x, y) + \psi_1(x, y)\varepsilon_1 + O(\varepsilon_1^2) \text{ as } \varepsilon_1 \to 0. \qquad (14.5.6)$$

Substituting (14.5.6) into (14.5.5),

$$\nabla^2\psi = \nabla^2\psi_0 + \nabla^2\psi_1\varepsilon_1 + O(\varepsilon_1^2) = 0; x > 0, \zeta(x; \varepsilon_1) < y < \infty; \qquad (14.5.7a)$$

$$\psi(x, y; \varepsilon_1) = \psi_0(x, y) + \psi_1(x, y)\varepsilon_1 + O(\varepsilon_1^2) \sim y \text{ as } y \to \infty; \qquad (14.5.7b)$$

$$\psi[x, \sqrt{2x}\beta_1\varepsilon_1; \varepsilon_1] = \psi_0(x, \sqrt{2x}\beta_1\varepsilon_1) + \psi_1(x, \sqrt{2x}\beta_1\varepsilon_1)\varepsilon_1 + O(\varepsilon_1^2) \qquad (14.5.7c)$$

$$= \psi_0(x, 0) + \psi_{0_y}(x, 0)\sqrt{2x}\beta_1\varepsilon_1 + (\varepsilon_1^2) + \varepsilon_1[\psi_1(x, 0) + O(\varepsilon_1^2)] + O(\varepsilon_1^2) \qquad (14.5.7d)$$

$$= \psi_0(x, 0) + [\psi_1(x, 0) + \psi_{0_y}(x, 0)\sqrt{2x}\beta_1]\varepsilon_1 + O(\varepsilon_1^2) = 0; \qquad (14.5.7e)$$

we obtain the following sequence of problems for $\psi_0(x, y)$ and $\psi_1(x, y)$, respectively:

$$O(1): \nabla^2\psi_0 = 0, 0 < y < \infty; \psi_0(x, 0) = 0; \psi_0(x, y) \sim y \text{ as } y \to \infty; \qquad (14.5.8a)$$

$$O(\varepsilon_1): \nabla^2\psi_1 = 0, 0 < y < \infty; \psi_1(x, 0) = -\psi_{0_y}(x, 0)\sqrt{2x}\beta_1; \psi_1(x, y) = o(y) \text{ as } y \to \infty. \qquad (14.5.8b)$$

Problem (14.5.8a) is identical to (14.3.3) in conjunction with boundary conditions (14.3.2b, 14.3.2c) and hence yields the same solution as (14.3.4) or

$$\psi_0(x, y) = y \qquad (14.5.9a)$$

while then (14.5.8b) is identical to (14.5.4a) and hence yields the same solution as (14.5.4b) or

$$\psi_1(x, y) = -\beta_1 \mathrm{Re}\left(2\sqrt{x + iy}\right). \qquad (14.5.9b)$$

Thus, for $\varepsilon_1 = \varepsilon^{1/2}$, (14.5.6) is equivalent to the free-stream solution of (14.5.2) through terms of $O(\varepsilon^{1/2})$ and $\zeta(x;\varepsilon^{1/2}) = \delta_1(x)$, the boundary-layer thickness, through terms of the same order. This implies that the outer solution never actually satisfied the inner boundary condition (14.3.1c) at the plate $y = 0$ but rather the condition (14.5.5d) at the top of the boundary layer $y = \delta_1(x) = \beta_1\sqrt{2x\varepsilon}$ instead. Note in this context that $\delta_1(x) = \beta_1\sqrt{2x\varepsilon} \to 0$ as $\varepsilon \to 0$ which accounts for the anomalous condition $\psi_0(x,0) = 0$. Physically this means that the free-stream solution can be interpreted as slipping inviscidly past an effective body consisting of the flat plate plus the boundary layer which it cannot penetrate while the boundary-layer solution satisfies both the no-slip and no-penetration conditions at that plate. D'Alembert's paradox occurs because this outer free-stream solution is being employed to calculate drag on a body rather than that inner boundary-layer solution.

Problems

14.1. Consider the following singularly perturbed boundary value problem for $y = y(x;\varepsilon)$:

$$\varepsilon\frac{d^2y}{dx^2} + (1+x)\frac{dy}{dx} + y = 0,\ 0 < x < 1,\ 0 < \varepsilon << 1;$$

$$y(0;\varepsilon) = 0,\ y(1;\varepsilon) = 1.$$

(a) Given that the boundary layer occurs at $x = 0$ and has thickness $\delta(\varepsilon) = \varepsilon$, find the first two terms in the outer and inner expansions for $y(x;\varepsilon)$ employing the intermediate limit technique of matched asymptotic expansions designating them by

$$y_0(x) + \varepsilon y_1(x) \text{ and } Y_0(\xi) + \varepsilon Y_1(\xi) \text{ where } \xi = \frac{x}{\varepsilon},$$

respectively.

(b) Use these results to construct a two-term uniformly valid additive composite defined by

$$y_u^{(1)}\left(x,\frac{x}{\varepsilon};\varepsilon\right) = y_0(x) + \varepsilon y_1(x) + Y_0\left(\frac{x}{\varepsilon}\right) + \varepsilon Y_1\left(\frac{x}{\varepsilon}\right) - y_0(0) - y_0'(0)x - \varepsilon y_1(0)$$

and represent it in the form

$$y_u^{(1)}\left(x,\frac{x}{\varepsilon};\varepsilon\right) = y_{u_0}^{(1)}\left(x,\frac{x}{\varepsilon}\right) + \varepsilon y_{u_1}^{(1)}\left(x,\frac{x}{\varepsilon}\right).$$

Compare $y_{u_0}^{(1)}$ to the one-term uniformly valid additive composite

$$y_u^{(0)}\left(x,\frac{x}{\varepsilon}\right) = y_0(x) + \varepsilon Y_0\left(x,\frac{x}{\varepsilon}\right) - y_0(0)$$

and note in this context that the term $\varepsilon\xi^2 e^{-\xi} = (x^2/\varepsilon)e^{-x/\varepsilon}$ belongs in $y_{u_0}^{(1)}$ rather than in $\varepsilon y_{u_1}^{(1)}$.

(c) Find a one-term uniformly valid asymptotic representation for $y(x;\varepsilon)$ by the following generalized method of multiple scales. Define

$$y(x;\varepsilon) = f(x,\zeta;\varepsilon) = f_0(x,\zeta) + \varepsilon f_1(x,\zeta) + O(\varepsilon^2) \text{ as } \varepsilon \to 0,\ \zeta = \frac{g(x)}{\varepsilon} > 0,\ 0 < x < 1;$$

where $g(x)$ is such that $g(x) \sim x$ as $x \to 0$ and the partial differential equation satisfied by f_0 with respect to ζ is of the same form as the ordinary differential equation satisfied by Y_0 with respect to ξ. Once $g(x)$ has been determined, calculate $f_0(x,\zeta)$ by proceeding exactly as in the regular method of multiple scales, with ζ assuming the same role as ξ, and compare it for $\zeta = g(x)/\varepsilon$ to $y_{u_0}^{(1)}(x, x/\varepsilon)$ from part (b). Hint: $g(x) = x + x^2/2$ and observe that unlike the regular method of multiple scales the change of variables involves an extra term proportional to $g''(x)$. Further $y_{u_0}^{(1)}(x, x/\varepsilon)$ will bear the same partial sum relationship to $f_0(x, [x + x^2/2]/\varepsilon)$ as it did to $f_0(x, x/\varepsilon)$ in the regular method of multiple scales. That is $e^{-x^2/(2\varepsilon)} \sim 1 - x^2/(2\varepsilon)$.

14.2. Consider Blasius-type flow past an infinite flat plate where the velocity profile satisfies the free-stream condition

$$\boldsymbol{v}^* = u^*\boldsymbol{i} + v^*\boldsymbol{j} \sim \frac{U_0}{\ell}(x^*\boldsymbol{i} - y^*\boldsymbol{j}) \text{ as } y^* \to \infty,\ x^* > 0.$$

This implies that the velocity components satisfy the far-field conditions

$$u^* \sim \frac{U_0 x^*}{\ell},\ v^* \sim -\frac{U_0 y^*}{\ell} \text{ as } y^* \to \infty,\ x^* > 0;\ u \sim x,\ v \sim -y \text{ as } y \to \infty,\ x > 0;$$

in dimensional and nondimensional variables, respectively, instead of (14.2.1d) and (14.2.2d) while all the other equations and boundary conditions of systems (14.2.1) and (14.2.2) remain the same.

(a) Defining the stream function

$$\psi = \psi(x,y;\varepsilon) \in C^2[\{(x,y) : x > 0, y \ge 0\}] \text{ such that } u = \frac{\partial\psi}{\partial y} \text{ and } v = -\frac{\partial\psi}{\partial x},$$

conclude that (14.3.1d) is replaced by

$$\psi(x,y;\varepsilon) \sim xy \text{ as } y \to \infty$$

while (14.3.1a, b, c) remain the same.

(b) Deduce that the outer or free-stream solution to lowest order of (14.3.4) is then given by

$$\psi_0(x,y) = \lim_{\varepsilon\to 0} \psi(x,y;\varepsilon) = xy \text{ for } y \text{ in the free stream.}$$

(c) Introducing the boundary-layer variables of system (14.3.7) concludes that the lowest order inner or boundary-layer solution

$$\varphi_0(x,\eta) = \lim_{\varepsilon\to 0}\varphi(x,\eta;\varepsilon)$$

then satisfies the matching condition

$$\varphi_{0_\eta}(x,\eta) \sim \psi_{0_y}(x,0) = x \text{ as } \eta \to \infty$$

rather than (14.3.7c) while (14.3.7a, 14.3.7b) remain the same.

(d) Hence conclude that (14.3.7a) in conjunction with the matching condition of part (c) implies

$$\varphi_{0_{\eta\eta\eta}} + \varphi_{0_x}\varphi_{0_{\eta\eta}} - \varphi_{0_\eta}\varphi_{0_{x\eta}} = g(x) = -\psi_{0_y}(x,0)\psi_{0_{yx}}(x,0) = -x$$

rather than (14.3.8). Thus denoting φ_0 by φ for ease of exposition the problem for it reduces to

$$\varphi_{\eta\eta\eta} + \varphi_x\varphi_{\eta\eta} - \varphi_\eta\varphi_{x\eta} = -x;\, 0 < \eta < \infty;$$

$$\varphi(x,0) = \varphi_\eta(x,0) = 0;\; \varphi_\eta(x,\eta) \sim x \text{ as } \eta \to \infty.$$

(e) Consider the transformation

$$x' = \frac{x}{c},\, \eta' = \eta,\, \varphi' = \frac{\varphi}{c} \text{ for } c > 0.$$

Show that then the system of part (d) becomes for $\varphi' = \varphi'(x',\eta')$:

$$\varphi'_{\eta'\eta'\eta'} + \varphi'_{x'}\varphi'_{\eta'\eta'} - \varphi'_{\eta'}\varphi'_{x'\eta'} = -x';\, 0 < \eta' < \infty;$$

$$\varphi'(x',0) = \varphi_{\eta'}(x',0) = 0;\; \varphi'_{\eta'}(x',\eta') \sim x' \text{ as } \eta' \to \infty.$$

In other words that system is invariant under this transformation.

(f) Since the transformed system of part (e) is identical in form to the original system of part (d) this tells one that the original system has a similarity solution such that φ/x must be a function of η which will convert its partial differential equation into an ordinary differential equation. Hence assume

$$\frac{\varphi(x,\eta)}{x} = f(\eta) \text{ or } \varphi(x,\eta) = xf(\eta)$$

and show that $f(\eta)$ then satisfies

$$f''' + ff'' - (f')^2 + 1 = 0,\, 0 < \eta < \infty;$$

$$f(0) = f'(0) = 0,\, f'(\eta) \sim 1 \text{ as } \eta \to \infty.$$

(g) Noting that

$$\psi(x,y;\varepsilon) \sim \varepsilon^{1/2}\varphi_0\left(x, \frac{y}{\varepsilon^{1/2}}\right) = \varepsilon^{1/2} x f\left(\frac{y}{\varepsilon^{1/2}}\right) \text{ as } \varepsilon \to 0 \text{ for } y \text{ in the boundary layer,}$$

where

$$\varepsilon = \frac{1}{R},\ f''(0) = 1.2326,\ \eta - f(\eta) \sim \beta_1 = 0.6480 \text{ as } \eta \to \infty;$$

compute the following parameters of the boundary layer:

$$\delta_1(x) = \int_0^\infty \left[1 - \frac{u(x,y)}{U(x)}\right] dy \equiv \frac{\beta_1}{\sqrt{R}} = \frac{0.6480}{\sqrt{R}} \text{ with } U(x) = x;$$

$$C_D = \frac{1}{R}\left[\frac{\partial u(x,0)}{\partial y} + \frac{\partial v(x,0)}{\partial x}\right] = \frac{1}{R}\frac{\partial u(x,0)}{\partial y} = \frac{x f''(0)}{\sqrt{R}};$$

$$D = \int_0^1 C_D\, dx = \frac{0.6163}{\sqrt{R}}.$$

Part III
Chapters 15–18

Chapter 15
Rayleigh–Bénard Natural Convection Problem

Rayleigh–Bénard buoyancy-driven convection of a layer of viscous Boussinesq fluid confined between two stress-free surfaces and heated from below is investigated in this chapter. The critical conditions for onset of a convective instability are obtained by a linear stability analysis of the pure conduction solution to the governing system of Boussinesq equations. The temperature difference between the plates must exceed a certain critical level in order to overcome the internal resistance to motion caused by viscosity before convection can occur in the form of rolls of a characteristic wavelength. These critical conditions are compared with those for the corresponding rigid–free and rigid–rigid lower–upper surface scenarios. The initial conditions are satisfied by means of Fourier series and integrals. Then the results of an air–cigarette smoke aerosol convective nonlinear stability analysis are presented and discussed in some detail to resolve a long-standing discrepancy between theory and experiment for thin layers of gases. These results are also compared with those involving carbon dioxide–smoke aerosols as well. There are four problems: The first two fill in some details assumed in the chapter involving the substantial derivative and the neglect of adiabatic gradient effects, while the last two consider the role played by eddies for Rayleigh–Bénard convection layers in planetary atmospheres and surface tension-driven Marangoni–Bénard convection, respectively.

15.1 Governing Boussinesq Equations of Motion

Rayleigh–Bénard convection (see Figs. 15.1 and 15.2) has to date provided perhaps the best studied example of nonlinear pattern selection (reviewed by Koschmieder, [61]). In the simplest mathematical version of this particular problem due to Lord Rayleigh [100], a layer of viscous fluid is confined between two horizontal surfaces of infinite planar extent located at $z = 0$ and $z = d$ on which the tangential components of stress are taken to be zero. These stress-free surfaces are assumed to be pure conductors and hence maintained at the temperatures $T = T_0$ and $T = T_1$,

© Springer International Publishing AG, part of Springer Nature 2017

D. J. Wollkind and B. J. Dichone, *Comprehensive Applied Mathematical Modeling in the Natural and Engineering Sciences*,
https://doi.org/10.1007/978-3-319-73518-4_15

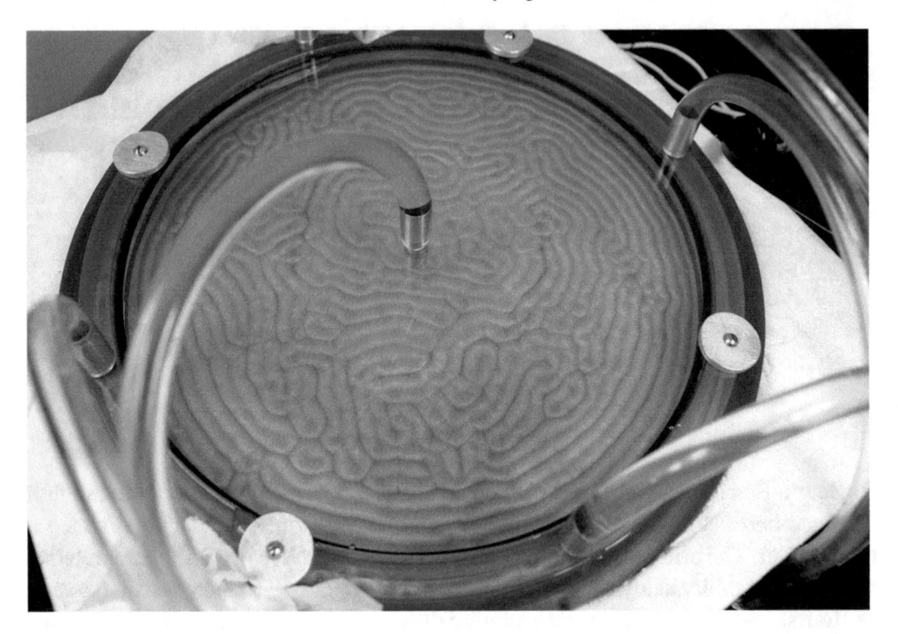

Fig. 15.1 A photograph of roll-type convection cells for the Rayleigh–Bénard problem with an experimental apparatus involving a silicone oil layer having a rigid upper surface, reproduced with permission from the original provided by O.J.E. Matsson and initially appearing in [79].

respectively, where $\Delta T = T_0 - T_1 > 0$ (see Fig. 15.3). This means that the fluid is either being heated from below or cooled from above or both. As the ΔT between those free surfaces is increased past a certain critical level $(\Delta T)_c$, convection begins to occur. The conditions for the onset of this instability can be determined analytically by examining the linear stability of a pure conduction solution to its governing system of equations under the Oberbeck [87]-Boussinesq [10] approximation. Implicit to this formulation is the assumption that the buoyancy effect due to the expansion of the hotter lower fluid serves as the driving mechanism for such convection. Although originally devised to explain the hexagonal cells (see Fig. 15.2) obtained experimentally by Bénard [7] in layers of spermaceti lying on a heated metallic plate that was maintained at a uniform temperature considerably higher than the mean temperature of the ambient air layer in contact with the upper surface of the melted whale oil, this model with its free–free boundary conditions has proven itself more useful for representing the roll-type convection cells occurring in closed containers (see Fig. 15.1), the isothermal boundary planes of which are rigid (no-slip) surfaces. The reasons for this are two-fold: The resulting flow is then purely gravity-driven as opposed to being generated by the variation of surface tension with temperature at the fluid–fluid interface (see Problem 10.6), which was the mechanism primarily responsible for the motion observed by Bénard in his thin oil layers exposed to the air (see Problem 15.4); subsequent investigations of convective instability problems demonstrated that the same qualitative results are obtained for rigid–rigid and rigid–free boundary conditions as for the free–free boundary conditions of the Rayleigh–Bénard model [61]. Specifically, Jeffreys [56], realizing that the boundary conditions employed by Lord Rayleigh [100] were somewhat artificial, extended the problem to take into account the possibility of rigid boundaries. The form of this rigid–rigid boundary condition model now necessitated that the critical conditions for the onset of convection be calculated numerically. Jeffreys [56] numerical results, which have been progressively improved upon over the years by a series of researchers using increasingly more accurate procedures, yielded only quantitative differences when compared with Rayleigh's [100] analytical ones [61]. The first part of this chapter is primarily concerned with the determination of that $(\Delta T)_c$ for the onset of convection and the particular critical wavelength characteristic of the emerging pattern as predicted by Rayleigh's linear stability analysis.

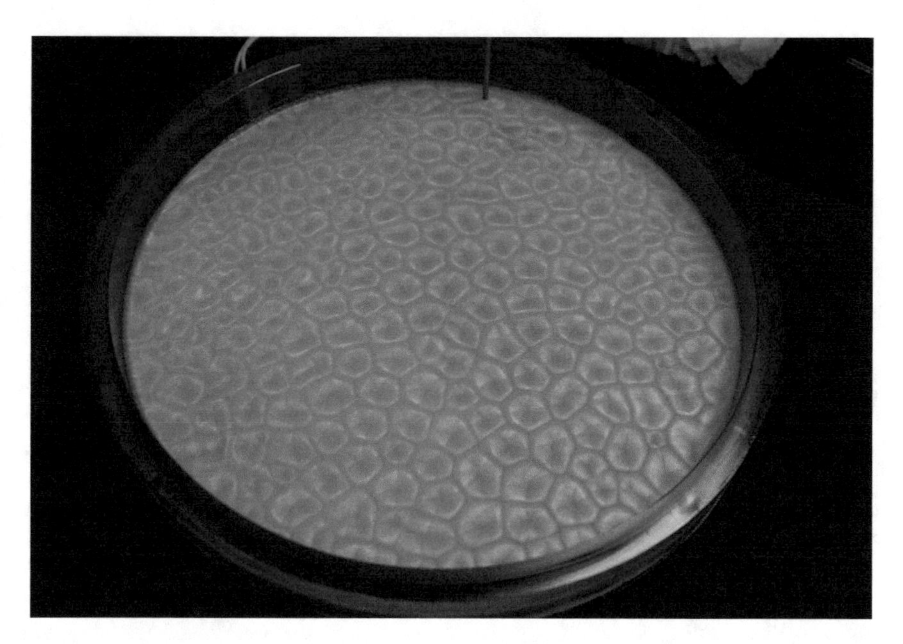

Fig. 15.2 A photograph of hexagonal-type convection cells for the Rayleigh–Bénard problem with an experimental apparatus involving a silicone oil layer having a free upper surface, reproduced with permission from the original provided by O.J.E. Matsson and initially appearing in [79].

We begin by considering the most general governing system of equations appropriate for this situation: Namely, the continuity equation (9.2.1), the momentum equations (9.4.20), and the energy equation (9.5.15b):

$$\frac{D\rho}{Dt} + \rho\theta = 0 \text{ with } \theta = \nabla \bullet \boldsymbol{v}; \tag{15.1.1a}$$

$$\rho\frac{D\boldsymbol{v}}{Dt} = -\nabla p + \rho\boldsymbol{f} + (\lambda_0 + \mu_0)\nabla(\nabla \bullet \boldsymbol{v}) + \mu_0\nabla^2\boldsymbol{v} \text{ for } \boldsymbol{f} = -g\boldsymbol{e}_3; \tag{15.1.1b}$$

$$\rho C_p \frac{DT}{Dt} = \frac{Dp}{Dt} + \Phi + r + \nabla \bullet (k\nabla T) \text{ with } k \equiv k_0 \text{ and } \Phi = \lambda_0\theta^2 + 2\mu_0\varepsilon_{ij}\varepsilon_{ij}. \tag{15.1.1c}$$

We next introduce the Boussinesq [10] approximation due originally to Oberbeck [87] which consists of the following two assumptions:

(i) The density satisfies the equation of state $\rho = \rho_0[1-\alpha_0(T-T_0)]$ where its variation with temperature is assumed to be affine with thermal expansivity constant $\alpha_0 > 0$ while its variation with pressure is assumed to be negligible.
(ii) The variation of ρ with T is only taken into account in the body force term $\rho \boldsymbol{f} = -\rho g \boldsymbol{e}_3$ of the momentum equations (15.1.1b). That is everywhere else ρ appears, it is assumed $\rho \equiv \rho_0$.

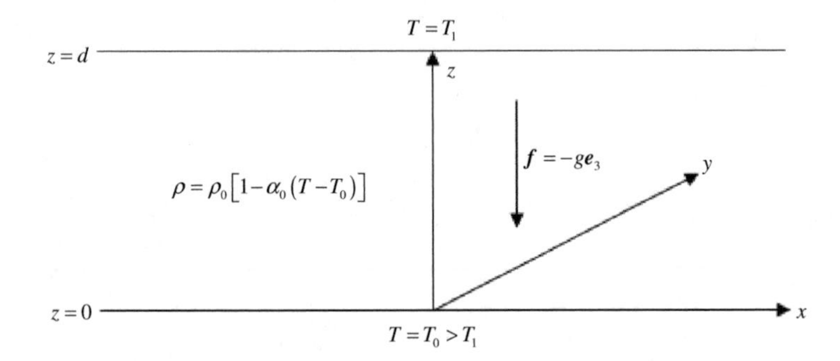

Fig. 15.3 A schematic plot of the buoyancy-driven Rayleigh–Bénard problem.

Under this approximation (15.1.1) reduces to the pseudo-incompressible set of governing equations:

$$\nabla \bullet \boldsymbol{v} = 0; \tag{15.1.2a}$$

$$\frac{D\boldsymbol{v}}{Dt} = -\frac{1}{\rho_0}\nabla p - g[1-\alpha_0(T-T_0)]\boldsymbol{e}_3 + \nu_0\nabla^2\boldsymbol{v} \text{ with } \nu_0 = \frac{\mu_0}{\rho_0}; \tag{15.1.2b}$$

$$\rho_0 C_p \frac{DT}{Dt} = \frac{Dp}{Dt} + \Phi + r + k_0\nabla^2 T \text{ with } \Phi = 2\mu_0\varepsilon_{ij}\varepsilon_{ij}. \tag{15.1.2c}$$

Note that although Rayleigh [100] introduced this reduction with no formal justification, Fife [34] placed it on a firm mathematical foundation with an appropriate asymptotic analysis. In particular, system (15.1.2) written in component form now becomes

$$\frac{\partial u}{\partial x} + \frac{\partial v}{\partial y} + \frac{\partial w}{\partial z} = 0; \tag{15.1.3a}$$

$$\frac{Du}{Dt} = -\frac{1}{\rho_0}\frac{\partial p}{\partial x} + \nu_0 \nabla^2 u; \tag{15.1.3b}$$

$$\frac{Dv}{Dt} = -\frac{1}{\rho_0}\frac{\partial p}{\partial y} + \nu_0 \nabla^2 v; \tag{15.1.3c}$$

$$\frac{Dw}{Dt} = -\frac{1}{\rho_0}\frac{\partial p}{\partial z} - g[1 - \alpha_0(T - T_0)] + \nu_0 \nabla^2 w; \tag{15.1.3d}$$

$$\underbrace{\rho_0 C_p \frac{DT}{Dt} - \frac{Dp}{Dt}}_{a} = \underbrace{r}_{b} + \underbrace{\Phi + k_0 \nabla^2 T}_{c} \text{ with } \Phi = 2\mu_0 \varepsilon_{ij}\varepsilon_{ij}; \tag{15.1.3e}$$

where

$$2\varepsilon_{ij} = \partial_j v_i + \partial_i v_j,\ \partial_i \equiv \frac{\partial}{\partial x_i},\ (x_1, x_2, x_3) \equiv (x, y, z),\ (v_1, v_2, v_3) \equiv (u, v, w);$$

$$\frac{D}{Dt} = \frac{\partial}{\partial t} + u\frac{\partial}{\partial x} + v\frac{\partial}{\partial y} + w\frac{\partial}{\partial z},\ \nabla^2 \equiv \nabla \bullet \nabla = \frac{\partial^2}{\partial x^2} + \frac{\partial^2}{\partial y^2} + \frac{\partial^2}{\partial z^2}.$$

15.2 Simplified Rayleigh Model

Although the simplifications of the energy equation (15.1.3e) to be introduced in this section are qualitatively valid for the natural convection of thin liquid layers such as the silicone oil ones depicted in Figs. 15.1 and 15.2, we shall assume that the fluid in question is a gas for the purpose of making the quantitative comparisons between its terms required to justify them. Our main reason for doing so is that ultimately we shall compare the stability predictions to be obtained with the experiments of Chandra [18] and Sutton [119] which involved the onset of convection in thin air or carbon dioxide layers. The (15.1.3e) terms to be simplified have been grouped by a, b, and c.

(a) Neglect the $-Dp/Dt$ term with respect to $\rho_0 C_p DT/Dt$ which is justified by making the following comparison between their sizes: From (15.1.3d) we can deduce that

$$-\frac{\partial p}{\partial z} \sim \rho_0 g \Rightarrow -\Delta p = \rho_0 g d \Rightarrow \left[-\frac{Dp}{Dt}\right] = \frac{-\Delta p}{\tau} = \rho_0 g \frac{d}{\tau}$$

and

$$\left[\rho_0 C_p \frac{DT}{Dt}\right] = \rho_0 C_p \frac{\Delta T}{\tau}$$

$$\Rightarrow \frac{[-DP/Dt]}{[\rho_0 C_p DT/Dt]} = \frac{\rho_0 g d/\tau}{\rho_0 C_p \Delta T/\tau} = \frac{g/C_p}{\Delta T/d} = \frac{\Gamma_{ad}}{\beta_0} = D_0 << 1$$

given that $D_0 \equiv$ the dissipation number is typically very small since $\beta_0 = \Delta T/d \equiv$ the pure conduction gradient greatly exceeds $\Gamma_{ad} = g/C_p \equiv$ the adiabatic gradient for this situation.

(b) Neglect this source term by taking $r \cong 0$ since such a Stefan–Boltzmann radiative effect given by $r(T) \sim T^4$ is negligible for the mean layer temperature characterized by T_0 and ΔT.

(c) Neglect the $\Phi \sim \mu_0(\partial_j v_i)^2$ term with respect to $k_0\nabla^2 T$ which is justified by making the following comparison between their sizes: From (15.1.3d) we can deduce that

$$w\frac{\partial w}{\partial z} \sim \alpha_0 g\Delta T \Rightarrow [w]^2 = \alpha_0 g(\Delta T)d \Rightarrow [\Phi] = \mu_0\frac{[w]^2}{d^2} = \mu_0\alpha_0 g\frac{\Delta T}{d}$$

and

$$[k_0\nabla^2 T] = k_0\frac{\Delta T}{d^2} \Rightarrow \frac{[\Phi]}{[k_0\nabla^2 T]} = \frac{\mu_0\alpha_0 g\Delta T/d}{k_0\Delta T/d^2} = \mu_0\alpha_0 g\frac{d}{k_0} << 1$$

given that when $d \cong 1$ cm, $\mu_0\alpha_0 gd/k_0$ is typically of the order 10^{-7} since

$$\frac{\mu_0\alpha_0 gd}{k_0} = \frac{\mu_0}{\rho_0}\frac{\Delta T}{\overline{T}_0}\frac{g}{C_p}\frac{\rho_0 C_p}{k_0}\frac{d}{\Delta T} = \frac{\nu_0}{\kappa_0}\frac{\Gamma_{ad}}{\beta_0}\frac{\Delta T}{\overline{T}_0} = PrD_0\frac{\Delta T}{\overline{T}_0}$$

where for gases $Pr \equiv$ Prandtl number $= \nu_0/\kappa_0 \cong 0.7$ with $\kappa_0 = k_0/(\rho_0 C_p)$ and $\alpha_0 = 1/\overline{T}_0$ with $\overline{T}_0 \equiv$ mean temperature while in general $\Delta T << \overline{T}_0$. These simplifications reduce (15.1.3e) to

$$\frac{DT}{Dt} = \kappa_0\nabla^2 T. \tag{15.2.1}$$

The appropriate boundary conditions to impose for this Rayleigh–Bénard model are as follows:

Kinematic Boundary Conditions:

Recall from (10.2.6b) that the kinematic boundary condition takes the form

$$w = \zeta_t + u\zeta_x + v\zeta_y \text{ at } z = \zeta(x, y, t).$$

For our Bénard problem the lower and upper surfaces are located at the fixed horizontal planes $z = 0$ and d, respectively, which correspond to $\zeta(x, y, t) \equiv 0$ and d. Since then, $\zeta(x, y, t)$ represents an undeviated planar interface, this reduces that boundary condition to

$$w = 0 \text{ at } z = 0 \text{ and } d. \tag{15.2.2a}$$

Dynamical Boundary Conditions:

Rayleigh assumed that these planar surfaces were stress-free in the sense that the tangential components of the stress vector $\boldsymbol{t}$ vanished upon them. That is for the tangent vectors $\boldsymbol{\tau}_{1,2}$

$$\boldsymbol{t} \bullet \boldsymbol{\tau}_1 = \boldsymbol{t} \bullet \boldsymbol{i} = t_x = \boldsymbol{t} \bullet \boldsymbol{\tau}_2 = \boldsymbol{t} \bullet \boldsymbol{j} = t_y = 0 \text{ at } z = 0 \text{ and } d$$

where

$$t_i = T_{ij} n_j \text{ for } i = x \text{ and } y \text{ with } \boldsymbol{n} = \mp \boldsymbol{k} \text{ at } z = 0 \text{ and } d, \text{ respectively,}$$

or

$$t_x = \mp T_{xz} = t_y = \mp T_{yz} = 0 \text{ at } z = 0 \text{ and } d, \text{ respectively,}$$

which implies that

$$T_{xz} = T_{yz} = 0 \text{ at } z = 0 \text{ and } d$$

or

$$\frac{\partial u}{\partial z} + \frac{\partial w}{\partial y} = 0 \text{ at } z = 0 \text{ and } d; \tag{15.2.2b}$$

$$\frac{\partial v}{\partial z} + \frac{\partial w}{\partial y} = 0 \text{ at } z = 0 \text{ and } d. \tag{15.2.2c}$$

Thermal Boundary Conditions:

These surfaces being pure conductors are maintained at the constant temperatures

$$T = T_0 \text{ at } z = 0 \text{ and } T = T_1 \text{ at } z = d \text{ where } \Delta T = T_0 - T_1 > 0. \tag{15.2.2d}$$

Far-Field Boundary Conditions:

The dependent variables must be finite in the whole extended x-y plane. Thus we assume that

$$u, v, w, p, \text{ and } T \text{ remain bounded as } x^2 + y^2 \to \infty. \tag{15.2.2e}$$

15.3 Pure Conduction Solution and its Perturbation System

There exists an exact solution to the governing system of simplified Boussinesq equations (15.1.3a, b, c, d) and (15.2.1) and boundary conditions (15.2.2a, b, c, d, e) that is static ($u = v = w \equiv 0$), steady-state ($\partial/\partial t \equiv 0$), and stratified in z ($\partial/\partial x = \partial/\partial y \equiv 0$). To deduce it, consider a solution to that system and boundary conditions of the form

$$u = v = w \equiv 0, T = T_0(z), p = p_0(z). \tag{15.3.1a}$$

Such a solution satisfies equations (15.1.3a, b, c) and boundary conditions (15.2.2a, b, c, e) identically. To determine $T_0(z)$ and $p_0(z)$ we shall examine this solution in conjunction with equations (15.1.3d) and (15.2.1) and boundary conditions (15.2.2d). Then equation (15.2.1) and boundary conditions (15.2.2d) yield the following problem for $T_0(z)$:

$$T_0''(z) = 0,\ 0 < z < d;\ T_0(0) = T_0,\ T_0(d) = T_1 < T_0;$$

which has solution

$$T_0(z) = T_0 - \beta_0 z \text{ where } \beta_0 = \frac{\Delta T}{d} > 0 \tag{15.3.1b}$$

and represents an adverse temperature gradient in the layer while given (15.3.1a, 15.3.1b) equation (15.1.3d) implies

$$p_0'(z) = -\rho_0 g(1 + \alpha_0 \beta_0 z) \text{ or } p_0(z) = -\rho_0 g \int (1 + \alpha_0 \beta_0 z)\, dz. \tag{15.3.1c}$$

It is the stability of this *pure conduction* solution (15.3.1) with which we are concerned. Hence we consider a solution to our basic vector system (15.1.2a, b, c, d) and (15.2.1) and boundary conditions (15.2.2) of the form of (15.3.1) plus a perturbation

$$\boldsymbol{v} = \boldsymbol{0} + \widetilde{\boldsymbol{v}} \text{ where } \widetilde{\boldsymbol{v}} \equiv (\widetilde{u}, \widetilde{v}, \widetilde{w}),\ T = T_0(z) + \widetilde{T},\ p = p_0(z) + \widetilde{p}; \tag{15.3.2}$$

which yields the associated system for its perturbation quantities (see Problem 15.1)

$$\nabla \bullet \widetilde{\boldsymbol{v}} = 0; \tag{15.3.3a}$$

$$\frac{\widetilde{D}\widetilde{\boldsymbol{v}}}{\widetilde{D}t} = -\frac{1}{\rho_0}\nabla\widetilde{p} + g\alpha_0\widetilde{T}\boldsymbol{e}_3 + \nu_0\nabla^2\widetilde{\boldsymbol{v}}; \tag{15.3.3b}$$

$$\frac{\widetilde{D}\widetilde{T}}{\widetilde{D}t} - \beta_0\widetilde{w} = \kappa_0\nabla^2\widetilde{T}; \tag{15.3.3c}$$

$$\widetilde{w} = \frac{\partial\widetilde{u}}{\partial z} + \frac{\partial\widetilde{w}}{\partial x} = \frac{\partial\widetilde{v}}{\partial z} + \frac{\partial\widetilde{w}}{\partial y} = \widetilde{T} = 0 \text{ at } z = 0 \text{ and } d; \tag{15.3.4a}$$

$$\widetilde{u}, \widetilde{v}, \widetilde{w}, \widetilde{p} \text{ and } \widetilde{T} \text{ remain bounded as } x^2 + y^2 \to \infty; \tag{15.3.4b}$$

where

$$\frac{\widetilde{D}}{\widetilde{D}t} = \frac{\partial}{\partial t} + \widetilde{\boldsymbol{v}} \bullet \nabla \text{ with } \widetilde{\boldsymbol{v}} \bullet \nabla \equiv \widetilde{u}\frac{\partial}{\partial x} + \widetilde{v}\frac{\partial}{\partial y} + \widetilde{w}\frac{\partial}{\partial z}.$$

15.4 Normal-Mode Linear Stability Analysis

We now transform (15.3.3) and (15.3.4) into a linear perturbation system by letting

$\widetilde{\boldsymbol{v}} = \varepsilon \boldsymbol{v}_1 + \boldsymbol{O}(\varepsilon^2)$ where $\boldsymbol{v}_1 = (u_1, v_1, w_1)$, $\widetilde{T} = \varepsilon T_1 + O(\varepsilon^2)$, $\widetilde{p} = \varepsilon p_1 + O(\varepsilon^2)$ for $|\varepsilon| << 1$; (15.4.1)

substituting (15.4.1) into it; neglecting terms of $O(\varepsilon^2)$; cancelling the resulting common ε-factor; and obtaining in component form

$$\frac{\partial u_1}{\partial x} + \frac{\partial v_1}{\partial y} + \frac{\partial w_1}{\partial z} = 0; \tag{15.4.2a}$$

$$\frac{\partial u_1}{\partial t} = -\frac{1}{\rho_0}\frac{\partial p_1}{\partial x} + \nu_0 \nabla^2 u_1; \tag{15.4.2b}$$

$$\frac{\partial v_1}{\partial t} = -\frac{1}{\rho_0}\frac{\partial p_1}{\partial y} + \nu_0 \nabla^2 v_1; \tag{15.4.2c}$$

$$\frac{\partial w_1}{\partial t} = -\frac{1}{\rho_0}\frac{\partial p_1}{\partial z} + g\alpha_0 T_1 + \nu_0 \nabla^2 w_1; \tag{15.4.2d}$$

$$\frac{\partial T_1}{\partial t} - \beta_0 w_1 = \kappa_0 \nabla^2 T_1; \tag{15.4.2e}$$

$$w_1 = \frac{\partial u_1}{\partial z} + \frac{\partial w_1}{\partial x} = \frac{\partial v_1}{\partial z} + \frac{\partial w_1}{\partial y} = T_1 = 0 \text{ at } z = 0 \text{ and } d; \tag{15.4.3a}$$

$$u_1, v_1, w_1, p_1 \text{ and } T_1 \text{ remain bounded as } x^2 + y^2 \to \infty; \tag{15.4.3b}$$

which essentially differs from the original perturbation system by the replacement of the substantial time derivative containing its only nonlinear terms with a normal partial time derivative instead.

We next seek a normal-mode solution of these equations and boundary conditions of the form

$$[u_1, v_1, w_1, p_1, T_1](x, y, z, t) = [U, V, W, P, \Theta](z)\exp(i\boldsymbol{k} \bullet \boldsymbol{r} + \omega t) \tag{15.4.4a}$$

where

$$\boldsymbol{k} = (k_1, k_2) \text{ with } k_{1,2} \in \mathbb{R},\ \boldsymbol{r} = (x, y), \text{ and } \omega \in \mathbb{C}. \tag{15.4.4b}$$

Analogous to our treatment of the slime mold aggregation and inviscid Couette model systems of Chapter 7 and Problem 12.2, respectively, this again represents an arbitrary component of a Fourier decomposable disturbance as will be demonstrated in the next section and thus is quite general. Hence we can determine the stability of the pure conduction solution to a relatively wide class of initially infinitesimal perturbations by examining the long-time behavior of this single mode. Further note in this context that the far-field boundary condition (15.4.3b) is satisfied automatically by virtue of $k_{1,2} \in \mathbb{R}$. Observe for $\alpha_1(x, y, z, t) = A(z)\exp(i\boldsymbol{k} \bullet \boldsymbol{r} + \omega t)$ where $\boldsymbol{k} \bullet \boldsymbol{r} = k_1 x + k_2 y$ that

$$\frac{\partial \alpha_1}{\partial x} = ik_1\alpha_1,\ \frac{\partial \alpha_1}{\partial y} = ik_2\alpha_1,\ \frac{\partial \alpha_1}{\partial z} = \frac{dA}{dz}\exp(i\boldsymbol{k} \bullet \boldsymbol{r} + \omega t),\ \frac{\partial \alpha_1}{\partial t} = \omega\alpha_1,$$

and

$$\nabla^2\alpha_1 = \left(\frac{d^2}{dz^2} - s^2\right)\alpha_1 \text{ where } s^2 = k_1^2 + k_2^2 \geq 0.$$

Substituting this normal-mode solution of (15.4.4) into equations (15.4.2) and boundary conditions (15.4.3a) and cancelling the common exponential factor, we obtain the following system of ordinary differential equations satisfied by the stratification functions for $0 < z < d$:

$$ik_1U + ik_2V + \frac{dW}{dz} = 0; \tag{15.4.5a}$$

$$\omega U = -\frac{ik_1}{\rho_0}P + \nu_0\left(\frac{d^2}{dz^2} - s^2\right)U; \tag{15.4.5b}$$

$$\omega V = -\frac{ik_2}{\rho_0}P + \nu_0\left(\frac{d^2}{dz^2} - s^2\right)V; \tag{15.4.5c}$$

$$\omega W = -\frac{1}{\rho_0}\frac{dP}{dz} + g\alpha_0\Theta + \nu_0\left(\frac{d^2}{dz^2} - s^2\right)W; \tag{15.4.5d}$$

$$\omega\Theta - \beta_0 W = \kappa_0\left(\frac{d^2}{dz^2} - s^2\right)\Theta; \tag{15.4.5e}$$

with boundary conditions:

$$W = \frac{dU}{dz} + ik_1W = \frac{dV}{dz} + ik_2W = \Theta = 0 \text{ at } z = 0 \text{ and } d. \tag{15.4.6}$$

Note that this reduces to

$$\frac{dU}{dz} = \frac{dV}{dz} = 0 \text{ at } z = 0 \text{ and } d. \tag{15.4.7}$$

Taking the derivative of (15.4.5a) with respect to z yields

$$i\left(k_1\frac{dU}{dz} + k_2\frac{dV}{dz}\right) + \frac{d^2W}{dz^2} = 0 \text{ for } 0 < z < d. \tag{15.4.8}$$

Assuming the requisite continuity at $z = 0$ and d and taking the limit of (15.4.8) as $z \to 0$ and d, we obtain, in conjunction with (15.4.7), that

$$\frac{d^2W}{dz^2} = 0 \text{ at } z = 0 \text{ and } d, \tag{15.4.9}$$

which, along with (15.4.6), comprise the boundary conditions for this problem. Since, those boundary conditions only involve W and Θ as does equation (15.4.5e), we proceed to eliminate U, V, and P from equations (15.4.5a, b, c, d) as follows: Solving (15.4.5a) for $i(k_1U + k_2V)$ yields

$$i(k_1U + k_2V) = -\frac{dW}{dz}. \tag{15.4.10}$$

Multiplying (15.4.5b) by ik_1; (15.4.5c) by ik_2; and adding these results; we obtain

$$\omega i(k_1 U + k_2 V) = \frac{s^2 P}{\rho_0} + \nu_0 \left(\frac{d^2}{dz^2} - s^2 \right) i(k_1 U + k_2 V). \tag{15.4.11}$$

Then substitution of (15.4.10) into (15.4.11) yields

$$\frac{s^2 P}{\rho_0} = \left[\nu_0 \left(\frac{d^2}{dz^2} - s^2 \right) - \omega \right] \frac{dW}{dz}. \tag{15.4.12}$$

Differentiating (15.4.12) with respect to z, we obtain

$$\left(\frac{s^2}{\rho_0} \right) \frac{dP}{dz} = \left[\nu_0 \left(\frac{d^2}{dz^2} - s^2 \right) - \omega \right] - \frac{d^2 W}{dz^2}. \tag{15.4.13}$$

Now, multiplying (15.4.5d) by $-s^2$ yields

$$-s^2 \omega W = \left(\frac{s^2}{\rho_0} \right) \frac{dP}{dz} - s^2 g \alpha_0 \Theta + \nu_0 \left(\frac{d^2}{dz^2} - s^2 \right) (-s^2 W). \tag{15.4.14}$$

Finally, substituting (15.4.13) into (15.4.14), we obtain

$$\nu_0 \left[\left(\frac{d^2}{dz^2} - s^2 \right)^2 - \omega \left(\frac{d^2}{dz^2} - s^2 \right) \right] W - s^2 g \alpha_0 \Theta = 0, \tag{15.4.15a}$$

which along with (15.4.5e) rewritten in the form

$$\beta_0 W + \left[\kappa_0 \left(\frac{d^2}{dz^2} - s^2 \right) - \omega \right] \Theta = 0, \tag{15.4.15b}$$

and boundary conditions (15.4.6) and (15.4.9)

$$W = \frac{d^2 W}{dz^2} = \Theta = 0 \text{ at } z = 0 \text{ and } d, \tag{15.4.15c}$$

constitutes the reduced system for $W = W(z)$ and $\Theta = \Theta(z)$.

Introducing the nondimensional independent variable and parameters

$$\xi = \frac{z}{d}, \; q = sd, \; \sigma = \frac{\omega d^2}{\nu_0}, \; Pr = \frac{\nu_0}{\kappa_0};$$

system (15.4.15) is transformed into the following system for $w(\xi) \equiv W(\xi d)$ and $\theta(\xi) \equiv \Theta(\xi d)$ where $D \equiv d/d\xi$ and hence $d/dz = D/d$:

$$\left.\begin{aligned}\left[(\nu_0/d^4)(D^2-q^2)^2-(\omega/d^2)(D^2-q^2)\right]w-(g\alpha_0 q^2/d^2)\theta=0;\\ \beta_0 w+[(\kappa_0/d^2)(D^2-q^2)-\omega]\theta=0;\end{aligned}\right\}\ 0<\xi<1; \tag{15.4.16a}$$

or

$$\left.\begin{aligned}\left[(D^2-q^2)^2-\sigma(D^2-q^2)\right]w-(g\alpha_0 d^2/\nu_0)q^2\theta=0;\\ (\beta_0 d^2/\kappa_0)w+(D^2-q^2-Pr\sigma)\theta=0;\end{aligned}\right\}\ 0<\xi<1; \tag{15.4.16b}$$

with boundary conditions

$$w=D^2w=\theta=0 \text{ at } \xi=0 \text{ and } 1. \tag{15.4.16c}$$

This is an eigenvalue problem for σ once q^2 and the material and experimental parameters of the problem have been fixed. Assuming the requisite continuity, (15.4.16c) in conjunction with (15.4.16a and 15.4.16b) implies that

$$D^4w=D^2\theta=0 \text{ at } \xi=0 \text{ and } 1. \tag{15.4.17a}$$

Then operating on (15.4.16a and 15.4.16b) by D^2 sequentially and proceeding in a similar manner, we can generalize this result to deduce that

$$D^{2N}w=D^{2N}\theta=0 \text{ for } N=0,1,2,3,\ldots \text{ at } \xi=0 \text{ and } 1. \tag{15.4.17b}$$

Alternatively, by solving (15.4.16a) for θ and substituting that result into (15.4.16b), we can reduce (15.4.16) to the following sixth-order differential equation boundary value problem for $w=w(\xi)$:

$$[(D^2-q^2)^3-(Pr+1)\sigma(D^2-q^2)^2+Pr\sigma^2(D^2-q^2)+Rq^2]w=0,\ 0<\xi<1; \tag{15.4.18a}$$

with the six boundary conditions

$$w=D^2w=D^4w=0 \text{ at } \xi=0 \text{ and } 1; \tag{15.4.18b}$$

where

$$R\equiv \text{Rayleigh number } = g\alpha_0\beta_0\frac{d^4}{\kappa_0\nu_0}=g\alpha_0\frac{d^3\Delta T}{\kappa_0\nu_0}. \tag{15.4.18c}$$

Assuming the requisite continuity (15.4.18b) in conjunction with (15.4.18a) implies that

$$D^6w=0 \text{ at } \xi=0 \text{ and } 1; \tag{15.4.19a}$$

and operating on (15.4.18a) by D^2 sequentially, we can generalize (15.4.19a) to deduce again that

$$D^{2N}w=0 \text{ for } N=0,1,2,3,\ldots \text{ at } \xi=0 \text{ and } 1. \tag{15.4.19b}$$

Proceeding in either manner we may conclude from the Fourier series Theorem and (15.4.17b) or (15.4.19b) that these problems have eigenvectors

$$[w,\theta](\xi) = [A_m, B_m]\sin(m\pi\xi) \text{ for } m = 1,2,3,\dots \text{ where } |A_m|^2 + |B_m|^2 \neq 0. \quad (15.4.20)$$

Then, observing that

$$[(D^2 - q^2)^p]\sin(m\pi\xi) = (-1)^p k_m^{2p}\sin(m\pi\xi) \text{ where } k_m^2 = m^2\pi^2 + q^2 > 0,$$

substituting either (15.4.20) into (15.4.16a and 15.4.16b) or (15.4.20) into (15.4.18a), and cancelling the common factor, we obtain that the corresponding eigenvalues σ_m for $m = 1,2,3,\dots$ satisfy either

$$\left.\begin{aligned}(k_m^4 + \sigma_m k_m^2)A_m - (g\alpha_0 d^2/\nu_0)q^2 B_m = 0;\\ (\beta_0 d^2/\kappa_0)A_m - (k_m^2 + Pr\sigma_m)B_m = 0;\end{aligned}\right\} \quad (15.4.21a)$$

or

$$Prk_m^2\sigma_m^2 + (Pr+1)k_m^4\sigma_m + k_m^6 - Rq^2 = 0; \quad (15.4.21b)$$

which are equivalent since the required vanishing determinantial condition for (15.4.21a)

$$\begin{vmatrix} k_m^4 + \sigma_m k_m^2 & -(g\alpha_0 d^2/\nu_0)q^2 \\ \beta_0 d^2/\kappa_0 & -(k_m^2 + Pr\sigma_m) \end{vmatrix} = 0$$

yields the quadratic secular equation of (15.4.21b).

Calculating the discriminant of this quadratic

$$\mathcal{D} = (Pr+1)^2 k_m^8 - 4Prk_m^2(k_m^6 - Rq^2) = (Pr-1)^2 k_m^8 + 4PrRk_m^2 q^2 = 4Pr^2 k_m^4 \Delta_m^2 \geq 0, \quad (15.4.21c)$$

we can conclude that $\sigma_m^{\pm} = -b_m \pm \Delta_m \in \mathbb{R}$ where $2b_m = (1 + Pr^{-1})k_m^2$ and $\Delta_m \geq 0$.

Recalling from Chapter 3 that a quadratic of this sort in σ_m has roots such that $\mathrm{Re}(\sigma_m^{\pm}) = \sigma_m^{\pm} < 0$ if and only if all its coefficients are positive and observing that Prk_m^2, $(Pr+1)k_m^4 > 0$, we can deduce the following stability criteria

$$k_m^6 - Rq^2 > 0 \text{ for } m = 1,2,3,\dots, \quad (15.4.22a)$$

since $\omega_m^{\pm} = \nu_0\sigma_m^{\pm}/d^2$. Noting that $\sigma_m^- < 0$ as defined these criteria can only have a substantive effect on σ_m^+. Hence

$$\sigma_m^+ < 0 \text{ if and only if } R < R_0(q^2;m) = \frac{k_m^6}{q^2} = \frac{(m^2\pi^2 + q^2)^3}{q^2}. \quad (15.4.22b)$$

The marginal stability curves on which these roots satisfy $\sigma_m^+ = 0$ are given by

$$R = R_0(q^2;m) = \frac{(m^2\pi^2 + q^2)^3}{q^2} \text{ for } m = 1,2,3,\dots. \quad (15.4.22c)$$

Observe from (15.4.22c) that $R_0(q^2; m+1) > R_0(q^2; m)$. Hence we can deduce that $\sigma_{m+1}^+ > \sigma_m^+$ and $m = 1$ correspond to the so-called most dangerous mode which becomes unstable first as R increases. For ease of exposition we denote the growth rate σ_1^+ corresponding to that most dangerous mode by σ and its associated marginal stability curve $R_0(q^2; 1)$ by $R_0(q^2)$ in what follows. Thus

$$R_0(q^2) \equiv \frac{(\pi^2 + q^2)^3}{q^2}. \tag{15.4.23}$$

Then

$$\frac{dR_0(q_c^2)}{dq^2} = 0 \Rightarrow 3(\pi^2 + q_c^2)^2 q_c^2 = (\pi^2 + q_c^2)^3 \Rightarrow 3q_c^2 = \pi^2 + q_c^2 \Rightarrow q_c^2 = \frac{\pi^2}{2} \tag{15.4.24a}$$

and

$$R_c = R_0(q_c^2) = R_0 \frac{(3\pi^2/2)^3}{\pi^2/2} = \frac{27\pi^4}{4} = 657.511. \tag{15.4.24b}$$

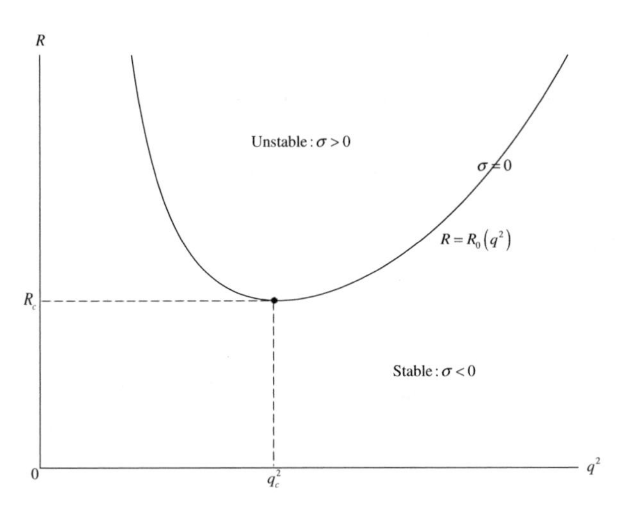

Fig. 15.4 A plot of the marginal stability curve $R = R_0(q^2)$ of (15.4.23).

Therefore the marginal stability curve $R = R_0(q^2)$ when plotted in the q^2-R plane of Fig. 15.4 has an absolute minimum at the point (q_c^2, R_c). From that figure we can see that for $R < R_c$ there exist no squared wavenumbers q^2 corresponding to $\sigma > 0$ while for $R > R_c$ there exists a band of such squared wavenumbers centered about q_c^2 characterized by a disturbance wavelength $\lambda_c = 2\pi/q_c = 2\sqrt{2}$. Hence we can conclude that there is linear instability provided

$$R = \frac{g\alpha_0 d^3 \Delta T}{\kappa_0 \nu_0} > R_c = \frac{27\pi^4}{4} \text{ or } \Delta T > (\Delta T)_c = \frac{27\pi^4 \kappa_0 \nu_0}{4g\alpha_0 d^3}. \tag{15.4.25}$$

15.5 Satisfaction of Initial Conditions

We are now ready to consider initial conditions for perturbation system (15.4.2). As in Section 7.7 we need only impose the following initial conditions on the perturbation quantity $w_1(x,y,z,t)$:

$$w_1(x,y,,z,0) = h_1(x,y,z), \ \frac{\partial w_1(x,y,z,0)}{\partial t} = h_2(x,y,z).$$

Synthesizing our approaches with Fourier series and integrals in Chapters 5 and 7, respectively, we next assume a superposition principle related to our normal-mode solution (15.4.4) and represent

$$w_1(x,y,z,t) = \int_{-\infty}^{\infty}\int_{-\infty}^{\infty}\sum_{m=1}^{\infty}(A_m^+ e^{\omega_m^+ t} + A_m^- e^{\omega_m^- t})\sin(m\pi\xi)e^{i(\mathbf{k}\bullet\mathbf{r})}\,dk_1\,dk_2.$$

Analogous to our approach in Section 7.7 we rewrite

$$\begin{aligned} A_m^+ e^{\omega_m^+ t} + A_m^- e^{\omega_m^- t} &= e^{-b_m\tau}(A_m^+ e^{\Delta_m\tau} + A_m^- e^{-\Delta_m\tau}) \\ &= e^{-b_m\tau}[c_m^+\cosh(\Delta_m\tau) + c_m^-\sinh(\Delta_m\tau)], \end{aligned}$$

where $\tau = \nu_0 t/d^2$, $c_m^+ = A_m^+ + A_m^-$, and $c_m^- = A_m^+ - A_m^-$. Hence

$$\begin{aligned} &w_1(x,y,z,t) \\ &= \int_{-\infty}^{\infty}\int_{-\infty}^{\infty}\sum_{m=1}^{\infty} e^{-b_m\tau}[c_m^+\cosh(\Delta_m\tau) + c_m^-\sinh(\Delta_m\tau)]\sin(m\pi\xi)e^{i(\mathbf{k}\bullet\mathbf{r})}\,dk_1\,dk_2. \end{aligned}$$

Thus

$$[h_1,h_2](x,y,z) = \int_{-\infty}^{\infty}\int_{-\infty}^{\infty}\sum_{m=1}^{\infty}\left[c_m^+, \nu_0\left(\frac{\Delta_m c_m^- - b_m c_m^+}{d^2}\right)\right]\sin(m\pi\xi)e^{i(\mathbf{k}\bullet\mathbf{r})}\,dk_1\,dk_2.$$

Applying the inverse Fourier transform to these integrals, we find the series satisfy

$$\sum_{m=1}^{\infty}\left[c_m^+, \nu_0\left(\frac{\Delta_m c_m^- - b_m c_m^+}{d^2}\right)\right]\sin(m\pi\xi) = \frac{1}{4\pi^2}\int_{-\infty}^{\infty}\int_{-\infty}^{\infty}[h_1,h_2](x,y,z)e^{-i(\mathbf{k}\bullet\mathbf{r})}\,dx\,dy.$$

Finally, applying the Fourier series Theorem to these results, we obtain

$$\left[c_m^+, v_0\left(\frac{\Delta_m c_m^- - b_m c_m^+}{d^2}\right)\right]$$

$$= \frac{1}{2\pi^2}\int_0^1 \left\{\int_{-\infty}^{\infty}\int_{-\infty}^{\infty}[h_1,h_2](x,y,\xi)e^{-i(\mathbf{k}\bullet\mathbf{r})}\,dx\,dy\right\}\sin(m\pi\xi)\,d\xi$$

completing the process and allowing us to synthesize arbitrary initial conditions for those functions satisfying the conditions of Sections 5.4 with respect to z and 7.6 with respect to both x and y. Such so-called Dirichlet conditions are required in order for a function to have Fourier series and integral representations.

15.6 Rayleigh Stability Criterion and Comparison with Experiment

Returning to the Bénard problem with rigid surfaces, as described in Section 15.1, the stress-free boundary conditions (15.2.2b, c) must be replaced by the no-slip conditions

$$u = v = 0 \text{ at } z = 0 \text{ and } d$$

while (15.2.2a) now represents the no-penetration rather than the kinematic boundary conditions at those surfaces. Then although the normal-mode analysis yields the same differential equations (15.4.5) the boundary conditions (15.4.6) become

$$W = U = V = \Theta = 0 \text{ at } z = 0 \text{ and } d;$$

which in conjunction with (15.4.5a) yield

$$W = \frac{dW}{dz} = \Theta = 0 \text{ at } z = 0 \text{ and } d.$$

The presence of that odd derivative of W in the boundary conditions rather than the even one of (15.4.9) requires the eigenvalue problem for σ to be solved numerically as mentioned earlier. From such computations R_c and $\lambda_c = 2\pi/q_c$ can be calculated. The most accurate values for these quantities are tabulated below for the cases of rigid–free and rigid–rigid surfaces along with the corresponding ones from the free–free case of the Bénard problem determined analytically in Section 15.4 for comparison purposes [61]:

Once the critical conditions for the onset of convective instability in the rigid–rigid case of the Bénard problem had been determined theoretically by Jeffreys [56], various experiments were devised to check the validity of these predictions. The experimental determination of the onset of convection agreed with theoretical predictions in liquid layers, but in gas layers there was a discrepancy when the visual method of adding an aerosol to the gas was employed. That technique, completely

Table 15.1 Critical conditions for free–free, rigid–free, and rigid–rigid surfaces.

$z = 0$-$z = d$	Free–Free	Rigid–Free	Rigid–Rigid
R_c	657.511	1100.650	1707.762
q_c	2.2214	2.682	3.117
λ_c	2.828	2.342	2.016

analogous to the method of making motion visible in silicone oils by means of the addition of aluminum powder, resulted in an observed onset of this instability occurring at lower temperature differences than those predicted theoretically which themselves were in agreement with experimental evidence for clean gases as determined through heat transfer or optical means. This phenomenon was first observed by Chandra [18] in air layers mixed with cigarette smoke (see Fig. 15.5). He found the nature of the instability to be crucially dependent upon layer depth.

For layers deeper than 10 mm, normal convection was observed at temperature gradients approximately 80% of Jeffrey's [56] predicted value, while if the layer depth was less than 7.5 mm, motion of an entirely different nature of a wavelength much shorter than anticipated occurred at temperature gradients a great deal lower than that prediction. Chandra termed these flow pattern convective rolls motion of Types I and II, respectively, the latter mode being subsequently described by Sutton [119] as a *columnar instability*. It is somewhat ironic that an experimental technique adopted to visualize a flow (via cigarette smoke) would in fact make such major changes in the flow itself that this innocent intervention turned out to have far reaching and interesting consequences.

Wollkind and Zhang [144] analyzed this phenomenon by introducing a two-phase flow model with the gas as the continuous phase and the cigarette smoke as the discrete one. They employed a particle–gas model with the following discrete-phase Boussinesq-type governing equations:

Balance of mass:

$$\nabla \bullet \boldsymbol{s} = 0.$$

Balance of momentum:

$$\frac{D_s\boldsymbol{s}}{D_st} = \left(\frac{\partial}{\partial t} + \boldsymbol{s} \bullet \nabla\right)\boldsymbol{s} = -\frac{1}{mN_0}\nabla\mathcal{P} - g[1 - \alpha_d(\theta_d - T_0)]\boldsymbol{e}_3 + \frac{K_0}{m}(\boldsymbol{v} - \boldsymbol{s}).$$

Balance of energy:

$$mN_0C_0\frac{D_s\theta_d}{D_st} = H_0(T - \theta_d) + k_d\nabla^2\theta_d.$$

Boussinesq equation of state:

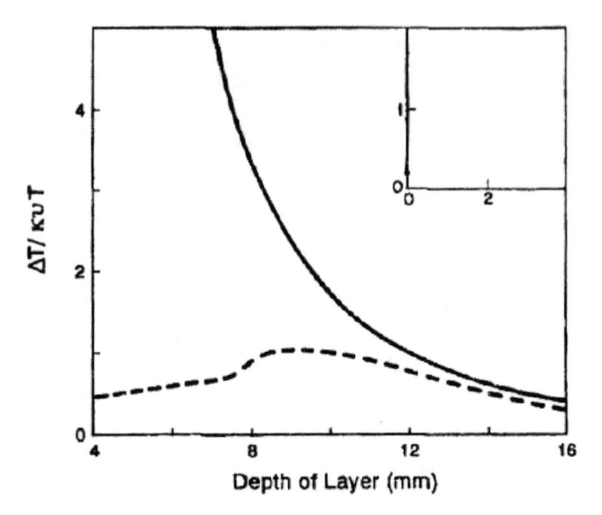

Fig. 15.5 A reproduction from [144] of Chandra's [18] graphical comparison between his observed critical conditions (dashed curve) for convective instability in a smoke–air layer of variable depth involving a rigid–rigid boundary experimental apparatus and Jeffrey's [56] theoretical predictions (solid curve) for a clean gas relevant to this situation. Here, ΔT and d denote the layer temperature difference (K) and depth (mm), respectively while T, κ, and ν are the clean gas mean temperature (K), thermal diffusivity (cm^2/sec), and kinematic viscosity (cm^2/sec). Chandra [18] employed $1709/(gd^3)$ for Jeffrey's [56] curve where g is the acceleration due to gravity at sea level (cm/sec^2), d has cm measure, and 1709 was the best available estimate of the associated critical Rayleigh number now determined more accurately to be 1708. In Fig. 15.9 Chandra's critical value will be adopted for comparison purposes with his experimental data. The inset to this figure was added in order to portray explicitly the extrapolated intercept of the columnar instability section of Chandra's curve.

$$N = N_0[1 - \alpha_d(\theta_d - T_0)].$$

Constitutive relations:

$$K_0 = 6\pi\mu_0 a_0,\ H_0 = 4\pi a_0 N_0 k_0 \varepsilon_0 \text{ where } \varepsilon_0 = \frac{\rho_0}{\rho_d},\ \alpha_d = \frac{1}{\bar{\theta}_0}.$$

Boundary conditions:

Dynamical: $\boldsymbol{s} \bullet \boldsymbol{e}_3 = s_3 = 0$ at $z = 0$ and d;

Thermal: $\theta_d = T_0$ at $z = 0$ and $\theta_d = T_1$ at $z = d$;

Far Field: $\boldsymbol{s}$, $\mathcal{P}$, and θ_d remain bounded as $x^2 + y^2 \to \infty$.

In addition they added a particle drag and thermal exchange interaction term given by

$$\left(\frac{K_0 N_0}{\rho_0}\right)(\mathbf{s}-\mathbf{v}) \text{ for } (15.1.2b,c,d) \text{ and } \left(\frac{H_0}{\rho_0 C_p}\right)(\theta_d - T) \text{ for } (15.2.1),$$

to the right-hand side of those gas balance of momentum and energy equations, respectively.

The spherical particles have radius $\equiv a_0$, mass $\equiv m$, material density $\equiv \rho_d$, number density $\equiv N_0$, specific heat $\equiv C_0$, and thermal conductivity $\equiv k_d$ with velocity $\equiv \mathbf{s}$, pressure $\equiv \mathcal{P}$, and temperature $\equiv \theta_d$, while mean temperature $\equiv \bar{\theta}_0$. Note, in this context, it is assumed that the gas–particle flow is dilute enough to justify the neglect of viscous particle effects in its balance of momentum equations but not so dilute as to invalidate the continuum hypothesis for these particles. It is also assumed that a_0 lies at the upper range over which the molecular process responsible for Brownian motion is of importance, and hence $\mathcal{P}$ represents the continuum effect of those particle–gas collisions. Further terms due to the substantial derivative of this pressure in the particle energy equation and Joule heating from particle drag in both energy equations have been neglected. The inclusion of the k_d term in the particle energy equation, however, accounts for the continuum mechanism by which heat radiates directly from the hotter particles and is reabsorbed by the cooler ones without first flowing through the gas. Finally, Wollkind and Zhang [144] retained the stress-free boundary conditions (15.2.2b, c) for the gas and in order to compare their predictions with Chandra's [18] experimental observations for the rigid–rigid case merely rescaled those critical conditions for the free–free case in accordance with Table 15.1 given that convective instability investigations of these two cases yield qualitatively similar results.

There exists a pure conduction solution of this particle–gas system given by (15.3.1) and

$$\mathbf{s} = \mathbf{0}, \theta_d = T_0(z), \mathcal{P} = \mathcal{P}_0(z) = -mN_d g \int (1 + \alpha_d \beta z)\, dz.$$

That state represents the physical situation of a quiescent gas layer with no particle settling across which an adverse affine temperature gradient is maintained. It was the stability of this solution to both three-dimensional linearly infinitesimal and two-dimensional nonlinear finite-amplitude disturbances with which Wollkind and Zhang [144] were concerned. Given that only roll-type convection can occur in the rigid–rigid case such a stratified longitudinal planform nonlinear stability analysis was sufficient for their comparison purposes with Chandra's results. Toward that end they first performed a normal-mode stability analysis of the pure conduction state for the particle–gas system by seeking a solution of its linearized version of the form of (15.4.4) with similar expansions for its particle variables – *e.g.*,

$$\widetilde{s}_3(x,y,z,t) \sim B_{11} \sin\left(\pi \frac{z}{d}\right) \exp(i[q_1 X + q_2 Y] + \sigma\tau) \text{ where } (X,Y) = \frac{(x,y)}{d};$$

finding critical conditions (q_c^2, R_c) for its most dangerous stationary mode with growth rate $\sigma = \sigma_s$ and then sought a solution for the full system of the form of

(8.5.1) with the scale factors to be introduced in part (c) of Problem 15.2 – *e.g.*,

$$\widetilde{w}(x,z,t) \sim A(\tau)\cos(q_c X)\sin\left(\pi\frac{z}{d}\right) + A^2(\tau)\left[w_{20}\left(\frac{z}{d}\right) + w_{22}\left(\frac{z}{d}\right)\cos(2q_c X)\right]$$
$$+A^3(\tau)\left[w_{33}\left(\frac{z}{d}\right)\cos(q_c X) + w_{33}\left(\frac{z}{d}\right)\cos(3q_c X)\right] \text{ where } \frac{dA}{d\tau} \sim \sigma_s A - a_1 A^3;$$

finding an expression for the Landau constant a_1 as a Fredholm-type solvability condition. When the other parameters of the problem were assigned typical values, $[R_c, q_c, a_1] = [R_c, q_c, a_1](d)$.

In this context, besides the nondimensional parameters already defined for the Bénard problem, Wollkind and Zhang [144] also employed

$$r = \frac{\alpha_d}{\alpha_0}, \xi = \frac{k_d}{k_0}, f = \frac{mN_0}{\rho_0}, h = \frac{C_0 f}{C_p}, F = h + rf(1+h).$$

They assigned the typical material parameter values for air at sea level with $\overline{T}_0 = 293$ K:

$$k_0 = 6.1\times 10^{-5}\frac{\text{cal}}{\text{cm sec K}}, C_p = 0.242\frac{\text{cal}}{\text{gm K}}, \rho_0 = 0.0012\frac{\text{gm}}{\text{cm}^3}, \nu_0 = 0.152\frac{\text{cm}^2}{\text{sec}};$$

and for cigarette smoke:

$$a_0 = 0.3\times 10^{-4} \text{ cm}, N_0 = \frac{2.35\times 10^4}{\text{cm}^3}, r = 1.0, \xi = 10^{-4}, \varepsilon_0 = \frac{1}{36}, f = 0.048, h = 4f;$$

respectively. Note that for these values of k_0, C_p, ρ_0, and ν_0:

$$\kappa_0 = \frac{k_0}{\rho_0 C_p} = 0.210\frac{\text{cm}^2}{\text{sec}} \Rightarrow Pr = \frac{\nu_0}{\kappa_0} = 0.72;$$

while for those values of f, h, and r:

$$F = 0.25. \tag{15.6.1}$$

After assigning these parameter values, Wollkind and Zhang [144] plotted $R_c = R_c(d)$ and $a_1 = a_1(d)$ as reproduced in Figs. 15.6 and 15.8, respectively. Before interpreting those plots and explaining Figs. 15.7 and 15.9 as well, in order to examine the behavior of this model with respect to layer depth, we shall describe Chandra's [18] experimental observations in more detail. His original apparatus consisted of a smoke–air layer of variable depth d contained between two rigid horizontal plates uniformly heated from below and cooled from above. Starting with the layer in motion at a particular depth, he lowered the temperature difference ΔT between the plates until that convective motion ceased. Chandra [18] reported the results of these experiments in both a tabular and graphical form, the latter by means of a plot reproduced in Fig. 15.5 of $\Delta T/(\kappa \nu T)$ versus $4\,\mathrm{mm} \le d \le 16\,\mathrm{mm}$ where $T = \overline{T}_0$, the mean layer temperature. Recalling for gases $\alpha_0 = 1/\overline{T}_0$, we can deduce from (15.4.18c) that

$$\frac{\Delta T}{\kappa_0 \nu_0 T} = \frac{R}{g d^3}. \tag{15.6.2}$$

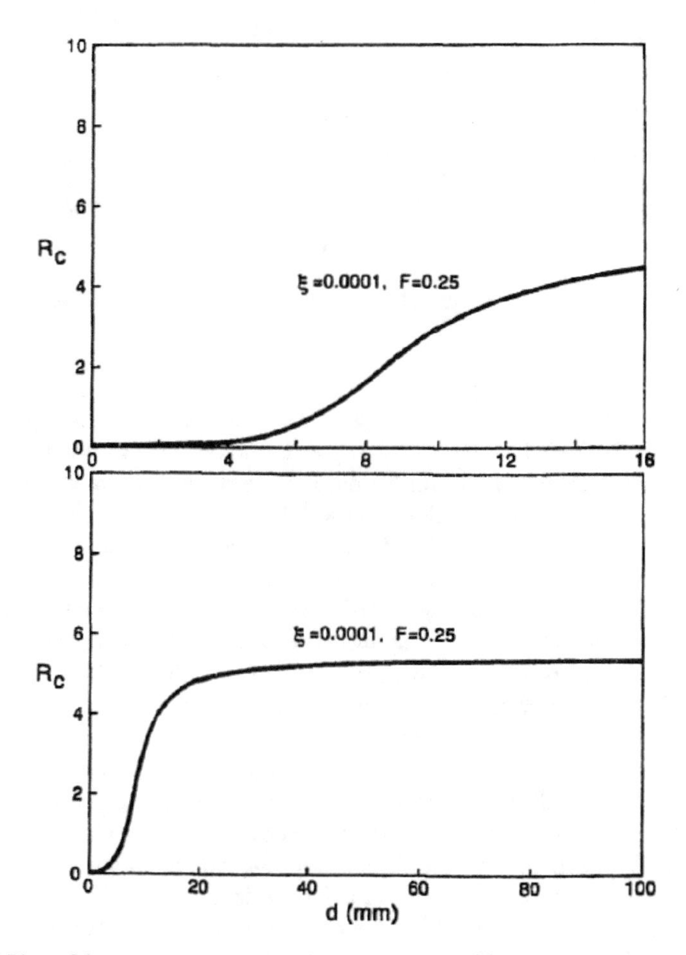

Fig. 15.6 Plots of the stationary marginal stability curve $R = R_c(d)$ where R is measured in units of π^4 and d in mm for (a) $0\ \mathrm{mm} \le d \le 16\ \mathrm{mm}$ and (b) $0\ \mathrm{mm} \le d \le 100\ \mathrm{mm}$ [144]. Here and in the figures which follow for Chandra's experiments the short-hand notation $\xi = 10^{-4}$ and $F = 0.25$ is being employed to represent all the particle and gas parameter value assignments.

Observe from this figure that Chandra's [18] experimentally determined marginal stability curve (dashed line) can be divided into three parts: Namely, a columnar instability section, a normal convective section, and a transitional region linking these two sections together. In particular, the normal convective section of that graph

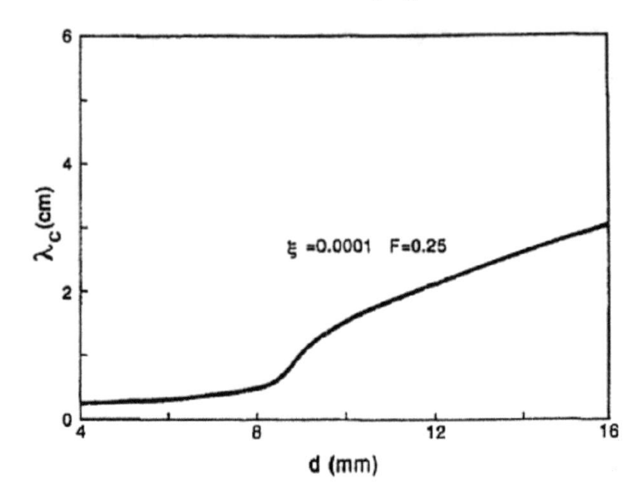

Fig. 15.7 A plot of the dimensional wavelength $\lambda_c(d)$ of (15.6.8) measured in cm versus d for 4 mm $\leq d \leq 16$ mm [144].

occurs for 12 mm $\leq d \leq 16$ mm and is closely approximated by 80% of the asymptotic curve (solid line) $1709/(gd^3)$ corresponding to Jeffrey's [56] theoretical prediction of the critical conditions for the onset of Rayleigh–Bénard instability in the case of rigid–rigid boundaries relevant to a clean gas, while the columnar instability section occurs for 4 mm $\leq d \leq 7.5$ mm and is a segment of a straight line with positive slope and intercept (see the inset of Fig. 15.5). We note, in this context, that the quantity $\Delta T/(\kappa\nu T)$ has units sec^2/cm^4 in Fig. 15.5 even though the depth of the layer appearing in that figure is being measured in mm. Finally, Chandra [18] also published photographs of the flow patterns for various depths which clearly indicated that the rolls associated with columnar or Type II instabilities were much more elongated in form when compared to those associated with normal convective or Type I instabilities than one would have anticipated from Jeffrey's [56] predicted critical wavelengths for such small differences in layer depth. Further, with a layer depth $d = 2$ mm, Chandra [18] found that stability had not yet been attained even for those ΔT which were so low as to be impossible to maintain uniformly over the whole layer. Hence, he was unable to observe the critical condition for this depth accurately.

Note from an examination of the plot of the stationary marginal stability curve $R = R_c(d)$ in Fig. 15.6 that it becomes virtually tangent to the d-axis at $d = 2$ mm consistent with Chandra's inability to attain a motionless state at this depth. Further observe from Fig. 15.8 that

$$a_1(d) > 0 \text{ when } d > 0. \tag{15.6.3}$$

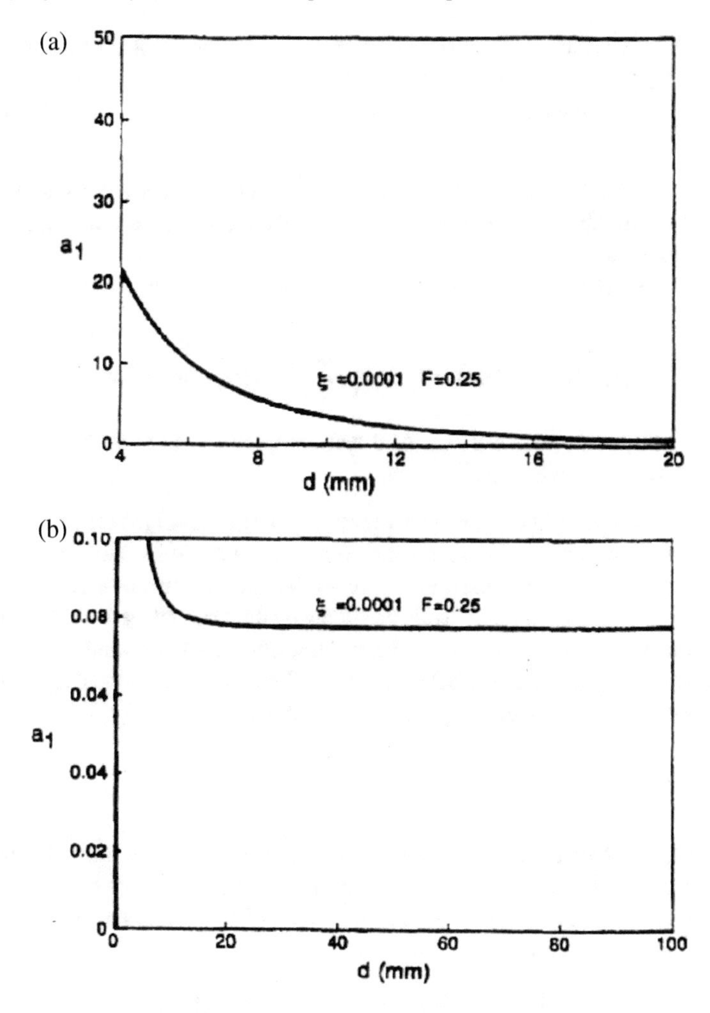

Fig. 15.8 Plots of $a_1(d)$ for (a) 0 mm $\leq d \leq$ 20 mm and (b) 0 mm $\leq d \leq$ 100 mm [144].

Then setting the wavenumber equal to its critical value q_c as has been done in the nonlinear expansion, one can represent σ_s in a Taylor series about $R = R_c$

$$\sigma_s(R) = \sigma_s(R_c) + \sigma_s'(R_c)(R - R_c) + O(R - R_c)^2;\ \sigma_s(R_c) = 0,\ \sigma_s'(R_c) = \sigma_1 > 0; \tag{15.6.4a}$$

and, taking R sufficiently close to its critical value so that $R - R_c = O(\varepsilon^2)$ where $|\varepsilon| << 1$, guarantee that

$$\sigma_s = O(\varepsilon^2). \tag{15.6.4b}$$

Under these conditions it can be shown that the implicit neglect of the fifth-order terms in the amplitude equation of the nonlinear expansion is valid (see the next chapter) and the long-time behavior of the solution to that equation can be catalogued as outlined in Section 8.5:

(i) For $R < R_c$, the undisturbed state $A = 0$, which corresponds to the pure conduction solution, is stable to both infinitesimal and finite-amplitude stationary disturbances.

(ii) For $R > R_c$, there exists a stable stationary equilibrium solution $A_e^2 = \sigma_s / a_1$ such that

$$\widetilde{w}(x, z, t) \to \widetilde{w}_e(x, z) \sim A_e \cos\left(q_c \frac{x}{d}\right) \sin\left(\pi \frac{z}{d}\right) \text{ as } t \to \infty, \tag{15.6.5}$$

which physically corresponds to a stationary roll-type convection pattern of dimensional wavelength $(2\pi/q_c)d$.

Wollkind and Zhang [144] were now ready to examine the behavior of their particle–gas model with respect to layer depth and compare those results with Chandra's [18] experimental evidence regarding the occurrence of columnar instabilities. They wished to demonstrate that the stationary rolls predicted by their model were compatible with his observations involving smoke–air layers described above. Toward that end, they first investigated the asymptotic behavior of $R_c(d)$ and $a_1(d)$ for large d as depicted in Figs. 15.6b and 15.8b, respectively. They deduced that

$$R_c(d) \sim \frac{1+\xi}{1+F} \frac{27}{4} \cong \frac{27}{4(1+F)} = \frac{27}{5} \text{ as } d \to \infty$$

for their choice of parameter values which is consistent with the normal convective mode being associated with 80% of the critical value of Jeffrey's [18] clean gas theoretical prediction once one realizes that the R_c of Fig. 15.6 is being measured in units of π^4. Similarly they deduced that

$$a_1(d) \sim \frac{(1+h)^2/(1+\xi)}{8[Pr^{-1}(1+f)(1+\xi)+1+h]} \cong \frac{(1+h)^2/8}{Pr^{-1}(1+f)+1+h} \text{ as } d \to \infty$$

in comparison with $a_1 = 1/[8(Pr^{-1} + 1)]$ for the clean gas Rayleigh–Bénard model which can be obtained by setting $f = h = 0$ in the above expression.

In order actually to provide quantitative agreement with Chandra's [18] experimental data, it was necessary for Wollkind and Zhang [144] to take into account the intrinsic difference between the critical conditions for the onset of Rayleigh–Bénard instability involving a clean gas when the boundaries are rigid–rigid rather than free–free. As mentioned earlier the simplest way of accomplishing this was merely to introduce the following rescaled marginal stability function appropriate for comparing their theoretical predictions with Chandra's [18] experimental observations summarized in Fig. 15.5:

$$1709\left(\frac{4}{27}\right)\frac{R_c(d)}{gd^3}, \tag{15.6.6}$$

which has the large-d asymptotic behavior

$$\left[\frac{1+\xi}{1+F}\right]\frac{1709}{gd^3} \tag{15.6.7a}$$

where

$$\frac{1+\xi}{1+F} \cong \frac{1}{1+F} = \begin{cases}1.0\\0.8\end{cases} \quad \text{for } F = \begin{cases}0.00\\0.25\end{cases}. \tag{15.6.7b}$$

Next in order to examine the correlation of their model's theoretically predicted results with Chandra's columnar instabilities they plotted the marginal stationary stability function of (15.6.6) versus d in Fig. 15.9a. Upon comparison with Chandra's columnar instability straight-line data points of $d = 4, 6, 7$, and 7.5 mm it can be seen from Fig. 15.9a that (15.6.6) provides an excellent fit to these points over this portion. They then plotted the asymptotic function of (15.6.7) versus d for both $F = 0$ and $F = 0.25$ in Fig. 15.9b. Note that the $F = 0$ curve, denoted by a dotted line in that figure, corresponds to Jeffrey's [56] theoretical prediction and the $F = 0.25$ one virtually coincides with the marginal stability function of (15.6.6) over its normal convective instability portion 12 mm $\leq d \leq$ 16 mm as well as also providing an excellent fit to the data end points of $d = 12$ and 16 mm for this interval. Thus this completes both the columnar and normal convective instability parts of a compatibility demonstration of Wollkind and Zhang's [144] theoretical predictions with Chandra's [18] experimental results. In fact, even the transition region data point $d = 8$ mm in Fig. 15.9a deviates just slightly and that of $d = 10$ mm in Fig. 15.9b just slightly more from their marginal stability curve. This completes the final part of the desired compatibility demonstration.

It only remained for Wollkind and Zhang [144] to demonstrate that the layer depth behavior of the critical wavelength of their theoretically predicted stationary rolls was compatible with the characteristic width of the convective flow patterns photographed by Chandra [18]. Using analogous reasoning to that employed in conjunction with the introduction of (15.6.6), Wollkind and Zhang [144] defined the dimensional wavelength

$$\lambda_c(d) = \left(\frac{2.016}{2\sqrt{2}}\right)\left[\frac{2\pi}{q_c(d)}\right]d, \tag{15.6.8a}$$

which corresponds to a similar rescaled quantity appropriate for rigid–rigid boundaries. When they plotted the critical wavelength function of (15.6.8) versus d in Fig. 15.7, the resultant curve had an inflection point which, occurring as it did at $d \cong 9$ mm, allowed for a much more dramatic shift in characteristic roll width between those depths associated with columnar and normal convective instabilities, respectively, than could be anticipated from the linear relation for a clean gas given by

$$\lambda_c(d) = 2.016d, \tag{15.6.8b}$$

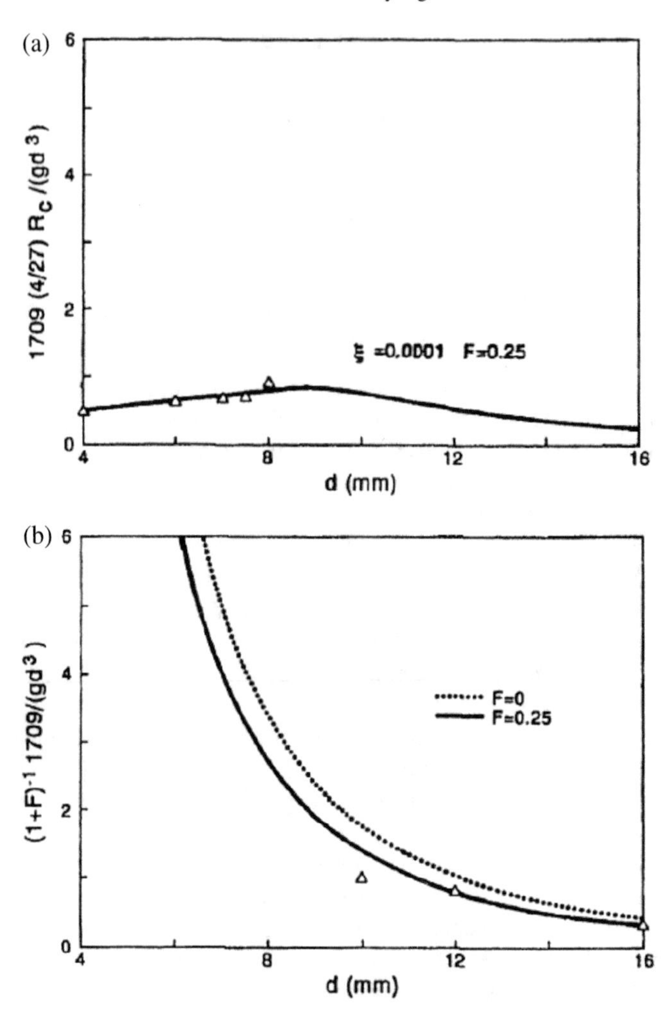

Fig. 15.9 Plots of (a) the marginal stability function of (15.6.6) versus 4 mm $\leq d \leq$ 16 mm for the $R_c(d)$ of Fig. 15.6 and (b) the asymptotic functions of (15.6.7) versus 4 mm $\leq d \leq$ 16 mm for $F = 0$ and 0.25 [144]. Here the small triangles represent Chandra's experimentally determined data points.

which serves as the large-d asymptote for (15.6.8a). This prediction, being in exact accordance with Chandra's [18] photographic experimental evidence, completed Wollkind and Zhang's [144] final compatibility demonstration.

Dassanayake continued Chandra's [18] work on columnar instabilities employing essentially the same apparatus and methodology but replacing the air layer involved in the latter's experiments with carbon dioxide as reported by Sutton [119]. Dassanayake's data set, denoted by encircled dots, is reproduced from Sutton [119] in Fig. 15.10. This is again a plot of $\Delta T/(\kappa\nu\overline{T}_0)$ measured in sec^2/cm^4 versus d in mm units which includes Chandra's [18] data, denoted here by crosses, for comparison purposes. In particular, Dassanayake's marginal stability curve consists of two parts: Namely, a columnar instability section occurring for 4 mm $\leq d \leq$ 7 mm which is a line segment with positive slope and 4-mm intercept and a normal convective section occurring for 7 mm $\leq d \leq$ 10 mm which is asymptotic to the dashed curve $1708/(gd^3)$ in Fig. 15.10 corresponding to Jeffrey's [56] theoretical prediction for a clean gas.

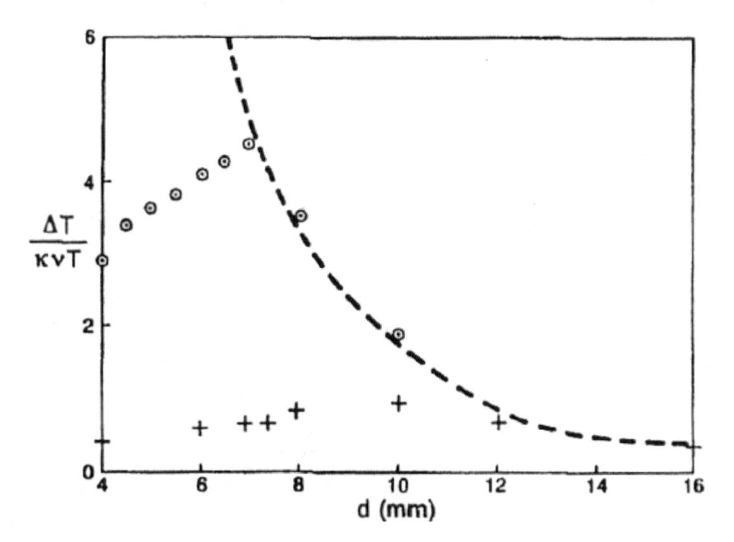

Fig. 15.10 A reproduction from [144] of Sutton's [119] graphical comparison between the experimental data points of Chandra [18] and Dassanayake, denoted by crosses and encircled dots, respectively, relevant to the occurrence of columnar instabilities in a smoke–gas layer of variable depth involving a rigid–rigid experimental apparatus where air was used as the carrier gas for the former series of experiments and carbon dioxide for the latter. Note that here the same notation is being employed as in Fig. 15.5 except that Jeffrey's [56] theoretical prediction (dashed curve) is now represented by $1708(gd^3)$ since Sutton [119] adopted the more accurate 1708 determination of the associated critical Rayleigh number as opposed to the 1709 used by Chandra [18]. In Fig. 15.11a Sutton's critical value will be adopted for comparison purposes with Dassanayake's experimental data.

In contrast, Chandra's marginal stability curve also contains a third part which is transitional in nature linking its columnar instability and normal convective sections. Upon comparing these two marginal stability curves, we see that the one for carbon dioxide has had both the 4-mm intercept and slope of its columnar instability section's line segment so markedly increased that it now intersects its normal convective section at a cusp-type maximum point possessing a much larger ordinate value than occurs with air while virtually eliminating the transitional region of the latter. The primary cause for this difference is that $\kappa_0 \nu_0$ for air is about four and one half times greater than that for carbon dioxide as we now demonstrate. Recall for air at $T = 293$ K that

$$\kappa_0 = 0.210 \frac{\text{cm}^2}{\text{sec}} \text{ and } \nu_0 = 0.152 \frac{\text{cm}^2}{\text{sec}} \Rightarrow \kappa_0 \nu_0 = 0.032 \frac{\text{cm}^4}{\text{sec}^2}$$

while the corresponding components of κ_0 for carbon dioxide are given by (Weast, [132])

$$k_0 = 3.8 \times 10^{-5} \frac{\text{cal}}{\text{cm sec K}}, \; C_p = 0.200 \frac{\text{cal}}{\text{gm K}}, \; \rho_0 = 0.0019 \frac{\text{gm}}{\text{cm}^3}$$

$$\Rightarrow \kappa_0 = \frac{k_0}{\rho_0 C_p} = 0.100 \frac{\text{cm}^2}{\text{sec}}.$$

Recalling that for gases

$$Pr = \frac{\nu_0}{\kappa_0} = 0.7 \Rightarrow \nu_0 = 0.7\kappa_0 = 0.070 \frac{\text{cm}^2}{\text{sec}} \Rightarrow \kappa_0 \nu_0 = 0.0007 \frac{\text{cm}^4}{\text{sec}^2}.$$

Hence the ratio of this quantity for air with that for carbon dioxide is 4.57 as was to be shown. As previously summarized in this section, Wollkind and Zhang [144] assigned parameter values for their particle–gas system appropriate for modeling Chandra's convective experiments involving air layers at sea level into which cigarette smoke had been mixed to provide visibility. They then provided a similar sort of correlation between their theoretical predictions and Dassanayake's data set. To do so, they varied the relevant particle parameters simultaneously about these reference values for a specific choice of the gas parameters particularized to smoke–carbon dioxide aerosols and selected the proper set to provide the desired correlation. Wollkind and Zhang [144] defined the following ratios in addition to r and ξ:

$$\eta = \frac{\rho_{0\text{CO}_2}}{\rho_{0\text{air}}}, \; n = \frac{N_d}{N_0}, \; 4\zeta = \left(\frac{C_0}{C_p}\right)_{\text{CO}_2/\text{smoke}};$$

and using these ratios represented the relevant model parameters in the form

$$\varepsilon_0 = \frac{\eta}{36}, \; f = 0.048\frac{n}{\eta}, \; h = 4\zeta f;$$

which reduced to their reference values for $\eta = n = \zeta = 1$. Here N_d represents the reference particle number density for the carbon dioxide experiments while taking

$h = 4f$ as was done for the air–smoke layers implicitly implied that $(C_0/C_p)_{\text{air/smoke}} = 4$ which, being the value of that ratio for aerosols of liquid water in air, assumes that the moisture content of air–cigarette smoke mixtures is significant enough to allow them to be treated as such aerosols.

In light of their values for $\rho_{0\text{CO}_2}$ and $\rho_{0\text{air}}$, Wollkind and Zhang [144] assigned

$$\eta = \frac{19}{12} = 1.6; \tag{15.6.9a}$$

retained

$$r = 1.0; \tag{15.6.9b}$$

and determined the proper selection for the particle ratios ζ, n, ξ that provided the best fit to Dassanayake's data of Fig. 15.10 by varying each component of that triad about the reference point $\zeta = n = 1$, $\xi = 10^{-4}$ and making the optimal selection visually which yielded

$$\zeta = 0.8, n = 0.3, \xi = 1.3 \times 10^{-4}. \tag{15.6.9c}$$

The result of such a process appears in Fig. 15.11a, while the associated $\lambda_c(d)$ and $a_1(d)$ are plotted in Figs. 15.11b, c. Note that Wollkind and Zhang's [144] theoretical predictions fit Dassanayake's data points extremely well except for the 4-mm intercept data point, which probably can be treated as an outlier due to the nonuniformity of ΔT at such a low depth since all the other columnar instability points lie on a straight line. Observe, in addition, that the intersection of the columnar instability and normal convective segments of their marginal stability curve of Fig. 15.11a at a cusp-type maximum point results in the concomitant precipitous behavior of $\lambda_c = \lambda_c(d)$ in the neighborhood of its inflection point (see Fig. 15.11b) while, given the positivity of $a_1 = a_1(d)$, the occurrence of a supercritical stationary roll convective pattern is again predicted (see Fig. 15.11c). We wish to make a few comments on the physical significance of the parameter values chosen in Fig. 15.11 to provide this fit. Taking $r = 1$ adopts the mean temperature equilibrium condition $\overline{T}_0 = \bar{\theta}_0$ which implies that the pointwise temperature differences $T \neq \theta$ caused by thermal disequilibrium tend to cancel out when spatially averaged over the whole layer. Taking $\zeta = 0.8$ implies $C_0 = 3.2C_p$ for carbon dioxide as opposed to $C_0 = 4C_p$ for air which is consistent with the fact that $C_{p\text{CO}_2} = 0.8C_{p\text{air}}$. Rewriting ξ for the carbon dioxide–smoke mixture in the form

$$\xi = \frac{k_{d\text{CO}_2}}{k_{0\text{CO}_2}} = \left(\frac{k_{d\text{CO}_2}}{k_{d\text{air}}}\right)\left(\frac{k_{d\text{air}}}{k_{0\text{air}}}\right)\left(\frac{k_{0\text{air}}}{k_{0\text{CO}_2}}\right) = 1.3 \times 10^{-4}$$

and identifying

$$\frac{k_{d\text{air}}}{k_{0\text{air}}} = \xi_0 = 10^{-4}, \ \frac{k_{0\text{air}}}{k_{0\text{CO}_2}} = \frac{6.1}{3.8} = 1.6 \Rightarrow \frac{k_{d\text{CO}_2}}{k_{d\text{air}}} = \frac{1.3}{1.6} = 0.8;$$

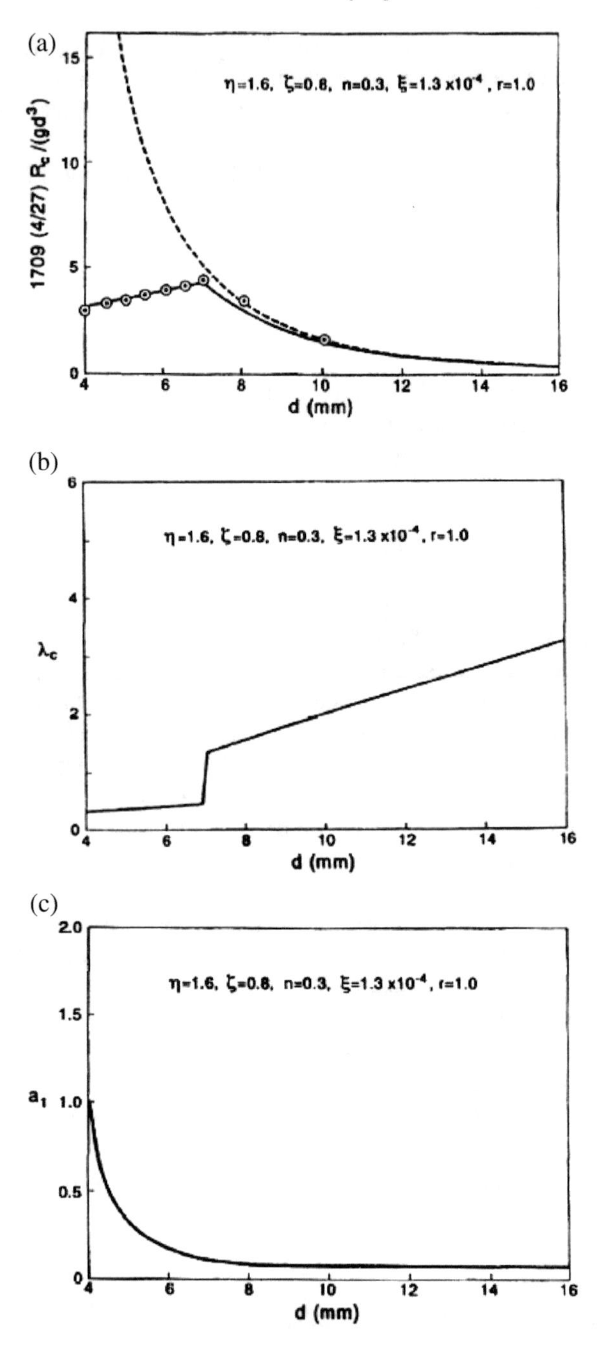

Fig. 15.11 Plots of (a) the marginal stability function of (15.6.6) with 1709 replaced by 1708, (b) $\lambda_c(d)$ of (15.6.8), and (c) $a_1(d)$; for $4\,\text{mm} \leq d \leq 16\,\text{mm}$ with $\eta = 1.6, r = 1.0, \zeta = 0.8, n = 0.3$, and $\xi = 1.3 \times 10^{-4}$ [144]. In (a), the encircled dots and dashed curve represent Dassanayake's experimentally determined carbon dioxide data points as reported by Sutton [119] and Jeffrey's [56] theoretically predicted marginal stability locus, respectively, from Fig. 15.10.

while $n = N_d/N_0 = 0.3 \Rightarrow N_d = 0.3N_0 = 7.05 \times 10^3/\text{cm}^3$; which correlates quite well with such particle to particle radiation effects measured by $k_{dCO_2,\text{air}}$ depending upon the relative closeness of the particles to each other as measured by $N_{d,0}$.

We close with a final commentary on exactly what contribution each of Wollkind and Zhang's [144] model system parameters makes to the correlation with experiment exhibited in Figs. 15.9a or 15.11a. In particular the slope and the 4-mm intercept of the straight-line segment columnar instability portion of their marginal stability curves depended crucially upon ε_0, ξ, and the product of a_0 with N_0 or N_d while the reduction factor $1/(1+F)$ characteristic of the normal convection portion depended crucially upon its constituent components r, f, and h. In this context, we note that for the carbon dioxide–smoke marginal curve of Fig. 15.11a $1/(1+F) = 0.96$ as opposed to that factor's value of 0.80 for the air–smoke marginal curve of Fig. 15.9a which was why the normal convective portion of the former virtually coincided with Jeffrey's [56] asymptotic curve which corresponds to $F = 0$ or $1/(1+F) = 1$.

Problems

15.1. For

$$\alpha(x,y,z,t) = \alpha_0(z) + \widetilde{\alpha}(x,y,z,t) \text{ and } \boldsymbol{v}(x,y,z,t) = \boldsymbol{0} + \widetilde{\boldsymbol{v}}(x,y,z,t),$$

which represents the pure conduction solution of (15.3.1) plus the perturbation of (15.3.2), show that

$$\frac{D\alpha}{Dt} = \frac{\partial \alpha}{\partial t} + (\boldsymbol{v} \bullet \nabla)\alpha = \frac{\partial \widetilde{\alpha}}{\partial t} + \left(\frac{d\alpha_0}{dz}\right)\widetilde{w} + (\widetilde{\boldsymbol{v}} \bullet \nabla)\widetilde{\alpha}$$

where $\widetilde{\boldsymbol{v}} = (\widetilde{u}, \widetilde{v}, \widetilde{w})$. Now deduce the governing equations (15.3.3) and boundary conditions (15.3.4).
Hint: Note that $\nabla \alpha_0 = (d\alpha_0/dz)\boldsymbol{e}_3$ and $\widetilde{\boldsymbol{v}} \bullet \boldsymbol{e}_3 = \widetilde{w}$.

15.2. Consider the governing equations for the Rayleigh–Bénard convection problem in the form

$$\nabla \bullet \boldsymbol{v} = 0, \tag{P2.1}$$

$$\rho_0 \frac{D\boldsymbol{v}}{Dt} = -\nabla p' - \rho_0(1 - \alpha_0 T)g\boldsymbol{e}_3 + \mu_0 \nabla^2 \boldsymbol{v}, \tag{P2.2}$$

$$\rho_0 C_p \frac{DT}{Dt} = \frac{Dp'}{Dt} + \Phi + k_0 \nabla^2 T; \tag{P2.3}$$

where $v = (u, w) \equiv$ velocity, $p' \equiv$ pressure, $T \equiv$ temperature, $\rho_0 \equiv$ density, $\alpha_0 \equiv$ thermal expansivity, $g \equiv$ gravitational acceleration at sea level, $\mu_0 \equiv$ shear viscosity, $C_p \equiv$ specific heat at constant pressure, $\Phi \equiv$ viscous dissipation, $k_0 \equiv$ thermal conductivity, $D/DT = \partial/\partial t + \boldsymbol{v} \bullet \nabla$ for $\nabla \equiv (\partial/\partial x, \partial/\partial z)$, $\nabla^2 \equiv \nabla \bullet \nabla$, $\boldsymbol{e}_3 \equiv (0, 1)$,

$t \equiv$ time, $(x, z) \equiv$ two-dimensional position; with boundary conditions

$$w = \frac{\partial u}{\partial z} + \frac{\partial w}{\partial x} = 0 \text{ at } z = 0 \text{ and } d;\ T = 0 \text{ at } z = 0,\ T = -\Delta T < 0 \text{ at } z = d. \qquad \text{(P2.4)}$$

(a) Introducing the reduced pressure

$$p = p' - p_0(z) \text{ where } p_0(z) = p_A - \rho_0 g z, \qquad \text{(P2.5)}$$

show that (P2.2) and (P2.3) are transformed into

$$\frac{D\boldsymbol{v}}{Dt} = -\frac{1}{\rho_0}\nabla p + \alpha_0 g T \boldsymbol{e}_3 + \nu_0 \nabla^2 \boldsymbol{v}, \qquad \text{(P2.6)}$$

$$\frac{DT}{Dt} + \frac{gw}{C_p} = \kappa_0 \nabla^2 T; \qquad \text{(P2.7)}$$

upon neglect of Dp/DT and Φ in the energy equation where $\nu_0 = \mu_0/\rho_0 \equiv$ kinematic viscosity and $\kappa_0 = k_0/(\rho_0 C_p) \equiv$ thermal diffusivity.

(b) Deduce that there exists a pure conduction solution of (P2.1), (P2.4), (P2.6), and (P2.7) given by

$$u = w = 0,\ T = T_0(z) = -\beta_0 z \text{ for } \beta_0 = \frac{\Delta T}{d},\ p = p_0(z) = -\frac{\rho_0 \alpha_0 \beta_0 g z^2}{2} + p_A. \qquad \text{(P2.8)}$$

(c) Seeking a solution of this system of the form

$$\boldsymbol{v} = \varepsilon \boldsymbol{v}_1 + \boldsymbol{O}(\varepsilon^2),\ p = p_0(z) + O(\varepsilon^2),\ T = T_0(z) + \varepsilon T_1 + O(\varepsilon^2), \qquad \text{(P2.9)}$$

where $\boldsymbol{v}_1 = (u_1, w_1)$, neglecting terms of $O(\varepsilon^2)$, and nondimensionalizing position, time, velocity, pressure, and temperature by d, d^2/κ_0, $\kappa_0 d$, $\kappa_0\mu_0/d^2$, and $\Delta\theta = \Delta T - gd/C_p > 0$, respectively, obtain

$$\nabla \bullet \boldsymbol{v}_1 = 0, \qquad \text{(P2.10)}$$

$$Pr^{-1}\frac{\partial \boldsymbol{v}_1}{\partial t} = -\nabla p_1 + R T_1 \boldsymbol{e}_3 + \nabla^2 \boldsymbol{v}_1, \qquad \text{(P2.11)}$$

$$\frac{\partial T_1}{\partial t} = w_1 + \nabla^2 T_1; \qquad \text{(P2.12)}$$

$$w_1 = \frac{\partial u_1}{\partial z} + \frac{\partial w_1}{\partial x} = T_1 = 0 \text{ at } z = 0 \text{ and } 1; \qquad \text{(P2.13)}$$

where $Pr = \nu_0/\kappa_0$, $R = g\alpha_0 d^3 \Delta\theta/(\kappa_0 \nu_0)$, and the same nomenclature has been employed for the dimensionless variables as that used for the original ones.

(d) Seeking a normal-mode solution of this system of the form

$$u_1(x, z, t) = A \sin(qx)\cos(\pi z)e^{\sigma t}, \qquad \text{(P2.14a)}$$

$$[w_1, T_1](x, z, t) = [B, C]\cos(qx)\sin(\pi z)e^{\sigma t}, \qquad \text{(P2.14b)}$$

$$p_1(x,z,t) = E\cos(qx)\cos(\pi z)e^{\sigma t}; \tag{P2.14c}$$

obtain the secular equation

$$k^2 Pr^{-1}\sigma^2 + k^4(Pr^{-1}+1)\sigma + k^6 - Rq^2 = 0 \text{ where } k^2 = \pi^2 + q^2 \tag{P2.15}$$

governing the linear stability of the pure conduction state of (P2.8).

(e) From the analysis of the related equation (15.4.21b) conclude that the onset of convective instability occurs should the temperature difference between the layers exceed the critical value

$$\Delta T > \frac{gd}{C_p} + \frac{27\pi^4\kappa_0\nu_0}{4g\alpha_0 d^3}. \tag{P2.16}$$

(f) Plot this instability criterion schematically in the d-ΔT plane and discuss its behavior with layer depth d paying particular attention to the limiting cases of thin and thick layers. Determine the minimum point of this curve and discuss its significance bearing in mind that the first and second terms on the right-hand side of the inequality of (P2.16) represent the adiabatic gradient and normal Rayleigh number effects, respectively, the former only important for gases.

(g) Finally represent the expression for the critical depth at which this minimum occurs as

$$d_c = 3\pi\sqrt[4]{\frac{k_0\nu_0\overline{T}_0}{4\rho_0 g^2}} \tag{P2.17}$$

by making use of the definition of κ_0 and the fact that $\alpha_0 = 1/\overline{T}_0$ for a gas where $\overline{T}_0 \equiv$ mean temperature in K. Calculate d_c (cm) for air given the parameter values introduced in Section 15.4:

$$k_0 = 2551\,\frac{\text{gm cm}}{\text{sec}^2\text{K}},\ \nu_0 = 0.152\,\frac{\text{cm}^2}{\text{sec}}, \overline{T}_0 = 293\text{ K},\ \rho_0 = 0.0012\,\frac{\text{gm}}{\text{cm}^3},\ g = 980\,\frac{\text{cm}}{\text{sec}^2}.$$

Note that this k_0 value is equivalent to the one introduced in Section 15.4 for air once the conversion factor cal $\equiv 4.182\times10^7$ (gm cm^2)/sec^2 at 293 K has been employed. What conclusions can be drawn from this result about the necessity of including the adiabatic gradient when modeling convective experiments involving air at sea level?

15.3. Consider the following nondimensionalized linear perturbation system relevant to the pure conduction solution for Rayleigh–Bénard convection in upper planetary atmospheres under the compressible gas version of the Boussinesq approximation (see Problem 15.2):

$$\frac{\partial u_1}{\partial x} + \frac{\partial w_1}{\partial z} = 0, \tag{P3.1a}$$

$$\frac{\partial u_1}{\partial t} = -\frac{\partial p_1}{\partial x} + \left(m\frac{\partial^2}{\partial x^2} + \frac{\partial^2}{\partial z^2}\right)u_1, \tag{P3.1b}$$

$$\frac{\partial w_1}{\partial t} = -\frac{\partial p_1}{\partial z} + RT_1 + \left(m\frac{\partial^2}{\partial x^2} + \frac{\partial^2}{\partial z^2}\right) w_1, \tag{P3.1c}$$

$$\frac{\partial T_1}{\partial t} = w_1 + \left(m\frac{\partial^2}{\partial x^2} + \frac{\partial^2}{\partial z^2}\right) T_1; \tag{P3.1d}$$

$$w_1 = \frac{\partial u_1}{\partial z} + \frac{\partial w_1}{\partial x} = T_1 = 0 \text{ at } z = 0 \text{ and } 1. \tag{P3.1e}$$

Here the gas has perturbation velocity components (u_1, w_1), reduced pressure p_1, and temperature T_1 which are functions of the spatial coordinates (x, z) and time t. Further R is the compressible fluid Rayleigh number defined in part (c) of Problem 15.2 and, to account for the role of eddies, an eddy kinematic viscosity and thermometric conductivity, which are assumed equal, have been introduced where these diffusivities are anisotropic with m being the ratio of its horizontal to vertical components.

(a) Seeking a normal-mode solution of this system of the form of (P2.14) obtain the secular equation

$$k_1^2\sigma^2 + 2k_1^2k_m^2\sigma + k_1^2k_m^4 - Rq^2 = 0 \tag{P3.2a}$$

where

$$k_m^2 = \pi^2 + mq^2 = \pi^2\left(1 + \frac{4m}{\lambda^2}\right) \text{ for } \lambda = \frac{2\pi}{q}. \tag{P3.2b}$$

(b) Upon analyzing (P3.2) conclude that $\sigma \in \mathbb{R}$ and the marginal curve for the onset of convective instability is given by

$$R = R_m(\lambda^2) = \pi^4\left(\frac{\lambda^2}{4} + 1\right)\left(1 + \frac{4m}{\lambda^2}\right)^2. \tag{P3.3}$$

(c) Demonstrate that the marginal curve of (P3.3) has a minimum point (λ_c^2, R_c) where $\lambda_c^2 = \lambda_c^2$ (m) can be represented implicitly or explicitly by

$$m = \frac{\lambda_c^2/4}{1 + 8/\lambda_c^2} \text{ or } \lambda_c^2(m) = 2m\left(1 + \sqrt{1 + \frac{8}{m}}\right). \tag{P3.4}$$

(d) Plot $\lambda_c(m)$ of (P3.4) in the m-λ plane identifying m_1 and m_2 such that

$$\lambda_c(m_1) = 10 \text{ and } \lambda_c(m_2) = 100. \tag{P3.5}$$

(e) Comment on the physical significance of these results with the respect to the aspect ratio $\lambda_c = (\lambda_c^*)/d$ where λ_c^* is the dimensional wavelength and d, the depth, characteristic of cells occurring in the upper atmosphere of Venus which tend to lie in the range $10 \leq \lambda_c \leq 100$. Show that, for the case of isotropic diffusivities where $m = 1$, (P3.4) reduces to

$$\lambda_c^2(1) = 8 \Rightarrow \lambda_c(1) = 2\sqrt{2} \tag{P3.6}$$

consistent with Table 15.1. Given (P3.6), compare the aspect ratios of convection cells typically observed in laboratory experiments with that of those occurring in planetary atmospheres.

15.4. The purpose of this problem is to examine an infinite Prandtl number surface tension-driven Marangoni-Bénard convection model system for the sake of completeness as well as to demonstrate how problems of this sort are usually analyzed when the methodology employed on the free–free boundary version of the Rayleigh–Bénard model cannot be used. The notation is the same as that introduced in this chapter and part (d) of Problem 10.6. Then the relevant linear perturbation system in dimensionless variables for its pure conduction solution is given by

$$\nabla^4 w_1 = 0, \tag{P4.1a}$$

$$\frac{\partial T_1}{\partial t} = w_1 + \nabla^2 T_1; \tag{P4.1b}$$

with boundary conditions

$$w_1 = 0, \tag{P4.2a}$$

$$\frac{\partial T_1}{\partial z} = 0, \tag{P4.2b}$$

$$\frac{\partial^2 w_1}{\partial z^2} = \mathcal{M}\nabla_2^2 T_1, \tag{P4.2c}$$

at the upper free surface $z = 0$;

$$w_1 = \frac{\partial w_1}{\partial z} = T_1 = 0 \text{ at the lower rigid plate } z = -1. \tag{P4.3}$$

Here (P4.2a) represents the kinematic condition, and (P4.2b) follows from an impedance-type condition $\partial T_1/\partial z + \mathcal{B}T_1 = 0$ with $\mathcal{B} \equiv$ Biot number $= 0$, and (P4.2c), from part (d) of Problem 10.6 with $\mathcal{M} = (\gamma_1\beta_0 d^2/(\rho_0 \nu_0 \kappa_0)) \equiv$ Marangoni number where $\gamma_1 \equiv$ surface entropy change and $\rho_0 \equiv$ constant density; while (P4.3) are the usual conditions for a pure-conducting rigid surface.

(a) Seeking a normal-mode solution of that system of the form

$$[w_1, T_1](x, y, z, t) = [w, \theta](z)\exp(i\boldsymbol{q} \bullet \boldsymbol{r} + \sigma t) \text{ where } \boldsymbol{q} = (q_1, q_1) \text{ and } \boldsymbol{r} = (x, y), \tag{P4.4}$$

show that

$$(D^2 - q^2)^2 w = 0,\ w + (D^2 - q^2 - \sigma)\theta = 0; \tag{P4.5a}$$

$$D \equiv \frac{d}{dz},\ q^2 = q_1^2 + q_2^2;\ 0 < z < -1; \tag{P4.5b}$$

$$w = D\theta = 0, \tag{P4.6a,b}$$

$$D^2 w = -q^2 \mathcal{M}\theta \text{ at } z = 0; \tag{P4.6c}$$

$$w = Dw = \theta = 0 \text{ at } z = -1. \tag{P4.7}$$

(b) Solve (P4.5a) to obtain

$$w(z) = (c_1 z + c_2)\cosh(qz) + (c_3 z + c_4)\sinh(qz). \tag{P4.8}$$

(c) Wishing to deduce a marginal stability curve, set $\sigma = 0$ in (P4.5b) and employ (P4.8) to yield

$$L[\theta] \equiv (D^2 - q^2)\theta = -(c_1 z + c_2)\cosh(qz) - (c_3 z + c_4)\sinh(qz) \tag{P4.9a}$$

and since $L[\theta]$ is linear seek a solution of (P4.9a) of the form $\theta(z) = \theta_c(z) + \theta_p(z)$ where

$$L[\theta_c] = 0 \tag{P4.9b}$$

and

$$L[\theta_p] = -(c_1 z + c_2)\cosh(qz) - (c_3 z + c_4)\sinh(qz). \tag{P4.9c}$$

(d) Deduce from (P4.9b) that

$$\theta_c(z) = c_5\cosh(qz) + c_6\sinh(qz) \tag{P4.10a}$$

and then, employing the method of undetermined coefficients, from (P4.9c) that θ_p is of the

$$\theta_p(z) = (A_1 z^2 + A_2 z)\cosh(qz) + (A_3 z^2 + A_4 z)\sinh(qz). \tag{P4.10b}$$

(e) Now operating on (P4.10b) by L obtain that

$$L[\theta_p] = (4qA_3 z + 2A_1 + 2qA_4)\cosh(qz) + (4qA_1 z + 2A_3 + 2qA_2)\sinh(qz). \tag{P4.11}$$

(f) Applying the boundary conditions of (P4.6) deduce from (P4.8) and (P4.10) that

$$c_2 = 0, \tag{P4.12a}$$

$$qc_6 + A_2 = 0, \tag{P4.12b}$$

$$2qc_3 = -q^2 M c_5. \tag{P4.12c}$$

(g) Comparing (P4.11) with (P4.9c) and employing the fact that $c_2 = 0$ from (P4.12a) conclude that

$$4qA_3 = -c_1,\ 2A_1 + 2qA_4 = -c_2,\ 4qA_1 = -c_3,\ 2A_3 + 2qA_2 = -c_4 \Rightarrow$$

$$A_1 = -\frac{c_3}{4q}, \tag{P4.13a}$$

$$A_3 = -\frac{c_1}{4q}, \tag{P4.13b}$$

$$A_2 = \frac{c_1}{4q^2} - \frac{c_4}{2q}, \tag{P4.13c}$$

$$A_4 = \frac{c_3}{4q^2}. \tag{P4.13d}$$

(h) Demonstrate from (P4.12b) and (P4.13c) that

$$c_1 = 2qc_4 - 4q^3 c_6 \tag{P4.14a}$$

while from (P4.12c) that

$$c_3 = -\frac{q\mathcal{M}c_5}{2}; \tag{P4.14b}$$

and thus, from (P4.13) and (P4.14), that

$$A_1 = \frac{\mathcal{M}c_5}{8}, A_2 = -qc_6, A_3 = q^2 c_6 - \frac{c_4}{2}, A_4 = -\frac{\mathcal{M}c_5}{8q}. \tag{P4.15}$$

(i) Observe from (P4.8), (P4.12a), and (P4.14) that $w(z)$ only depends on c_4, c_5, and c_6. Further observe from (P4.10) and (P4.15) that $\theta(z)$ only depends on these three coefficients as well. Then, upon substitution of those solutions into the boundary conditions of (P4.7), one obtains a linear homogeneous system of three equations in c_4, c_5, and c_6. Hence, assuming that $c_4^2 + c_5^2 + c_6^2 \neq 0$, the vanishing of the determinant of its coefficients yields the marginal stability curve [61]

$$\mathcal{M} = \mathcal{M}_0(q) = 8q^2 \frac{[q - \cosh(q)\sinh(q)]\cosh(q)}{q^3\cosh(q) - \sinh^3(q)} \tag{P4.16}$$

for the onset of convective instability which occurs when $M > M_0(q)$. Plot (P4.16) in the q-M plane and determine that it has an absolute minimum point at (q_c, M_c) given by [61]

$$q_c = 1.993 \text{ and } \mathcal{M}_c = 79.607. \tag{P4.17}$$

(j) Show that the critical temperature difference $(\Delta T)_{\text{crit}}$ associated with (P4.17) for the onset of surface tension-driven convection is given by

$$(\Delta T)_{\text{crit}} = 79.607 \frac{\rho_0 \nu_0 \kappa_0}{\gamma_1 d} \tag{P4.18a}$$

while the corresponding critical temperature difference $(\Delta T)_c$ associated with the rigid–free case for the onset of buoyancy-driven convection from Table 15.1 is given by

$$(\Delta T)_c = 1100.650 \frac{\kappa_0 \nu_0}{\alpha_0 g d^3}. \tag{P4.18b}$$

(k) Finally conclude by comparing (P4.18a) with (P4.18b) that the onset of surface tension-driven as opposed to buoyancy-driven convection will occur for those layer depths such that

$$d^2 < d^2_{\text{equal}} = 13.826 \frac{\gamma_1}{\rho_0 \alpha_0 g}. \tag{P4.19}$$

and calculate d_{equal} (mm) for a Dow Corning silicone oil layer having material properties

$$\gamma_1 = 0.058 \frac{\text{gm}}{\text{sec}^2\,\text{K}}, \rho_0 = 0.968 \frac{\text{gm}}{\text{cm}^3}, \alpha_0 = \frac{0.960 \times 10^{-3}}{\text{K}}. \tag{P4.20}$$

Chapter 16
Heat Conduction in a Finite Bar with a Nonlinear Source

Although various overviews of nonlinear stability theory and its results have been presented in sections of previous chapters, specifically 8.5 and 15.6, a definitive study of the so-called Stuart [118]-Watson [131] method of weakly nonlinear stability theory, as applied to partial differential equations, has been deferred until this chapter, which, in particular, concentrates on a longitudinal-planform application of that method to a nonlinear extension of the one-dimensional heat conduction equation treated in Chapter 5 while Chapters 17 and 18 will be concerned with its hexagonal- and rhombic-planform applications to two-dimensional partial differential equations, respectively. That model reaction–diffusion equation for temperature introduced in the chapter has a nonlinear source term containing odd power terms explicitly through fifth-order, which is an extension of the linear source one treated in Chapter 5. This is equivalent to the interaction–dispersion equation for population density originally analyzed by Wollkind et al. [142] through terms of third order in its supercritical parameter range. That analysis is extended through terms of fifth order to examine the behavior in its subcritical regime. It is shown that, under the proper conditions, the two subcritical cases behave in exactly the same manner as the two supercritical ones, unlike the outcome for the truncated system. Further, there also exists a region of metastability, allowing for the possibility of population outbreaks discussed in Chapter 3. These results are then used to offer an explanation for the occurrence of isolated vegetative patches, where linear theory predicts instability, and sparse homogeneous distributions, where linear theory predicts stability, in the relevant ecological parameter range, where there is subcriticality, for a plant-groundwater model to be treated in Chapter 18. Finally, these results are discussed in the context of Matkowsky's [78] two-time nonlinear stability analysis, which is related to multiple scales singular perturbation theory introduced in Section 14.1. The problem applies this Stuart-Watson nonlinear stability analysis through terms of third order to a particular reaction-long-range diffusion model equation [142].

© Springer International Publishing AG, part of Springer Nature 2017

D. J. Wollkind and B. J. Dichone, *Comprehensive Applied Mathematical Modeling in the Natural and Engineering Sciences*,
https://doi.org/10.1007/978-3-319-73518-4_16

16.1 Nondimensional Governing Equation

We consider the following prototype reaction–diffusion heat conduction problem in a finite bar for $T = T(x, \tau) \equiv$ temperature in the bar where $x \equiv$ its one-dimensional spatial variable and $\tau \equiv$ time:

$$\rho_0 c_0 \frac{\partial T}{\partial \tau} = k_0 \frac{\partial^2 T}{\partial x^2} + r_0 T_e r(\theta), 0 < x < L; \tag{16.1.1a}$$

$$T(0, \tau) = T(L, \tau) = T_e; \tag{16.1.1b}$$

with

$$r(\theta) = \theta + \alpha\theta^3 + \gamma\theta^5 + O(\theta^7) \text{ for } \theta = \frac{T - T_e}{T_e}. \tag{16.1.1c}$$

Here $\rho_0 \equiv$ density, $c_0 \equiv$ specific heat, $k_0 \equiv$ thermal conductivity, $r_0 \equiv$ reaction rate, $T_e \equiv$ ambient temperature, and $L \equiv$ bar length while α and γ represent dimensionless reaction coefficients. Note that

$$T(x, \tau) \equiv T_e \tag{16.1.1d}$$

is an exact solution to this boundary value problem. Observe that if $|(T - T_e)/T_e| << 1$, then $r(\theta) \cong \theta = (T - T_e)/T_e$ which implies $r_0 T_e r(\theta) \cong r_0 T_e (T - T_e)/T_e = r_0 (T - T_e)$ and this problem reduces to the linear source one treated in Chapter 5. Dividing (16.1.1a) by $\rho_0 c_0$, we obtain

$$\frac{\partial T}{\partial \tau} = \kappa_0 \frac{\partial^2 T}{\partial x^2} + s_0 T_e r(\theta) \text{ where } \kappa_0 = \frac{k_0}{\rho_0 c_0} \text{ and } s_0 = \frac{r_0}{\rho_0 c_0}. \tag{16.1.2}$$

Introducing the nondimensional variables and parameter

$$z = \frac{\pi x}{L}, t = \frac{\kappa_0 \pi^2 \tau}{L^2}, \theta(z, t) = \frac{T - T_e}{T_e}, \beta = \frac{s_0 L^2}{\kappa_0 \pi^2} = \frac{r_0 L^2}{k_0 \pi^2}, \tag{16.1.3}$$

our original problem transforms into

$$\frac{\partial \theta}{\partial t} - \frac{\partial^2 \theta}{\partial z^2} = \beta r(\theta), 0 < z < \pi; \theta(0, t) = \theta(\pi, t) = 0. \tag{16.1.4a}$$

Note that the exact solution (16.1.1d) corresponds in this dimensionless boundary value problem to

$$\theta(z, t) \equiv \theta_e = 0. \tag{16.1.4b}$$

16.2 Linear Stability Analysis

It is the stability of the equilibrium solution (16.1.4b) with which we are concerned in this chapter. We begin by examining its linear stability: Consider

$$\theta(z,t;\varepsilon_1) = \theta_e + \varepsilon_1\theta_1(z,t) + O(\varepsilon_1^2) \text{ where } \theta_e = 0 \text{ and } |\varepsilon_1| << 1. \quad (16.2.1a)$$

Then, as usual, upon substituting (16.2.1a) into (16.1.4a), neglecting terms of $O(\varepsilon_1^2)$, and cancelling the resulting common ε_1 factor, we obtain the following linear perturbation system for $\theta_1 = \theta_1(z,t)$:

$$\frac{\partial\theta_1}{\partial t} - \frac{\partial^2\theta_1}{\partial z^2} = \beta\theta_1, 0 < z < \pi; \theta_1(0,t) = \theta_1(\pi,t) = 0. \quad (16.2.1b)$$

Seeking a separation of variables normal-mode solution of (16.2.1) of the form

$$\theta_1(z,t) = \Theta(z)e^{\sigma t}, \quad (16.2.2a)$$

we find that $\Theta = \Theta(z)$ satisfies the linear eigenvalue problem for σ given by

$$[D^2 - (\sigma - \beta)]\Theta = 0, 0 < z < \pi \text{ where } D \equiv \frac{d}{dz}; \Theta(0) = \Theta(\pi) = 0. \quad (16.2.2b)$$

As with the Rayleigh–Bénard problem of the previous chapter assuming the requisite continuity at $z = 0$ and π, we can deduce that

$$D^{2N}\Theta = 0 \text{ at } z = 0 \text{ and } \pi \text{ for } N = 0, 1, 2, 3, \ldots \quad (16.2.2c)$$

and thus conclude the eigenvectors are given by

$$\Theta(z) = \theta_{1m}\sin(mz) \text{ where } \theta_{1m} \neq 0 \text{ for } m = 1, 2, 3, \ldots \quad (16.2.2d)$$

with corresponding eigenvalues

$$\sigma_m = \beta - m^2 \text{ for } m = 1, 2, 3, \ldots. \quad (16.2.2e)$$

Hence by the superposition principle

$$\theta_1(z,t) = \sum_{m=1}^{\infty} \theta_{1m}e^{\sigma_m t}\sin(mz). \quad (16.2.2f)$$

Observing that this problem is actually equivalent to the one treated in Chapter 5, we can now reprise our argument of Section 5.6. Therefore $m = 1$ again represents its most dangerous mode since $\sigma_m > \sigma_{m+1}$ from (16.2.2e) and our stability criterion becomes

$$\sigma_1 = \beta - 1 < 0 \text{ or } \beta = \frac{s_0 L^2}{\kappa_0\pi^2} = \frac{r_0 L^2}{k_0\pi^2} < 1 \quad (16.2.2g)$$

which, not surprisingly, is identical to that of (5.6.5). Physically this means that the bar can conduct heat away faster than it is being generated and the ambient temperature solution T_e is linearly stable while should $\beta > 1$ the reverse is true and thermal runaway results.

16.3 One-Dimensional Planform Stuart-Watson Method of Nonlinear Stability Theory

This is an extension to fifth order of a model equation introduced by Wollkind et al. [142] to illustrate the Stuart-Watson method of weakly nonlinear stability analysis of prototype reaction–diffusion equations. Here we shall be following the approach of Davis et al. [26]. Asymptotic analyses of this sort are very useful for predicting pattern formation in such nonlinear systems. That analysis requires the expansion of θ in powers of an unknown function $A(t)$ with spatially dependent coefficients that at $O(A)$ is proportional to the most dangerous mode of linear stability theory. The pattern formational aspect of this system can be predicted from the behavior of that amplitude function which is governed by its Landau ordinary differential equation

$$\frac{dA}{dt} \sim \sigma A - a_1 A^3 - a_3 A^5 = F(A) \tag{16.3.1}$$

where σ is the growth rate of linear stability theory and $a_{1,3}$ are the Landau constants. Its long-time behavior is crucially dependent upon the signs of these Landau constants. Wollkind et al. [142] concentrated on the case for which $r(\theta) = \sin(\theta)$ to satisfy those conditions employed by Matkowsky [78] to develop his two-time method of weakly nonlinear stability theory since their main concern was to compare the results obtained from the application of the Stuart-Watson method with those he deduced. Then $a_1 > 0$ identically (see below) and it is only necessary to include terms through third order in $r(\theta)$ to make pattern formation predictions for this problem. In that event there are two solutions of the truncated system: The first a homogeneous one that is stable for $\sigma < 0$ and the second a supercritical re-equilibrated pattern forming one that exists and is stable for $\sigma > 0$. These results can be directly applied to our problem for its generalized $r(\theta)$ in the parameter range where $a_1 > 0$. In the range where $a_1 < 0$ and there is so-called subcriticality, the solutions to the truncated problem can grow without bound, and one must take the fifth-order terms into account in order to determine the long-time behavior of the system. Then we shall show that if there is a parameter range over which the other Landau constant $a_3 > 0$, the pattern formation properties of our system can be ascertained without having to resort to considering even higher-order terms in $r(\theta)$. That requires the development of a formula for this Landau constant and an examination of its sign as a function of α and γ.

In order to motivate the form of this Stuart-Watson expansion as well as its amplitude equation (16.3.1), we shall proceed sequentially in two stages:

(i) Let

$$\theta(z,t) = A(t)\sin(z) + \varphi(z,t),\ \varphi(z,t) = o(A) \text{ as } A \to 0; \tag{16.3.2a}$$

and

$$\frac{dA(t)}{dt} = \sigma A(t) + B(t),\ B(t) = o(A) \text{ as } A \to 0; \qquad (16.3.2b)$$

where $A(t) \equiv$ amplitude function and $\sin(z) \equiv$ most dangerous mode of linear stability theory. To deduce the forms of φ and B we need only examine the $\alpha\theta^3$ term in (16.1.1c) as follows:

$$\alpha\theta^3 = \alpha A^3(t)\sin^3(z) + o(A^3). \qquad (16.3.3)$$

Let us calculate $\sin^3(z)$ by developing a formula for

$$\sin^2(z)\sin(y) = \frac{1}{2}[1-\cos(2z)]\sin(y) = \frac{1}{2}[\sin(y) - \sin(y)\cos(2z)]. \qquad (16.3.4a)$$

Recall

$$\sin(y)\cos(2z) = \frac{1}{2}[\sin(y+2z) + \sin(y-2z)]. \qquad (16.3.4b)$$

Then substituting (16.3.4b) into (16.3.4a), we obtain

$$\sin^2(z)\sin(y) = \frac{1}{4}[2\sin(y) - \sin(y+2z) - \sin(y-2z)]. \qquad (16.3.4c)$$

Now letting $y = z$ in (16.3.4c) we find that

$$\sin^3(z) = \frac{1}{4}[2\sin(z) - \sin(3z) - \sin(-z)] = \frac{1}{4}[3\sin(z) - \sin(3z)]. \qquad (16.3.4d)$$

Thus substituting (16.3.4d) into (16.3.3), we obtain

$$\alpha\theta^3 = \alpha A^3(t)\frac{3\sin(z) - \sin(3z)}{4} + o(A^3). \qquad (16.3.5)$$

Hence from (16.3.5) we can deduce that

$$\varphi(z,t) = A^3(t)[\theta_{31}\sin(z) + \theta_{33}\sin(3z)] + O(A^5),\ B(t) = -a_1A^3(t) + O(A^5).$$

Therefore from (16.3.2) we obtain through terms of third order that

$$\theta(z,t) = A(t)\sin(z) + A^3(t)[\theta_{31}(\beta)\sin(z) + \theta_{33}(\beta)\sin(3z)] + O(A^5) \qquad (16.3.6a)$$

and

$$\frac{dA(t)}{dt} = \sigma(\beta)A(t) - a_1A^3(t) + O(A^5); \qquad (16.3.6b)$$

where θ_{nm} denotes the coefficient of the term proportional to $A^n(t)\sin(mz)$ appearing in this expansion. Note that whereas these coefficients as well as the growth rate σ in the amplitude equation may be functions of β, a_1 is assumed to be independent of that bifurcation parameter and called the Landau constant after Lev Landau [63] who first proposed the equivalent truncated form of this equation in his work on the transition from laminar to turbulent flow.

(ii) Let

$$\theta(z,t) = A(t)\sin(z) + A^3(t)[\theta_{31}\sin(z) + \theta_{33}\sin(3z)] + \psi(z,t); \quad (16.3.7a)$$

$$\frac{dA(t)}{dt} = \sigma A(t) - a_1 A^3(t) + C(t);\ \psi(z,t),\ C(t) = o(A^3) \text{ as } A \to 0. \quad (16.3.7b)$$

To deduce the forms of $\psi(z,t)$ and $C(t)$ we must now examine both the $\alpha\theta^3$ and $\gamma\theta^5$ terms in (16.1.1c) as follows:

$$\alpha\theta^3 = \alpha A^3(t)\sin^3(z) + 3\alpha A^5(t)[\theta_{31}\sin^3(z) + \theta_{33}\sin^2(z)\sin(3z)] + o(A^5); \quad (16.3.8a)$$

now letting $y = 3z$ in (16.3.4c) we find that

$$\sin^2(z)\sin(3z) = \frac{1}{4}[2\sin(3z) - \sin(5z) - \sin(z)], \quad (16.3.8b)$$

and hence upon substitution of (16.3.4d) and (16.3.8b) into (16.3.8a) obtain

$$\alpha\theta^3 = \alpha A^3(t)\frac{3\sin(z) - \sin(3z)}{4}$$
$$+ 3\alpha A^5(t)\frac{(3\theta_{31} - \theta_{33})\sin(z) + 2\theta_{33}\sin(3z) - \theta_{33}\sin(5z)}{4} + o(A^5); \quad (16.3.8c)$$

while

$$\gamma\theta^5 = \gamma A^5(t)\sin^5(z) + o(A^5); \quad (16.3.9a)$$

where

$$\sin^5(z) = \sin^3(z)\sin^2(z) = \frac{3\sin^3(z) - \sin^2(z)\sin(3z)}{4} \quad (16.3.9b)$$

which upon substitution of (16.3.4c) and (16.3.8b) into (16.3.9b) yields

$$\sin^5(z) = \frac{9\sin(z) - 3\sin(3z) + \sin(z) + \sin(5z) - 2\sin(3z)}{16}$$
$$= \frac{10\sin(z) - 5\sin(3z) + \sin(5z)}{16} \quad (16.3.9c)$$

and thus upon substitution of (16.3.9c) into (16.3.9a) we obtain

$$\gamma\theta^5 = \gamma A^5(t)\frac{[10\sin(z) - 5\sin(3z) + \sin(5z)]}{16} + o(A^5). \quad (16.3.9d)$$

Hence from (16.3.8) and (16.3.9d) we can deduce that

$$\psi(z,t) = A^5(t)[\theta_{51}\sin(z) + \theta_{53}\sin(3z) + \theta_{55}\sin(5z)] + O(A^7),$$
$$C(t) = -a_3 A^5(t) + O(A^7).$$

Therefore from (16.3.7) we finally obtain through terms of fifth order the Stuart-Watson expansion

$$\theta(z,t) \sim A(t)\sin(z) + A^3(t)[\theta_{31}(\beta)\sin(z) + \theta_{33}\sin(3z)] \\ + A^5(t)[\theta_{51}(\beta)\sin(z) + \theta_{53}(\beta)\sin(3z) + \theta_{55}(\beta)\sin(5z)] \quad (16.3.10a)$$

along with the Landau amplitude equation (16.3.1)

$$\dot{A}(t) \equiv \frac{dA(t)}{dt} \sim \sigma(\beta)A(t) - a_1A^3(t) - a_3A^5(t). \quad (16.3.10b)$$

Note that to this order (16.3.10a) satisfies the boundary conditions $\theta(0,t) = \theta(\pi,t) = 0$ identically.
In what follows we shall rewrite (16.3.4a) in the form

$$\underbrace{\frac{\partial\theta}{\partial t}}_{①} + \underbrace{\left(-\frac{\partial^2\theta}{\partial z^2}\right)}_{②} = \underbrace{\beta r(\theta) \sim \beta(\theta + \alpha\theta^3 + \gamma\theta^5)}_{③}. \quad (16.3.11)$$

Then substituting the expansion of (16.3.10a) into each of the terms in (16.3.11) while employing (16.3.10b) in ①; the fact that $-D^2\sin(mz) = m^2\sin(mz)$ in ②; and (16.3.8), (16.3.9d) in ③; we obtain:

$$\begin{aligned} ① \sim\ & \dot{A}(t)\sin(z) + 3A^2(t)\dot{A}(t)[\theta_{31}\sin(z) + \theta_{33}\sin(3z)] \\ & + 5A^4(t)\dot{A}(t)[\theta_{51}(\beta)\sin(z) + \theta_{53}\sin(\beta)\sin(3z) + \theta_{55}(\beta)\sin(5z)] \\ & = [\sigma A(t) - a_1A^3(t) - a_3A^5(t)]\sin(z) \\ & + 3A^2(t)[\sigma A(t) - a_1A^3(t)][\theta_{31}\sin(z) + \theta_{33}\sin(3z)] \\ & + 5A^4(t)[\sigma A(t)][\theta_{51}(\beta)\sin(z) + \theta_{53}(\beta)\sin(3z) + \theta_{55}(\beta)\sin(5z)] \\ & = \sigma A(t)\sin(z) + A^3(t)[(3\sigma\theta_{31} - a_1)\sin(z) + 3\sigma\theta_{33}\sin(3z)] \\ & + A^5(t)[(5\sigma\theta_{51} - a_3 - 3a_1\theta_{31})\sin(z) + (5\sigma\theta_{53} - 3a_1\theta_{33})\sin(3z) \\ & + 5\sigma\theta_{55}\sin(5z)]; \end{aligned} \quad (16.3.12a)$$

$$\begin{aligned} ② \sim\ & 1A(t)\sin(z) + A^3(t)[\theta_{31}\sin(z) + 9\theta_{33}\sin(3z)] \\ & + A^5(t)[\theta_{51}\sin(z) + 9\theta_{53}\sin(3z) + 25\theta_{55}\sin(5z)]; \end{aligned} \quad (16.3.12b)$$

$$\textcircled{3} \sim \beta A(t)\sin(z) + A^3(t)\left[\beta\left(\theta_{31} + \frac{3\alpha}{4}\right)\sin(z) + \beta\left(\theta_{33} - \frac{\alpha}{4}\sin(3z)\right)\right]$$
$$+ A^5(t)\beta\left(\theta_{51} + \theta_{31}\frac{9\alpha}{4} - \theta_{33}\frac{3\alpha}{4} + \frac{5\gamma}{8}\right)\sin(z)$$
$$+ A^5(t)\left[\beta\left(\theta_{53} + \frac{3\alpha\theta_{33}}{2} - \frac{5\gamma}{16}\right)\sin(3z) + \beta\left(\theta_{55} + \frac{\gamma}{16} - \frac{3\alpha\theta_{33}}{4}\right)\sin(5z)\right]. \tag{16.3.12c}$$

Since $\textcircled{1} + \textcircled{2} = \textcircled{3}$, this yields a series of problems, one for each term appearing explicitly in our expansion of the form $A^n(t)\sin(mz)$, given by

$$A(t)\sin(z): \sigma + 1 = \beta; \tag{16.3.13a}$$

$$A^3(t)\sin(z): 3\sigma\theta_{31} - a_1 + \theta_{31} = \beta\left(\theta_{31} + \frac{3\alpha}{4}\right); \tag{16.3.13b}$$

$$A^3(t)\sin(3z): 3\sigma\theta_{33} + 9\theta_{33} = \beta\left(\theta_{33} - \frac{\alpha}{4}\right); \tag{16.3.13c}$$

$$A^5(t)\sin(z): 5\sigma\theta_{51} - a_3 - 3a_1\theta_{31} + \theta_{51} = \beta\left(\theta_{51} + \theta_{31}\frac{9\alpha}{4} - \theta_{33}\frac{3\alpha}{4} + \frac{5\gamma}{8}\right); \tag{16.3.13d}$$

$$A^5(t)\sin(3z): 5\sigma\theta_{53} - 3a_1\theta_{33} + 9\theta_{53} = \beta\left(\theta_{53} + \theta_{33}\frac{3\alpha}{2} - \frac{5\gamma}{16}\right); \tag{16.3.13e}$$

$$A^5(t)\sin(5z): 5\sigma\theta_{55} + 25\theta_{55} = \beta\left(\theta_{55} + \frac{\gamma}{16} - \theta_{33}\frac{3\alpha}{4}\right). \tag{16.3.13f}$$

Although we have catalogued all three $O(A^5)$ problems (16.3.13d–f) in the interest of completeness, only (16.3.13d) which contains the Landau constant a_3 is actually required for our present purposes.

We now solve the remaining problems (16.3.13a–d) sequentially. Then, from (16.3.13a, 16.3.13c), we obtain in a straightforward manner that

$$\sigma(\beta) = \beta - 1 = \sigma_1 \tag{16.3.14a}$$

and

$$\theta_{33}(\beta) = -\frac{\alpha\beta}{8(\beta+3)}, \tag{16.3.14b}$$

while (16.3.13b, 16.3.13d) which contain the Landau constants yield

$$2\sigma(\beta)\theta_{31}(\beta) = a_1 + \frac{3\alpha\beta}{4}, \tag{16.3.14c}$$

$$4\sigma(\beta)\theta_{51}(\beta) = a_3 + 3\theta_{31}(\beta)\left(a_1 + \frac{3\alpha\beta}{4}\right) - \theta_{33}(\beta)\frac{3\alpha\beta}{4} + \frac{5\gamma\beta}{8}. \tag{16.3.14d}$$

Note that the growth rate of (16.3.14a) not surprisingly is equal to that associated with the most dangerous mode of linear theory and that (16.3.13e, 16.3.13f) could be solved for θ_{53} and θ_{55} in a similar straightforward manner as (16.3.13c) was for θ_{33} of (16.3.14b) should those solutions have been desired.

(i) Assuming that $\theta_{31}(\beta)$ is well behaved at the critical bifurcation value of $\beta = 1$ and taking the limit of (16.3.14c) as $\beta \to 1$ while noting that $\sigma(\beta) = \beta - 1 \to 0$ in this limit, we obtain the solvability condition

$$a_1 = -\frac{3\alpha}{4} \tag{16.3.15a}$$

and, upon substitution of (16.3.14a) and (16.3.15a) into (16.3.14c), the solution

$$\theta_{31}(\beta) \equiv \theta_{31} = \frac{3\alpha}{8}. \tag{16.3.15b}$$

Hence we can deduce that

$$a_1 > 0 \text{ for } \alpha < 0 \text{ and } a_1 < 0 \text{ for } \alpha > 0. \tag{16.3.16a}$$

Thus, when particularized to Wollkind et al.'s [142] specific choice of

$$r(\theta) = \sin(\theta) = \theta - \frac{\theta^3}{6} + O(\theta^5) \text{ then } \alpha = -\frac{1}{6} \Rightarrow a_1 = \frac{1}{8}. \tag{16.3.16b}$$

Now, in this case, defining $0 < \varepsilon << 1$ such that

$$\varepsilon^2 = \frac{\sigma(\beta)}{a_1} \text{ or } \beta = 1 + \frac{\varepsilon^2}{8} \tag{16.3.17a}$$

and introducing the rescaled variables

$$\eta = \sigma t, \mathcal{A}(\eta) = \frac{A(t)}{\varepsilon}, \tag{16.3.17b}$$

into amplitude equation (16.3.17b) we obtain

$$\frac{d\mathcal{A}}{d\eta} = \mathcal{A} - \mathcal{A}^3 + O(\varepsilon^2), \tag{16.3.17c}$$

which justifies the truncation procedure

$$\frac{dA}{dt} = \sigma A - a_1 A^3. \tag{16.3.17d}$$

Now multiplying (16.3.17d) by $A(t)$ and rewriting it as

$$\frac{1}{2}\frac{dA^2}{dt} = \sigma A^2 - a_1 A^4 = \sigma A^2 \left[1 - \frac{A^2}{\sigma/a_1}\right] = f_3(A^2), \tag{16.3.17e}$$

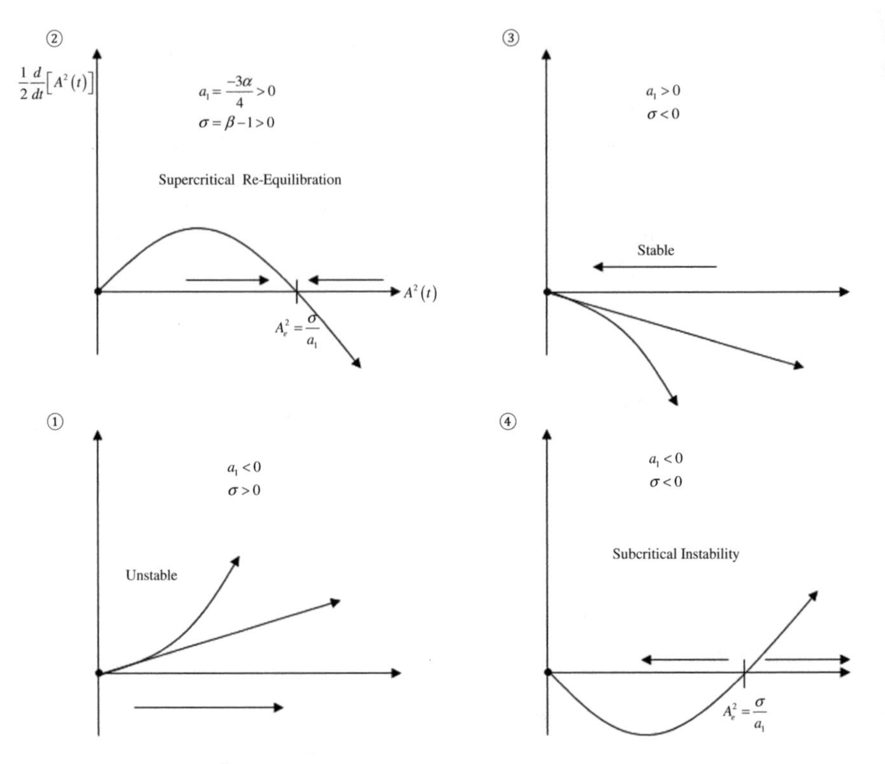

Fig. 16.1 Plots of $f_3(A^2)$ for the third-order truncated amplitude equation with $\sigma = \beta - 1$ and $a_1 = -3\alpha/4$. Here the circled numbers correspond to the quadrants in the α-β plane of Fig. 16.5 with horizontal axis $\beta = 1$ and vertical axis $\alpha = 0$.

we can easily deduce its long-time behavior by means of the four phase-plane plots of $(1/2)dA^2/dt = f_3(A^2)$ that constitute Fig. 16.1 which catalogues the four qualitatively different cases corresponding to the possibility of σ and a_1 being either positive or negative. These serve as graphical representations of the cases discussed earlier for amplitude equation (16.3.17d) truncated through terms of $O(A^3)$. For the supercritical re-equilibration case ② when σ, $a_1 > 0$, we have

$$\lim_{t\to\infty} A(t) = A_e = \varepsilon \text{ and hence } \lim_{t\to\infty} \theta(z,t) \sim \theta_e(z) = \delta \sin(z) \text{ as } \delta \to 0,$$

since

$$\lim_{t\to\infty} \theta(z,t) = \varepsilon \sin(z) + \varepsilon^3[\theta_{31} \sin(z) + \theta_{33}(\beta) \sin(3z)] + O(\varepsilon^5)$$
$$= (\varepsilon + \theta_{31}\varepsilon^3) \sin(z) + \varepsilon^3\theta_{33}(1) \sin(3z) + O(\varepsilon^5)$$
$$= \delta \sin(z) + \delta^3 \frac{\sin(3z)}{192} + O(\delta^5) \sim \delta \sin(z) \text{ as } \delta \to 0$$

where $\delta = \varepsilon + \varepsilon^3\theta_{31} > 0$. This equilibrium state, plotted in Fig. 16.2, is an arch-type pattern formed from one-cycle of a sine curve with its maximum amplitude δ occurring at $z = \pi/2$.

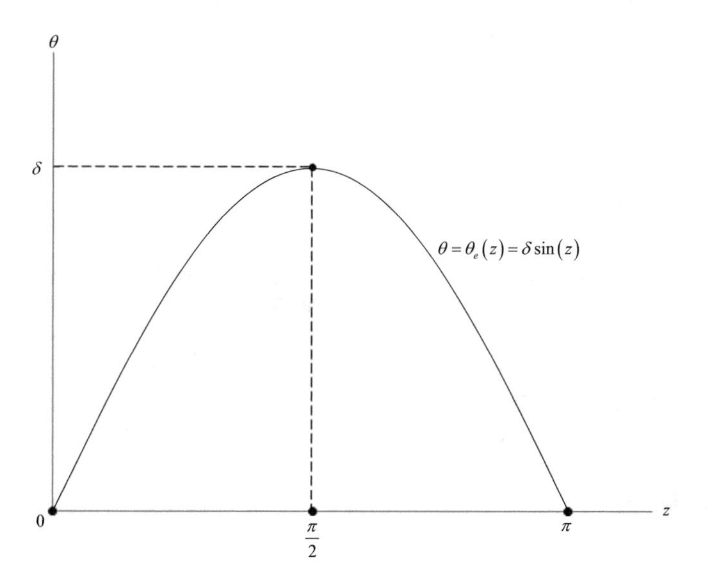

Fig. 16.2 Plot of the arch solution $\theta_e(z)$ for $0 \le z \le \pi$.

(ii) We next proceed to analyze (16.3.14d) involving a_3 and θ_{51} in an analogous manner to that just employed to evaluate a_1 and θ_{31}. Thus, assuming $\theta_{51}(\beta)$ to be well-behaved at $\beta = 1$ and taking the limit of this relation as $\beta \to 1$, we obtain the solvability condition

$$a_3 = -\frac{5\gamma}{8} - 3\theta_{31}\left(a_1 + \frac{3\alpha}{4}\right) + \theta_{33}(1)\frac{3\alpha}{4} = -\frac{5\gamma}{8} - \frac{3\alpha^2}{128} \tag{16.3.18a}$$

and, upon substitution of (16.3.14a) and (16.3.18a) into (16.3.14d), the solution

$$\theta_{51}(\beta) = \frac{5\gamma}{32} + \theta_{31}\frac{9\alpha}{16} + \frac{3\alpha^2(4\beta+3)}{512(\beta+3)}. \tag{16.3.18b}$$

Observe that, by virtue of the value of a_1, a_3 is independent of θ_{31}. Also observe that unlike this quantity θ_{51} is a function of β. Finally note, in addition, should we have assumed that the Stuart-Watson expansion for $\theta(z,t)$ and the Landau equation for dA/dt contained even powers of $A(t)$ as well as odd ones then the solvability conditions and solutions for those coefficients would have shown them to be zero [142]. Hence the fact that these quantities only contain odd but not even powers follows as a direct consequence of the form of $r(\theta)$. Matkowsky [78] employed his specific form for $r(\theta)$ to simplify calculations and our generalization of (16.1.1c) was chosen to preserve the rationale of this selection. Such a choice is also reminiscent of the fact that prospective nonlinear extensions of the restorative force $\boldsymbol{F}$ for the one-dimensional linear spring problem (see Chapter 19), given originally by $\boldsymbol{F} = -kx(t)\boldsymbol{i}$ with $k \equiv$ spring constant and $x(t) \equiv$ its displacement from equilibrium, involve only odd but never even powers of this displacement.

16.4 Truncated Landau Equation

Having determined its coefficients, we shall examine amplitude equation (16.3.1) truncated through terms of $O(A^5)$, *i.e.*, $dA/dt = F(A)$ and defer until after this examination has been completed a justification for that truncation. We seek conditions under which the inclusion of fifth-order terms will re-equilibrate the growing solutions predicted through third order when $a_1 < 0$, *i.e.*, cases ① and ④ of Fig. 16.1. Hence we assume a parameter range in which $a_1 < 0$ or $\alpha > 0$. Further, anticipating our results to be demonstrated below, we assume that $a_3 > 0$ while as always $\sigma \in \mathbb{R}$. This equation has three equilibrium points

$$A(t) \equiv A_e \text{ such that } F(A_e) = 0 \tag{16.4.1a}$$

satisfying either

$$A_e = 0 \text{ or } 2a_3 A_e^{\pm 2} = \pm\sqrt{a_1^2 + 4a_3\sigma} - a_1. \tag{16.4.1b}$$

Observe that, since they must be real and positive, A_e^{+2} exists for $\sigma \geq \sigma_{-1} = -a_1^2/(4a_3)$ while A_e^{-2} only exists for $\sigma_{-1} \leq \sigma < 0$. Multiplication of our truncated amplitude equation by A yields

$$\frac{1}{2}\frac{dA^2}{dt} = \sigma A^2 - a_1 A^4 - a_3 A^6 = A^2(A_e^{-2} - A^2)(A^2 - A_e^{+2}) = f_5(A^2). \tag{16.4.1c}$$

Then we can determine the global stability properties of these equilibrium points by plotting $(1/2)dA^2/dt = f_5(A^2)$ for $\sigma < \sigma_{-1} < 0$, $\sigma_{-1} < \sigma < 0$, and $\sigma > 0$, respectively, in the phase-plane plots of Fig. 16.3. From this figure we can see that 0 is globally stable for $\sigma < \sigma_{-1} < 0$; A_e^{+2}, for $\sigma > 0$; and in the overlap region where either can be stable depending on initial conditions 0 is stable for $0 < A^2(0) < A_e^{-2}$ and A_e^{+2}, for

$A^2(0) > A_e^{-2}$, while A_e^{-2} which only exists in that bistability region is not stable there.

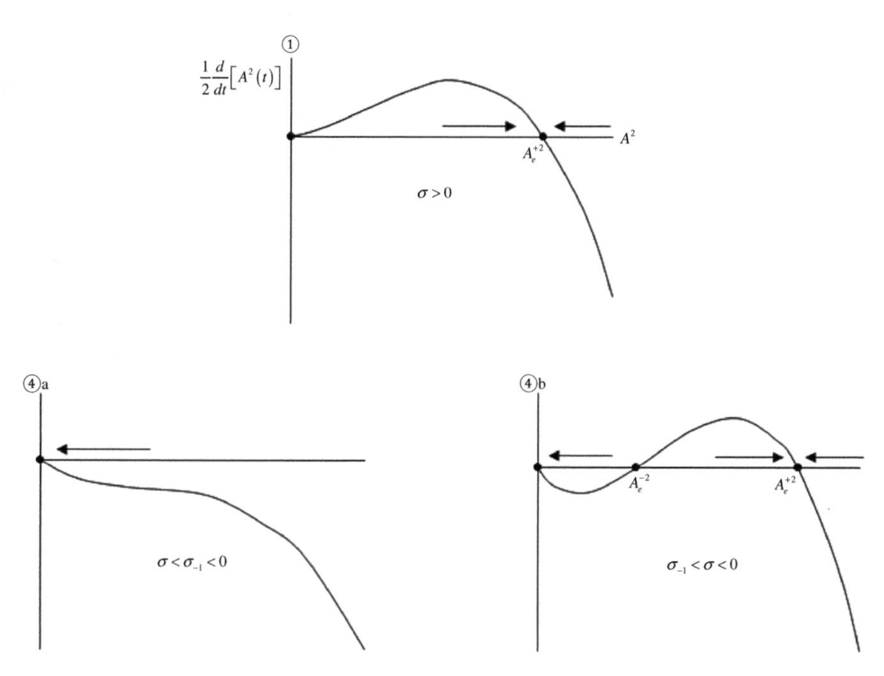

Fig. 16.3 Plots of $f_5(A^2)$ for the fifth-order truncated amplitude equation with $a_1 < 0$; $a_3 > 0$; and $\sigma < \sigma_{-1} = -a_1^2/(4a_3) < 0$, $\sigma_{-1} < \sigma < 0$, and $\sigma > 0$, respectively. Here the circled numbers correspond to the quadrants in the α-β plane of Fig. 16.5.

To justify this truncation procedure we next consider our Landau equation (16.4.1) in the form $dA/dt = F(A) + O(A^7)$, define $\varepsilon^2 = -a_1$, assume $a_3 = O(1)$ as $\varepsilon \to 0$, and let $\sigma = O(\varepsilon^4)$. Then $A_e^{+2} = O(\varepsilon^2)$ which implies that $A_e^+ = O(\varepsilon)$. Note $\alpha = 10^{-2}$, $\gamma = -2$ yield Landau constants satisfying these conditions. Now, analogous to our approach at third order, we introduce the rescaled variables $\eta = \sigma t$, $\mathcal{A}(\eta) = A(t)/A_e^+$ where $\mathcal{A}$, $d\mathcal{A}/d\eta = O(1)$. Since $dA/dt = \sigma A_e^+ d\mathcal{A}/d\eta$, $\sigma A = \sigma A_e^+ \mathcal{A}$, $a_1 A^3 = a_1 A_e^{+3} \mathcal{A}^3$, and $a_3 A^5 = a_3 A_e^{+5} \mathcal{A}^5$ are all of $O(\varepsilon^5)$ while $O(A^7) = O(A_e^{+7} \mathcal{A}^7) = O(\varepsilon^7)$ under these conditions, this justifies our truncation procedure at fifth order. Finally, when $\sigma > 0$, we have the same type of equilibrium solution as depicted in Fig. 16.2 except in this case

$$\delta = \varepsilon_0 + \theta_{31}(1)\varepsilon_0^3 + \theta_{51}(1)\varepsilon_0^5 \text{ where } A_e^+ = A_0\varepsilon = \varepsilon_0 \text{ with } A_0 = O(1) \text{ as } \varepsilon \to 0. \tag{16.4.2}$$

These results depend upon

$$a_3 > 0 \Rightarrow \gamma < -\frac{3\alpha^2}{80}. \tag{16.4.3}$$

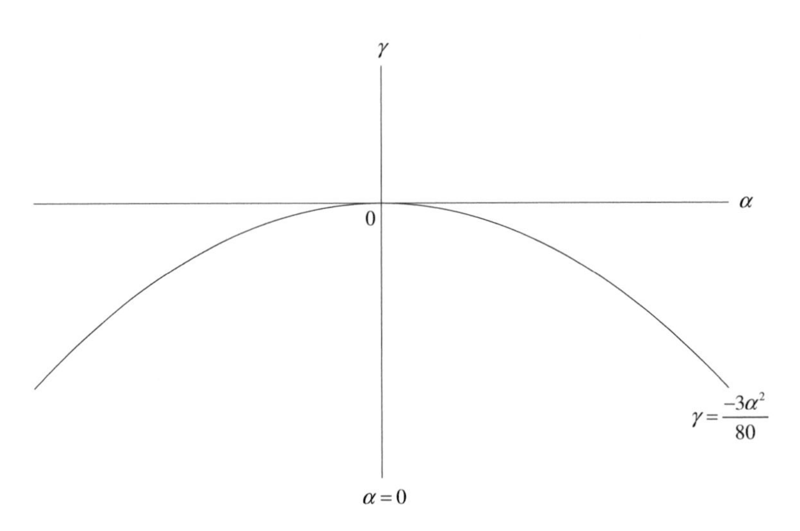

Fig. 16.4 Plot of the region in the α-γ plane where $a_3 > 0$.

Recall in addition we have already taken $\alpha > 0$ to guarantee that $a_1 = -3\alpha/4 < 0$. This region is plotted in the fourth quadrant of the α-γ plane of Fig. 16.4. Note from Fig. 16.3 that unlike the situation depicted in Fig. 16.1 for $\alpha > 0$, *i.e.*, cases (1) and (4), in this event the solutions remain bounded for all such parameter values when the fifth-order terms in $r(\theta)$ are retained.

16.5 Pattern Formation Results

Should there exist a parameter range in a dynamical systems model of a given phenomenon for which the third-order Landau constant $a_1 < 0$ and hence the bifurcation is subcritical, the weakly nonlinear stability analysis must be pushed to fifth order as originally pointed out by DiPrima et al. [28]. This has been standard operating procedure particularly over the last five years when practitioners of the Palermo nonlinear stability theory group began considering fifth-order terms in the Landau equation during their investigation of subcritical bifurcation for a variety of two-component reaction–diffusion systems [38, 39, 122]. By necessity such calculations are long and technically complicated. Thus, when surveying the theory, there is some merit in introducing a simple model equation that preserves all the salient features of a more complex system but considerably reduces the labor involved in determining the Landau constants. This was our rationale for considering the generalized Matkowsky equation under investigation. That was also the rationale for Drazin and Reid's [31] employment of their specific version of the Matkowsky equation in order to develop weakly nonlinear theory relevant to hydrodynamic stability. Matkowsky [78] regarded his problem as a mathematical model for temperature distribution in a finite bar with a nonlinear source term, the ends of which were maintained at the ambient, while Drazin and Reid [31] offered their corresponding version as a phenomenological model of parallel flow in a channel. Hence they both envisioned their instabilities to be rate driven by considering the bifurcation parameter $\beta \sim s_0$ or r_0. For ecological applications in which T, T_e, κ_0, s_0, and L are replaced by N, N_e, D_0, R_0, and $\mathcal{L}$, respectively, in our basic equation where $N \equiv$ population density, $N_e \equiv$ equilibrium population density, $D_0 \equiv$ dispersion constant, $R_0 \equiv$ interaction rate, and $\mathcal{L} \equiv$ favorable territory size, it is often more relevant to envision these instabilities to be territory size driven by considering $\beta \sim \mathcal{L}^2$ and then the instability criterion describes the evolution of spatially heterogeneous structure in a specific domain.

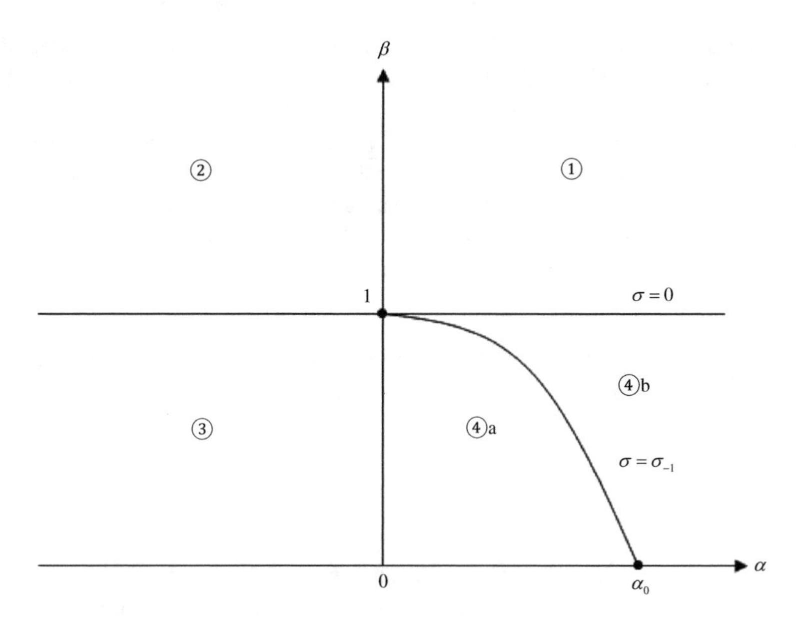

Fig. 16.5 Bifurcation diagram in the α-β plane with $\sigma_{-1} = -a_1^2/(4a_3)$ and $\sigma = \beta - 1$ for $a_1 = -3\alpha/4$ and $a_3 = -5\gamma/8 - 3\alpha^2/128 > 0$ where the circled numbers correspond to the quadrants denoted in Figs. 16.1 and 16.3.

Given that the fifth-order extensions referenced above primarily concentrated only on the subcritical regime, we begin this section by synthesizing our fifth-order results of Fig. 16.3 valid for $a_1 < 0$, or equivalently $\alpha > 0$, and $a_3 > 0$, or equivalently $3\alpha^2 + 80\gamma < 0$, with those valid for $a_1 > 0$, or equivalently $\alpha < 0$, and $a_3 > 0$, as well, depicted in the third quadrant of Fig. 16.4. Note that under these conditions $A_e^{+2} > 0$ for $\sigma > 0$ and $A_e^{-2} < 0$, identically. If we plot a figure analogous to the supercritical cases (2) and (3) of Fig. 16.1, it is obvious that the qualitative morphological behavior of those cases is preserved at fifth order with the only change being now $A_e^2 = A_e^{+2}$. We accomplish this synthesis by means of Fig. 16.5, a bifurcation diagram in α-β space, where the relevant regions associated with these predicted morphological identifications are represented graphically. Since those results also depend on the behavior of σ while $\sigma = 0$ and $\sigma = \sigma_{-1}$ are the critical loci for that quantity in this regard, it is necessary for us to generate loci equivalent to them in α-β space. In this context, using our previous solvability conditions and definitions, we can deduce the following equivalences:

$$\sigma = \beta - 1 = 0 \iff \beta = 1, \tag{16.5.1a}$$

$$\sigma = \beta - 1 = \sigma_{-1} = -\frac{a_1^2}{4a_3} = \frac{18\alpha^2}{3\alpha^2 + 80\gamma} \iff \beta = 1 + \frac{18\alpha^2}{3\alpha^2 + 80\gamma}, \tag{16.5.1b}$$

which are plotted in Fig. 16.5. Here that first locus (16.5.1a) is a horizontal line parallel to the α-axis which divides our α-β space into the four quadrants formed by it and the β-axis while the second (16.5.1b) is a concave downward decreasing curve having a horizontal tangent at its β-intercept of 1 and an α-intercept of $\alpha_0 > 0$ where $\alpha_0^2 = -80\gamma/21$ which separates the fourth quadrant of that space into two parts. From an examination of the modification of the supercritical cases of Fig. 16.1 described above and the subcritical cases of Fig. 16.3, we construct Table 16.1 cataloguing the stable equilibrium points for A^2 in each of the quadrants of Fig. 16.5.

Table 16.1 Stable equilibrium points for A^2 in the quadrants of Fig. 16.5.

Quadrant	1	2	3	4a	4b
Stable equilibrium point	A_e^{+2}	A_e^{+2}	0	0	$\begin{cases} 0 \\ A_e^{+2} \end{cases}$

Note that these fifth-order results for our model equation are much more self-consistent than those obtained in the case of its third-order truncation in that the behavior for the subcritical quadrants 1 and 4a now exactly resemble the behavior for the supercritical quadrants 2 and 3, respectively. In the subcritical quadrant 4b we have what biologists refer to as metastability in that the 0 equilibrium point is stable to initially small disturbances but the model will switch to the equilibrium point A_e^{+2} for sufficiently large ones. The existence of such a region of metastability allows the ecological version of our model equation to exhibit outbreak behavior for which the maximum population level increases several fold upon a sufficient initial perturbation in amplitude. For that ecological formulation, the fact that $A^2 = 0$ represents a globally stable equilibrium point implies that

$$\lim_{\tau \to \infty} N(x, \tau) = N_e.$$

Hence this solution represents a homogeneous population. In many actual biological systems such as the interaction–diffusion plant-groundwater one employed by Chaiya et al. [17] to model vegetative pattern formation in a flat arid environment (see Chapter 18), the homogeneous patterns in the subcritical parameter range correspond to relatively sparse distributions while most of those patterns in the supercritical range correspond to much denser distributions, where the threshold between these two types of distributions occurs at some N_C. We can induce this sort of behavior in the ecological version of our model equation by adopting the relationship

$$N_e = N_e(\alpha) = N_C e^{-\alpha} \tag{16.5.2}$$

which is plotted in Fig. 16.6. Then from this relation and Table 16.1 in conjunction with Fig. 16.2 we can deduce the stable pattern predictions given in Table 16.2 for the quadrants of Fig. 16.5.

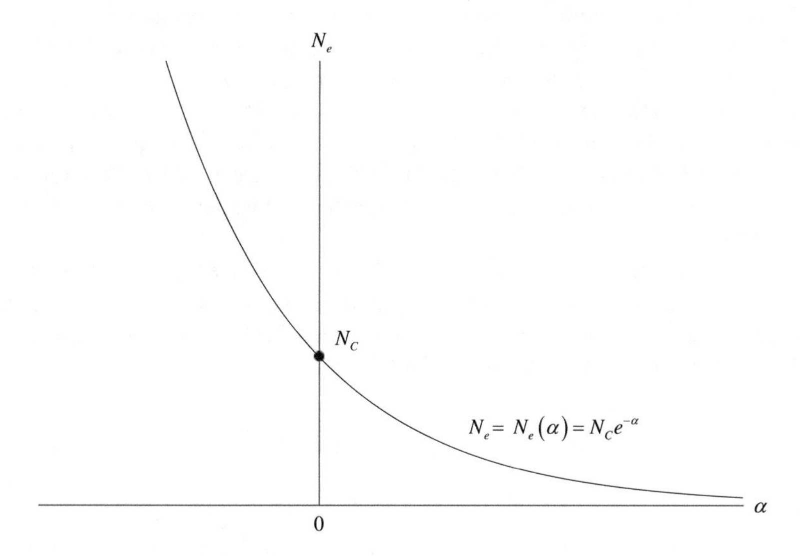

Fig. 16.6 Plot of the population equilibrium density N_e versus α.

In Chaiya et al. [17] it was conjectured that the region of parameter space of subcriticality where $a_1 < 0$ corresponded to isolated vegetative patches when $\sigma > 0$ and low-density homogeneous distributions when $\sigma < 0$ as opposed to the occurrence of periodic patterns for $\sigma > 0$ and high-density homogeneous distributions when $\sigma < 0$ where $a_1 > 0$ which were already predicted by their rhombic-planform two-dimensional nonlinear stability analysis. Such isolated patches are a compromise between periodic patterns and homogeneous stable states that are sparse enough to resemble bare ground. They then associated equilibrium points 0 and A_e^{+2} of quad-

Table 16.2 Morphological stability predictions for Table 16.1.

1	Arch
2	Arch
3	Dense Homogeneous
4a	Sparse Homogeneous
4b	{ Sparse Homogeneous Arch

rants 1 and 4 of Table 16.1 with the sparse homogeneous state and the isolated patch, respectively, that would occur in a postulated fifth-order extension should $a_3 > 0$ for this parameter range. Our fifth-order results summarized in Table 16.2 represent the first step in a conclusive demonstration of the validity of this conjecture. We conclude by noting that although these results are only strictly asymptotically valid in a neighborhood of the marginal stability curve $\beta = 1$, Boonkorkuea et al. [9] by comparing their theoretical predictions of this sort with existing numerical simulations of vegetative pattern formation for a model evolution equation recently showed that the former can often be extrapolated to those regions of parameter space relatively far from the marginal curve. These theoretical predictions also associated that region of parameter space where numerical simulation generated isolated patches with $\sigma > 0$ and $a_1 < 0$.

Finally, for the sake of definiteness, we offer a closed-form representation of $r(\theta)$, composed of combinations of common functions, that produce Landau constants consistent in sign with our subcriticality assumptions. Recall the following Maclaurin polynomials truncated through terms of fifth order:

$$\sinh(z) \sim z + \frac{z^3}{6} + \frac{z^5}{120}, \ \arctan(z) \sim z - \frac{z^3}{3} + \frac{z^5}{5}. \tag{16.5.3a}$$

Then

$$4\sinh\left(\frac{\theta}{2}\right) \sim 4\left(\frac{\theta}{2} + \frac{\theta^3}{48} + \frac{\theta^3}{32[120]}\right) = 2\theta + \frac{\theta^3}{12} + \frac{\theta^5}{960}. \tag{16.5.3b}$$

$$2\arctan\left(\frac{\theta}{2}\right) \sim 2\left(\frac{\theta}{2} - \frac{\theta^3}{24} + \frac{\theta^5}{160}\right) = \theta - \frac{\theta^3}{12} + \frac{\theta^5}{80}. \tag{16.5.3c}$$

Now defining $r(\theta)$ to be the difference between these two functions we obtain

$$r(\theta) = 2\left[2\sinh\left(\frac{\theta}{2}\right) - \arctan\left(\frac{\theta}{2}\right)\right] \sim \theta + \frac{\theta^3}{6} - \frac{11\theta^5}{960} = \theta + \alpha\theta^3 + \gamma\theta^5 \tag{16.5.4a}$$

which implies

$$\alpha = \frac{1}{6} > 0, \ \gamma = -\frac{11}{960} \text{ such that } 80\gamma + 3\alpha^2 = -\frac{11}{12} + \frac{1}{12} = -\frac{5}{6} < 0. \tag{16.5.4b}$$

We close by returning to a topic introduced at the start of Section 16.3. Matkowsky [78] analyzed a model equation which when placed in the form of (16.1.4a) required

$$r(0) = r''(0) = 0, r'(0) > 0, r'''(0) < 0. \tag{16.5.5}$$

Note that $r(\theta) = \sin(\theta)$ as employed by Drazin and Reid [31] and Wollkind et al. [142] satisfies these conditions. Instead of being saddled with the usual deficiencies of linear stability theory which, although producing a solution satisfying a given initial condition at $t = 0$ (see Chapter 5), is only valid in a boundary-layer interval of that time, or nonlinear theory, which, although correctly predicting the long-time behavior of the most dangerous mode, cannot satisfy a general initial condition consisting as it does of only this single mode, Matkowsky's [78] two-time multiple scales analysis produced a uniformly valid solution in t which both satisfied the initial condition and predicted long-time behavior (see Section 22.1). In particular he posed the initial condition

$$\theta(x, 0; \varepsilon_1) \sim \varepsilon_1 h(x) = \varepsilon_1 \sum_{m=1}^{\infty} h_m \sin(mz), 0 < x < \pi,$$

and sought a solution of his model equation of the form of (16.2.1)

$$\theta(x, t; \varepsilon_1) = \zeta(z, t, \eta_1; \varepsilon_1) \sim \varepsilon_1 \zeta_1(z, t, \eta_1) + \varepsilon_1^3 \zeta_3(x, t, \eta_1),$$

with $\eta_1 = \varepsilon_1^2 t$ where $\varepsilon_1^2 = \beta - \beta_c$ for $\beta_c = 1/r'(0)$. Matkowsky [78] found that

$$\zeta_1(x, t, \eta_1) = \mathcal{A}_1(\eta_1) \sin(x) + \sum_{m=2}^{\infty} \mathcal{A}_m(\eta_1) e^{\sigma_m(\beta_c)t} \sin(mx)$$

where $\sigma_m(\beta) = r'(0)\beta - m^2$ and hence $\sigma_m(\beta_c) = 1 - m^2 < 0$ for $m \geq 2$. Then the multiple scales uniformity condition at the next order (see Chapter 14) required the following amplitude equation for $\mathcal{A}_1 = \mathcal{A}_1(\eta_1)$ [78]:

$$\dot{\mathcal{A}}_1 = r'(0)\mathcal{A}_1 + \beta_c r'''(0) \frac{\mathcal{A}_1^3}{8}. \tag{16.5.6a}$$

Now given that

$$\zeta_1(x, 0, 0) = \sum_{m=1}^{\infty} \mathcal{A}_m(0) \sin(mx),$$

the initial condition will be satisfied provided

$$\mathcal{A}_m(0) = h_m \text{ for } m \geq 1, \tag{16.5.6b}$$

where, as usual, the h_m's are related to $h(x)$ by

$$h_m = \frac{2}{\pi} \int_0^{\pi} h(x) \sin(mx)\, dx.$$

Note in this context that should we have defined $\sigma(\beta) = \varepsilon_1^2$ in (16.3.17a) then amplitude equation (16.3.17d) would be equivalent to (16.5.6a) since for $r(\theta) \sim \theta + \alpha\theta^3$; $r'(0) = 1$, $r'''(0) = 6\alpha$; and $\beta_c r'''(0)/8 = 3\alpha/4 = -a_1$. Thus, for $a_1 > 0$ or $\alpha < 0$,

$$\lim_{\eta_1 \to \infty} \mathcal{A}_1^2(\eta_1) = \mathcal{A}_e^2 = \frac{1}{a_1}$$

and hence

$$\lim_{t \to \infty} \theta(x, t; \varepsilon_1) = \lim_{t, \eta_1 \to \infty} \zeta(x, t, \eta_1; \varepsilon_1) \sim \varepsilon_1 \lim_{t, \eta_1 \to \infty} \zeta_1(x, t, \eta_1) = \varepsilon_1 \mathcal{A}_e \sin(x).$$

Observe that under these conditions

$$\varepsilon_1 \mathcal{A}_e = \sqrt{\frac{\sigma(\beta)}{a_1}} \sim \delta,$$

which is consistent with the arch-type equilibrium solution depicted in Fig. 16.2. This completes our sketch of Matkowsky's [78] demonstration that the long-time predictions of the Stuart-Watson method of nonlinear stability theory and its corresponding Landau equation may be reproduced by the employment of the two-time multiple scales technique which also preserves the satisfaction of the initial condition formerly restricted to a linear stability theory approach.

Problem

16.1. Consider the model interaction–long-range diffusion equation [142] for $f = f(s, \tau)$ with $s \equiv$ the one-dimensional spatial variable and $\tau \equiv$ time:

$$\frac{\partial f}{\partial \tau} + \frac{\partial}{\partial s}\left[D(f)\frac{\partial f}{\partial s}\right] + D_2\frac{\partial^4 f}{\partial s^4} + r_0 f_e \sinh\left(\frac{f}{f_e}\right) = 0, \ |s| < \infty; \tag{P1.1a}$$

where

$$D(f) = D_0 + D_1 f \tag{P1.1b}$$

and

$$f \text{ remains bounded as } s^2 \to \infty. \tag{P1.1c}$$

Here f may be interpreted as the deviation from a homogeneous distribution f_e for the density of a population over an unbounded flat environment where r_0 is the reference scale of its interaction rate while D_2 and $D(f)$ represent its long- and short-range diffusion effects, respectively. In addition an affine-type density dependent constitutive relation of (P1.1b) has been assumed for the latter with intercept $D_0 > 0$ and slope D_1. The sign of this intercept implies that when (P1.1a) is placed in the normal interaction–diffusion equation canonical form its diffusion coefficient is negative. For an activation-inhibition type system the possibility of such an occurrence depends on its lateral inhibition mechanism being much more severe and acting over a much longer range than is usual as surveyed in detail by Murray [84]. Finally, the interaction source term is also opposite in sign from that contained in this canonical interaction–diffusion equation form.

(a) Introducing the nondimensional variables and parameters

$$x = \frac{s}{\sqrt[4]{\frac{D_2}{r_0}}},\ t = r_0\tau,\ \zeta = \frac{f}{f_e},\ \mathcal{R} = \frac{D_0}{\sqrt{D_2 r_0}},\ \alpha = \frac{f_e D_1}{D_0}; \tag{P1.2}$$

transform (P1.1) into the following equation for $\zeta = \zeta(x,t)$:

$$\frac{\partial \zeta}{\partial t} + \mathcal{R}\frac{\partial}{\partial x}\left[\mathcal{D}(\zeta)\frac{\partial \zeta}{\partial x}\right] + \frac{\partial^4 \zeta}{\partial x^4} + \sinh(\zeta) = 0,\ -\infty < x < \infty; \tag{P1.3a}$$

where

$$\mathcal{D}(\zeta) = 1 + \alpha\zeta \tag{P1.3b}$$

and

$$\zeta \text{ remains bounded as } x \to \pm\infty. \tag{P1.3c}$$

(b) Show that

$$\zeta \equiv \zeta_e = 0, \tag{P1.4}$$

which represents a homogeneous density distribution, is a solution to equation (P1.3).

(c) Recalling

$$\cos(y \pm z) = \cos(y)\cos(z) \mp \sin(y)\sin(z), \tag{P1.5a}$$

deduce that

$$\cos(y)\cos(z) = \frac{1}{2}[\cos(y+z) + \cos(y-z)] \tag{P1.5b}$$

and

$$\sin(y)\sin(z) = \frac{1}{2}[\cos(y-z) - \cos(y+z)]. \tag{P1.5c}$$

(d) In particular, use (P1.5b) to conclude that

$$\cos^2(\omega x) = \frac{1}{2}[1 + \cos(2\omega x)] \tag{P1.6a}$$

$$\cos(\omega x)\cos(2\omega x) = \frac{1}{2}[\cos(\omega x) + \cos(3\omega x)]; \tag{P1.6b}$$

and, then by employing (P1.6a, P1.6b), that

$$\cos^3(\omega x) = \frac{1}{4}[\cos(\omega x) + \cos(3\omega x)]. \tag{P1.6c}$$

(e) Similarly use (P1.5c) to conclude that

$$\sin^2(\omega x) = \frac{1}{2}[1 - \cos(2\omega x)], \tag{P1.7a}$$

$$\sin(\omega x)\sin(2\omega x) = \frac{1}{2}[\cos(\omega x) - \cos(3\omega x)]. \quad \text{(P1.7b)}$$

(f) Examine the stability of the solution (P1.4) by considering a Stuart-Watson cosine expansion of equation (P1.3) of the form

$$\zeta(x,t) \sim A(t)\cos(\omega x) + A^2(t)[\zeta_{20} + \zeta_{22}\cos(2\omega x)] + A^3(t)[\zeta_{31}\cos(\omega x) + \zeta_{33}\cos(3\omega x)] \quad \text{(P1.8a)}$$

where

$$\dot{A}(t) \equiv \frac{dA(t)}{dt} \sim \sigma A(t) - a_1 A^3(t) \quad \text{(P1.8b)}$$

and $\omega = 2\pi/\lambda$, λ being the wavelength of the class of periodic perturbations under investigation. Substituting (P1.8) into (P1.3), noting that $\sinh(\zeta) \sim \zeta + \zeta^3/6$ while $\partial[\mathcal{D}(\zeta)\partial\zeta/\partial x]/\partial x = (1 + \alpha\zeta)\partial^2\zeta/\partial x^2 + \alpha(\partial\zeta/\partial x)^2$, and employing the identities of (P1.6) and (P1.7), obtain a sequence of problems, one for each pair of values of n and m which corresponds to a term of the form $A^n(t)\cos(m\omega x)$ appearing explicitly in (P1.8a).

(g) From the linear stability problem for $n = m = 1$, arrive at the secular equation

$$\sigma = \sigma_{\omega^2}(\mathcal{R}) = [\mathcal{R} - \mathcal{R}_0(\omega^2)]\omega^2 \text{ where } \mathcal{R}_0(\omega^2) = \omega^2 + \frac{1}{\omega^2}. \quad \text{(P1.10)}$$

(h) Plot the marginal stability curve for (P1.10) in the ω^2-R plane

$$\mathcal{R} = \mathcal{R}_0(\omega^2) \quad \text{(P1.11a)}$$

and conclude that it has a minimum value of $\mathcal{R}_c$ at $\omega^2 = \omega_c^2$ where

$$\omega_c^2 = 1 \text{ and } \mathcal{R}_c = \mathcal{R}_0(\omega_c^2) = 2. \quad \text{(P1.11b)}$$

(i) After DiPrima et al. [28], take $\omega \equiv \omega_c > 0$ in (P1.8a). Then show (P1.10) becomes

$$\sigma = \sigma_1(\mathcal{R}) = \mathcal{R} - 2 \quad \text{(P1.12)}$$

and hence observe that the homogenous solution is linearly stable or unstable depending upon whether $\mathcal{R} < \mathcal{R}_c = 2$ or $\mathcal{R} > \mathcal{R}_c = 2$, respectively.

(j) Under this condition solve the problems for ζ_{20}, ζ_{22}, and ζ_{33} in a straightforward manner to obtain that

$$\zeta_{20}(\mathcal{R},\alpha) \equiv 0, \quad \text{(P1.13a)}$$

$$\zeta_{22}(\mathcal{R},\alpha) = \frac{\alpha\mathcal{R}}{13 - 2\mathcal{R}}, \quad \text{(P1.13b)}$$

$$\zeta_{33}(\mathcal{R},\alpha) = \frac{9\alpha\mathcal{R}\zeta_{22}(\mathcal{R},\alpha) - 1/12}{4(28 - 3\mathcal{R})}. \quad \text{(P1.13c)}$$

(k) Demonstrate that the problem for ζ_{31} now gives rise to the following relation

$$a_1(\alpha) = 2\sigma_1(\mathcal{R})\zeta_{31}(\mathcal{R},\alpha) + \frac{1}{8} - \frac{\alpha\mathcal{R}\omega_c^2\zeta_{22}(\mathcal{R},\alpha)}{2}. \quad \text{(P1.14)}$$

(l) Taking the limit of (P1.14) as $\mathcal{R} \to \mathcal{R}_c = 2$ deduce the solvability condition

$$a_1(\alpha) = \frac{1}{8}\left[1 - \left(\frac{4\alpha}{3}\right)^2\right] \Rightarrow \zeta_{31}(\mathcal{R}, \alpha) = \frac{\alpha^2(9\mathcal{R} + 26)}{36(13 - 2\mathcal{R})}. \qquad \text{(P1.15)}$$

(m) Finally defining $\beta = 1/\mathcal{R}$ which implies $\sigma_1(1/\beta) = (1 - 2\beta)/\beta)$ plot its horizontal marginal stability line $\beta = \beta_c = 1/2$ and the loci $\alpha = \alpha_0^{\pm} = \pm 3/4$ in the α-β plane identifying those regions corresponding to the four cases depicted in Fig. 16.1 by examining the signs of $\sigma_1(1/\beta)$ and $a_1(\alpha)$.

Chapter 17
Nonlinear Optical Ring-Cavity Model Driven by a Gas Laser

In this chapter, the development of spontaneous stationary equilibrium patterns, induced by the injection of a laser (an acronym for *l*ight *a*mplification by *s*timulated *e*mission of *r*adiation) pump field into a purely absorptive two-level atomic sodium vapor ring cavity, is investigated by means of a hexagonal planform nonlinear stability analysis applied to the appropriate governing evolution equation for this optical phenomenon. In the quasi-equilibrium limit for its atomic variables, the mathematical system modeling that phenomenon can be reduced to a single modified Swift–Hohenberg nonlinear partial differential time-evolution equation describing the real part of the deviation of the intracavity field envelope function from its uniform solution on an unbounded two-dimensional spatial domain. Diffraction of radiation can induce transverse patterns consisting of stripes and hexagonal arrays of bright spots or honeycombs in an initially uniform plane-wave configuration. A threshold-dependent paradigm is introduced to interpret these stable patterns. Then, those theoretical predictions are compared with both relevant experimental evidence and existing numerical simulations from some recent nonlinear optical pattern formation studies. This approach follows that of Wollkind et al. [137]. There are four problems: The first two fill in some details of this analysis, while the last two examine bistability for a related nonlinear optical phenomenon and hexagonal pattern formation for the relevant amplitude–phase equations with a hypothetical growth rate and set of Landau constants.

17.1 Maxwell–Bloch Governing Equations

Consider the injection of a laser pump field into a ring cavity containing a purely absorptive two-level atomic sodium vapor medium (see Fig. 17.1). Pattern formation in this phenomenon is modeled by coupling Maxwell's equation for the intracavity field with the Schrödinger-type Bloch equations for the atomic variables of polarization and population difference. That system may be reduced to a single nonlinear modified Swift–Hohenberg model equation describing the real part of the deviation

© Springer International Publishing AG, part of Springer Nature 2017

D. J. Wollkind and B. J. Dichone, *Comprehensive Applied Mathematical Modeling in the Natural and Engineering Sciences*,
https://doi.org/10.1007/978-3-319-73518-4_17

of the intracavity field envelope function from its uniform solution on an unbounded two-dimensional transverse spatial domain provided the decay rate for the field is negligible when compared with the dephasing rate for the polarization and the decay rate for the population difference. We begin with a sketch of the reduction procedure required to obtain the governing modified Swift–Hohenberg model equation.

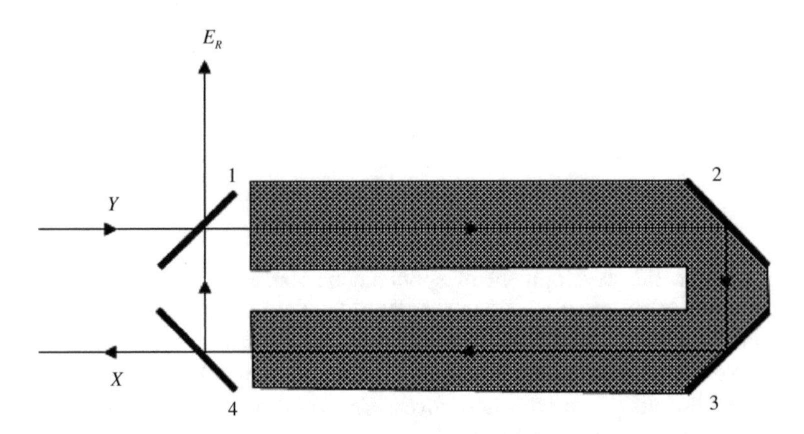

Fig. 17.1 Schematic diagram of a ring cavity filled with atomic sodium vapor (shaded region) where Y and X are related to the envelopes of the injected pump and internal cavity fields, respectively, while E_R bears the same relationship to the corresponding reflected field. Here mirrors 1 and 4 have very small transmissivity coefficients while mirrors 2 and 3 are 100% reflective.

The nondimensional complex system of coupled Maxwell–Bloch equations for this phenomenon is given by [35, 76, 77, 83]

$$\frac{\partial X}{\partial t} = -(1+i\theta)X + Y - \beta P + i\chi\nabla_2^2 X, \tag{17.1.1a}$$

$$\varepsilon_1 \frac{\partial P}{\partial t} = fX - (1+i\Delta)P, \tag{17.1.1b}$$

$$\varepsilon_2 \frac{\partial f}{\partial t} = 1 - f - \frac{X^*P + XP^*}{2}. \tag{17.1.1c}$$

Here, $t \equiv$ time, $(x, y) \equiv$ transverse coordinates, $\nabla_2 \equiv (\partial/\partial x, \partial/\partial y)$, and $\nabla_2^2 \equiv \nabla_2 \bullet \nabla_2$. Further, X and Y are related to the internal cavity and injected pump fields; P and f, to the atomic polarization and population difference between the upper and lower energy levels; while θ and Δ, are the cavity mistuning and atomic detuning parameters; β and χ, the coefficients of absorption and diffraction; and $\varepsilon_{1,2} = r/r_{1,2}$, where r is the dimensional decay rate associated with X and $r_{1,2}$ bear a similar relationship to P and f, being the dephasing and decay rates for the polarization and population difference. In particular, $\theta = (\Omega_c - \Omega_0)/r$ and $\Delta = (\Omega_a - \Omega_0)/r$, where Ω_c is the frequency of the longitudinal cavity mode nearest to the resonant mode Ω_0 and Ω_a is the atomic transition frequency. Finally, the oscillation $e^{-i\Omega_0 t/r}$ has been factored out of the cavity, pump, and polarization fields in (17.1.1a, b), where $i = \sqrt{-1}$. Hence the dynamical variables X and P represent the complex-valued envelopes of these fields, while f, measured in occupation probabilities, satisfies $0 < f \leq 1$ and Y, the pump field envelope function, is assumed to be a real positive constant. In this context, the asterisked quantities of (17.1.1c) denote complex conjugates. In addition, it has been assumed that the dependent variables are initially independent of the cavity longitudinal spatial dimension. Then, as demonstrated directly by Lugiato and Oldano [76], these envelope functions retain that independence for all later time and thus need only be defined on the cavity two-dimensional transverse spatial domain.

Should $\varepsilon_{1,2} << 1$ as is typically the case [83], one can employ a steady-state assumption on (17.1.1b, c) to obtain the quasi-equilibrium conditions satisfied implicitly by the atomic variables

$$P = \frac{fX}{1 + i\Delta} \tag{17.1.2a}$$

and

$$f = 1 - \frac{X^*P + XP^*}{2} = 1 - \mathrm{Re}(X^*P). \tag{17.1.2b}$$

These equations can be solved simultaneously to yield explicit representations for the atomic variables as follows: Upon substituting (17.1.2a) into (17.1.2b) and taking the real part of the result, we find that

$$f = 1 - \frac{f|X|^2}{1 + \Delta^2} \Rightarrow f = \frac{1 + \Delta^2}{1 + \Delta^2 + |X|^2} \tag{17.1.3a}$$

which, in conjunction with (17.1.2a), yields

$$P = \frac{(1 - i\Delta)X}{1 + \Delta^2 + |X|^2}, \tag{17.1.3b}$$

where $|X|^2 = XX^*$ and $1 + \Delta^2 = (1 + i\Delta)(1 - i\Delta)$. Hence substitution of (17.1.3b) into (17.1.1a) reduces system (17.1.1) to a single nonlinear evolution equation for the intracavity field envelope function X:

$$\frac{\partial X}{\partial t} = -X\left[1 + i\theta + \frac{\beta(1 - i\Delta)}{1 + \Delta^2 + |X|^2}\right] + Y + i\chi\nabla_2^2 X. \tag{17.1.4}$$

Given that the pump field envelope function Y is a real positive constant, there exists a steady-state uniform equilibrium solution $X \equiv X_e$ to (17.1.4) satisfying

$$Y = X_e\left[1 + \frac{\beta}{D} + \left(\theta - \frac{\beta\Delta}{D}\right)i\right] \text{ with } D = 1 + \Delta^2 + \alpha \text{ for } \alpha = |X_e|^2 \tag{17.1.5a}$$

or

$$Y^2 = YY^* = \alpha\left[\left(1+\frac{\beta}{D}\right)^2 + \left(\theta - \frac{\beta\Delta}{D}\right)^2\right], \tag{17.1.5b}$$

which is single-valued in Y provided $0 < \beta < \beta_{\text{crit}}$, where [77]

$$27\beta_{\text{crit}}(1+\Delta^2)(1+\theta^2) = (\beta_{\text{crit}} - 2 + 2\theta\Delta)^3. \tag{17.1.5c}$$

The implicit representation for β_{crit} of (17.1.5c) can most easily be derived as follows: Expanding (17.1.5b), we rewrite it in the form

$$Y^2 = \alpha\left[1+\theta^2 + \frac{2\beta(1-\theta\Delta)}{D} + \frac{\beta^2(1+\Delta^2)}{D^2}\right]. \tag{17.1.5d}$$

Since ultimately we wish to identify β_{crit} with that value of β for which $dY^2/d\alpha = 0$ has a unique solution, it is necessary as a preliminary step in this determination to compute

$$\frac{d}{d\alpha}\left(\frac{\alpha}{D}\right) = \frac{D-\alpha D'}{D^2} = \frac{D-\alpha}{D^2} = \frac{1+\Delta^2}{D^2} \tag{17.1.6a}$$

and

$$\begin{aligned}\frac{d}{d\alpha}\left(\frac{\alpha}{D^2}\right) &= \frac{1}{D}\frac{d}{d\alpha}\left(\frac{\alpha}{D}\right) + \frac{\alpha}{D}\frac{d}{d\alpha}\left(\frac{1}{D}\right)\\ &= \frac{1+\Delta^2}{D^3} + \frac{\alpha}{D}\frac{-D'}{D^2} = \frac{1+\Delta^2-\alpha}{D^3}.\end{aligned} \tag{17.1.6b}$$

Now, upon differentiating (17.1.5d) with respect to α and employing (17.1.6), we find that

$$\frac{dY^2}{d\alpha} = 1+\theta^2 + \frac{2\beta(1-\theta\Delta)(1+\Delta^2)}{D^2} + \frac{\beta^2(1+\Delta^2)(1+\Delta^2-\alpha)}{D^3}. \tag{17.1.6c}$$

Recalling that $-\alpha = 1+\Delta^2 - D$, substituting this value into (17.1.6c), and multiplying the result by D^3, we obtain that

$$D^3\frac{dY^2}{d\alpha} = (1+\theta^2)D^3 + \beta[2(1-\theta\Delta)-\beta](1+\Delta^2)D + 2\beta^2(1+\Delta^2)^2. \tag{17.1.6d}$$

Hence we can conclude that $dY^2/d\alpha > 0$ for $0 < \beta < \beta_{\text{crit}}$ where D satisfies (see Fig. 17.2)

$$D^3 + bD + c > 0 \tag{17.1.7a}$$

with

$$b = \frac{\beta[2(1-\theta\Delta)-\beta](1+\Delta^2)}{1+\theta^2},\ c = \frac{2\beta^2(1+\Delta^2)^2}{1+\theta^2}; \tag{17.1.7b}$$

and β_{crit} corresponds to that β-value such that the discriminant of the related cubic equation $D^3 + bD + c = 0$ is zero or [124]

$$\mathcal{D} = 4b^3 + 27c^2 = 0. \tag{17.1.7c}$$

Therefore, since

$$\begin{aligned}\mathcal{D} &= \frac{4\beta^3(1+\Delta^2)^3[2(1-\theta\Delta)-\beta]^3}{(1+\theta^2)^3} + \frac{27[4\beta^4(1+\Delta^2)^4]}{(1+\theta^2)^2} \\ &= \frac{4\beta^3(1+\Delta^2)^3[(2-2\theta\Delta-\beta)^3 + 27\beta(1+\Delta^2)(1+\theta^2)]}{(1+\theta^2)^3},\end{aligned} \tag{17.1.7d}$$

we can deduce from (17.1.7c, d) that β_{crit} is defined implicitly by (17.1.5c). Observe from Fig. 17.2 that for $\beta = \beta_{\text{crit}}$ there then exists an α-value where $dY^2/d\alpha = 0$ while for $\beta > \beta_{\text{crit}}$ there exists an α-interval where $dY^2/d\alpha < 0$ as implied by this derivation (see Problem 17.3).

This equilibrium solution X_e satisfies the transverse far-field condition

$$|X|^2 \text{ remains bounded as } x^2 + y^2 \to \infty \tag{17.1.8}$$

and represents an initial uniform plane-wave configuration. It is the stability of this solution to various periodic transverse-type disturbances with which we shall be concerned in what follows.

Toward that end, we now consider a solution to (17.1.4) of the form

$$X = X_e(1+A) \tag{17.1.9}$$

and retaining terms through third order in A obtain

$$\frac{\partial A}{\partial t} \equiv A_t \sim \sum_{n=1}^{3}\sum_{\ell=0}^{n} \frac{1}{(n-\ell)!\ell!}\frac{\partial^n F(0,0)}{\partial A^{n-\ell}\partial A^{*\ell}} A^{n-\ell}A^{*\ell} + i\chi\nabla_2^2 A$$

where

$$F(A, A^*) = -(1+A)\left[1 + i\theta + \frac{\beta(1-i\Delta)}{1+\Delta^2+\alpha(1+A)(1+A^*)}\right].$$

Finally, after Firth and Scroggie [35], we shall concentrate on the special resonant excitation case of $\Delta = 0$ given that this assumption allows us to simplify our computations enormously while still preserving the salient pattern formation features of the more general case of $\Delta \neq 0$. Then, in particular, examining the linear problem which corresponds to retaining only first-order terms in the expansion introduced above, taking its complex conjugate, and performing a stability analysis on that two-equation system by, after Lugiato and Lefever [75], seeking a normal-mode solution of its perturbation quantities of the form (see Problem 17.1)

$$[\mathcal{A}_1, \mathcal{A}_1^*](x,t) = [k_1, k_2]e^{\sigma t}\cos(qx), \tag{17.1.10a}$$

where

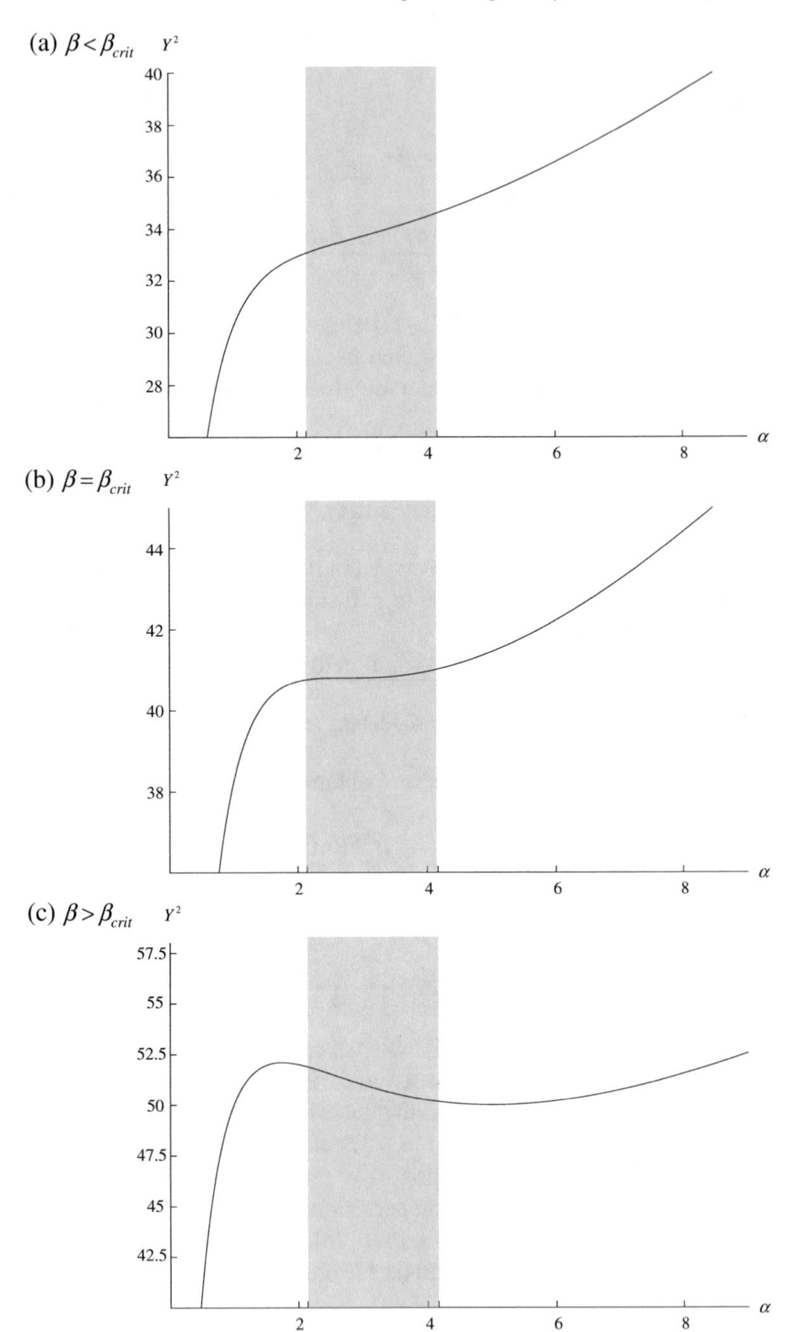

Fig. 17.2 Plots of Y^2 of (17.1.5b) versus α for $\Delta = 0$, $\theta = -1$, and β equals (a) $8.8 < \beta_{\text{crit}} = 10.2$, (b) β_{crit}, and (c) $12 > \beta_{\text{crit}}$.

$$|k_1|^2 + |k_2|^2 \neq 0 \tag{17.1.10b}$$

and $\lambda = 2\pi/q \equiv$ wavelength of the periodic disturbance under investigation, results in the quadratic secular equation for the growth rate σ:

$$\sigma^2 + 2\left[1 + \frac{\beta}{(\alpha+1)^2}\right]\sigma + \left[1 + \frac{\beta(1-\alpha)}{(\alpha+1)^2}\right]\left[1 + \frac{\beta}{\alpha+1}\right] + (\theta + \chi q^2)^2 = 0. \tag{17.1.11a}$$

The marginal stability curve for (17.1.11a) on which $\sigma = 0$ has its lowest threshold in (α, β) space when $\theta + \chi q^2 = 0$. Hence we can conclude that the critical wavenumber q_c satisfies

$$q_c^2 = -\frac{\theta}{\chi} > 0 \tag{17.1.11b}$$

and, for $q^2 \equiv q_c^2$, (17.1.11a) has the corresponding roots

$$\sigma_R(\alpha, \beta) = -1 + \frac{\beta(\alpha-1)}{(\alpha+1)^2}, \sigma_I(\alpha, \beta) = -1 - \frac{\beta}{\alpha+1}, \tag{17.1.11c}$$

where σ_R, the growth rate of the most dangerous mode, yields the marginal stability curve

$$\beta = \beta_0(\alpha) = \frac{(\alpha+1)^2}{\alpha-1} \text{ for } \alpha > 1, \tag{17.1.12}$$

plotted in Fig. 17.3 while σ_I is strongly stabilizing. In this context, we note that under these conditions a similar linear stability analysis of the uniform plane-wave solution to the original system (17.1.1) of governing equations would yield a fifth-order secular equation having the following asymptotic behavior for $\varepsilon_2 = O(\varepsilon_1)$:

$$\sigma_{1,2} = O(1), \sigma_3 \sim -\frac{1}{\varepsilon_1}, \sigma_{4,5} \sim \frac{\Sigma_{4,5}}{\varepsilon_1} \text{ where } \Sigma_{4,5} = O(1) \text{ as } \varepsilon_1 \to 0 \tag{17.1.13a}$$

such that to lowest order $\Sigma_{4,5}$ satisfy

$$r_0\Sigma^2 + (r_0+1)\Sigma + 1 + \alpha = 0 \text{ where } r_0 = \frac{\varepsilon_2}{\varepsilon_1} = \frac{r_1}{r_2} = O(1), \tag{17.1.13b}$$

which, since all its coefficients are positive, implies that

$$\text{Re}(\Sigma_{4,5}) < 0 \text{ as } \varepsilon_1 \to 0. \tag{17.1.13c}$$

Specifically for $\varepsilon_1 = \varepsilon_2$ or $r_0 = 1$ these reduce to

$$\Sigma_{4,5} \sim -1 \pm i\alpha^{1/2} \text{ as } \varepsilon_1 \to 0. \tag{17.1.13d}$$

Further, we make the observation that $\sigma_{1,2} \sim \sigma_{R,I}$ as $\varepsilon_1 \to 0$, this correspondence being consistent with our steady-state assumption. Thus, with no loss of generality for the nonlinear optical instabilities under examination, we may consider (17.1.1) in the limit as $\varepsilon_1 \to 0$ should that quantity be very small, since the neglected roots are highly stabilizing. Next, writing $A = R + iI$, substituting into our third-order equation for A with $\Delta = 0$ and, after Firth and Scroggie [35], taking

$$\chi q_c^2 = -\theta = 1 \text{ or } \theta = -1, \tag{17.1.13e}$$

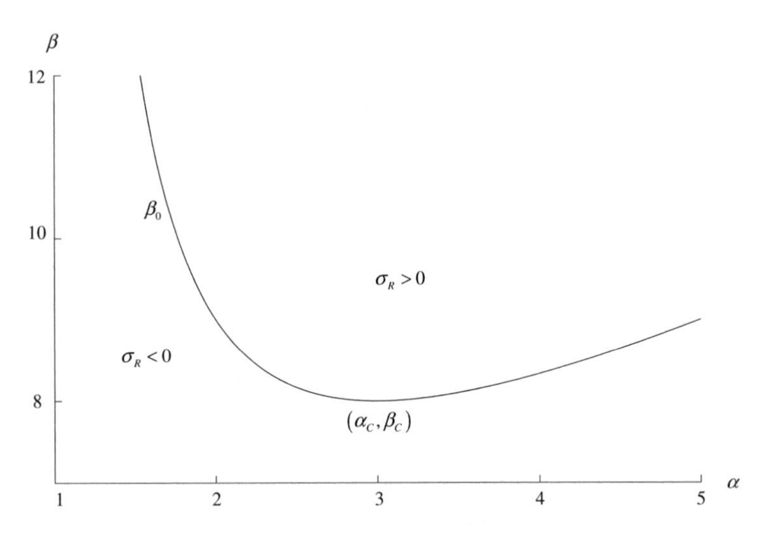

Fig. 17.3 Plot in the α-β plane of the marginal stability curve $\beta = \beta_0(\alpha)$ of (17.1.12) with minimum point $(\alpha_c, \beta_c) = (3, 8)$ on which the growth rate $\sigma_R(\alpha, \beta) = 0$.

and separating real and imaginary parts, we derive equations for R_t and I_t which have growth rates σ_R and σ_I, respectively. Making use of these facts to deduce the asymptotic quasi-equilibrium condition $I \sim [\chi/\sigma_I(\alpha,\beta)](\nabla_2^2 + q_c^2)R$ from the latter and employing this to eliminate I from the former, we obtain the modified Swift–Hohenberg equation (see Problem 17.2)

$$R_t \sim \sigma_R(\alpha,\beta)R - \omega_0(\alpha,\beta)R^2 - \omega_1(\alpha,\beta)R^3 + \left[\frac{\chi^2}{\sigma_I(\alpha,\beta)}\right](\nabla_2^2 + q_c^2)^2 R, \tag{17.1.14a}$$

where

$$\omega_0(\alpha,\beta) = \frac{\beta\alpha(\alpha-3)}{(\alpha+1)^3}, \; \omega_1(\alpha,\beta) = \frac{\beta\alpha[8\alpha - (\alpha+1)^2]}{(\alpha+1)^4}. \tag{17.1.14b}$$

17.2 Striped Planform Stuart-Watson Expansion

We wish to analyze the modified Swift–Hohenberg equation (reviewed by Cross and Hohenberg, [24]) of (17.1.14) for the type of nonlinear optical patterns mentioned in the opening paragraph. We note that this evolution equation admits the trivial zero solution

$$R \equiv 0, \tag{17.2.1a}$$

which corresponds to the steady-state uniform plane-wave equilibrium solution $X \equiv X_e$ of (17.1.4) and hence satisfies the companion far-field condition to (17.1.8)

$$R \text{ remains bounded as } x^2 + y^2 \to \infty. \tag{17.2.1b}$$

In our development of (17.1.14), we have paralleled and synthesized the approaches of both Lugiato and Oldano [76], who performed a linear stability analysis on (17.1.1), and Firth and Scroggie [35], who numerically integrated (17.1.4) while using (17.1.14) in conjunction with (17.1.11c) as a guide to select the appropriate parameter range favoring pattern formation. We further note that these numerical simulations were of necessity performed on a square grid with periodic boundary conditions. Given that the characteristic wavelength of the nonlinear optical patterns to be investigated is very small with respect to the territorial length scale of the cavity cross-sectional area, it again seems reasonable as a first approximation for us to consider our evolution equation on an unbounded transverse spatial domain since under this condition the actual cross-sectional boundary does not significantly influence the pattern [43]. In this context, the boundedness property of the transverse far-field conditions of (17.1.8) and (17.1.1b) physically represents the analytical analogue of the periodic boundary conditions adopted for numerical simulation. It is the nonlinear stability of the plane-wave solution (17.1.1a) to one-dimensional striped planform and two-dimensional hexagonal planform perturbations with which we shall be concerned in this and the next section, respectively. As a necessary prelude to that hexagonal planform analysis of (17.1.14), we perform a nonlinear stability analysis of its zero state in this section along the lines of that developed in Problem 16.1 by seeking a one-dimensional real Stuart-Watson expansion of it through third-order terms of the form

$$\begin{aligned} R(x,t) \sim{}& A_1(t)\cos(q_c x) + A_1^2(t)[R_{20} + R_{22}\cos(2q_c x)] \\ &+ A_1^3(t)[R_{31}\cos(q_c x) + R_{33}\cos(3q_c x)], \end{aligned} \tag{17.2.2a}$$

where the amplitude function $A_1(t)$ satisfies the Landau equation

$$\frac{dA_1(t)}{dt} \sim \sigma A_1(t) - a_1 A_1^3(t) \tag{17.2.2b}$$

and q_c is the critical wavenumber of linear stability theory defined by (17.1.13e). Substituting the solution of (17.2.2) into (17.1.14a), we obtain a sequence of problems, one for each pair of n and m values that corresponds to a term of the form $A_1^n(t)\cos(mq_c x)$ appearing explicitly in (17.1.2a).

Then, the linear stability problem for $n = m = 1$ yields the secular equation

$$\sigma = \sigma_R(\alpha, \beta) = \frac{\beta}{\beta_0(\alpha)} - 1, \tag{17.2.3a}$$

where σ_R and β_0 are as defined in (17.1.11c) and (17.1.12), respectively. Thus, the locus $\beta = \beta_0(\alpha)$ of Fig. 17.3 serves as its marginal stability curve in the α-β plane with a minimum at (α_c, β_c) for

$$\alpha_c = 3, \beta_c = 8. \tag{17.2.3b}$$

Hence,

$$\sigma_R < 0 \text{ for } 0 < \beta < \beta_0(\alpha), \sigma_R = 0 \text{ for } \beta = \beta_0(\alpha), \sigma_R > 0 \text{ for } \beta > \beta_0(\alpha) \geq \beta_c. \tag{17.2.3c}$$

Therefore, $R \equiv 0$ is linearly stable for $0 < \beta < \beta_0(\alpha)$, neutrally stable for $\beta = \beta_0(\alpha)$, and unstable for $\beta > \beta_0(\alpha) \geq \beta_c = 8$. Continuing our description of the results of this one-dimensional expansion procedure, the second-order problems corresponding to $n = 2$ and $m = 0$ or 2 can be solved in a straightforward manner to yield

$$R_{20} = \left(\frac{\omega_0}{2}\right)\left(\frac{1}{\sigma_I} - \sigma_R\right), R_{22} = \left(\frac{\omega_0}{2}\right)\left(\frac{9}{\sigma_I} - \sigma_R\right). \tag{17.2.4}$$

Although there are also two third-order problems, it is permissible for us to concentrate our attention exclusively on the one corresponding to $n = 3$ and $m = 1$ which contains the Landau constant a_1 for the Fredholm-type solvability method to be employed. That problem is given by

$$a_1 - 2\sigma_R R_{31} = \omega_0(2R_{20} + R_{22}) + \frac{3\omega_1}{4}. \tag{17.2.5a}$$

Now, taking the limit of (17.2.5) as $\beta \to \beta_0(\alpha)$, making use of (17.2.3a) and (17.2.4) as well as (17.1.11c) and (17.1.14b), and assuming the requisite continuity at $\beta = \beta_0(\alpha)$, we obtain the solvability condition

$$a_1 = \frac{\alpha\beta_0(\alpha)}{(\alpha+1)^4}\left\{\frac{-19}{9}\left[\frac{\alpha(\alpha-3)}{\alpha-1}\right]^2 + \frac{3}{4}[8\alpha - (\alpha+1)^2]\right\} = a_1(\alpha). \tag{17.2.5b}$$

As demonstrated in Chapter 16, the stability behavior of the Landau equation (17.2.2b) truncated through terms of third order and thus the pattern formation aspect of our model system is crucially dependent upon the sign of a_1. Hence in order to determine this behavior we need to examine the formula for a_1 of (17.2.5b). Toward that end, we plot $a_1(\alpha)$ of (17.2.5b) versus α in Fig. 17.4. From this figure, we observe that, as in Chapter 8, a_1 has two zeroes at $\alpha = \alpha_{1,2}$ such that

$$a_1 < 0 \text{ for } 1 < \alpha < \alpha_1 \text{ or } \alpha > \alpha_2, a_1 > 0 \text{ for } \alpha_1 < \alpha < \alpha_2; \tag{17.2.6a}$$

where

$$\alpha_1 = 2.143 \text{ and } \alpha_2 = 4.167. \tag{17.2.6b}$$

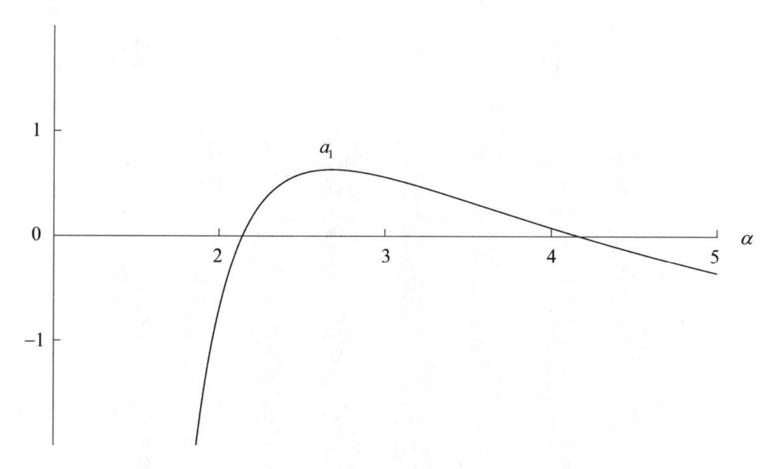

Fig. 17.4 Plot of a_1 of (17.2.5b) versus α.

Again, given these conditions of (17.2.3c) for $\sigma_R(\alpha, \beta)$ and (17.2.6) for $a_1(\alpha)$, we note that the amplitude function $A_1(t)$ of (17.2.2b) undergoes a pitchfork bifurcation at $\beta = \beta_0(\alpha)$ when $\alpha_1 < \alpha < \alpha_2$ from which we may conclude, as in Chapter 8 and Fig. 16.1, that

1. For $0 < \beta < \beta_0(\alpha)$ and $\alpha_1 < \alpha < \alpha_2$, the undisturbed state $A_1 = 0$ is stable since $\sigma_R < 0$ and $a_1 > 0$ yielding a uniform plane-wave configuration $R(x, t) \to 0$ as $t \to \infty$.
2. For $\beta > \beta_0(\alpha)$ and $\alpha_1 < \alpha < \alpha_2, A_1 = A_e = \sqrt{\frac{\sigma_R}{a_1}}$ is stable since $\sigma_R, a_1 > 0$ yielding a periodic one-dimensional pattern consisting of stationary parallel stripes

$$R(x, t) \to R_e(x) \sim A_e \cos\left(\frac{2\pi x}{\lambda_c}\right) \text{ as } t \to \infty \tag{17.2.7a}$$

with characteristic wavelength, exactly fitting the cavity, of

$$\lambda_c = \frac{2\pi}{q_c} = 2\pi\chi^{1/2}. \tag{17.2.7b}$$

These supercritical stripes are represented in the contour plot of Fig. 17.5 where, after Firth and Scroggie [35], regions of higher intensity ($R > 0$ in this case) appear bright and those of lower intensity ($R < 0$) dark. Here the axes are being measured in units of λ_c. When $\alpha < \alpha_1$ or $\alpha > \alpha_2$, that bifurcation is subcritical. In the next section we shall concentrate on the behavior of the modified Swift–Hohenberg equation (17.1.14) in its supercritical regime where $\alpha_1 < \alpha < \alpha_2$.

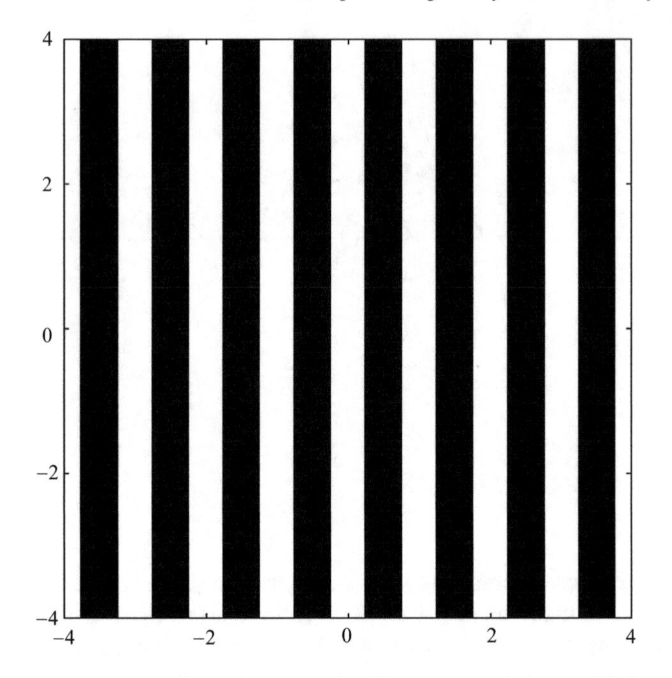

Fig. 17.5 Contour plot of the supercritical stripes of (17.2.7).

17.3 Hexagonal Planform Stuart-Watson Expansion

In order to investigate the possibility of occurrence for our modified Swift–Hohenberg equation of the type of two-dimensional optical patterns mentioned earlier, we consider a hexagonal planform solution of (17.1.14) that to lowest order satisfies (as outlined in Chapter 8)

$$R(x,y,t) \sim A_1(t)\cos[q_c x + \varphi_1(t)] + A_2(t)\cos\left[\frac{q_c(x-\sqrt{3}y)}{2} - \varphi_2(t)\right] + A_3(t)\cos\left[\frac{q_c(x+\sqrt{3}y)}{2} - \varphi_3(t)\right] \tag{17.3.1a}$$

where, for $(j,k,\ell) \equiv$ even permutation of (1,2,3),

$$\frac{dA_j}{dt} \sim \sigma A_j - 4a_0 A_k A_\ell \cos(\varphi_j + \varphi_k + \varphi_\ell) - A_j[a_1 A_j^2 + 2a_2(A_k^2 + A_\ell^2)], \tag{17.3.1b}$$

$$A_j \frac{d\varphi_j}{dt} \sim 4a_0 A_k A_\ell \sin(\varphi_j + \varphi_k + \varphi_\ell). \tag{17.3.1c}$$

Here we have represented the hexagonal expansion of (17.3.1) in its real form. Given that many nonlinear stability analyses employ its equivalent complex form, we shall now examine that equivalence for the sake of completeness. Consider the complex hexagonal planform expansion

$$R(x,y,t) \sim R_1\mathcal{A}_1(t)e^{i\boldsymbol{Q}_1\bullet\boldsymbol{r}} + R_1^*\mathcal{A}_1^*(t)e^{-i\boldsymbol{Q}_1\bullet\boldsymbol{r}} + R_2\mathcal{A}_2(t)e^{i\boldsymbol{Q}_2\bullet\boldsymbol{r}} + R_2^*\mathcal{A}_2^*(t)e^{-i\boldsymbol{Q}_2\bullet\boldsymbol{r}} + R_3\mathcal{A}_3(t)e^{i\boldsymbol{Q}_3\bullet\boldsymbol{r}} + R_3^*\mathcal{A}_3^*(t)e^{-i\boldsymbol{Q}_3\bullet\boldsymbol{r}} \tag{17.3.2a}$$

where

$$\boldsymbol{r} = x\boldsymbol{e}_1 + y\boldsymbol{e}_2;\ \boldsymbol{Q}_j = q_c[\cos(\theta_j)\boldsymbol{e}_1 + \sin(\theta_j)\boldsymbol{e}_2],\ j = 1,2,3; \tag{17.3.2b}$$

with

$$\theta_1 = 0,\ \theta_2 = \frac{2\pi}{3},\ \theta_3 = \frac{4\pi}{3}; \tag{17.3.2c}$$

while for $(j,k,l) \equiv$ even permutation of (1,2,3),

$$\frac{d\mathcal{A}_j}{dt} \sim \sigma\mathcal{A}_j - 4a_0\mathcal{A}_k^*\mathcal{A}_\ell^* - \mathcal{A}_j[a_1|\mathcal{A}_j|^2 + 2a_2(|\mathcal{A}_k|^2 + |\mathcal{A}_\ell|^2)]. \tag{17.3.2d}$$

Now noting that

$$\boldsymbol{Q}_1\bullet\boldsymbol{r} = q_c x,\ \boldsymbol{Q}_2\bullet\boldsymbol{r} = \frac{q_c(-x+\sqrt{3}y)}{2},\ \boldsymbol{Q}_3\bullet\boldsymbol{r} = \frac{-q_c(x+\sqrt{3}y)}{2}; \tag{17.3.3a}$$

writing the complex amplitudes in polar form

$$\mathcal{A}_j(t) = A_j(t)e^{i\varphi_j(t)},\ j = 1,2,3; \tag{17.3.3b}$$

taking

$$R_j = \frac{1}{2},\ j = 1,2,3; \tag{17.3.3c}$$

and recalling

$$\cos(\theta) = \frac{e^{i\theta} + e^{-i\theta}}{2}; \tag{17.3.3d}$$

the complex representation of (17.3.2) reduces to the real one of (17.3.1). Further substituting the complex amplitude functions of (17.3.3b) into (17.3.2d), computing the indicated derivatives, noting that

$$\mathcal{A}_j^* = A_j(t)e^{-i\varphi_j(t)}, \tag{17.3.3e}$$

and multiplying the result by $e^{-i\varphi_j(t)}$, we obtain

$$\frac{dA_j}{dt} + iA_j\frac{d\varphi_j}{dt} \sim \sigma A_j - 4a_0A_kA_\ell e^{-i(\varphi_j+\varphi_k+\varphi_\ell)} - A_j[a_1A_j^2 + 2a_2(A_k^2 + A_\ell^2)], \tag{17.3.3f}$$

the real and imaginary parts of which yield the amplitude–phase equations of (17.3.1b, c), respectively, thus completing our equivalency demonstration. The nonlinear stability behavior of these amplitude–phase equations to be described below depends only on the values of their growth rate and Landau constants. We can determine the solvability conditions for that growth rate and these Landau constants most easily by introducing the transformation

$$A_2(t) = A_3(t) = \frac{B_1(t)}{2}, \varphi_1(t) = \varphi_2(t) = \varphi_3(t) \equiv 0, \tag{17.3.4a}$$

which reduces (17.3.1) to

$$R(x,y,t) \sim h_0(x,y,t) = A_1(t)\cos(q_c x) + B_1(t)\cos\left(\frac{q_c x}{2}\right)\cos\left(\frac{\sqrt{3}q_c y}{2}\right), \tag{17.3.4b}$$

where

$$\frac{dA_1}{dt} \sim \sigma A_1 - a_0 B_1^2 - A_1(a_1 A_1^2 + a_2 B_1^2), \tag{17.3.5a}$$

$$\frac{dB_1}{dt} \sim \sigma B_1 - 4a_0 A_1 B_1 - B_1\left[2a_2 A_1^2 + \frac{(a_1 + 2a_2)B_1^2}{4}\right]. \tag{17.3.5b}$$

Hence, we consider solutions of (17.1.14) of the form [136, 143]

$$\begin{aligned} R(x,y,t) \sim\ & h_0(x,y,t) + h_{20}(x)A_1^2(t) + h_{11}(x,y)A_1(t)B_1(t) + h_{02}(x,y)B_1^2(t) \\ & + h_{30}(x)A_1^3(t) + h_{21}(x,y)A_1^2(t)B_1(t) + h_{12}(x,y)A_1(t)B_1^2(t) + h_{03}(x,y)B_1^3(t) \end{aligned} \tag{17.3.6a}$$

with

$$h_{20}(x) = R_{2000} + R_{2040}\cos(2q_c x),\ h_{30}(x) = R_{3020}\cos(q_c x) + R_{3060}\cos(3q_c x),$$

$$h_{11}(x,y) = R_{1111}\cos\left(\frac{q_c x}{2}\right)\cos\left(\frac{\sqrt{3}q_c y}{2}\right) + R_{1131}\cos\left(\frac{3q_c x}{2}\right)\cos\left(\frac{\sqrt{3}q_c y}{2}\right),$$

$$h_{02}(x,y) = R_{0200} + R_{0220}\cos(q_c x) + R_{0202}\cos(\sqrt{3}q_c y) + R_{0222}\cos(q_c x)\cos(\sqrt{3}q_c y),$$

$$\begin{aligned} h_{21}(x,y) = {} & R_{2111}\cos\left(\frac{q_c x}{2}\right)\cos\left(\frac{\sqrt{3}q_c y}{2}\right) + R_{2131}\cos\left(\frac{3q_c x}{2}\right)\cos\left(\frac{\sqrt{3}q_c y}{2}\right) \\ & + R_{2151}\cos\left(\frac{5q_c x}{2}\right)\cos\left(\frac{\sqrt{3}q_c x}{2}\right), \end{aligned}$$

$$\begin{aligned} h_{12}(x,y) = {} & R_{1200} + R_{1220}\cos(q_c x) + R_{1240}\cos(2q_c x) + R_{1202}\cos(\sqrt{3}q_c y) \\ & + R_{1222}\cos(q_c x)\cos(\sqrt{3}q_c y) + R_{1242}\cos(2q_c x)\cos(\sqrt{3}q_c y), \end{aligned}$$

$$h_{03}(x,y) = R_{0311}\cos\left(\frac{q_c x}{2}\right)\cos\left(\frac{\sqrt{3}q_c y}{2}\right) + R_{0331}\cos\left(\frac{3q_c x}{2}\right)\cos\left(\frac{\sqrt{3}q_c y}{2}\right) +$$

$$R_{0313}\cos\left(\frac{q_c x}{2}\right)\cos\left(\frac{3\sqrt{3}q_c y}{2}\right) + R_{0333}\cos\left(\frac{3q_c x}{2}\right)\cos\left(\frac{\sqrt{3}q_c y}{2}\right). \tag{17.3.6b}$$

We note that the forms of the second- and third-order terms in the expansions of (17.3.6) can be deduced by examining the functional dependence of $h_0^2(x,y,t)$ and $h_0^3(x,y,t)$, respectively. Observe in addition that the terms in the amplitude equations (17.3.5a, b) are those with coefficients containing components proportional to $\cos(q_c x/2)$ and $\cos(\sqrt{3}q_c y/2)$, respectively. Further note in his context that by taking $B_1(t) \equiv 0$ these expansions can be reduced to those of (17.2.2) once we make the identification $R_{nm} = R_{n0M0}$ for $M = 2m$ where the notation R_{njMk} is being employed to represent the coefficient of each higher-order term appearing in (17.3.6) of the form $A_1^n(t)B_1^j(t)\cos(Mq_c x/2)\cos(k\sqrt{3}q_c y/2)$. Thus σ and a_1 are given by (17.2.3a) and (17.2.5b), respectively, and we need only evaluate the remaining Landau constants a_0 and a_2.

Then, substituting (17.3.6) into (17.1.14), expanding R^2 and R^3 in powers of $A_1(t)$ and $B_1(t)$, employing the relevant trigonometric identities for the resulting products of cosine functions contained in its coefficients, making use of the Landau amplitude equations (17.3.5), and computing the required Laplacian, we obtain a sequence of problems, one for each term appearing explicitly in our expansion. Solving these problems through terms of second order, we find specifically that

$$R_{2000} = R_{20},\ R_{1131} = \frac{\omega_0}{4/\sigma_1 - \sigma_R} \tag{17.3.7}$$

while, in particular, the other two Landau constants satisfy

$$4a_0 - \sigma_R R_{1111} = \omega_0, \tag{17.3.8a}$$

$$a_2 - \sigma_R R_{2111} + \frac{(\omega_0 - 4a_0)R_{1111}}{2} = \omega_0\left(R_{2000} + \frac{R_{1131}}{2}\right) + \frac{3\omega_1}{4}. \tag{17.3.8b}$$

Finally, taking the limit of (17.3.8) as $\beta \to \beta_0(\alpha)$ and employing (17.3.7) as well as (17.1.11c), (17.1.12), (17.1.14b), (17.2.3a), and (17.2.4), we obtain the solvability conditions

$$4a_0 = \frac{\beta_0(\alpha)\alpha(\alpha-3)}{(\alpha+1)^3} = \frac{\alpha(\alpha-3)}{\alpha^2-1} = 4a_0(\alpha), \tag{17.3.8c}$$

$$a_2 = \frac{\alpha\beta_0(\alpha)}{4(\alpha+1)^4}\left\{-5\left[\frac{\alpha(\alpha-3)}{\alpha-1}\right]^2 + 3[8\alpha - (\alpha+1)^2]\right\} = a_2(\alpha). \tag{17.3.8d}$$

Note from (17.3.8b) that the expression $a_2(\alpha)$ of (17.3.8d) does not explicitly contain the component $\lim_{\beta\to\beta_0(\alpha)} R_{1111}$ since the coefficient of the latter quantity, namely $\lim_{\beta\to\beta_0(\alpha)}(\omega_0 - 4a_0)$, is identically equal to zero by virtue of (17.3.8c). Although, as pointed out by Wollkind et al. [142] and demonstrated in Chapter 16, such independence can be expected in single equation models, thus eliminating the necessity of determining R_{1111} in this instance, we can conclude, for the sake of completeness, from (17.2.3a) and (17.3.8a, c) that

$$R_{1111} = \frac{-\beta_0(\alpha)\alpha(\alpha-3)}{(\alpha+1)^3} = -4a_0(\alpha). \tag{17.3.9}$$

Having determined formulae for their growth rate and Landau constants, we now return to the six-disturbance hexagonal planform amplitude–phase equations (17.3.1b, c). In cataloguing the critical points of these equations and summarizing their orbital stability behavior, it is necessary to employ the quantities

$$\sigma_{-1} = \frac{-4a_0^2}{a_1+4a_2}, \sigma_1 = \frac{16a_1a_0^2}{(2a_2-a_1)^2}, \sigma_2 = \frac{32(a_1+a_2)a_0^2}{(2a_2-a_1)^2}. \tag{17.3.10}$$

There exist equivalence classes of critical points of (17.3.1b, c) of the form $(A_0, B_0, B_0, 0, 0, 0)$ corresponding to $\phi_1 = \phi_2 = \phi_3 = 0$ with

$$\mathrm{I}: A_0 = B_0 = 0; \mathrm{II}: A_0^2 = \frac{\sigma}{a_1}, B_0 = 0;$$

$$\mathrm{III}^{\pm}: A_0 = B_0 = A_0^{\pm} = \frac{-2a_0 \pm [4a_0^2 + (a_1+4a_2)\sigma]^{1/2}}{a_1+4a_2}; \tag{17.3.11}$$

$$\mathrm{IV}: A_0 = \frac{-4a_0}{2a_2-a_1}, B_0^2 = \frac{\sigma-\sigma_1}{a_1+2a_2},$$

where it is assumed that $a_1, a_1+4a_2 > 0$. In order to investigate the stability of these critical points of (17.3.11), we seek solutions of the amplitude–phase equations (17.3.1b, c) of the form

$$A_1(t) = A_0 + \varepsilon c_1 e^{pt} + O(\varepsilon^2); A_j(t) + B_0 + \varepsilon c_j e^{pt} + O(\varepsilon^2), j = 2 \text{ and } 3; \tag{17.3.12a}$$

$$\varphi_j(t) = 0 + \varepsilon c_{j+3} e^{pt} + O(\varepsilon^2), j = 1, 2, \text{ and } 3 \text{ with } |\varepsilon| << 1. \tag{17.3.12b}$$

Then, upon examining the signs of $\mathrm{Re}(p)$ for each of the critical points of (17.3.8), one can deduce orbital stability criteria for these critical points that are posed in terms of σ [136]. Thus, critical point I is stable in this sense for $\sigma < 0$, while the stability behavior of II and $\mathrm{III}^{\pm}$, which depends upon the signs of a_0 and $2a_2 - a_1$ as well, has been summarized in Table 17.1 under the further assumption that $a_1 + a_2 > 0$ [136]. In this parameter range, $A_0^+ > 0$ and $A_0^- < 0$.

Finally, critical IV, which reduces to II for $\sigma = \sigma_1$ and to $\mathrm{III}^{\pm}$ for $\sigma = \sigma_2$, and hence, called a generalized cell, is not stable for any value of σ [136]. Here, we use the term orbital stability of pattern formation to mean a family of solutions in the plane that may interchange with each other but do not grow or decay into a solution type from a different family. Such an interpretation depends upon the translational and rotational symmetries inherent to the amplitude–phase equations, these invariancies also limiting each equivalence class of critical points to a single

Table 17.1 Orbital stability behavior for critical points II and III$^{\pm}$.

a_0	$2a_2 - a_1$	Stable structures
+	$-, 0$	III$^-$ for $\sigma > \sigma_{-1}$
+	+	III$^-$ for $\sigma_{-1} < \sigma < \sigma_2$, II for $\sigma > \sigma_1$
0	$-$	III$^{\pm}$ for $\sigma > 0$
0	+	II for $\sigma > 0$
$-$	+	III$^+$ for $\sigma_{-1} < \sigma < \sigma_2$, II for $\sigma > \sigma_1$
$-$	$-, 0$	III$^+$ for $\sigma > \sigma_{-1}$

member that must be considered explicitly. Also, note that by stability in the third row of this table, we merely mean neutral stability.

We next offer a morphological interpretation of the potentially stable critical points catalogued above relative to the transverse optical patterns under investigation. Then, critical points I and II represent the uniform plane-wave and supercritical striped states, respectively, described in the last section. Observing that the equilibrium plane-wave deviation associated with critical points III$^{\pm}$, $R(x, y, t) \to R_e^{\pm}(x, y)$ as $t \to \infty$, satisfies to lowest order

$$R_e^{\pm}(x, y) \sim A_0^{\pm} g_0(x, y) \text{ as } t \to \infty \text{ where } g_0(x, y) = \cos\left(\frac{2\pi x}{\lambda_c}\right) + 2\cos\left(\frac{\pi x}{\lambda_c}\right)\cos\left(\frac{\sqrt{3}\pi y}{\lambda_c}\right) \tag{17.3.13}$$

and, noting from Fig. 17.6 that the function $g_0(x, y)$ exhibits hexagonal symmetry, we can deduce that these critical points represent hexagonal arrays possessing individual hexagons with bright circular regions at their centers and dark boundaries for III$^+$ and with dark circular regions at their centers and bright boundaries for III$^-$. The contour plots relevant to these critical points are depicted in Fig. 17.7 where (a) is for III$^+$ and (b), for III$^-$. Hence, in what follows, we shall identify transverse hexagonal optical arrays of bright spots or honeycombs with critical points III$^+$ or III$^-$, respectively.

Having summarized those stability criteria and morphological identifications, we now return to our expressions for the Landau constants of (17.3.1b, c) given by (17.2.5b) and (17.3.8a, b). First, we examine the signs of a_0, $2a_2 - a_1$, $a_1 + 4a_2$, and $a_1 + a_2$ as functions of α by plotting these quantities versus α in Fig. 17.8. From the results of this examination, we observe that besides α_c, α_1, and α_2 defined in (17.2.3b) and (17.2.6) there exist the following other significant values of α which are given in Table 17.2:

$$\alpha_7 < \alpha_5 < \alpha_3 < \alpha_1 < \alpha_c < \alpha_2 < \alpha_4 < \alpha_6 < \alpha_8 \tag{17.3.14a}$$

such that

$$a_1 + a_2 = 0 \text{ for } \alpha = \alpha_3 \text{ or } \alpha_4,\ a_1 + a_2 > 0 \text{ for } \alpha_3 < \alpha < \alpha_4; \tag{17.3.14b}$$

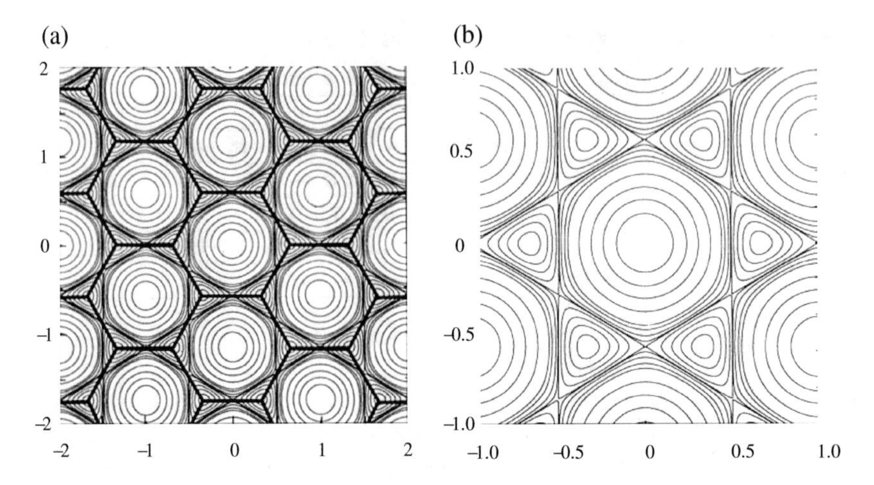

Fig. 17.6 Contour plots for $g_0(x, y)$ of (17.3.13). In this context, the axes are again being measured in units of λ_c. Here (a) exhibits the relevant hexagonal symmetry while (b), an enlargement of its central region, emphasizes the level curve behavior of an individual cell, each of which has an elevation ranging from 3 at its center to −3/2 at its vertices. The third, fourth, and fifth rings from that center are at 3/4, zero, and −3/5 elevation, respectively, while the boundary of each cell is of variable depth with its high points at elevation −1 occurring at the midpoint of each edge.

$$a_1 + 4a_2 = 0 \text{ for } \alpha = \alpha_5 \text{ or } \alpha_6,\ a_1 + 4a_2 > 0 \text{ for } \alpha_5 < \alpha < \alpha_6; \tag{17.3.14c}$$

$$2a_2 - a_1 = 0 \text{ for } \alpha = \alpha_7 \text{ or } \alpha_8,\ 2a_2 - a_1 > 0 \text{ for } \alpha_7 < \alpha < \alpha_8; \tag{17.3.14d}$$

$$a_0 = 0 \text{ for } \alpha = \alpha_c,\ a_0 < 0 \text{ for } 1 < \alpha < \alpha_c,\ a_0 > 0 \text{ for } \alpha > \alpha_c. \tag{17.3.14e}$$

Table 17.2 The α-values of (17.3.14).

α_7	α_5	α_3	α_1	α_c	α_2	α_4	α_6	α_8
1.646	2.024	2.075	2.143	3	4.167	4.290	4.383	5.107

We shall next restrict our attention to those $\alpha \in (\alpha_3, \alpha_4)$ and observe from (17.3.14) that then

$$2a_2 - a_1 > 0. \tag{17.3.15}$$

Thus, under this additional assumption the relevant entries of Table 17.1 which now only involve the sign of a_0 may be summarized by Table 17.3.

In order to compare our theoretical predictions with numerical simulations and relevant experimental evidence, we must represent the stability results of Table 17.3 graphically in the α-β plane. To do so, it is now necessary for us to generalize the approach which produced the $\sigma = \sigma_0 = 0$ marginal stability curve $\beta = \beta_0(\alpha)$ of (17.1.12) from (17.2.3a) in order that we may generate the analogous loci associated

Table 17.3 Orbital stability behavior of critical points II and III$^{\pm}$ when $2a_2 - a_1 > 0$.

a_0	Stable structures
+	III$^-$ for $\sigma_{-1} < \sigma < \sigma_2$, II for $\sigma > \sigma_1$
0	II for $\sigma > 0$
−	III$^+$ for $\sigma_{-1} < \sigma < \sigma_2$, II for $\sigma > \sigma_1$

with $\sigma = \sigma_j$ for $j = -1, 1$, and 2, respectively. That is, employing (17.2.3a), (17.2.5b), (17.3.8c, d), and (17.3.10), we solve

$$\sigma_R(\alpha, \beta) = \sigma_j[a_0(\alpha), a_1(\alpha), a_2(\alpha)], \text{ with } j = -1, 1, \text{ and } 2, \tag{17.3.16a}$$

for β to obtain

$$\beta = \beta_0(\alpha)\{1 + \sigma_j[a_0(\alpha), a_1(\alpha), a_2(\alpha)]\} = \beta_j(\alpha) \text{ with } j = -1, 1, \text{ and } 2, \tag{17.3.16b}$$

respectively. Since all the quantities required for the transverse optical patterns of Table 17.3 have been evaluated, we can represent graphically the regions corresponding to these patterns in the α-β plane of Fig. 17.9, where the loci of (17.1.12) and (17.3.16) are denoted by β_j with $j = -1, 0, 1$, and 2, in that figure, and identify this correspondence in Table 17.4. From the latter, we see that stable spots can only occur for $\alpha < \alpha_c = 3$ and stable honeycombs for $\alpha > \alpha_c = 3$. In this context we observe from Fig. 17.9a that all of the loci intersect each other at the point $(\alpha_c, \beta_c) = (3, 8)$ and note that in the left half of that figure for $\alpha < \alpha_c$ (see the top half of Table 17.4 and Fig. 17.9b) the α-component of the intersection points of β_1 and β_2 with β_0 corresponds to α_1 and α_3, respectively, while in its right half for $\alpha > \alpha_c$ (see the bottom half of Table 17.4 and Fig. 17.9c) the α-components of these intersection points correspond to α_2 and α_4.

Although the stability behavior summarized in Table 17.3 is strictly for $a_1 > 0$ or $\alpha \in (\alpha_1, \alpha_2)$, we have extended these results to the case of $a_1 + a_2 > 0$ or $\alpha \in (\alpha_3, \alpha_4)$ in Table 17.4. This extrapolation has been accomplished by realizing in the relevant α-intervals of $(\alpha_3, \alpha_1]$ and $[\alpha_2, \alpha_4)$ only III$^{\pm}$ structures can exist since $a_1 < 0$ and these hexagonal structures are stable for

$$\beta_{-1}(\alpha) < \beta < \beta_2(\alpha). \tag{17.3.17}$$

Note that in those α-intervals there are no stable patterns for $\beta > \beta_2$ and metastability between the plane-wave and subcritical striped states for $\beta < \beta_{-1}$.

Again considering $\alpha \in (\alpha_1, \alpha_2)$, we observe from Fig. 17.9, in conjunction with Tables 17.3 and 17.4, that part $(\beta_0 < \beta < \beta_2)$ of the region $(\beta > \beta_0)$, where the one-dimensional analysis of Section 17.2 predicted supercritical striped patterns, is further divided into two subregions characterized by hexagonal patterns consisting of either spots (when $\alpha < \alpha_c$) or honeycombs (when $\alpha > \alpha_c$), respectively. In the overlap regions satisfying $\sigma_1 < \sigma < \sigma_2$ or

Table 17.4 Pattern formation predictions for Fig. 17.9.

α-interval	β-range	Stable pattern
(α_3, α_c)	$\beta < \beta_{-1}$	Plane wave
	$\beta_{-1} < \beta < \beta_0$	Plane wave and spots
$(\alpha_3, \alpha_1]$	$\beta_0 < \beta < \beta_2$	Spots
(α_1, α_c)	$\beta_0 < \beta < \beta_1$	Spots
	$\beta_1 < \beta < \beta_2$	Spots and stripes
	$\beta > \beta_2$	Stripes
α_c	$\beta > \beta_c$	Stripes
(α_c, α_2)	$\beta_0 < \beta < \beta_1$	Honeycombs
	$\beta_1 < \beta < \beta_2$	Honeycombs and stripes
	$\beta > \beta_2$	Stripes
$[\alpha_2, \alpha_4)$	$\beta_0 < \beta < \beta_2$	Honeycombs
(α_c, α_4)	$\beta < \beta_{-1}$	Plane wave
	$\beta_{-1} < \beta < \beta_0$	Plane wave and honeycombs

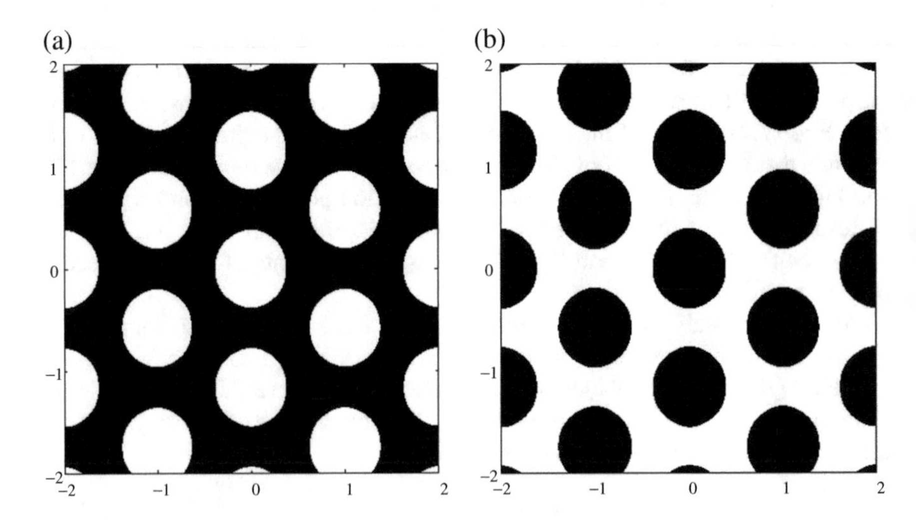

Fig. 17.7 Contour plots of (17.3.13) for critical points (a) III$^+$ and (b) III$^-$ of (17.3.11).

$$\beta_1(\alpha) < \beta < \beta_2(\alpha), \tag{17.3.18}$$

where stripes and spots ($\alpha_1 < \alpha < \alpha_c = 3$) or stripes and honeycombs ($3 = \alpha_c < \alpha < \alpha_2$) are predicted initial conditions determine which stable equilibrium pattern of each pair will be selected. Returning to the extended interval (α_3, α_4), there also exists a region of bistability corresponding to $\sigma_{-1} < \sigma < 0$ or

$$\beta_{-1}(\alpha) < \beta < \beta_0(\alpha), \tag{17.3.19}$$

the plane wave being stable for $\sigma < 0$ or $\beta < \beta_0$ and hexagons for (17.3.17). Given that $\sigma_{-1} < 0$ for $a_0 \neq 0$ or $\alpha \neq \alpha_c = 3$, the hexagons predicted in this overlap region would be subcritical in nature. Finally to justify the truncation procedure inherent to the asymptotic representation of (17.3.1b, c) it is necessary that the Landau constants of these amplitude–phase equations satisfy the size constraint [142]

$$\frac{|a_0|}{(a_1 + 4a_2)^2} << 1. \tag{17.3.20}$$

Observing from Fig. 17.8c that this inequality is satisfied in the parameter range of interest, we can conclude that such a truncation procedure is valid for our hexagonal planform nonlinear stability analysis of the modified Swift–Hohenberg equation (17.1.14).

17.4 Pattern Formation Results

We are finally ready to compare these theoretical predictions developed in Section 17.3 with relevant numerical simulations and experimental evidence as well as place them in the context of some recent nonlinear optical pattern formation studies. We begin naturally by comparing our predictions with the numerical simulations of Firth and Scroggie [35]. As mentioned earlier, these authors numerically integrated (17.1.4) on a square grid with periodic boundary conditions for $\Delta = 0$, $\theta = -1$, and χ normalized to unity while using (17.1.14) and (17.1.11c), respectively, as a guide to select the appropriate values of α and β that favor pattern formation. Implicitly exploiting our relationship (17.3.8a) and assuming that α_c could be determined from the expression for ω_0 in (17.1.14b), they concluded that $\alpha_c = 3$ and constructed contour plots of Re(X) in the x-y plane by selecting a value of β greater than $\beta_c = 8$ and increasing α across the instability region. In this event Firth and Scroggie [35] observed a transition from stable III$^+$- to II- to III$^-$-type patterns for $\alpha < 3$, $\alpha \cong 3$, and $\alpha > 3$, respectively.

To compare that simulation result with our theoretical predictions, we now select the typical fixed value of $\beta = 8.8$ employed by Firth and Scroggie [35] in the three parts of the contour plots depicting their results reproduced in Fig. 17.10. Upon

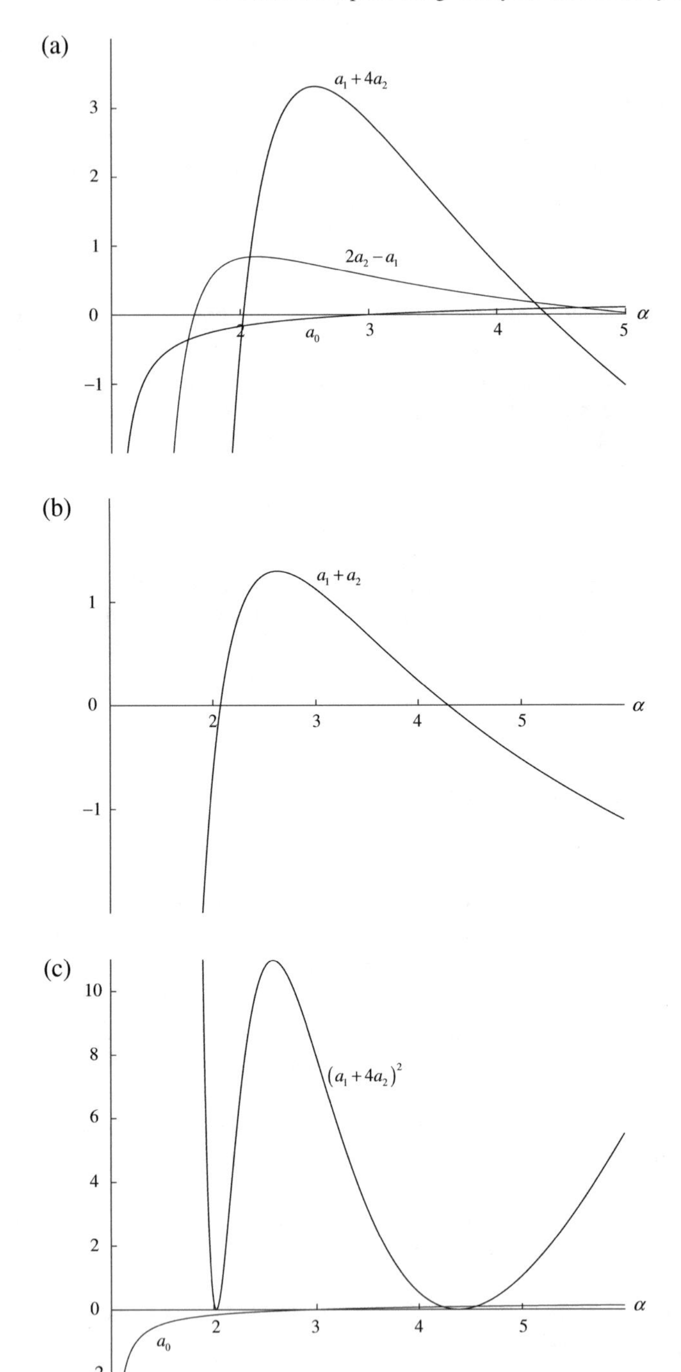

Fig. 17.8 Plots of (a) a_0, $2a_2 - a_1$, and $a_1 + 4a_2$; (b) $a_1 + a_2$; (c) a_0 and $(a_1 + 4a_2)^2$; of (17.2.5b) and (17.3.8c, d) versus α.

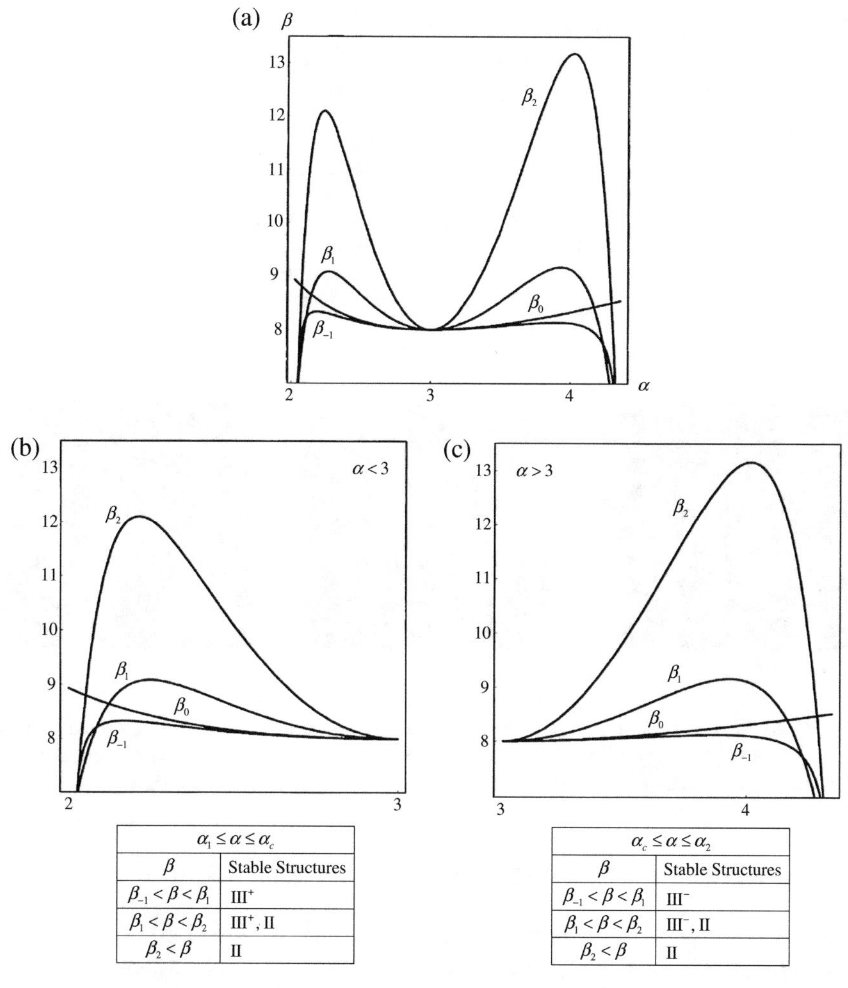

$\alpha_1 \leq \alpha \leq \alpha_c$	
β	Stable Structures
$\beta_{-1} < \beta < \beta_1$	III$^+$
$\beta_1 < \beta < \beta_2$	III$^+$, II
$\beta_2 < \beta$	II

$\alpha_c \leq \alpha \leq \alpha_2$	
β	Stable Structures
$\beta_{-1} < \beta < \beta_1$	III$^-$
$\beta_1 < \beta < \beta_2$	III$^-$, II
$\beta_2 < \beta$	II

Fig. 17.9 (a) Stability diagram in the α-β plane for the modified Swift–Hohenberg equation (17.1.14) with the predicted nonlinear optical patterns summarized in Table 17.4 denoted by the enlargements for (b) $\alpha_1 < \alpha < \alpha_c$ and (c) $\alpha_c < \alpha < \alpha_2$. Here $\beta_j = \beta_0(1 + \sigma_j)$ for $j = -1, 0, 1, 2$.

examination of Fig. 17.9 and Table 17.4, we can deduce the predicted morphological sequence given in Table 17.5 as α is increased along that horizontal transit line.

In order to interpret these results with respect to Firth and Scroggie's [35] simulations depicted in Fig. 17.10, we shall assume the maximal possible domain for hexagons when considering regions of bistability in Table 17.5. This assumption is consistent with the fact that those authors were unable to observe the stable coexistence of stripes and hexagonal patterns. Making such an interpretation for $\beta = 8.8$ yields the predicted sequence of morphologies

Table 17.5 Morphological stability predictions versus α for $\beta = 8.8$.

α-interval	Stable pattern
(2.08, 2.14)	Spots
(2.14, 2.16)	Spots and stripes
(2.16, 2.45)	Spots
(2.45, 2.73)	Spots and stripes
(2.73, 3.30)	Stripes
(3.30, 3.62)	Honeycombs and stripes
(3.62, 4.10)	Honeycombs
(4.10, 4.17)	Honeycombs and stripes
(4.17, 4.28)	Honeycombs

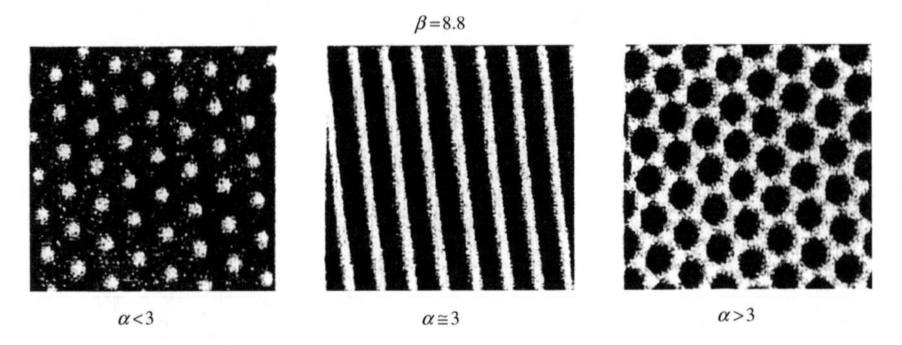

Fig. 17.10 Plots relevant to the spot-stripe-honeycomb transition as α is increased for the simulations of Firth and Scroggie [35] of (17.1.4) with $\Delta = 0$, $\theta = -1$, $\chi = 1$, and $\beta = 8.8$, adapted by the permission of the authors from [35] for which Euro. Phys. Lett. holds the copyright.

$$\text{III}^+ \text{ for } \alpha < 2.73, \text{ II for } 2.73 < \alpha < 3.30, \text{ III}^- \text{ for } \alpha > 3.30, \tag{17.4.1}$$

in qualitative agreement with the results of Firth and Scroggie [35]. Recall that in plotting Figs. 17.5 and 17.7a, b for our striped and hexagonal optical patterns, respectively, the standard operating procedure for selecting its bright and dark regions has been employed: Namely, we used $R = 0$ as the critical threshold separating these regions. Upon comparison of those plots with the striped and hexagonal optical patterns depicted in Fig. 17.10 observe that they do not replicate Firth and Scroggie's [35] simulations exactly. In order to improve the quantitative accuracy of our plots in comparison with these simulation results it is necessary for us to re-examine that methodology. In effect such a zero-threshold protocol requires this color change to coincide with the equilibrium intensity state characterized by α. That is, we implicitly assumed a critical intensity $\alpha_{\text{crit}} = |X_{\text{crit}}|^2 = |X_e|^2 = \alpha$. Since α varies this means a threshold of that sort is based upon a sliding scale. Note that for this zero-deviation threshold the boundaries of the spots of Fig. 17.7 correspond to the fourth curve of Fig. 17.6b counting outward from its center. This being the case it is instructive to examine precisely what effect the adoption of a different protocol based on a fixed value of α_{crit} would have on the appearance of our optical patterns. For the sake of definiteness let us select $\alpha_{\text{crit}} \geq \alpha_4$. Then all $\alpha \in (\alpha_3, \alpha_4)$ in the patterned interval of Fig. 17.9 would satisfy $\alpha < \alpha_{\text{crit}}$ or $|X_e| < |X_{\text{crit}}|$. Hence this would give rise to what Wollkind and Stephenson [143] termed higher threshold patterns in which the dark regions now predominate. These higher threshold patterns of stripes and hexagonal arrays of spots or honeycombs are plotted in the three parts of Fig. 17.11a which upon comparison with the simulations of Firth and Scroggie [35] depicted in Fig. 17.10 are seen to provide the desired quantitative agreement. In particular, the boundaries of the spots in that figure have been taken to correspond to the third curve in Fig. 17.6b; the boundaries of the honeycombs, the fifth curve in Fig. 17.6b; and the width of the interstripes and stripes, to be in a 2:1 ratio.

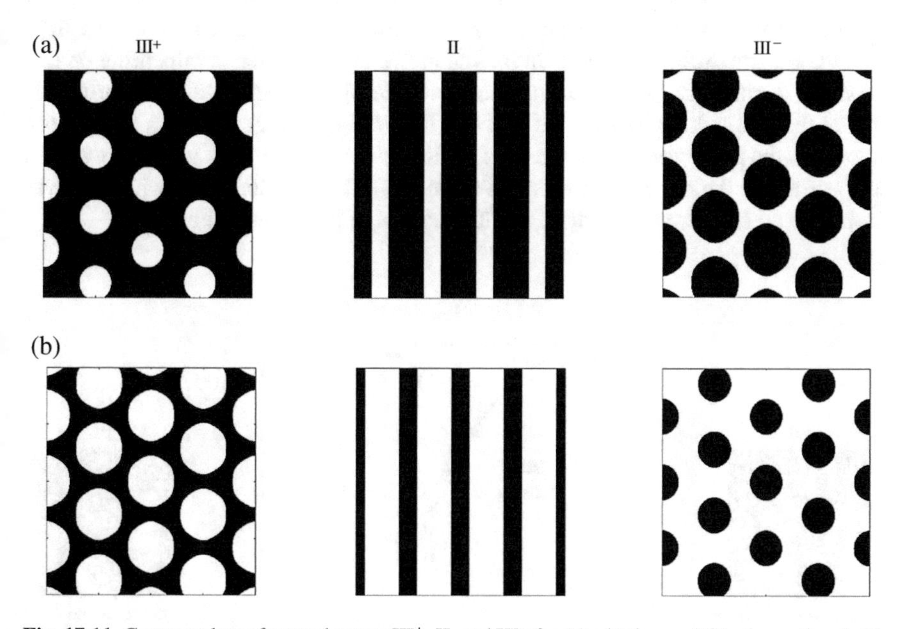

Fig. 17.11 Contour plots of critical points III$^+$, II, and III$^-$ for (a) a higher and (b) a lower threshold protocol.

Having completed an investigation of this higher threshold protocol there is also merit in examining the morphological consequences of the adoption of a lower threshold one. Toward that end we select $\alpha_{\text{crit}} \leq \alpha_3 < \alpha \in (\alpha_3, \alpha_4)$ giving rise to the corresponding lower threshold patterns plotted in the three parts of Fig. 17.11b. Here the bright regions predominate instead. Observe that for the zero-threshold protocol if the light and dark regions are interchanged the stripes of Fig. 17.5 are preserved and the spots and honeycombs of Fig. 17.7 are interchanged. Although this is no longer true for the higher and lower threshold protocols such an interchange in the higher threshold patterns of Fig. 17.11a will result in a similar correspondence with the lower threshold patterns of Fig. 17.11b and *vice versa*.

Finally if $\alpha_3 < \alpha_{\text{crit}} < \alpha_4$, that region is partitioned into three subregions with higher threshold optical patterns occurring for $\alpha_3 < \alpha < \alpha_{\text{crit}}$ and lower threshold ones, for $\alpha_{\text{crit}} < \alpha < \alpha_4$, while zero-threshold patterns are retained for $\alpha = \alpha_{\text{crit}}$. Specifically, all three types of striped patterns can occur should $\alpha_{\text{crit}} = \alpha_c$; all three III^+ patterns, should $\alpha_3 < \alpha_{\text{crit}} < \alpha_c$; and all three III^- patterns, should $\alpha_c < \alpha_{\text{crit}} < \alpha_4$. This completes a demonstration of the capability of our predictions to reproduce numerically or experimentally observed optical patterns upon adoption of an appropriate threshold protocol. That the adoption of such a protocol to replicate patterns of this sort may not be as well known a theoretical pattern generation mechanism as one might have suspected is attested to by the following: Kondo and Asai [60] investigated the formation of stripes in marine angelfish by employing an activator–inhibitor model which incorporated the kinetics of Turing [121]. Analogous to our contour plots high and low concentrations of the activator species were represented by light and dark colors, respectively, in their simulated patterns. When commenting on this work in conjunction with the actual appearance of the angelfish, Meinhardt [81] stated that the light stripes on the fish still required further explanation since those shown in the photographs contained in Kondo and Asai [60] were very narrow with respect to the dark spaces in between them while all the models of which he was aware could only produce stripes and interstripes of the same width.

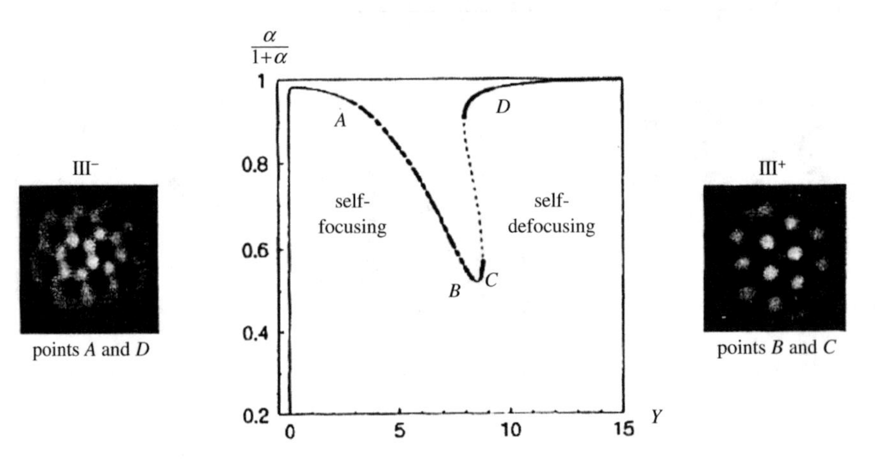

Fig. 17.12 Plot relevant to the transition between spots (III^+) and honeycombs (III^-) as Y is increased for the experiments of Ackemann et al. [2], adapted by the permission of the authors from [2] for which the American Physical Society holds the copyright. See Problem 17.3 in the exercises for a definition of self-focusing and self-defocusing in nonlinear optical media.

Lugiato and Oldano [76] formulated their ring-cavity laser model of Fig. 17.1 to demonstrate that a nonlinear optical system could produce stationary patterns analogous to the chemical reaction–diffusion Turing structures considered in Chapter 8 but with diffraction in the optical system taking the place of diffusion in the chemical one. Since the experimental motivation for the numerical simulations Firth and Scroggie [35] performed on this ring-cavity model was the sort of transition between the two types of hexagonal patterns that Ackemann et al. [2] observed in a sodium vapor cell with a feedback mirror where optical pattern formation was induced by external pumping, we next compare our theoretical predictions with the latter authors' primary experimental results. Fig. 17.12 adapted from those results represents a plot of a steady-state plane-wave configuration measure $\alpha/(1+\alpha)$ versus the rate of external pumping Y. In particular, for the portion of this curve having positive slope they found spots occurred at the left-hand end of the patterned interval and honeycombs at the right-hand end while for the portion having negative slope the locations of occurrence of these two types of hexagonal patterns were interchanged. Observe from the photographic inserts in Fig. 17.12 that both the spots and honeycombs generated during those experiments were of the higher threshold variety. To compare these experimental results with our theoretical predictions, we must first re-examine Fig. 17.2 which is a plot of relationship (17.1.5b) between α and Y^2 with $\Delta = 0$ and $\theta = -1$ for various fixed values of β. Specifically, in this instance, β_{crit}, as defined implicitly by (17.1.5c), satisfies

$$\beta_{\text{crit}} = 10.2 \tag{17.4.2a}$$

and the three parts of that figure have been plotted for (a) $\beta = 8.8$, (b) $\beta = 10.2$, and (c) $\beta = 12$, respectively. Observe from these plots that

$$\frac{dY^2}{d\alpha} > 0 \text{ for } \beta = 8.8, \quad \frac{dY^2}{d\alpha} \geq 0 \text{ for } \beta = 10.2, \quad \frac{dY^2}{d\alpha} < 0 \text{ for } \beta = 12, \tag{17.4.2b}$$

over the patterned region which has been denoted by shading in Fig. 17.2. Next, we must determine the morphological sequence analogous to (17.4.1) for $\beta = 8.8$ when β is assigned the other two values employed in (17.4.2b) instead. From Fig. 17.9 and Table 17.4, we obtain the predicted morphological sequences associated with the horizontal transit lines $\beta = 10.2$

$$\text{III}^+ \text{ for } 2.1 < \alpha < 2.5, \text{ II for } 2.5 < \alpha < 3.55, \text{ III}^- \text{ for } 3.55 < \alpha < 4.25 \tag{17.4.3a}$$

and $\beta = 12$

$$\text{III}^+ \text{ for } 2.2 < \alpha < 2.3, \text{ II for } 2.3 < \alpha < 3.8, \text{ III}^- \text{ for } 3.8 < \alpha < 4.2, \tag{17.4.3b}$$

respectively. Finally combining these results from (17.4.2) and (17.4.3) while identifying the sign of the slope of the curve in Fig. 17.12 with that of the reciprocal of $dY^2/d\alpha$ in Fig. 17.2b, c, we can deduce that for $\beta = 10.2$ the slope is positive, spots occur at the left-hand end of the patterned interval and honeycombs at its right-hand end as the external pump rate increases and for $\beta = 12$ the slope is negative and this behavior reverses. Hence, in spite of the fact that the optical device employed by Ackemann et al. [2] differs from a ring-cavity configuration, we have demonstrated the potential of our theoretical predictions for the proper choice of parameter values to provide a reasonable correlation with their experimental results. Indeed, we have selected the precise β-values of (17.4.3b) in order to improve the qualitative accuracy of that correlation. These selections produce patterned Y-interval lengths for both values of β and a point of vertical tangency for $\beta = 10.2$ corresponding to $\alpha = 2.73$ where $dY^2/d\alpha = 0$ in agreement with the plot of Fig. 17.12.

So far we have been concentrating on the supercritical behavior of our model system and thus have left for last a discussion of its subcritical behavior. We begin by considering the onset of subcritical hexagonal optical patterns. From Fig. 17.9 and Table 17.4, for a fixed value of $\alpha = \alpha_0$ such that $\alpha_0 \in (\alpha_3, \alpha_c) \cup (\alpha_c, \alpha_4)$ and $\beta < \beta_1$, we obtain the morphological sequence

$$\text{I for } \beta < \beta_{-1}, \text{ I/III for } \beta_{-1} < \beta < \beta_0, \text{ III for } \beta_0 < \beta < \beta_1. \tag{17.4.4}$$

where III in (17.4.4) represents III$^+$ or III$^-$ depending upon whether $\alpha_0 \in (\alpha_3, \alpha_c)$ or $\alpha_0 \in (\alpha_c, \alpha_4)$, respectively. Employing a morphological persistence argument, we can conclude that the transition from a plane-wave configuration to a hexagonal pattern would occur supercritically at $\beta = \beta_0$ for increasing β but that the reverse transition from a hexagonal pattern to a plane-wave configuration would occur subcritically at $\beta = \beta_{-1}$ for decreasing β thus resulting in a region of hysteresis. Although Firth and Scroggie [35] saw the formation of III$^+$ or III$^-$ patterns close to threshold when varying β while holding α fixed, they did not mention the occurrence of such hysteresis possibly because of the difficulty in distinguishing numerically between β_{-1} and β_0 for this instance (see Fig. 17.9). There does, however, exist a general theoretical result relevant to our prediction. Aranson et al. [3] analyzed a modified Swift–Hohenberg equation of the form of (17.1.14) but with the idealized parameter values

$$\sigma_R = -1, \omega_0 = -\beta, \omega_1 = 1, \sigma_I = \chi^2 \tag{17.4.5}$$

and numerically found a subcritical bifurcation that gave rise to localized solutions in two spatial dimensions over a range of real β. Here, β is the modification parameter which when 0 reduces (17.1.14) with parameters chosen by (17.4.5) to an ordinary Swift–Hohenberg equation in normal form. We finally turn to the possibility of occurrence of one-dimensional stable stripes even when $a_1 < 0$. Firth and Scroggie [35] asserted that the sign of ω_1 in (17.1.14b) determined whether such one-dimensional bifurcations occurred supercritically ($a_1 > 0$) or subcritically ($a_1 < 0$) for their system with $\Delta = 0$ and $\theta = -1$. From (17.2.5) we see that this assertion is different than our conclusion of (17.2.6) since $\omega_1 > 0$ whenever

$$\alpha^- < \alpha < \alpha^+ \tag{17.4.6a}$$

where

$$\alpha^\pm = 3 \pm 2\sqrt{2} \text{ or } \alpha^- \cong 0.172 \text{ and } \alpha^+ \cong 5.828. \tag{17.4.6b}$$

Firth and Scroggie [35] then performed a numerical simulation of that system for $\alpha = 5$ and $\beta = 10$ obtaining stable stripes. That choice of parameters yields a $\sigma > 0$

since $\beta_0(5) = 9$ and an $a_1 < 0$ since $\alpha_2 = 4.167$. This was precisely the unstable case for the amplitude equation (17.2.2b) truncated through terms of third order treated in the last chapter that could give rise to a stable re-equilibrated solution should a term of the proper sign be retained at fifth-order. For our model such a solution would represent a striped pattern. This behavior is reminiscent of that occurring for the analysis and simulation performed by Geddes et al. [40] on a nonlinear optical model system involving a so-called Kerr medium where pattern formation was driven by a pump field. Although squares did not saturate theoretically at third order (see Chapter 18), these authors nonetheless generated stable square patterns by numerical simulation and suggested such saturation would occur theoretically at quintic order (see Chapter 16).

We close with a few additional observations about this problem. Firth and Scroggie [35] were concerned with the spontaneous formation of stationary patterns relevant to optical bistability rather than the occurrence of pulsed, oscillatory, or chaotic temporal or spatiotemporal patterns as is often the major focus for problems in nonlinear optics [83]. For their $\Delta = 0$ analysis, they pointed out that since (17.1.14) was based on a perturbation expansion any results from it would only be strictly valid in the vicinity of the marginal curve $\beta = \beta_0(\alpha)$. Thus, although one could not then formally predict a spot-stripe-honeycomb transition at $\alpha = \alpha_c = 3$ when one increased α for constant $\beta > \beta_c$, Firth and Scroggie [35] stated that this was nevertheless essentially the observed behavior from their simulations. Besides these $\Delta = 0$ simulations already described in this section, they also numerically integrated (17.1.4) with $\Delta \neq 0$ and obtained results broadly similar to those with $\Delta = 0$. For example, upon increasing α with β held constant, Firth and Scroggie [35] saw the spot-stripe-honeycomb transition now occurring at a minimum threshold $\alpha = \alpha_c(\Delta)$ where $\alpha_c(0) = \alpha_c$. That being the case the range of validity of our results is much wider than could be expected otherwise, an extrapolation implicitly exploited earlier.

In conclusion, our theoretical predictions when compared with relevant numerical simulations and experimental evidence from existing nonlinear optical pattern formation studies provide consistency in the former case and very good agreement when parameter values are chosen appropriately in the latter case. Hence this laser-injected atomic sodium optical ring-cavity problem, involving a single evolution equation for the real part of the intracavity field, is compatible with our long-range aim of employing the simplest reasonable natural science models that preserve the essential features of pattern formation and are still consistent with observation.

Problems

Problems 17.1 and 17.2 depend upon the function $F(A, A^*)$ defined in Section 17.1 when $\Delta = 0$ given by

$$G(A,A^*) = -(1+A)\left[1+i\theta+\frac{\beta}{1+\alpha(1+A)(1+A^*)}\right].$$

Define

$$\gamma_{n\ell} = \frac{1}{(n-\ell)!\ell!}\frac{\partial^n G(0,0)}{\partial A^{n-\ell}\partial A^{*\ell}}.$$

Then

$$\gamma_{10} = -\left(1+\frac{\beta}{D^2}+i\theta\right),\ \gamma_{11} = \frac{\alpha\beta}{D^2} \text{ with } D = 1+\alpha;$$

$$\gamma_{20} = \frac{\alpha\beta}{D^3},\ \gamma_{21} = \frac{2\alpha\beta}{D^3},\ \gamma_{22} = -\frac{\beta\alpha^2}{D^3};$$

$$\gamma_{30} = -\frac{\beta\alpha^2}{D^4},\ \gamma_{31} = \frac{\alpha\beta(1-2\alpha)}{D^4},\ \gamma_{32} = -\frac{3\beta\alpha^2}{D^4},\ \gamma_{33} = \frac{\beta\alpha^3}{D^4}.$$

17.1. Let A of (17.1.9) be of the form

$$A(x,y,t) = \varepsilon_1\mathcal{A}_1(x,y,t)+O(\varepsilon_1^2) \text{ where } |\varepsilon_1| << 1.$$

(a) Show that the linear problem discussed in Section 17.1 satisfies

$$\frac{\partial\mathcal{A}_1}{\partial t} = \gamma_{10}\mathcal{A}_1+\gamma_{11}\mathcal{A}_1^*+i\chi\nabla_2^2\mathcal{A}_1. \tag{P1.1a}$$

(b) Taking the complex conjugate of (P1.1a) obtain

$$\frac{\partial\mathcal{A}_1^*}{\partial t} = \gamma_{10}^*\mathcal{A}_1^*+\gamma_{11}\mathcal{A}_1-i\chi\nabla_2^2\mathcal{A}_1^*. \tag{P1.1b}$$

(c) Conclude that seeking a normal-mode solution of (P1.1) of the form (17.1.10a) yields the following linear homogeneous system of equations for k_1 and k_2:

$$\left[\sigma+1+\frac{\beta}{D^2}+i(\theta+\chi q^2)\right]k_1-\frac{\alpha\beta k_2}{D^2} = 0, \tag{P1.2a}$$

$$-\frac{\alpha\beta k_1}{D^2}+\left[\sigma+1+\frac{\beta}{D^2}-i(\theta+\chi q^2)\right]k_2 = 0. \tag{P1.2b}$$

(d) Deduce that the vanishing of the determinant of the matrix of coefficients of (P1.2) required by the nontriviality condition of (17.1.10b) results in the secular equation

$$\sigma^2+2\left[1+\frac{\beta}{(\alpha+1)^2}\right]\sigma+\left[1+\frac{\beta}{(\alpha+1)^2}\right]^2-\left[\frac{\alpha\beta}{(\alpha+1)^2}\right]^2+(\theta+\chi q^2)^2 = 0. \tag{P1.3a}$$

(e) Finally, demonstrate that (P1.3a) is equivalent to (17.1.11a) by employing the difference of squares

$$\left[1+\frac{\beta}{(\alpha+1)^2}\right]^2-\left[\frac{\alpha\beta}{(\alpha+1)^2}\right]^2=\left[1+\frac{\beta(1-\alpha)}{(\alpha+1)^2}\right]\left[1+\frac{\beta}{\alpha+1}\right]. \quad \text{(P1.3b)}$$

17.2. Rewrite the equation for A in Section 17.1 when $\Delta = 0$ as

$$A_t \sim \gamma_{10}A+\gamma_{11}A^*+\gamma_{20}A^2+\gamma_{21}AA^*+\gamma_{22}A^{*2}+\gamma_{30}A^3+\gamma_{31}A^2A^*+\gamma_{32}AA^{*2} \quad \text{(P2.1)}$$

$$+\gamma_{33}A^{*3}+i\chi\nabla_2^2A. \quad (17.4.1)$$

(a) Letting $A = R+iI$ with $\theta = -\chi q_c^2 = -1$ and representing $\gamma_{10} = \gamma_{10}^{(r)}+i\chi q_c^2$ where $\gamma_{10}^{(r)} = -(1+\beta/D^2)$ in (P2.1) obtain

$$R_t+iI_t \sim \gamma_{10}^{(r)}(R+iI)+\gamma_{11}(R-iI)+\gamma_{20}(R^2+2iRI-I^2)+\gamma_{21}(R^2+I^2) \quad \text{(P2.2)}$$

$$+\gamma_{22}(R^2-2iRI-I^2)+\gamma_{30}(R^3+3iR^2I-3RI^2-iI^3)+\gamma_{31}(R^3+iR^2I+RI^2+iI^3)$$
$$+\gamma_{32}(R^3-iR^2I+RI^2-iI^3)+\gamma_{33}(R^2-3iR^2I-3RI^2+iI^3)+i\chi(\nabla_2^2+q_c^2)(R+iI). \quad (17.4.2)$$

(b) Separating the real and imaginary parts of (P2.2) and retaining terms through third order on the right-hand side of the former and through second order on the corresponding side of the latter while noting I, in this context, actually represents a second-order effect (see below) show that

$$R_t \sim [\gamma_{10}^{(r)}+\gamma_{11}]R+(\gamma_{20}+\gamma_{21}+\gamma_{22})R^2+(\gamma_{30}+\gamma_{31}+\gamma_{32}+\gamma_{33})R^3-\chi(\nabla_2^2+q_c^2)I, \quad \text{(P2.3a)}$$

$$I_t \sim [\gamma_{10}^{(r)}-\gamma_{11}]I+\chi(\nabla_2^2+q_c^2)R. \quad \text{(P2.3b)}$$

(c) Upon simplifying the coefficients in these equations make the following identifications:

$$\gamma_{10}^{(r)}+\gamma_{11} = -1+\frac{\beta(\alpha-1)}{(\alpha+1)^2} = \sigma_R(\alpha,\beta), \quad \text{(P2.4a)}$$

$$\gamma_{10}^{(r)}-\gamma_{11} = -1-\frac{\beta}{\alpha+1} = \sigma_I(\alpha,\beta), \quad \text{(P2.4b)}$$

$$\gamma_{20}+\gamma_{21}+\gamma_{22} = \frac{\beta\alpha(3-\alpha)}{(\alpha+1)^3} = -\omega_0(\alpha,\beta), \quad \text{(P2.4c)}$$

$$\gamma_{30}+\gamma_{31}+\gamma_{32}+\gamma_{33} = \frac{\beta\alpha[(\alpha+1)^2-8\alpha]}{(\alpha+1)^4} = -\omega_1(\alpha,\beta). \quad \text{(P2.4d)}$$

(d) Finally, employing (P2.4) in (P2.3), assuming in addition that (P2.3b) satisfies the quasi-equilibrium condition

$$I \sim -\frac{\chi}{\sigma_I(\alpha,\beta)}(\nabla_2^2+q_c^2)R, \quad \text{(P2.5)}$$

and using this asymptotic relation to eliminate I from (P2.3a), deduce the modified Swift–Hohenberg equation (17.1.14)

$$R_t \sim \sigma_R(\alpha,\beta)R - \omega_0(\alpha,\beta)R^2 - \omega_1(\alpha,\beta)R^3 + \frac{\chi^2}{\sigma_I(\alpha,\beta)}(\nabla_2^2 + q_c^2)^2 R, \quad \text{(P2.6a)}$$

where

$$\omega_0(\alpha,\beta) = \frac{\beta\alpha(\alpha-3)}{(\alpha+1)^3}, \; \omega_1(\alpha,\beta) = \frac{\beta\alpha[8\alpha^2 - (\alpha+1)^2]}{(\alpha+1)^4}. \quad \text{(P2.6b)}$$

17.3. In their study of pattern formation driven by a pump field in a unidirectional ring cavity containing a so-called Kerr medium Scroggie et al. [113] deduced the following relationship between X_e and Y analogous to (17.1.5a) but with $\theta > 0$

$$Y = X_e[1 + i\eta(|X_e|^2 - \theta)^2] \quad \text{(P3.1a)}$$

where the parameter η equals $+1$ for a self-focusing Kerr medium and -1 for a self-defocusing one. Hence $|\eta| = 1$. A Kerr medium is a material the refractive index of which changes with an applied electric field. This change is proportional to the square of the intensity of the applied field and is positive for a self-focusing medium while it is negative for a self-defocusing one.

(a) Taking the complex conjugate of (P1.1a) deduce that

$$Y^* = Y = X_e^*[1 - i\eta(|X_e|^2 - \theta)^2]. \quad \text{(P3.1b)}$$

(b) Computing the product of (P3.1a) and (P3.1b) derive the analogous relationship to (17.1.5b)

$$YY^* = Y^2 = \alpha[1 + (\alpha - \theta)^2] = \alpha^3 - 2\theta\alpha^2 + (1 + \theta^2)\alpha \text{ where } \alpha = X_e X_e^* = |X_e|^2. \quad \text{(P3.2)}$$

Note that this relationship is independent of η by virtue of $|\eta| = 1$.

(c) Taking the derivative of (P3.2) with respect to α obtain that

$$\frac{dY^2}{d\alpha} = 3\alpha^2 - 4\theta\alpha + 1 + \theta^2. \quad \text{(P3.3)}$$

(d) Conclude from (P3.3) that $dY^2/d\alpha > 0$ for $0 < \theta < \theta_{\text{crit}}$ where α satisfies

$$3\alpha^2 - 4\theta\alpha + 1 + \theta^2 > 0, \quad \text{(P3.4a)}$$

and θ_{crit} corresponds to that θ-value such that the discriminant of the related quadratic equation

$$a\alpha^2 + b\alpha + c = 0 \text{ with } a = 3, b = -4\theta, c = 1 + \theta^2 \quad \text{(P3.4b)}$$

vanishes or

$$\mathcal{D} = b^2 - 4ac = 0. \tag{P3.4c}$$

(e) Calculating the discriminant of (P3.4) show that

$$\mathcal{D} = 4(\theta_{\text{crit}}^2 - 3) = 0 \text{ or } \theta_{\text{crit}} = \sqrt{3}. \tag{P3.5}$$

(f) Finally, from these results deduce that (P3.2) is single-valued in Y^2 for $0 < \theta < \sqrt{3}$ and S-shaped (exhibits optical bistability) for $\theta > \sqrt{3}$.

17.4. Consider an interfacial perturbation $\zeta(x, y, t)$ of an originally planar surface where $t \equiv$ time and $(x, y) \equiv$ a transverse Cartesian coordinate system, all of which are dimensionless. Upon seeking a real hexagonal planform solution for $\zeta(x, y, t)$ of the form of (17.3.1), assume it is found that the coefficients of the amplitude–phase equations satisfy

$$\sigma = 2(1 - \beta),\ a_0 = -\alpha,\ a_1 = a_2 = 4(1 + \alpha^2), \tag{P4.1a}$$

for $\beta > 0$ and $\alpha \in \mathbb{R}$, an experimentally controllable and a material parameter, respectively. Here the equivalence classes of potentially stable critical points of these equations defined in (17.3.11) have the following morphological identifications: I represents a planar interface; II, parallel ridges; and III$^{\pm}$, hexagonal arrays of elevated dots or circular holes, respectively. The orbital stability conditions for those critical points can be posed in terms of σ. Thus critical point I is stable in this sense for $\sigma < 0$ while the orbital stability behavior of II and III$^{\pm}$ which depends only on the sign of a_0 (since $2a_2 - a_1 = a_1 > 0$) and on the quantities σ_j for $j = -1, 1$, and 2 defined in (17.3.10) has been summarized in Table 17.3. Construct a morphological stability diagram analogous to Fig. 17.9 in the physically relevant portion of the α-β plane consistent with these predictions identifying those regions corresponding to stable dots, ridges, and holes, respectively.

Chapter 18
Vegetative Flat Dryland Rhombic Pattern Formation Driven by Root Suction

In this chapter, a rhombic planform nonlinear cross-diffusive instability analysis is applied to a particular interaction–diffusion plant-groundwater model system in an arid flat environment to investigate the formation of stationary two-dimensional vegetative patterns consisting of periodic arrays of spots (leopard bush), gaps (pearled bush), or bicontinuous mazes (labyrinthine tiger bush) observed to occur in such environments. That model contains a root suction effect, as a cross-diffusion term in the groundwater equation, and is deduced from a three-component system, containing surface water as well, by making a hydrological assumption to eliminate the latter dependent variable, similar in nature to the reduction procedure introduced in Chapter 7, to obtain the simplified slime mold aggregation model. In addition, the threshold-dependent paradigm introduced in Chapter 17 is employed to interpret stable rhombic patterns. These patterns are driven by root suction since the plant equation does not yield the required positive feedback necessary for the generation of standard Turing-type self-diffusive instabilities discussed in Chapter 8. The results of that analysis can be represented by plots in a root suction coefficient versus rainfall rate dimensionless parameter space. From those plots, regions corresponding to bare ground and vegetative patterns, consisting of isolated patches, rhombic arrays of pseudo-spots or -gaps separated by an intermediate rectangular state, and homogeneous distributions from low to high density, may be identified in this parameter space. Then, a morphological sequence of stable vegetative states is produced upon traversing an experimentally-determined root suction characteristic curve as a function of rainfall through these regions. Finally, that predicted sequence along a rainfall gradient is compared with observational evidence relevant to the occurrence of leopard, pearled, or labyrinthine tiger bush vegetative patterns, used to motivate an aridity classification scheme, and placed in the context of some recent biological nonlinear pattern formation studies. This approach follows that of Chaiya et al. [17]. There are four problems: The first two fill in some details of this analysis, while the last two examine critical conditions for the onset of instability for a related vegetation model and rhombic pattern formation for an ion-sputtered solid surface erosion model.

© Springer International Publishing AG, part of Springer Nature 2017

D. J. Wollkind and B. J. Dichone, *Comprehensive Applied Mathematical Modeling in the Natural and Engineering Sciences*,
https://doi.org/10.1007/978-3-319-73518-4_18

18.1 Basic Governing Equations and a Simplified Model

We begin with a brief sketch of the reduction procedure required to obtain our governing plant-groundwater model system. Recently, von Hardenberg et al. [126] devised a plant-groundwater (sometimes called soil water) interaction–diffusion system to model self-organized vegetative pattern formation in arid environments (reviewed by Rietkerk et al., [104]). Here the positive feedback for an activator consumer (*e.g.*, plants) in the plant equation and the self-diffusivity advantage for an inhibitory limiting resource (*e.g.*, groundwater) provided the necessary conditions for the onset of Turing [121] pattern formation. von Hardenberg et al.'s [126] model also included the effect of plant root suction by adding a cross-diffusion term in their groundwater equation. Rietkerk et al. [103] performed numerical simulations using reflecting boundary conditions on a similar interaction–diffusion model system consisting of three partial differential equations describing the spatiotemporal behavior of plant density, soil water, and surface water, respectively, but excluding the root suction cross-diffusion term in their soil water equation. That model had been carefully developed by HilleRisLambers et al. [48] for a flat semiarid grazing system. We wish to formulate an interaction–diffusion model system for $N(X,Y,\tau) \equiv$ plant biomass density gm/m^2 and $W(X,Y,\tau) \equiv$ ground (soil) water content (mm of depth), where $(X,Y) \equiv$ two-dimensional coordinate system (m, m) and $\tau \equiv$ time (d), based upon the interaction terms of Rietkerk et al. [103] and the diffusion terms of von Hardenberg et al. [126], defined on a flat unbounded arid environment.

Toward that end, we first introduce the auxiliary-dependent variable $O(X,Y,\tau) \equiv$ surface water content (mm of depth) and the coupled interaction–diffusion model system given by

$$\frac{\partial N}{\partial \tau} = Q^{(N)} - \nabla_2 \bullet \boldsymbol{J}^{(N)}, \; \frac{\partial W}{\partial \tau} = Q^{(W)} - \nabla_2 \bullet \boldsymbol{J}^{(W)}, \; \frac{\partial O}{\partial \tau} = Q^{(O)} - \nabla_2 \bullet \boldsymbol{J}^{(O)}, \quad (18.1.1a)$$

where [103, 126]

$$Q^{(N)} = \frac{c g_M WN}{W+k_1} - dN, \; Q^{(W)} = i_M O \frac{N+k_2 f}{N+k_2} - \frac{g_M WN}{W+k_1} - rW,$$
$$Q^{(O)} = R - i_M O \frac{N+k_2 f}{N+k_2}, \quad (18.1.1b)$$

$$\boldsymbol{J}^{(N)} = -D_N \nabla_2 N, \; \boldsymbol{J}^{(W)} = -D_W \nabla_2 (W - \rho N), \; \boldsymbol{J}^{(O)} = -D_O \nabla_2 O, \quad (18.1.1c)$$

with $\nabla_2 \equiv (\partial/\partial X, \partial/\partial Y)$ and $\nabla_2 \bullet \boldsymbol{J} = (\partial J_1/\partial X, \partial J_2/\partial Y)$ for $\boldsymbol{J} = (J_1, J_2)$. Here $g_M \equiv$ maximum specific water uptake rate by the plants, $c \equiv$ conversion of water uptake by the plants to plant growth, $d \equiv$ specific loss rate of plant density due to mortality, $k_1 \equiv$ half-saturation soil water constant relevant to specific plant growth and water uptake, $r \equiv$ specific loss rate of soil water due to evaporation and drainage, $R \equiv$ rainfall rate, $i_M \equiv$ maximum specific water infiltration rate, $k_2 \equiv$ plant saturation shaping constant for water infiltration, $f \equiv$ fraction of maximum specific water infiltration rate in the absence of plants, $D_N \equiv$ dispersal coefficient for plants, $D_W \equiv$ diffusion coefficient for ground water, $D_O \equiv$ diffusion coefficient for surface water, $D_S \equiv$ coefficient of plant root suction, and $\rho \equiv D_S/D_W$.

For the sake of model analysis, HilleRisLambers et al. [48] introduced a quasi-steady-state approximation in (18.1.1a) by taking $\partial O/\partial \tau \equiv 0$. We next, after Keller and Segel's [59] employment of Haldane's assumption in their slime mold problem (see Chapter 7), simplify our model system (18.1.1) by assuming that the surface water is in hydrological equilibrium and making the quasi-*stationary* approximation that

$$Q^{(O)} \equiv 0 \Rightarrow R = i_M O \frac{N + k_2 f}{N + k_2}. \tag{18.1.2}$$

Then, introducing (18.1.2) into (18.1.1) by replacing the rate of infiltration term in $Q^{(W)}$ with R, we obtain the final formulation of our interaction–diffusion plant-groundwater model system

$$\frac{\partial N}{\partial \tau} = F(W,N) + D_N \nabla_2^2 N, \; \frac{\partial W}{\partial \tau} = G(N,W) + D_W(\nabla_2^2 W - \rho \nabla_2^2 N) \tag{18.1.3a}$$

where

$$\nabla_2^2 \equiv \nabla_2 \bullet \nabla_2, \; F(N,W) = \frac{c g_M W N}{W + k_1} - dN, \; G(N,W) = R - \frac{g_M W N}{W + k_1} - rW. \tag{18.1.3b}$$

Our main purpose in doing so is to devise a model system of this sort that demonstrates root suction alone can generate the two-dimensional vegetative patterns (*e.g.*, leopard, pearled, and labyrinthine tiger bush) occurring in arid flat environments as described by Rietkerk et al. [104]. Here arid or semiarid refers to environments where the yearly average rate of rainfall is less than the corresponding rate of evaporation and water is a limiting resource for plant growth.

18.2 Equilibrium Points and their Linear Stability

There exist two equilibrium points of model system (18.1.3)

$$N \equiv N_0, \; W \equiv W_0 \tag{18.2.1a}$$

satisfying

$$F(N_0, W_0) = G(N_0, W_0) = 0 \tag{18.2.1b}$$

given by

$$N_0 = 0, \; W_0 = \frac{R}{r}; \tag{18.2.1c}$$

and

$$N_0 = N_e = \frac{c}{d}\left(R - \frac{rk_1}{\delta - 1}\right), W_0 = W_e = \frac{k_1}{\delta - 1} \text{ with } \delta = \frac{cg_M}{d}. \tag{18.2.1d}$$

Note that (18.2.1c) which exists for all parameter values corresponds to a bare ground or no vegetation situation while (18.2.1d) which only exists for $N_e, W_e > 0$ or, equivalently,

$$\delta > 1 + \frac{rk_1}{R}, \tag{18.2.2}$$

corresponds to a community equilibrium point or a state exhibiting a nontrivial homogenous vegetative distribution. In this context, we adopt the far-field condition that

$$N, W \text{ remain bounded as } X^2 + Y^2 \to \infty. \tag{18.2.3}$$

We next wish to examine the linear stability behavior of these critical points and shall proceed sequentially. That is we begin with (18.2.1c) by considering a normal mode solution to system (18.1.3) and boundary condition (18.2.3) of the form

$$N(X,Y,\tau) = 0 + \varepsilon_1 N_{11}\cos(QX)e^{\Sigma\tau} + O(\varepsilon_1^2), \tag{18.2.4a}$$

$$W(X,Y,\tau) = \frac{R}{r} + \varepsilon_1 W_{11}\cos(QX)e^{\Sigma\tau} + O(\varepsilon_1^2), \tag{18.2.4b}$$

$$|\varepsilon_1| << 1, N_{11}^2 + W_{11}^2 \neq 0, Q \geq 0, \tag{18.2.4c}$$

and find that

$$\Sigma = \Sigma_1 = \frac{(\delta - 1)R - rk_1}{R + rk_1} - D_N Q^2 \text{ or } \Sigma = \Sigma_2 = -r - D_W Q^2. \tag{18.2.4d}$$

Upon examination of (18.2.4d) it follows that $\Sigma_2 < 0$ identically and $\Sigma_1 < 0$ provided

$$\delta < 1 + \frac{rk_1}{R}. \tag{18.2.4e}$$

Hence we can conclude that (18.2.4e) represents the linear stability criterion for this bare ground equilibrium point. Since the community equilibrium point of (18.2.1d) does not exist for these parameter values by virtue of (18.2.2), an exchange of stabilities between the two critical points of system (18.1.3) occurs at $\delta = 1 + rk_1/R$. The primary focus of our research is on the stability behavior of this community equilibrium point. Now introducing the nondimensional variables and parameters

$$(x,y) = \left(\frac{d}{D_W}\right)^{1/2}(X,Y), t = \tau d, n = \frac{N}{N_e}, w = \frac{W}{W_e}, \tag{18.2.5a}$$

$$\gamma = \frac{r}{d}, \alpha = \frac{(\delta - 1)R - rk_1}{k_1 d}, \mu = \frac{D_N}{D_W}, \beta = c\rho, \tag{18.2.5b}$$

we transform system (18.1.3) and far-field condition (18.2.3) into

$$\frac{\partial n}{\partial t} = \Theta(n,w) + \mu\nabla^2 n, \quad \frac{\partial w}{\partial t} = \Psi(n,w) + \nabla^2 w - \alpha\beta\nabla^2 n, \tag{18.2.6a}$$

where $\nabla^2 \equiv \partial^2/\partial x^2 + \partial^2/\partial y^2$ and

$$\Theta(n,w) = \frac{\delta wn}{w+\delta-1} - n, \quad \Psi(n,w) = \alpha + \gamma(1-w) - \frac{\alpha\delta wn}{w+\delta-1}; \tag{18.2.6b}$$

while

$$n, w \text{ remain bounded as } x^2 + y^2 \to \infty. \tag{18.2.6c}$$

Observe that the equilibrium point in question corresponds to

$$n = w \equiv 1 \tag{18.2.7a}$$

in our dimensionless formulation since

$$\Theta(1,1) = \Psi(1,1) = 0. \tag{18.2.7b}$$

Here we are concerned with the stability of this solution to initially infinitesimal one-dimensional perturbations. Toward that end, we consider a reduced form of our basic system with $\nabla^2 \equiv \partial^2/\partial x^2$; seek a normal mode solution to it satisfying

$$n(x,t) = 1 + \varepsilon_1 n_{11}\cos(qx)e^{\sigma t} + O(\varepsilon_1^2), \quad w(x,t) = 1 + \varepsilon_1 w_{11}\cos(qx)e^{\sigma t} + O(\varepsilon_1^2), \tag{18.2.8a}$$

where $q \geq 0$ and σ are the wavenumber and growth rate of the linear perturbation quantities (*i.e.*, $|\varepsilon_1| << 1$) with $|n_{11}|^2 + |w_{11}|^2 \neq 0$ for the constants n_{11} and w_{11}; and obtain

$$\mathcal{D}(\sigma;q^2) = \sigma^2 + [(1+\mu)q^2 - \psi_{01}]\sigma + \mu q^4 - (\mu\psi_{01} + \alpha\beta\theta_{01})q^2 - \theta_{01}\psi_{10} = 0 \tag{18.2.8b}$$

where

$$\theta_{ps} = \frac{1}{p!s!}\frac{\partial^{p+s}}{\partial n^p \partial w^s}\Theta(1,1), \quad \psi_{ps} = \frac{1}{p!s!}\frac{\partial^{p+s}}{\partial n^p \partial w^s}\Psi(1,1) \tag{18.2.8c}$$

which are tabulated below for the relevant values of p and s.

Table 18.1 Interaction expansion coefficients.

$\theta_{10} = \theta_{20} = \theta_{21} = \theta_{30} = 0$, $\theta_{01} = \theta_{11} = \Delta = 1 - \delta^{-1}$, $\theta_{02} = \theta_{12} = -\Delta\delta^{-1}$, $\theta_{03} = \Delta\delta^{-2}$
$\psi_{10} = -\alpha$, $\psi_{01} = -\gamma - \alpha\Delta$, $\psi_{20} = \psi_{21} = \psi_{30} = 0$, $\psi_{02} = \psi_{12} = \alpha\Delta\delta^{-1}$, $\psi_{03} = -\alpha\Delta\delta^{-2}$

Note that in (18.2.8b) we have implicitly made use of the fact that $\theta_{10} = 0$. Upon substitution of these expansion coefficients from Table 18.1 into (18.2.8b), we obtain the explicit secular equation for σ

$$\sigma^2 + [(1+\mu)q^2 + \gamma + \alpha\Delta]\sigma + \mu q^4 + [\mu\gamma + \alpha(\mu-\beta)\Delta]q^2 + \alpha\Delta = 0, \tag{18.2.8d}$$

where $\Delta > 0$ for those δ satisfying (18.2.2). Thus, since quadratics of the form of (18.2.8d) have roots with negative real parts provided their coefficients are positive, we can conclude that the community equilibrium point is stable to linear homogeneous perturbations for which $q^2 = 0$. Further observe that in the absence of plant root suction ($\beta = 0$) this equilibrium point would be linearly stable to heterogeneous perturbations, for which $q^2 > 0$, as well. When root suction is considered, to guarantee the onset of such a cross-diffusive instability we require that

$$\beta > \beta_0(q^2;\alpha,\gamma,\Delta,\mu) = \frac{\mu}{\alpha\Delta}q^2 + \frac{1}{q^2} + \frac{\mu\gamma}{\alpha\Delta} + \mu. \tag{18.2.9}$$

For fixed values of α, γ, Δ, and μ, the curve $\beta = \beta_0(q^2;\alpha,\gamma,\Delta,\mu)$ in the $q^2 \geq 0$ and $\beta \geq 0$ portion of the q^2-β plane is marginal in the sense that it separates the linearly stable region where $0 \leq \beta < \beta_0(q^2;\alpha,\gamma,\Delta,\mu)$ from the unstable region of (18.2.9). The marginal stability curve, the linearly stable region, and the unstable region can be characterized by $\sigma_0 = 0$, $\mathrm{Re}(\sigma_0) < 0$, and $\sigma_0 > 0$, respectively, where σ_0 corresponds to that root of (18.2.8d) having the largest real part. This marginal stability curve has a minimum point at (q_c^2, β_c) given by (see Problem 18.1)

$$q_c^2 = \sqrt{\frac{\Delta}{\mu}}\sqrt{\alpha},\ \beta_c = 2\sqrt{\frac{\mu}{\Delta}}\left(\frac{1}{\sqrt{\alpha}}\right) + \left(\frac{\gamma\mu}{\Delta}\right)\left(\frac{1}{\alpha}\right) + \mu. \tag{18.2.10}$$

Hence, when $0 < \beta < \beta_c$ there exist no squared wavenumbers q^2 corresponding to growing disturbances while when $\beta > \beta_c$ there exists a band of such wavenumbers centered about q_c^2 for which $\sigma_0 > 0$ (see Fig. 18.1).

In particular, taking $q^2 \equiv q_c^2$, as we shall do in our weakly nonlinear stability analyses of the next two sections, it is possible assuming α, γ, μ, and Δ are fixed, to represent $\sigma_0 = \sigma_c(\beta)$ such that, for β sufficiently close to β_c, σ_c is still real when $\beta < \beta_c$. Then $\sigma_c(\beta)$ satisfies the conditions that $\sigma_c(\beta) < 0$ when $\beta < \beta_c$, $\sigma_c(\beta_c) = 0$, and $\sigma_c(\beta) > 0$ when $\beta > \beta_c$ (see Fig. 18.1). Hence, the locus $\beta = \beta_c(\alpha;\gamma,\Delta,\mu)$ with $\beta_c(\alpha;\gamma,\Delta,\mu)$ defined by (18.2.10), is a marginal stability curve in the α-β plane, when γ, Δ, and μ are fixed. We plot that locus in Fig. 18.2 for $0 < \alpha \leq 3.5$, $\gamma = 1$, $\Delta = 0.6$, and $\mu = 0.01$, corresponding to the typical parameter values [103, 126]

$$g_M = 0.05\frac{\text{mm/d}}{\text{gm/m}^2},\ c = 10\frac{\text{gm/m}^2}{\text{mm}},\ d = r = \frac{0.2}{\text{d}},\ k_1 = 5\text{ mm}, \tag{18.2.11a}$$

$$\frac{2}{3}\frac{\text{mm}}{\text{d}} < R \leq 3\frac{\text{mm}}{\text{d}},\ D_N = 0.1\frac{\text{m}^2}{\text{d}},\ D_W = 10\frac{\text{m}^2}{\text{d}}. \tag{18.2.11b}$$

Note for these values $\delta = 2.5$, $\gamma = 1$, and $\alpha = 1.5R - 1$.

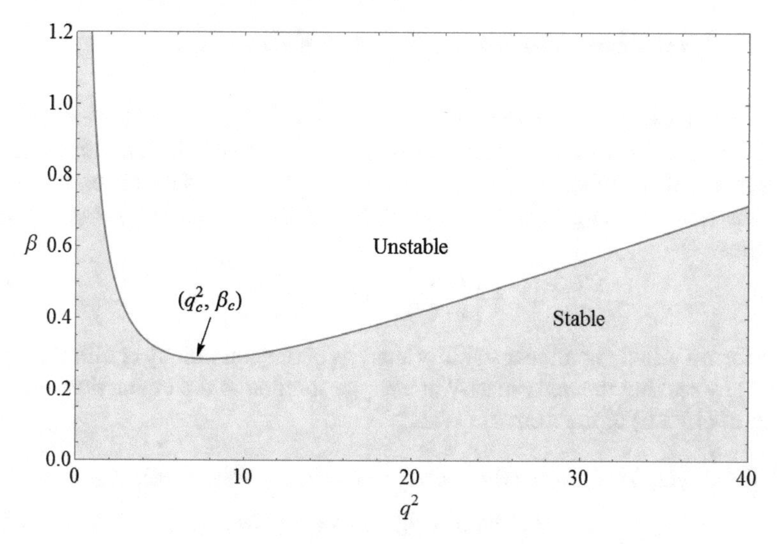

Fig. 18.1 Plot of $\beta = \beta_0(q^2; \alpha, \gamma, \Delta, \mu)$ of (18.2.9) for $\alpha = \gamma = 1$, $\Delta = 0.6$, $\mu = 0.01$.

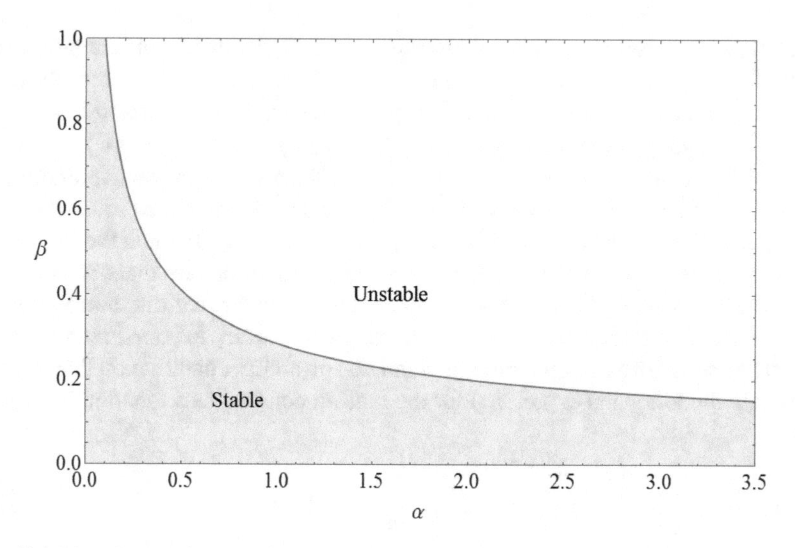

Fig. 18.2 Plot of $\beta = \beta_c(\alpha; \gamma, \Delta, \mu)$ of (18.2.10) for $0 < \alpha \leq 3.5$, $\gamma = 1$, $\Delta = 0.6$, $\mu = 0.01$.

18.3 Striped Planform Stuart-Watson Expansion

In the previous section we determined the critical conditions for the onset of cross-diffusive instabilities that can occur provided $\alpha > 0$. To predict the long-time behavior and spatial pattern of such growing disturbances it is necessary for us to take the nonlinear terms of our governing model equations into account. Defining the vector quantities

$$\mathbf{v}(x,t) = \begin{bmatrix} n(x,t) \\ w(x,t) \end{bmatrix} \text{ and } \mathbf{v}_{jk} = \begin{bmatrix} n_{jk} \\ w_{jk} \end{bmatrix}, \tag{18.3.1a}$$

we perform a weakly nonlinear stability analysis of our community equilibrium point (18.1.1) by seeking the real Stuart-Watson type solution to the interaction–diffusion system of (18.2.6) of the form of (17.2.2)

$$\mathbf{v}(x,t) \sim \mathbf{v}_{00} + A_1(t)\mathbf{v}_{11}\cos(q_c x) + A_1^2(t)[\mathbf{v}_{20} + \mathbf{v}_{22}\cos(2q_c x)] \\ + A_1^3(t)[\mathbf{v}_{31}\cos(q_c x) + \mathbf{v}_{33}\cos(3q_c x)], \tag{18.3.1b}$$

where $n_{00} = w_{00} = 1$, and the amplitude function $A_1(t)$ satisfies the Landau equation

$$\frac{dA_1}{dt} \sim \sigma A_1 - a_1 A_1^3. \tag{18.3.1c}$$

Then, after substituting this solution into (18.2.6) and expanding its interaction terms in a Taylor series about $A_1 \equiv 0$, we obtain a sequence of vector systems, one for each pair of values of j and k which corresponds to a nonzero term of the form $A_1^j(t)\cos(kq_c x)$ appearing explicitly in (18.3.1). The system for $j = k = 0$ is satisfied identically since $\mathbf{v}_{00}$ represents the uniform homogeneous solution. The $O(A_1)$ system for $j = k = 1$ is equivalent to the linear problem of the previous section with $q \equiv q_c$, $\theta_{01}w_{11} = (\sigma\theta_{10} + \mu q_c^2)n_{11}$, $n_{11} = 1$, and $\sigma = \sigma_c(\beta)$ while the two $O(A_1^2)$ systems for $j = 2$ and $k = 0$ or 2 can be solved in a straightforward manner. Although there are also two $O(A_1^3)$ systems, we need only consider that one corresponding to $j = 3$ and $k = 1$ which contains the Landau constant a_1 for our Fredholm-type method of solvability. Employing this standard solvability condition on that system, we obtain the following expression for the Landau constant as a function of α, γ, δ, and μ

$$a_1 = \left.\frac{(\psi_{01} - q_c^2)b_{31}^{(1)} - \theta_{01}b_{31}^{(2)}}{(\mu+1)q_c^2 - \theta_{10} - \psi_{01}}\right|_{\beta=\beta_c} = a_1(\alpha;\gamma,\delta,\mu), \tag{18.3.2a}$$

where the components of

$$\boldsymbol{b}_{jk} = \begin{bmatrix} b_{jk}^{(1)} \\ b_{jk}^{(2)} \end{bmatrix}, \tag{18.3.2b}$$

for $j = 3$ and $k = 1$ along with the solutions for the two $O(A_1^2)$ systems are given by

$$n_{2k} = \frac{(2\sigma_c - \psi_{01} + k^2 q_c^2) b_{2k}^{(1)} + \theta_{01} b_{2k}^{(2)}}{\mathcal{D}(2\sigma_c; k^2 q_c^2)}, \quad w_{2k} = \frac{\psi_{10} b_{2k}^{(1)} + (2\sigma_c - \theta_{10} + \mu k^2 q_c^2) b_{2k}^{(2)}}{\mathcal{D}(2\sigma_c; k^2 q_c^2)}, \tag{18.3.2c}$$

where

$$\boldsymbol{b}_{2k} = \frac{1}{2}(n_{11}^2 \boldsymbol{v}_{20} + n_{11} w_{11} \boldsymbol{v}_{11} + w_{11}^2 \boldsymbol{v}_{02}) \text{ for } k = 0 \text{ and } 2 \tag{18.3.2d}$$

with

$$\boldsymbol{v}_{jk} = \begin{bmatrix} \theta_{jk} \\ \psi_{jk} \end{bmatrix}; \tag{18.3.2e}$$

and

$$\boldsymbol{b}_{31} = n_{11}(2n_{20} + n_{22})\boldsymbol{v}_{20} + \left[n_{11}\left(w_{20} + \frac{w_{22}}{2} \right) + w_{11}\left(n_{20} + \frac{n_{22}}{2} \right) \right] \boldsymbol{v}_{11}$$
$$+ w_{11}(2w_{20} + w_{22})\boldsymbol{v}_{02} + \frac{3}{4}(n_{11}^3 \boldsymbol{v}_{30} + n_{11}^2 w_{11} \boldsymbol{v}_{21} + n_{11} w_{11}^2 \boldsymbol{v}_{12} + w_{11}^3 \boldsymbol{v}_{30}). \tag{18.3.2f}$$

As usual the stability behavior of the Landau equation (18.3.1c) is dependent upon the sign of a_1. Thus, to ascertain this behavior we must analyze the formula for $a_1(\alpha; \gamma, \delta, \mu)$ given by (18.3.2). Hence, we plot $a_1(\alpha; \gamma, \delta, \mu)$ in Fig. 18.3 for the same α-domain and choice of parameter values as used in Fig. 18.2 (note $\Delta = 0.6$ corresponds to $\delta = 2.5$).

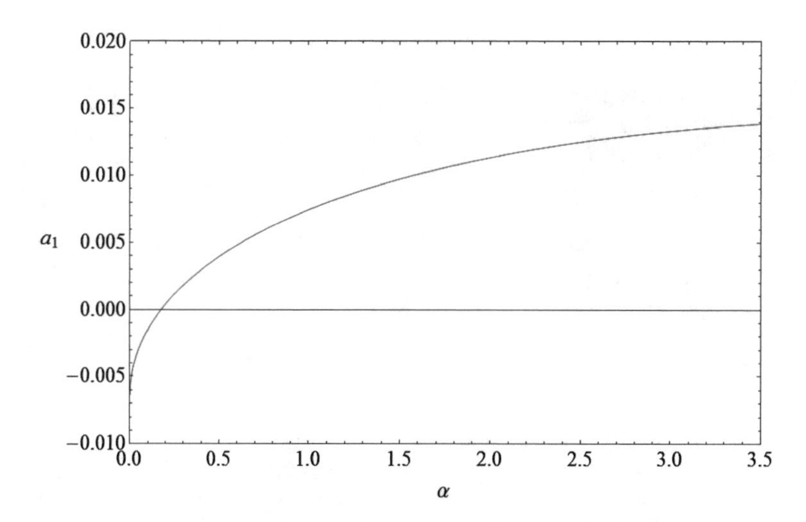

Fig. 18.3 Plot of $a_1(\alpha; \gamma, \delta, \mu)$ of (18.3.2) for $0 < \alpha \leq 3.5$, $\gamma = 1$, $\delta = 2.5$, $\mu = 0.01$.

From this figure we observe that $a_1(\alpha; 1, 2.5, 0.01)$ has a zero at $\alpha = 0.172$ such that

$$\begin{cases} a_1 < 0, & \text{for } 0 < \alpha < \alpha_1 \equiv 0.172, \\ a_1 > 0, & \text{for } \alpha > \alpha_1 \equiv 0.172. \end{cases} \tag{18.3.2g}$$

Given these conditions, we may conclude that:

(i) For $0 < \beta < \beta_c$ and $\alpha > \alpha_1$, the undisturbed state $A_1 = 0$ is stable, yielding a uniform homogeneous vegetative pattern $n(x,t) \to 1$ as $t \to \infty$.

(ii) For $\beta > \beta_c$ and $\alpha > \alpha_1$, $A_1 = A_e = (\sigma_c/a_1)^{1/2} > 0$ is stable, yielding a periodic one-dimensional vegetative pattern consisting of stationary parallel stripes

$$n(x,t) \to n_e(x) \sim 1 + A_e \cos\left(\frac{2\pi x}{\lambda_c}\right) \text{ as } t \to \infty \tag{18.3.3a}$$

of characteristic wavelength

$$\lambda_c = \frac{2\pi}{q_c} \text{ and } \lambda_c^* = \left(\frac{D_W}{d}\right)^{1/2} \lambda_c \tag{18.3.3b}$$

in dimensionless and dimensional variables, respectively. These supercritical stripes are plotted in Fig. 18.4a where regions of higher density ($n > 1$) appear dark and those of lower density ($n < 1$) appear light. When $0 < \alpha < \alpha_1$, that bifurcation is subcritical.

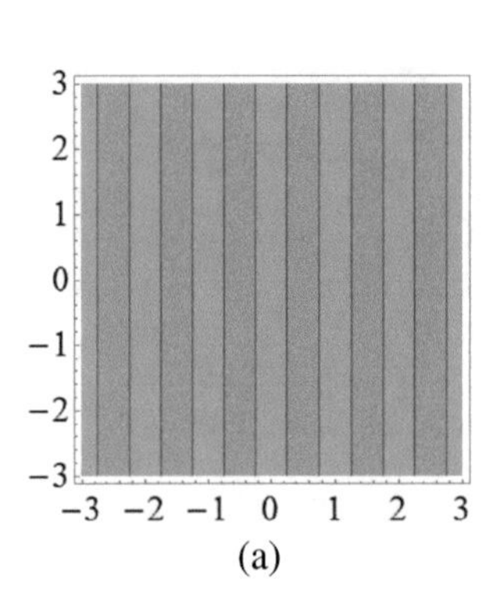

(a)

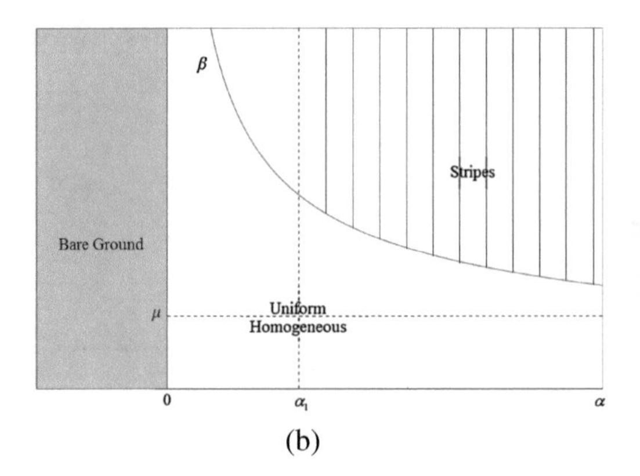

(b)

Fig. 18.4 (a) Contour plot of the supercritical stripes of (18.3.3) where the x-variable is measured in units of λ_c. (b) Schematic stability diagram in the α-β plane for our one-dimensional interaction–diffusion model system denoting the predicted vegetative patterns. Here, the lower and upper bounds on α correspond to $R = 0$ and 3, respectively, measured in units of mm/d.

Finally we synthesize the one-dimensional pattern formation results of this and the previous section in the α-β plane of Fig. 18.4b. Here, we have plotted simultaneously the cross-diffusive instability boundary curve $\beta = \beta_c$ of Fig. 18.2 and the vertical lines $\alpha = 0$, $\alpha = \alpha_1$, with the regions $\alpha < 0$; $\alpha > \alpha_1$, $0 < \beta < \beta_c$; and $\alpha > \alpha_1$, $\beta > \beta_c$;

corresponding to bare ground, uniform homogeneous distributions of vegetation, and stationary striped vegetative patterns, respectively, being identified in that parameter space. In this context, observe that the line $\beta = \mu$ serves as a horizontal asymptote for $\beta = \beta_c$ while $\delta \left\{ \begin{matrix} < \\ = \\ > \end{matrix} \right\} 1 + rk_1/R$ corresponds to $\alpha \left\{ \begin{matrix} < \\ = \\ > \end{matrix} \right\} 0.$

18.4 Rhombic Planform Stuart-Watson Expansion

Wishing to refine our one-dimensional predictions summarized in Fig. 18.4b and to investigate the possibility of occurrence of the two-dimensional vegetative patterns mentioned earlier, we next consider a rhombic planform solution of system (18.2.6) of the form [143]

$$\begin{aligned} n(x,y,t) \sim\ & n_{0000} + A_1(t)n_{1010}\cos(q_c x) + B_1(t)n_{0101}\cos(q_c z) \\ & + A_1^2(t)[n_{2000} + n_{2020}\cos(2q_c z)] \\ & + A_1(t)B_1(t)[n_{1111}\cos(q_c\{x+z\}) + n_{111(-1)}\cos(q_c\{x-z\})] \\ & + B_1^2(t)[n_{0200} + n_{0202}\cos(2q_c z)] \\ & + A_1^3(t)[n_{3010}\cos(q_c x) + n_{3030}\cos(3q_c x)] \\ & + A_1^2(t)B_1(t)[n_{2101}\cos(q_c z) + n_{2121}\cos(q_c\{2x+z\}) + n_{212(-1)}\cos(q_c\{x-z\})] \\ & + A_1(t)B_1^2(t)[n_{1210}\cos(q_c x) + n_{1212}\cos(q_c\{x+2z\}) + n_{121(-2)}\cos(q_c\{x-2z\})] \\ & + B_1^3(t)[n_{0301}\cos(q_c z) + n_{0303}\cos(3q_c z)], \end{aligned} \quad (18.4.1a)$$

where

$$n_{0000} = 1,\ z = x\cos(\varphi) + y\sin(\varphi) \quad (18.4.1b)$$

with an analogous expansion for $w(x,y,t)$, such that

$$\frac{dA_1}{dt} \sim \sigma A_1 - A_1(a_1A_1^2 + b_1B_1^2),\ \frac{dB_1}{dt} \sim \sigma B_1 - B_1(b_1A_1^2 + a_1B_1^2). \quad (18.4.1c)$$

Here we are employing the notation $n_{j\ell km}$ for the coefficient of each term in (18.4.1a) of the form $A_1^j(t)B_1^\ell(t)\cos(q_c\{kx+mz\})$. Then substituting this rhombic planform solution of (18.4.1) into system (18.2.6), we again obtain a sequence of problems, each of which corresponds to one of these terms. Solving those problems we find that

$$\sigma = \sigma_c(\beta),\ a_1 = a_1(\alpha;\gamma,\delta,\mu), \quad (18.4.2)$$

while applying the same method of analysis, as employed for deducing (18.3.2), to the $j = 2$, $\ell = m = 1$, $k = 0$ system yields the Fredholm-type solvability condition for the second rhombic planform third-order Landau constant

$$b_1 = \left.\frac{(\psi_{01} - q_c^2)b_{2101}^{(1)} - \theta_{01}b_{2101}^{(2)}}{(\mu+1)q_c^2 - \theta_{10} - \psi_{01}}\right|_{\beta=\beta_c} = b_1(\alpha, \varphi; \gamma, \delta, \mu), \tag{18.4.3a}$$

where the components of

$$\boldsymbol{b}_{2101} = \begin{bmatrix} b_{2101}^{(1)} \\ b_{2101}^{(2)} \end{bmatrix}, \tag{18.4.3b}$$

as well as the solutions for the relevant second-order systems are given by

$$n_{j0k0} = n_{0j0k} = n_{jk},\ w_{j0k0} = w_{0j0k} = w_{jk}; \tag{18.4.3c}$$

$$n_{111m} = \frac{\{2\sigma_c - \psi_{01} + [1 + m^2 + 2m\cos(\varphi)]q_c^2\}b_{111m}^{(1)} + \theta_{01}b_{111m}^{(2)}}{\mathcal{D}(2\sigma_c; [1 + m^2 + 2m\cos(\varphi)]q_c^2)}, \tag{18.4.3d}$$

$$w_{111m} = \frac{\psi_{10}b_{111m}^{(1)} + \{2\sigma_c - \theta_{10} + \mu[1 + m^2 + 2m\cos(\varphi)]q_c^2\}b_{111m}^{(2)}}{\mathcal{D}(2\sigma_c; [1 + m^2 + 2m\cos(\varphi)]q_c^2)}, \tag{18.4.3e}$$

where

$$\boldsymbol{b}_{111m} = \begin{bmatrix} b_{111m}^{(1)} \\ b_{111m}^{(2)} \end{bmatrix} = n_{1010}n_{0101}\boldsymbol{v}_{20} + \frac{1}{2}(n_{1010}w_{0101} + n_{0101}w_{1010})\boldsymbol{v}_{11} + w_{1010}w_{0101}\boldsymbol{v}_{22} \tag{18.4.3f}$$

for $m = \pm 1$; and

$$\begin{aligned} \boldsymbol{b}_{2101} &= [n_{1010}\{n_{1111} + n_{111(-1)}\} + 2n_{0101}n_{2000}]\boldsymbol{v}_{20} \\ &+ \left[\frac{1}{2}n_{1010}\{w_{1111} + w_{111(-1)}\} + n_{0101}w_{2000} + w_{0101}n_{2000} + \frac{1}{2}w_{1010}\{n_{1111} + n_{111(-1)}\}\right]\boldsymbol{v}_{11} \\ &+ [w_{1010}\{w_{1111} + w_{111(-1)}\} + 2w_{0101}w_{2000}]\boldsymbol{v}_{02} \\ &+ \frac{3}{2}n_{1010}^2 n_{0101}\boldsymbol{v}_{30} + \left[\frac{1}{2}n_{1010}^2 w_{0101} + n_{1010}n_{0101}w_{1010}\right]\boldsymbol{v}_{21} \\ &+ \left[n_{1010}w_{1010}w_{0101} + \frac{1}{2}n_{0101}w_{1010}^2\right]\boldsymbol{v}_{12} + \frac{3}{2}w_{1010}^2 w_{0101}\boldsymbol{v}_{03}. \end{aligned} \tag{18.4.3g}$$

Having developed these formulae for its growth rate and Landau constants, we now turn our attention to the rhombic planform amplitude equations (18.4.1c), which possess the following equivalence classes of critical points (A_0, B_0) where $A_0, B_0 \geq 0$:

$$\text{I: } A_0 = B_0 = 0; \text{ II: } A_0^2 = \frac{\sigma}{a_1}, B_0 = 0; \text{V: } A_0 = B_0 \text{ with } A_0^2 = \frac{\sigma}{a_1 + b_1}. \tag{18.4.4a}$$

Assuming that $a_1, a_1 + b_1 > 0$ and investigating the stability of these critical points one finds that (see Problem 18.2):

$$\text{I is stable for } \sigma < 0; \text{ II, for } \sigma > 0, b_1 > a_1; \text{ and V, for } \sigma > 0, a_1 > b_1. \tag{18.4.4b}$$

Note that I and II, as in the one-dimensional analysis of the previous section, represent the uniform homogeneous and supercritical striped states, respectively, while V can be identified with a rhombic pattern possessing characteristic angle φ (see Fig. 18.7 and [143]). In the next section we shall use these criteria to refine our one-dimensional predictions of Fig. 18.4b relevant to the former states due to the presence of the latter. Toward that end, we examine the sign of $a_1 \pm b_1$. We illustrate this procedure by defining the ratio of Landau constants [40]

$$\eta(\alpha, \varphi; \gamma, \delta, \mu) = \frac{b_1(\alpha, \varphi; \gamma, \delta, \mu)}{a_1(\alpha; \gamma, \delta, \mu)} \tag{18.4.5}$$

and plotting that quantity versus φ in Fig. 18.5 for a fixed value of α, namely $\alpha = 3$, and with the other parameters taking on their values of Fig. 18.3.

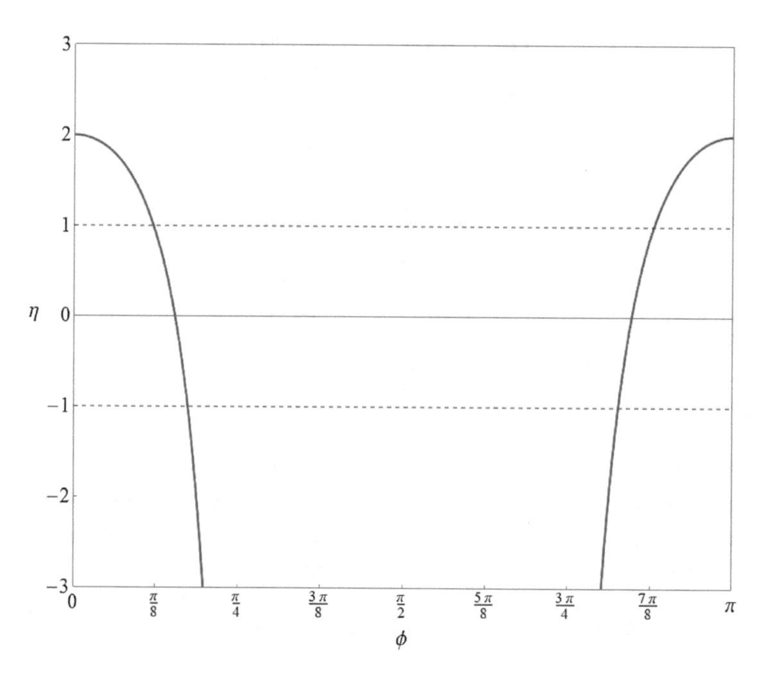

Fig. 18.5 Plot of η of (18.4.5) for $0 \le \varphi \le \pi$, $\gamma = 1$, $\delta = 2.5$, $\mu = 0.01$, $\alpha = 3$. Here the dashed horizontal lines denote $\eta = \pm 1$.

Restricting ourselves to the interval of interest $\varphi \in [0, \pi/2]$, we see from this figure that there exists a band of stable rhombic patterns, where $a_1 \pm b_1 > 0$ or equivalently $-1 < \eta < 1$, given by $\varphi \in (\varphi_m, \varphi_M)$ with $0 < \varphi_m < \varphi_M < \pi/2$ when $\beta > \beta_c$. Observe from Fig. 18.5 the intercept and symmetry properties

$$\eta(3, 0; 1, 2.5, 0.01) = 2, \tag{18.4.6}$$

$$\eta(3, \pi - \varphi; 1, 2.5, 0.01) = \eta(3, \varphi; 1, 2.5, 0.01). \tag{18.4.7}$$

Here, these properties of (18.4.7) are a consequence of mode interference occurring exactly at $\varphi = 0$ and modal interchange, respectively [24]. Repeating the process used to produce Fig. 18.5 but for other $\alpha_1 < \alpha \leq 3.5$, we find the same generic behavior as for $\alpha = 3$ and summarize these results for selected values in Table 18.2.

Table 18.2 Angle range for stable rhombic patterns.

α	φ_m	φ_M
0.2	0.109632	0.185916
0.4	0.243290	0.384288
0.5	0.268140	0.416588
0.6	0.285482	0.438294
0.8	0.308829	0.466429
1.0	0.324280	0.484419
1.2	0.335538	0.497225
1.4	0.344251	0.506968
1.6	0.351279	0.514728
1.8	0.357124	0.521116
2.0	0.362099	0.526508
2.2	0.366411	0.531150
2.4	0.370204	0.535210
2.6	0.373583	0.538809
2.8	0.376622	0.542033
3.0	0.379381	0.544949
3.2	0.381903	0.547906
3.4	0.385319	0.550045

We have deferred until now a detailed morphological interpretation of the stable rhombic patterns that can be identified with critical point V for the values of the characteristic angle relevant to Table 18.2. Then, to lowest order, the equilibrium vegetative pattern associated with that critical point satisfies

$$n(x, y, t) \to n_e(x, y) \sim 1 + A_0 g(x, z) \text{ as } t \to \infty \text{ for } z = x\cos(\varphi) + y\sin(\varphi), \tag{18.4.8a}$$

$$g(x, z) = \cos\left(\frac{2\pi x}{\lambda_c}\right) + \cos\left(\frac{2\pi z}{\lambda_c}\right). \tag{18.4.8b}$$

The three parts of Fig. 18.6 are threshold contour plots of (18.4.8b) for $\varphi = 0.5$ with threshold values of $1, 0$, and -1, respectively. Hence from left to right the parts of this figure correspond to what we termed higher, zero, and lower threshold patterns, respectively, in the last chapter. In this context note that $|g(x, z)| \leq 2$. Traditionally, most pattern formation analyses of this type have used the dimensional homogeneous vegetative solution value N_e of (18.2.1d) as the threshold to trigger the color change from light to dark (see Fig. 18.5a). Thus all spatial regions characterized by $N = N_e n \geq N_e$ appear dark and those characterized by $N < N_e$, light, where again

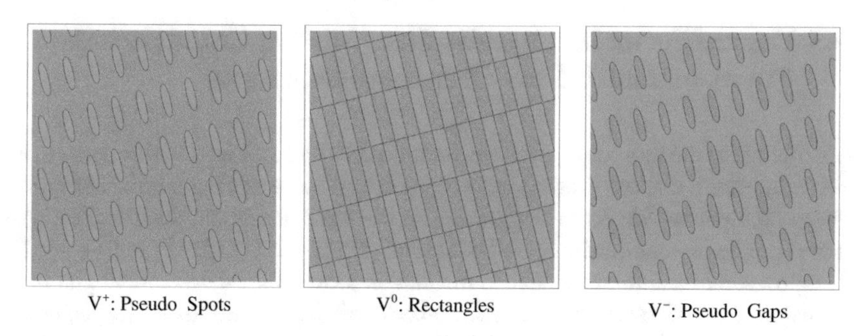

Fig. 18.6 Rhombic patterns relevant to $g(x, z)$ of (18.4.8) for $\varphi = 0.5$ with threshold values from left to right of 1, 0, and −1. Here, the spatial variables are being measured in units of λ_c and regions exceeding that threshold in each part appear dark while those below it appear light.

dark regions correspond to high plant biomass density and light ones to low plant biomass density or bare ground. This is equivalent to our zero threshold case of Fig. 18.6. Note from (18.2.1d) and (18.2.5b) that

$$\alpha = \frac{(\delta - 1)N_e}{k_1 c}. \tag{18.4.9a}$$

For fixed values of the other parameters and δ satisfying (18.2.2) we may consider α and N_e to be increasing straight line functions of R alone given by $\alpha(R)$ and $N_e(R)$, respectively, where

$$\alpha(R) = \frac{(\delta - 1)N_e(R)}{k_1 c} \tag{18.4.9b}$$

is a dimensionless measure of the rainfall rate R. We now wish to select a particular $R = R_c$ and adopt the protocol that

$$N_c = N_e(R_c) \tag{18.4.10}$$

represents this threshold instead. Then, when $\alpha_1 < \alpha < \alpha_c = \alpha(R_c)$ or $N_e < N_c$, a higher threshold pattern of the type depicted in Fig. 18.6 would occur while, when $\alpha > \alpha_c$ or $N_e > N_c$, a lower threshold type would occur. Given their appearance in Fig. 18.6 we label these higher, zero, and lower threshold-type rhombic vegetative arrays as pseudo-spots, rectangles, and pseudo-gaps and denote them by V^+, V^0, and V^-, respectively, in what follows. We shall defer the specific choice for R_c, our rationale for making that selection, and its morphological interpretations until the comparison of these results with some recent vegetative pattern formation studies included in the next section.

18.5 Pattern Formation Results, Root Suction Characteristic Curve, and an Aridity Classification Scheme

As a prelude to the morphological interpretations to be developed in this section, we first demonstrate that our model system does not generate any theoretical hexagonal pattern predictions. We do so by performing the same hexagonal planform analysis on the community equilibrium point of system (18.2.6) as introduced in Chapter 17. If we proceed in a similar manner to the one employed for the hexagonal planform expansion of that chapter, the Fredholm-type solvability conditions for $[n_{0220}, w_{0220}]$ and $[n_{1220}, w_{1220}]$, respectively, yield [17]

$$a_0 = \left.\frac{(\psi_{01} - q_c^2)b_{0220}^{(1)} - \theta_{01}b_{0220}^{(2)}}{(\mu+1)q_c^2 - \theta_{10} - \psi_{01}}\right|_{\beta=\beta_c} = a_0(\alpha;\gamma,\delta,\mu), \tag{18.5.1a}$$

$$a_2 = -\left.\frac{8a_0\theta_{01}a_{0220} + (q_c^2 - \psi_{01})b_{1220}^{(1)} + \theta_{01}b_{1220}^{(2)}}{(1+\mu)q_c^2 - \theta_{10} - \psi_{01}}\right|_{\beta=\beta_c} = a_2(\alpha;\gamma,\delta,\mu); \tag{18.5.1b}$$

where the components of

$$\boldsymbol{b}_{j220} = \begin{bmatrix} b_{j220}^{(1)} \\ b_{j220}^{(2)} \end{bmatrix}, \tag{18.5.1c}$$

for $j = 0$ and 1 as well as the solutions for the relevant second-order systems are given by

$$n_{0200}|_{\beta=\beta_c} = \left.\frac{-\psi_{01}b_{0220}^{(2)} + \theta_{01}b_{0220}^{(2)}}{\theta_{10}\psi_{01} - \theta_{01}\psi_{10}}\right|_{\beta=\beta_c}, \tag{18.5.1d}$$

$$w_{0200}|_{\beta=\beta_c} = \left.\frac{-\theta_{10}b_{0220}^{(2)} + \psi_{10}b_{0220}^{(1)}}{\theta_{10}\psi_{01} - \theta_{01}\psi_{10}}\right|_{\beta=\beta_c}; \tag{18.5.1e}$$

$$n_{1331}|_{\beta=\beta_c} = \left.\frac{(3q_c^2 - \psi_{01})b_{1131}^{(1)} + \theta_{01}b_{1131}^{(2)}}{(\theta_{10} - 3\mu q_c^2)(\psi_{01} - 3q_c^2) - \theta_{01}\psi_{10}}\right|_{\beta=\beta_c}, \tag{18.5.1f}$$

$$w_{1131}|_{\beta=\beta_c} = \left.\frac{(3\mu q_c^2 - \theta_{01})b_{1131}^{(2)} + \psi_{01}b_{1131}^{(1)}}{(\theta_{10} - 3\mu q_c^2)(\psi_{01} - 3q_c^2) - \theta_{01}\psi_{10}}\right|_{\beta=\beta_c}, \tag{18.5.1g}$$

where

$$\boldsymbol{b}_{1131} = \begin{bmatrix} b_{1131}^{(1)} \\ b_{1131}^{(2)} \end{bmatrix} = n_{1020}n_{0111}\boldsymbol{v}_{20} + w_{1020}w_{0111}\boldsymbol{v}_{02} + \frac{1}{2}(n_{1020}w_{0111} + n_{0111}w_{1020})\boldsymbol{v}_{11}; \tag{18.5.1h}$$

$$\theta_{01}w_{0220}|_{\beta=\beta_c} = -(a_0n_{1020} + b_{0220}^{(1)})\Big|_{\beta=\beta_c}; \tag{18.5.1i}$$

and

$$b_{0220} = \frac{1}{4}(n_{0111}^2 v_{20} + n_{0111}w_{0111}v_{11} + w_{0111}^2 v_{02}), \tag{18.5.1j}$$

$$\begin{aligned} \boldsymbol{b}_{1220} &= \left(2n_{1020}n_{0220} + \frac{1}{2}n_{0111}n_{1131}\right)\boldsymbol{v}_{20} \\ &+ \left[n_{1020}w_{0200} + n_{0111}w_{0220} + \frac{1}{4}(n_{0111}w_{1131} + n_{1131}w_{0111}) + n_{0200}w_{1020}\right]\boldsymbol{v}_{11} \\ &+ \left[2w_{1020}w_{0200} + w_{0111}\left(2w_{0220} + \frac{1}{2}w_{1131}\right)\right]\boldsymbol{v}_{02} \\ &+ \frac{3}{4}n_{1020}w_{0111}^2\boldsymbol{v}_{30} + \frac{1}{4}(2n_{1020}w_{0111} + n_{0111}w_{1020})\boldsymbol{v}_{21} \\ &+ \frac{1}{4}w_{0111}(2n_{0111}w_{1020} + n_{1020}w_{0111})\boldsymbol{v}_{12} + \frac{3}{4}w_{1020}w_{0111}^2\boldsymbol{v}_{03}. \end{aligned} \tag{18.5.1k}$$

Observe, as pointed out by Wollkind and Stephenson [143], that the expression for a_2 of (18.5.1b) does not contain the component $n_{0220}|_{\beta=\beta_c}$ since its coefficient vanishes identically in this limit by virtue of (18.5.1a) and hence is often referred to as a free mode which is why that component is not catalogued above. Having determined formulae for these Landau constants, we now return to the six disturbance hexagonal planform amplitude phase equations (17.3.1b, c) and its critical points (17.3.11). Critical points I and II represent the uniform homogeneous and supercritical striped states described in the previous two sections; III$^{\pm}$, hexagonal close-packed arrays of spots and gaps, respectively (see Chapter 17); and IV, a generalized cell. Employing our expressions for the Landau constants of (18.3.2) and (18.5.1), we examine the signs of a_0, $a_1 + 4a_2$, and $2a_2 - a_1$ by plotting those quantities as well as a_2 versus $\alpha_1 < \alpha \leq 3.5$ for $\gamma = 1$, $\delta = 2.5$, and $\mu = 0.01$ in Fig. 18.7 and observe that they are all identically negative.

Recall from Chapter 17 that although I is stable for $\sigma < 0$, II is not stable for $2a_2 - a_1 < 0$ and III$^{\pm}$ does not exist for $a_1 + 4a_2 < 0$ while IV is not stable for any set of parameter values. Hence, we have demonstrated that this hexagonal planform analysis does not yield any additional stable stationary heterogeneous vegetative patterns for our model system.

Having established this fact, let us return to the subject with which we ended the last section: Namely, the selection of the proper value to be assigned for R_c and hence, N_c. In order to motivate our specific choice, we first summarize the simulation results of Rietkerk et al. [103]. Their two-dimensional numerical simulations of model system (18.1.1), with $\rho = 0$ and its other parameters set at values consistent with (18.2.11), yielded close-packed vegetative patterns of spots or gaps depending upon whether R was less than or greater than 1 mm/d, respectively (see Fig. 18.8).

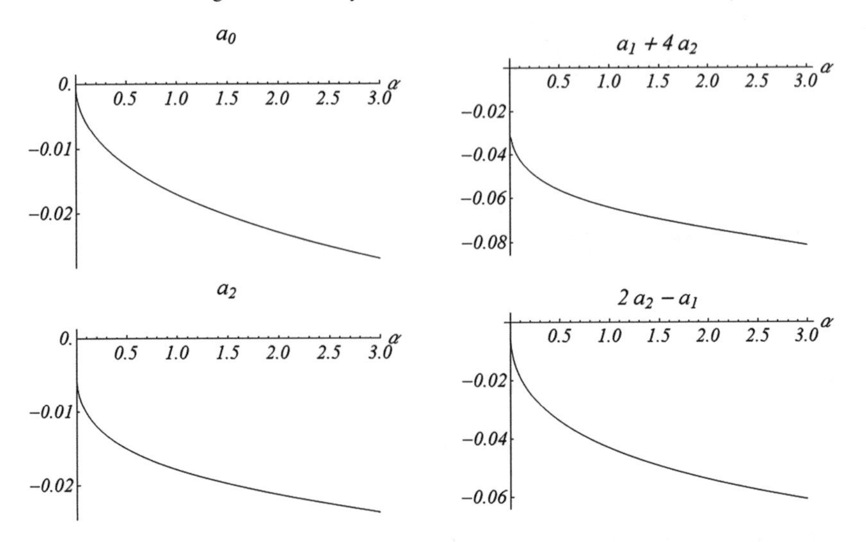

Fig. 18.7 Plots of a_0, $a_1 + 4a_2$, and $2a_2 - a_1$ versus $\alpha_1 < \alpha \leq 3.5$ for $\gamma = 1$, $\delta = 2.5$, and $\mu = 0.01$. Here, an analogous plot of a_2 has also been included for the sake of completeness.

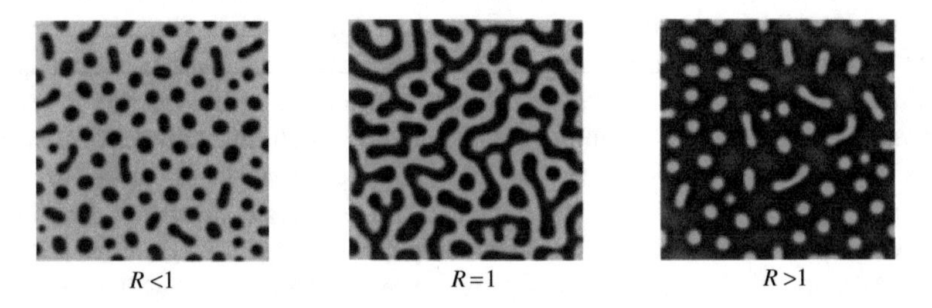

Fig. 18.8 The critical threshold value of $R_c = 1$ was selected to correspond with the simulated patterns of Rietkerk et al. [103] given above from left to right for R less than, equal to, or greater than 1, respectively. Adapted by permission of the authors from [103] for which the The University of Chicago Press holds the copyright.

Motivated by the desire to replicate this behavior and given the similarity in appearance between these two types of patterns and the left- or right-hand parts of Fig. 18.6, respectively, we then select

$$R_c = 1 \text{ mm/d} \Rightarrow \alpha_c = 0.5. \tag{18.5.2a}$$

Thus, from our rhombic planform analysis of the previous section, we can make the prediction that V^+ patterns will occur for $R < R_c$ or $\alpha < \alpha_c = 0.5$ and V^- ones, for $R > R_c$ or $\alpha > 0.5$ when $\beta > \beta_c$. Influenced by that resemblance in appearance just cited, we have referred to these periodic rhombic arrays of $V^{\pm}$ as vegetative pseudo-spots or pseudo-gaps, respectively. We now incorporate these two-dimensional rhombic planform morphological stability results for $\gamma = 1$, $\delta = 2.5$, and $\mu = 0.01$ in the α-β plane of Fig. 18.9 and identify regions corresponding to the predicted vegetative patterning. Note in this context that (18.4.9), (18.4.10), and (18.5.2a) imply

$$N_c = \frac{50}{3} \frac{\text{gm}}{\text{m}^2} \tag{18.5.2b}$$

while the plant component of the equilibrium point for the model system employed by Rietkerk et al. [103] was identical with N_e of (18.2.1d).

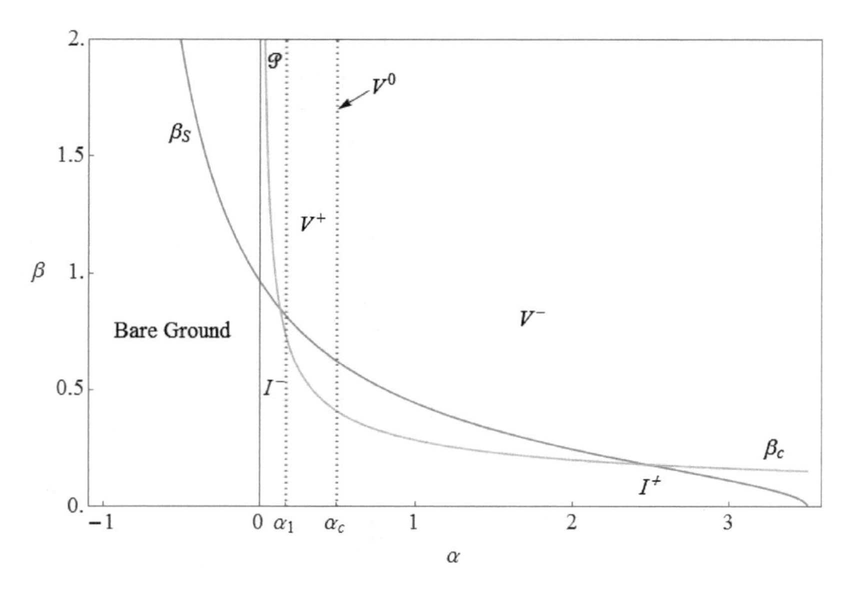

Fig. 18.9 Stability diagram in the α-β plane for our two-dimensional interaction–diffusion model system with $\gamma = 1$, $\delta = 2.5$, and $\mu = 0.01$, identifying the predicted vegetative patterns. Here β_S denotes the root suction characteristic (18.5.8) as a function of saturation with $\beta^{(0)} = 0.22$.

Fig. 18.9 represents the two-dimensional refinement of our one-dimensional predictions of Fig. 18.4b. Observe that the occurrence of striped and rhombic patterns is mutually exclusive by virtue of stability criteria (18.4.4b). Hence since rhombic patterns occur for all $\alpha > \alpha_1$ in the patterned $\beta > \beta_c$ region this precludes the occurrence of any striped patterns there. Thus our major two-dimensional refinement of those one-dimensional predictions is the replacement of the whole region of striped vegetative patterns (II) appearing in Fig. 18.4b with a rhombic (V) one instead. Specifically, these are identified in Fig. 18.9 where $\beta > \beta_c$ as rhombic arrays of pseudo-spots (V^+) for $\alpha_1 < \alpha < \alpha_c$ and of pseudo-gaps (V^-) for $\alpha > \alpha_c$ in accordance with our morphological threshold introduced in (18.5.2a). Further, note that the $0 < \beta < \beta_c$ region of Fig. 18.9 which can be identified with a uniform homogeneous vegetative pattern varies from a relatively sparse distribution ($N_e < N_c = 50/3$

gm/m^2) for $0 < \alpha < \alpha_c$ to a relatively dense one ($N_e > N_c = 50/3$ gm/m^2) for $\alpha > \alpha_c$ and hence have been designated by $\mathrm{I}^{\mp}$, respectively.

So far, with the implicit exception of the above-mentioned I^-, we have been concerned with the morphological stability behavior of our model system for $a_1 > 0$. We next consider its behavior for $a_1 < 0$. Wollkind et al. [142] showed a particular partial differential evolution equation containing fourth-order spatial derivatives could be used to mimic pattern formation in reaction–diffusion systems. A comparison of the simulation results of Lejeune et al. [64] with the weakly nonlinear stability ones of Boonkorkuea et al. [9] for a strongly related evolution equation describing vegetative pattern formation in arid isotropic environments led to the conjecture that when $a_1 < 0$ localized structures would occur where $\sigma > 0$, characterized by isolated patches of vegetation at low densities that were a spatial compromise between the periodic patchy vegetation and bare ground stable states. Chen and Ward [20] found local structures occurring in conjunction with such subcriticality for the Gray-Scott reaction–diffusion chemical model system. Cangelosi et al. [14] employed the same argument to identify a region of their relevant parameter space with isolated clusters for a mussel-algae interaction–diffusion model system. The resulting morphological sequence deduced from that identification provided close agreement with mussel bed patterning observations both in the field [128] and laboratory [70]. Given the similarity of behavior among all these phenomena we conjecture with some confidence that isolated patches of vegetation would occur for $0 < \alpha < \alpha_1$ where $\beta > \beta_c$ and identify that region graphically in Fig. 18.9 using the designation $\mathcal{P}$.

Taking into account as in Chapter 16 terms through fifth-order in the expansion and amplitude equation of (18.3.1) when $a_1 < 0$, by, in particular, considering

$$\frac{dA_1}{dt} \sim \sigma A_1 - a_1 A_1^3 - a_3 A_1^5$$

and calculating a_3, we might enhance our understanding of the morphological stability of this system in the subcritical regime. As demonstrated in Chapter 16 should $a_3 > 0$ that equation will have three equilibrium points: Namely, 0 and $2a_3 A_e^{\pm 2} = \pm\sqrt{a_1^2 + 4a_3\sigma} - a_1$ such that 0 is globally stable for $\sigma < \sigma_{-1} = -a_1^2/(4a_3) < 0$; A_e^+, for $\sigma > 0$; and in the overlap region $\sigma_{-1} < \sigma < 0$ where either can be stable depending on initial conditions 0 is stable for $A_1^2(0) < A_e^{-2}$ and A_e^+, for $A_1^2(0) > A_e^{-2}$; while A_e^- which only exists in that bistability region is not stable there. Here the potentially stable critical points 0 and A_e^+ would correspond to I^- and $\mathcal{P}$, respectively.

Determining the generalized marginal curve for our problem, analogous to (18.2.9) but with $\sigma \neq 0$, we would find that (see Problem 18.1)

$$\beta_\sigma(\alpha;\gamma,\Delta,\mu) = \frac{2\mu^{1/2}}{\alpha\Delta}[\alpha\Delta + (\gamma+\alpha\Delta)\sigma + \sigma^2]^{1/2} + \frac{\mu\gamma + (1+\mu)\sigma}{\alpha\Delta} + \mu,$$

where $\beta_{\sigma=0}(\alpha;\gamma,\Delta,\mu) = \beta_c(\alpha;\gamma,\Delta,\mu)$. Plotting the marginal curve associated with $\sigma = \sigma_{-1}$,

$$\beta = \beta_{-1}(\alpha;\gamma,\Delta,\mu) = \beta_{\sigma_{-1}}(\alpha;\gamma,\Delta,\mu) < \beta_c(\alpha;\gamma,\Delta,\mu),$$

in Fig. 18.9, we could make the proper morphological identifications. These would differ from those already appearing there only due to the presence of the overlap region $\beta_{-1} < \beta < \beta_c$ where $\mathcal{P}$ as well as I^- patterns could now occur. Should $a_1^2/(4a_3) << 1$, as was the case in Kealy-Dichone et al. [58] involving hexagonal patterns, these marginal curves would almost coincide or $\beta_{-1} \cong \beta_c$, causing the overlap region virtually to disappear and resulting in the exact same identifications as the ones appearing in Fig. 18.9.

Traditionally, morphological sequences of the sort referred to above have been generated from stability diagrams such as Fig. 18.9 by traversing appropriate horizontal (*e.g.*, see Table 17.5) or vertical lines (*e.g.*, see [58]) in that two-component parameter space. A procedure of this sort is inherently dependent upon the implicit assumption that these two components are independent of each other. In the case of Fig. 18.9, however, α and β, being nondimensional measures of rainfall and the coefficient of plant root suction, respectively, are actually related. To obtain the proper morphological sequence of vegetative states along a rainfall gradient predicted from Fig. 18.9, it is first necessary for us to deduce that relationship. Toward this end, employing our basic definitions and the parameter values of (18.2.11), we find that

$$\beta = c\rho = c\frac{D_S}{D_W} = D_S\frac{\text{gm d}}{\text{mm m}^4}. \tag{18.5.3}$$

Note in (18.5.3) the units for D_S, as indicated below, are mm m^4/(gm d) consistent with β being a dimensionless parameter. Adopting the root suction characteristic of Roose and Fowler [105], we take

$$D_S = \beta^{(0)} f_m(S)\frac{\text{mm m}^4}{\text{gm d}} \tag{18.5.4a}$$

where

$$f_m(S) = (S^{-1/m} - 1)^{1-m} \text{ for } 0 < m < 1 \text{ and } 0 < S \leq 1. \tag{18.5.4b}$$

Here $S \equiv$ the relative water saturation in the soil while the parameters $\beta^{(0)}$ and m are determined from experimental data for different soils. To complete our formulation we let

$$S = \frac{R}{R_M} \text{ for } 0\frac{\text{mm}}{\text{d}} < R \leq R_M \tag{18.5.5a}$$

where, specifically,

$$R_M = 3\frac{\text{mm}}{\text{d}}. \tag{18.5.5b}$$

Upon recalling that

$$R = \frac{k_1 d(\alpha + \gamma)}{\delta - 1} \tag{18.5.6}$$

and substituting (18.5.6) into (18.5.5) yields

$$S = \frac{\alpha + \gamma}{\alpha_M + \gamma} \text{ for } \alpha_0 < \alpha \leq \alpha_M \tag{18.5.7a}$$

where, specifically,

$$\alpha_0 = -\gamma = -1, \alpha_M = 3.5. \tag{18.5.7b}$$

Hence,

$$S = \frac{\alpha + 1}{4.5} \text{ for } -1 < \alpha \leq 3.5. \tag{18.5.7c}$$

Finally, selecting $m = 0.5$ after Roose and Fowler [105], and incorporating (18.5.7c) and (18.5.4) into (18.5.3), we obtain the one-parameter family of root suction characteristic curves

$$\beta = \beta_S(\alpha) = \beta^{(0)} f_{0.5}\left(\frac{\alpha + 1}{4.5}\right) \text{ for } -1 < \alpha \leq 3.5 \tag{18.5.8a}$$

where

$$f_{0.5}(S) = \frac{(1 - S^2)^{1/2}}{S}. \tag{18.5.8b}$$

We plot this curve with $\beta^{(0)} = 0.22$ in Fig. 18.9, where the assignment of that parameter has been made both for the purpose of definiteness and to be consistent with our silt loam soil choice for m [105]. Representing the stability boundary in that figure by $\beta_c(\alpha)$, for ease of exposition, we observe that there exist two points of intersection between it and $\beta_S(\alpha)$ satisfying

$$\beta_S(p_{0,2}) = \beta_c(p_{0,2}) \tag{18.5.9a}$$

where, specifically,

$$p_0 = 0.13217, p_2 = 2.47622. \tag{18.5.9b}$$

Here, from (18.5.6), these α-values of (18.5.9b) correspond to

$$R_0 = 0.75478 \, \frac{\text{mm}}{\text{d}}, R_2 = 2.31748 \, \frac{\text{mm}}{\text{d}}, \tag{18.5.9c}$$

respectively. In this context, we define

$$p_1 = \alpha_c, R_1 = 0.78133 \, \frac{\text{mm}}{\text{d}} \tag{18.5.9d}$$

where $\alpha(R_1) = \alpha_1$. The morphological sequence of predicted stable vegetative states along a rainfall gradient obtained upon traversing the curve $\beta = \beta_S(\alpha)$ in the α-β plane of Fig. 18.9 is tabulated in Table 18.3. Note, in general, that

$$p_0 \begin{Bmatrix} < \\ = \\ > \end{Bmatrix} \alpha_1 \text{ if } \beta^{(0)} \begin{Bmatrix} > \\ = \\ < \end{Bmatrix} 0.20. \quad (18.5.10a)$$

Thus, isolated patches are only predicted for transit curves of the form of (18.5.8) provided

$$\beta^{(0)} > 0.20. \quad (18.5.10b)$$

Table 18.3 Morphological stability predictions along a rainfall gradient for $\beta = \beta_S(\alpha)$ in Fig. 18.9. Here $p_0 = 0.13$, $\alpha_1 = 0.17$, $p_1 = 0.50 = \alpha_c$, $p_2 = 2.48$, $R_0 = 0.75$, $R_1 = 0.78$, and $R_2 = 2.32$.

α-range	R-range (mm/d)	Stable pattern
$-1 < \alpha < 0$	$0 < R < \frac{2}{3}$	Bare ground
$0 < \alpha < p_0$	$\frac{2}{3} < R < R_0$	Sparse homogeneous
$p_0 < \alpha < \alpha_1$	$R_0 < R < R_1$	Isolated patches
$\alpha_1 < \alpha < \alpha_c = p_1$	$R_1 < R < R_c = 1$	Pseudo-spots
$\alpha = \alpha_c = p_1$	$R = R_c = 1$	Rectangles
$\alpha_c = p_1 < \alpha < p_2$	$R_c = 1 < R < R_2$	Pseudo-gaps
$p_2 < \alpha < 3.5$	$R_2 < R < 3$	Dense homogeneous

We next compare our results with those from some recent biological pattern formation studies. We begin with the work of Gowda et al. [42]. These authors examined the standard sequence of patterned states (gaps $\longrightarrow$ labyrinth $\longrightarrow$ spots) generated in a general activator-inhibitor reaction–diffusion system as a bifurcation parameter was varied and then applied their results to the particular von Hardenberg et al. [126] plant-groundwater model as its precipitation parameter was decreased. They employed both numerical simulation and analytical weakly nonlinear hexagonal planform bifurcation methods. Gowda et al. [42] found, for the default set of parameter values von Hardenberg et al. [126] used in their numerical integration, that, although the simulation method reproduced the latter's standard sequence, the hexagonal planform analysis as in our problem failed to predict vegetative patterns. These calculations, in accordance with (18.2.11), were performed for $1/\mu = 100$. When those calculations were repeated for $1/\mu = 27$, they found that the same standard sequence of vegetative patterns was produced for both the simulation and weakly nonlinear stability methods with the transition between the two hexagonal states occurring exactly where the second-order Landau constant changed sign (see Boonkorkuea et al., [9]). Gowda et al. [42] concluded that weakly nonlinear stability theory failed to produce the correct results in the first instance because the simulated morphological sequence of vegetative patterns occurred for large amplitudes as the precipitation parameter was decreased. Given our results, we wish to suggest another possible explanation for this discrepancy: Namely, that a rhombic planform weakly nonlinear stability analysis might yield a predicted morphological sequence involving pseudo-gaps $\longrightarrow$ rectangles $\longrightarrow$ pseudo-spots as the precipitation parameter was decreased in this case. As supporting evidence for such a conjecture we offer the following observation. The gap- and spot-type simulated patterns for $1/\mu = 100$ appearing in both Gowda et al. [42] and von Hardenberg et al. [126] were much less regular in nature than were the corresponding simulated hexagonal patterns for $1/\mu = 27$ appearing in Gowda et al. [42]. The simulated transition states between these two types of patterns were also different consisting of labyrinths in the former instance but of parallel stripes in the latter case. Since for each value of α we predict multistable rhombic states with an interval of characteristic angles (see Fig. 18.5 and Table 18.2) and as initial conditions vary point by point over a flat environment these states can be selected quite randomly, it is possible to generate simulated patterns resembling those appearing in von Hardenberg et al. [126] from families of pseudo-gaps and pseudo-spots including labyrinths from families of rectangles. In this context, the numerically simulated two-dimensional patterns of Rietkerk *et al.* [103] used earlier to motivate our choice for R_c also bore a strong resemblance to those of von Hardenberg et al. [126] including a labyrinthine transition state at $R = 1$ mm/d (see Fig. 18.8). Unlike the von Hardenberg et al. [126] model ours is extremely robust to variations in μ. We performed additional rhombic and hexagonal planform nonlinear stability analyses on system (18.2.6) and found identical qualitative behavior for all $0.001 \le \mu \le 1$.

We end this phase of our discussion by restating von Hardenberg et al.'s [126] claim that the power of model systems such as ours of (18.1.3) is their predicted sequence of stable states along a rainfall gradient such as the one summarized in Table 18.3 can be used to motivate an aridity classification scheme which is characterized by the three rainfall thresholds

$$p_0 = 0.13217 < p_1 = 0.5 < p_2 = 2.47622. \tag{18.5.11}$$

Here we are employing the notation of von Hardenberg et al. [126] for these dimensionless rainfall (precipitation) rate thresholds and use them to introduce the following possible aridity classes based upon the inherent vegetative states of our system:

- *Dry-subhumid* ($p_2 < \alpha < 3.5$): The only vegetative state the system supports corresponds to a dense homogeneous distribution.
- *Semiarid* ($p_1 < \alpha < p_2$): The only vegetative state the system supports corresponds to pseudo-gaps of low threshold type.
- *Arid* ($p_0 < \alpha < p_1$): The only vegetative states the system supports correspond to either pseudo-spots of high threshold type or isolated patches.
- *Hyperarid* ($-1 < \alpha < p_0$): The only possible stable states the system supports correspond to either a sparse homogeneous vegetative distribution or bare ground.

As pointed out by von Hardenberg et al. [126] the advantage of the proposed aridity classification scheme pertains to the information it contains about dynamical aspects of drylands. Regions whose aridity classes imply the occurrence of upper threshold vegetative patterns, isolated patches, or a sparse homogeneous distribution are vulnerable to desertification. The mere knowledge of that threat, however, allows land managers to reverse this process for those regions by implementing crust disturbance, seed augmentation, or irrigation strategies. Meron et al. [82] suggested a cycling mechanism between plants and water to account for the formation of bare patches characteristic of vegetative patterning along such a precipitation gradient. Note that a process of this sort occurs in all directions for two-dimensional vegetative patterns (pseudo-spots or gaps, rectangles, and isolated patches) but only in two directions for one-dimensional ones (stripes).

It remains for us to compare our theoretical predictions with relevant observational evidence. Striking periodic or localized self-organized vegetative patterns covering widespread areas of arid or semiarid flat regions of Africa, Australia, the Americas, and the Near East became noticeable through aerial photography more than 70 years ago. In this instance, an arid or semiarid flat region refers to a plateau-like environment that is characterized by an extended dry season where yearly potential evaporation exceeds yearly rainfall and water availability is a limiting factor for plant growth. The periodic patterns reported consisted of spots (leopard bush), labyrinths, or gaps (pearled bush) and the localized ones, isolated spots. These included bushy vegetation punctuated by bare gaps in Senegal, labyrinthine mazes in Niger, and periodic or localized spots of trees or shrubs in Sudan or French Guiana, respectively, as depicted in the photographs included in [42] and [64]. Incidentally labyrinthine patterns have been associated with certain types of tiger bush or banded thicket vegetative distributions found in arid or semiarid flat environments [23]. Given the similarity of appearance between the periodic observed patterns depicted in the photographs included in [42] and the theoretical patterns of Fig. 18.6, we now identify our rhombic arrays of pseudo-spots (V^+) or pseudo-gaps (V^-) with these leopard or pearled bush vegetative patterns, respectively, and our rhombic arrays of rectangles (V^0) with the tiger bush labyrinthine vegetative patterns. In light of those identifications we then investigate the predicted wavelengths of these vegetative patterns. From (18.2.10) and (18.3.3b), we can deduce that

$$\lambda_c = \frac{2\pi\sqrt[4]{\mu/\Delta}}{\sqrt[4]{\alpha}} = \lambda_c(\alpha). \tag{18.5.12}$$

Designating the α's associated with $V^{\pm}$ by $\alpha^{\pm}$, respectively, it follows from Table 18.3 that

$$\alpha_1 < \alpha^+ < \alpha_c = p_1 < \alpha^- < p_2. \tag{18.5.13a}$$

Then, we can see from (18.5.12) and (18.5.13a) that

$$\lambda_1 = \lambda_c(\alpha_1) > \lambda_c^+ = \lambda_c(\alpha^+) > \lambda_c^0 = \lambda_c(\alpha_c) > \lambda_c^- = \lambda_c(\alpha^-) > \lambda_2 = \lambda_c(p_2), \tag{18.5.13b}$$

and hence conclude that the vegetative distributions of spots in leopard bush have a tendency to be more widely spaced than the labyrinthine components of the mazes which in turn are more widely spaced than the bare patches that regularly punctuate the vegetation cover in pearled bush [65]). Employing the length scale of (18.3.3b) and the parameter values of (18.2.11) in (18.5.13b) yields the corresponding dimensional wavelength relationships

$$\lambda_1^* = 25 \text{ m} > \lambda_c^{*+} > \lambda_c^{*0} = 19 \text{ m} > \lambda_c^{*-} > \lambda_2^* = 12.7 \text{ m} \tag{18.5.13c}$$

consistent with the photographs of [42] and in agreement with Boonkorkuea et al. [9], who interestingly enough found for the evolution equation of Lejeune et al. [65] an identical power law relationship to (18.5.12) between their pattern wavelength and plant biomass (see Problem 18.4). Finally it is obvious that our isolated patches $\mathcal{P}$ should be identified with the localized vegetative spots depicted in the photographs included in [64].

We close with a more detailed commentary on the role played by cross-diffusion in generating pattern formation instabilities for our two-component model system. Given that $\theta_{10} = 0$, our system violates the activator positive feedback necessary condition for the occurrence of a Turing *self-diffusive* instability which requires $\theta_{10} > 0$. Hence the cross-diffusive effect of plant root suction on groundwater generates our instability since as noted earlier if $\beta = 0$ its community equilibrium point would be identically linearly stable. Indeed, the other requirement of $\mu < 1$ for a Turing *self-diffusive* instability to occur might also be violated should $\mu = 1$, and a cross-diffusive instability of this type could still be generated although, in our actual parameter range, it is not violated. Recently, Stancevic et al. [116] considered a reaction–chemotaxis–diffusion three-component in-host viral dynamics model system for the concentrations of uninfected or infected cells and the virus. They found that the cross-diffusive effect of chemotaxis toward the infected cells by the uninfected ones generated their pattern formation instability in a similar manner as for our two-component system. Since the community equilibrium point of their system was linearly stable in the absence of diffusion and chemotaxis, Stancevic et al. [116] referred to this as a Turing instability. To distinguish between these two cases, we shall refer to ours as a Turing *cross-diffusive* instability instead. The von Hardenberg et al. [126] two-component nondimensional model system also included the cross-diffusive effect of plant root suction on groundwater. Since that system's interaction terms satisfied the activator positive feedback condition for its community equilibrium point while $1/\mu = 100$ in their default set of parameter values, the presence of this cross-diffusive effect mediated rather than generated their Turing self-diffusive instability. In particular, von Hardenberg et al. [126] took the coefficient of that term $b = 3$ in this default set. Upon inspection of (18.2.6) we can see that this coefficient is related to our parameters by

$$b = \alpha\beta. \tag{18.5.14a}$$

Thus that assignment would yield the root suction characteristic curve

$$\beta = \frac{3}{\alpha}, \tag{18.5.14b}$$

which, as a decreasing function of α, is in qualitative accord with our formulation of (18.5.8). The three-component model systems of Rietkerk et al. [103] and HilleRis-Landers et al. [48] by explicitly including surface water were able to generate Turing instabilities where none would have occurred for our simplified two-component version of that model without root suction. Finally, Wang et al. [129] conducted a definitive analysis of Turing instabilities for their predator–prey model system by including both self- and cross-diffusion terms in the prey and predator equations and performing weakly nonlinear hexagonal planform bifurcations and numerical simulations on its community equilibrium point. Since the Allee positive feedback effect for the activator prey and the self-diffusivity advantage for the inhibitory predator were satisfied for their specific model, this was an investigation of cross-diffusion mediated rather than generated instabilities.

Implicit to our continuum formulation were the assumptions that the pattern wavelength was relatively large when compared with the mean coverage diameter of an individual plant but quite small when compared with the territorial length scale characteristic of the arid environment which allowed us to have considered our interaction–diffusion equations on an unbounded spatial domain [42]. We conclude by noting that despite the fact these results of our weakly nonlinear stability analyses are only strictly valid in the vicinity of the marginal stability curve and the occurrence of the rhombic vegetative arrays along our specific coefficient of root suction characteristic curve satisfied this particular constraint, recent numerical simulations of pattern formation in a variety of partial differential equation model systems have demonstrated that such theoretical predictions can often be extrapolated to regions of the relevant parameter space substantially removed from the marginal curve [9, 41]. Given that a weakly nonlinear hexagonal planform analysis of our system did not predict any stable patterns, we offered an alternative explanation involving these stable rhombic patterns for why numerical simulations analyses of similar systems have in the past yielded periodic pattern formation over a parameter range where theoretical weakly nonlinear hexagonal ones did not. We finish by reiterating for the purpose of emphasis the fact that all of our pattern formation results for this model have been generated by the cross-diffusion process of root suction as opposed to the mediating effect it has often had on self-diffusion generated Turing patterns.

Problems

18.1. The purpose of this problem is to derive the generalized ($\sigma \neq 0$) marginal stability curve introduced in conjunction with the subcriticality discussion of Section 18.5 and then examine its behavior in the neighborhood of the neutral stability curve of Fig. 18.2.

(a) Solve the secular equation of (18.2.8d) for β and obtain

$$\beta = \beta_0(q^2;\alpha,\gamma,\Delta,\mu,\sigma) = \frac{\mu}{\alpha\Delta}q^2 + \frac{\alpha\Delta + (\gamma+\alpha\Delta)\sigma + \sigma^2}{\alpha\Delta}\frac{1}{q^2} + \frac{\mu\gamma + (1+\mu)\sigma}{\alpha\Delta} + \mu.$$

(b) Demonstrate that

$$\frac{d\beta_0}{dq^2}(q_\sigma^2;\alpha,\gamma,\Delta,\mu,\sigma)=0 \Rightarrow q_\sigma^2=\frac{[\alpha\Delta+(\gamma+\alpha\Delta)\sigma+\sigma^2]^{1/2}}{\mu^{1/2}}.$$

(c) Show that

$$\beta_0(q_\sigma^2;\alpha,\gamma,\Delta,\mu,\sigma)=\beta_\sigma(\alpha;\gamma,\mu,\Delta)$$
$$=\frac{2\mu^{1/2}}{\alpha\Delta}[\alpha\Delta+(\gamma+\alpha\Delta)\sigma+\sigma^2]^{1/2}+\frac{\mu\gamma+(1+\mu)\sigma}{\alpha\Delta}+\mu,$$

and hence derive the generalized marginal stability function of Section 18.5.

(d) By setting $\sigma=0$ in the results of the previous two parts of this problem, deduce that these quantities then reduce to the critical conditions of (18.2.10) or

$$q_{\sigma=0}^2=q_c^2=\sqrt{\frac{\Delta}{\mu}}\sqrt{\alpha};$$

$$\beta_{\sigma=0}(\alpha;\gamma,\Delta,\mu)=\beta_c(\alpha;\gamma,\Delta,\mu)=2\sqrt{\frac{\mu}{\Delta}}\left(\frac{1}{\sqrt{\alpha}}\right)+\left(\frac{\mu\gamma}{\Delta}\right)\left(\frac{1}{\alpha}\right)+\mu.$$

(e) Taking the derivative of $\beta_\sigma(\alpha;\gamma,\mu,\Delta)$ with respect to σ, obtain

$$\frac{d\beta_\sigma}{d\sigma}(\alpha;\gamma,\Delta,\mu)=\frac{\mu^{1/2}(\gamma+\alpha\Delta+2\sigma)}{\alpha\Delta[\alpha\Delta+(\gamma+\alpha\Delta)\sigma+\sigma^2]^{1/2}}+\frac{1+\mu}{\alpha\Delta}.$$

(f) Evaluating this derivative at $\sigma=0$ determine that

$$\left.\frac{d\beta_\sigma}{d\sigma}(\alpha;\gamma,\Delta,\mu)\right|_{\sigma=0}=\frac{\mu^{1/2}\gamma}{(\alpha\Delta)^{3/2}}+\frac{1+\mu}{\alpha\Delta}+\frac{\mu^{1/2}}{(\alpha\Delta)^{1/2}}>0,$$

and thus conclude that in the neighborhood of the neutral stability curve of Fig. 18.2

$$\beta_{\sigma<0}(\alpha;\gamma,\Delta,\mu)<\beta_{\sigma=0}(\alpha;\gamma,\Delta,\mu)=\beta_c(\alpha;\gamma,\Delta,\mu)<\beta_{\sigma>0}(\alpha;\gamma,\mu,\Delta).$$

18.2. Consider the rhombic planform amplitude equations (18.4.1c)

$$\frac{dA_1}{dt}\sim\sigma A_1-A_1(a_1A_1^2+b_1B_1^2)=F(A_1,B_1),\ \frac{dB_1}{dt}\sim\sigma B_1-B_1(b_1A_1^2+a_1B_1^2)=G(A_1,B_1).$$

(a) Show that these equations possess the following equivalence classes of critical points

$$A_1(t)\equiv A_0,\ B_1(t)\equiv B_0 \text{ where } F(A_0,B_0)=G(A_0,B_0)=0 \text{ and } A_0,B_0\geq 0$$

given by (18.4.4a)

$$\text{I: } A_0=B_0=0;\ \text{II: } A_0^2=\frac{\sigma}{a_1},\ B_0=0;\ \text{V: } A_0^2=B_0^2=\frac{\sigma}{a_1+b_1}.$$

Here we are assuming that a_1, $a_1 + b_1 > 0$. Hence critical points II and V will only occur provided $\sigma > 0$.

(b) Begin an investigation of the stability of these critical points by first seeking a solution of the amplitude equations of the form

$$A_1(t) = A_0 + \varepsilon_1 \mathcal{A}(t) + O(\varepsilon_1^2),\ B_1(t) = B_0 + \varepsilon_1 \mathcal{B}(t) + O(\varepsilon_1^2) \text{ with } |\varepsilon_1| << 1$$

and showing that the perturbation quantities satisfy the linear homogeneous system

$$\frac{d\mathcal{A}}{dt} = c_{11}\mathcal{A} + c_{12}\mathcal{B},\ \frac{d\mathcal{B}}{dt} = c_{21}\mathcal{A} + c_{22}\mathcal{B},$$

where

$$c_{11} = \sigma - 3a_1A_0^2 - b_1B_0^2,\ c_{22} = \sigma - 3a_1B_0^2 - b_1A_0^2,$$

$$c_{12} = c_{21} = -2b_1A_0B_0.$$

(c) Continue that investigation by letting $[\mathcal{A},\mathcal{B}](t) = [C_1, C_2]e^{pt}$ where $C_1^2 + C_2^2 \neq 0$ and obtaining the quadratic satisfied by p

$$(p - c_{11})(p - c_{22}) - c_{12}^2 = 0.$$

(d) Particularizing that quadratic to the specific (A_0, B_0) values of the critical points and noting that it then has the associated roots $p_1 = c_{11} + c_{12}$ and $p_2 = c_{22} - c_{12}$ since either $c_{11} = c_{22}$ or $c_{12} = 0$ in this event, conclude that

$$\text{I: } p_{1,2} = \sigma;\ \text{II: } p_1 = -2\sigma,\ p_2 = \frac{(a_1 - b_1)\sigma}{a_1}\ \text{V: } p_1 = -2\sigma,\ p_2 = \frac{2(b_1 - a_1)\sigma}{a_1 + b_1};$$

(e) Finally, requiring $p_{1,2} < 0$, deduce the stability criteria given by (18.4.4b)

I is stable for $\sigma < 0$; II, for $\sigma > 0$, $b_1 > a_1$; and V, for $\sigma > 0$, $a_1 > b_1$.

18.3. Let $h(x, y, t)$ represent a dimensionless interfacial deviation from a planar interface for the governing damped Kuramoto–Sivashinsky evolution equation describing a solid surface during ion-sputtered erosion studied by Pansuwan et al. [92]. Here $t \equiv$ time and $(x, y) \equiv$ a two-dimensional Cartesian coordinate system scaled with the wavelength of a disturbance to that planar interface. In order to investigate the possibility of occurrence of rhombic arrays of mounds or pits etched on that solid surface during this erosion process, they considered a nonlinear perturbation solution to their governing equation which to lowest order was of the form

$$h(x, y, t) \sim A_1(t)\cos(2\pi x) + B_1(t)\cos(2\pi z) \text{ with } z = x\cos(\varphi) + y\sin(\varphi).$$

while $A_1(t)$ and $B_1(t)$ satisfied the amplitude equations of (18.4.1c) examined in Problem 18.2, and found that

$$\sigma(\beta) = 2(1 - \beta),\ a_1(\alpha) = \frac{1}{4} + \frac{\alpha^2}{18},\ b_1(\alpha, \varphi) = \frac{1}{2} + \alpha^2\frac{\cos^2(\varphi)[4\cos^2(\varphi) - 3]}{[4\cos^2(\varphi) - 1]^2}$$

where $\alpha \in \mathbb{R}$ and $\beta > 0$ were nondimensional parameters proportional to the tilt-dependent coefficient of the erosion rate and the ion-bombardment flux, respectively.

(a) Show that these Landau constants satisfy the properties of (18.4.7)

$$b_1(\alpha, 0) = 2a_1(\alpha) = \frac{1}{2} + \frac{\alpha^2}{9} \text{ and } b_1(\alpha, \pi - \varphi) = b_1(\alpha, \varphi).$$

(b) Represent the stability criteria of Problem 18.2 for square arrays when $\varphi = \pi/2$ (see below) in terms of α and β. Can any of these critical points coexist – *i.e.*, can there be bistability?

(c) In the parameter ranges of interest for α and β just determined where each of the critical points are stable when $\varphi = \pi/2$ demonstrate that there exists a steady-state equilibrium solution of the governing equation which to lowest order satisfies

$$\lim_{t\to\infty} h(x, y, t) \sim h_0(x, y) = A_0 \cos(2\pi x) + B_0 \cos(2\pi y).$$

Particularizing this result to critical points II and V use level curves to construct zero-threshold contour plots of $h_0(x, y)$ and hence conclude that these two states can be identified with a striped and checkerboard pattern, respectively. In the latter case, make use of the trigonometric identity

$$\cos(\gamma - \delta) + \cos(\gamma + \delta) = 2\cos(\gamma)\cos(\delta)$$

with γ and δ chosen appropriately to deduce the relevant level curves.

18.4. The linear stability analysis performed by Boonkorkuea et al. [9] on the Lejeune et al. [65] nondimensional evolution equation for plant biomass in an arid environment yielded the following parabolic dispersion relationship between σ, the growth rate, and q^2, the wavenumber squared, of their initially infinitesimal disturbance to its equilibrium density α

$$\sigma = -2\alpha^2 + \frac{(\alpha - \beta)q^2}{2} - \frac{\alpha q^4}{8}$$

for $\beta = L^2$ and $\Lambda = 1$ where L and Λ were the facilitation-to-competition range and interaction ratios, respectively.

(a) Show from this secular equation that if $\beta \geq \alpha$ then $\sigma < 0$ and there is stability identically while if $\beta < \alpha$ then this parabola has a maximum value at its vertex (q_c^2, σ_c) where

$$q_c^2(\alpha, \beta) = 2\left(1 - \frac{\beta}{\alpha}\right) > 0 \text{ and } \sigma_c(\alpha, \beta) = -2\alpha^2 + \frac{(\alpha - \beta)^2}{2\alpha}.$$

(b) Demonstrate from part (a) that

$$\sigma_c < 0 \text{ for } \beta > \beta_c, \sigma_c = 0 \text{ for } \beta = \beta_c, \sigma_c > 0 \text{ for } \beta < \beta_c;$$

where

$$\beta_c(\alpha) = \alpha - 2\alpha^{3/2} = \alpha(1 - 2\alpha^{1/2}) \text{ for } 0 < \alpha < \frac{1}{4}.$$

(c) Hence conclude from part (b) that $\beta = \beta_c(\alpha)$ serves as a marginal stability curve in the α-β plane for this problem and plot it as well as its tangent at the origin $\beta = \alpha$ in that plane.

(d) Determine from part (c) that this marginal curve has a maximum value at $\alpha = \alpha_0$ where

$$\alpha_0 = \frac{1}{9} \cong 0.111 \text{ and } \beta_c(\alpha_0) = \frac{1}{27} \cong 0.037.$$

(e) Defining the critical wavelength corresponding to $q_c > 0$ by

$$\lambda_c(\alpha, \beta) = \frac{2\pi}{q_c(\alpha, \beta)},$$

deduce from part (a) that, for $\beta < \alpha$,

$$\lambda_c(\alpha, \beta) = \sqrt{2}\pi \left(1 - \frac{\beta}{\alpha}\right)^{-1/2}.$$

(f) Evaluating the function of part (e) on the marginal stability curve $\beta = \beta_c(\alpha)$ of part (b), obtain that

$$\lambda_c(\alpha, \beta_c) = \frac{\pi}{\alpha^{1/4}} = \lambda_c(\alpha)$$

which is identical in form to the power law relationship between pattern wavelength and plant biomass of (18.5.12).

Part IV
Chapters 19–22

Chapter 19
Calculus of Variations Revisited Plus the Gamma and Bessel Functions

In this chapter, a variety of topics that use the Calculus of Variations, introduced in Chapters 4 and 12, are explored. After developing general conditions for constrained optimization in a pastoral interlude, they are applied to Queen Dido's isoperimetric problem. Then the Euler–Lagrange equations of motion for conservative dynamical systems are derived in a pastoral interlude and applied to a linear spring, a double plane pendulum, and a vibrating string, respectively. Finally, Bessel functions, needed in the problems, are introduced and examined, which requires a similar preliminary examination of the gamma function as well, since the former are defined in terms of the latter. In this context, the Method of Stationary Phase and Laplace's Method are developed and employed to derive asymptotic representations for the Bessel functions and Stirling's formula for $n!$, respectively. There are nine problems: The first seven and the last deal with a variety of Calculus of Variations applications and asymptotic representations for the special functions, as well as a complete examination of the Bessel equation of order one-half. The eighth problem requires a Calculus of Variations approach to derive the wave equation governing the motion of a vibrating circular membrane which, when solved by separation of variables, gives rise to the eigenvalue example involving Bessel's equation of order zero handled in the final section of the chapter.

19.1 Pastoral Interlude: Euler–Lagrange Equations for Constrained Optimization

We have considered the Calculus of Variations method of extremalization of functionals for integrands involving first derivatives of a function of one and several independent variables in Sections 4.4 and 12.2, respectively. We now wish to continue our development of this topic by generalizing these approaches to include the following three cases as well:

© Springer International Publishing AG, part of Springer Nature 2017

D. J. Wollkind and B. J. Dichone, *Comprehensive Applied Mathematical Modeling in the Natural and Engineering Sciences*,
https://doi.org/10.1007/978-3-319-73518-4_19

(i) More Derivatives

We wish to characterize a function y that extremalizes $I(Y) = \int_{x_1}^{x_2} f(x,Y,Y',Y'')\,dx$ among all sufficiently smooth functions Y that satisfy $Y(x_1) = y_1$, $Y'(x_1) = v_1$, $Y(x_2) = y_2$, $Y'(x_2) = v_2$. Introducing the arbitrary function $s(x)$ which satisfies $s(x_1) = s'(x_1) = s(x_2) = s'(x_2) = 0$, and considering the admissible functions $Y(x;\varepsilon) = y(x) + \varepsilon s(x)$, we require that the function

$$\mathcal{I}(\varepsilon) = \int_{x_1}^{x_2} f(x, y+\varepsilon s, y'+\varepsilon s', y''+\varepsilon s'')\,dx$$

has an extremum at $\varepsilon = 0$. Then $\mathcal{I}'(0) = 0$ implies that $\int_{x_1}^{x_2} (f_2 s + f_3 s' + f_4 s'')\,dx = 0$, which upon integration by parts and using the boundary conditions that s satisfies, yields

$$\int_{x_1}^{x_2} \left(f_2 - \frac{df_3}{dx} + \frac{d^2 f_4}{dx^2} \right) s\,dx = 0.$$

Finally, applying the fundamental lemma of the Calculus of Variations, we obtain

$$\frac{\partial f}{\partial y} - \frac{d}{dx}\left(\frac{\partial f}{\partial y'}\right) + \frac{d^2}{dx^2}\left(\frac{\partial f}{\partial y''}\right) = 0 \text{ or, more generally, } \frac{\partial f}{\partial y} + \sum_{j=1}^{n} (-1)^j \frac{d^j}{dx^j}\left[\frac{\partial f}{\partial y^{(j)}}\right] = 0.$$

(ii) More Dependent Variables

We wish to find functions y_1 and y_2 that extremalize $I(Y_1,Y_2) = \int_{x_1}^{x_2} f(x, Y_1, Y_1', Y_2, Y_2')\,dx$ among all sufficiently smooth functions Y_i that satisfy $Y_i(x_1) = u_i$, $Y_i(x_2) = \ell_i$ for $i = 1,2$. We consider admissible functions $Y_i(x)$ of the form $Y_i(x;\varepsilon_i) = y_i(x) + \varepsilon_i s_i(x)$ for $i = 1,2$, where the $s_i(x)$ are arbitrary functions which satisfy $s_i(x_1) = s_i(x_2) = 0$ for $i = 1,2$. Then define

$$\mathcal{I}(\varepsilon_1,\varepsilon_2) = \int_{x_1}^{x_2} f(x, y_1+\varepsilon_1, y_1'+\varepsilon_1 s_1', y_2+\varepsilon_2 s_2, y_2'+\varepsilon_2 s_2')\,dx.$$

Since $\mathcal{I}(\varepsilon_1,\varepsilon_2)$ has an extreme value at $\varepsilon_1 = \varepsilon_2 = 0$, it is necessary that

$$\frac{\partial \mathcal{I}}{\partial \varepsilon_i}(0,0) = 0 \text{ for } i = 1,2.$$

These lead to the simultaneous ordinary differential equations

$$\frac{\partial f}{\partial y_i} - \frac{d}{dx}\left(\frac{\partial f}{\partial y_i'}\right) = 0 \text{ for } i = 1,2; \text{ or, more generally, } i = 1,2,\ldots,n.$$

(iii) Integral Constraint

As a prelude to this topic, we develop the concept of a Lagrange multiplier as follows. We wish to extremalize a function $f(x,y)$ subject to the side constraint that $g(x,y) \equiv C$ or $y = G(x)$. Then consider $F(x) = f[x, G(x)]$ and $g[x, G(x)] \equiv C$. Thus

$$F'(x) = f_x[x, G(x)] + f_y[x, G(x)]G'(x) \text{ and } g_x[x, G(x)] + g_y[x, G(x)]G'(x) \equiv 0.$$

If (x_0, y_0) is a critical point where $y_0 = G(x_0)$, it is necessary that $F'(x_0) = 0$ or

$$f_x(x_0, y_0) + f_y(x_0, y_0)G'(x_0) = 0 \text{ and } g_x(x_0, y_0) + g_y(x_0, y_0)G'(x_0) = 0.$$

Hence

$$\frac{f_x(x_0, y_0)}{f_y(x_0, y_0)} = \frac{g_x(x_0, y_0)}{g_y(x_0, y_0)} = -G'(x_0)$$

$$\Rightarrow f_x(x_0, y_0) = -\lambda g_x(x_0, y_0), f_y(x_0, y_0) = -\lambda g_y(x_0, y_0)$$

or

$$\nabla f + \lambda \nabla g = \nabla(f + \lambda g) = \mathbf{0} \text{ at } x = x_0,\ y = y_0.$$

Now, returning to the Calculus of Variations, we wish to extremalize $I(Y) = \int_{x_1}^{x_2} f(x, Y, Y')\,dx$ among all sufficiently smooth functions $Y = Y(x)$ satisfying $Y(x_1) = y_1$, $Y(x_2) = y_2$, and the constraint $J(Y) = \int_{x_1}^{x_2} g(x, Y, Y')\,dx \equiv C$, where C is a prescribed constant. Denoting this extremalizing function by y, we can no longer proceed by considering admissible functions of the form $y + \varepsilon s$. If s is a fixed function, we cannot expect $y + \varepsilon s$ to satisfy $J(y + \varepsilon s) \equiv C$ for an interval of ε values containing $\varepsilon = 0$. Only when such an interval exists can we proceed as before to require a vanishing derivative with respect to ε at $\varepsilon = 0$. Instead we define

$$Y(x; \varepsilon_1, \varepsilon_2) = y(x) + \varepsilon_1 s_1(x) + \varepsilon_2 s_2(x)$$

where $s_i(x)$ are smooth functions such that $s_i(x_1) = s_i(x_2) = 0$ for $i = 1, 2$. Regarding $s_1(x)$ and $s_2(x)$ as fixed functions our problem is to deduce a condition implied by the fact that

$$\mathcal{I}(\varepsilon_1, \varepsilon_2) = \int_{x_1}^{x_2} f(x, y + \varepsilon_1 s_1 + \varepsilon_2 s_2, y' + \varepsilon_1 s_1' + \varepsilon_2 s_2')\,dx$$

has an extreme value at $\varepsilon_1 = \varepsilon_2 = 0$ among all nearby $(\varepsilon_1, \varepsilon_2)$ satisfying

$$\mathcal{J}(\varepsilon_1, \varepsilon_2) = \int_{x_1}^{x_2} g(x, y + \varepsilon_1 s_1 + \varepsilon_2 s_2, y' + \varepsilon_1 s_1' + \varepsilon_2 s_2')\,dx \equiv C.$$

Using a Lagrange multiplier approach for constrained optimization leads to

$$\frac{\partial}{\partial \varepsilon_i}(\mathcal{I} + \lambda \mathcal{J})(0, 0) = 0 \text{ for } i = 1, 2.$$

From these we can deduce, in the usual way, that

$$\int_{x_1}^{x_2} \left[f_2 - \frac{df_3}{dx} + \lambda \left(g_2 - \frac{dg_3}{dx} \right) \right] s_i(x)\,dx = 0 \text{ for } i = 1, 2.$$

Since the above integral relations were deduced by assuming s_1 and s_2 fixed, λ may depend on them. We observe, however, that the integral relation with $i = 1$ implies λ is independent of s_2. Then applying the fundamental lemma of the Calculus of Variations to the $i = 2$ relation yields

$$f_y + \lambda g_y - \frac{d}{dx}(f_{y'} + \lambda g_{y'}) = 0$$

as the differential equation $y(x)$ and λ must satisfy. Note that this is equivalent to the unconstrained extremalization of $\int_{x_1}^{x_2} f^* \, dx$ where $f^* = f + \lambda g$. In fact we may use that observation to generalize this result to integrals containing more derivatives, more dependent variables, and more independent variables. For example, the problem of finding the partial differential equation satisfied by a function $w(x, y)$, prescribed on $\partial \mathcal{D}$, that extremalizes

$$\iint_{\mathcal{D}} f(x, y, W, W_x, W_y) \, dx \, dy \text{ subject to the constraint}$$

$$\iint_{\mathcal{D}} g(x, y, W, W_x, W_y) \, dx \, dy \equiv C$$

yields the equation (see Problem 19.1)

$$\frac{\partial f^*}{\partial w} - \frac{\partial}{\partial x}\left(\frac{\partial f^*}{\partial w_x}\right) - \frac{\partial}{\partial y}\left(\frac{\partial f^*}{\partial w_y}\right) = 0 \text{ where } f^* = f + \lambda g$$

while that of finding the ordinary differential equations satisfied by the functions $y_1(x)$ and $y_2(x)$ prescribed at $x = x_1$ and x_2 that extremalizes

$$\int_{x_1}^{x_2} f(x, Y_1, Y_1', Y_2, Y_2') \, dx \text{ subject to the constraint } \int_{x_1}^{x_2} g(x, Y_1, Y_1', Y_2, Y_2') \, dx \equiv C$$

yields the equations (see the next section)

$$\frac{\partial f^*}{\partial y_i} - \frac{d}{dx}\left(\frac{\partial f^*}{\partial y_i'}\right) = 0 \text{ for } i = 1, 2 \text{ where } f^* = f + \lambda g.$$

19.2 Queen Dido's Problem

We illustrate the approach just developed by introducing the following isoperimetric problem:

Of all closed non-self-intersecting plane curves for which the total length has the given value L, we seek the one for which the enclosed area is a maximum (see Fig. 19.1).

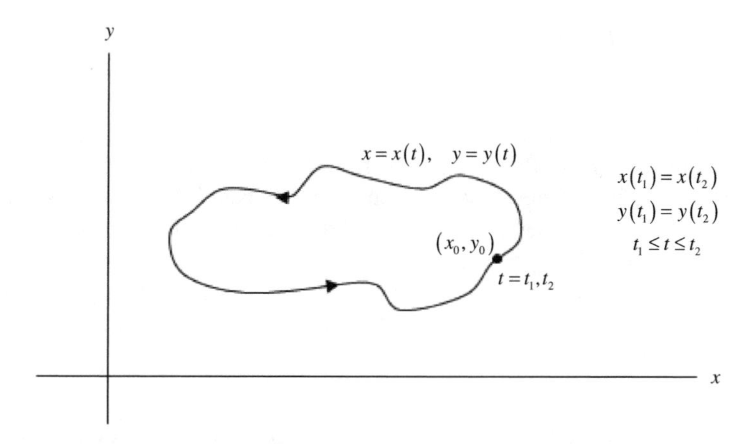

Fig. 19.1 The isoperimetric or Queen Dido's problem.

Consider a closed non-self-intersecting family of rectifiable curves described parametrically by $x = x(t)$, $y = y(t)$ for $t_1 \le t \le t_2$ where $x(t_1) = x(t_2) = x_0$ and $y(t_1) = y(t_2) = y_0$. Then

$$I = \frac{1}{2}\int_{t_1}^{t_2} (x\dot{y} - y\dot{x})\,dt \equiv \text{ enclosed area and } J = \int_{t_1}^{t_2} \sqrt{\dot{x}^2 + \dot{y}^2}\,dt = L.$$

Hence, using our Euler–Lagrange equations for constrained optimization with

$$x = t,\ y_1 = x,\ y_2 = y,\ f = \frac{1}{2}(x\dot{y} - y\dot{x}),\ \text{and } g = \sqrt{\dot{x}^2 + \dot{y}^2},$$

we find that

$$\frac{\partial f^*}{\partial x} - \frac{d}{dt}\frac{\partial f^*}{\partial \dot{x}} = 0 \text{ and } \frac{\partial f^*}{\partial y} - \frac{d}{dt}\frac{\partial f^*}{\partial \dot{y}} = 0 \text{ for } f^* = \frac{1}{2}(x\dot{y} - y\dot{x}) + \lambda\sqrt{\dot{x}^2 + \dot{y}^2}$$

or

$$\frac{1}{2}\dot{y} - \frac{d}{dt}\left(-\frac{1}{2}y + \frac{\lambda\dot{x}}{\sqrt{\dot{x}^2 + \dot{y}^2}}\right) = 0 \text{ and } -\frac{1}{2}\dot{x} - \frac{d}{dt}\left(\frac{1}{2}x + \frac{\lambda\dot{y}}{\sqrt{\dot{x}^2 + \dot{y}^2}}\right) = 0.$$

Upon integrating these equations with respect to t, we obtain

$$y - \frac{\lambda\dot{x}}{\sqrt{\dot{x}^2 + \dot{y}^2}} = k \text{ and } x + \frac{\lambda\dot{y}}{\sqrt{\dot{x}^2 + \dot{y}^2}} = h \Rightarrow (x-h)^2 + (y-k)^2 = \lambda^2.$$

Thus we have the well-known result that the closed smooth curve of given perimeter that maximizes its enclosed area is a circle. Here the Lagrange multiplier $\lambda = L/(2\pi)$ is the radius of that circle. This is generally referred to today as Queen Dido's problem after the first queen of Carthage. According to the Roman poet Virgil in his *Aeneid*, Princess Elissa of Tyre (later Queen Dido) when arriving at Cyprus in 825 BC offered to buy as much land as could be encompassed by the hide of an ox from the Berber chieftain Iarbas, who controlled that portion of the North African coast. He agreed and Dido cut the oxhide into a very long thin strip so that she had enough to include an entire nearby hill when marking the boundary of a semicircular region against the straight shoreline which afterward became her city of Carthage [94].

19.3 Hamilton's Principle for Conservative Forces of Particle Mass Systems

(a) Particle mass:

This concerns the motion of a particle of mass m due to a conservative force field $\boldsymbol{F} = -\nabla V$ where $V = V(x_1, x_2, x_3) \equiv$ potential energy. Let $x_i(t)$, where t is time, describe the trajectory of the particle that leaves point $\boldsymbol{\alpha}$ at $t = t_1$ and arrives at point $\boldsymbol{\beta}$ when $t = t_2$ so that $x_i(t_1) = \alpha_i$, $x_i(t_2) = \beta_i$ for $i = 1, 2, 3$. Denote its kinetic energy by T. Then

$$T = \frac{m}{2} \sum_{i=1}^{3} \dot{x}_i^2.$$

Now define its Lagrangian function by

$$L(x_1, x_2, x_3, \dot{x}_1, \dot{x}_2, \dot{x}_3) = T(\dot{x}_1, \dot{x}_2, \dot{x}_3) - V(x_1, x_2, x_3).$$

Hamilton's Principle asserts that the particle will move on a path which renders a minimum the action quantity

$$I = \int_{t_1}^{t_2} L(x_1, x_2, x_3, \dot{x}_1, \dot{x}_2, \dot{x}_3)\, dt.$$

This is referred to as the Principle of Least Action. Applying the Euler–Lagrange equations to this integral we obtain

$$\frac{\partial L}{\partial x_i} - \frac{d}{dt}\frac{\partial L}{\partial \dot{x}_i} = 0 \text{ for } i = 1, 2, 3.$$

To illustrate this technique let us consider the simple example of a so-called one-dimensional linear spring-mass system acting in the absence of any dissipative forces – *e.g.*, friction or air resistance (see Fig. 19.2). Here $\boldsymbol{F} = -kx\boldsymbol{i}$ where $x \equiv$ displacement of the mass m from its equilibrium position, $k \equiv$ linear spring restorative constant, and $\boldsymbol{i} \equiv$ unit vector in the x-direction. Hence

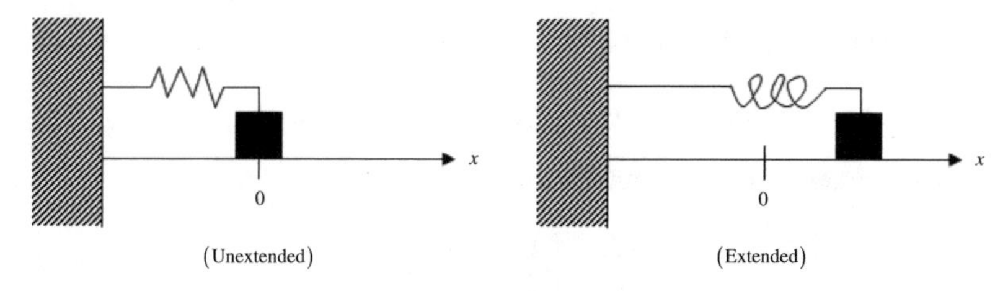

Fig. 19.2 The linear spring problem.

$$V'(x) = kx \Rightarrow V(x) = \frac{k}{2}x^2 \text{ and } T = \frac{m}{2}\dot{x}^2 \Rightarrow L = T - V = \frac{1}{2}(m\dot{x}^2 - kx^2).$$

Thus
$$\frac{d}{dt}\frac{\partial L}{\partial \dot{x}} - \frac{\partial L}{\partial x} = 0 \text{ yields } \frac{d}{dt}(m\dot{x}) + kx = 0 \text{ or } m\ddot{x} + kx = 0$$
as the equation of motion.

(b) Generalized Coordinates for Particle Mass Systems:
If rigid constraints exist in a particular problem, a mechanical system consisting of N point masses m_k, $k = 1,2,\ldots,N$, can be described by coordinates whose number is equal to the number of degrees of freedom of that system. These independent parameters, which define the position in space of this mechanical system at any time t, are called generalized coordinates, denoted by $q_j = q_j(t)$, $j = 1,2,\ldots,n \equiv$ degrees of freedom, and selected for convenience. Since, for the kth particle,

$$x_i^{(k)} = x_i^{(k)}(q_1,q_2,\ldots,q_n) \text{ and } \dot{x}_i^{(k)} = \sum_{j=1}^{n} \frac{\partial x_i^{(k)}}{\partial q_j}\dot{q}_j = \dot{x}_i^{(k)}(q_1,q_2,\ldots,q_n,\dot{q}_1,\dot{q}_2,\ldots,\dot{q}_n),$$

we see that although its potential energy written in terms of these generalized coordinates

$$\mathcal{V}_k = \mathcal{V}_k(q_1,q_2,\ldots,q_n)$$

is still a function of position alone, its kinetic energy

$$\mathcal{T}_k = \frac{m_k}{2}\sum_{i=1}^{3}\left[\dot{x}_i^{(k)}\right]^2 = \mathcal{T}_k(q_1,q_2,\ldots,q_n,\dot{q}_1,\dot{q}_2,\ldots,\dot{q}_n)$$

is now a function of position as well as of the velocities

$$\dot{q}_j = \dot{q}_j(t); j = 1,2,\ldots,n.$$

Then the Lagrangian function for the kth particle, in generalized coordinates, is given by

$$\mathcal{L}_k = \mathcal{T}_k - \mathcal{V}_k = \mathcal{L}_k(q_1, q_2, \ldots, q_n, \dot{q}_1, \dot{q}_2, \ldots, \dot{q}_n)$$

and our Euler–Lagrange equations of motion satisfy

$$\frac{d}{dt}\frac{\partial \mathcal{L}}{\partial \dot{q}_j} - \frac{\partial \mathcal{L}}{\partial q_j} = 0 \text{ for } j = 1, 2, \ldots, n$$

where

$$\mathcal{L} = \sum_{k=1}^{N} \mathcal{L}_k.$$

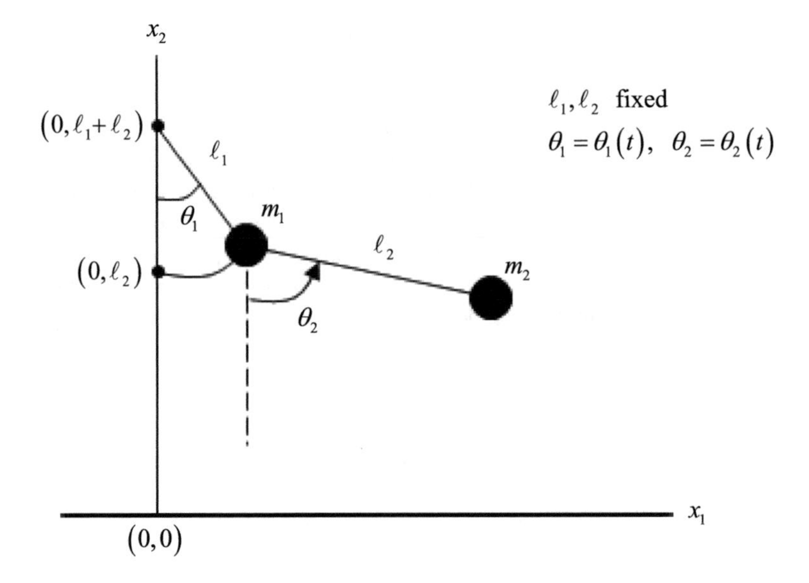

Fig. 19.3 The double plane pendulum problem.

Let us use this approach to deduce the equations of motion for a double plane pendulum (see Fig. 19.3). The first pendulum has mass m_1 and length ℓ_1, while the second one, which is attached to it, has mass m_2 and length ℓ_2. Considering $\ell_{1,2}$ as fixed, we define the generalized coordinates $\theta_1 = \theta_1(t)$ and $\theta_2 = \theta_2(t)$, which are measures of the angles in radians that ℓ_1 and ℓ_2 make with the vertical, respectively. Then the first particle mass has coordinates

$$x_1^{(1)} = \ell_1 \sin(\theta_1), x_2^{(1)} = \ell_1[1 - \cos(\theta_1)] + \ell_2$$

$$\Rightarrow \dot{x}_1^{(1)} = \ell_1 \cos(\theta_1)\dot{\theta}_1, \dot{x}_2^{(1)} = \ell_1 \sin(\theta_1)\dot{\theta}_1;$$

hence

$$\mathcal{T}_1 = \frac{m_1}{2}\ell_1^2\dot{\theta}_1^2 \text{ and } \mathcal{V}_1 = m_1 g\ell_1[1-\cos(\theta_1)];$$

while the second particle mass has coordinates

$$x_1^{(2)} = \ell_1\sin(\theta_1)+\ell_2\sin(\theta_2),\, x_2^{(2)} = \ell_1[1-\cos(\theta_1)]+\ell_2[1-\cos(\theta_2)]$$

$$\Rightarrow \dot{x}_1^{(2)} = \ell_1\cos(\theta_1)\dot{\theta}_1+\ell_2\cos(\theta_2)\dot{\theta}_2,\, \dot{x}_2^{(2)} = \ell_1\sin(\theta_1)\dot{\theta}_1+\ell_2\sin(\theta_2)\dot{\theta}_2;$$

hence

$$\mathcal{T}_2 = \frac{m_2}{2}[\ell_1^2\dot{\theta}_1^2+\ell_2^2\dot{\theta}_2^2+2\ell_1\ell_2\cos(\theta_1-\theta_2)\dot{\theta}_1\dot{\theta}_2]$$

and

$$\mathcal{V}_2 = m_2 g\{\ell_1[1-\cos(\theta_1)]+\ell_2[1-\cos(\theta_2)]\}.$$

Thus

$$\mathcal{L} = \mathcal{L}_1+\mathcal{L}_2 = \mathcal{T}_1-\mathcal{V}_1+\mathcal{T}_2-\mathcal{V}_2 = \mathcal{T}_1+\mathcal{T}_2-(\mathcal{V}_1+\mathcal{V}_2) = \mathcal{T}-\mathcal{V}$$

or

$$\begin{aligned}\mathcal{L} = {}& \frac{m_1+m_2}{2}\ell_1^2\dot{\theta}_1^2+\frac{m_2}{2}\ell_2^2\dot{\theta}_2^2+m_2\ell_1\ell_2\cos(\theta_1-\theta_2)\dot{\theta}_1\dot{\theta}_2\\ &-(m_1+m_2)g\ell_1[1-\cos(\theta_1)]-m_2 g\ell_2[1-\cos(\theta_2)]\end{aligned}$$

and we have the Euler–Lagrange equations of motion (see Problem 19.2)

$$\frac{\partial\mathcal{L}}{\partial\theta_j}-\frac{d}{dt}\frac{\partial\mathcal{L}}{\partial\dot{\theta}_j} = 0 \text{ for } j = 1,2.$$

Finally,

$$\mathcal{L}(\theta_1,\theta_2,\dot{\theta}_1,\dot{\theta}_2) = \mathcal{T}(\theta_1,\theta_2,\dot{\theta}_1,\dot{\theta}_2)-\mathcal{V}(\theta_1,\theta_2).$$

Now define

$$\mathcal{E}(t) = \sum_{j=1}^{2}\dot{\theta}_j\frac{\partial\mathcal{L}}{\partial\dot{\theta}_j}-\mathcal{L}.$$

Then

$$\dot{\mathcal{E}}(t) = \sum_{j=1}^{2}\left(\ddot{\theta}_j\frac{\partial\mathcal{L}}{\partial\dot{\theta}_j}+\dot{\theta}_j\frac{d}{dt}\frac{\partial\mathcal{L}}{\partial\dot{\theta}_j}\right)-\sum_{j=1}^{2}\left(\frac{\partial\mathcal{L}}{\partial\theta_j}\dot{\theta}_j+\frac{\partial\mathcal{L}}{\partial\dot{\theta}_j}\ddot{\theta}_j\right) = 0.$$

Therefore

$$\mathcal{E}(t) = \sum_{j=1}^{2}\dot{\theta}_j\frac{\partial\mathcal{L}}{\partial\dot{\theta}_j}-\mathcal{L} \equiv \mathcal{E}_0.$$

Observe that

$$\frac{\partial \mathcal{L}}{\partial \dot{\theta}_j} = \frac{\partial \mathcal{T}}{\partial \dot{\theta}_j} \Rightarrow \mathcal{E}(t) = \sum_{j=1}^{2} \dot{\theta}_j \frac{\partial \mathcal{T}}{\partial \dot{\theta}_j} - \mathcal{L} \equiv \mathcal{E}_0.$$

Noting that

$$\mathcal{T}(\theta_1, \theta_2, \lambda\dot{\theta}_1, \lambda\dot{\theta}_2) = \lambda^2 \mathcal{T}(\theta_1, \theta_2, \dot{\theta}_1, \dot{\theta}_2),$$

taking the partial derivative of this relationship with respect to λ, and setting $\lambda = 1$, yields

$$\sum_{j=1}^{2} \dot{\theta}_j \frac{\partial \mathcal{T}}{\partial \dot{\theta}_j} = 2\mathcal{T}.$$

Hence the so-called Hamiltonian function satisfies

$$\mathcal{E}(t) = \sum_{j=1}^{2} \dot{\theta}_j \frac{\partial \mathcal{T}}{\partial \dot{\theta}_j} - \mathcal{L} = 2\mathcal{T} - (\mathcal{T} - \mathcal{V}) = \mathcal{T} + \mathcal{V} \equiv \mathcal{E}_0$$

and there is conservation of kinetic plus potential energy.

19.4 Derivation of the One-Dimensional Elastic String Equation

In this section we apply Hamilton's Principle to a system involving a continuous distribution of mass as distinguished from a discrete set of mass particles. The means of affecting this application is the simple devise of replacing sums over discrete particles by integrals over the continuous mass distribution. As an illustration of this procedure, we consider a perfectly flexible elastic string stretched under a constant tension force τ_0 along the x-axis with its end points fixed at $x = 0$ and $x = \ell$ (see Fig. 19.4). This undisturbed state is called the equilibrium configuration. The string is then permitted to vibrate freely in a plane containing the x-axis in such a fashion that each point on it moves in a straight line perpendicular to this axis with an amplitude of vibration assumed small enough so its slope will be negligible when compared to unity. Further, assume that there are no dissipative forces acting on the string and hence we are again dealing with a conservative system. Finally, we shall also neglect the weight of the string.

We denote the transverse displacement of the string at position x and time t by $w = w(x, t)$ such that

$$w(0, t) = w(\ell, t) = 0.$$

Then, assuming a linear mass density of $\rho = \rho(w)$ and velocity of $\dot{w} \equiv \partial w / \partial t$ for the string, its kinetic energy is given by

$$T = \frac{1}{2} \int_0^{\ell} \rho(w) \dot{w}^2 \, dx.$$

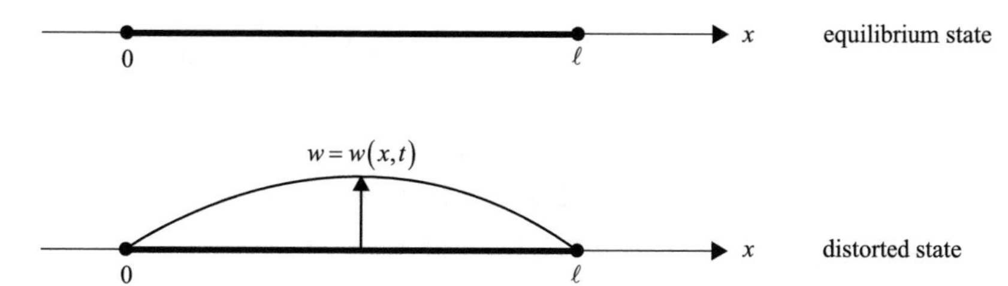

Fig. 19.4 The elastic string problem.

Since the string is perfectly flexible, the amount of work which must be done upon it to affect a given configuration is only dependent on the increase of the length of the string relative to its equilibrium length ℓ. Therefore in order to compute the potential energy of the string at any time we must merely determine the amount of work required to stretch it from the length ℓ to its total length in the configuration exhibited at that time. Thus, since the stretching force is equal to τ_0, its potential energy is given by

$$V = \tau_0 \int_0^\ell \left(\sqrt{1 + w_x^2} - 1\right) dx \text{ where } w_x \equiv \frac{\partial w}{\partial x}.$$

Now, applying Hamilton's Principle with $L = T - V$ to the vibrating string, we find that the function which describes the motion of the vibrating string is the one which renders

$$I = \int_{t_1}^{t_2} L\,dt = \int_{t_1}^{t_2} \int_0^\ell \left\{ \frac{\rho(w)}{2} \dot{w}^2 - \tau_0 \left(\sqrt{1 + w_x^2} - 1\right) \right\} dx\,dt$$

a minimum among all functions $w(x, t)$ that coincide with the actual configuration at the arbitrary times $t = t_1, t_2$ and vanish at $x = 0, \ell$. The minimization of this action integral is accomplished by using the Euler–Lagrange equation for two independent variables where we replace y by t and consider $\mathcal{D}$ to be the rectangle $0 \le x \le \ell, t_1 \le t \le t_2$ in the x-t plane. That is our equation of motion for

$$f(x, t, w, w_x, \dot{w}) = \frac{\rho(w)}{2} \dot{w}^2 - \tau_0 \left(\sqrt{1 + w_x^2} - 1\right)$$

becomes

$$\frac{\partial f}{\partial w} - \frac{\partial}{\partial x}\left(\frac{\partial f}{\partial w_x}\right) - \frac{\partial}{\partial t}\left(\frac{\partial f}{\partial \dot{w}}\right) = 0.$$

Now since

$$\frac{\partial f}{\partial w} = \frac{\rho'(w)}{2}\dot{w}^2, \; -\frac{\partial f}{\partial w_x} = \tau_0 w_x(1+w_x^2)^{-1/2}, \; \frac{\partial f}{\partial \dot{w}} = \rho(w)\dot{w},$$

this yields

$$\frac{\partial}{\partial t}[\rho(w)\dot{w}] = \tau_0 \frac{\partial}{\partial x}[w_x(1+w_x^2)^{-1/2}] + \frac{\rho'(w)}{2}\dot{w}^2.$$

Then noting that

$$\frac{\partial}{\partial t}[\rho(w)\dot{w}] = \rho'(w)\dot{w}^2 + \rho(w)\ddot{w}$$

and

$$\frac{\partial}{\partial x}[w_x(1+w_x^2)^{-1/2}] = w_{xx}(1+w_x^2)^{-1/2} - w_x^2 w_{xx}(1+w_x^2)^{-3/2}$$

$$= \frac{[w_{xx}(1+w_x^2) - w_x^2 w_{xx}]}{(1+w_x^2)^{3/2}} = \frac{w_{xx}}{(1+w_x^2)^{3/2}} \equiv \kappa(w),$$

we obtain

$$\rho(w)\ddot{w} + \frac{\rho'(w)}{2}\dot{w}^2 = \tau_0 \kappa(w).$$

For $\rho(w) \cong \rho(0) = \rho_0$ and $|w_x| << 1$, this reduces to the one-dimensional wave equation

$$\ddot{w} = c_0^2 w_{xx} \text{ where } c_0^2 \equiv \frac{\tau_0}{\rho_0}.$$

Note: $[\tau_0] = ML/\tau^2$, $[\rho_0] = M/L$, $[\tau_0/\rho_0] = (ML/\tau^2)/(M/L) = (L/\tau)^2 \Rightarrow [c_0] = L/\tau$.

In Problem 19.3 this methodology is applied to the two-dimensional situation of a vibrating circular membrane in order to deduce the relevant governing equation of motion.

19.5 Pastoral Interlude: Gamma Function

The gamma function is defined by

$$\Gamma(z) = \int_0^\infty t^{z-1} e^{-t}\, dt \text{ for } \mathrm{Re}(z) > 0$$

and is plotted in Fig. 19.5a for $z \in \mathbb{R}$. Then upon integration by parts we can deduce that

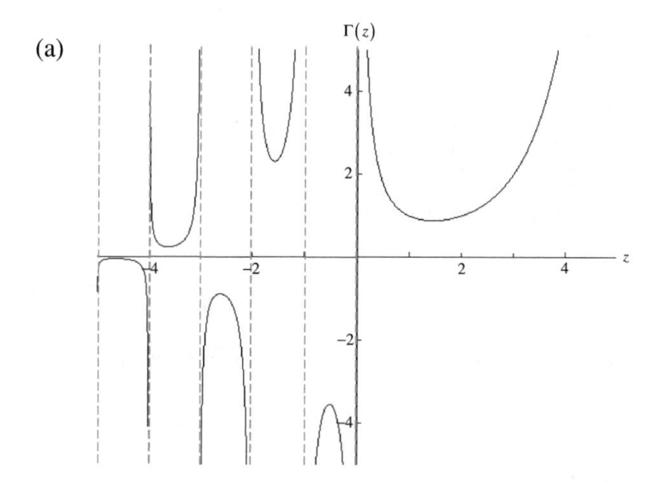

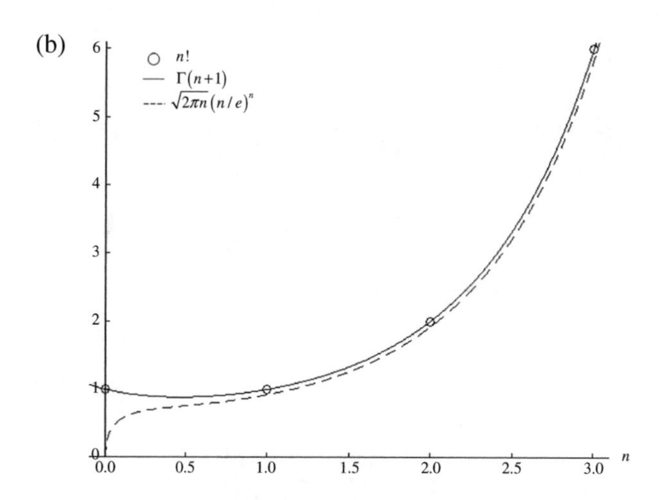

Fig. 19.5 Plots of (a) $\Gamma(z)$ for real z and (b) Stirling's formula for $\Gamma(n+1) = n!$.

$$\Gamma(z+1) = \int_0^\infty \underbrace{t^z}_{u} \underbrace{e^{-t}\,dt}_{dv} = \underbrace{t^z}_{u} \underbrace{(-e^{-t})}_{v}\Bigg|_{t=0}^{t\to\infty} - \int_0^\infty \Bigg(\underbrace{-e^{-t}}_{v}\Bigg) \underbrace{zt^{z-1}\,dt}_{du}$$

$$= z\int_0^\infty t^{z-1} e^{-t}\,dt = z\Gamma(z)$$

since

$$t^z = 0 \text{ at } t = 0 \text{ and } t^z e^{-t} \to 0 \text{ as } t \to \infty \text{ for } \operatorname{Re}(z) > 0.$$

Further employing this identity successively n-times yields

$$\Gamma(z+1) = z\Gamma\left(\underbrace{z-1}_{w}+1\right) = z\left(\underbrace{z-1}_{w}\right)\Gamma\left(\underbrace{z-1}_{w}\right) = z(z-1)\cdots(z-n+1)\Gamma(z-n+1).$$

For $z = n$, a positive integer, this reduces to

$$\Gamma(n+1) = n(n+1)\cdots(1)\Gamma(1) = n! \text{ since } \Gamma(1) = \int_0^\infty e^{-t}\,dt = -e^{-t}\Big|_0^\infty = 1.$$

We use analytic continuation in conjunction with the identity

$$\Gamma(z+1) = z\Gamma(z)$$

written in the form

$$\frac{1}{\Gamma(z)} = \frac{z}{\Gamma(z+1)}$$

to define this function for $\text{Re}(z) \leq 0$. Then

$$\frac{1}{\Gamma(0)} = \frac{0}{\Gamma(1)} = 0, \frac{1}{\Gamma(-1)} = \frac{-1}{\Gamma(0)} = 0, \cdots, \frac{1}{\Gamma(-N)} = 0 \text{ for } N = 0, 1, 2, 3, \ldots.$$

Consider

$$\underbrace{\int_0^\infty e^{-(r^n)}\,dr}_{\substack{t=r^n,\\ r=t^{1/n},\\ dr=(1/n)t^{(1/n)-1}\,dt}} = \frac{1}{n}\int_0^\infty t^{(1/n)-1}e^{-t}\,dt = \frac{1}{n}\Gamma\left(\frac{1}{n}\right).$$

Hence, for $n = 2$, we obtain

$$\int_0^\infty e^{-r^2}\,dr = \frac{1}{2}\Gamma\left(\frac{1}{2}\right)$$

and recalling from Chapter 6 that

$$\int_0^\infty e^{-r^2}\,dr = \frac{\sqrt{\pi}}{2},$$

this implies

$$\Gamma\left(\frac{1}{2}\right) = \sqrt{\pi}.$$

Finally taking $z = m - 1/2$ in our recursion relation yields

$$\Gamma\left(m+\frac{1}{2}\right) = \Gamma\left(m-\frac{1}{2}+1\right) = \left(m-\frac{1}{2}\right)\left(m-\frac{3}{2}\right)\cdots\left(m-n+\frac{1}{2}\right)\Gamma\left(m-n+\frac{1}{2}\right).$$

Thus setting $n = m$ in the above relation we find that

$$\Gamma\left(m+\frac{1}{2}\right)=\frac{(2m-1)(2m-3)\cdots(1)}{2^m}\Gamma\left(\frac{1}{2}\right)=\frac{(2m)!}{2^{2m}m!}\sqrt{\pi}.$$

19.6 Laplace's Method and Stirling's Formula

We first wish to obtain a one-term asymptotic representation for Laplace integrals of the form

$$I(k)=\int_a^b f(x)e^{kg(x)}\,dx$$

in the limit as $k\longrightarrow\infty$ where $f(x)\in C[a,b]$ and $g(x)\in C^\infty[a,b]$ and then employ it to derive Stirling's formula for $n!$ as $n\longrightarrow\infty$. We shall consider the following two cases:

(i) Let $g(x)$ be such that $g'(x)\neq 0$ for any $x\in[a,b]$. Then integrating $I(k)$ by parts we obtain

$$\begin{aligned}I(k)&=\int_a^b \frac{f(x)}{kg'(x)}kg'(x)e^{kg(x)}\,dx=\frac{f(x)}{kg'(x)}e^{kg(x)}\Big|_{x=a}^{b}-\int_a^b\frac{d}{dx}\left[\frac{f(x)}{kg'(x)}\right]e^{kg(x)}\,dx\\&=\underbrace{\frac{f(x)}{kg'(x)}e^{kg(x)}\Big|_{x=a}^{b}}_{\substack{\text{Either the contribution}\\\text{at } x=a \text{ or } b \text{ dominates}}}+O\left(\frac{e^{Mk}}{k^2}\right)\ \text{as } k\longrightarrow\infty \text{ where } M=\max\{g(a),g(b)\}.\end{aligned}$$

We now examine a typical example of this type where $g(x)=x$ which implies $g'(x)\equiv 1$:

$$\begin{aligned}I(k)&=\int_0^1\frac{e^{kx}}{1+x^2}\,dx=\frac{e^{kx}}{k(1+x^2)}\Big|_{x=0}^{1}-\int_0^1\frac{2x}{(1+x^2)^2}\frac{e^{kx}}{k}\,dx\\&=\frac{e^k}{2k}-\frac{1}{k}+O\left(\frac{e^k}{k^2}\right)\sim\frac{e^k}{2k}\ \text{as } k\longrightarrow\infty.\end{aligned}$$

(ii) We next consider $g(x)$ such that $g'(\alpha)=0$ where $a<\alpha<b$ and $g''(x)<0$ for $x\in[a,b]$. Let $\delta>0$ such that $a<\alpha-\delta<\alpha<\alpha+\delta<b$. Hence

$$I(k)=\underbrace{\int_a^{\alpha-\delta}f(x)e^{kg(x)}\,dx}_{①}+\underbrace{\int_{\alpha-\delta}^{\alpha+\delta}f(x)e^{kg(x)}\,dx}_{②}+\underbrace{\int_{\alpha+\delta}^{b}f(x)e^{kg(x)}\,dx}_{③}.$$

We shall examine each of these integrals separately. Note that in ① and ③ $|x-\alpha|>\delta$.

(1) Here $a \le x \le \alpha - \delta$. Then

$$\left|(1)\right| \le \int_a^{\alpha-\delta} |f(x)| e^{kg(x)}\, dx.$$

Making use of the Taylor polynomial for

$$g(x) = g(\alpha) + g''(x_1)\frac{(x-\alpha)^2}{2} \text{ where } x_1 \in (x, \alpha),$$

and applying the generalized integral mean value theorem (see Section 20.2), we obtain

$$\left|(1)\right| \le \frac{e^{kg(\alpha)}}{e^{k\Delta_1}} \int_a^{\alpha-\delta} |f(x)|\, dx \text{ where } \Delta_1 = \frac{-g''(c_1)\delta^2}{2} > 0 \text{ for } c_1 \in (a, \alpha-\delta).$$

(3) Here $\alpha + \delta \le x \le b$. Then in a similar manner to (1) we can deduce that

$$\left|(3)\right| \le \frac{e^{kg(\alpha)}}{e^{k\Delta_2}} \int_{\alpha+\delta}^{b} |f(x)|\, dx \text{ where } \Delta_2 = \frac{-g''(c_2)\delta^2}{2} > 0 \text{ for } c_2 \in (\alpha+\delta, b).$$

(2) Here $\alpha - \delta \le x \le \alpha + \delta$. Then, assuming δ to be "small enough" (see below) so that

$$g(x) \cong g(\alpha) - A(x-\alpha)^2 \text{ where } A = \frac{-g''(\alpha)}{2} > 0,$$

$$\begin{aligned}
(2) &\cong f(\alpha)e^{kg(\alpha)} \int_{\alpha-\delta}^{\alpha+\delta} e^{-kA(x-\alpha)^2}\, dx = 2f(\alpha)e^{kg(\alpha)} \int_{\alpha}^{\alpha+\delta} e^{-kA(x-\alpha)^2}\, dx \\
&= \frac{2f(\alpha)}{(kA)^{1/2}} e^{kg(\alpha)} \int_0^{(kA)^{1/2}\delta} e^{-z^2}\, dz \sim \frac{2f(\alpha)}{(kA)^{1/2}} e^{kg(\alpha)} \int_0^{\infty} e^{-z^2}\, dz \\
&= \sqrt{\frac{-2\pi}{g''(\alpha)}} f(\alpha) \frac{e^{kg(\alpha)}}{k^{1/2}} \text{ as } k \to \infty,
\end{aligned}$$

where δ is such that $(kA)^{12}\delta \to \infty$ as $k \to \infty$ which implies that $\delta = o(k^{-1/2})$ as $k \to \infty$.

Finally comparing these three results and observing that the numerators of each are proportional to $e^{kg(\alpha)}$ while $k^{1/2}$, the denominator of (2), is less than $e^{k\Delta_{1,2}}$, the denominators appearing in the inequalities of (1) and (3), as $k \to \infty$ for such δ, we can conclude that the dominant contribution to $I(k)$ comes from (2) which yields the asymptotic representation of Laplace's Method (reviewed by Copson [22])

$$I(k) = \int_a^b f(x)e^{kg(x)}\, dx \sim \sqrt{\frac{-2\pi}{g''(\alpha)}} f(\alpha) \frac{e^{kg(\alpha)}}{k^{1/2}} \text{ as } k \to \infty,$$

when

$$g'(\alpha) = 0 \text{ where } a < \alpha < b \text{ and } g''(x) < 0 \text{ for } x \in [a, b].$$

We are now ready to find an asymptotic representation for $n!$ as $n \to \infty$. Recall from Section 19.5

$$\Gamma(n+1) = n! = \int_0^\infty t^n e^{-t}\, dt = \int_0^\infty e^{-t} e^{n \ln(t)}\, dt.$$

As a first attempt identify this as a Laplace integral with

$$x = t,\ k = n,\ f(t) = e^{-t},\ g(t) = \ln(t),\ a = 0,\ b \to \infty.$$

Then since $g'(t) = 1/t \neq 0$ case (i) applies and thus we try integration by parts obtaining

$$n! = \int_0^\infty \underbrace{e^{-t}}_{u} \underbrace{t^n\, dt}_{dv} = \underbrace{e^{-t}}_{u} \underbrace{\frac{t^{n+1}}{n+1}}\Bigg|_{t=0}^{t\to\infty} - \int_0^\infty \underbrace{\frac{t^{n+1}}{n+1}}_{v} \underbrace{(-e^{-t})\, dt}_{du} = \int_0^\infty e^{-t} \frac{t^{n+1}}{n+1}\, dt,$$

due to the facts that $t^{n+1} = 0$ at $t = 0$ and $t^{n+1}e^{-t} \to 0$ as $t \to \infty$ for $n > 0$, which clearly does not work. Instead we introduce the following change of variables

$$n! = \underbrace{\int_0^\infty t^n e^{-t}\, dt}_{\substack{t=nx,\\ dt=n\,dx}} = n^{n+1} \int_0^\infty x^n e^{-nx}\, dx = n^{n+1} \int_0^\infty e^{n[\ln(x) - x]}\, dx = n^{n+1} I(n).$$

Now identifying $I(n)$ as a Laplace integral with

$$k = n,\ f(x) \equiv 1,\ g(x) = \ln(x) - x,\ a = 0,\ b \to \infty;$$

we find that

$$g'(x) = \frac{1}{x} - 1,\ g'(1) = 1,\ g''(x) = -\frac{1}{x^2} < 0;$$

and thus case (ii) applies with $\alpha = 1$. Employing the asymptotic representation of Laplace's Method and noting that $g(1) = g''(1) = -1$ while $f(1) = 1$, we obtain Stirling's Formula

$$n! = n^{n+1} I(n) \sim n^{n+1} \sqrt{2\pi} \frac{e^{-n}}{n^{1/2}} = \sqrt{2\pi} n^{n+1/2} e^{-n} \text{ as } n \to \infty.$$

We examine its accuracy in Table 19.1 by defining [1]

$$\Gamma(z+1) = \sqrt{2\pi z} z^z e^{-z} f_1(z) \text{ as } z \to \infty$$

where

$$f_1(z) = 1 + \frac{1}{12z} + \frac{1}{288z^2} - \frac{139}{51{,}840z^3} + O(z^{-4}).$$

Table 19.1 Asymptotic accuracy of Stirling's Formula $\sqrt{2\pi n}\, n^n e^{-n} = n!/f_1(n)$.

n	$n!$	$f_1(n)$
1	1	1.08444
5	120	1.01678
10	3.62880×10^6	1.00837
50	3.04141×10^{64}	1.00167
100	9.33262×10^{157}	1.00083

From this table we can see that Stirling's Formula represents a very good approximation to $n!$ for large n. Indeed it is even a reasonably good approximation when $n = 1$! (here the "!" should be interpreted as an exclamation point rather the factorial notation). In this context see Fig. 19.5b.

19.7 Pastoral Interlude: Bessel Functions

(i) The Bessel Function of the First Kind of order $\nu \geq 0$ is denoted by $J_\nu(x)$ and satisfies the Bessel ordinary differential equation

$$L[J_\nu(x)] = x^2 J_\nu''(x) + xJ_\nu'(x) + (x^2 - \nu^2)J_\nu(x) = 0, x > 0.$$

Here $x = 0$ is a so-called regular singular point for this equation. Recall $x = 0$ is called an ordinary point for the equation

$$y''(x) + p(x)y'(x) + q(x)y(x) = 0, |x| < R;$$

if

$$p(x) = \sum_{n=0}^{\infty} p_n x^n \text{ and } q(x) = \sum_{n=0}^{\infty} q_n x^n \text{ for } |x| < R;$$

and one seeks a series solution of such an equation of the form (see Section 12.7)

$$y(x) = \sum_{n=0}^{\infty} a_n x^n.$$

Consider the Euler equation

$$x^2 y''(x) + p_0 xy'(x) + q_0 y(x) = 0, x > 0;$$

one seeks a solution of this equation of the form $y(x) = x^r$. Finally consider

$$L[y(x)] = x^2 y''(x) + [xp(x)]xy'(x) + [x^2 q(x)]y(x) = 0, x > 0.$$

If

$$xp(x) = \sum_{n=0}^{\infty} p_n x^n \text{ and } x^2 q(x) = \sum_{n=0}^{\infty} q_n x^n \text{ for } 0 < x < R,$$

then $x = 0$ is said to be a regular singular point of this equation and one seeks a solution of it of the form

$$y(x) = \sum_{n=0}^{\infty} a_n(r) x^{n+r}$$

by the method of Frobenius. In particular for the Bessel equation

$$y(x) = J_\nu(x),\ xp(x) \equiv 1, \text{ and } x^2 q(x) = x^2 - \nu^2;$$

which implies that

$$p_0 = 1,\ p_n = 0 \text{ for } n \geq 1;\ q_0 = -\nu^2,\ q_1 = 0,\ q_2 = 1,\ q_n = 0 \text{ for } n \geq 3.$$

Hence we seek a solution of the Bessel equation of the form

$$J_\nu(x) = \sum_{n=0}^{\infty} a_n x^{n+r}.$$

Then

$$J_\nu'(x) = \sum_{n=0}^{\infty} (n+r) a_n x^{n+r-1} \text{ and } J_\nu''(x) = \sum_{n=0}^{\infty} (n+r)(n+r-1) a_n x^{n+r-2}.$$

Thus

$$L[J_\nu(x)] = \sum_{n=0}^{\infty} \underbrace{[(n+r)(n+r-1) + (n+r) - \nu^2]}_{(n+r)^2-\nu^2} a_n x^{n+r} + \underbrace{\sum_{n=0}^{\infty} a_n x^{n+r+2}}_{\substack{j=n+2,\\ n=j-2}}$$

$$= \sum_{n=0}^{\infty} (n+r+\nu)(n+r-\nu) a_n x^{n+r} + \sum_{n=2}^{\infty} a_{n-2} x^{n+r} = 0$$

or

$$(r^2 - \nu^2) a_0 + [(r+1)^2 - \nu^2] a_1 x + \sum_{n=2}^{\infty} [(n+r+\nu)(n+r-\nu) a_n + a_{n-2}] x^n = 0,$$

which, most generally, for $a_0 \neq 0$ implies that

$$r = r_{1,2} = \pm\nu,\ a_1 = 0, \text{ and } a_n = -\frac{a_{n-2}}{(n+r+\nu)(n+r-\nu)} \text{ for } n \geq 2.$$

Observe that for the regular singular point problem with arbitrary coefficient functions this procedure would yield the so-called indicial equation to determine r

$$F(r) = r^2 + (p_0 - 1)r + q_0 = 0$$

and the recursion relation to determine the a_n's for each r

$$F(n+r)a_n + D_n(r) = 0 \text{ for } n \geq 1 \text{ where } D_n(r) = \sum_{k=0}^{n-1} [(k+r)p_{n-k} + q_{n-k}]a_k,$$

the form of the sum in $D_n(r)$ following from the employment of the Cauchy product. When particularized to the coefficient functions appropriate for the Bessel equation these relations reduce to the results just obtained since

$$F(r) = r^2 - \nu^2 \text{ and } D_n(r) = \begin{cases} 0, & \text{for } n = 1 \\ a_{n-2}, & \text{for } n \geq 2 \end{cases}.$$

Note that given this second-order difference equation then the a_{2m}'s and a_{2m+1}'s are proportional to a_0 and a_1, respectively. Therefore, from the latter relationship and the fact that $a_1 = 0$, we can conclude

$$a_{2m+1} = 0 \text{ for } m = 0, 1, 2, \ldots.$$

We shall now determine the a_{2m}'s and hence the solution for each of the two values of $r = \pm\nu$.

(1) $r = \nu$: For the Bessel function of the first kind it is traditional to take

$$a_0 = \frac{2^{-\nu}}{\Gamma(\nu+1)}.$$

Then from the second-order difference equation for the a_{2m}'s and the recursion relation for $\Gamma(z)$

$$a_{2m} = (-1)^m \frac{(1/2)^{2m+\nu}}{m!\Gamma(\nu+m+1)} \text{ for } m = 0, 1, 2, \ldots,$$

and hence

$$J_\nu(x) = \sum_{m=0}^{\infty} \frac{(-1)^m (x/2)^{2m+\nu}}{m!\Gamma(\nu+m+1)}.$$

(2) $r = -\nu$: In an identical manner to the previous case, it is now traditional to take

$$a_0 = \frac{2^{\nu}}{\Gamma(-\nu+1)};$$

from which it follows in exactly the same way that

$$a_{2m} = \frac{(-1)^m (1/2)^{2m-v}}{m!\Gamma(-v+m+1)} \text{ for } m = 0, 1, 2, \ldots;$$

and hence

$$J_{-v}(x) = \sum_{m=0}^{\infty} \frac{(-1)^m (x/2)^{2m-v}}{m!\Gamma(-v+m+1)}.$$

We now wish to particularize these results to Bessel functions of the first kind of integer order by letting $v = n$, where n is a nonnegative integer. That is

① :

$$J_n(x) = \sum_{m=0}^{\infty} \frac{(-1)^m (x/2)^{2m+n}}{m!\Gamma(n+m+1)}.$$

Recalling that $\Gamma(N+1) = N!$ when N is a positive integer and defining $0! \equiv \Gamma(1) = 1$, we can conclude that

$$\Gamma(n+m+1) = (n+m)!$$

and thus

$$J_n(x) = \sum_{m=0}^{\infty} \frac{(-1)^m (x/2)^{2m-n}}{m!(n+m)!};$$

while

② :

$$J_{-n}(x) = \sum_{m=0}^{\infty} \frac{(-1)^m (x/2)^{2m-n}}{m!\Gamma(-n+m+1)}.$$

Recalling that $1/\Gamma(-N+1) = 0$ for $N = 0, 1, 2, 3, \ldots$, we can conclude that

$$\frac{1}{\Gamma(-n+m+1)} = 0 \text{ for } m = 0, 1, \ldots, n-1$$

and therefore

$$J_{-n}(x) = \underbrace{\sum_{m=n}^{\infty} \frac{(-1)^m (x/2)^{2m-n}}{m!\Gamma(-n+m+1)}}_{\substack{k=m-n,\\ m=n+k,\\ 2m-n=2k+n}} = \sum_{k=0}^{\infty} \frac{(-1)^{k+n} (x/2)^{2k+n}}{(k+n)!\Gamma(k+1)}$$

$$= (-1)^n \sum_{k=0}^{\infty} \frac{(-1)^k (x/2)^{2k+n}}{(k+n)!k!} = (-1)^n J_n(x)$$

and thus $J_{\pm v}(x)$ are not linearly independent if v is a nonnegative integer.

There is a generating function for $J_{\pm n}(x)$ given by

$$e^{(z-1/z)(x/2)} = \sum_{n=-\infty}^{\infty} J_n(x)z^n,$$

which is demonstrated as follows:

$$e^{(z-1/z)(x/2)} = e^{(x/2)z}e^{-(x/2)z^{-1}} = \left[\sum_{k=0}^{\infty} \frac{(x/2)^k z^k}{k!}\right]\left[\frac{(-1)^m (x/2)^m z^{-m}}{m!}\right].$$

Then the coefficient of each z^n can be determined by letting $k = m + n$ for all m with n, a fixed nonnegative integer:

$$\sum_{m=0}^{\infty} \frac{(x/2)^{m+n}}{(m+n)!}\frac{(-1)^m (x/2)^m}{m!} = \sum_{m=0}^{\infty} \frac{(-1)^m (x/2)^{2m+n}}{m!(n+m)!} = J_n(x);$$

while the coefficient of each z^{-n} can be determined by letting $m = k + n$ for all k with n, a fixed nonnegative integer:

$$\sum_{k=0}^{\infty} \frac{(x/2)^k}{k!}\frac{(-1)^{k+n}(x/2)^{k+n}}{(k+n)!} = (-1)^n \sum_{k=0}^{\infty} \frac{(-1)^k (x/2)^{2k+n}}{(k+n)!k!}$$
$$= (-1)^n J_n(x) = J_{-n}(x).$$

(ii) Bessel Functions of the Second Kind of Order $\nu \geq 0$.
Given the linear dependence between $J_{\pm n}(x)$ just demonstrated, we wish to deduce a second linearly independent solution of the Bessel equation of order ν written in the form

$$L[y(x)] = x^2 y''(x) + xy'(x) + (x^2 - \nu^2)y(x) = 0$$

when $\nu = n$, a nonnegative integer. We begin by solving this problem completely from first principles for $\nu = 0$. As usual, we seek a solution of the form

$$y(x) = \varphi(x,r) = \sum_{n=0}^{\infty} a_n(r)x^{n+r}$$

where $a_0(r) \equiv a_0 = 1$ and find

$$L[\varphi(x,r)] = a_0(r)r^2 x^r + a_1(r)(r+1)^2 x^{r+1} + \sum_{n=2}^{\infty}[(n+r)^2 a_n(r) + a_{n-2}(r)]x^{n+r}$$
$$= 0$$

which implies that

$$r^2 = 0 \text{ or } r = r_{1,2} = 0;\ a_1(0) = 0;\ a_n(r) = \frac{-a_{n-2}(r)}{(n+r)^2} \text{ for } n \geq 2.$$

Hence

$$a_{2m+1}(0) = 0 \text{ for } m = 1, 2, 3, \ldots;$$

and

$$a_{2m}(r) = \frac{-a_{2m-2}(r)}{(2m+r)^2} = \frac{(-1)^2 a_{2m-4}(r)}{(2m+r)^2(2m-2+r)^2}$$
$$= \frac{(-1)^m a_0}{(2m+r)^2(2m-2+r)^2 \cdots (4+r)^2(2+r)^2} \text{ for } m = 1, 2, 3, \ldots.$$

Thus

$$a_{2m}(0) = \frac{(-1)^m}{(2m)^2(2m-2)^2 \cdots (4)^2(2)^2} = \frac{(-1)^m}{2^{2m}(m!)^2} \text{ for } m = 0, 1, 2, 3, \ldots.$$

Therefore

$$y_1(x) = \varphi(x, 0) = \sum_{m=0}^{\infty} a_{2m}(0)x^{2m} = \sum_{m=0}^{\infty} \frac{(-1)^m x^{2m}}{2^{2m}(m!)^2} = J_0(x).$$

To seek a second linearly independent solution, we consider

$$L[\varphi(x, r)] = a_0 r^2 x^r$$

by taking

$$a_1(r)(r+1)^2 \equiv 0 \text{ and } (n+r)^2 a_n(r) + a_{n-2}(r) \equiv 0 \text{ for } n \geq 2.$$

Then

$$\frac{\partial}{\partial r} L[\varphi(x, r)] = L\left[\frac{\partial \varphi(x, r)}{\partial r}\right] = a_0[2r + r^2 \ln(x)]x^r.$$

Thus

$$L\left[\frac{\partial \varphi(x, 0)}{\partial r}\right] = 0.$$

Hence

$$y_2(x) = \ln(x)\varphi(x, 0) + \sum_{n=1}^{\infty} a_n'(0)x^n.$$

Differentiating $a_1(r)(r+1)^2 \equiv 0$ with respect to r, we obtain

$$a_1'(r)(r+1)^2 + 2a_1(r)(r+1) \equiv 0,$$

which when evaluated at $r = 0$ yields

$$a_1'(0) = 0$$

since $a_1(0) = 0$. Similarly, since $a_{2m+1}(0) = 0$, by differentiating the recursion relation we can deduce that

$$a'_{2m+1}(0) = 0 \text{ for } m = 1, 2, 3, \ldots.$$

Recall that for

$$f(x) = c_0 (x+b_1)^{p_1} (x+b_2)^{p_2} \cdots (x+b_n)^{p_n},$$

$$\ln[f(x)] = \ln(c_0) + p_1 \ln(x+b_1) + p_2 \ln(x+b_2) + \cdots + p_n \ln(x+b_n)$$

which upon logarithmic differentiation implies

$$\frac{f'(x)}{f(x)} = \frac{p_1}{x+b_1} + \frac{p_2}{x+b_2} + \cdots + \frac{p_n}{x+b_n}.$$

Therefore, upon applying this result to the formula for $a_{2m}(r)$,

$$\frac{a'_{2m}(r)}{a_{2m}(r)} = -2\left(\frac{1}{2m+r} + \frac{1}{2m-2+r} + \cdots + \frac{1}{4+r} + \frac{1}{2+r}\right)$$

and evaluating that expression at $r = 0$, we can deduce

$$\frac{a'_{2m}(0)}{a_{2m}(0)} = -\left(\frac{1}{m} + \frac{1}{m-1} + \cdots + \frac{1}{2} + 1\right) = -H_m \text{ where } H_m = \sum_{k=1}^{m} \frac{1}{k}$$

or

$$a'_{2m}(0) = -H_m a_{2m}(0) = \frac{(-1)^{m+1} H_m}{2^{2m}(m!)^2}.$$

Finally,

$$y_2(x) = \ln(x) J_0(x) + \sum_{m=1}^{\infty} a'_{2m}(0) x^{2m} = \ln(x) J_0(x) + \sum_{m=1}^{\infty} \frac{(-1)^{m+1} H_m}{2^{2m}(m!)^2} x^{2m}.$$

By convention the Bessel Function of the Second Kind of Order Zero, denoted by $Y_0(x)$, is defined as [62]

$$\begin{aligned} Y_0(x) &= \frac{2}{\pi}[y_2(x) + \{\gamma - \ln(2)\} J_0(x)] \\ &= \frac{2}{\pi}\left[\left\{\gamma + \ln\left(\frac{x}{2}\right)\right\} J_0(x) \sum_{m=1}^{\infty} \frac{(-1)^{m+1} H_m}{2^{2m}(m!)^2} x^{2m}\right] \end{aligned}$$

where [1]

$$\gamma = \lim_{m \to \infty} [H_m - \ln(m)] = 0.57721566491053286060651 2\ldots.$$

Functions $J_0(x)$ and $Y_0(x)$ are plotted in Fig. 19.6. For order ν, we define [62]

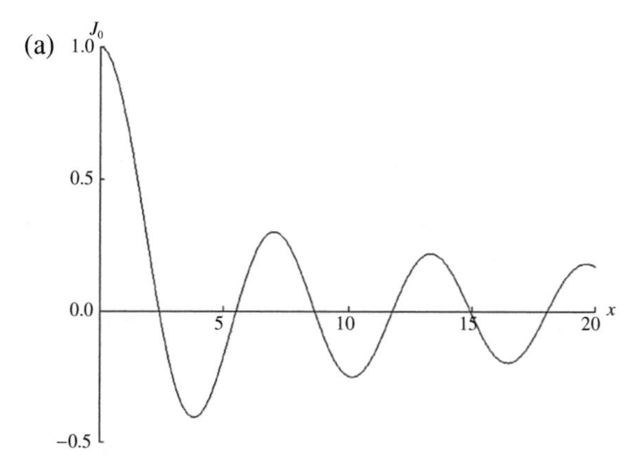

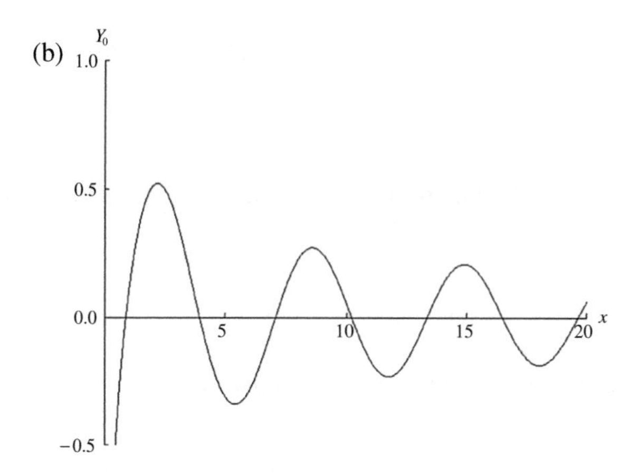

Fig. 19.6 Plots of (a) $J_0(x)$ and (b) $Y_0(x)$ versus x.

$$Y_\nu(x) = \frac{\cos(\pi\nu)J_\nu(x) - J_\nu(x)}{\sin(\pi\nu)} \text{ for } \nu \neq n;$$

$$Y_n(x) = \lim_{\nu\to n} Y_\nu(x) = \frac{2}{\pi}\left[\gamma + \ln\left(\frac{x}{2}\right)\right]J_n(x) - \frac{1}{\pi}\sum_{m=0}^{n-1}\frac{(n-m-1)!}{m!}\left(\frac{x}{2}\right)^{2m-n}$$
$$+\frac{1}{\pi}\sum_{m=0}^{\infty}(-1)^{m+1}\frac{H_m + H_{m+n}}{m!(m+n)!}\left(\frac{x}{2}\right)^{2m+n} \quad \text{for } n \geq 1$$

where $H_0 \equiv 0$ and $Y_{n=0}(x) = Y_0(x)$. Then the Bessel equation of order ν has the general solution composed of a linear combination of these two functions

$J_\nu(x)$ and $Y_\nu(x)$ given by

$$y(x) = c_1 J_\nu(x) + c_2 Y_\nu(x).$$

In particular, these two functions have asymptotic representations [62]

$$J_\nu(x) \sim \sqrt{\frac{\pi}{2x}} \cos\left(x - \frac{\nu\pi}{2} - \frac{\pi}{4}\right), \; Y_\nu(x) \sim \sqrt{\frac{\pi}{2x}} \sin\left(x - \frac{\nu\pi}{2} - \frac{\pi}{4}\right) \text{ as } x \to \infty;$$

which is demonstrated explicitly in the next section for $\nu = n$.

19.8 Method of Stationary Phase and Asymptotic Representation of Bessel Functions

In this section we shall first develop a one-term asymptotic representation for an integral of the form

$$I(k) = \int_a^b f(x)e^{ikg(x)}\,dx \text{ as } k \to \infty,$$

where f, which may be a complex function, and g, which is a real one, are such that

$$f(x) \in C[a,b] \text{ and } g(x) \in C^\infty[a,b] \text{ with } g'(\alpha) = 0 \text{ for } a < \alpha < b,$$

by the Method of Stationary Phase (reviewed by Copson, [22]) and then use it to deduce the asymptotic behavior of $J_n(x)$ as $x \to \infty$.

In order to motivate this procedure we now consider

$$I(k) = \int_a^b f(x)\cos[kg(x)]\,dx + i\int_a^b f(x)\sin[kg(x)]\,dx$$

and introduce the following special function for $g(x)$:

$$g(x) = \begin{cases} g_1(x), & \text{for } a \le x \le x_1 \\ g_0, & \text{for } x_1 \le x \le x_2 \\ g_2(x), & \text{for } x_2 \le x \le b \end{cases}.$$

Let us examine the behavior for such a $g(x)$ of

$$\int_a^b f(x)\cos[kg(x)]\,dx$$

when k is large and positive. In the intervals $a \le x \le x_1$ and $x_2 \le x \le b$ the function $y = g(x)$ satisfies

$$g(x+\Delta x) = y + \Delta y \cong g(x) + g'(x)\Delta x \text{ for } \Delta x \text{ sufficiently small}$$

or

$$\Delta y \cong g'(x)\Delta x;$$

and in those intervals $\cos[kg(x)]$ exhibits periodicity of the form

$$\cos[kg(x+\Delta x)] = \cos(ky + k\Delta y) = \cos(ky + 2\pi) = \cos(ky) = \cos[kg(x)]$$

which implies that

$$2\pi = k\Delta y \cong kg'(x)\Delta x \text{ or } \Delta x \cong \frac{2\pi}{g'(x)k}.$$

Note that for k large, Δx will be small and hence $\cos[kg(x)]$ will oscillate rapidly in these intervals while over each such period $f(x)$ will not. Thus the integrals in question are approximately zero by cancellation and therefore

$$\int_a^b f(x)\cos[kg(x)]\,dx \cong \int_{x_1}^{x_2} f(x)\cos[kg(x)]\,dx$$

for k large with the same argument holding for $\sin[kg(x)]$. Observe that $g\prime(x) \equiv 0$ for $x \in [x_1, x_2]$. Now given a $g(x)$ the derivative of which has a single zero at $x = \alpha$ in the interval (a, b) it is plausible that the major contribution to our original $I(k)$ would be

$$I(k) \sim \int_{\alpha-\delta}^{\alpha+\delta} f(x)e^{ikg(x)}\,dx \text{ as } k \to \infty$$

for sufficiently small δ as in Laplace's Method of Section 19.6. There are three possible cases for such a situation depending upon whether $g''(\alpha)$ is less than, equal to, or greater than 0. In this section we shall assume that $g''(\alpha) \neq 0$ and handle in Problem 19.7 the case for which $g''(\alpha) = 0$ but $g'''(\alpha) \neq 0$. We begin below with the case of $g'(\alpha) = 0$, $g''(\alpha) < 0$.

(i) $g'(\alpha) = 0, g''(\alpha) < 0$: Proceeding exactly as in Section 19.6 we obtain

$$I(k) \sim 2f(\alpha)\sqrt{\frac{-2}{g''(\alpha)}\frac{e^{ikg(\alpha)}}{k^{1/2}}}\int_0^\infty e^{-iz^2}\,dz \text{ as } k \to \infty.$$

(ii) $g'(\alpha) = 0, g''(\alpha) > 0$: Again proceeding as in Section 19.6 but not making use of the evenness of the integrand we would obtain

$$I(k) \sim f(\alpha)\sqrt{\frac{2}{g''(\alpha)}\frac{e^{ikg(\alpha)}}{k^{1/2}}}\int_{-\infty}^\infty e^{iz^2}\,dz \text{ as } k \to \infty.$$

Then

$$\int_{-\infty}^{\infty} e^{iz^2}\,dz = \int_{-\infty}^{0} e^{iz^2}\,dz + \int_{0}^{\infty} e^{iz^2}\,dz = 2\int_{0}^{\infty} e^{iz^2}\,dz$$

since

$$\underbrace{\int_{-\infty}^{0} e^{iz^2}\,dz}_{\substack{w=-z,\\ dw=-dz}} = -\int_{\infty}^{0} e^{iw^2}\,dw = \int_{0}^{\infty} e^{iw^2}\,dw = \int_{0}^{\infty} e^{iz^2}\,dz.$$

Therefore

$$I(k) \sim 2f(\alpha)\sqrt{\frac{2}{g''(\alpha)}}\frac{e^{ikg(\alpha)}}{k^{1/2}}\int_{0}^{\infty} e^{iz^2}\,dz \text{ as } k \to \infty.$$

Although this result could have been obtained in exactly the same manner as was that for case (i) by means of symmetry, we chose to use an alternate approach because the latter must be employed in the corresponding Method of Stationary Phase Problem 19.7 which lacks this property.

Both cases (i) and (ii) can be combined in the single formula

$$I(k) \sim 2f(\alpha)\sqrt{\frac{2}{|g''(\alpha)|}}\frac{e^{ikg(\alpha)}}{k^{1/2}}\int_{0}^{\infty} e^{isgm[g''(\alpha)]z^2}\,dz \text{ as } k \to \infty$$

where

$$sgm(\gamma) = \begin{cases} 1, & \text{for } \gamma > 0 \\ -1, & \text{for } \gamma < 0 \end{cases}.$$

Given that

$$\int_{0}^{\infty} e^{-iz^2}\,dz = \overline{\int_{0}^{\infty} e^{iz^2}\,dz},$$

where the overbar denotes the complex conjugate, we need only evaluate

$$\int_{0}^{\infty} e^{iz^2}\,dz$$

to deduce the desired asymptotic representation. Hence we perform the complex contour integration about the closed curve Γ_R depicted in Fig. 19.7a with $n = 2$ and then take the limit as $R \to \infty$ where since the integrand is analytic within that region and on its boundary

$$\oint_{\Gamma_R}^{z=re^{i\theta}} e^{iz^2}\,dz = \int_{①}^{\theta=0} e^{iz^2}\,dz + \int_{②}^{r=R} e^{iz^2}\,dz + \int_{③}^{\theta=\pi/4} e^{iz^2}\,dz = 0.$$

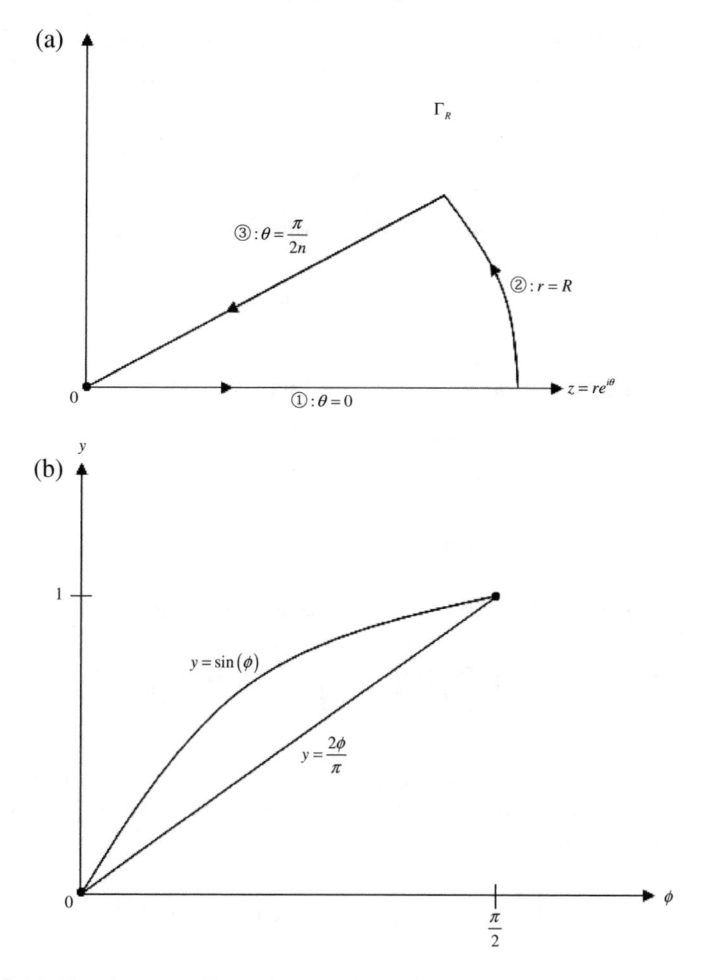

Fig. 19.7 (a) Closed contour Γ_R in the complex z-plane and (b) plots of $y = \sin(\varphi)$, $2\varphi/\pi$ for $0 \leq \varphi \leq \pi/2$.

We shall now evaluate each of these line integrals separately in the order ①, ③, and ②:

① $\theta = 0$: $z = r$ where $0 \leq r \leq R$ and $dz = dr$. Thus

$$\int_{①}^{\theta=0} e^{iz^2}\, dz = \int_0^R e^{ir^2}\, dr$$

which is the integral we wish to evaluate in the limit as $R \to \infty$.

③ $\theta = \pi/4$: $z = re^{i\pi/4}$ where r ranges from R to 0, $dz = e^{i\pi/4}\,dr$, and $z^2 = e^{i\pi/2}r^2 = ir^2$. Thus

$$\int_{③}^{\theta=\pi/4} e^{iz^2}\,dz = e^{i\pi/4}\int_R^0 e^{-r^2}\,dr = e^{-i\pi/4}\int_0^R e^{-r^2}\,dr.$$

Note that for the general integrand e^{iz^n} where n is a positive integer we would select

$$\int_{③}^{\theta=\pi/(2n)} e^{iz^n}\,dz$$

as depicted in Fig. 19.7a since then $z^n = ir^n$ and

$$\int_{③}^{\theta=\pi/(2n)} e^{iz^n}\,dz = e^{-i\pi/(2n)}\int_0^R e^{-r^n}\,dr$$

which reduces to our result for $n = 2$. Observe that in Problem 19.7 it is necessary to take $n = 3$.

② $r = R$: $z = Re^{i\theta}$ where $0 \le \theta \le \pi/4$ and $dz = iRe^{i\theta}\,d\theta$. Thus

$$\int_{②}^{r=R} e^{iz^2}\,dz = iR\int_0^{\pi/4} e^{iR^2e^{2i\theta}}\,e^{i\theta}\,d\theta.$$

Since, unlike the two other line integrals, our ultimate goal here is to demonstrate that this one goes to zero in the limit as $R \to \infty$, we proceed as follows. Recall for complex functions

$$\left|\int_C f(z)\,dz\right| \le \int_C |f(z)||dz| \text{ and } |e^z| = e^{\mathrm{Re}(z)}.$$

Then, noting that $ie^{2i\theta} = -\sin(2\theta) + i\cos(2\theta)$,

$$\left|\int_{②}^{r=R} e^{iz^2}\,dz\right| \le R\underbrace{\int_0^{\pi/4} e^{-R^2\sin(2\theta)}\,d\theta}_{\varphi=2\theta,d\varphi=2d\theta}$$

$$= \frac{R}{2}\int_0^{\pi/2} e^{-R^2\sin(\varphi)}\,d\varphi \le \frac{R}{2}\int_0^{\pi/2} e^{-2R^2\varphi/\pi}\,d\varphi$$

since (see Fig. 19.7b) $0 \le 2\varphi/\pi \le \sin(\varphi)$ for $0 \le \varphi \le \pi/2$. Hence

$$\left|\int_{②}^{r=R} e^{iz^2}\,dz\right| \le \frac{\pi}{4R}e^{-2R^2\varphi/\pi}\Big|_{\pi/2}^{0} = \frac{\pi}{4R}(1 - e^{-R^2}).$$

Observe from this result it follows that

$$\left| \lim_{R\to\infty} \int_{②}^{r=R} e^{iz^2}\, dz \right| \le \frac{\pi}{4} \lim_{R\to\infty} \frac{1-e^{-R^2}}{R} = 0,$$

which implies that

$$\lim_{R\to\infty} \int_{②}^{r=R} e^{iz^2}\, dz = 0.$$

We can therefore conclude that

$$\lim_{R\to\infty} \oint_{\Gamma_R}^{z=re^{i\theta}} e^{iz^2}\, dz = \lim_{R\to\infty} \int_{①}^{\theta=0} e^{iz^2}\, dz$$

$$+ \lim_{R\to\infty} \int_{②}^{r=R} e^{iz^2}\, dz + \lim_{R\to\infty} \int_{③}^{\theta=\pi/4} e^{iz^2}\, dz = 0,$$

which yields the desired result

$$\int_0^\infty e^{ir^2}\, dr = e^{i\pi/4} \underbrace{\int_0^\infty e^{-r^2}\, dr}_{\sqrt{\pi}/2} = e^{i\pi/4}\frac{\sqrt{\pi}}{2}.$$

So

$$\int_0^\infty e^{-ir^2}\, dr = e^{-i\pi/4}\frac{\sqrt{\pi}}{2}.$$

Returning to our stationary phase asymptotic formula we have demonstrated that

$$\int_0^\infty e^{isgm[g''(\alpha)]z^2}\, dz = e^{isgm[g''(\alpha)]\pi/4}\frac{\sqrt{\pi}}{2}.$$

Thus finally substituting this result into that formula we obtain

$$I(k) = \int_a^b f(x)e^{ikg(x)}\, dx \sim F(\alpha,k) \text{ as } k \to \infty$$

where

$$F(\alpha,k) = \sqrt{\frac{2\pi}{|g''(\alpha)|}}\frac{f(\alpha)}{k^{1/2}} e^{i\{kg(\alpha)+\frac{\pi}{4}sgm[g''(\alpha)]\}}.$$

More generally, if $g'(\alpha_j) = 0$ and $g''(\alpha_j) \neq 0$ for $j = 1, 2, \ldots, N$, and these are the only zeroes of $g'(x)$ in the interval $x \in [a, b]$, then

$$I(k) \sim \sum_{j=1}^{N} \beta_j F(\alpha_j, k) \text{ as } k \to \infty,$$

where

$$\beta_j = \begin{cases} 1, & \text{when } a < \alpha_j < b \\ 1/2, & \text{when } \alpha_j = a \text{ or } b \end{cases}.$$

Here the presence of $\beta_j = 1/2$ if the α_j occurs at either end point of the interval being dependent upon the fact that in this event the relevant integrals under examination would only be over the half-intervals

$$\int_a^{a+\delta} f(x)e^{ikg(x)}\,dx \text{ or } \int_{b-\delta}^{b} f(x)e^{ikg(x)}\,dx$$

depending upon whether

$$\alpha_j = a \text{ or } b,$$

respectively.

In order to use the Method of Stationary Phase to deduce the one-term asymptotic representation for $J_n(x)$ as $x \to \infty$, we must first develop an integral formula for these functions.

To do so it is next necessary to reformulate the Fourier series Theorem of Section 5.4 in complex form. Recall this theorem in real form stated that

$$f(x) = \frac{a_0}{2} + \sum_{n=1}^{\infty}\left[a_n \cos\left(\frac{n\pi x}{L}\right) + b_n \sin\left(\frac{n\pi x}{L}\right)\right]$$

where

$$a_n = \frac{1}{L}\int_{-L}^{L} f(u)\cos\left(\frac{n\pi u}{L}\right) du \text{ and } b_n = \frac{1}{L}\int_{-L}^{L} f(u)\sin\left(\frac{n\pi u}{L}\right) du;$$

or

$$f(x) = \frac{1}{2L}\int_{-L}^{L} f(u)\,du + \frac{1}{L}\sum_{n=1}^{\infty}\int_{-L}^{L} f(u)\cos\left[\frac{n\pi}{L}(x-u)\right] du$$

upon employment of the trigonometric identity for the cosine of the sum of two angles. Now since $\cos(z) = (e^{iz} + e^{-iz})/2$, this yields the result

$$f(x) = \frac{1}{2L}\int_{-L}^{L} f(u)\,du + \frac{1}{2L}\sum_{n=1}^{\infty}\int_{-L}^{L} f(u)\left[e^{i\frac{n\pi}{L}(x-u)} + e^{-i\frac{n\pi}{L}(x-u)}\right] du.$$

Hence, observing that

$$\underbrace{\sum_{n=1}^{\infty}\int_{-L}^{L} f(u)e^{-i\frac{n\pi}{L}(x-u)}\,du}_{n=-m} = \underbrace{\sum_{m=-1}^{-\infty}\int_{-L}^{L} f(u)e^{i\frac{m\pi}{L}(x-u)}\,du}_{m=n}$$

$$= \sum_{n=-\infty}^{-1}\int_{-L}^{L} f(u)e^{i\frac{n\pi}{L}(x-u)}\,du,$$

we obtain the desired complex form of the Fourier series Theorem

$$f(x) = \underbrace{\frac{1}{2L}\int_{-L}^{L} f(u)\,du}_{n=0} + \frac{1}{2L}\sum_{n=1}^{\infty}\int_{-L}^{L} f(u)e^{i\frac{n\pi}{L}(x-u)}\,du$$

$$+ \frac{1}{2L}\sum_{n=-\infty}^{-1}\int_{-L}^{L} f(u)e^{i\frac{n\pi}{L}(x-u)}\,du$$

$$= \frac{1}{2L}\sum_{n=-\infty}^{\infty}\int_{-L}^{L} f(u)e^{i\frac{n\pi}{L}(x-u)}\,du = \sum_{n=-\infty}^{\infty} c_n e^{i\frac{n\pi x}{L}}$$

where

$$c_n = \frac{1}{2L}\int_{-L}^{L} f(u)e^{-i\frac{n\pi u}{L}}\,du.$$

Returning to the generating function for $J_n(x)$ deduced in Section 19.7

$$e^{(z-1/z)(x/2)} = \sum_{n=-\infty}^{\infty} J_n(x)z^n,$$

letting $z = e^{i\varphi}$, and noting that in this event

$$z - \frac{1}{z} = e^{i\varphi} - e^{-i\varphi} = 2i\sin(\varphi),$$

we can find that

$$e^{ix\sin(\varphi)} = \sum_{n=-\infty}^{\infty} J_n(x)e^{in\varphi}.$$

Then taking $x = \varphi$, $L = \pi$, and $u = \theta$ in the complex form of the Fourier series Theorem yields

$$f(\varphi;x) = \sum_{n=-\infty}^{\infty} c_n(x)e^{in\varphi} \text{ where } c_n(x) = \frac{1}{2\pi}\int_{-\pi}^{\pi} f(\theta;x)e^{-in\theta}\,d\theta$$

where here x is being treated as a parameter. Thus making the identifications

$$f(\varphi;x) = e^{ix\sin(\varphi)} \text{ and } c_n(x) = J_n(x),$$

we obtain the required integral formula for $J_n(x)$ as follows:

$$J_n(x) = \frac{1}{2\pi}\int_{-\pi}^{\pi} e^{i[x\sin(\theta)-n\theta]}\,d\theta$$

$$= \frac{1}{2\pi}\left\{ \underbrace{\int_{-\pi}^{\pi} \cos\left[\underbrace{x\sin(\theta)-n\theta}_{\text{odd}}\right]}_{\text{even}} d\theta + i\underbrace{\int_{-\pi}^{\pi} \sin\left[\underbrace{x\sin(\theta)-n\theta}_{\text{odd}}\right]}_{\text{odd}} d\theta \right\}$$

$$= \frac{1}{\pi}\int_{0}^{\pi} \cos[x\sin(\theta)-n\theta]\,d\theta = \frac{1}{\pi}\mathrm{Re}\left\{\int_{0}^{\pi} e^{i[x\sin(\theta)-n\theta]}\,d\theta\right\}.$$

We next consider

$$I(x;n) = \int_{a=0}^{b=\pi} e^{-in\theta}e^{ix\sin(\theta)}\,d\theta$$

and determine its asymptotic behavior as $x \longrightarrow \infty$ by means of the Method of Stationary Phase formula upon associating x with k and θ with x and making the identifications $f(\theta;n) = e^{-in\theta}$ and $g(\theta) = \sin(\theta)$. Then since $g'(\theta) = \cos(\theta) = 0$ for $0 \leq \theta \leq \pi$ only at $\theta = \pi/2$ and $g''(\pi/2) = -\sin(\pi/2) = -1$, we select $\alpha = \pi/2$ and obtain the one-term asymptotic representation

$$I(x;n) \sim \sqrt{\frac{2\pi}{|g''(\pi/2)|}}\frac{f(\pi/2;n)}{x^{1/2}}e^{i\{xg(\pi/2)+\frac{\pi}{4}sgm[g''(\pi/2)]\}} \text{ as } x \longrightarrow \infty$$

or

$$I(x;n) \sim \sqrt{\frac{2\pi}{x}}e^{i(x-n\pi/2-\pi/4)} \text{ as } x \longrightarrow \infty.$$

Hence

$$J_n(x) = \frac{1}{\pi}\mathrm{Re}\{I(x;n)\} \sim \sqrt{\frac{2}{\pi x}}\mathrm{Re}\{e^{i(x-n\pi/2-\pi/4)}\} = \sqrt{\frac{2}{\pi x}}\cos\left(x-\frac{n\pi}{2}-\frac{\pi}{4}\right) \text{ as } x \longrightarrow \infty,$$

which completes our demonstration. Further it can be shown that [130]

$$Y_n(x) = \frac{1}{\pi}\int_{0}^{\pi} \sin[x\sin(\theta)-n\theta]\,d\theta = \frac{1}{\pi}\mathrm{Im}\left\{\int_{0}^{\pi} e^{i[x\sin(\theta)-n\theta]}\,d\theta\right\}.$$

Thus

$$Y_n(x) = \frac{1}{\pi}\mathrm{Im}\{I(x;n)\} \sim \sqrt{\frac{2}{\pi x}}\mathrm{Im}\{e^{i(x-n\pi/2-\pi/4)}\} = \sqrt{\frac{2}{\pi x}}\sin\left(x-\frac{n\pi}{2}-\frac{\pi}{4}\right) \text{ as } x \longrightarrow \infty.$$

Specifically, for $n = 0$,

(a)

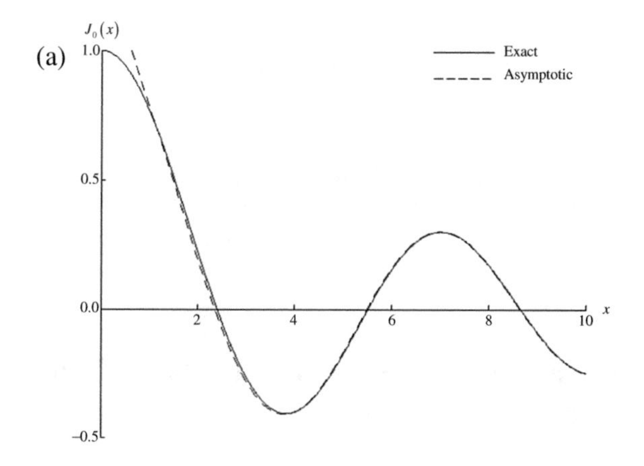

(b)

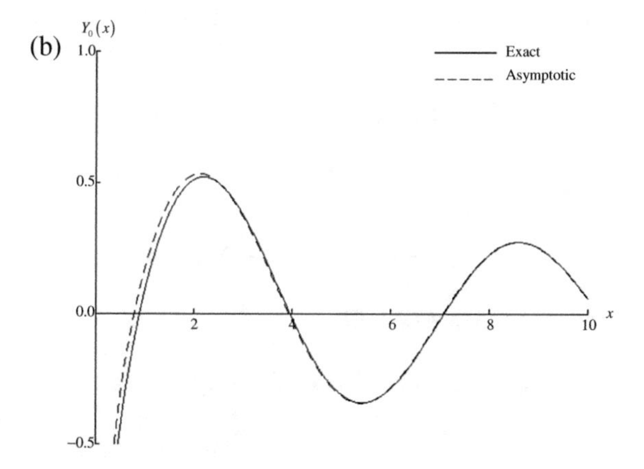

(c)

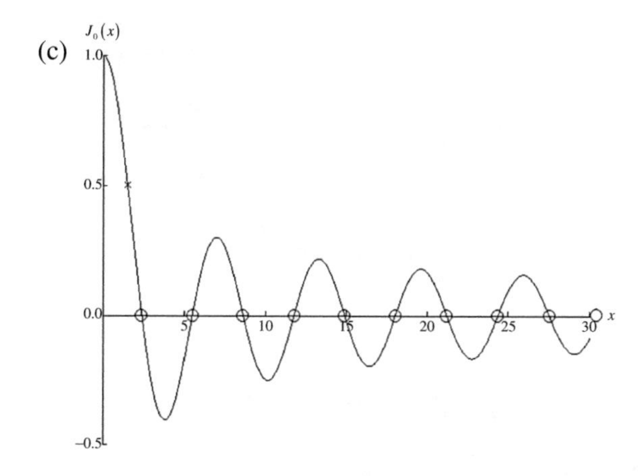

Fig. 19.8 Plots of the one-term asymptotic representations for (a) $J_0(x)$ and (b) $Y_0(x)$ as well as (c) the first ten zeroes $J_0(\alpha_n) = 0$ for $n = 1, 2, 3, \ldots, 10$, where the cross denotes the reference point $J_0(1.5211) = 0.5$.

$$J_0(x) \sim \sqrt{\frac{2}{\pi x}} \cos\left(x - \frac{\pi}{4}\right) \text{ as } x \to \infty$$

and

$$Y_0(x) \sim \sqrt{\frac{2}{\pi x}} \sin\left(x - \frac{\pi}{4}\right) \text{ as } x \to \infty.$$

These one-term asymptotic representations are compared with $J_0(x)$ and $Y_0(x)$ in Fig. 19.8a, b.

19.9 An Eigenvalue Problem Involving the Bessel Function of the First Kind of Order Zero

Consider the canonical Sturm–Liouville eigenvalue problem for $y = y(x)$ of Section 12.7:

$$\mathcal{L}[y] = -\frac{1}{r(x)}[p(x)y']' + q(x)y = \lambda y \text{ for } a < x < b;$$

with the identifications

$$r(x) = p(x) = x,\ q(x) \equiv 0,\ a = 0, \text{ and } b = 1;$$

where $y(x)$ satisfies the boundary conditions

$$y, y' \text{ are bounded as } x \to 0 \text{ and } y(1) = 0.$$

We shall first show that $\lambda > 0$ as follows: Multiplying our original equation by $x\overline{y(x)}$ and integrating on x from 0 to 1, we obtain

$$-\int_0^1 \underbrace{\overline{y(x)}}_{u}\ \underbrace{[xy'(x)]'}_{v'}\, dx = -\underbrace{\underbrace{\overline{y(x)}}_{u}\ \underbrace{[xy'(x)]}_{v}\Bigg|_{x=0}^{1}}_{0}$$

$$+\int_0^1 \underbrace{xy'(x)}_{v}\ \underbrace{\overline{y'(x)}}_{u'}\, dx = \int_0^1 x|y'(x)|^2\, dx$$

and thus

$$\int_0^1 x|y'(x)|^2\, dx = \lambda \int_0^1 |y(x)|^2\, dx;$$

which implies that

$$\lambda > 0$$

since $y(x) \not\equiv 0$ being an eigenfunction and $\lambda = 0 \Rightarrow y(x) \equiv C = 0$ by virtue of continuity. Observe that $\mathcal{L}[y] = \lambda y$ is equivalent to

$$[xy'(x)]' + \lambda xy(x) = xy''(x) + y'(x) + \lambda xy(x) = 0, 0 < x < 1;$$

or

$$x^2y''(x) + xy'(x) + \lambda x^2 y(x) = 0.$$

We next introduce the following change of variables:

$$s = \sqrt{\lambda}x, \ u(s) = y(x) = y\left(\frac{s}{\sqrt{\lambda}}\right).$$

Then since

$$y'(x) = \frac{dy}{dx}(x) = \frac{du}{dx}(s) = \frac{du}{ds}(s)\frac{ds}{dx} = \sqrt{\lambda}\frac{du}{ds}(s)$$

and

$$y''(x) = \frac{d^2y}{dx^2}(x) = \frac{d}{dx}\left[\sqrt{\lambda}\frac{du}{ds}(s)\right] = \lambda\frac{d^2u}{ds^2}(s);$$

this transforms our equation into

$$s^2\frac{d^2u}{ds^2}(s) + s\frac{du}{dx}(s) + s^2u(s) = 0$$

or the Bessel equation of order 0. Hence we can conclude that

$$u(s) = c_1J_0(s) + c_2Y_0(s) \text{ or } y(x) = c_1J_0(\sqrt{\lambda}x) + c_2Y_0(\sqrt{\lambda}x).$$

Since y must remain bounded as $x \to 0$ and $Y_0 \to -\infty$ as its argument goes to 0 (see Fig. 19.7), we require that $c_2 = 0$ or

$$y(x) = c_1J_0(\sqrt{\lambda}x) \text{ where } c_1 \neq 0.$$

Further

$$y(1) = 0 \Rightarrow J_0(\sqrt{\lambda}) = 0 \Rightarrow \sqrt{\lambda_n} = \alpha_n \Rightarrow \lambda_n = \alpha_n^2 > 0, \ y_n(x) = J_0(\alpha_n x) \text{ for } n = 1, 2, 3, \ldots,$$

where the α_n's represent the infinite number of zeroes of J_0, the first ten of which are depicted in Fig. 19.8c and tabulated below to four significant figures [62]:

$$\alpha_1 = 2.4048, \alpha_2 = 5.5201, \alpha_3 = 8.6537, \alpha_4 = 11.7915, \alpha_5 = 14.9309,$$

$$\alpha_6 = 18.0711, \alpha_7 = 21.2116, \alpha_8 = 24.3525, \alpha_9 = 27.4935, \alpha_{10} = 30.6346.$$

Note in this context we can deduce from the one-term asymptotic representation of J_0 that

$$\alpha_n - \frac{\pi}{4} \sim (2n-1)\frac{\pi}{2} \text{ or } \alpha_n \sim \beta_n = \left(n - \frac{1}{4}\right)\pi \text{ as } n \to \infty.$$

Observe for comparison purposes that $\beta_{10} = 30.6305$.

Recall for the general Sturm–Liouville eigenvalue problem of Section 12.7 that

$$\int_a^b r(x)y_n(x)y_m(x)\,dx = 0 \text{ for } n \neq m \text{ provided } [p(x)y_n y_m' - y_m y_n']\big|_a^b = 0;$$

the latter condition being satisfied for our problem since $p(x) = x$ disappears at $x = a = 0$ while $y_n(x), y_m(x)$ are zero at $x = b = 1$. Therefore since $r(x) = x$, we have demonstrated the orthogonality condition

$$\int_0^1 xJ_0(\alpha_n x)J_0(\alpha_m x)\,dx = 0 \text{ for } n \neq m.$$

It remains to deduce the normalization constant when $m = n$. We proceed to do so in the following indirect manner. Considering our problem in the form

$$[xy_n'(x)]' = -\lambda_n x y_n(x)$$

and multiplying this by $2xy_n'(x)$, we obtain

$$2xy_n'(x)[xy'(x)]' = ([xy_n'(x)]^2)' = -\lambda_n x^2[y_n^2(x)]' = -2\lambda_n x^2 y_n(x)y_n'(x).$$

Now integration of this result on x from 0 to 1 yields

$$\int_0^1 ([xy_n'(x)]^2)'\,dx = [xy_n'(x)]^2\big|_0^1 = [y_n'(1)]^2 = \alpha_n^2[J_0'(\alpha_n)]^2 = \lambda_n[J_0'(\alpha_n)]^2$$

$$= -\lambda_n \int_0^1 \underbrace{x^2}_{u}\,\underbrace{[y_n^2(x)]'}_{v'}\,dx$$

$$= -\lambda_n \left\{ \underbrace{\left[\underbrace{x^2}_{u}\,\underbrace{y_n^2(x)}_{v}\right]\Bigg|_0^1}_{0} - \int_0^1 \underbrace{2x}_{u'}\,\underbrace{y_n^2(x)}_{v}\,dx \right\}$$

$$= 2\lambda_n \int_0^1 xy_n^2(x)\,dx$$

which implies

$$\int_0^1 xy_n^2(x)\,dx = \int_0^1 xJ_0^2(\alpha_n x)\,dx = \frac{1}{2}[J_0'(\alpha_n)]^2.$$

Observe for

$$J_0(s) = \sum_{m=0}^{\infty} \frac{(-1)^m s^{2m}}{2^{2m}(m!)^2},$$

$$J_0'(s) = \sum_{m=1}^{\infty} \frac{(-1)^m 2m s^{2m-1}}{2^{2m}(m!)^2} = \underbrace{\sum_{m=1}^{\infty} \frac{(-1)^m s^{2m-1}}{2^{2m-1} m!(m-1)!}}_{\substack{j=m-1,\\ m=j+1,\\ 2m-1=2j+1}} = -\sum_{j=0}^{\infty} \frac{(-1)^j s^{2j+1}}{2^{2j+1}(j+1)!j!} = -J_1(s);$$

or, more generally, that

$$[s^{-\nu} J_\nu(s)]' = -s^{-\nu} J_{\nu+1}(s).$$

Thus

$$J_0'(\alpha_n) = -J_1(\alpha_n)$$

and

$$\int_0^1 x J_0^2(\alpha_n)\, dx = \frac{1}{2} J_1^2(\alpha_n).$$

Therefore we have deduced the orthonormality relationship

$$\int_0^1 x J_0(\alpha_n) J_0(\alpha_m)\, dx = \frac{1}{2} J_1^2(\alpha_n) \delta_{nm}.$$

Hence any function $f(x)$ that is piecewise continuous and has a piecewise continuous derivative for $x = r/R \in [0, 1]$ can be represented in both this dimensionless and its associated dimensional form by

$$f(x) = \sum_{n=0}^{\infty} c_n J_0(\alpha_n x) = \sum_{n=0}^{\infty} c_n J_0\left(\frac{\alpha_n r}{R}\right) = f\left(\frac{r}{R}\right) = F(r)$$

where

$$c_n = \frac{2}{J_1^2(\alpha_n)} \underbrace{\int_0^1 x f(x) J_0(\alpha_n x)\, dx}_{\substack{x=r/R,\\ dx=dr/R}} = \frac{2}{J_1^2(\alpha_n) R^2} \int_0^R r F(r) J_0\left(\frac{\alpha_n r}{R}\right) dr.$$

Problems

19.1. Consider the surface area functional for $W = W(x, y) > 0$ in $\mathcal{D}$ and prescribed to be 0 on $\partial\mathcal{D}$

$$I(W) = \iint_{\mathcal{D}} \underbrace{\sqrt{1 + W_x^2 + W_y^2}}_{f}\, dx\, dy$$

subject to the fixed volume constraint that

$$J(W) = \iint_{\mathcal{D}} \underbrace{W}_{g} \, dx\,dy \equiv C,$$

where C is a given positive constant.

(a) Using a constrained optimization Calculus of Variations approach in conjunction with the introduction of a Lagrange multiplier λ, employ the governing partial differential equation

$$\frac{\partial f^*}{\partial w} - \frac{\partial}{\partial x}\left(\frac{\partial f^*}{\partial w_x}\right) - \frac{\partial}{\partial y}\left(\frac{\partial f^*}{\partial w_y}\right) = 0 \text{ where } f^* = f + \lambda g,$$

and hence show that (see Problem 10.1)

$$\lambda = \kappa(w) = \frac{w_{xx}(1+w_y^2) - 2w_x w_y w_{xy} + w_{yy}(1+w_x^2)}{(1+w_x^2+w_y^2)^{3/2}}$$

for the minimizing function $w = w(x, y)$.

(b) The result of part (a) implies that this surface $z = w(x, y)$ of fixed volume which minimizes its area must be one of constant curvature λ. Excluding the trivial planar solution $w(x, y) \equiv 0$ since $C > 0$, the only function satisfying this condition is a hemisphere. Demonstrate that

$$\lambda = -\frac{2}{a} = -2\left(\frac{2\pi}{3C}\right)^{1/3}$$

for the hemisphere of radius a

$$w(x, y) = (a^2 - x^2 - y^2)^{1/2} \text{ with } \mathcal{D} = \{(x, y) | 0 \leq x^2 + y^2 \leq a^2\}$$

by computing its curvature $\kappa(w) = -2/a$ and using the integral constraint to find a relationship between that radius a and the fixed volume parameter C. In doing so, the volume of this hemisphere may assumed to be $(2/3)\pi a^3$ without derivation.

19.2. Deduce the explicit equations of motion for the double plane pendulum problem from the Euler–Lagrange equations

$$\frac{d}{dt}\frac{\partial \mathcal{L}}{\partial \dot{\theta}_j} - \frac{\partial \mathcal{L}}{\partial \theta_j} = 0 \text{ for } j = 1 \text{ and } 2$$

where

$$\mathcal{L} = \frac{m_1+m_2}{2}\ell_1^2\dot{\theta}_1^2 + \frac{m_2}{2}\ell_2^2\dot{\theta}_2^2 + m_2\ell_1\ell_2\cos(\theta_1-\theta_2)\dot{\theta}_1\dot{\theta}_2$$
$$-(m_1+m_2)g\ell_1[1-\cos(\theta_1)] - m_2 g\ell_2[1-\cos(\theta_2)].$$

19.3. Consider a Laplace integral of the form

$$I(\nu) = \int_a^b \varphi(\theta)e^{\nu h(\theta)}\,d\theta$$

where $\varphi \in C[a,b]$ and $h \in C^\infty[a,b]$. In this problem any results for Laplace's Method of Section 19.6 and any arguments appearing in Section 19.8 on the Method of Stationary Phase may be employed.

(a) For an h such that $h'(\alpha) = 0$ and $h''(\theta) < 0$, as in Section 19.6, conclude

$$I(\nu) \sim \varphi(\alpha)e^{\nu h(\alpha)}\left[\frac{-\beta(\alpha)\pi}{\nu h''(\alpha)}\right]^{1/2} \quad \text{as } \nu \to \infty$$

where

$$\beta(\alpha) = \begin{cases} 2, & \text{if } a < \alpha < b \\ 1/2, & \text{if } \alpha = a \text{ or } b \end{cases}.$$

Explain why this asymptotic formula is different when α occurs at either end point of the interval as opposed to at an interior point.

(b) We now consider

$$I(\nu) = \int_{a=0=\alpha}^{b=\pi} \varphi(\theta)e^{\nu h(\theta)}\,d\theta$$
$$= \underbrace{\int_{a=0=\alpha}^{\delta} \varphi(\theta)e^{\nu h(\theta)}\,d\theta}_{①} + \underbrace{\int_{\delta}^{\theta_c} \varphi(\theta)e^{\nu h(\theta)}\,d\theta}_{②} + \underbrace{\int_{\theta_c}^{b=\pi} \varphi(\theta)e^{\nu h(\theta)}\,d\theta}_{③}$$

where h is such that $h'(0) = h'(\pi) = 0$, $h'(\theta) < 0$ for $0 < \theta < \pi$; and $h''(\theta_c) = 0$, $h''(\theta) < 0$ for $0 \le \theta < \theta_c$, $h''(\theta) > 0$ for $\theta_c < \theta \le \pi$, as depicted in Fig. 19.9. Deduce from part (a) and the results of Section 19.6 that:

$$① \sim \varphi(0)e^{\nu h(0)}\left[\frac{-\pi}{2\nu h''(0)}\right]^{1/2} \quad \text{as } \nu \to \infty;$$

$$|②| \le \frac{e^{\nu h(0)}}{e^{\nu \Delta_1}}\int_{\delta}^{\theta_c} |\varphi(\theta)|\,d\theta \text{ where } \Delta_1 = \frac{-h''(\theta_1)\delta^2}{2} > 0;$$

$$|③| \le e^{\nu h(\theta_c)}\int_{\theta_c}^{\pi} |\varphi(\theta)|\,d\theta;$$

and thus, employing the reasoning of Section 19.6 and the fact that $h(0) > h(\theta_c)$, conclude that the contribution from ① dominates those from ② and ③, respectively. Hence again obtain that

$$I(\nu) \sim \varphi(0)e^{\nu h(0)}\left[\frac{-\pi}{2\nu h''(0)}\right]^{1/2} \quad \text{as } \nu \to \infty.$$

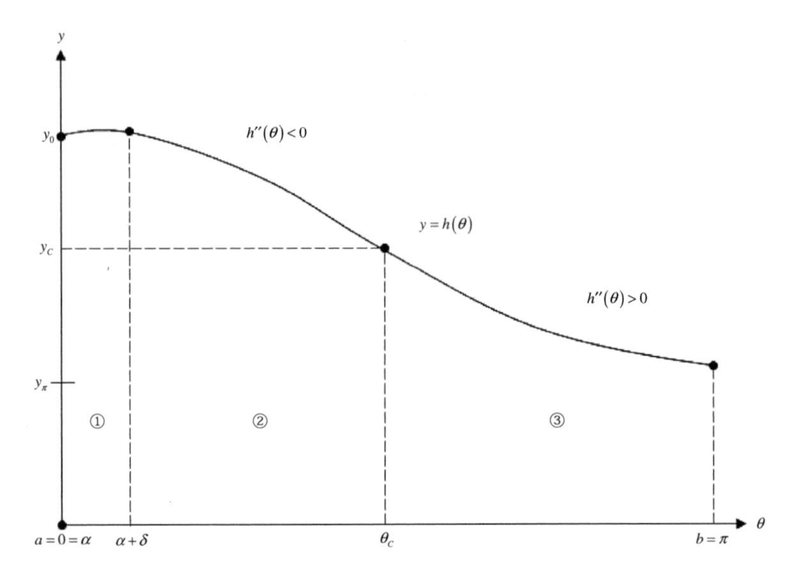

Fig. 19.9 Plot of $y = h(\theta)$ relevant to Problem 19.3(b).

(c) Defining $f(\theta) = e^{h(\theta)}$, show that under the conditions of part (b),

$$\int_{a=0=\alpha}^{b=\pi} \varphi(\theta)[f(\theta)]^{\nu}\, d\theta \sim \varphi(\alpha)[f(\alpha)]^{\nu+1/2}\left[\frac{-\pi}{2\nu f''(\alpha)}\right]^{1/2} \quad \text{as } \nu \to \infty;$$

where $f'(\alpha) = 0$ and $f''(\alpha) < 0$.

(d) Recalling from Section 12.7 the integral relationship for the Legendre polynomial

$$P_\nu(\mu) = \frac{1}{\pi}\int_0^{\pi}\left[\underbrace{[\mu + (\mu^2 - 1)^{1/2}\cos(\theta)]^n}_{f(\theta)}\right]^{\nu} d\theta$$

for $\mu > 1$, employ the one term asymptotic representation of part (c) to deduce that

$$P_\nu(\mu) \sim \frac{1}{\sqrt{2\pi\nu}}\,\frac{[\mu + (\mu^2 - 1)^{1/2}]^{\nu+1/2}}{(\mu^2 - 1)^{1/4}} \quad \text{as } \nu \to \infty.$$

19.4. In the Saddle Point Method (see Copson, [22]) the locus in the complex z-plane where

$$\frac{1}{2}\psi''(\zeta)(z - \zeta)^2$$

is real and negative plays a pivotal role. Here ζ is the saddle point and ψ an analytic function. Demonstrate that this locus is a straight line through the pole by representing both $z-\zeta$ and $\psi''(\zeta)/2$ in polar form. That is let

$$z-\zeta = re^{i\theta} \text{ and } \frac{\psi''(\zeta)}{2} = R_0 e^{i\varphi_0} \text{ where } r, R_0 > 0.$$

19.5. Consider the Bessel equation of order $\nu = 1/2$ for $y = y(x)$:

$$L[y] = x^2 y'' + xy' + \left(x^2 - \frac{1}{4}\right) y = 0,\ x > 0.$$

(a) Show directly by the method of Frobenius that this equation has a general solution given by

$$y(x) = c_1 \frac{\cos(x)}{x^{1/2}} + c_2 \frac{\sin(x)}{x^{1/2}}.$$

That is let

$$y(x) = \sum_{n=0}^{\infty} a_n(r) x^{n+r},$$

find the two values of $r = r_{1,2}$ satisfying the indicial equation, and determine the $a_n(r_{1,2})$ from the recursion relation. Do not automatically take the coefficients with odd indices to be zero as is traditionally done for deriving the Bessel function formulae and deduce the $a_n(r_{1,2})$ from a bottom-up approach by examining $n = 2, 4, \ldots$, which depend upon a_0 and $n = 3, 5, \ldots$, which depend on a_1 separately and in that order until its general form can be ascertained and identified with the desired Maclaurin series for $\cos(x)$ and $\sin(x)$ given by

$$\cos(x) = \sum_{n=0}^{\infty} (-1)^n \frac{x^{2n}}{(2n)!} \text{ and } \sin(x) = \sum_{n=0}^{\infty} (-1)^n \frac{x^{2n+1}}{(2n+1)!}.$$

(b) Now introduce the transformation

$$y(x) = x^{-1/2} v(x)$$

and by obtaining and solving the differential equation resulting from $L[x^{-1/2}v] = 0$ for $v(x)$, demonstrate in this alternate manner that the general solution in part (a) is correct.

(c) Finally show, upon substitution of $\nu = 1/2$ into the series representations for $J_{\pm\nu}(x)$ of Section 19.7, that

$$J_{1/2}(x) = \sqrt{\frac{2}{\pi x}} \sin(x) \text{ and } J_{-1/2}(x) = \sqrt{\frac{2}{\pi x}} \cos(x).$$

Here the result from Section 19.5

$$\Gamma\left(m+\frac{1}{2}\right) = \frac{(2m)!}{2^{2m}m!}\sqrt{\pi}$$

may be employed directly and used in conjunction with $\Gamma(z+1) = z\Gamma(z)$ to deduce that

$$\Gamma\left(m+\frac{3}{2}\right) = \Gamma\left(m+\frac{1}{2}+1\right) = \left(m+\frac{1}{2}\right)\Gamma\left(m+\frac{1}{2}\right) = \frac{(2m+1)!}{2^{2m+1}m!}\sqrt{\pi}.$$

19.6. Consider the integral representation for $J_0(x)$

$$J_0(x) = \frac{1}{\pi}\int_0^{\pi} \cos[x\sin(\theta)]\,d\theta$$

which follows from that for $J_n(x)$ of Section 19.8 with $n = 0$.

(a) The purpose of this part is to show that

$$J_0(x) = \frac{1}{\pi}\int_0^{\pi} e^{ix\cos(\varphi)}\,d\varphi$$

by justifying each step in the following transformation procedure:

$$\underbrace{\int_0^{\pi} \cos[x\sin(\theta)]\,d\theta}_{\substack{\theta=\pi/2-\varphi \\ d\theta=-d\varphi}} = \int_{-\pi/2}^{\pi/2} \cos[x\cos(\varphi)]\,d\varphi = 2\int_0^{\pi/2} \cos[x\cos(\varphi)]\,d\varphi$$

$$= \int_0^{\pi} \cos[x\cos(\theta)]\,d\theta$$

$$= \int_0^{\pi} e^{ix\cos(\varphi)}\,d\varphi \text{ since } \int_0^{\pi} \sin[x\cos(\theta)]\,d\theta = 0.$$

(b) The purpose of this part is to use part (a) and the Method of Stationary Phase to deduce the one-term asymptotic representation for $J_0(x)$ of Section 19.8 in an alternate manner. Justify

$$J_0(x) = \int_{0=\alpha_1}^{\pi=\alpha_2} \underbrace{\frac{1}{\pi}}_{f(\varphi)} \exp[ix\underbrace{\cos(\varphi)}_{g(\varphi)}]\,d\varphi \sim \frac{1}{2}\sum_{j=1}^{2} F(\alpha_j, x) \text{ as } x \to \infty$$

where

$$F(\alpha, x) = \sqrt{\frac{2\pi}{|g''(\alpha)|}}\frac{f(\alpha)}{x^{1/2}}e^{i\{xg(\alpha)+\frac{\pi}{4}sgm[g''(\alpha)]\}}$$

or

$$J_0(x) \sim \sqrt{\frac{2}{\pi x}}\frac{e^{i(x-\pi/4)}+e^{-i(x-\pi/4)}}{2} = \sqrt{\frac{2}{\pi x}}\cos\left(x-\frac{\pi}{4}\right) \text{ as } x \to \infty;$$

by filling in the details.

(c) The purpose of this part is to use the integral relations for the Legendre polynomial and the Bessel function of the first kind of order zero to deduce that

$$\lim_{\nu\to\infty} P_\nu\left[\cos\left(\frac{\theta}{\nu}\right)\right] = J_0(\theta).$$

Recall from Section 12.7 that the integral representation for $P_\nu(\mu)$ can be extended by analytic continuation to those $|\mu| < 1$ and taking $\mu = \cos(\theta/\nu)$ show that

$$P_\nu\left[\cos\left(\frac{\theta}{\nu}\right)\right] = \frac{1}{\pi}\int_0^\pi \left[\cos\left(\frac{\theta}{\nu}\right) + i\sin\left(\frac{\theta}{\nu}\right)\cos(\varphi)\right]^\nu d\varphi.$$

Rewriting

$$\left[\cos\left(\frac{\theta}{\nu}\right) + i\sin\left(\frac{\theta}{\nu}\right)\cos(\varphi)\right]^\nu = \exp\left(\frac{\ln[\cos(\theta\alpha) + i\sin(\theta\alpha)\cos(\varphi)]}{\alpha}\right),$$

where $\alpha = \frac{1}{\nu}$, and using L'Hospital's Rule to calculate

$$\begin{aligned}\lim_{\alpha\to 0}\frac{\ln[\cos(\theta\alpha) + i\sin(\theta\alpha)\cos(\varphi)]}{\alpha} &= \lim_{\alpha\to 0}\frac{-\theta\sin(\theta\alpha) + i\theta\cos(\theta\alpha)\cos(\varphi)}{\cos(\theta\alpha) + i\sin(\theta\alpha)\cos(\varphi)} \\ &= i\theta\cos(\varphi),\end{aligned}$$

conclude that

$$\begin{aligned}\lim_{\nu\to\infty} P_\nu\left[\cos\left(\frac{\theta}{\nu}\right)\right] &= \frac{1}{\pi}\int_0^\pi \lim_{\nu\to\infty}\left[\cos\left(\frac{\theta}{\nu}\right) + i\sin\left(\frac{\theta}{\nu}\right)\cos(\varphi)\right]^\nu d\varphi \\ &= \frac{1}{\pi}\int_0^\pi e^{i\theta\cos(\varphi)}\, d\varphi = J_0(\theta).\end{aligned}$$

19.7. (a) Consider

$$I(\nu) = \int_0^\pi f(\theta)e^{i\nu g(\theta)}\, d\theta$$

where $f(\theta)$ is an analytic function of the real variable θ and $g(\theta)$ is a real-valued function possessing only one point $\theta = \alpha$ such that $g'(\alpha) = 0$ for $\theta \in [0, \pi]$. Given that, $g''(\alpha) = 0$ and $g'''(\alpha) \neq 0$, show

$$I(\nu) \sim \frac{\Gamma(1/3)}{\sqrt{3}}\beta(\alpha)f(\alpha)\left[\frac{6}{|g'''(\alpha)|\nu}\right]^{1/3} e^{i\nu g(\alpha)} \text{ as } \nu \to \infty$$

$$\text{where } \beta(\alpha) = \begin{cases} 1, & \text{for } 0 < \alpha < \pi \\ 1/2, & \text{for } \alpha = 0 \text{ or } \pi \end{cases}.$$

(b) Use the result of part (a) to deduce that

$$J_\nu(\nu) = \mathrm{Re}\left\{\int_0^\pi \frac{1}{\pi} e^{i\nu[\sin(\theta)-\theta]}\, d\theta\right\} \sim \frac{\Gamma(1/3)}{2^{2/3}3^{1/6}\pi\nu^{1/3}} \text{ as } \nu \to \infty.$$

Hints:

For (a): For $0 < \alpha < \pi$,

$$I(\nu) \sim \int_{\alpha-\delta}^{\alpha+\delta} f(\theta)e^{i\nu g(\theta)}\, d\theta \text{ as } \nu \to \infty.$$

Let $g(\theta) \cong g(\alpha) + g'''(\alpha)(\theta - \alpha)^3/6$ and $z=(\nu A)^{1/3}(\theta - \alpha)$ where $A = |g'''(\alpha)|/6$. Thus

$$I(\nu) \sim f(\alpha)\left[\frac{6}{|g'''(\alpha)|\nu}\right]^{1/3} e^{i\nu g(\alpha)} \int_{-\infty}^{\infty} e^{i\,\mathrm{sgm}[g'''(\alpha)]z^3}\, dz \text{ as } \nu \to \infty.$$

Note:

$$\int_{-\infty}^{\infty} e^{iz^3}\, dz = \int_{-\infty}^{0} e^{iz^3}\, dz + \int_0^{\infty} e^{iz^3}\, dz$$
$$= \int_0^{\infty} (e^{iz^3} + e^{-iz^3})\, dz = 2\mathrm{Re}\left\{\int_0^{\infty} e^{iz^3}\, dz\right\}$$

and

$$\int_{-\infty}^{\infty} e^{-iz^3}\, dz = 2\mathrm{Re}\left\{\int_0^{\infty} e^{-iz^3}\, dz\right\} = 2\mathrm{Re}\left\{\int_0^{\infty} e^{iz^3}\, dz\right\} \text{ as well.}$$

Consider the contour Γ_R depicted in Fig. 19.7a with $n = 3$. Then

$$\oint_{\Gamma_R} e^{iz^3}\, dz = 0$$

and, in the limit as $R \to \infty$, this implies

$$\int_0^{\infty} e^{iz^3}\, dz = e^{i\pi/6}\int_0^{\infty} e^{-r^3}\, dr = e^{i\pi/6}\frac{\Gamma(1/3)}{3} = (\sqrt{3}+i)\frac{\Gamma(1/3)}{6}.$$

For $\alpha = 0$ or π, one obtains $1/2$ the result for $0 < \alpha < \pi$.

For (b): Considering

$$I(\nu) = \int_0^\pi \frac{1}{\pi} e^{i\nu[\sin(\theta)-\theta]}\, d\theta;$$

make the identifications $f(\theta) \equiv 1/\pi$ and $g(\theta) = \sin(\theta) - \theta$. Hence $\alpha = 0$.

19.8. Consider a thin elastic membrane extended over a closed non-self-intersecting curve C. The planar domain $\mathcal{D}$ enclosed by C coincides with the equilibrium configuration of the membrane and is of area D. Assume the membrane is permitted to vibrate so that each of its points undergoes a motion in a direction perpendicular to this x-y

plane and its stretch-resistant restorative force per unit length or surface tension has a constant value τ_0. Denote that vertical displacement from equilibrium by $w = w(x, y, t)$, and assume a surface mass density $\rho = \rho(w)$ for the membrane. Then its kinetic and potential energies at any time t are given by

$$T = \frac{1}{2} \iint_{\mathcal{D}} \rho(w)\dot{w}^2 \, dx\, dy \text{ and } V = \tau_0 \iint_{\mathcal{D}} [(1 + w_x^2 + w_y^2)^{1/2} - 1] \, dx\, dy$$

where $\dot{w} \equiv \partial w/\partial t$, $w_x \equiv \partial w/\partial x$, and $w_y \equiv \partial w/\partial y$ while $w(x, y, t) = 0$ for $(x, y) \in C$.

(a) Noting that D is equal to

$$\iint_{\mathcal{D}} dx\, dy,$$

explain the physical significance of the integral appearing in V.

(b) Construct the Lagrangian for this system

$$L = T - V = \iint_{\mathcal{D}} f(w, \dot{w}, w_x, w_y) \, dx\, dy$$

where

$$f(w, \dot{w}, w_x, w_y) = \rho(w)\frac{\dot{w}^2}{2} - \tau_0[(1 + w_x^2 + w_y^2)^{1/2} - 1].$$

Then according to Hamilton's Principle, the action integral

$$I = \int_{t_1}^{t_2} \iint_{\mathcal{D}} f(w, \dot{w}, w_x, w_y) \, dx\, dy\, dt$$

is minimized by the function $w(x, y, t)$ that describes the actual motion of the membrane over all such functions which have continuous second partial derivatives, vanish on C, and coincide with the actual membrane configuration at $t = t_1$ and t_2, where the latter times are completely arbitrary. Show that the Euler–Lagrange equation for this integral

$$\frac{\partial f}{\partial w} - \frac{\partial}{\partial x}\left(\frac{\partial f}{\partial w_x}\right) - \frac{\partial}{\partial y}\left(\frac{\partial f}{\partial w_y}\right) - \frac{\partial}{\partial t}\left(\frac{\partial f}{\partial \dot{w}}\right) = 0$$

yields the equation of motion (see Problem 19.1 and Section 19.4)

$$\rho(w)\ddot{w} + \rho'(w)\frac{\dot{w}^2}{2} = \tau_0 \kappa(w)$$

where

$$\kappa(w) = \frac{w_{xx}(1 + w_y^2) - 2w_x w_y w_{xy} + w_{yy}(1 + w_x^2)}{(1 + w_x^2 + w_y^2)^{3/2}}$$

(c) Assuming that w, the deviation of the membrane from its equilibrium configuration, is small enough to allow $\rho(w) \cong \rho(0) = \rho_0$, $|w_x| << 1$, and $|w_y| << 1$, deduce that this equation of motion reduces to the two-dimensional wave equation

$$\ddot{w} = c_0^2 \nabla_2^2 w \text{ where } c_0^2 \equiv \frac{\tau_0}{\rho_0} \text{ and } \nabla_2^2 w \equiv w_{xx} + w_{yy}.$$

Further, since $[\tau_0] = (ML/\tau^2)/L = M/\tau^2$ and $[\rho_0] = M/L^2$, conclude that $[c_0] = L/\tau$.

(d) We now examine the problem of a circular membrane of radius R vibrating under the conditions already stipulated with the additional stipulation that its initial vibration at time $t = 0$ is axisymmetric in that it only depends on the radial distance from the center of this circle. Under these conditions, upon introduction of polar coordinates (r, θ), it can be deduced that

$$w(x, y, t) = W(r, t);$$

i.e., the displacement is then independent of θ for all later times $t > 0$ as well. If $F(r)$ and $K(r)$ represent the initial displacement and velocity functions of this membrane, respectively, explain why the governing partial differential equation (PDE), boundary conditions (BC's), and initial conditions (I.C.'s) for this circular membrane undergoing axisymmetric vibrations are given by (the form of ∇_2^2 in polar coordinates developed in Section 12.2 may be assumed)

$$\text{PDE: } \frac{\partial^2 W}{\partial t^2} = c_0^2 \left(\frac{\partial^2 W}{\partial r^2} + \frac{1}{r}\frac{\partial W}{\partial r} \right), 0 < r < R, t > 0;$$

$$\text{BC's: } W(r, t) \text{ remain bounded as } r \to 0 \text{ and } W(R, t) = 0 \text{ for } t > 0;$$

$$\text{I.C.'s: } W(r, 0) = F(r) \text{ and } \frac{\partial W}{\partial t}(r, 0) = K(r) \text{ for } 0 < r < R.$$

(e) Solve this mixed initial-boundary value problem by first seeking a separation of variables solution to the PDE and BC's of the form

$$W(r, t) = H(r)G(t)$$

obtaining

$$\frac{\ddot{G}(t)}{c_0^2 G(t)} = \frac{1}{H(r)} \left(\frac{d^2 H}{dr^2}(r) + \frac{1}{r}\frac{dH}{dr}(r) \right) \equiv -\lambda, \text{ the separation constant;}$$
$$\text{for } 0 < r < R \text{ and } t > 0;$$

$$H(r) \text{ remain bounded as } r \to 0 \text{ and } H(R) = 0.$$

(f) Introducing the change of variables $s = \sqrt{\lambda} r$ and employing the results of Section 19.9, transform the boundary value problem for $H(r) = u(s)$ into

$$s^2 \frac{d^2u}{ds^2}(s) + s\frac{du}{ds}(s) + s^2 u(s) = 0; \text{ for } 0 < s < \sqrt{\lambda}R;$$

$$u(s) \text{ remain bounded as } s \to 0 \text{ and } u(\sqrt{\lambda}R) = 0;$$

and hence conclude that

$$\lambda_n = \frac{\alpha_n^2}{R^2} \text{ and } H_n(r) = J_0\left(\frac{\alpha_n r}{R}\right) \text{ for } n = 1, 2, 3, \ldots;$$

while $G_n(t)$ satisfies

$$\ddot{G}_n(t) = \omega_n^2 G_n(t) = 0 \text{ where } \omega_n^2 = c_0^2 \lambda_n = \frac{c_0^2 \alpha_n^2}{R^2} \text{ for } t > 0$$

and thus

$$G_n(t) = A_n \cos(\omega_n t) + B_n \sin(\omega_n t) \text{ where } \omega_n = \frac{c_0 \alpha_n}{R} \text{ for } n = 1, 2, 3, \ldots.$$

(g) Employing the superposition principle show that the most general solution emerging from the separation of variables approach of the previous two parts is given by

$$W(r,t) = \sum_{n=1}^{\infty} W_n(r,t) = \sum_{n=1}^{\infty} H_n(r)G_n(t) = \sum_{n=1}^{\infty} J_0\left(\frac{\alpha_n r}{R}\right)[A_n \cos(\omega_n t) + B_n \sin(\omega_n t)].$$

(h) Finally making use of the orthonormality property of the $J_0(\alpha_n r/R)$ as demonstrated in Section 19.9 determine that

$$A_n = \frac{2}{J_1^2(\alpha_n)R^2} \int_0^R rF(r)J_0\left(\frac{\alpha_n r}{R}\right) dr$$

and

$$B_n = \frac{2}{J_1^2(\alpha_n)c_0\alpha_n R} \int_0^R rK(r)J_0\left(\frac{\alpha_n r}{R}\right) dr;$$

in order to satisfy the I.C.'s.

19.9. The purpose of this problem is to use a Calculus of Variations approach to deduce the so-called elastic beam equations for a particular prototype situation related to the behavior of two-ply laminated architectural glass plates used in the automotive industry. Consider a laminated composite beam of length L and width b composed of two layers of glass of thicknesses $h_{1,2}$ separated by a thin plastic interlayer of thickness t. The total potential energy of such a beam subjected to a distributed load q and a point load P at its center can be written as [5]

$$\Pi = \int_0^L f(x, w, u_1, u_2, w', u_1', u_2', w'')\, dx$$

where

$$f(x,w,u_1,u_2,w',u_1',u_2',w'') = \sum_{i=1}^{2} \frac{E}{2}\left\{A_i\left[\frac{du_i}{dx}+\frac{1}{2}\left(\frac{dw}{dx}\right)^2\right]^2 + I_i\left(\frac{d^2w}{dx^2}\right)^2\right\}$$
$$+\frac{Gb}{2t}\left(u_1-u_2-h\frac{dw}{dx}\right)^2 - qw - \frac{Pw_L}{2L^2}x.$$

Here x is the axial coordinate of the beam; w, its common lateral displacement with $w_L \equiv w|_{x=L}$; E, the elastic modulus of the glass; G, the shear modulus of the plastic; and $h = \frac{h_1}{2}+\frac{h_2}{2}+t$ while $u_{1,2}, A_{1,2}$, and $I_{1,2}$ are the axial displacements, cross-sectional areas, and second moments of area of the top and bottom glass layers, respectively. Assume that the actual displacements are the ones that render the total potential energy a minimum [5].

(a) Employing a Calculus of Variations approach conclude from the appropriate Euler–Lagrange equation for w

$$\frac{\partial f}{\partial w} - \frac{d}{dx}\frac{\partial f}{\partial w'} + \frac{d^2}{dx^2}\frac{\partial f}{\partial w''} = 0,$$

that these displacements satisfy the equation

$$\left[\frac{d^2}{dx^2}\left(EI\frac{d^2}{dx^2}\right) - \frac{Gbh^2}{t}\frac{d^2}{dx^2}\right]w$$
$$= q + \left(\frac{dN_1}{dx}+\frac{dN_2}{dx}\right)\frac{dw}{dx} + (N_1+N_2)\frac{d^2w}{dx^2} - \frac{Gbh}{t}\left(\frac{du_1}{dx}-\frac{du_2}{dx}\right),$$

where $I = I_1 + I_2$ and $N_{1,2}$, the axial forces in the top and bottom glass layers, are given by

$$N_{1,2} = EA_{1,2}\left[\frac{du_{1,2}}{dx}+\frac{1}{2}\left(\frac{dw}{dx}\right)^2\right].$$

(b) Continuing this approach conclude from the appropriate Euler–Lagrange equations for $u_{1,2}$

$$\frac{d}{dx}\frac{\partial f}{\partial u_{1,2}'} = \frac{\partial f}{\partial u_{1,2}},$$

that these displacements, in addition, satisfy

$$\frac{dN_1}{dx} = \frac{Gb}{t}\left(u_1-u_2-h\frac{dw}{dx}\right) = -\frac{dN_2}{dx}.$$

(c) Deduce from (b) that

$$N_1 + N_2 \equiv N_0, \text{ a constant,}$$

and hence the differential equation of (a) reduces to

$$\left[\frac{d^2}{dx^2}\left(EI\frac{d^2}{dx^2}\right)-\frac{Gbh^2}{t}\frac{d^2}{dx^2}\right]w=q+N_0\frac{d^2w}{dx^2}-\frac{Gbh}{t}\left(\frac{du_1}{dx}-\frac{du_2}{dx}\right).$$

Chapter 20
Alternate Methods of Solution for Heat and Wave Equation Problems

In this chapter, the Laplace transform (first introduced in Chapter 7) method of solution is employed to solve the problems of heat conduction in a laterally insulated semi-infinite or infinite bar that had been solved by a similarity solution method in Chapter 6 and Problem 6.1, respectively. The Laplace transform method of solving the latter problem involves the use of a Dirac delta function which is introduced in a pastoral interlude. Then the sound wave problem, the solution of which had been obtained in Chapter 12 by employing D'Alembert's characteristic coordinate method, is solved by using a Fourier integral approach. There are two problems: The first dealing with heat conduction in a laterally insulated semi-infinite bar in contact with a time-dependent reservoir, which can only be solved by the Laplace transform method, and the second, with the first-order linear partial differential equation example, solved by the method of characteristics in Chapter 12, which now is to be solved by the Fourier integral method introduced in this chapter.

20.1 Laplace Transform Method of Solution for Heat Conduction in a Semi-Infinite Bar

Recall from Chapter 6 that when investigating impulsive heat conduction in a laterally insulated semi-infinite bar, we deduced the following problem for its temperature function $u(x,t;\kappa_0)$:

$$\text{PDE: } \frac{\partial u}{\partial t} = \kappa_0 \frac{\partial^2 u}{\partial x^2}, \; 0 < x < \infty, \, t > 0;$$

$$\text{BC's: } u(0,t;\kappa_0) = 1, \; u(x,t;\kappa_0) \to 0 \text{ as } x \to \infty \text{ for } t > 0;$$

$$\text{I.C.: } u(x,0;\kappa_0) = 0 \text{ for } 0 < x < \infty;$$

and solved it by a similarity solution method to obtain

© Springer International Publishing AG, part of Springer Nature 2017
D. J. Wollkind and B. J. Dichone, *Comprehensive Applied Mathematical Modeling in the Natural and Engineering Sciences*,
https://doi.org/10.1007/978-3-319-73518-4_20

$$u(x,t;\kappa_0) = \text{erf}_c\left(\frac{x}{2\sqrt{\kappa_0 t}}\right).$$

We indicated at the time that this method of solution was crucially dependent upon the BC's being independent of t and if that were not the case then a Laplace transform method of solution would be required instead. Toward that end we now solve this problem by using such a Laplace transform method.

We introduced the concept of Laplace transforms in Chapter 7. Recall that the Laplace transform of a real function $f(t)$ is defined by

$$\mathcal{L}\{f(t)\} = \int_0^\infty e^{-st} f(t)\,dt = F(s) \text{ for } \text{Re}(s) > \alpha.$$

In Chapter 7 we showed that $\mathcal{L}\{e^{at}\} = 1/(s-a)$ for $\text{Re}(s) > \text{Re}(a)$ and employed this result to deduce that $\mathcal{L}\{\sin(\omega t)\} = \omega/(s^2+\omega^2)$. Similarly, $\mathcal{L}\{\cos(\omega t)\} = s/(s^2+\omega^2)$ can be deduced in the same manner. We shall now develop some additional properties of Laplace transforms required for this chapter and its problems. Consider the Laplace transform of the derivative of a function $y(t)$:

$$\mathcal{L}\left\{\frac{dy}{dt}(t)\right\} = \int_0^\infty e^{-st}\frac{dy}{dt}\,dt;$$

which, upon integration by parts, yields

$$\mathcal{L}\left\{\frac{dy}{dt}(t)\right\} = e^{-st}y(t)\Big|_{t=0}^{\infty} + s\int_0^\infty e^{-st}y(t)\,dt.$$

Then we can conclude that

$$\lim_{t\to\infty} e^{-st}y(t) = 0$$

provided there exists an $M, T > 0$ such that $|y(t)| \le Me^{\alpha t}$ for $t > T$ which are also the conditions for the existence of

$$\mathcal{L}\{y(t)\} = \int_0^\infty e^{-st}y(t)\,dt = Y(s).$$

Hence, we have deduced that for such a $y(t)$

$$\mathcal{L}\left\{\frac{dy}{dt}(t)\right\} = sY(s) - y(0).$$

Sometimes $F(s)$ is known and we wish to find its inverse Laplace transform $f(t) = \mathcal{L}^{-1}\{F(s)\}$. This can be accomplished by means of complex contour integration and from such computations an inverse Laplace transform table may be constructed as in Table 20.1 [115].

Table 20.1 Inverse Laplace transforms of some selected functions.

$F(s)$	$f(t)$
$\dfrac{e^{-a\sqrt{s}}}{s}$	$\text{erf}_c\left(\dfrac{a}{2\sqrt{t}}\right)$
$\dfrac{e^{-a\sqrt{s}}}{\sqrt{s}}$	$\dfrac{e^{-a^2/(4t)}}{\sqrt{\pi t}}$
$\dfrac{1}{s\sqrt{s}}$	$2\sqrt{\dfrac{t}{\pi}}$
$\dfrac{e^{-a\sqrt{s}}}{s\sqrt{s}}$	$2\sqrt{\dfrac{t}{\pi}}e^{-a^2/(4t)} - a\text{erf}_c\left(\dfrac{a}{2\sqrt{t}}\right)$

Let us now return to the problem at hand for $u(x,t;\kappa_0)$. In what follows we shall designate that function by $u(x,t)$ for ease of exposition and define its Laplace transform by

$$U(x,s) = \mathcal{L}\{u(x,t)\} = \int_0^\infty e^{-st}u(x,t)\,dt.$$

First, multiplying its PDE by e^{-st} and integrating on t from 0 to ∞, we obtain

$$\int_0^\infty e^{-st}\frac{\partial u}{\partial t}(x,t)\,dt = \kappa_0 \int_0^\infty e^{-st}\frac{\partial^2 u}{\partial x^2}(x,t)\,dt.$$

Now, assuming the requisite conditions on $u(x,t)$, this yields

$$\mathcal{L}\left\{\frac{\partial u}{\partial t}(x,t)\right\} = sU(x,s) - \underbrace{u(x,0)}_{=0} = \kappa_0 \frac{\partial^2 U}{\partial x^2}(x,s)$$

which, upon substitution of the I.C., implies

$$\frac{\partial^2 U}{\partial x^2}(x,s) - \frac{s}{\kappa_0}U(x,s) = 0,\ 0 < x < \infty;$$

while, employing the BC's,

$$U(0,s) = \int_0^\infty e^{-st}\underbrace{u(0,t)}_{=1}\,dt = \mathcal{L}\{1\} = \mathcal{L}\{e^{-at}\big|_{a=0}\} = \left.\frac{1}{s-a}\right|_{a=0} = \frac{1}{s}$$

and

$$\lim_{x\to\infty} U(x,s) = \int_0^\infty e^{-st}\underbrace{\lim_{x\to\infty} u(x,t)}_{=0}\,dt = 0.$$

Seeking a solution of this differential equation of the form $U(x,s) = e^{m(s)x}$, we find that

$$m^2(s) = \frac{s}{\kappa_0} \Rightarrow m_{1,2}(s) = \pm\frac{\sqrt{s}}{\sqrt{\kappa_0}} \text{ where } \mathrm{Re}(\sqrt{s}) > 0$$

and hence

$$U(x,s) = A(s)e^{(x/\sqrt{\kappa_0})\sqrt{s}} + B(s)e^{-(x/\kappa_0)\sqrt{s}}.$$

Applying the far-field boundary condition as $x \to \infty$, implies $A(s) = 0$ while that at $x = 0$ yields $U(0,s) = B(s) = 1/s$. Therefore

$$U(x,s) = \frac{e^{-(x/\sqrt{\kappa_0})\sqrt{s}}}{s}.$$

Finally, inverting $U(x,s)$ from Table 20.1,

$$\begin{aligned} u(x,t) &= \mathcal{L}^{-1}\{U(x,s)\} = \mathcal{L}^{-1}\left\{\frac{e^{-(x/\sqrt{\kappa_0})\sqrt{s}}}{s}\right\} \\ &= \mathcal{L}^{-1}\left\{\frac{e^{-a\sqrt{s}}\Big|_{a=x/\sqrt{\kappa_0}}}{s}\right\} = \mathrm{erf}_c\left(\frac{a}{2\sqrt{t}}\right)\Bigg|_{a=x/\sqrt{\kappa_0}} \\ &= \mathrm{erf}_c\left(\frac{x}{2\sqrt{\kappa_0 t}}\right) \end{aligned}$$

as was to be demonstrated. Problem 20.1 consists of an example where the BC at $x = 0$ is a function of t and hence can only be solved by this Laplace transform method.

20.2 Pastoral Interlude: Dirac Delta Function

We begin this section with the concept of symbolic functions which are defined in terms of a functional acting on a member of a class of testing functions as follows:

$$\mathcal{F}(\varphi) = \int_{-\infty}^{\infty} \sigma(x)\varphi(x)\,dx$$

where $\sigma(x) \equiv$ the symbolic function while $\varphi(x) \in C^{\infty}(-\infty,\infty)$, the testing function, is of bounded support – *i.e.*, it goes to 0 identically outside of a finite interval. Further, equality of symbolic functions is defined by

$$\mathcal{F}_1(\varphi) = \mathcal{F}_2(\varphi) \Leftrightarrow \sigma_1(x) = \sigma_2(x).$$

As a specific example of a symbolic function we introduce the Dirac Delta Function $\delta(x)$

$$\int_{-\infty}^{\infty} \delta(x)\varphi(x)\,dx = \varphi(0).$$

In order to develop a feel for this symbolic function, consider a delta sequence $w_k(x)$ given by

$$w_k(x) = kw(kx) \text{ for } k > 0 \text{ where } w(x) > 0 \text{ and } \int_{-\infty}^{\infty} w(x)\,dx = 1.$$

Then

$$\lim_{k\to\infty} w_k(x) = \delta(x),$$

which we shall demonstrate as follows: Examine

$$\int_{-\infty}^{\infty} w_k(x)\varphi(x)\,dx = \underbrace{\int_{-\infty}^{\infty} w(kx)\varphi(x)k\,dx}_{\substack{u=kx \\ du=k\,dx}} = \int_{-\infty}^{\infty} w(u)\varphi\left(\frac{u}{k}\right)du.$$

Applying the generalized integral mean value theorem for $f(u), g(u) \in C[a,b]$ and $g(u) > 0$,

$$\int_a^b f(u)g(u)\,du = f(c)\int_a^b g(u)\,du \text{ where } a < c < b;$$

which reduces to the integral mean value theorem if $g(u) \equiv 1$, we obtain

$$\int_{-\infty}^{\infty} w_k(x)\varphi(x)\,dx = \int_{-\infty}^{\infty} \underbrace{w(u)}_{g(u)}\underbrace{\varphi\left(\frac{u}{k}\right)}_{f(u)}du = \varphi\left(\frac{c}{k}\right)\int_{-\infty}^{\infty} w(u)\,du = \varphi\left(\frac{c}{k}\right).$$

Now taking the limit of both sides of this expression as $k \to \infty$, yields

$$\lim_{k\to\infty}\int_{-\infty}^{\infty} w_k(x)\varphi(x)\,dx = \int_{-\infty}^{\infty}\left[\lim_{k\to\infty} w_k(x)\right]\varphi(x)\,dx = \lim_{k\to\infty}\varphi\left(\frac{c}{k}\right) = \varphi(0).$$

Thus, from the definition of $\delta(x)$ and the equality condition for symbolic functions, we conclude

$$\lim_{k\to\infty} w_k(x) = \delta(x)$$

as was to be demonstrated. We offer three examples of such delta sequences:

$$w(x) = \begin{cases} \frac{1}{2} & \text{for } |x| < 1 \\ 0 & \text{for } |x| > 1 \end{cases},\ w_k(x) = \begin{cases} \frac{k}{2} & \text{for } |x| < \frac{1}{k} \\ 0 & \text{for } |x| > \frac{1}{k} \end{cases};$$

$$w(x) = \frac{1}{\pi(x^2+1)},\ w_k(x) = \frac{k}{\pi(k^2x^2+1)};$$

$$w(x) = \frac{e^{-x^2}}{\sqrt{\pi}},\ w_k(x) = \frac{ke^{-k^2x^2}}{\sqrt{\pi}}.$$

Observe in the third example if we take $k = 1/\sqrt{t}$, then

$$W_t(x) = \frac{e^{-x^2/t}}{\sqrt{\pi t}} \text{ such that } \lim_{t\to 0^+} W_t(x) = \delta(x).$$

Note that this, in effect, implies $\delta(x) = 0$ for $x \neq 0$ while since

$$\int_{-\infty}^{\infty} w_k(x)\,dx = \int_{-\infty}^{\infty} w(kx)k\,dx = \int_{-\infty}^{\infty} w(u)\,du = 1,$$

this also implies that

$$\int_{-\infty}^{\infty} \delta(x)\,dx = 1.$$

As a final property of $\delta(x)$, we introduce the concept of a symbolic derivative defined by

$$\int_{-\infty}^{\infty} \sigma'(x)\varphi(x)\,dx = -\int_{-\infty}^{\infty} \sigma(x)\varphi'(x)\,dx$$

and examine the symbolic derivative $f'_\sigma(x)$ of a function $f(x)$ which is differentiable for every x except at the origin where it has a jump discontinuity such that $f(0^+) - f(0^-) = a > 0$. Then

$$\begin{aligned}\int_{-\infty}^{\infty} f'_\sigma(x)\varphi(x)\,dx &= -\int_{-\infty}^{\infty} f(x)\varphi'(x)\,dx \\ &= -\left[\int_{-\infty}^{0} f(x)\varphi'(x)\,dx + \int_{0}^{\infty} f(x)\varphi'(x)\,dx\right].\end{aligned}$$

Upon integrating both integrals on the right-hand side of this relation by parts and employing the bounded support of $\varphi(x)$ to conclude that

$$\lim_{x\to\pm\infty} \varphi(x) = 0,$$

we obtain

$$\begin{aligned}\int_{-\infty}^{\infty} f'_\sigma(x)\varphi(x)\,dx = &-\left[f(x)\varphi(x)|_{-\infty}^{0^-} - \int_{-\infty}^{0} f'(x)\varphi(x)\,dx\right] \\ &-\left[f(x)\varphi(x)|_{0^+}^{\infty} - \int_{0}^{\infty} f'(x)\varphi(x)\,dx\right]\end{aligned}$$

$$= f(0^+)\underbrace{\varphi(0^+)}_{=\varphi(0)} - f(0^-)\underbrace{\varphi(0^-)}_{=\varphi(0)} + \int_{-\infty}^{0} f'(x)\varphi(x)\,dx$$

$$+ \int_{0}^{\infty} f'(x)\varphi(x)\,dx$$

$$= \underbrace{[f(0^+) - f(0^-)]}_{=a}\varphi(0) + \int_{-\infty}^{\infty} f'(x)\varphi(x)\,dx$$

$$= a\varphi(0) + \int_{-\infty}^{\infty} f'(x)\varphi(x)\,dx$$

$$= \int_{-\infty}^{\infty} [a\delta(x) + f'(x)]\varphi(x)\,dx.$$

Thus, from the equality condition for symbolic functions, we conclude

$$f'_{\sigma}(x) = f'(x) + a\delta(x),$$

which means that the presence of a term of the form $a\delta(x)$ in an equation involving symbolic derivatives of a particular quantity implies that its symbolic antiderivative undergoes a jump discontinuity of value a at $x = 0$. We shall use this fact in the next section dealing with the Laplace transform method of solution for Problem 6.1 involving heat conduction in a laterally insulated infinite bar to derive a jump condition satisfied by the spatial derivative of the transform of its temperature function at the midpoint of the bar.

20.3 Laplace Transform Method of Solution for Heat Conduction in an Infinite Bar

Recall that, for Problem 6.1 dealing with heat conduction in a laterally insulated infinite bar, we defined the amount of heat in this bar between its axial coordinate $s = \pm x$ by

$$Q(x,t) = \rho_0 c_0 \mathcal{A} \int_{-x}^{x} T(s,t)\,ds + Q_e,\ 0 < x < \infty,\ t > 0;$$

where

$$Q(0,t) = Q_e,\ Q(x,t) \to Q_e + Q_0 \text{ as } x \to \infty;$$

and found by a similarity solution method for $Q(x,t)$ that its temperature function

$$T(x,t) = \frac{Q_0 e^{-x^2/(4\kappa_0 t)}}{2\mathcal{A}\sqrt{\rho_0 c_0 k_0 \pi t}}.$$

In this section we shall pose a mixed initial-boundary value problem for that function and solve it by means of a Laplace transform method.

Given that there is no constantly replenished source when $t > 0$, $T(x,t)$ satisfies the heat conduction equation (P1.12) of Problem 6.1:

$$\text{PDE: } \frac{\partial T}{\partial t} = \kappa_0 \frac{\partial^2 T}{\partial x^2} \text{ where } \kappa_0 = \frac{k_0}{\rho_0 c_0}; -\infty < x < \infty, t > 0;$$

with

$$\text{BC's: } T(x,t) \to 0 \text{ as } x \to \pm\infty;$$

the boundary conditions following directly from the far-field condition for $Q(x,t)$. We must still deduce an I.C. for $T(x,t)$ consistent with our initial introduction of a heat source Q_0 at the midpoint of the bar and proceed as follows: The far-field condition for $Q(x,t)$ implies

$$\int_{-\infty}^{\infty} T(s,t)\, ds = \frac{Q_0}{\rho_0 c_0 \mathcal{A}}.$$

Taking the limit of this integral as $t \to 0^+$, yields

$$\int_{-\infty}^{\infty} \left[\lim_{t \to 0^+} T(s,t) \right] ds = \frac{Q_0}{\rho_0 c_0 \mathcal{A}}.$$

Now letting

$$\lim_{t \to 0^+} T(s,t) = T_0 \delta(s),$$

$$\underbrace{T_0 \int_{-\infty}^{\infty} \delta(s)\, ds}_{=1} = T_0 = \frac{Q_0}{\rho_0 c_0 \mathcal{A}}.$$

Therefore, we have deduced the following initial condition for $T(x,t)$:

$$\text{I.C.: } \lim_{t \to 0^+} T(x,t) = \frac{Q_0 \delta(x)}{\rho_0 c_0 \mathcal{A}}.$$

Having completing the formulation of this mixed initial-boundary value problem for $T(x,t)$, we solve it by a Laplace transform method. Toward that end define

$$\theta(x,s) = \mathcal{L}\{T(x,t)\} = \lim_{\varepsilon \to 0} \int_{\varepsilon}^{\infty} e^{-st} T(x,t)\, dt.$$

Then, proceeding exactly as in Section 20.1, we find from the PDE and the I.C. that

$$s\theta(x,s) - \lim_{\varepsilon\to 0} T(x,\varepsilon) = s\theta(x,s) - \frac{Q_0}{\rho_0 c_0 \mathcal{A}}\delta(x) = \kappa_0 \frac{\partial^2\theta}{\partial x^2}(x,s)$$

or

$$\frac{\partial^2\theta}{\partial x^2}(x,s) - \frac{s}{\kappa_0}\theta(x,s) = -\frac{Q_0}{k_0\mathcal{A}}\delta(x), \ -\infty < x < \infty;$$

while the BC's imply that

$$\theta(x,s) \to 0 \text{ as } x \to \pm\infty.$$

Defining

$$\theta(x,s) = \begin{cases} \theta_1(x,s) & \text{for } x < 0 \\ \theta_2(x,s) & \text{for } x > 0 \end{cases}$$

and assuming a continuity condition at $x = 0$

$$\theta_1(0,s) = \theta_2(0,s),$$

the presence of $\delta(x)$ in the differential equation, as mentioned at the end of Section 20.2, implies a jump condition on $\partial\theta/\partial x$ at $x = 0$

$$\frac{\partial\theta_1}{\partial x}(0,s) - \frac{\partial\theta_2}{\partial x}(0,s) = \frac{Q_0}{k_0\mathcal{A}};$$

which basically is a Green's function type formulation [45] for $\theta(x,s)$.

Recalling that $\delta(x) = 0$ for $x \neq 0$, the differential equation and far-field conditions yield the following problems for $\theta_{1,2}(x,s)$:

$$\frac{\partial^2\theta_1}{\partial x^2}(x,s) - \frac{s}{\kappa_0}\theta_1(x,s) = 0 \text{ for } x < 0, \ \theta_1(x,s) \to 0 \text{ as } x \to -\infty;$$

and

$$\frac{\partial^2\theta_2}{\partial x^2}(x,s) - \frac{s}{\kappa_0}\theta_2(x,s) = 0 \text{ for } x > 0, \ \theta_2(x,s) \to 0 \text{ as } x \to \infty.$$

Solving these problems, we obtain that

$$\theta_1(x,s) = c_1(s)e^{\sqrt{s/\kappa_0}x} \text{ and } \theta_2(x,s) = c_2(s)e^{-\sqrt{s/\kappa_0}x} \text{ where } \mathrm{Re}(\sqrt{s}) > 0;$$

and, applying the continuity and jump conditions at $x = 0$, determine that

$$c_1(s) = c_2(s) = \frac{Q_0}{2k_0\mathcal{A}}\sqrt{\frac{\kappa_0}{s}} = \frac{Q_0}{2\mathcal{A}\sqrt{\rho_0 c_0 k_0 s}}.$$

Hence our solution becomes

$$\theta(x,s) = \frac{Q_0}{2\mathcal{A}\sqrt{\rho_0 c_0 k_0}} \frac{e^{-a\sqrt{s}}\Big|_{a=x/\sqrt{\kappa_0}}^{a=-x/\sqrt{\kappa_0}} \text{ for } \begin{matrix} x \le 0 \\ x \ge 0 \end{matrix}}{\sqrt{s}}.$$

Finally, inverting $\theta(x,s)$ from Table 20.1,

$$\begin{aligned}
T(x,t) &= \frac{Q_0}{2\mathcal{A}\sqrt{\rho_0 c_0 k_0}} \mathcal{L}^{-1}\{\theta(x,s)\} \\
&= \frac{Q_0}{2\mathcal{A}\sqrt{\rho_0 c_0 k_0}} \mathcal{L}^{-1}\left\{ \frac{e^{-a\sqrt{s}}\Big|_{a=x/\sqrt{\kappa_0}}^{a=-x/\sqrt{\kappa_0}} \text{ for } \begin{matrix} x \le 0 \\ x \ge 0 \end{matrix}}{\sqrt{s}} \right\} \\
&= \frac{Q_0}{2\mathcal{A}\sqrt{\rho_0 c_0 k_0 \pi t}} e^{-a^2/(4t)}\Big|_{a^2=x^2/\kappa_0} \\
&= \frac{Q_0}{2\mathcal{A}\sqrt{\rho_0 c_0 k_0 \pi t}} e^{-x^2/(4\kappa_0 t)}
\end{aligned}$$

as was to be demonstrated. Observe that this is of the form of the delta sequence $W_t(x)$ as $t \to 0^+$ offered in Section 20.2 consistent with our I.C. when t is replaced by $4\kappa_0 t$.

20.4 Fourier Integral Method of Solution for the Sound Wave Problem

We begin this section with a restatement of the Fourier integral theorem developed in Chapter 7. Consider a real-valued function $f(x)$ of a real variable such that f and f' are piecewise continuous on any finite interval and

$$\int_{-\infty}^{\infty} |f(u)|^2\, du \text{ converges.}$$

Then the Fourier integral theorem states that

$$\frac{f(x+0)+f(x-0)}{2} \doteq f(x) = \int_{-\infty}^{\infty} C(\alpha)e^{i\alpha x}\, d\alpha \text{ where } C(\alpha) = \frac{1}{2\pi}\int_{-\infty}^{\infty} f(u)e^{-i\alpha u}\, du$$

or, synthesizing these results,

$$f(x) = \int_{-\infty}^{\infty} \left[\frac{1}{2\pi}\int_{-\infty}^{\infty} f(u)e^{-i\alpha u}\, du \right] e^{i\alpha x}\, d\alpha.$$

Recall in the sound wave problem of Chapter 11 we deduced the following mixed initial-boundary value problem for its condensation function $s(x,t)$:

$$\text{PDE: } \frac{\partial^2 s}{\partial t^2} = c_0^2 \frac{\partial^2 s}{\partial x^2}, -\infty < x < \infty, t > 0;$$

$$\text{BC's: } s(x,t) \text{ remains bounded as } x^2 \to \infty \text{ for } t > 0;$$

$$\text{I.C.'s: } s(x,0) = f(x) \text{ and } \frac{\partial s}{\partial t}(x,0) = h'(x) \text{ for } -\infty < x < \infty,$$

where

$$\int_{-\infty}^{\infty} |f(u)|^2\, du \text{ and } \int_{-\infty}^{\infty} |h(u)|^2\, du \text{ both converge;}$$

and solving it by D'Alembert's Method of characteristic coordinates found that

$$s(x,t) = \frac{1}{2}[f(x+c_0t) + f(x-c_0t)] + \frac{1}{2c_0}[h(x+c_0t) - h(x-c_0t)].$$

We shall resolve this problem by means of a Fourier integral method of solution as follows: Let

$$s(x,t) = Ce^{i\alpha x + \sigma t} \text{ for } C \neq 0$$

where $\alpha \in \mathbb{R}$ to satisfy the BC's. Then, since

$$\frac{\partial s}{\partial t} = \sigma s \text{ and } \frac{\partial s}{\partial x} = i\alpha s \text{ where } i^2 = -1,$$

substitution of this solution into our PDE results in the secular equation

$$\sigma^2 = -\alpha^2 c_0^2 \Rightarrow \sigma = \sigma_{1,2} = \pm i\alpha c_0.$$

Thus, invoking a superposition principal, this yields the solution

$$\begin{aligned} s(x,t) &= \int_{-\infty}^{\infty} [C_1(\alpha)e^{i\alpha x + \sigma_1 t} + C_2(\alpha)e^{i\alpha x + \sigma_2 t}]\, d\alpha \\ &= \int_{-\infty}^{\infty} [C_1(\alpha)e^{i\alpha(x+c_0t)} + C_2(\alpha)e^{i\alpha(x-c_0t)}]\, d\alpha. \end{aligned}$$

Applying the I.C.'s, we obtain

$$s(x,0) = \int_{-\infty}^{\infty} [C_1(\alpha) + C_2(\alpha)]e^{i\alpha x}\, d\alpha = f(x)$$

and

$$\frac{\partial s}{\partial t}(x,0) = \int_{-\infty}^{\infty} i\alpha c_0[C_1(\alpha) - C_2(\alpha)]e^{i\alpha x}\, d\alpha = h'(x).$$

Now, employing the Fourier integral theorem to invert these relations, we find that

$$C_1(\alpha) + C_2(\alpha) = \frac{1}{2\pi}\int_{-\infty}^{\infty} f(u)e^{-i\alpha u}\, du$$

and

$$C_1(\alpha)-C_2(\alpha)=\frac{1}{2\pi c_0}\int_{-\infty}^{\infty}h'(u)\frac{e^{-i\alpha u}}{i\alpha}\,du=\frac{1}{2\pi c_0}\int_{-\infty}^{\infty}h(u)e^{-i\alpha u}\,du;$$

since, upon integration by parts,

$$\int_{-\infty}^{\infty}h'(u)\frac{e^{-i\alpha u}}{i\alpha}\,du=-h(u)e^{-i\alpha u}\Big|_{u=-\infty}^{\infty}+\int_{-\infty}^{\infty}h(u)e^{-i\alpha u}\,du=\int_{-\infty}^{\infty}h(u)e^{-i\alpha u}\,du;$$

given that

$$\lim_{u\to\pm\infty}h(u)=0,$$

which follows directly from the convergence property of $|h(u)|^2$. Next, solving these equations simultaneously,

$$C_1(\alpha)=\frac{1}{4\pi}\int_{-\infty}^{\infty}\left[f(u)+\frac{1}{c_0}h(u)\right]e^{-i\alpha u}\,du$$

and

$$C_2(\alpha)=\frac{1}{4\pi}\int_{-\infty}^{\infty}\left[f(u)-\frac{1}{c_0}h(u)\right]e^{-i\alpha u}\,du.$$

Then substitution of those values into our solution yields,

$$s(x,t)=\frac{1}{2}\int_{-\infty}^{\infty}\left[\frac{1}{2\pi}\int_{-\infty}^{\infty}f(u)e^{-i\alpha u}\,du\right]\{e^{i\alpha(x+c_0t)}+e^{i\alpha(x-c_0t)}\}\,d\alpha$$
$$+\frac{1}{2c_0}\int_{-\infty}^{\infty}\left[\frac{1}{2\pi}\int_{-\infty}^{\infty}h(u)e^{-i\alpha u}\,du\right]\{e^{i\alpha(x+c_0t)}-e^{-i\alpha(x-c_0t)}\}\,d\alpha;$$

which, upon employment of the Fourier integral theorem in its synthesized form, finally reduces this expression to

$$s(x,t)=\frac{1}{2}[f(x+c_0t)+f(x-c_0t)]+\frac{1}{2c_0}[h(x+c_0t)-h(x-c_0t)],$$

as was to be demonstrated for our second-order characteristic coordinate problem. In Problem 20.2, it is shown that a first-order characteristic coordinate problem, namely, the example presented in Section 11.3(i), can be solved in a similar manner as well.

Problems

20.1. In this problem we examine heat conduction in a laterally insulted semi-infinite bar driven by a time-dependent heat reservoir in contact with its $x=0$ end in the absence of any bulk sources. Thus we consider heat conduction in such a bar, originally at some ambient temperature T_e, which, from time $t\geq 0$, is placed in contact with a heat reservoir of elevated temperature $T_e+T_1\sqrt{t/\tau}$. Heat conduction for this scenario is then governed by the following boundary-initial value partial differential diffusion equation problem for $T = T(x,t) \equiv$ temperature distribution in the bar:

$$\text{PDE: } \frac{\partial T}{\partial t} = \kappa_0 \frac{\partial^2 T}{\partial x^2}, \, 0 < x < \infty; \, t > 0;$$

$$\text{BC's: } T(0,t) = T_e + T_1\sqrt{\frac{t}{\tau}}, \, T(x,t) \to T_e \text{ as } x \to \infty \text{ for } t > 0;$$

$$\text{I.C.: } T(x,0) = T_e \text{ for } 0 < x < \infty.$$

(a) Introducing $u = u(x,t)$ defined by $T(x,t) = T_e + T_1 u(x,t)$, transforms this system into

$$\text{PDE: } \frac{\partial u}{\partial t} = \kappa_0 \frac{\partial^2 u}{\partial x^2}, \, 0 < x < \infty, \, t > 0;$$

$$\text{BC's: } u(0,t) = \sqrt{\frac{t}{\tau}}, \, u(x,t) \to 0 \text{ as } x \to \infty \text{ for } t > 0;$$

$$\text{I.C.: } u(x,0) = 0 \text{ for } 0 < x < \infty.$$

(b) Since the BC at $x = 0$ is a function of t, a Laplace transform method of solution must be employed to find $u(x,t)$. Toward that end define its Laplace transform by

$$U(x,s) = \mathcal{L}\{u(x,t)\} = \int_0^\infty e^{-st} u(x,t)\, dt,$$

and, using the methodology employed in this chapter as well as Table 20.1, show that

$$\frac{\partial^2 U}{\partial x^2}(x,s) - \frac{s}{\kappa_0} U(x,s) = 0, \, 0 < x < \infty;$$

$$U(0,s) = \frac{\sqrt{\pi}}{2s\sqrt{\tau s}}, \, U(x,s) \to 0 \text{ as } x \to \infty.$$

(c) Solving this problem, obtain that

$$U(x,s) = \left(\frac{\sqrt{\pi/\tau}}{2}\right) \frac{e^{-(x/\sqrt{\kappa_0})\sqrt{s}}}{s\sqrt{s}}.$$

(d) Finally, inverting $U(x,s)$ by means of Table 20.1, find that

$$u(x,t) = \mathcal{L}^{-1}\{U(x,s)\} = \sqrt{\frac{t}{\tau}} e^{-x^2/(4\kappa_0 t)} - \sqrt{\pi}\left(\frac{x}{2\sqrt{\kappa_0 \tau}}\right) \operatorname{erf}_c\left(\frac{x}{2\sqrt{\kappa_0 t}}\right).$$

20.2. The purpose of this problem is to demonstrate that a first-order characteristic coordinate-type linear partial differential equation can also be solved by the Fourier integral method employed in Section 20.4. Toward that end we reconsider the example introduced in Section 11.3(i) for $\rho(x,t)$:

$$\text{PDE: } a_0 \frac{\partial \rho}{\partial x} + \frac{\partial \rho}{\partial t} = m_0 \rho, \ -\infty < x < \infty, \ t > 0;$$

$$\text{BC's: } \rho(x,t) \text{ remains bounded as } x \to \pm\infty \text{ for } t > 0;$$

$$\text{I.C.: } \rho(x,0) = f(x) \text{ for } -\infty < x < \infty,$$

where

$$\int_{-\infty}^{\infty} |f(x)|^2 \, dx \text{ converges.}$$

Recall from Section 11.3(i) that solving this system by the method of characteristics yields

$$\rho(x,t) = e^{m_0 t} f(x - a_0 t).$$

(a) Let

$$\rho(x,t) = Ce^{i\alpha x + \sigma t} \text{ for } C \neq 0,$$

where $\alpha \in \mathbb{R}$ to satisfy the BC's, and from the PDE obtain the secular equation

$$\sigma = m_0 - i\alpha a_0.$$

(b) Then, invoking a superposition principle in conjunction with the results of part (a), conclude

$$\rho(x,t) = e^{m_0 t} \int_{-\infty}^{\infty} C(\alpha) e^{i\alpha(x - a_0 t)} \, d\alpha.$$

(c) Applying the I.C., obtain the Fourier integral relationship

$$\rho(x,0) = \int_{-\infty}^{\infty} C(\alpha) e^{i\alpha x} \, d\alpha = f(x).$$

(d) Finally, employing this Fourier integral relationship in the solution of part (b), deduce that

$$\rho(x,t) = e^{m_0 t} \int_{-\infty}^{\infty} C(\alpha) e^{i\alpha(x - a_0 t)} \, d\alpha = e^{m_0 t} f(x - a_0 t),$$

as was to be demonstrated.

Chapter 21
Finite Mathematical Models

So far, we have predominantly been concentrating on differential equation models for natural and engineering science phenomena. In this chapter, for the sake of completeness, we shall examine three phenomena, one each from the biological, political, and financial sciences, which, of necessity, are modeled by finite mathematical techniques instead. The first deals with the discrete-time rabbit reproduction population dynamics model, posed by Leonardo of Pisa, which gives rise to the finite difference equation, the solution of which produces the Fibonacci sequence. In the chapter, this is solved both directly in scalar form and by placing it in a system formulation that is then solved by a Jordan canonical form method. Two other methods of solution are introduced in the problems for the system formulation: Namely, a Cayley–Hamilton Theorem approach and a direct eigenvalue–eigenvector expansion method. The second model deals with the minimum fraction of the popular vote that can elect the President of the United States posed by George Pólya. The 1960 and 1996 presidential elections are examined in the chapter, while the 2008 election is considered in a problem. The third model deals with the financial mathematics problem of the repayment of a loan or mortgage. A loan shark example with 100% interest rate per pay period is examined in the chapter and a similar one with only a 50% interest rate per period is examined in the last problem.

21.1 Discrete-Time Population Dynamics: Fibonacci Sequence

As mentioned in Section 3.4, difference equations are more appropriate to model biological population interactions for which the generations are nonoverlapping [47]. Interest in studying the population dynamics of such a species adhering to a constant reproductive schedule dates back at least to Leonardo of Pisa (Fibonacci) who posed the following idealized problem in his *Liber abaci* (1202):

How many pairs of rabbits can be produced from a single pair in one year if it is assumed that every month each pair begets a new pair which from the second month

© Springer International Publishing AG, part of Springer Nature 2017

D. J. Wollkind and B. J. Dichone, *Comprehensive Applied Mathematical Modeling in the Natural and Engineering Sciences*,
https://doi.org/10.1007/978-3-319-73518-4_21

becomes reproductive?

Fibonacci solved this problem and obtained the sequence of new pairs produced per month for one year given by

$$1, 1, 2, 3, 5, 8, 13, 21, 34, 55, 89, 144,$$

which is known today as the Fibonacci sequence.

We wish to find an explicit formula for x_k where $k \equiv$ month and $x_k \equiv$ number of new pairs produced that month. Observe that this sequence satisfies the recursion relation

$$x_{k+2} = x_k + x_{k+1} \text{ for } k \geq 0; \text{ where } x_1 = 1 \text{ and } x_0 = 0 \text{ (defined).}$$

Before solving this second-order constant coefficient difference equation directly by a scalar method, we shall reformulate it as a first-order 2×2 matrix system and solve that by a canonical Jordan form method. Let $y_k = x_{k+1}$. Then

$$x_{k+1} = 0 \cdot x_k + 1 \cdot y_k \text{ for } k \geq 0, \text{ where } x_0 = 0;$$

$$y_{k+1} = x_{k+2} = x_k + x_{k+1} = 1 \cdot x_k + 1 \cdot y_k \text{ for } k \geq 0, \text{ where } y_0 = x_1 = 1.$$

Thus, defining the vector quantity

$$X_k = \begin{bmatrix} x_k \\ y_k \end{bmatrix},$$

this problem is equivalent to the first-order 2×2 matrix system

$$X_{k+1} = AX_k \text{ for } k \geq 0, \text{ where } A = \begin{pmatrix} 0 & 1 \\ 1 & 1 \end{pmatrix} \text{ and } X_0 = \begin{bmatrix} 0 \\ 1 \end{bmatrix}.$$

Hence

$$X_1 = AX_0, X_2 = AX_1 = A^2X_0, \ldots, X_k = A^kX_0.$$

We shall next use a Jordan canonical form matrix method to represent A^k and in Problem 21.1 develop an alternate Cayley–Hamilton Theorem approach to deduce that representation. Toward this end we first compute the eigenvalues and eigenvectors of A as follows: Consider

$$AZ = \lambda Z \text{ where } Z = \begin{bmatrix} z_1 \\ z_2 \end{bmatrix} \text{ with } z_2 = \lambda z_1 \neq 0$$

$$\Rightarrow \det(A - \lambda I) = 0 \text{ where } I = \begin{pmatrix} 1 & 0 \\ 0 & 1 \end{pmatrix} \text{ or } \begin{vmatrix} -\lambda & 1 \\ 1 & 1-\lambda \end{vmatrix} = \lambda^2 - \lambda - 1 = 0$$

$$\Rightarrow \lambda = \lambda_{1,2} = \frac{1}{2}(1 \pm \sqrt{5}) \text{ and } Z = Z^{(1,2)} = \begin{bmatrix} 1 \\ \lambda_{1,2} \end{bmatrix}.$$

Therefore $(A)_{ij} = a_{ij}$ for $i,j = 1$ and 2 has the Jordan canonical form

$$\sum_{k=1}^{2} a_{ik} z_k^{(j)} = \lambda_j z_i^{(j)} = \sum_{k=1}^{2} z_i^{(k)} \lambda_j \delta_{kj}$$

where $z_i^{(j)} = (Q)_{ij}$ and $\lambda_j \delta_{ij} = (J)_{ij}$ for $i,j = 1$ and 2

or

$$AQ = QJ \text{ where } Q = \begin{pmatrix} 1 & 1 \\ \lambda_1 & \lambda_2 \end{pmatrix} \text{ and } J = \begin{pmatrix} \lambda_1 & 0 \\ 0 & \lambda_2 \end{pmatrix};$$

which implies

$$A = QJQ^{-1} \text{ where } Q^{-1} = \frac{1}{\lambda_2 - \lambda_1} \begin{pmatrix} \lambda_2 & -1 \\ -\lambda_1 & 1 \end{pmatrix}.$$

Now

$$A^2 = (QJQ^{-1})(QJQ^{-1}) = QJ^2Q^{-1}, \ldots, A^k = QJ^kQ^{-1};$$

or

$$\begin{aligned} A^k &= \frac{1}{\lambda_2 - \lambda_1} \begin{pmatrix} 1 & 1 \\ \lambda_1 & \lambda_2 \end{pmatrix} \begin{pmatrix} \lambda_1^k & 0 \\ 0 & \lambda_2^k \end{pmatrix} \begin{pmatrix} \lambda_2 & -1 \\ -\lambda_1 & 1 \end{pmatrix} \\ &= \frac{1}{\lambda_2 - \lambda_1} \begin{pmatrix} \lambda_1^k & \lambda_2^k \\ \lambda_1^{k+1} & \lambda_2^{k+1} \end{pmatrix} \begin{pmatrix} \lambda_2 & -1 \\ -\lambda_1 & 1 \end{pmatrix} \\ &= \frac{1}{\lambda_2 - \lambda_1} \begin{pmatrix} \lambda_2\lambda_1^k - \lambda_1\lambda_2^k & \lambda_2^k - \lambda_1^k \\ \lambda_2\lambda_1^{k+1} - \lambda_1\lambda_2^{k+1} & \lambda_2^{k+1} - \lambda_1^{k+1} \end{pmatrix} \\ &= \begin{pmatrix} \alpha & \beta \\ \beta & \alpha + \beta \end{pmatrix} \end{aligned}$$

where

$$\alpha = \frac{\lambda_2\lambda_1^k - \lambda_1\lambda_2^k}{\lambda_2 - \lambda_1} \text{ and } \beta = \frac{\lambda_1^k - \lambda_2^k}{\lambda_1 - \lambda_2};$$

given that, since $\lambda_1\lambda_2 = -1$ and $\lambda_1 + \lambda_2 = 1$,

$$\lambda_2\lambda_1^{k+1} - \lambda_1\lambda_2^{k+1} = \lambda_1\lambda_2(\lambda_1^k - \lambda_2^k) = \lambda_2^k - \lambda_1^k$$

and

$$\begin{aligned} \alpha + \beta &= \frac{\lambda_2\lambda_1^k - \lambda_1^k + \lambda_2^k - \lambda_1\lambda_2^k}{\lambda_2 - \lambda_1} \\ &= \frac{(\lambda_2 - 1)\lambda_1^k + (1 - \lambda_1)\lambda_2^k}{\lambda_2 - \lambda_1} = \frac{\lambda_2^{k+1} - \lambda_1^{k+1}}{\lambda_2 - \lambda_1}. \end{aligned}$$

Finally, having deduced this representation for A^k, it follows that

$$X_k = \begin{bmatrix} x_k \\ y_k \end{bmatrix} = A^k X_0 = \begin{pmatrix} \alpha & \beta \\ \beta & \alpha + \beta \end{pmatrix} \begin{bmatrix} 0 \\ 1 \end{bmatrix}$$

which yields the explicit formula for x_k, as desired, and y_k:

$$x_k = \beta = \frac{\lambda_1^k - \lambda_2^k}{\lambda_1 - \lambda_2} = \frac{(1+\sqrt{5})^k - (1-\sqrt{5})^k}{2^k\sqrt{5}}$$

and (equivalently)

$$y_k = x_{k+1} = \alpha + \beta = \frac{\lambda_2^{k+1} - \lambda_1^{k+1}}{\lambda_2 - \lambda_1} = \frac{(1+\sqrt{5})^{k+1} - (1-\sqrt{5})^{k+1}}{2^{k+1}\sqrt{5}}.$$

We shall now derive this result directly from $x_{k+2} = x_k + x_{k+1}$ for $k \geq 0$ by letting $x_k = \lambda^k$, analogous to seeking an exponential solution for a differential equation, and obtaining

$$\lambda^{k+2} = \lambda^k + \lambda^{k+1} \Rightarrow \lambda^2 - \lambda - 1 = 0 \Rightarrow \lambda = \lambda_{1,2} = \frac{1}{2}(1 \pm \sqrt{5}),$$

which not surprisingly yields the eigenvalues from our Jordan canonical form analysis. This means that our difference equation has a general solution composed of the linear combination

$$x_k = c_1\lambda_1^k + c_2\lambda_2^k,$$

where the constants c_1 and c_2 are selected to satisfy the initial conditions. That is

$$x_0 = c_1 + c_2 = 0 \text{ and } x_1 = \lambda_1 c_1 + \lambda_2 c_2 = 1 \Rightarrow c_1 = -c_2 = \frac{1}{\lambda_1 - \lambda_2}.$$

Thus we again find that

$$x_k = \frac{\lambda_1^k - \lambda_2^k}{\lambda_1 - \lambda_2} = \frac{(1+\sqrt{5})^k - (1-\sqrt{5})^k}{2^k\sqrt{5}}.$$

In the last half of Problem 21.1 a direct matrix method of this sort is also suggested.

21.2 Minimum Fraction of Popular Votes Necessary to Elect the American President

In this section we shall present George Pólya's [98] very clever model for determining the minimum fraction of popular votes necessary to elect John F. Kennedy, President of the USA in 1960. Pólya [97] proposed a four-step process for modeling complex phenomena by first understanding the problem; then devising a plan of attack; next implementing that plan to obtain a solution; and finally looking back at the phenomenological implications of that solution within the framework of the model. For this model he made the following assumptions:

1. There are only two presidential candidates and each voter casts a ballot for one of them.
2. Each state has a voting population proportional to its number of representatives r in the US House of Representatives with a proportionality constant N, a large even positive integer.
3. Each state has an electoral vote $e = r + 2$, equal to the number of its members in the US Congress, *i.e.*, the sum of its representatives plus its two senators, the US legislature being a bicameral body with every state having two members in the US Senate.
4. The candidate who wins the popular vote in a state receives all that states electoral votes while the losing candidate receives none.
5. The candidate receiving the majority of the total number of electoral votes wins the election.

We tabulate the 1960 e values for each state in Table 21.1. Designating the multiplicity of those states with exactly $r = e - 2$ representatives by m_r,

$$\sum rm_r = 437 \equiv \text{ total number of representatives in 1960}$$

while the total number of electoral votes in 1960 is then given by

$$\sum em_r = \sum (r+2)m_r = \sum rm_r + 2\sum m_r = 437 + 2(50) = 437 + 100 = 537,$$

where $\sum m_r = 50 \equiv$ the total number of states in the 1960 election. In this context we note that for every election since 1912 there has always been 435 members in the US House except in 1960 since Alaska and Hawaii had just achieved statehood and with no reapportionment having yet taken place, each was given one representative. Further, the District of Columbia (D.C.) did not participate in presidential elections until 1964 which is why its number of electoral votes is listed as 0 in Table 21.1. Given this situation observe that in order to win the presidency, John F. Kennedy who was elected US President in 1960 needed to receive at least 269 electoral votes, which was the majority value for this election. Note that for all following presidential elections, with D.C. having been assigned 3 electoral votes and the House returned to 435 members, the total number of electoral votes is 538 and the winning candidate must receive at least 270 of them (see Problem 21.2 which deals with the 2008 election). Assume that the winning candidate carries s states. Then

$$\sum_{i=1}^{s} e_i = \sum_{i=1}^{s} (r_i + 2) = \sum_{i=1}^{s} r_i + 2s \geq 269 \Rightarrow \sum_{i=1}^{s} r_i \geq 269 - 2s.$$

If W denotes the total popular vote for the winning candidate, then

$$W \geq \sum_{i=1}^{s} \left(\frac{r_i N}{2} + 1 \right) = \frac{N}{2} \sum_{i=1}^{s} r_i + s$$

Table 21.1 The states and their electoral votes $e = r + 2$ for the 1960 Presidential election [54].

Alabama	11	Kentucky	10	North Dakota	4
Alaska	3	Louisiana	10	Ohio	25
Arizona	4	Maine	5	Oklahoma	8
Arkansas	8	Maryland	9	Oregon	6
California	32	Massachusetts	16	Pennsylvania	32
Colorado	6	Michigan	20	Rhode Island	4
Connecticut	8	Minnesota	11	South Carolina	8
Delaware	3	Mississippi	8	South Dakota	4
D.C.	0	Missouri	13	Tennessee	11
Florida	10	Montana	4	Texas	24
Georgia	12	Nebraska	6	Utah	4
Hawaii	3	Nevada	3	Vermont	3
Idaho	4	New Hampshire	4	Virginia	12
Illinois	27	New Jersey	16	Washington	9
Indiana	13	New Mexico	4	West Virginia	8
Iowa	10	New York	45	Wisconsin	12
Kansas	8	North Carolina	14	Wyoming	3

where the minimum total on the right-hand side of this inequality represents a scenario in which that candidate gets two more votes than the opponent in any state carried and no votes at all in those states lost. Combining these inequalities yields

$$W \geq \frac{N}{2}(269 - 2s) + s.$$

Further the total popular vote is given by $T = 437N \cong 68.8$ million for the 1960 election. Thus

$$\frac{W}{T} \geq \frac{269 - 2s}{874} + \frac{s}{437N} = \frac{269 - 2s}{874} + \frac{s}{68.8} \times 10^{-6}.$$

Here the expression on the right-hand side of this inequality represents the minimum fraction of the popular vote necessary to elect a candidate in the 1960 US presidential election as a function of $s \equiv$ the number of states won by that candidate. Since its second term is so small being at most 7.27×10^{-7} for $s = 50$, Pólya [98] neglected it and concentrated on the first term in determining his minimum fraction. Given that in one of the cases handled in Problem 21.2 this second term actually determines which scenario provides the desired minimum fraction we shall retain it for the sake of definiteness in what follows while using the first term in our argument for calculating that minimum fraction. Due to the presence of the minus sign in its numerator, it is clear that the minimum fraction will occur when s takes on the largest allowable value and thus the strategy to achieve this result is to win as many of the states having the lowest number of electoral votes or equivalently representatives as possible. In other words this is a classical minimax problem. Thus arranging the states in increasing order of representatives, we start counting them in Table 21.2 until the sum of their electoral votes just reaches 269.

Table 21.2 Calculations to determine the s value for the minimum fraction in the 1960 election.

r	m_r	$\sum m_r$	e	em_r	$\sum em_r$	m_e	$\sum m_e$	em_e	$\sum em_4$
1	6	6	3	18	18	6	6	18	18
2	9	15	4	36	54	8	14	32	50
3	1	16	5	5	59	1	15	5	55
4	3	19	6	17	77	3	18	18	73
6	7	26	8	56	133	7	25	56	129
7	2	28	9	18	151	2	27	18	147
8	4	32	10	40	191	4	31	40	187
9	3	35	11	33	224	3	34	33	220
10	3	38	12	36	260	3	37	36	256
11	2	40	13	26	286	1	$s = 38$	13	269
12	1	41	14	14	300				
14	2	43	16	32	332				
18	1	44	20	20	352				
22	1	45	24	24	376				
23	1	46	25	25	401				
25	1	47	27	27	428				
30	2	49	32	64	492				
43	1	50	45	45	537				

Observe from the left-hand side of this table that winning the 38 states having 12 or fewer electoral votes yields a total of 260 electoral votes. Then adding one of the two states with 13 electoral votes would yield 273 total electoral votes or 4 over the desired limit of 269. Hence dropping one of the states with 4 electoral votes and replacing it with this state, we would just reach that total of 269 electoral votes which is reflected in the right-hand side of Table 21.3 where the relevant changes are indicated by highlighting. Here m_e denotes the multiplicity of states with exactly e electoral votes *actually* carried by the candidate employing this optimal strategy. Therefore, as indicated on the right-hand side of Table 21.2, $s = 38$ and

$$\left(\frac{W}{T}\right)_{\min} = \frac{\overbrace{269 - 76}^{=193}}{874} + \frac{38}{69{,}000{,}000} = 0.220823799 + 0.000000552 = 0.220824351,$$

which means that a candidate could have won the 1960 election for the US presidency with only about 22% of the popular vote. Although in the 1960 election this choice of optical strategy was unique, it does not have to be. To demonstrate this we consider the relevant data for the 1996 presidential election compiled in Table 21.3 with its e and m_r values obtained from Israel [54].

Here D.C. is one of the 8 "states" with 3 electoral votes and since its population would only warrant 1 representative if it were a state that electoral vote is being assumed to consist of 1 "virtual" representative and 2 "virtual" senators just the

Table 21.3 Calculations to determine the s value for the minimum fraction in the 1996 election.

					①				②			
e	m_r	em_r	$\sum m_r$	$\sum em_r$	m_e	em_e	$\sum m_e$	$\sum em_e$	m_e	em_e	$\sum m_e$	$\sum em_e$
3	8	24	8	24	3	8	8	24	3	8	8	24
4	6	24	14	48	6	24	14	48	6	24	14	48
5	4	20	18	68	4	20	18	68	4	20	18	68
6	2	12	20	80	2	12	20	80	2	12	20	80
7	3	21	23	101	3	21	23	101	3	21	23	101
8	6	48	29	149	6	48	29	149	6	48	29	149
9	2	18	31	167	2	18	31	167	2	18	31	167
10	2	20	33	187	2	20	33	187	2	20	33	187
11	4	44	37	231	3	33	36	220	4	44	37	231
12	2	24	39	255	2	24	38	244	2	24	39	255
13	2	26	41	281	2	26	$s = 40$	270	0	0	39	255
14	1	14	42	295	-	-	-	-	0	0	39	255
15	1	15	43	310	-	-	-	-	1	15	$s = 40$	270

same as all the other 7 states for the purpose of computing the minimum fraction. Thus $(270 - 2s)/874$ is the relevant quantity to consider (see Problem 21.2). As indicated in the highlighted rows of Table 21.3, there are two different scenarios for achieving this optical strategy when reaching the states with 13 electoral votes by either ① including them and dropping a state with 11 or ② adding the state with 15 instead of them, both of which yield $s = 40$. The combinatorics of the 1960 data set precludes an option analogous to scenario ② which would require taking only two states with 12 electoral votes and adding a nonexistent one with 21 as a replacement for the third state with 12.

21.3 Financial Mathematics: Compound Interest, Annuities, and Mortgages

In this section we derive a formula for the repayment of a debt or mortgage payment on a loan from first principles and then apply it to various situations. We begin with the concept of the sum of a *geometric series*. Consider the series for $R \neq 1$:

$$S_n = a + aR + aR^2 + \cdots + aR^{n-1} = a\sum_{k=0}^{n-1} R^k;$$

$$RS_n = aR + aR^2 + \cdots + aR^{n-1} + aR^n = a\underbrace{\sum_{k=0}^{n-1} R^{k+1}}_{j=k+1} = a\sum_{j=1}^{n} R^j = a\sum_{k=1}^{n} R^k;$$

and taking the difference between these two series obtained by telescoping

$$S_n - RS_n = (1-R)S_n = a\left(\sum_{k=0}^{n-1} R^k - \sum_{k=1}^{n} R^k\right) = a\left[1 + \sum_{k=1}^{n-1}(R^k - R^k) - R^n\right] = a(1-R^n)$$

which implies that

$$S_n = a\frac{R^n - 1}{R-1}.$$

Next we develop a formula for *compound interest*. Letting $A \equiv$ principal, $n \equiv$ number of periods, and $r \equiv$ rate of interest per period, consider the difference equation for its value

$$C_{n+1} = C_n + rC_n = C_n(1+r) \text{ for } n \geq 0 \text{ with } C_0 = A.$$

Then

$$C_1 = C_0(1+r) = A(1+r);\ C_2 = C_1(1+r) = A(1+r)^2;\ldots;\ C_n = A(1+r)^n,\ n \geq 0.$$

Finally we introduce the concept of an *annuity*. Consider n pay periods denoted by $j = 1, 2, \ldots, n-1, n$ and for each period put down a payment P. Each payment will compound from the period it is put down until $j = n$ with an interest rate r per period as follows:

$$c_1 = P(1+r)^{n-1},\ c_2 = P(1+r)^{n-2}, \ldots, c_{n-1} = P(1+r),\ c_n = P.$$

The sum of these $c_j = P(1+r)^{n-j}$, a geometric series, is the value of the *annuity* given by

$$\begin{aligned} V_n &= \sum_{j=1}^{n} c_j = P\sum_{j=1}^{n}(1+r)^{n-j} = \underbrace{P}_{a}\sum_{k=0}^{n-1}\underbrace{(1+r)}_{R}{}^{k} \\ &= P\frac{(1+r)^n - 1}{(1+r)-1} = P\frac{(1+r)^n - 1}{r},\ n \geq 0. \end{aligned}$$

Now we are ready to consider the repayment of a debt or *mortgage* payment on a loan. Suppose that A, the principal, is borrowed from a lending institution to be repaid with a fixed mortgage payment P over each of n periods at an interest rate per period of r. In effect the borrower is setting up an annuity for the lending institution which at the end of the loan period will have a value of

$$V_n = P\frac{(1+r)^n - 1}{r}.$$

If that institution kept the principal of the loan A and let it compound at the same interest rate r over those n periods, its worth would be

$$C_n = A(1+r)^n,$$

at the end of this time. Hence the institution wants the value of the annuity set up by the borrower with the mortgage repayment schedule to be equal to that potential worth or

$$V_n = P\frac{(1+r)^n - 1}{r} = A(1+r)^n = C_n$$

and solving this relation for P yields the mortgage payment formula

$$P = rA\frac{(1+r)^n}{(1+r)^n - 1},$$

or equivalently, upon division of both the numerator and denominator of the fraction appearing in this formula by $(1+r)^n$,

$$P = \frac{rA}{1-(1+r)^{-n}},$$

the usual representation for a mortgage payment.

Let us use this formula to calculate a typical mortgage payment. Suppose \$479,992 is borrowed from a bank at an interest rate of 3.25% per annum to be repaid each month for 15 years. Usually referred to as a 15-year mortgage, its number of periods or months will be $n = 180$ and the interest rate per month then satisfies $r = 0.0325/12 = 0.002308\overline{3}$ while $A = 479{,}992$. Employing these values in our formula for P, we obtain

$$P = 3{,}372.75$$

rounded to the nearest cent, as the monthly payment for such a 15-year mortgage. To examine how much of such a mortgage payment represents interest and how much is actually a repayment of the principal, we shall finally pose an idealized so-called *loan shark* example that emphasizes this point in a dramatic fashion.

The Loan Shark Example: Consider a loan of \$1500 which is to be repaid over a 4-month period at an interest rate of 100% per month. Then $A = 1500$, $n = 4$, and $r = 1$. Hence

$$(1+r)^{-4} = 2^{-4} = \frac{1}{16}, \; 1-(1+r)^{-4} = \frac{15}{6} \text{ and } P = \frac{rA}{1-(1+r)^{-4}} = \frac{1500}{15/16} = 1600.$$

We now construct Table 21.4 which describes the repayment schedule for this loan. Here the notation introduced to designate column ① has been chosen to indicate the updating process used to obtain it from the previous period. Observe that for our example of Table 21.4 in which the payment actually exceeds the original loan more than half of the latter is repaid during the last period! In Problem 21.3 it is shown that for a loan shark example also over 4 pay periods but with the interest rate only a half of this one that effect is much less dramatic.

As a check on these computations let us demonstrate that the equality which was used to derive the formula for the mortgage payment P is satisfied for the example described in Table 21.4. Hence toward that end we calculate

Table 21.4 Mortgage repayment schedule for the loan shark example with $r = 1$.

	①			
	① - ④ from previous period	②	③ = r ① = ①	④ = ② - ③
Period	Principal Outstanding	Payment	Interest Paid	Principal Repaid
1	1500	1600	1500	100
2	1400	1600	1400	200
3	1200	1600	1200	400
4	800	1600	800	800
Total	0	6400	4900	1500

$$V_4 = P\frac{(1+r)^4 - 1}{r} = 1600(2^4 - 1) = 1600(16-1) = 1600(15) = 24000$$

and

$$C_4 = A(1+r)^4 = 1500(2^4) = 1500(16) = 24000.$$

From these calculations we can conclude that

$$V_4 = C_4,$$

as was to be expected from our derivation.

Problems

21.1. Consider the Fibonacci rabbit reproduction example of Section 21.1 in its 2×2 matrix form

$$X_{k+1} = AX_k = \begin{pmatrix} a & b \\ c & d \end{pmatrix} \begin{bmatrix} x_k \\ y_k \end{bmatrix} \Rightarrow X_k = A^k X_0 \text{ with } X_0 = \begin{bmatrix} x_0 \\ y_0 \end{bmatrix} \text{ for } k \geq 0,$$

where $a = x_0 = 0$ and $b = c = d = y_0 = 1$. The purpose of this problem is two-fold: Namely, to develop the alternate Cayley–Hamilton Theorem approach for deducing the representation of A^k and to solve the original problem by a direct matrix method.

(a) The Cayley–Hamilton Theorem states that a matrix satisfies its own characteristic polynomial. Since for a general 2×2 matrix that characteristic polynomial is given by

$$\det(A - \lambda I) = \begin{vmatrix} a-\lambda & b \\ c & d-\lambda \end{vmatrix} = \lambda^2 - (a+d)\lambda + ad - bc = g(\lambda),$$

this implies

$$g(A) = A^2 - (a+b)A + (ad-bc)I = \begin{pmatrix} 0 & 0 \\ 0 & 0 \end{pmatrix} \text{ where } A^2 = \begin{pmatrix} a^2+bc & ab+bd \\ ca+dc & cb+d^2 \end{pmatrix}.$$

Show that result by direct computation.

(b) The division algorithm for polynomials states that given another polynomial $p(\lambda) = \lambda^k$ with $\deg(p) = k > \deg(g) = 2$ there exists polynomials $q(\lambda)$ and $r(\lambda)$ such that

$$p(\lambda) = \lambda^k = q(\lambda)g(\lambda) + r(\lambda) \text{ where } \deg(r) < \deg(g).$$

Hence, conclude that $r(\lambda)$ is of the form

$$r(\lambda) = \alpha + \beta\lambda.$$

(c) Given that $\lambda_{1,2} = (1 \pm \sqrt{5})/2$, the eigenvalues of A, satisfy $g(\lambda_{1,2}) = 0$, deduce from part (b)

$$\lambda_1^k = \alpha + \beta\lambda_1 \text{ and } \lambda_2^k = \alpha + \beta\lambda_2.$$

(d) Solving these equations of part (c) simultaneously determines that

$$\alpha = \frac{\lambda_2\lambda_1^k - \lambda_1\lambda_2^k}{\lambda_2 - \lambda_1} \text{ and } \beta = \frac{\lambda_1^k - \lambda_2^k}{\lambda_1 - \lambda_2}.$$

(e) Finally, invoking the Cayley–Hamilton Theorem, obtain that

$$p(A) = A^k = q(A)g(A) + r(A) = r(A) = \alpha I + \beta A = \begin{pmatrix} \alpha & \beta \\ \beta & \alpha+\beta \end{pmatrix} \text{ since } g(A) = \begin{pmatrix} 0 & 0 \\ 0 & 0 \end{pmatrix},$$

as was to be demonstrated.

(f) We now wish to use a direct matrix method analogous to the direct scalar method employed in Section 21.1 to solve the original Fibonacci matrix difference equation problem. That is, consider

$$X_{k+1} = AX_k = \begin{pmatrix} 0 & 1 \\ 1 & 1 \end{pmatrix} \begin{bmatrix} x_k \\ x_{k+1} \end{bmatrix} \text{ for } k \geq 0 \text{ with } X_0 = \begin{bmatrix} 0 \\ 1 \end{bmatrix};$$

and, analogous to the method for solving a first-order matrix system of differential equations, show that seeking a vector solution of it of the form

$$X_k = Z\lambda^k \text{ where } Z = \begin{bmatrix} z_1 \\ z_2 \end{bmatrix}$$

yields the eigenvalue problem

$$AZ = \lambda Z.$$

(g) Recalling from the eigenvalues and eigenvectors computed in Section 21.1 for A that

$$\lambda = \lambda_{1,2} = \frac{1}{2}(1 \pm \sqrt{5}) \text{ and } Z = Z^{(1,2)} = \begin{bmatrix} 1 \\ \lambda_{1,2} \end{bmatrix},$$

conclude from part (f) that this difference equation admits the solutions

$$X_k^{(1)} = Z^{(1)}\lambda_1^k = \begin{bmatrix} 1 \\ \lambda_1 \end{bmatrix}\lambda_1^k \text{ and } X_k^{(2)} = Z^{(2)}\lambda_2^k = \begin{bmatrix} 1 \\ \lambda_2 \end{bmatrix}\lambda_2^k.$$

(h) Hence deduce from part (g) that this difference equation has a general solution composed of the linear combination

$$X_k = c_1 X_k^{(1)} + c_2 X_k^{(2)} = c_1 \begin{bmatrix} 1 \\ \lambda_1 \end{bmatrix}\lambda_1^k + c_2 \begin{bmatrix} 1 \\ \lambda_2 \end{bmatrix}\lambda_2^k.$$

(i) Applying the initial condition to the general solution of part (h), obtain that

$$X_0 = c_1 \begin{bmatrix} 1 \\ \lambda_1 \end{bmatrix} + c_2 \begin{bmatrix} 1 \\ \lambda_2 \end{bmatrix} = \begin{bmatrix} 0 \\ 1 \end{bmatrix} \Rightarrow c_1 + c_2 = 0$$

$$\text{and } \lambda_1 c_1 + \lambda_2 c_2 = 1 \Rightarrow c_1 = -c_2 = \frac{1}{\lambda_1 - \lambda_2}.$$

(j) Finally synthesizing the results of parts (h) and (j) again find that

$$x_k = \frac{\lambda_1^k - \lambda_2^k}{\lambda_1 - \lambda_2} = \frac{(1+\sqrt{5})^k - (1-\sqrt{5})^k}{2^k\sqrt{5}}$$

and (equivalently)

$$x_{k+1} = \frac{\lambda_1^{k+1} - \lambda_2^{k+1}}{\lambda_1 - \lambda_2} = \frac{(1+\sqrt{5})^{k+1} - (1-\sqrt{5})^{k+1}}{2^{k+1}\sqrt{5}},$$

as was to be demonstrated.

21.2. The purpose of this problem is to find the minimum fraction of popular votes necessary to elect a President of the US under the conditions prevailing in the 2008 election. The states and their corresponding electoral votes are listed in Table 21.5. Proceed in a manner similar to that developed in section 2 for the 1960 election by employing the same model and assumptions. First assume that D.C.'s electoral votes consist of 1 "virtual" representative r and 2 "virtual" senators just as all the other states that have an electoral vote $e = r + 2$. Then consider a second case in which D.C.'s electoral votes are assumed to consist of 3 "virtual" representatives but no "virtual" senators. Since there are 538 total electoral votes, a candidate must receive at least 270 of them to win in a two-candidate election. Note that for the second case the subcases in which D.C. is included as a winning "state" and the one in which it is not must be treated separately. Further take the total number of votes cast in this election to be 131.4 million.

Table 21.5 The states and their electoral votes $e = r + 2$ for the 2008 Presidential election.

Alabama	9	Kentucky	8	North Dakota	3
Alaska	3	Louisiana	9	Ohio	20
Arizona	10	Maine	4	Oklahoma	7
Arkansas	6	Maryland	10	Oregon	7
California	55	Massachusetts	12	Pennsylvania	21
Colorado	9	Michigan	17	Rhode Island	4
Connecticut	7	Minnesota	10	South Carolina	8
Delaware	3	Mississippi	6	South Dakota	3
D.C.	3	Missouri	11	Tennessee	11
Florida	27	Montana	3	Texas	34
Georgia	15	Nebraska	5	Utah	5
Hawaii	4	Nevada	5	Vermont	3
Idaho	4	New Hampshire	4	Virginia	13
Illinois	21	New Jersey	15	Washington	11
Indiana	11	New Mexico	5	West Virginia	5
Iowa	7	New York	31	Wisconsin	10
Kansas	6	North Carolina	15	Wyoming	3

Hints: Assume the winning candidate carries s states. Consider two cases: (1) in which D.C.'s electoral votes consist of 1 "virtual" representative and 2 "virtual" senators just the same as all the other states with $e = r + 2$; (2) in which D.C.'s electoral votes e_k are assumed to consist of 3 "virtual" representatives r_k but no "virtual" senators. The latter case has subcases (2a) when D.C. is not included as a winning "state" and (2b) when it is. The number of voters in each state is proportional to the r_i value for that state with proportionality constant N_j, a large positive even integer, where $j = 1$ and 2, for the two cases respectively. Let W denote the number of popular votes obtained by the winning candidate and T, the total number of votes cast in the election.

Then

$$W \geq \begin{cases} \sum_{i=1}^{s} \left(\frac{r_i N_1}{2} + 1 \right) = \frac{N_1}{2} \sum_{i=1}^{s} r_i + s & \text{for Case (1)} \\ \sum_{i=1}^{s} \left(\frac{r_i N_2}{2} + 1 \right) = \frac{N_2}{2} \sum_{i=1}^{s} r_i + s & \text{for Case (2)} \end{cases},$$

$$T = \begin{cases} 436 N_1 & \text{for Case (1)} \\ 438 N_2 & \text{for Case (2)} \end{cases} \approx 131.4 \text{ million}.$$

Cases (1) & (2a):

$$\sum_{i=1}^{s} e_i = \sum_{i=1}^{s} (r_i + 2) = \sum_{i=1}^{s} r_i + 2s \geq 270 \Rightarrow \sum_{i=1}^{s} r_i \geq 270 - 2s;$$

$$\Rightarrow \frac{W}{T} \geq \begin{cases} \frac{1}{872}\sum_{i=1}^{s} r_i + \frac{s}{436N_1} & \text{for Case (1)} \\ \frac{1}{876}\sum_{i=1}^{s} r_i + \frac{s}{438N_2} & \text{for Case (2a)} \end{cases} \geq \begin{cases} \frac{270-2s}{872} + \frac{s}{436N_1} & \text{for Case (1)} \\ \frac{270-2s}{876} + \frac{s}{438N_2} & \text{for Case (2a)} \end{cases}.$$

Case (2b):

$$\sum_{i=1}^{s} e_i = \sum_{i=1}^{k-1}(r_i+2) + r_k + \sum_{i=k+1}^{s}(r_i+2) = \sum_{i=1}^{s} r_1 + 2(s-1) \geq 270 \Rightarrow \sum_{i=1}^{s} r_i \geq 272 - 2s;$$

$$\Rightarrow \frac{W}{T} \geq \frac{1}{876}\sum_{i=1}^{s} r_i + \frac{s}{438N_2} \geq \frac{272-2s}{876} + \frac{s}{438N_2}.$$

21.3. The purpose of this problem is to consider a loan shark example in which \$1300 is borrowed and repaid over a 4 month period with a payment each month consistent with an interest rate of 200% per the whole period.

(a) Conclude that then

$$A = 1300,\ n = 4,\ \text{and } r = \frac{2}{4} = \frac{1}{2}.$$

(b) Show that under the conditions of part (a) the monthly mortgage payment $P = 810$.

(c) Now construct a table analogous to Table 21.4 for this situation.

(d) Finally demonstrate that $V_4 = C_4 = 6{,}581.25$.

Chapter 22
Concluding Capstone Problems

In closing, six capstone problems are offered in this chapter. They deal with the propagation of a noncytopathic virus in a target cell population, Jeans' criterion for the occurrence of a self-gravitational instability in a rotating inviscid adiabatic gas cloud of infinite extent, the critical conditions for the onset of a Rayleigh–Bénard instability in a Boussinesq dissociating gas layer at chemical quasi-equilibrium being heated from below or above, a complex nonlinear stability expansion of the model interaction long-range diffusion equation (which was analyzed by a real expansion approach in Problem 16.1), an exact solution of the Black–Scholes heat-type partial differential equation of mathematical finance, and an age-structured discrete-time American dipper population model. Each of these problems synthesize several concepts presented in earlier chapters, and as such they serve as a fitting conclusion to this book on comprehensive applied mathematical modeling given that the first three and the last one are data-driven, the fourth allows a comparison between two different methods of analysis, and the fifth exhibits how techniques developed for the natural and engineering sciences can be applied to an example from quantitative finance.

22.1 Viral Dynamics

We begin with a target cell population dynamics model of a noncytopathic viral interaction. Tuckwell and Wan [120] introduced a three-component model for a basic viral interaction of a general type. The components of that model were uninfected target cells, infected such cells, and the free virus, with spatially homogeneous densities at time t denoted by $M(t)$, $I(t)$, and $V(t)$, respectively, which satisfied the dimensional nonlinear coupled system of first-order ordinary differential equations:

$$\frac{dM}{dt} = \lambda - \rho M - \beta MV, \ \frac{dI}{dt} = \beta MV - \delta I, \ \frac{dV}{dt} = bI - \gamma V.$$

© Springer International Publishing AG, part of Springer Nature 2017

D. J. Wollkind and B. J. Dichone, *Comprehensive Applied Mathematical Modeling in the Natural and Engineering Sciences*,
https://doi.org/10.1007/978-3-319-73518-4_22

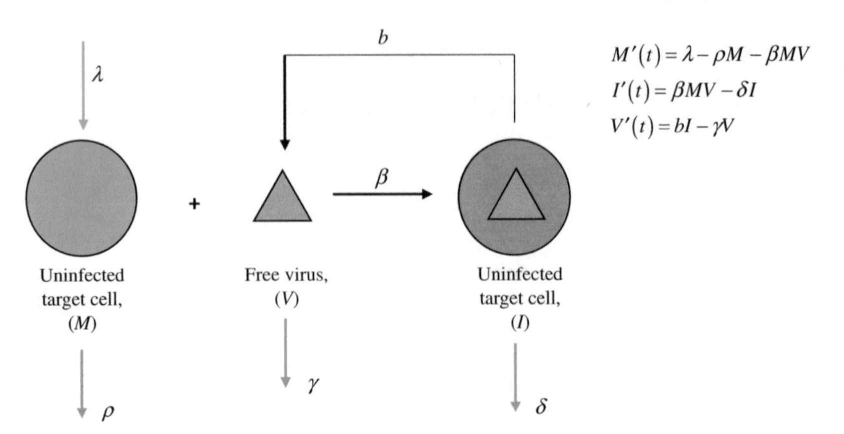

Fig. 22.1 Schematic diagram of the basic target cell–virus general interaction. Should this viral interaction be noncytopathic $\rho = \delta$.

A schematic diagram of this basic target cell–virus general interaction is depicted in Fig. 22.1. Since for our specific case the viral interaction is noncytopathic – *i.e.*, the virus does not kill its host cells, we assume that, in addition, the death rates for M and I are equal or $\rho = \delta$. Further $\lambda \equiv$ source rate for M and $\beta \equiv$ infection rate for M and I while b and γ are the production and clearance rates for V, respectively.

(a) Introducing the following dimensionless parameters $\varepsilon = \rho/\gamma$, $R_0 = \lambda\beta b/(\rho^2\gamma)$; and nondimensional "quasi-steady-state time" variables

$$\tau = \rho t,\ m(\tau;\varepsilon) = \frac{\rho M(t)}{\lambda},\ i(\tau;\varepsilon) = \frac{\rho I(t)}{\lambda},\ v(\tau;\varepsilon) = \frac{\beta V(t)}{\rho};$$

show that the basic system is transformed into

$$\frac{dm}{d\tau} = 1 - m - mv,\ \frac{di}{d\tau} = mv - i,\ \varepsilon\frac{dv}{d\tau} = R_0 i - v.$$

(b) Since for the typical noncytopathic equine infectious anemia virus $\varepsilon = O(10^{-3})$, seek a solution of (a) of the form $[m, i, v](\tau;\varepsilon) = [m_0, i_0, v_0](\tau) + \boldsymbol{O}(\varepsilon)$ and deduce that

$$\frac{dm_0}{d\tau} = 1 - m_0 - m_0 v_0,\ \frac{di_0}{d\tau} = m_0 v_0 - i_0 \text{ where } v_0 = R_0 i_0.$$

(c) Introducing the "transient time" variables

$$\eta = \frac{\tau}{\varepsilon} = \gamma t,\ \mathcal{M}(\eta;\varepsilon) = m(\tau;\varepsilon),\ \mathcal{I}(\eta;\varepsilon) = i(\tau;\varepsilon),\ \mathcal{V}(\eta;\varepsilon) = v(\tau;\varepsilon);$$

show that the system of (a) is transformed into the "boundary-layer equations"

$$\frac{d\mathcal{M}}{d\eta} = \varepsilon(1 - \mathcal{M} - \mathcal{M}\mathcal{V}),\ \frac{d\mathcal{I}}{d\eta} = \varepsilon(\mathcal{M}\mathcal{V} - \mathcal{I}),\ \frac{d\mathcal{V}}{d\eta} = R_0\mathcal{I} - \mathcal{V}.$$

(d) Seeking a solution of (c) of the form $[\mathcal{M}, \mathcal{I}, \mathcal{V}](\eta; \varepsilon) = [\mathcal{M}_0, \mathcal{I}_0, \mathcal{V}_0](\eta) + \boldsymbol{O}(\varepsilon)$, deduce that

$$\mathcal{M}_0(\eta) \equiv m^{(0)},\ \mathcal{I}_0(\eta) \equiv i^{(0)},\ \mathcal{V}_0(\eta) = R_0 i^{(0)} + [v^{(0)} - R_0 i^{(0)}]e^{-\eta};$$

where $m^{(0)}$, $i^{(0)}$, and $v^{(0)}$ are the $O(1)$ values as $\varepsilon \to 0$ of the designated initial conditions

$$\mathcal{M}(0; \varepsilon) = m^{(0)},\ \mathcal{I}(0; \varepsilon) = i^{(0)},\ \mathcal{V}(0; \varepsilon) = v^{(0)}.$$

(e) Determine the proper initial conditions to impose for the one-term outer solution functions of part (b) by using the one-term matching rule

$$m_0(0) = \lim_{\eta\to\infty} \mathcal{M}_0(\eta),\ i_0(0) = \lim_{\eta\to\infty} \mathcal{I}_0(\eta),\ v_0(0) = \lim_{\eta\to\infty} \mathcal{V}_0(\eta);$$

in conjunction with the results of (d) for the one-term inner solution functions to conclude that

$$m_0(0) = m^{(0)},\ i_0(0) = i^{(0)},\ v_0(0) = R_0 i^{(0)}$$

where the target cell initial values can be normalized to satisfy

$$m^{(0)} + i^{(0)} = 1.$$

(f) Constructing the one-term uniformly valid additive composites defined by

$$m_u^{(0)}(\tau) = m_0(\tau) + \mathcal{M}_0\left(\frac{\tau}{\varepsilon}\right) - m^{(0)},\ i_u^{(0)}(\tau) = i_0(\tau) + \mathcal{I}_0\left(\frac{\tau}{\varepsilon}\right) - i^{(0)},$$
$$v_u^{(0)}(\tau) = v_0(\tau) + \mathcal{V}_0\left(\frac{\tau}{\varepsilon}\right) - R_0 i^{(0)};$$

obtain, from the results of parts (b), (d), and (e), that

$$m_u^{(0)}(\tau) = m_0(\tau),\ i_u^{(0)}(\tau) = i_0(\tau),\ v_u^{(0)}(\tau) = v_0(\tau) + [v^{(0)} - R_0 i^{(0)}]e^{-\tau/\varepsilon}.$$

Observe, for the target cell variables, the outer solution is actually uniformly valid to this order.

(g) Returning to part (b) and taking the sum of its two differential equations find that

$$\frac{d}{d\tau}(m_0 + i_0) + (m_0 + i_0) = 1$$

with initial condition from part (e) of

$$m_0(0) + i_0(0) = 1.$$

(h) Solving the differential equation problem of part (g) for $m_0 + i_0$ demonstrate that

$$m_0(\tau) + i_0(\tau) \equiv 1 \text{ or } i_0 = 1 - m_0.$$

(i) From the results of parts (b) and (h) conclude that

$$v_0 = R_0 i_0 = R_0(1 - m_0).$$

(j) Finally, substituting the result of part (h) into the differential equation for $dm_0/d\tau$ of part (b), obtain the following Riccati equation for $m_0 = m_0(\tau; R_0)$:

$$\frac{dm_0}{d\tau} = 1 - (R_0 + 1)m_0 + R_0 m_0^2 = f(m_0; R_0),\ \tau > 0;\ 0 \le m_0(0; R_0) = m^{(0)} \le 1;$$

where the initial condition follows from the results of part (e).

(k) Demonstrate that $m_e(R_0) = 1$ or $1/R_0$ are the equilibrium points of this equation of part (j) satisfying $f(m_e; R_0) = 0$, and examine their linear stability behavior in terms of the bifurcation parameter $R_0 \equiv$ basic reproductive ratio. That is let

$$m_0(\tau; \varepsilon_1) = m_e(R_0) + \varepsilon_1 m_1(\tau; \varepsilon_1) + O(\varepsilon_1^2) \text{ where } |\varepsilon_1| << 1,$$

substitute this solution into that equation, neglect terms of $O(\varepsilon_1^2)$, cancel the common ε_1 factor to yield $dm_1/d\tau = f'(m_e; R_0)m_1$, and determine the long-time behavior of $m_1(\tau; \varepsilon_1)$ as a function of R_0 for each of these two equilibrium points finding $m_e = 1$ is stable for $0 < R_0 < 1$ and $m_e = 1/R_0$ is stable for $R_0 > 1$.

(ℓ) Make the change of dependent variable $m_0 = 1 + y$ in the differential equation of part (j) and obtain the Bernoulli equation for $y = y(\tau; R_0)$:

$$\frac{dy}{d\tau} + (1 - R_0)y = R_0 y^2.$$

(m) After first noting that $y(\tau; R_0) \equiv 0$ is a solution to this Bernoulli equation, divide that equation by $-y^2$, and introducing the new dependent variable $u = 1/y$ (see Section 3.7), obtain the linear constant coefficient first-order ordinary differential equation for $u = u(\tau; R_0)$:

$$\frac{du}{d\tau} + (R_0 - 1)u = -R_0.$$

(n) Solving this equation of part (m) for $R_0 \neq 1$, determine from that solution in conjunction with the changes of dependent variables of parts (ℓ) and (m) that $m_0(\tau; m^{(0)}, R_0)$ of part (j) has solution

$$m_0(\tau; m^{(0)}, R_0) = 1 + \frac{(1 - R_0)(m^{(0)} - 1)}{R_0(m^{(0)} - 1) + (1 - R_0 m^{(0)})e^{(1-R_0)\tau}} \text{ for } R_0 \neq 1.$$

(o) Noting that the differential equation of part (j) reduces to $dm_0/d\tau = (m_0 - 1)^2$ for $R_0 = 1$, show that

$$m_0(\tau; m^{(0)}, R_0) = 1 + \frac{m^{(0)} - 1}{1 + (1 - m^{(0)})\tau} \text{ for } R_0 = 1.$$

(p) From the results of parts (n) and (o) determine the long-time behavior of $m_0(\tau; m^{(0)}, R_0)$ and compare this global prediction with the linear stability one of part (k). Finally, recalling that $i_0 = 1 - m_0$ and $v_0 = R_0 i_0$, employ the one-term uniformly valid solutions of part (f) to make a phenomenological interpretation of the spread of the virus as a function of R_0. Now particularizing these results to the special case of $m^{(0)} = 0.9999$ and $v^{(0)} = 1.5R_0 i^{(0)}$, plot the relevant solutions for $\lambda = 2 \times 10^3$ cells/(ml d), $\beta = 3.33 \times 10^{-7}$ ml/(copies d), $b = 893$ copies/(cells d), $\rho = \delta = 0.056/\text{d}$, and $\gamma = 11.1/\text{d}$ or for $R_0 = 17.1$ and $\varepsilon = 5 \times 10^{-3}$ [112].

22.2 Self-Gravitational Instabilities

We continue with an inviscid fluid model of a self-gravitating infinite expanse of a rotating adiabatic gas. The governing equations for this situation are given by [19]:

$$\text{Continuity Equation: } \frac{D\rho}{Dt} + \rho \nabla \bullet \boldsymbol{v} = 0;$$

$$\text{Momentum Equations: } \frac{D\boldsymbol{v}}{Dt} + 2\boldsymbol{\Omega} \times \boldsymbol{v} + \boldsymbol{\Omega} \times (\boldsymbol{\Omega} \times \boldsymbol{r}) = -\frac{1}{\rho}\mathcal{P}'(\rho)\nabla\rho + \boldsymbol{g};$$

$$\text{Poisson's Equation: } \nabla \bullet \boldsymbol{g} = -4\pi G_0 \rho.$$

Here $t \equiv$ time, $\boldsymbol{r} = (x, y, z) \equiv$ position vector, $D/Dt \equiv$ material derivative, $\rho \equiv$ mass density, $\boldsymbol{\Omega} = (0, 0, \Omega_0) \equiv$ rotation vector, $\boldsymbol{v} = (u, v, w) \equiv$ velocity vector with respect to the rotating frame, $\mathcal{P}(\rho) = p_0(\rho/\rho_0)^\gamma \equiv$ adiabatic pressure, $\boldsymbol{g} = -\nabla\varphi \equiv$ gravitational acceleration vector with $\varphi \equiv$ self-gravitating potential, and $G_0 \equiv$ universal gravitational constant. The continuity and momentum equations follow from their derivations for an inviscid fluid of Chapter 9 with the addition of the extra second and third terms on the left-hand side of the latter which represent the Coriolis effect and centrifugal force, respectively, due to the rotation [46].

(a) Sketch the derivation of Poisson's equation from the divergence theorem for $\boldsymbol{g}$

$$\iint_S \boldsymbol{g} \bullet \boldsymbol{e}_a \, d\sigma = \iiint_R \nabla \bullet \boldsymbol{g} \, d\tau,$$

where $R \equiv$ a small spherical region of radius a centered about a point in space with bounding surface S possessing unit outward-pointing normal $\boldsymbol{e}_a$. Since by Newton's universal law of gravitation $\boldsymbol{g} = -(G_0 M_a/a^2)\boldsymbol{e}_a$ on S, where M_a denotes the mass in this spherical region, conclude that

$$\iint_S \boldsymbol{g} \bullet \boldsymbol{e}_a \, d\sigma = -\frac{G_0 M_a}{a^2} \iint_S \boldsymbol{e}_a \bullet \boldsymbol{e}_a \, d\sigma = -\frac{G_0 M_a}{a^2} \iint_S d\sigma$$

$$= -\frac{G_0 M_a}{a^2} 4\pi a^2 = -4\pi G_0 M_a.$$

Given the relationship between M_a and the density ρ

$$M_a = \iiint_R \rho \, d\tau,$$

obtain that

$$\iiint_R (\nabla \bullet \boldsymbol{g} + 4\pi G_0 \rho) \, d\tau = 0$$

from the divergence theorem. Now, employing the integral mean value theorem and cancelling the common spherical volume factor of $(4\pi a^3/3)$, deduce that

$$(\nabla \bullet \boldsymbol{g})^* = -4\pi G_0 \rho^*$$

where the asterisk denotes the quantity in question being evaluated at an intermediate point in R. Finally, taking the limit as $a \to 0$, derive Poisson's equation for any point in space

$$\nabla \bullet \boldsymbol{g} = -4\pi G_0 \rho.$$

(b) Show that with $\boldsymbol{\Omega}$, $\boldsymbol{v}$, $\boldsymbol{r}$, and $\boldsymbol{g}$ as defined

$$\boldsymbol{\Omega} \times \boldsymbol{v} = \Omega_0(-v, u, 0), \; \boldsymbol{\Omega} \times (\boldsymbol{\Omega} \times \boldsymbol{r}) = -\Omega_0^2(x, y, 0), \text{ and } \nabla \bullet \boldsymbol{g} = -\nabla^2 \varphi.$$

Thus conclude from part (a) that the momentum and Poisson's equations become

$$\text{Momentum Equations: } \frac{D\boldsymbol{v}}{Dt} + 2\Omega_0(-v, u, 0) - \Omega_0^2(x, y, 0) = -\frac{1}{\rho} \mathcal{P}'(\rho) \nabla \rho - \nabla \varphi;$$

$$\text{Poisson's Equation: } \nabla^2 \varphi = 4\pi G_0 \rho;$$

respectively. Hint: Recall from part (c) of Problem 12.1 that

$$\boldsymbol{\Omega} \times (\boldsymbol{\Omega} \times \boldsymbol{r}) = (\boldsymbol{\Omega} \bullet \boldsymbol{r})\boldsymbol{\Omega} - (\boldsymbol{\Omega} \bullet \boldsymbol{\Omega})\boldsymbol{r}.$$

(c) Demonstrate that

$$\boldsymbol{v} \equiv \boldsymbol{0} = (0, 0, 0), \; \rho = \rho_0, \; \varphi = \varphi_0(x, y) = \Omega_0^2 \frac{x^2 + y^2}{2} \text{ where } \Omega_0^2 = 2\pi G_0 \rho_0 > 0;$$

represents an exact static homogeneous solution of our basic equations.

(d) Now seeking a linear perturbation solution of these basic equations of the form

$$\boldsymbol{v} = \varepsilon \boldsymbol{v}_1 + \boldsymbol{O}(\varepsilon^2),\ \rho = \rho_0[1 + \varepsilon s + O(\varepsilon^2)],$$
$$\varphi = \varphi_0 + \varepsilon\varphi_1 + O(\varepsilon^2) \text{ where } \boldsymbol{v}_1 = (u_1, v_1, w_1);$$

with $|\varepsilon| << 1$, deduce that the perturbation quantities satisfy

$$\frac{\partial s}{\partial t} + \frac{\partial u_1}{\partial x} + \frac{\partial v_1}{\partial y} + \frac{\partial w_1}{\partial z} = 0;$$

$$\frac{\partial u_1}{\partial t} - 2\Omega_0 v_1 + c_0^2\frac{\partial s}{\partial x} + \frac{\partial \varphi_1}{\partial x} = 0 \text{ where } c_0^2 = \mathcal{P}'(\rho_0) = \frac{\gamma p_0}{\rho_0} > 0;$$

$$\frac{\partial v_1}{\partial t} + 2\Omega_0 u_1 + c_0^2\frac{\partial s}{\partial y} + \frac{\partial \varphi_1}{\partial y} = 0;$$

$$\frac{\partial w_1}{\partial t} + c_0^2\frac{\partial s}{\partial z} + \frac{\partial \varphi_1}{\partial z} = 0;\ 2\Omega_0^2 s - \nabla^2\varphi_1 = 0.$$

(e) Assuming a normal-mode solution for these perturbation quantities of the form

$$[u_1, v_1, w_1, s, \varphi_1](x, y, z, t) = [A, B, C, E, F]e^{i(k_1x + k_2y + k_3z) + \sigma t},$$

where $|A|^2 + |B|^2 + |C|^2 + |E|^2 + |F|^2 \neq 0$ and $k_{1,2,3} \in \mathbb{R}$ to implicitly satisfy the far-field boundedness property for those quantities, derive the following equations for $[A, B, C, E, F]$:

$$ik_1A + ik_2B + ik_3C + \sigma E = 0;$$

$$\sigma A - 2\Omega_0 B + ic_0^2k_1E + ik_1F = 0;\ \sigma B + 2\Omega_0 A + ic_0^2k_2E + ik_2F = 0;$$

$$\sigma C + ic_0^2k_3E + ik_3F = 0;\ 2\Omega_0^2E + k^2F = 0 \text{ where } k^2 = k_1^2 + k_2^2 + k_3^2.$$

(f) Setting the determinant of the 5×5 coefficient matrix for the system of constants of part (e) equal to zero to satisfy their nontriviality property, obtain upon cancellation of k^2 the equation

$$\sigma^4 + (c_0^2k^2 + 2\Omega_0^2)\sigma^2 + 4\Omega_0^2(c_0^2k^2 - 2\Omega_0^2)\cos^2(\theta) = 0 \text{ where } k_3 = k\cos(\theta),$$

θ being the azimuthal angle between the wavenumber vector $\boldsymbol{k} = (k_1, k_2, k_3)$ and $\boldsymbol{\Omega}$.

(g) Since this secular equation of part (f) is a quadratic in σ^2, conclude that $\sigma^2 \in \mathbb{R}$ by showing that its discriminant

$$\mathcal{D} = (c_0^2k^2 + 2\Omega_0^2)^2 - 16\Omega_0^2(c_0^2k^2 - 2\Omega_0^2)\cos^2(\theta) \geq 0.$$

Hint: Consider the two cases of $c_0^2k^2 - 2\Omega_0^2 \leq 0$ and $c_0^2k^2 - 2\Omega_0^2 > 0$ separately. For the latter case note that $\mathcal{D} \geq (c_0^2k^2 - 6\Omega_0^2)^2$.

(h) For $\theta = \pi/2$ conclude from part (f) that $\sigma^2 \leq 0$, while for $\theta \neq \pi/2$ deduce from parts (f) and (g) and the stability criteria governing such quadratics that $\sigma^2 < 0$ if and only if

$$c_0^2k^2 - 2\Omega_0^2 > 0.$$

(i) Make the interpretation from the results of parts (c) and (h) that there can only be $\sigma^2 > 0$ and hence unstable behavior should

$$c_0^2 k^2 - 4\pi G_0 \rho_0 < 0,$$

usually referred to as Jeans' gravitational instability criterion after Sir James Jeans [55] who first proposed it. Since Jeans' original analysis was for a non-rotating system with $\Omega_0 = 0$ which implies $\rho_0 = 0$ as well, that is often called Jeans' swindle. This problem demonstrates that incongruity can be rectified by adding rotation to the system and such a model also has the added advantage of being more astrophysically realistic.

Fig. 22.2 A Galaxy Evolution Explorer image of the Andromeda galaxy M31.

(j) Finally this instability criterion can be posed in terms of wavelength $\lambda = 2\pi/k$. Show from the result of part (i) that it then takes the form

$$\lambda > \lambda_J = c_0 \sqrt{\frac{\pi}{G_0 \rho_0}} \equiv \text{Jeans' length},$$

where $c_0 \equiv$ speed of sound in an adiabatic gas of density ρ_0. This formula is of fundamental importance in astrophysics and cosmology where many significant deductions concerning the formation of galaxies and stars have been based upon it. In particular Jeans' interpretation of the criterion now bearing his name was that a gas cloud of characteristic dimension much greater than λ_J would tend to form condensations with mean distance of separation comparable to λ_J that then developed into those protostars observable in the outer arms of spiral galaxies such as Andromeda M31 (see Fig. 22.2).

(k) Using the formula λ_J of part (j) with the parameters c_0 and ρ_0 assigned the values

$$c_0 = \frac{2}{3} \times 10^4 \text{ cm/sec and } \rho_0 = 10^{-22} \text{ gm/cm}^3$$

employed by Jeans [55] for this purpose, while taking

$$G_0 = 6.67 \times 10^{-8} \text{ cm}^3/(\text{gm sec}^2)$$

in cgs units, demonstrate that

$$\lambda_J = 4.58 \times 10^{18} \text{ cm } = 1.48 \text{ pc},$$

where 1 pc $\equiv 3.09 \times 10^{18}$ cm, which compares quite favorably with the mean distance between actual adjacent condensations originally formed in the outer arms of Andromeda since, in those parts of M31, the averaged observed distance between protostars in such chains is about 1.4 pc or somewhat more if allowances are made for foreshortening [55].

(ℓ) In addition, using the formula for the rate of rotation from part (c)

$$\Omega_0 = \sqrt{2\pi\rho_0 G_0}$$

and the values for ρ_0 and G_0 from part (k), conclude that

$$\Omega_0 = 6.47 \times 10^{15}/\text{sec},$$

and then show the corresponding rotational velocity $V_0 = r_0\Omega_0$ for the reference radial distance

$$r_0 = 1\,\text{kpc} = 10^3\,\text{pc} = 3.09 \times 10^{21}\,\text{cm} = 3.09 \times 10^{16}\,\text{km}$$

is given by

$$V_0 = 200\,\text{km/sec},$$

consistent with the spectroscopic measurements of the Andromeda nebula [107].

22.3 Chemically Driven Convection

The purpose of this problem is to examine the linear stability of an initially quiescent horizontal layer of a dissociating gas confined between two stress-free surfaces being heated from below or above. An extension of the Rayleigh–Bénard Boussinesq fluid model of Chapter 15 will be employed that is appropriate for considering a gas undergoing the reversible dissociation reaction

$$A_2 \rightleftharpoons 2A.$$

This requires the introduction of an additional dependent variable defined in two-dimensions by

$$\alpha = \alpha(\boldsymbol{r}, t) = \frac{N_A}{N_A + 2N_{A_2}} \text{ for } \boldsymbol{r} = (x, z) \equiv \text{ position vector and } t \equiv \text{ time,}$$

where

$$N_A \equiv \text{ number of } A \text{ atoms and } N_{A_2} \equiv \text{ number of } A_2 \text{ molecules,}$$

which represents the mass fraction of dissociation. The same notation is being used as that of Chapter 15 in what follows except that the thermal expansivity of the density will be designated by a rather than α_0, α having been reserved by tradition for this new dependent variable.

Our two-dimensional Boussinesq equations of motion for $\boldsymbol{v} = (u, w)$ are then of the form

$$\nabla_2 \bullet \boldsymbol{v} = 0; \; \frac{D\boldsymbol{v}}{Dt} = -\left(\frac{1}{\rho_0}\right)\nabla_2 p - g[1 - a(T - T_0) - b(\alpha - \alpha_0)]\boldsymbol{e}_3 + \nu_0 \nabla_2^2 \boldsymbol{v};$$

with conservation of energy and species given by

$$\rho_0\left(C_p\frac{DT}{Dt} + c_0\frac{D\alpha}{Dt}\right) = k_0\nabla_2^2 T + \rho_0 c_0 D_0 \nabla_2^2 \alpha;$$

$$\frac{D\alpha}{Dt} + \frac{\alpha - \alpha_e(T)}{\tau} = D_0 \nabla_2^2 \alpha.$$

Here $b \equiv$ dissociation expansivity of the density, $c_0 \equiv$ specific dissociation energy, $D_0 \equiv$ diffusion coefficient of the binary mixture, $\tau \equiv$ dissociation relaxation time, and $\alpha_e(T) \equiv$ equilibrium dissociation as an explicit function of temperature.

This dissociating fluid layer of depth d is maintained at the constant temperatures of T_0 below ($z = 0$) and T_1 above ($z = d$). Both of these boundaries are considered to be free surfaces on which the tangential component of stress vanishes and furthermore they are assumed to be in chemical equilibrium. Hence the following boundary conditions are adopted at $z = 0$ and d:

$$w = \frac{\partial u}{\partial z} + \frac{\partial w}{\partial x} = 0 \text{ at } z = 0, d;$$

and

$$T = T_0, \alpha = \alpha_e(T_0) = \alpha_0 \text{ at } z = 0; T = T_1, \alpha = \alpha_e(T_1) \text{ at } z = d.$$

(a) As a simplification retain only the first two terms in the Taylor series for $\alpha_e(T)$ about $T = T_0$ yielding [138]:

$$\alpha_e(T) = \alpha_e(T_0) + \alpha_e'(T_0)(T - T_0) = \alpha_0 + m_0(T - T_0) \text{ where } m_0 = \alpha_e'(T_0) > 0,$$

consistent with the Boussinesq equation of state for density. Further assume that the system under investigation is in a state of chemical quasi-equilibrium – *i.e.*, there is only a slight departure from a dissociating fluid in chemical equilibrium. Then it is possible to replace all of the partial derivatives of α in the governing equations with their corresponding partial derivatives with respect to T by introducing $d\alpha = m_0 dT$ [140]. Show that these modifications transform the conservation of energy and species equations into

$$\frac{DT}{Dt} = \kappa_e \nabla^2 T = \frac{\alpha_0 - \alpha}{m_0 \tau} + \frac{T - T_0}{\tau} + D_0 \nabla^2 T,$$

where $\kappa_e = k_e/(\rho_0 c_e)$, $k_e = k_0 + \rho_0 c_0 D_0 m_0$, and $c_e = C_p + c_0 m_0$ are the effective thermometric conductivity, thermal conductivity, and specific heat of the binary mixture, respectively.

(b) Initially the layer of dissociating fluid is quiescent, steady-state, and stratified in z. Demonstrate that there exists a solution to the modified system of governing PDE's and BC's given by

$$u = w = 0, T = T_0(z) = T_0 - \beta z \text{ where } \beta = \frac{T_0 - T_1}{d}, \alpha = \alpha_0(z) = \alpha_0 - m_0 \beta z,$$

$$p = p_0(z) = -\rho_0 g \int [1 + (a + b m_0)\beta z]\, dz = -\rho_0 g \left[1 + \frac{(a + b m_0)\beta z}{2}\right] z + p_A;$$

where $p_A \equiv$ atmospheric pressure, which corresponds to the situation described above.

(c) In order to examine the stability of this state of part (b) to linear perturbations consider a solution to the modified system of governing PDE's and BC's of the form

$$\boldsymbol{v}(\boldsymbol{r},t) = \varepsilon_1 \boldsymbol{v}_1(\boldsymbol{r},t) + \boldsymbol{O}(\varepsilon_1^2) \text{ for } \boldsymbol{v}_1 = (u_1, w_1), p(\boldsymbol{r},t) = p_0(z) + \varepsilon_1 p_1(\boldsymbol{r},t) + O(\varepsilon_1^2),$$

$$T(\boldsymbol{r},t) = T_0(z) + \varepsilon_1 T_1(\boldsymbol{r},t) + O(\varepsilon_1^2), \alpha(\boldsymbol{r},t) = \alpha_0(z) + \varepsilon_1 \alpha_1(\boldsymbol{r},t) + O(\varepsilon_1^2);$$

neglect terms of $O(\varepsilon_1^2)$, cancel the resulting ε_1 common factor, and obtain

$$\nabla_2 \bullet \boldsymbol{v}_1 = 0, \frac{\partial \boldsymbol{v}_1}{\partial t} = -\left(\frac{1}{\rho_0}\right) \nabla_2 p_1 + g(aT_1 + b\alpha_1)\boldsymbol{e}_3 + \nu_0 \nabla_2^2 \boldsymbol{v}_1;$$

$$\frac{\partial T_1}{\partial t} - \beta w_1 = \kappa_e \nabla_2^2 T_1, \alpha_1 = m_0[\tau(D_0 - \kappa_e)\nabla_2^2 T_1 + T_1];$$

$$w_1 = \frac{\partial u_1}{\partial z} + \frac{\partial w_1}{\partial x} = T_1 = \alpha_1 = 0 \text{ at } z = 0, d.$$

(d) Deduce from part (c) by eliminating α_1 between the momentum and species perturbation equations that

$$\frac{\partial \boldsymbol{v}_1}{\partial t} = -\left(\frac{1}{\rho_0}\right)\nabla_2 p_1 + g[(a + bm_0)T_1 + bm_0\tau(D_0 - \kappa_e)\nabla_2^2 T_1]\boldsymbol{e}_3 + \nu_0\nabla_2^2\boldsymbol{v}_1.$$

(e) Now considering the perturbation momentum equations from part (d) and the perturbation continuity and energy equations from part (c) in conjunction with the appropriate BC's, nondimensionalize position, time, velocity, pressure, and temperature by the scale factors d, d^2/ν_0, ν_0/d, $\rho_0\nu_0^2/d^2$, and βd, respectively, to obtain

$$\nabla_2 \bullet \boldsymbol{v}_1 = 0,$$

$$\frac{\partial \boldsymbol{v}_1}{\partial t} = -\nabla_2 p_1 + RPr^{-1}[(1+\delta)T_1 + \varepsilon B\nabla_2^2 T_1]\boldsymbol{e}_3 + \nabla_2^2\boldsymbol{v}_1,$$

$$\frac{\partial T_1}{\partial t} - w_1 = Pr^{-1}\nabla_2^2 T_1;$$

$$w_1 = \frac{\partial u_1}{\partial z} + \frac{\partial w_1}{\partial x} = T_1 = 0 \text{ at } z = 0, 1;$$

where

$$Pr = \frac{\nu_0}{\kappa_e},\ R = \frac{g\alpha\beta d^4}{\kappa_e\nu_0},\ \delta = \frac{bm_0}{a},\ S = \frac{\nu_0}{D_0},\ \varepsilon = \frac{\tau\nu_0\delta}{d^2},\ B = S^{-1} - Pr^{-1};$$

and the same nomenclature has been employed for the dimensionless variables as that used for the original ones.

(f) Seeking a normal-mode solution of this system of part (e) of the form

$$u_1(x,z,t) = A\sin(qx)\cos(\pi z)e^{\sigma t},$$

$$[w_1, T_1](x,z,t) = [C, E]\cos(qx)\sin(\pi z)e^{\sigma t},$$

$$p_1(x,z,t) = F\cos(qx)\cos(\pi z)e^{\sigma t};$$

where $|A|^2 + |C|^2 + |E|^2 + |F|^2 \neq 0$, obtain the secular equation

$$Prk^2\sigma^2 + k^4(Pr+1)\sigma + k^6 - Rq^2(1+\delta-\varepsilon Bk^2) = 0 \text{ where } k^2 = \pi^2 + q^2$$

governing the linear stability of the pure conduction state of part (b).

(g) In what follows it will be assumed that $0 < \varepsilon B << 1$ with a detailed examination of this assumption being deferred until the last parts of the section. Under this assumption conclude that the marginal stability curve on which $\sigma = 0$

$$R = R_0(q^2; \varepsilon B) = \frac{(q^2+\pi^2)^3}{q^2[1+\delta-\varepsilon B(q^2+\pi^2)]}$$

has a vertical asymptote at $q_0^2 = (1+\delta)/(\varepsilon B) - \pi^2 > 0$ with a normal convective instability occurring for $R > R_0 > 0$ when $0 < q^2 < q_0^2$ and a chemically driven one occurring for $R < R_0 < 0$ when $q^2 > q_0^2$. These two branches shall be denoted by "+" and "−", respectively.

(h) Designating by q_c^2 those q^2 such that $dR_0(q_c^2;\varepsilon B)/dq^2 = 0$, show that $Q_c = q_c^2$ satisfies

$$\overbrace{\varepsilon B Q_c^2}^{①} - \overbrace{2(1+\delta)Q_c}^{②} + \overbrace{\pi^2(1+\delta-\varepsilon B\pi^2)}^{③} = 0.$$

(i) Let $Q = q^2$. Then each of the branches identified in part (g) has a critical point in the Q-R plane with the "+" branch having a relative minimum at (Q_c^+, R_c^+) and the "−" branch, a relative maximum at (Q_c^-, R_c^-). Using the pair-wise balancing technique introduced in Section 3.2 deduce the one-term asymptotic representations

$$Q_c^+ \sim \frac{\pi^2}{2} \text{ and } Q_c^- \sim \frac{2(1+\delta)}{\varepsilon B} \text{ as } \varepsilon B \to 0,$$

by balancing the ②-term denoted in the quadratic of part (h) with its ③- and ①-terms, respectively.

(j) Defining $R_c^\pm = R_0(Q_c^\pm;\varepsilon B)$ deduce from parts (g) and (i) the one-term asymptotic representations

$$R_c^+ \sim \frac{27\pi^4}{4(1+\delta)} \text{ and } R_c^- \sim -\frac{4(1+\delta)}{(\varepsilon B)^2} \text{ as } \varepsilon B \to 0.$$

(k) Make a schematic plot of the marginal stability curve $R = R_0(Q;\varepsilon B)$ in the Q-R plane identifying regions of linear stability and instability. Explain the physical significance of the chemically driven instability region for $R < 0$, and compare the results for $R > 0$ with the ordinary Rayleigh–Bénard problem for a nondissociating fluid.

(ℓ) The chemically-driven instabilities deduced above depend upon the parameters $B > 0$ and $0 < \varepsilon << 1$. Given the definition of B, the former inequality means that the effective Prandtl number Pr must exceed the Schmidt number S. Wollkind and Frisch [140] derived a necessary and sufficient condition for that occurrence in this case to be

$$S < \frac{8(4+\alpha_0)}{15(3+\alpha_0)}.$$

Determine the plausibility of this condition by recalling from Chapter 15 that Pr typically has a value of 0.7 for gases and examining the variation of the right-hand side of this inequality versus α_0 over its domain of definition $0 \le \alpha_0 \le 1$. Since S is known to range from 0.5 to 1.0 as α_0 ranges over its domain [68], assume a linear interpolation for S; namely,

$$S = \frac{1+\alpha_0}{2};$$

and show this inequality is satisfied when $0 < \alpha_0 < \alpha_0^+ = [\sqrt{769} - 22]/15 \cong 0.38$. Further, demonstrate in this context, that if $\delta = 1.33$, as deduced in part (n) below, then $\varepsilon = \tau\nu_0\delta/d^2$ has the typically small value of 0.535×10^{-3} for nitrogen with $\alpha_0 = 0.33 \in (0, \alpha_0^+)$, $T_0 = 8,452$ K, $\tau = 10^{-5}$ sec, and $d = 1.0$ cm, and $\nu_0 = 5.4 \times 10^{-6}(T_0/\text{K})^{1.75}$ cm^2/sec $= 40.23$ cm^2/sec [140].

(m) Finally, α_0 and T_0 are related by [140]

$$\frac{\alpha_0^2}{1-\alpha_0} = \frac{\rho_d}{\rho_0}\exp\left(-\frac{T_d}{T_0}\right)$$

where $\rho_d = 130$ g/cm^3 and $T_d = 113,000$ K. Taking $\rho_0 = 0.00125$ g/cm^3 for nitrogen, show that the values for α_0 and T_0 of part (ℓ) are compatible with this relation.

(n) Generalizing the relation of part (m), it follows that

$$\frac{\alpha_e^2(T)}{1-\alpha_e(T)} = \frac{\rho_d}{\rho_0}\exp\left(-\frac{T_d}{T}\right).$$

By implicitly differentiating this relationship, evaluating the result at $T = T_0$, and recalling that $a = 1/T_0$ while noting $b = 1/(1+\alpha_0)$ since $p = \rho(\mathcal{R}/M)T(1+\alpha)$ deduce both

$$m_0 = \alpha_e'(T_0) = \frac{\alpha_0(1-\alpha_0)}{2-\alpha_0}\frac{T_d}{T_0^2} \text{ and } \delta = \frac{bm_0}{a} = \frac{\alpha_0(1-\alpha_0)}{(2-\alpha_0)(1+\alpha_0)}\frac{T_d}{T_0} = 1.33.$$

22.4 Complex Form of Nonlinear Stability Expansions

We next revisit the nonlinear stability Problem 16.1 and examine it by employing a complex expansion of the form suggested in 17.3 during our explanation of the real version of the hexagonal planform expansion treated in that chapter. Toward this end consider the nondimensional evolution equation from part (a) of that problem for $\zeta = \zeta(x,t)$:

$$\mathcal{L}[\zeta] = \frac{\partial \zeta}{\partial t} + \mathcal{R}\frac{\partial}{\partial x}\left[\mathcal{D}(\zeta)\frac{\partial \zeta}{\partial x}\right] + \frac{\partial^4 \zeta}{\partial x^4} + \sinh(\zeta) = 0, \; -\infty < x < \infty;$$

where $\mathcal{D}(\zeta) = 1 + \alpha\zeta$, $\sinh(\zeta) \sim \zeta + \zeta^3/6$, ζ remains bounded as $x \to \pm\infty$, $\mathcal{L}[0] = 0$, and $\partial[\mathcal{D}(\zeta)\partial\zeta/\partial x]/\partial x = (1+\alpha\zeta)\partial^2\zeta/\partial x^2 + \alpha(\partial\zeta/\partial x)^2$. The stability of the solution $\zeta \equiv 0$ to this model equation shall be investigated by introducing the complex Stuart-Watson expansion that to lowest order is given by

$$\zeta(x,t) \sim \eta(x,t) = \eta_{10}A(t)e^{i\omega x} + \eta_{01}A^*(t)e^{-i\omega x}$$

where $\omega \in \mathbb{R}$ and $A(t)$ is a complex amplitude function while $A^*(t)$ denotes its conjugate.

(a) Show that

$$\eta^2(x,t) = \eta_{10}^2 A^2(t)e^{2i\omega x} + 2\eta_{10}\eta_{01}|A(t)|^2 + \eta_{01}A^{*2}(t)e^{-2i\omega x} \text{ with } |A(t)|^2 \equiv A(t)A^*(t)$$

and

$$\eta^3(x,t) = \eta_{10}^3 A^3(t)e^{3i\omega x} + 3\eta_{10}^2\eta_{01}A(t)|A(t)|^2 e^{i\omega x} + 3\eta_{10}\eta_{01}^2 A^*(t)|A(t)|^2 e^{-i\omega x} + \eta_{01}^3 A^{*3}(t)e^{-3i\omega x}.$$

Motivated by the form of these results, conclude that through third-order terms this expansion should be given by

$$\zeta(x,t) \sim \eta_{10}A(t)e^{i\omega x} + \eta_{01}A^*(t)e^{-i\omega x} + \eta_{20}A^2(t)e^{2i\omega x} + \eta_{11}|A(t)|^2 + \eta_{02}A^{*2}(t)e^{-2i\omega x} + \eta_{30}A^3(t)e^{3i\omega x} + \eta_{21}A(t)|A(t)|^2 e^{i\omega x} + \eta_{12}A^*(t)|A(t)|^2 e^{-i\omega x} + \eta_{03}A^{*3}(t)e^{-3i\omega x}$$

with complex amplitude equation

$$\frac{dA(t)}{dt} \sim \sigma A(t) - a_1 A(t)|A(t)|^2.$$

(b) From the amplitude equation in part (a) deduce that

$$\frac{dA^*(t)}{dt} \sim \sigma^* A^*(t) - a_1^* A^*(t)|A(t)|^2$$

and then, from these two amplitude equations, that

$$\frac{d|A(t)|^2}{dt} = A(t)\frac{dA^*(t)}{dt} + A^*(t)\frac{dA(t)}{dt} \sim 2\mathrm{Re}(\sigma)|A(t)|^2 - 2\mathrm{Re}(a_1)|A(t)|^4.$$

(c) Substituting the expansion of part (a) in the evolution equation and employing the amplitude equations derived in part (b), obtain the following sequence of problems, each one of which is proportional to a term appearing explicitly in that expansion:

(i) $A(t)e^{i\omega x}$: $(\sigma - \mathcal{R}\omega^2 + \omega^4 + 1)\eta_{10} = 0$;
(ii) $A^*(t)e^{-i\omega x}$: $(\sigma^* - \mathcal{R}\omega^2 + \omega^4 + 1)\eta_{01} = 0$;
(iii) $A^2(t)e^{2i\omega x}$: $(2\sigma - 4\mathcal{R}\omega^2 + 16\omega^4 + 1)\eta_{20} = 2\alpha\mathcal{R}\omega^2\eta_{10}^2$;
(iv) $|A(t)|^2$: $[2\mathrm{Re}(\sigma) + 1]\eta_{11} = 2\alpha\mathcal{R}(\omega^2 - \omega^2)\eta_{10}\eta_{01} = 0$;
(v) $A^{*2}(t)e^{-2i\omega x}$: $(2\sigma^* - 4\mathcal{R}\omega^2 + 16\omega^4 + 1)\eta_{02} = 2\alpha\mathcal{R}\omega^2\eta_{01}^2$
(vi) $A(t)|A(t)|^2 e^{i\omega x}$: $[\sigma + 2\mathrm{Re}(\sigma) - \mathcal{R}\omega^2 + \omega^4 + 1]\eta_{21} = a_1\eta_{10} + \alpha\mathcal{R}\omega^2(\eta_{11}\eta_{10} + \eta_{20}\eta_{01}) - \eta_{10}^2\eta_{01}/2$.

Although there are also three other third-order problems, only the one appearing above need be considered for the purposes of this section.

(d) Solve the problems catalogued in part (c) sequentially and find that, at first order:

$$\sigma = \sigma^* = \sigma_{\omega^2}(\mathcal{R}) = [\mathcal{R} - \mathcal{R}_0(\omega^2)]\omega^2 \text{ where } \mathcal{R}_0(\omega^2) = \omega^2 + \frac{1}{\omega^2} \Rightarrow \omega_c = 1, \mathcal{R}_c = 2.$$

In what follows take $\omega \equiv \omega_c = 1$. Then $\sigma = \sigma_1(\mathcal{R}) = \mathcal{R} - 2$. Further take $\eta_{10} = \eta_{01} = 1/2$. Note under these conditions that $A^*(t) = A(t)$ and $\eta(x,t) = A(t)\cos(\omega x)$; at second order:

$$\eta_{20} = \eta_{02} = \frac{\alpha\mathcal{R}}{2(13 - 2\mathcal{R})}, \; \eta_{11} \equiv 0;$$

and finally at third-order:

$$4(\mathcal{R} - 2)\eta_{21}(\mathcal{R}, \alpha) = a_1 + \alpha\mathcal{R}\eta_{20}(\mathcal{R}, \alpha) - \frac{1}{8} = a_1 + \frac{\alpha^2\mathcal{R}^2}{2(13 - 2\mathcal{R})} - \frac{1}{8}.$$

(e) Taking the limit of the result of part (d) as $\mathcal{R} \to 2$ obtain the solvability conditions

$$a_1 = \frac{1}{8}\left[1 - \left(\frac{4\alpha}{3}\right)^2\right] \text{ and } \eta_{21}(\mathcal{R}, \alpha) = \frac{\alpha^2(9\mathcal{R} + 26)}{72(13 - 2\mathcal{R})}.$$

(f) Compare the results of parts (d) and (e) with those of Problem 16.1, and observe that the expressions for a_1 are identical while

$$\eta_{20} = \eta_{02} = \frac{\zeta_{22}}{2}, \; \eta_{11} = \zeta_{20} \equiv 0, \; \eta_{21} = \frac{\zeta_{31}}{2}.$$

Discuss the import of these relationships.

(g) Comment on the computational efficacy of combining exponentials when using this complex expansion method as opposed to the employment of trigonometric identities required earlier.

22.5 The Black–Scholes Equation

Having examined a discrete mathematical finance model in Section 21.3, we now consider a continuous one as represented by the Black–Scholes equation for the value of the European call option $V = V(S,t)$ of a stock with price S at time t [123] (the amazing thing about this formulation is that of being deterministic even though its derivation employs stochastic effects which disappear from the final form [134])

$$\frac{\partial V}{\partial t} + \frac{1}{2}\sigma^2 S^2 \frac{\partial^2 V}{\partial S^2} + rS\frac{\partial V}{\partial S} = rV, \; 0 < S < \infty, \; 0 \le t < T;$$

$$V(S,T) = \Psi(S;K);$$

where $r \equiv$ interest rate on a risk-free bond, $\sigma \equiv$ volatility of the stock, $K \equiv$ the strike price (the amount for which the stock can be sold at the terminal time T), and

$$\Psi(S;K) = \begin{cases} 0 & \text{for } S < K \\ S-K & \text{for } S \geq K \end{cases}.$$

It is the purpose of this problem to solve the Black–Scholes equation exactly by exploiting its similarity to heat-type PDE's and Euler-type ODE's.

(a) Motivated by the form of the Black–Scholes equation, first introduce the dimensionless variables

$$x = \ln\left(\frac{S}{K}\right),\ \tau = \frac{1}{2}\sigma^2(T-t),\ v(x,\tau) = \frac{V(S,t)}{K};$$

then show that

$$\frac{\partial V}{\partial t} = K\frac{\partial v}{\partial \tau}\frac{d\tau}{dt} = -\frac{K\sigma^2}{2}\frac{\partial v}{\partial \tau},\ \frac{\partial V}{\partial S} = K\frac{\partial v}{\partial x}\frac{dx}{dS} = \frac{K}{S}\frac{\partial v}{\partial x},$$

$$\frac{\partial^2 V}{\partial S^2} = \frac{\partial}{\partial S}\left(\frac{\partial V}{\partial S}\right) = \frac{\partial}{\partial S}\left(\frac{K}{S}\frac{\partial v}{\partial x}\right) = \frac{K}{S^2}\left(\frac{\partial^2 v}{\partial x^2} - \frac{\partial v}{\partial x}\right);$$

and hence demonstrate that the Black–Scholes equation transforms into the constant coefficient PDE

$$\frac{\partial v}{\partial \tau} = \frac{\partial^2 v}{\partial x^2} + (R-1)\frac{\partial v}{\partial x} - Rv = 0 \text{ for } R = \frac{2r}{\sigma^2},\ -\infty < x < \infty,\ 0 < \tau \leq \frac{\sigma^2 T}{2};$$

with initial condition

$$v(x,0) = \psi(x) = \begin{cases} 0 & \text{for } x < 0 \\ e^x - 1 & \text{for } x \geq 0 \end{cases};$$

given that

$$V(S,T) = \Psi(S;K) = \Psi(Ke^x;K) = Kv(x,0).$$

(b) Next, let

$$v(x,\tau) = e^{\gamma x + \delta\tau}u(x,\tau)$$

and show that

$$v_\tau = e^{\gamma x+\delta\tau}(\delta u + u_\tau),\ v_x = e^{\gamma x+\delta\tau}(\gamma u + u_x),\ v_{xx} = e^{\gamma x+\delta\tau}(\gamma^2 u + 2\gamma u_x + u_{xx}).$$

(c) Substituting the results of part (b) into the PDE for v of part (a) and cancelling the common exponential factor, demonstrate that this equation is transformed into

$$u_\tau = u_{xx} + (2\gamma + R - 1)u_x + [\gamma^2 + (R-1)\gamma - R - \delta]u = 0.$$

(d) Show that by selecting the following values of γ and δ

$$\gamma = \frac{1-R}{2},\ \delta = \gamma^2 + (R-1)\gamma - R = -\frac{(R+1)^2}{4};$$

the PDE of part (d) reduces to the canonical heat-type equation for u

$$u_\tau = u_{xx}, \ -\infty < x < \infty, \ 0 < \tau \leq \frac{\sigma^2 T}{2};$$

while the initial condition becomes

$$u(x,0) = e^{-\gamma x} v(x,0) = e^{(R-1)x/2} \psi(x) = u_0(x)$$
$$= \begin{cases} 0 & \text{for } x < 0 \\ e^{(R+1)x/2} - e^{(R-1)x/2} & \text{for } x \geq 0 \end{cases}.$$

Further adopt the implicit far-field conditions that u is of bounded support which implies

$$u \to 0 \textit{ identically} \text{ as } |x| \to \infty.$$

(e) Defining the Fourier integral transform of $u(x,t)$ in the usual way

$$U(k,\tau) = \int_{-\infty}^{\infty} u(x,\tau) e^{ikx} \, dx,$$

multiplying the canonical heat-type equation for u of part (d) by e^{ikx}, and integrating the result on x from $-\infty$ to ∞, obtain

$$\begin{aligned}\int_{-\infty}^{\infty} u_\tau(x,\tau)\, dx &= \frac{\partial U}{\partial \tau}(k,\tau) \\ &= \int_{-\infty}^{\infty} u_{xx}(x,\tau) e^{ikx} \, dx \\ &= -\int_{-\infty}^{\infty} u_x(x,\tau)(ik) e^{ikx} \, dx \\ &= (ik)^2 \int_{-\infty}^{\infty} u(x,\tau) e^{ikx} \, dx = -k^2 U(x,\tau),\end{aligned}$$

upon integrating by parts twice and imposing the far-field conditions.

(f) Show that the result of part (e) implies

$$U(x,\tau) = U_0(k) e^{-k^2 \tau}$$

where

$$U_0(k) = \int_{-\infty}^{\infty} u_0(x) e^{ikx} \, dx.$$

(g) To proceed with this derivation it is necessary to find the inverse Fourier transform $f(x,\tau)$ of $e^{-k^2\tau}$ or

$$f(x,\tau) = \frac{1}{2\pi} \int_{-\infty}^{\infty} e^{-k^2\tau - ixk} \, dk.$$

Show that, upon completing the square,

$$\tau k^2 + ikx = \tau\left(k + \frac{ix}{2\tau}\right)^2 + \frac{x^2}{4\tau};$$

and thus

$$f(x,\tau) = \frac{e^{-x^2/4\tau}}{2\pi}\int_{-\infty}^{\infty} e^{-\tau[k+ix/(2\tau)]^2}\,dk.$$

Making the change of variables $z = \sqrt{\tau}[k + ix/(2\tau)]$ in this integral, finally obtain

$$f(x,\tau) = \frac{e^{-x^2/(4\tau)}}{2\pi\sqrt{\tau}}\underbrace{\int_{-\infty}^{\infty} e^{-z^2}\,dz}_{=\sqrt{\pi}} = \frac{e^{-x^2/(4\tau)}}{2\sqrt{\pi\tau}}.$$

(h) To complete the derivation, the convolution theorem for the inverse Fourier transform of products is required. Toward that end let

$$F(k) = \int_{-\infty}^{\infty} f(x)e^{ikx}\,dx,\ G(k) = \int_{-\infty}^{\infty} g(x)e^{ikx}\,dx;$$

define the convolution

$$h(x) = f(x) * g(x) \equiv \int_{-\infty}^{\infty} f(x-s)g(s)\,ds;$$

examine its Fourier integral transform

$$H(k) = \int_{-\infty}^{\infty} h(x)e^{ikx}\,dx = \int_{-\infty}^{\infty}\left[\int_{-\infty}^{\infty} f(x-s)g(s)\,ds\right]e^{ikx}\,dx;$$

and show that the sequential interchange of the order of integration and the introduction of the change of variables $y = x - s$ yields

$$\begin{aligned} H(k) &= \int_{-\infty}^{\infty} g(s)\left[\int_{-\infty}^{\infty} f(x-s)e^{ikx}\,dx\right]ds \\ &= \int_{-\infty}^{\infty} g(s)\left[\int_{-\infty}^{\infty} f(y)e^{iky}\,dy\right]e^{iks}\,ds = F(k)G(k). \end{aligned}$$

(i) Applying this convolution theorem to the formula for $U(k,\tau)$ of part (f) and employing the result of part (g) obtain

$$u(x,\tau) = \int_{-\infty}^{\infty} f(x-s,\tau)u_0(s)\,ds = \frac{1}{2\sqrt{\pi\tau}}\int_{-\infty}^{\infty} u_0(s)e^{-(x-s)^2/(4\tau)}\,ds.$$

(j) Since it is more convenient to represent the analytical solution of part (i) in terms of tabulated functions, make the change of variables $y = (s-x)/\sqrt{2\tau}$ and convert that integral into

$$u(x,\tau) = \frac{1}{\sqrt{2\pi}} \int_{-\infty}^{\infty} u_0(x+\sqrt{2\tau}y)e^{-y^2/2}\,dy.$$

Noting that $u_0(x+\sqrt{2\tau}y) = 0$ for $y < -x/\sqrt{2\tau}$, show this integral truncates to

$$u(x,\tau) = \frac{1}{\sqrt{2\pi}} \int_{-x/\sqrt{2\tau}}^{\infty} u_0(x+\sqrt{2\tau}y)e^{-y^2/2}\,dy,$$

which upon substitution of the explicit form of $u_0(x+\sqrt{2\tau}y)$ becomes

$$u(x,\tau) = \overbrace{\frac{1}{\sqrt{2\pi}} \int_{-x/\sqrt{2\tau}}^{\infty} e^{(R+1)(x+\sqrt{2\tau}y)/2} e^{-y^2/2}\,dy}^{I_1} - \overbrace{\frac{1}{\sqrt{2\pi}} \int_{-x/\sqrt{2\tau}}^{\infty} e^{(R-1)(x+\sqrt{2\tau}y)/2} e^{-y^2/2}\,dy}^{I_2}.$$

(k) Consider I_1 in the result of part (j). Completing the square to obtain

$$y^2 - (R+1)\sqrt{2\tau}y = \left[y - (R+1)\sqrt{\frac{\tau}{2}}\right]^2 - (R+1)^2\frac{\tau}{2}$$

and making the change of variables $z = (R+1)\sqrt{\tau/2} - y$, demonstrate that I_1 becomes

$$I_1 = \frac{e^{(R+1)x/2+(R+1)^2\tau/4}}{\sqrt{2\pi}} \int_{-\infty}^{\eta_1} e^{-z^2/2}\,dz = e^{(R+1)x/2+(R+1)^2\tau/4}\mathrm{Erf}(\eta_1)$$

where $\mathrm{Erf}(\eta)$ represents the cumulative normal distribution function defined in Chapter 6 as

$$\mathrm{Erf}(\eta) = \frac{1}{\sqrt{2\pi}} \int_{-\infty}^{\eta} e^{-z^2/2}\,dz = \frac{1}{2}\left[1 + \mathrm{erf}\left(\frac{\eta}{\sqrt{2}}\right)\right] \text{ and } \eta_1 = \frac{x}{\sqrt{2\tau}} + (R+1)\sqrt{\frac{\tau}{2}}.$$

(ℓ) Since I_2 only differs from I_1 by the presence of the factor $R-1$ rather than $R+1$, deduce that

$$I_2 = e^{(R-1)x/2+(R-1)^2\tau/4}\mathrm{Erf}(\eta_2) \text{ where } \eta_2 = \frac{x}{\sqrt{2\tau}} + (R-1)\sqrt{\frac{\tau}{2}}.$$

(m) Therefore from parts (b), (d), (j), (k), and (ℓ), conclude that

$$v(x,\tau) = e^{(1-R)x/2-(R+1)^2\tau/4}u(x,\tau) = e^x\mathrm{Erf}(\eta_1) - e^{-R\tau}\mathrm{Erf}(\eta_2).$$

(n) Recalling that $x = \ln(S/K)$, $\tau = \sigma^2(T-t)/2$, and $R = 2r/\sigma^2$, finally obtain the solution

$$V(S,t) = Kv(x,\tau) = S\,\mathrm{Erf}(\eta_1) - Ke^{-r(T-t)}\mathrm{Erf}(\eta_2)$$

where

$$\eta_{1,2} = \frac{\ln(S/K)}{\sigma\sqrt{T-t}} + \left(\frac{2r}{\sigma^2} \pm 1\right)\frac{\sigma}{2}\sqrt{T-t} = \frac{\ln(S/K) + (r \pm \sigma^2/2)(T-t)}{\sigma\sqrt{T-t}},$$

which is called the Black–Scholes formula for this call option.

(o) Show that

$$\text{Erf}(\eta) \to 0 \text{ as } \eta \to -\infty, \ \text{Erf}(\eta) \to 1 \text{ as } \eta \to \infty;$$

$$\eta_{1,2} \to -\infty \text{ as } S \to 0, \ \eta_{1,2} \to \infty \text{ as } S \to \infty;$$

and hence deduce the expected asymptotic behavior for the Black–Scholes formula given by

$$V(S,t) \sim 0 \text{ as } S \to 0 \text{ and } V(S,t) \sim S \text{ as } S \to \infty.$$

22.6 Age-Structured Discrete-Time American Dipper Population Model

Fig. 22.3 A lithograph made in 1867 by Joseph Wolf of the American dipper (*Cinclus mexicanus*) subspecies *C.m. ardesiacus*, also known as a water ouzel. Here the adult is in the foreground and the juvenile, in the background.

We close by considering the discrete-time age-structured model for the American dipper population in the Black Hills of South Dakota (see Fig. 22.3) of the form of the system of Problem 21.1 where $x_k \equiv$ number of juvenile females (inexperienced breeders) and $y_k \equiv$ number of adult females (experienced breeders) at time k years given by [91]

$$X_{k+1} = AX_k \equiv \begin{pmatrix} s_j\rho_j & s_j\rho_a \\ s_a & s_a \end{pmatrix} \begin{bmatrix} x_k \\ y_k \end{bmatrix},$$

with $s_{j,a}$ denoting the annual survival rates and $\rho_{j,a}$, the annual reproductive rates of the juvenile and adult populations, respectively. Here, it is being assumed that there exists a 1:1 male to female sex ratio in this population or, more realistically, that there are at least a sufficient number of males to pair and mate with all available females.

(a) Show that the eigenvalues λ and eigenvectors Z of the matrix A defined by

$$AZ = \lambda Z$$

satisfy the quadratic characteristic polynomial

$$g(\lambda) = \lambda^2 - (s_j\rho_j + s_a)\lambda + s_j s_a(\rho_j - \rho_a) = 0$$

with corresponding eigenvectors

$$Z = \begin{bmatrix} z_1 \\ z_2 \end{bmatrix} = \begin{bmatrix} \lambda - s_a \\ s_a \end{bmatrix}.$$

(b) Demonstrate that the discriminant of this quadratic

$$\mathcal{D} = (s_j\rho_j + s_a)^2 + 4s_j s_a(\rho_a - \rho_j) = (s_j\rho_j - s_a)^2 + 4s_j s_a\rho_a > 0$$

and reduces to

$$\mathcal{D} = (s_j\rho + s_a)^2 \text{ when } \rho_j = \rho_a = \rho.$$

(c) Conclude from (a) and (b) that in general these eigenvalues and their corresponding eigenvectors are given by

$$2\lambda_{1,2} = s_j\rho_a + s_a \pm \sqrt{\mathcal{D}} \text{ and } Z^{(1,2)} = \begin{bmatrix} \lambda_{1,2} - s_a \\ s_a \end{bmatrix}.$$

(d) Deduce from the results of (g) and (h) of Problem 21.1 that this difference equation admits the solutions

$$X_k^{(1)} = Z^{(1)}\lambda_1^k = \begin{bmatrix} \lambda_1 - s_a \\ s_a \end{bmatrix} \lambda_1^k \text{ and } X_k^{(2)} = Z^{(2)}\lambda_2^k = \begin{bmatrix} \lambda_2 - s_a \\ s_a \end{bmatrix} \lambda_2^k$$

and hence has general solution composed of the linear combination

$$X_k = c_1 X_k^{(1)} + c_2 + X_k^{(2)} = c_1 \begin{bmatrix} \lambda_1 - s_a \\ s_a \end{bmatrix} \lambda_1^k + c_2 \begin{bmatrix} \lambda_2 - s_a \\ s_a \end{bmatrix} \lambda_2^k.$$

(e) In order to satisfy the initial conditions

$$X_0 = \begin{bmatrix} x_0 \\ y_0 \end{bmatrix},$$

determine that the constants $c_{1,2}$ must be chosen as

$$c_1 = \frac{x_0 s_a - y_a(\lambda_2 - s_a)}{s_a\sqrt{\mathcal{D}}} \text{ and } c_2 = \frac{y_0(\lambda_1 - s_a) - x_0 s_a}{s_a\sqrt{\mathcal{D}}}.$$

(f) For the special case $\rho_j = \rho_a = \rho$, conclude from (a) that the eigenvalues are now given by

$$\lambda_1 = s_j\rho + s_a \text{ and } \lambda_2 = 0;$$

and, as a partial check on these calculations, demonstrate from (b) and (c) that the general eigenvalue formulas then reduce to values consistent with those results.

(g) Population growth occurs if the dominant eigenvalue of this system λ_1 is greater than 1. The magnitude of that eigenvalue then represents the long-term growth rate of the population and the corresponding eigenvector gives its steady-state age distribution. For the special case introduced in (f) show that this requirement for population growth is given by

$$s_j\rho + s_a > 1$$

and its corresponding eigenvector by

$$\begin{bmatrix} s_j\rho \\ s_a \end{bmatrix}.$$

(h) If it is further assumed that $s_j = s_a = s$, show that the results of (g) simplify to

$$s(\rho + 1) > 1 \text{ and } s\begin{bmatrix} \rho \\ 1 \end{bmatrix}.$$

(i) First, for the case of (h) that there is no age-structured difference in either the reproductive or survival rates, Palmer and Javed (2014), from 2002 field data, estimated an annual reproductive rate of $\rho = 0.6973$ and annual survival rate of $s = 0.5931$ (calculated from all individuals). Conclude from the simplified results of (h) that this indicates an annual population growth of less than 1% with a stable age structure in that population of 41% juveniles and 59% adults.

(j) Since the juvenile survival rate may be less than that of the adults, Palmer and Javed [91] next considered the case where $s_j \neq s_a$. They used an estimated annual survival rate of $s_j = 0.4615$ pooled for all juveniles and $s_a = 0.6374$ for adults from the 2002 cohort while retaining $\rho = 0.6973$. Conclude from the more general results of (g) that this indicates an approximate annual loss in the population of 4% with a stable age structure in that population of 34% juveniles and 66% adults.

(k) Palmer and Javed [91] felt that the best way to reverse this trend would be to raise the reproductive rate by employing various conservation strategies. Find what percentage increase of ρ would be required to produce an annual growth in the population of 4%.

References

[1] Abramowitz, M., Stegun, I.A.: Handbook of Mathematical Functions. Dover, Mineola (1965)
[2] Ackemann, T., Logvin, Y.A., Heur, A., Lange, W.: Transition between positive and negative hexagons in optical pattern formation. Phys. Rev. Lett. **75**, 3450–3453 (1995)
[3] Aranson, I.S., Gorshkov, K.A., Lomov, A.S., Rabinovich, M.I.: Stable particle-like solutions of multidimensional nonlinear fields. Physica D **43**, 435–453 (1990)
[4] Arrowsmith, D.K., Place, C.M.: Ordinary Differential Equations. Chapman and Hall, London (1982)
[5] Aşic, M.Z., Tezcan, S.: A mathematical model for the behavior of laminated glass beams. Comput. Struct. **83**, 1742–1753 (2005)
[6] Batchelor, G.K.: An Introduction to Fluid Dynamics. Cambridge University Press, Cambridge (1967)
[7] Bénard, H.: Les tourbillons cellulaires dans une nappe liquide transportant de la chaleur par convection en regime permanent. Ann. De Chimie et de Physique **23**, 62–144 (1901)
[8] Bliss, G.A.: Calculus of Variations. Mathematical Association of America, Washington, DC (1944)
[9] Boonkorkuea, N., Lenbury, Y., Alvarado, F.J., Wollkind, D.J.: Nonlinear stability analyses of vegetative pattern formation in an arid flat environment. J. Biol. Dyn. **4**, 346–380 (2010)
[10] Boussinesq, J.: Théorie analytique de la chaleur: mice en harmonie avec la thermodynamique et avec la théorie mécanique de la lumiére, vol. 2. Gauthier-Villars, Paris (1903)
[11] Boyce, W.E., DiPrima, R.C.: Elementary Differential Equations and Boundary Value Problems. Wiley, New York (2012)
[12] Brand, L.: Vector and Tensor Analysis. Wiley, NewYork (1947)
[13] Briggs, G.E., Haldane, J.B.: A note on the kinetics of enzyme action. Biochem. J. **19**, 338–339 (1925)
[14] Cangelosi, R.A., Wollkind, D.J., Kealy-Dichone, B.J., Chaiya, I.: Nonlinear stability analyses of Turing patterns for a mussel-algae model. J. Math. Biol. **70**, 1249–1294 (2015)
[15] Castets, V., Dulos, E., Boissonade, J., DeKepper, P.: Experimental evidence of sustained standing Turing-type nonequilibrium chemical patterns. Phys. Rev. Lett. **64**, 2953–2956 (1990)
[16] Caughley, G.: Plant-herbivore systems. In: May, R. (ed.) Theoretical Ecology: Principles and Applications, pp. 94–113. W.B. Saunders, Philadelphia (1975)
[17] Chaiya, I., Wollkind, D.J., Cangelosi, R.A., Kealy-Dichone, B.J., Rattanakul, C.: Vegetative rhombic pattern formation driven by root suction for an interaction-diffusion plant-ground water model system in an arid flat environment. Am. J. Plant Sci. **6**, 1278–1300 (2015)

© Springer International Publishing AG, part of Springer Nature 2017

D. J. Wollkind and B. J. Dichone, *Comprehensive Applied Mathematical Modeling in the Natural and Engineering Sciences*,
https://doi.org/10.1007/978-3-319-73518-4

[18] Chandra, K.: Instability of fluids heated from below. Proc. Roy. Soc. A **164**, 231–242 (1938)

[19] Chandrasekhar, S.: Hydrodynamic and Hydromagnetic Stability. Oxford University Press, Oxford (1961)

[20] Chen, W., Ward, M.J.: The stability and dynamics of localized spot patterns in the two-dimensional Gray-Scott model. SIAM J. Dyn. Syst. **10**, 586–666 (2011)

[21] Collings, J.B., Wollkind, D.J., Moody, M.E.: Outbreaks and oscillations in a temperature dependent model for a mite predator-prey interaction. Theor. Pop. Biol. **38**, 159–190 (1990)

[22] Copson, E.T.: Asymptotic Expansions. Cambridge, London (1967)

[23] Couteron, P., Mahamane, A., Ouedraogo, P., Seghieri, J.: Differences between banded thickets (tiger bush) in two sites in West Africa. J. Veg. Sci. **11**, 321–328 (2000)

[24] Cross, M.C., Hohenberg, P.C.: Pattern formation outside of equilibrium. Rev. Mod. Phys. **65**, 851–112 (1993)

[25] Danby, J.M.A.: Computing Applications to Differential Equations. Reston Publishing Co., Reston (1985)

[26] Davis, M.G., Wollkind, D.J., Cangelosi, R.A., Kealy-Dichone, B.J.: The behavior of a population interaction-diffusion equation in its subcritical regime. Involve **11**, 297–309 (2018)

[27] DeKepper, P., Castets, V., Dulos, E., Boissonade, J.: Turing-type chemical patterns in the chlorite-iodide-malonic acid reaction. Physica D **49**, 161–169 (1991)

[28] DiPrima, R.C., Eckhaus, W., Segel, L.A.: Nonlinear wavenumber interaction in near-critical two-dimensional flow. J. Fluid Mech. **49**, 705–744 (2010)

[29] Dobrushkin, V.A.: Applied Differential Equations: An Introduction. CRC Press, Boca Raton (2014)

[30] Doedel, E.J., Pattenroth, R., Champneys, A.R., Fairgrieve, T.F., Kuznetsov, Y.A., Olderman, B.E., Sandstede, B., Wang, X.: Auto 2000: continuation and bifurcation software for ordinary differential equations. Technical Report, Concordia University, Montreal (2002)

[31] Drazin, P.G., Reid, W.H.: Hydrodynamic Instability. Cambridge University Press, Cambridge (1981)

[32] Durand, L.: Stability and oscillations of a soap film. Am. J. Phys. **49**, 334–343 (1981)

[33] Edwards, C.H.: Advanced Calculus of Several Variable. Dover, Mineola (1994)

[34] Fife, P.C.: The Bénard problem for general fluid dynamical equations and remarks on the Boussinesq approximation. Indiana Univ. Math. J. **20**, 303–326 (1970/1971)

[35] Firth, W.J., Scroggie, A.J.: Spontaneous pattern formation in an absorptive system. Euro. Phys. Lett. **26**, 521–526 (1994)

[36] Fransz, H.G.: The Functional Response to Prey Density in an Acarine System (Simulation Monographs). Pudoc, Wageningen (1974)

[37] Friedman, B.: Principles and Techniques of Applied Mathematics. Dover, Mineola (2011)

[38] Gambino, G., Greco, A.M., Lombardo, M.C., Sammartino, M.: A subcritical bifurcation for a nonlinear reaction-diffusion system. In: Greco, A.M., Rionero, S., Ruggeri, T. (eds.) Waves and Stability in Continuous Media, pp. 163–172. World Scientific Publishing Comp, Singapore (2010)

[39] Gambino, G., Lombardo, M.C., Sammartino, M.: Turing instability and traveling fronts for a nonlinear reaction-diffusion system with cross-diffusion. Math. Comput. Simulat. **82**, 1112–1132 (2012)

[40] Geddes, J.B., Indie, R.A., Moloney, J.V., Firth, W.J.: Hexagons and squares in a passive nonlinear optical system. Phys. Rev. **50**, 3471–3485 (1994)

[41] Golovin, A.A., Matkowsky, B.J., Volpert, V.A.: Turing pattern formation in the Brusselator model with superdiffusion. SIAM J. Appl. Math. **69**, 251–272 (2008)

[42] Gowda, K., Riecke, H., Silber, M.: Transitions between patterned states in vegetation models for semiarid ecosystems, Phys. Rev. E **8**, 022701-1–022701-8 (2014)

[43] Graham, M.D., Kevrekidis, I.G., Asakura, K., Lauterbach, J., Krischer, K., Rottermund, H.H., Ertl, G.: Effects of boundaries on pattern formation: catalytic oxidation of CO on platinum. Science **264**, 80–82 (1994)
[44] Granville, W.A.: Elements of Differential and Integral Calculus, revised edn. Forgotten Books, London (2016)
[45] Greenberg, M.D.: Applications of Greens Functions in Science and Engineering. Dover, Mineola (2015)
[46] Greenspan, H.P.: The Theory of Rotating Fluids. Cambridge University Press, Cambridge (1968)
[47] Hassell, M.P.: The Dynamics of Arthropod predator-prey Systems. Princeton University Press, Princeton (1978)
[48] HilleRisLambers, R., Rietkerk, M., van den Bosch, F., Prins, H.T.H., de Kroon, H.: Vegetation pattern formation in semi-arid grazing systems. Ecology **82**, 50–61 (2001)
[49] Holling, C.S.: The functional response of predators to prey density and its role in mimicry and population regulation. Mem. Entomol. Soc. Can. **45**, 1–60 (1965)
[50] Hoyt, S.C.: Integrated chemical control of insects and biological control of mites on apples in Washington. J. Econ. Entomol. **62**, 74–86 (1969)
[51] Huffaker, C.B.K., Shea, K.P., Herman, S.G.: Experimental studies on predation (III), complex dispersion and levels of food in an acarine predator-prey interaction. Hilgardia **34**, 305–330 (1963)
[52] Hughes, W.F., Gaylord, E.W.: Basic Equations of Engineering Science. McGraw-Hill, New York (1964)
[53] Ince, E.L.: Ordinary Differential Equations. Dover, Mineola (1956)
[54] Israel, F.L.: Student's atlas of American presidential elections 1789–1996. Congressional Quarterly Inc., Washington, D.C. (1997)
[55] Jeans, J.H.: Astronomy and Cosmology. Cambridge University Press, Cambridge (1929)
[56] Jeffreys, H.: Some cases of instability in fluid motion. Proc. Roy. Soc. A **118**, 195–208 (1928)
[57] Kareiva, P., Odell, G.M.: Swarms of predators exhibit "preytaxis" if individual predators use area search. Am. Nat. **130**, 233–270 (1987)
[58] Kealy-Dichone, B.J., Wollkind, D.J., Cangelosi, R.A.: Rhombic analysis extension of a plant-surface water interaction-diffusion model for hexagonal pattern formation in an arid flat environment. Am. J. Plant Sci. **6**, 1256–1277 (2015)
[59] Keller, E.F., Segel, L.A.: Initiation of slime mold aggregation viewed as an instability. J. Theor. Biol. **26**, 399–415 (1970)
[60] Kondo, S., Asai, R.: A reaction-diffusion wave on the skin of the marine angelfish Pomacanthus. Nature **376**, 765–768 (1995)
[61] Koschmieder, E.L.: Bénard Cells and Taylor Vortices. Cambridge, London (1993)
[62] Kreyszig, E.: Advanced Engineering Mathematics. Wiley, NY (20114)
[63] Landau, L.D.: On the problem of turbulence. Doklady Akademii Nauk SSR **44**, 339–342 (1944)
[64] Lejeune, O., Tlidi, M., Couteron, P.: Localized vegetation patches: a self-organized response to resource scarcity. Phys. Rev. E **66**, 010901-1–01901-4 (2002)
[65] Lejeune, O., Tlidi, M., Lefever, R.: Vegetation spots and stripes in arid landscapes. Int. J. Quantum Chem. **98**, 261–271 (2004)
[66] Lengyel, I., Epstein, I.R.: A chemical approach to designing Turing patterns in reaction-diffusion systems. Natl. Acad. Sci. USA **89**, 3977–3979 (1992)
[67] Leslie, P.H.: Some further notes on the use of matrices in population mathematics. Biometrica **35**, 213–245 (1948)
[68] Lighthill, M.J.: Dynamics of a dissociating gas Part 2. Quasi-equilibrium transfer theory. J. Fluid Mech. **8**, 161–182 (1960)

[69] Lin, C.C., Segel, L.A.: Mathematics Applied to Deterministic Problems in the Natural Sciences. SIAM Classics Series. SIAM, Philadelphia (1988)
[70] Liu, Q.X., Doelman, A., Rottschäfer, V., de Jager, M., Herman, P.M.J., Rietkerk, M., van de Koppel, J.: Phase separation explains a new class of self-organized spatial patterns in ecological systems. PNAS **10**, 11905–11910 (2013)
[71] Logan, J.A.: Population model of the association of Tetranychus mcdanieli (Acarina: Tetranychidae) with Metasieulus occidentalis (Acarina: Phytoseiidae) in the apple ecosystem. Ph.D. Dissertation. Washington State University, Pullman (1977)
[72] Logan, J.A.: Recent advances and new directions in Phytoseiid population models. In: Hoyt, S.C. (ed.) Recent Advances in the Knowledge of the Phytoseiidae, pp. 49–71. Division of Agricultural Sciences Publications, University of California, Berkeley (1982)
[73] Logan, J.A., Hilbert, D.W.: Modeling the effect of temperature on arthropod population systems. In: Lauenroth, W.K., Skogerboe, G.V., Flug, M. (eds.) Analysis of Ecological Systems: State of the Art in Ecological Modeling, pp. 113–122. Elsevier, Amsterdam (1983)
[74] Logan, J.A., Wollkind, D.J., Hoyt, S.C., Tanigoshi, L.K.: An analytic model for description of temperature dependent rate phenomena in arthropods. Envir. Entomol. **5**, 1130–1140 (1976)
[75] Lugiato, L.A., Lefever, R.: Spatial dissipative structures in passive optical systems. Phys. Rev. Lett. **58**, 2209–2211 (1987)
[76] Lugiato, L.A., Oldano, C.: Stationary spatial patterns in passive optical systems. Phys. Rev. A **37**, 3896–3908 (1988)
[77] Mandel, P., Georgio, M., Erneux, T.: Transverse effects in coherently driven nonlinear cavities. Phys. Rev. A **47**, 4277–4286 (1993)
[78] Matkowsky, B.J.: A simple nonlinear dynamic stability problem. Bull. Am. Math. Soc. **76**, 646–649 (1970)
[79] Matsson, O.J.E.: A Student Project for Rayleigh-Bénard Convection. American Society for Engineering Education, Washington, D.C. (2008)
[80] May, R.A.: Stability and Complexity in Model Ecosystems. Princeton University Press, Princeton (1973)
[81] Meinhardt, H.: Dynamics of stripe formation. Nature **376**, 722–723 (1995)
[82] Meron, E., Gilad, E., von Hardenberg, J., Shachuk, M., Zarmi, Y.: Vegetation patterns along a rainfall gradient. Chaos, Solitons Fractals **19**, 367–376 (2004)
[83] Moloney, J.V., Newell, A.C.: Nonlinear Optics. Westview Press, Boulder (2004)
[84] Murray, J.D.: Mathematical Biology. Springer, Berlin (1989)
[85] Murray, J.D.: Mathematical Biology II: Spatial Models and Biomedical Applications. Springer, Berlin (2003)
[86] Noble, B.: Applications of Undergraduate Mathematics in Engineering. Macmillan, New York (1967)
[87] Oberbeck, A.: Ueber die Wärmeleitung der Flüssigkeiten bei Berücksichtigung infolge von Temperaturdifferenzen. Annalen der Physik **243**, 271–292 (1879)
[88] O'Malley, R.E.: Singular Perturbation Methods in Ordinary Differential Equations. Springer, Berlin (1991)
[89] Ouyang, Q., Swinney, H.L.: Transition to chemical turbulence. Chaos **1**, 411–420 (1991)
[90] Ouyang, Q., Li, R., Li, G., Swinney, H.L.: Dependence of Turing pattern wavelength on diffusion rate. J. Chem. Phys. **102**, 2551–2555 (1995)
[91] Palmer, J.S., Javed, J.: An age-structured model for the American dipper population in the Black Hills of South Dakota. Proc. S. D. Acad. Sci. **93**, 79–88 (2014)
[92] Pansuwan, A., Rattanakul, C., Lenbury, Y., Wollkind, D.J., Harrison, L., Rajapakse, I., Cooper, K.: Nonlinear stability analyses of pattern formation during ion-sputtered erosion. Math. Comput. Model. **41**, 939–964 (2005)

[93] Pearson, J.E.: Pattern formation in a (2+1) activator-inhibitor-immobilizer system. Phys. A **188**, 178–189 (1992)

[94] Pesch, H.J.: The princess and infinite-dimensional optimization. Documenta Mathematica **Extra Volume ISMP**, 345–356 (2012)

[95] Plant, R.E., Mangel, M.: Modeling and simulation in agricultural pest management. SIAM Rev. **29**, 235–261 (1987)

[96] Poincaré, H.: Sur les integrals irréguliéres des équations linéaires. Acta Mathematica **8**, 295–344 (1886)

[97] Pólya, G.: How to Solve It. Doubleday, Garden City (1957)

[98] Pólya, G.: The minimum fraction of the popular vote that can elect the President of the United States. Math. Teach. **54**, 130–133 (1961)

[99] Prigogine, I., Lefever, R.: On symmetry breaking instabilities in dissipative systems II. J. Chem. Phys. **48**, 1695–1700 (1968)

[100] Rayleigh, Lord: On convection in a horizontal layer of fluid, when the higher temperature is on the underside. Phil. Mag. **32**, 529–546 (1916)

[101] Reid, W.H., Harris, D.L.: Some further results on the Bénard problem. Phys. Fluids **1**, 102–110 (1958)

[102] Riedel, V.: Gerisch: Regulation of extra cellular cyclic-AMP-phosphodiesterase activity during the development of Dictyostelium discoideum. Biochem. Biophys. Res. Comm. **42**, 119–124 (1971)

[103] Rietkerk, M., Boerlijst, M.C., van Langevelde, F., HilleRisLambers, R., van de Koppel, J., Kumar, L., Prins, H.H.T., de Roos, A.M.: Self-organization of vegetation in arid ecosystems. Am. Nat. **160**, 524–530 (2002)

[104] Rietkerk, M., Dekker, S.C., de Ruiter, P.C., van de Koppel, J.: Self-organized patchiness and catastrophic shift in ecosystems. Science **305**, 1926–1929 (2004)

[105] Roose, T., Fowler, A.C.: A model for water uptake by plant roots. J. Theor. Biol. **228**, 155–171 (2004)

[106] Rosenzweig, M.L.: Paradox of enrichment: destabilization of exploitation ecosystems in ecological time. Science **171**, 385–387 (1971)

[107] Rubin, V.C., Ford, W.K.J.: Rotation of the Andromeda nebula from a spectroscopic survey of emission. Astrophys. J **159**, 379 (1970)

[108] Sagan, C.: The solar system. In: Flanagan, D. (ed.) The Solar System, A Scientific American Book, pp. 1–11. W. H. Freeman and Co., San Francisco (1975)

[109] Schlichting, H.: Boundary-Layer Theory. McGraw-Hill, New York (1968)

[110] Schnackenberg, J.: Simple chemical reaction system with limit cycle behavior. J. Theor. Biol. **81**, 389–400 (1979)

[111] Schnell, S.: Validity of the Michaelis-Menten equation-steady-state or reaction stationary assumption: that is the question. FEBS J. **281**, 464–472 (2014)

[112] Schwartz, E.J., Vaidya, N.K., Dorman, H.S., Carpenter, S., Mealey, R.N.: Dynamics of lentiviral infection in vivo in the absence of adaptive immune responses. Virology **513**, 108–113 (2018)

[113] Scroggie, A.J., Firth, W.J., MacDonald, G.S., Tlidi, M., Lefever, R., Lugiato, L.A.: Pattern formation in a passive Kerr cavity. Chaos, Solitons Fractals **4**, 1323–1354 (1994)

[114] Segel, L.A.: Mathematics Applied to Continuum Mechanics. SIAM Classics Series. SIAM, Philadelphia (2007)

[115] Spiegel, M.R.: Laplace Transforms. McGraw Hill, New York (1965)

[116] Stancevic, O., Angstmann, C.N., Murray, J.M., Henry, B.I.: Turing patterns from dynamics of early HIV infection. Bull. Math. Biol. **75**, 774–795 (2013)

[117] Stephenson, L.E., Wollkind, D.J.: Weakly nonlinear stability analyses of one-dimensional Turing pattern formation in activator-inhibitor/immobilizer model systems. J. Math. Biol. **33**, 771–815 (1995)

[118] Stuart, J.T.: On the nonlinear mechanics of wave disturbances in stable and unstable flows, 1. The basic behavior in plane Poiseuille flow. J. Fluid Mech. **9**, 353–370 (1960)

[119] Sutton, O.G.: On the stability of a fluid heated from below. Proc. Roy. Soc. A **204**, 297–309 (1950)

[120] Tuckwell, H.C., Wan, Y.M.: On the behavior of solutions in viral dynamical models. BioSyst. **73**, 157–161 (2004)

[121] Turing, A.M.: The chemical basis of morphogenesis. Phil. Trans. Roy. Soc. London B **237**, 37–72 (1952)

[122] Tulumello, E., Lombardo, M.C., Sammartino, M.: Cross-diffusion driven instability in a predator-prey system with cross-diffusion. Acta. Appl. Math. **132**, 621–633 (2014)

[123] Ugur, O.: An Introduction to Computational Finance. Imperial College Press, London (2009)

[124] Uspensky, J.V.: Theory of Equations. McGraw Hill, New York (1963)

[125] Van Dyke, M.D.: Perturbation Methods in Fluid Mechanics. Parabolic Press, Palo Alto (1975)

[126] von Hardenberg, J., Meron, E., Shachak, M., Zarmi, Y.: Diversity of vegetation patterns and desertification. Phys. Rev. Letts. **87**, 198101-1–198101-4 (2001)

[127] Walgraef, D.: Spatio-Temporal Pattern Formation. Springer, New York (1997)

[128] Wang, R.H., Liu, Q.X., Sun, G.Q., Zhen, J., van de Koppel, J.: Nonlinear dynamic and pattern bifurcation in a model for spatial patterns in young mussel beds. J. R. Soc. Interface **6**, 705–718 (2009)

[129] Wang, W.M., Wang, W.J., Lin, Y.Z., Tan, Y.J.: Pattern selection in a predation model with self and cross diffusion. Chin. Phys. B **20**(034702–1), 034702–8 (2011)

[130] Watson, G.N.: Treatise on the Theory of Bessel Functions. Cambridge, London (1922)

[131] Watson, J.: On the nonlinear mechanics of wave disturbances in stable and unstable flows, 2. The development of a solution for plane Poiseuille flow and plane Couette flow. J. Fluid Mech. **9**, 371–389 (1960)

[132] Weast, R.C.: Handbook of Chemistry and Physics, 49th edn. Chem. Rubber Co., Cleveland (1968)

[133] Weinstock, R.: Calculus of Variations with Applications to Physics and Engineering. Dover, Mineola (1974)

[134] Wilmott, P., Hewison, S., Dewynne, J.: The Mathematics of Financial Derivatives. Cambridge University Press, Cambridge (1995)

[135] Wollkind, D.J.: Singular perturbation techniques: a comparison of the Method of Matched Asymptotic Expansions with that of multiple scales. SIAM Rev. **19**, 502–516 (1977)

[136] Wollkind, D.J.: Rhombic and hexagonal weakly nonlinear stability analyses: theory and application. In: Debnath, L. (ed.) Nonlinear Instability Analyses II, pp. 221–272. WIT Press, Southampton (2001)

[137] Wollkind, D.J., Alvarado, F.J., Edmeade, D.E.: Non-linear stability analyses of optical pattern formation in an atomic sodium vapour ring cavity. IMA J. Appl. Math. **73**, 902–935 (2008)

[138] Wollkind, D.J., Bdzil, J.: Comments on chemical instabilities. Phys. Fluids **14**, 1813–1814 (1971)

[139] Wollkind, D.J., Collings, J.B., Logan, J.A.: Metastability in a temperature-dependent model system for apredator-prey outbreak interaction on fruit trees. Bull. Math. Biol. **50**, 379–409 (1988)

[140] Wollkind, D.J., Frisch, H.L.: Chemical instabilities I: a heated horizontal layer of a dissociating fluid. Phys. Fluids **14**, 13–18 (1971)

[141] Wollkind, D.J., Logan, J.A.: Temperature-dependent predator-prey mite ecosystem on apple tree foliage. J. Math. Biol. **6**, 265–283 (1978)

[142] Wollkind, D.J., Manoranjan, V.S., Zhang, L.: Weakly nonlinear stability analyses of prototype reaction-diffusion model equations. SIAM Rev. **36**, 176–214 (1994)

[143] Wollkind, D.J., Stephensen, L.E.: Chemical Turing pattern formation analyses: comparison of theory with experiment. SIAM J. Appl. Math. **61**, 387–431 (2000)
[144] Wollkind, D.J., Zhang, L.: The effect of suspended particles on Rayleigh-Bénard convection II: a nonlinear stability analysis of a thermal disequilibrium model. Mathl. Comput. Model. **19**, 43–74 (1994)

Index

© Springer International Publishing AG, part of Springer Nature 2017

D. J. Wollkind and B. J. Dichone, *Comprehensive Applied Mathematical Modeling in the Natural and Engineering Sciences*,
https://doi.org/10.1007/978-3-319-73518-4

Printed in the United States
By Bookmasters